THE BLACK BOOK OF QUANTUM CHROMODYNAMICS

The Black Book of Quantum Chromodynamics

A Primer for the LHC Era

John Campbell

Theoretical Physics Department, Fermilab, Batavia, Illinois, USA

Joey Huston

Department of Physics and Astronomy, Michigan State University, East Lansing, Michigan, USA

Frank Krauss

Institute for Particle Physics Phenomenology, Durham University, Durham, UK

Great Clarendon Street, Oxford, OX2 6DP,
United Kingdom

Oxford University Press is a department of the University of Oxford.
It furthers the University's objective of excellence in research, scholarship,
and education by publishing worldwide. Oxford is a registered trade mark of
Oxford University Press in the UK and in certain other countries

First published 2018
First published in paperback 2022

Published in the United States of America by Oxford University Press
198 Madison Avenue, New York, NY 10016, United States of America

British Library Cataloguing in Publication Data

Data available

Library of Congress Cataloging in Publication Data

Data available

ISBN 978-0-19-965274-7 (Hbk.)
ISBN 978-0-19-287196-1 (Pbk.)

DOI: 10.1093/oso/9780192871961.001.0001

Printed and bound by
CPI Group (UK) Ltd, Croydon, CR0 4YY

To our families, with love.

To colleagues and friends,
who shaped our understanding of
particle physics, with gratitude.

Contents

Acknowledgements

We are greatly indebted to a large number of people, who inspired us to do particle physics, who did their best to teach us something, who collaborated with us, and who, by far and large, shaped our view of QCD at the LHC and other collider experiments: we truly are standing on the shoulders of giants.

We also owe a debt of gratitude to our friends and colleagues from CTEQ and MCnet who put up with our antics while further helping our understanding of science in countless night-time discussions during graduate schools and other meetings. We would like to thank Stefano Catani, Daniel de Florian, Keith Hamilton, Stefan Höche, Simon Plätzer, Stefan Prestel, and Marek Schönherr for patiently answering our many questions concerning some of the more specialized aspects in this book. You have been a huge help! Any inaccuracy or error is not a reflection of your explanations but of our limited understanding of them.

We are extremely grateful to Josh Isaacson, Pavel Nadolsky, Petar Petrov, and Alessandro Tricoli for carefully reading parts of the book while it was being written, pointing out conceptual shortcomings and misunderstandings from our side, the typical errors with factors of two, unfortunate formulations, and much more. However good, for all mistakes left in the book, the buck stops with us. A list of updates, clarifications, and corrections is maintained at the following website:

http://www.ippp.dur.ac.uk/BlackBook

We would like to thank the following for useful conversations and for providing figures for the book: Simon Badger, Johannes Bellm, Keith Ellis, Steve Ellis, Rick Field, Jun Gao, Nigel Glover, Stefan Höche, Silvan Kuttimalai, Gionata Luisoni, Matthew Mondragon, Ivan Pogrebnyak, Stefan Prestel, Marek Schönherr, Ciaran Williams, Jan Winter, and Un-ki Yang.

The genesis of this book was a review article co-written by two of us [320] and we thank both our co-author, James Stirling, and the IOP for their collaboration and support. Some parts of this book benefitted tremendously from us being allowed to test them on unsuspecting students in lectures during regular courses, at graduate schools, or summer institutes, and on our colleagues during talks at conferences and workshops. Thank you, for your patience with us and your feedback!

Many of the plots in this book have been created using the wonderful tools apfel-web [331], JaxoDraw [249], RIVET [291], MATPLOTLIB [638], xmGRACE, and XFIG. On that occasion we would also like to thank Andy Buckley, Holger Schulz and David Grellscheid for their continuous support and ingenious help with some of the finer issues with RIVET and LaTeX.

We are also grateful for the great support by our publisher, and in particular by the team taking care of this manuscript: Sonke Adlung, Harriet Konishi and Hemalatha Thirunavukkarasu.

Finally, of course we would like to thank our families for putting up with us while we assembled this manuscript. Surely, we were not always the most pretty to watch or the easiest people to have around.

1
Introduction

1.1 The physics of the LHC era

1.1.1 Particle physics in the LHC era

The turn-on of the LHC in 2008 culminated an almost 20-year design and construction effort, resulting in the largest particle accelerator (actually the largest machine) ever built. At its inception a competition still existed with the TEVATRON which, although operating at a much lower energy, had a data sample with a large integrated luminosity and well-understood detectors and physics-analysis software. The TEVATRON had discovered the top quark and was continuing its search for the Higgs boson. As is well known, the LHC suffered considerable damage from a cryogenic quench soon after turn-on that resulted in a shut-down for about 1.5 years. Its (re)turn-on in 2010 was at a much lower energy (7 TeV rather than 14 TeV) and at much lower intensities. The small data sample at the lower energy can be considered in retrospect as a blessing in disguise. There was not enough data to even consider a search for the Higgs boson (or even for much in the way of new physics), but there was enough data to produce W and Z bosons, top quarks, photons, leptons and jets — in other words, all of the particles of the Standard Model except for the Higgs boson. The result was the *re-discovery of the Standard Model* (a coinage for which one of the authors takes credit) and the development of the analysis tools and the detailed understanding of the detectors that allowed for the discovery of the Higgs boson on July 4, 2012, with data from 7 TeV in 2011 and 8 TeV in 2012. The LHC turned off again in early 2013 for repairs and upgrades (to avoid the type of catastrophic quench that occurred in 2008). The LHC detectors also used this two-year period for repairs and upgrades. The LHC ran again in 2015, at an energy much closer to design (13 TeV). The increased energy allowed for more detailed studies of the Higgs boson, but more importantly offered a much greater reach for the discovery of possible new physics. At the time of completion of this book, a great deal of physics has been measured at the operating energy of 13 TeV. Given the new results continually pouring out at this new energy, the decision was made to concentrate in this book on results from 7 and 8 TeV running. This is sufficient for the data comparisons needed to illustrate the theoretical machinery developed here.

The Black Book of Quantum Chromodynamics. John Campbell, Joey Huston and Frank Krauss, Oxford University Press. © John Campbell, Joey Huston, and Frank Krauss 2018, First published in paperback 2022. DOI: 10.1093/oso/9780192871961.003.0001

1.1.2 The quest for the Higgs boson — and beyond

1.1.2.1 Finding the Higgs boson

The LHC was designed as a discovery machine, with a design centre-of-mass energy a factor of seven larger than that of the TEVATRON. This higher collision energy opened up a wide phase space for searches for new physics, but there was one discovery that the LHC was *guaranteed* to make; that of the Higgs boson, or an equivalent mechanism for preventing WW scattering from violating unitarity at high masses.

The Higgs boson couples directly to quarks, leptons and to W and Z bosons, and indirectly (through loops) to photons and gluons. Thus the Higgs boson final states are just the building blocks of the SM with which we have much experience, both at the TEVATRON and the LHC. The ATLAS and CMS detectors were designed to find the Higgs boson and to measure its properties in detail.

The cross-section for production of a Higgs boson is not small. However, the final states for which the Higgs boson branching ratio is large (such as $b\bar{b}$) have backgrounds which are much larger from other more common processes. The final states with low backgrounds (such as $ZZ^* \rightarrow \ell^+\ell^-\ell^+\ell^-$) suffer from poor statistics, primarily due to the Z branching ratio to leptons. The Higgs$\rightarrow \gamma\gamma$ final state suffers from a small branching ratio and a large SM background. Thus one might not expect this final state to be promising for a Higgs boson search. However, due to the intrinsic narrow width of the Higgs boson, a diphoton signal can be observable if the experimental resolution of the detector is good enough that the signal stands out over the background.

The measurable final states of the Higgs boson decays were further subdivided into different topologies so that optimized cuts could be used to improve on the signal-to-background ratio for each topology (for example, in ATLAS the diphoton channel was divided into 12 topologies). The extracted signal was further weighted by the expectations of the SM Higgs boson in those topologies. In this sense, the Higgs boson that was discovered in 2012 was indeed the Standard Model Higgs boson. However, as will be discussed in Chapter 9, detailed studies have determined the properties of the new particle to be consistent with this assumption.

1.1.2.2 The triumph of the Gauge Principle

The discovery of the Higgs boson by the ATLAS and CMS collaboration, reported in July 2012 and published in [15, 368], is undoubtedly the crowning achievement of the LHC endeavour so far. It is hard to overestimate the importance of this discovery for the field of particle physics and beyond.

The Higgs boson is the only fundamental scalar particle ever found, which in itself makes it unique; all other scalars up to now were bound states, and the fundamental particles found so far have been all either spin-1/2 fermions or spin-1 vector bosons. This discovery is even more significant as it marks a triumph of the human mind: the Higgs boson is the predicted visible manifestation of the Brout–Englert–Higgs (BEH) mechanism [516, 601, 619–621, 675], which allows the generation of particle masses in a gauge-invariant way [580, 835, 888]. Ultimately, this discovery proves the paradigm of gauge invariance as the governing principle of the sub-nuclear world at the smallest distances and largest energies tested in a laboratory so far. With this discovery a 50-year-old prediction concerning the character of nature has been proven

The question now is not whether the Higgs boson exists but instead what are its properties? Is the Higgs boson perhaps a portal to some new phenomena, new particles, or even new dynamics? There are some hints from theory and cosmology that the discovery of the Higgs boson is not the final leg of the journey.

1.1.2.3 Beyond the Standard Model

By finding the last missing particle and thereby completing the most accurate and precise theory of nature at the sub-nuclear ever constructed, the paradigms by which it has been constructed have proved overwhelmingly successful. Despite this there are still fundamental questions left unanswered. These questions go beyond the realm of the SM, but they remain of utmost importance for an even deeper understanding of the world around us.

Observations of matter — Earth, other planets in the Solar System or beyond, other stars, or galaxies — suggest that the symmetry between matter and anti-matter is broken. This is a universe filled by matter and practically devoid of anti-matter. While naively there is no obvious reason why one should be preferred over the other, at some point in the history of the Universe — and presumably very early — this asymmetry had to emerge from what is believed to have been a symmetric initial state. In order for this to happen, a set of conditions, the famous **Sakharov conditions** [710, 834] had to be met. One of these intricate conditions is the violation of **CP**, which demands that the symmetry under the combined parity and charge–conjugation (CP) transformation must be broken. Experimentally, the existence of CP violation has been confirmed and is tightly related to the existence of at least three generations of matter fields in the SM. Due to the BEH mechanism, particles acquire masses, and their mass and electroweak interaction eigenstates are no longer aligned after EWSB. The existence of a complex phase in the CKM matrix, which parametrizes the interrelation between these two set of eigenstates, ultimately triggers CP violation in the quark sector. However, the amount of CP violation established is substantially smaller than necessary to explain how the universe evolved from an initial symmetric configuration to the matter-dominated configuration seen today [358].

Likewise, the existence of dark matter (DM) is now well established, first evidenced by the rotational curves of galaxies [831]. DM denotes matter which interacts only very weakly with normal matter (described by the SM) and therefore certainly does not interact through electromagnetism or the strong nuclear force. Despite numerous attempts it has not been directly detected. DM interacts through gravity and thereby has influenced the formation of large-scale structures in the Universe. Cosmological precision measurements by the WMAP and PLANCK collaborations [125, 623, 862] conclude that dark matter provides about 80% of the total matter content of the Universe. This in turn contributes about 25% of the overall energy balance, with the rest of the energy content of the Universe provided by what is known as dark energy (DE), which is even more mysterious than DM. The only thing known is that the interplay of DM and DE has been crucial in shaping the Universe as observed today and will continue to determine its future. One possible avenue in searches for DM particles at collider experiments is that they have no coupling to ordinary matter through gauge interactions but instead couple through the Higgs boson.

These examples indicate that the SM, as beautiful as it is, will definitely not provide the ultimate answer to the questions concerning the fundamental building blocks of the world around us and how they interact at the shortest distances. The SM will have to be extended by a theory encompassing at least enhanced CP violation, dark matter, and dark energy. Any such extension is already severely constrained by the overwhelming success of the gauge principle: the gauge sector of the SM has been scrutinized to incredibly high precision, passing every test up to now with flying colours. See for example [179] for a recent review, combining data from e^-e^+ and hadron collider experiments. The Higgs boson has been found only recently, and it is evident that this discovery and its implications will continue to shape our understanding of the micro–world around us. The discovery itself, and even more so the mass of the new particle and our first, imprecise measurements of its properties, already rule out or place severe constraints on many new physics models going beyond the well-established SM [515].

Right now, we are merely at the beginning of an extensive programme of precision tests in the Higgs sector of the SM or the theory that may reveal itself beyond it. It can be anticipated that at the end of the LHC era, either the SM will have prevailed completely, with new physics effects and their manifestation as new particles possibly beyond direct human reach, or alternatively, we will have forged a new, even more beautiful model of particle physics.

1.1.3   LHC: Accelerator and detectors

1.1.3.1   LHC, the machine

The LHC not only is the world's largest particle accelerator but it is also the world's largest machine, at 27 km in circumference. The LHC is a proton-proton collider (although it also operates with collisions of protons on nuclei, and nuclei on nuclei), located approximately 100 m underground and straddling the border between France and Switzerland. The LHC occupies the tunnel formerly used for the LEP accelerator in which electrons and positrons collided at centre-of-mass energies up to 209 GeV. The LHC contains 9593 magnets, including 1232 superconducting dipole magnets, capable of producing magnetic fields of the order of 8.3 T, and a maximum proton beam energy of 7 TeV (trillion electron-volts), leading to a maximum collision energy of 14 TeV. Thus far, the LHC has run at collision energies of 7 TeV (2010, 2011), 8 TeV (2012) and 13 TeV (2015,2016), greatly exceeding the previous record of the Fermilab TEVATRON of 1.96 TeV.[1] The large radius of the LHC is necessitated because of the desire to reach as high a beam energy as possible (7 TeV) using dipoles with the largest magnetic fields possible (in an accelerator). Running at full energy, the power consumption (including the experiments) is 750 GWh per year. At full power, the LHC will collide 2808 proton bunches, each approximately 30 cm long and 16 microns in diameter and containing 1.15×10^{11} protons, leading to a luminosity of $10^{34}\text{cm}^{-2}/\text{s}$ and a billion proton-proton collisions per second. The spacing between the bunches is 25 ns leading to collisions occurring every 25 ns; thus, at full **luminosity** there will

---

[1]Unlike the LHC, the TEVATRON was a proton-antiproton collider.

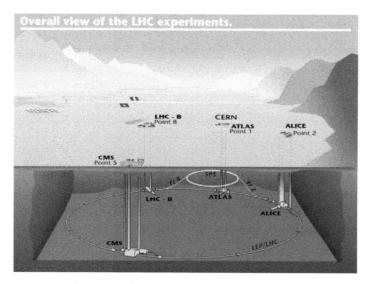

Fig. 1.1 A 3D layout of the LHC, showing the location of the four major experiments. Reprinted with permission from CERN.

be on average 25 interactions every beam crossing, most of which will be relatively uninteresting. The high luminosity for the machine is needed to produce events from processes with small cross-sections, for example involving physics at the TeV scale.

There are seven experiments running at the LHC (ATLAS, CMS, LHCB, ALICE, TOTEM, LHCf and MoEDAL), with ATLAS and CMS being the two general-purpose detectors. A schematic drawing of the LHC, indicating the position of the four larger experiments is shown in Fig. 1.1.

1.1.3.2   The detectors

It seems paradoxical that the largest devices are needed to probe the smallest distance scales. The ATLAS detector, for example, is 46 m long, 25 m in diameter and weighs 7000 tonnes. The CMS detector, although smaller than ATLAS at 15 m in diameter and 21.5 m in length, is twice as massive, at 14,000 tonnes. This can be compared to the CDF detector at the TEVATRON which was *only* 12m×12m×12m (and 5000 tonnes). The key to the size and complexity of the LHC detectors is the need to measure the four-vectors of the large number of particles present in LHC events, whose momenta can extend to the TeV range. The large particle multiplicity requires very fine segmentation; the ATLAS detector, for example, has 160 million channels to read out, half of which are in the pixel detector. The large energies/momenta require, in addition to fine segmentation, large magnetic fields and tracking volumes and thick calorimetry.

Both ATLAS and CMS are what are known as general-purpose 4π detectors, meaning that they attempt to cover as much of the solid angle around the collision point as possible, in order to reconstruct as much information about each event as possible.[2] There is a *universal* cylindrically symmetric configuration for a 4π detector, embodied, for example, in the ATLAS detector, as shown in Fig. 1.2. Collisions take place in the centre of the detector. Particles produced in each collision first encounter the pixel detector (6) and the silicon tracking detector (5). The first layer of the pixel detector is actually mounted on the beam-pipe in order to be as close to the interaction point as possible. The beam-pipe itself, in the interaction region, is composed of beryllium in order to present as little material as possible to the particles produced in the collision. The proximity of the pixel and silicon detectors to the collision point and the very fine segmentation (50×400 μm for the pixel detector and 70 μm for the silicon detector) allow for the reconstruction of secondary vertices from bottom and charm particles, which can travel distances of a few mm from the interaction point before decaying. The next tracking device (4), the transition radiation detector, is a straw-tube detector that provides information not only on the trajectory of the charged particle but also on the likelihood of the particle being an electron. All three tracking devices sit inside the central magnetic field of 2T produced by the solenoid (3).

The energies of the particles produced in the collision (both neutral and charged) are measured by the ATLAS calorimeters, the lead-liquid argon electromagnetic calorimeter (7) and the iron-scintillator hadronic calorimeter (Tilecal) (8). Both the ATLAS and CMS electromagnetic calorimeter designs emphasized good resolution for the measurement of the energies of photons and electrons, primarily to be able to distinguish the Higgs boson to $\gamma\gamma$ signal from the much larger diphoton background. The width of a light Higgs boson is much less than the experimental resolution, so any improvement in the resolution will lead to a better discrimination over the background.

Energetic muons can pass through the calorimetry, while other particles are absorbed. The toroidal magnets (2), in both the central and forward regions, produce an additional magnetic field (4 T) in which a second measurement of the muon momentum can be carried out using the muon tracking chambers (1), using several different technologies. One of the unique characteristics of the ATLAS detector (and part of its acronym) is the presence of the air-core toroidal muon system. The relatively small amount of material in the tracking volume leads to less multiple scattering and thus a more precise measurement of the muon's momentum. The muon momentum can be measured to a precision of 10% at a transverse momentum value of 1 TeV.

1.1.3.3 Challenges

To use a popular analogy, sampling the physics at the LHC is similar to trying to drink from a fire hose. Over 1 billion proton-proton collisions occur each second, but the limit of practical data storage is on the order of hundreds of events per second only. Thus, the experimental triggers have to provide a reduction capability of a factor of the order of 10^7, while still recording *bread-and-butter* signatures such as W and Z boson production. This requires a high level of sophistication for the on-detector hardware triggers and access to large computing resources for the higher-level triggering. Timing

---

[2]The main limitation for the solid-angle coverage is in the forward/backward directions, where the instrumentation is cut off by the presence of the beam pipe.

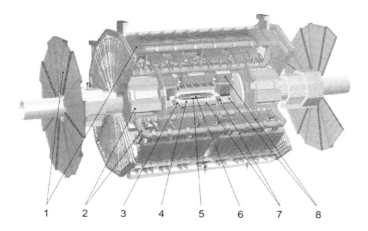

1   2   3   4   5   6   7   8

Fig. 1.2 A layout of the ATLAS detector, showing the major detector components, from `en.wikipedia.org/wiki/ATLAS_experiment`. Original image from CERN. Reprinted with permission from CERN.

is also an important issue. The ATLAS detector is 25m in diameter. With a bunch-crossing time of 25ns, this means that as new interactions are occurring in one bunch crossing, the particles from the previous bunch crossing are still passing through the detector. Each crossing produces 25 interactions. Experimental analyses thus face both in-time pileup and out-of-time pileup. The latter can be largely controlled through the readout electronics (modulo substantial variations in the population of the individual bunches), while the former requires sophisticated treatment in the physics analyses.

The dynamic ranges at the LHC are larger than at the TEVATRON. Leptons from W boson decays on the order of tens of GeV are still important, but so are multi-TeV leptons. Precise calibration and the maintenance of linearity are both crucial. To some extent, the TEVATRON has served as a boot camp, providing a learning experience for physics at the LHC, albeit at lower energies and intensities. Coming later, the LHC has benefited from advances in electronics, in computing, and perhaps most importantly, in physics analysis tools. The latter comprise both tools for theoretical predictions at higher orders in perturbative QCD and tools for the simulation of LHC final states.

Despite the difficulties, the LHC has had great success during its initial running, culminating in the discovery of the Higgs boson, but, alas, not in the discovery of new physics. The results obtained so far comprise a small fraction of the total data taking planned for the LHC. New physics may be found with this much larger data sample, but discovering it may require precise knowledge of SM physics, including QCD.

1.2 About this book

The reader is assumed to be already familiar with textbook methods for the calculation of simple Feynman diagrams at tree level, the evaluation of cross-sections through phase-space integration with analytic terms, and the ideas underlying the regularization and renormalization of ultraviolet divergent theories; however, for a short review, readers are referred to Appendix B.1, and for a more pedagogical introduction to these issues to a wealth of outstanding textbooks on various levels, including the books by Peskin and Schröder [803], Halzen and Martin [606], Ramond [822], Field [525] and others. For a review of QCD at collider experiments, the reader is referred to the excellent books by Ellis, Stirling, and Webber [504] and by Dissertori, Knowles, and Schmelling [467]. Of course, for a real understanding of various aspects it is hard to beat the original literature, and readers are encouraged to use the references in this book as a starting point for their journey through particle physics.

This book aims to provide an intuitive approach as to how to apply the framework of perturbative theory in the context of the strong interaction towards predictions at the LHC and ultimately towards an understanding of the signals and backgrounds at the LHC. Thus, even without the background discussed at the beginning of this section, this book should be useful for anyone wishing for a better understanding of QCD at the LHC.

The ideas for this book have been developed over various lecture series given at graduate level lectures or at advanced schools on high-energy physics by the authors. The authors hope that this book turns out to be useful in supporting the self-study of young researchers in particle physics at the beginning of their career as well as more advanced researchers as a resource for their actual research and as material for a graduate course on high-energy physics.

1.2.1 Contents

Chapter 2 provides a first overview of the content of this book and aims at putting various techniques and ideas into some coherent perspective. First of all, a physical picture underlying hadronic interactions, and especially scattering reactions at hadron colliders, is developed. To arrive at this picture, the ideas underlying the all-important factorization formalism are introduced which, in the end, allows the use of perturbative concepts in the discussion of the strong interaction at high energies and the calculation of cross-sections and other related observables. These concepts are then used in a specific example, namely the inclusive production of W bosons at hadron colliders. There, their production cross-section is calculated at leading and at next-to-leading order in the strong coupling constant, thereby reminding the reader of the ingredients of such calculations and fixing the notation and conventions used in this book. This part also includes a first discussion of observables relevant for the phenomenology of strong interactions at hadron colliders. In addition, some generic features and issues related to such fixed-order calculations are sketched. In a second part, the perturbative concepts already employed in the fixed-order calculations are extended to also include dominant terms to all orders through the resummation formalism. Generic features of analytical resummation are introduced there and some first practical applications for W production at hadron colliders are briefly discussed. As a somewhat alternative use of resummation techniques, jet production in electron–positron annihilations and in hadronic collisions is also discussed and, especially in the latter, some characteristic patterns are developed.

The next chapter, Chapter 3, is fairly technical, as it comprises a presentation of most of the sometimes fairly sophisticated technology that is being used in order to evaluate cross-section at leading and next-to leading order in the perturbative expansion of QCD. It also includes a brief discussion of emerging techniques for even higher order corrections in QCD. In addition, the interplay between QCD and electroweak corrections is touched upon in this chapter. Starting with a discussion of generic features, such as a meaningful definition of perturbative orders for various calculations, the corresponding technology is introduced, representing the current state of the art. As simple illustrative examples for the methods employed in such calculations, again inclusive W boson production and its production in association with a jet are employed. The calculations are worked out in some detail at both leading and next-to leading order in the perturbative expansion in the strong coupling.

The overall picture and phenomena encountered in hadron–hadron collisions, developed in Chapter 2, is discussed in the context of specific processes in Chapter 4. The processes discussed here range from the commonplace (e.g. jet production) to some of the most rare (e.g. production of Higgs bosons). In each case the underlying theoretical description of the process is described, typically at next-to leading order precision. Special emphasis is placed on highlighting phenomenologically relevant observables and issues that arise in the theoretical calculations. The chapter closes with a summary of what is achievable with current technology and an outlook of what may become important and relevant in the future lifetime of the LHC experiments.

Following the logic outlined in Chapter 2, in Chapter 5 the discussion of fixed-order technology is extended to the resummation of dominant terms, connected to large logarithms, to all orders. After reviewing in more detail standard analytic resummation techniques, and discussing their systematic improvement to greater precision by the inclusion of higher-order terms, the connection to other schemes is highlighted. In the second part of this chapter, numerical resummation as encoded in parton showers is discussed in some detail. The physical picture underlying their construction is introduced, some straightforward improvements by introducing some generic high–order terms are presented and different implementations are discussed. Since the parton showers are at the heart of modern event simulation, bridging the gap between fixed-order perturbation theory at high scales and phenomenological models for hadronization and the like at low scales, their improvement has been in the focus of actual research in the past decade. Therefore, some space is devoted to the discussion of how the simple parton shower picture is systematically augmented with fixed-order precision from the corresponding matrix elements in several schemes.

In Chapter 6, an important ingredient for the success of the factorization formalism underlying the perturbative results in the previous two chapters is discussed in more detail, namely the parton distribution functions. Having briefly introduced them, mostly at leading order, in Chapter 2, and presented some simple properties, in this chapter the focus shifts on their scaling behaviour at various orders and how this can be employed to extract them from experimental data. Various collaborations perform such fits with slightly different methodologies and slightly different biases in how data are selected and treated, leading to a variety of different resulting parton distributions. They are compared for some standard candles in this chapter as well, with a special emphasis on how the intrinsic uncertainties in experimental data and the more theoretical fitting procedure translates into systematic errors.

The tour of ingredients for a complete picture of hadronic interactions terminates in Chapter 7, where different non-perturbative aspects are discussed. Most of the ideas to address them are fairly qualitative and can be embedded in phenomenological models only. Therefore, rather than presenting in detail all developments in this field, the book focuses more on generic features and basic strategies underlying their treatment in different contexts. Issues discussed there include hadronization, the transition from the partons of perturbation theory of the strong interaction, quarks and gluons, to the experimentally observable hadrons, and their decays into stable ones, the underlying event, which is due to softer further interactions between the hadronic structures of the incident particles, and its connection to very inclusive observables such as total and elastic cross-sections.

In Chapters 8 and 9 theoretical results from analytic calculations and simulation tools are compared with a host of experimental data. Chapter 8 focuses on data especially from the TEVATRON,[3] where the foundations of our current understanding of the SM and in particular the dynamics of the strong interaction have been shaped. In Chapter 9 the most sophisticated calculations and simulations are compared with the most recent, most precise and most challenging data so far, taken at the LHC during Run I. This comparison ranges from inclusive particle production over event shape observables to data testing the dynamics of the SM — and potentially beyond — over scales ranging over two order of magnitude in the same process. This is the most challenging test of our understanding of nature at its most fundamental level ever performed. It is fair to state that while our most up-to-date tools, analytical calculations and simulations fare amazingly well in this comparison, some first cracks are showing that will motivate the community to push even further in the years to come.

1.2.2 A user's guide

This book is meant to provide PhD students in experimental particle physics working at the LHC who have a keen interest in theoretical issues, as well as PhD students working in particle theory with an emphasis on phenomenology at colliders, a starting point for their research. It is meant to introduce and expose the reader to all relevant concepts in current collider phenomenology, introduce and explain the technology that by now is routinely used in the perturbative treatment of the strong interaction, and provide an integrated perspective on the results of such calculations and simulations and the corresponding data.

The book consists of three parts. The first part is an overview of the relevant terminology and technology, worked out through one standard example and providing a coherent perspective on hadronic interactions at high energies. Readers and teachers, using this book for lectures, are invited to study Chapter 2 first before embarking on a more in-depth discussion of various theoretical or experimental aspects. The other two parts consist of a more detailed discussion of various aspects of the perturbative treatment of the strong interaction in hadronic reactions in the second part of the book, in Chapters 3–7. While these chapters frequently refer back to the overview chapter, Chapter 2, they are fairly independent from each other and could in principle be used in

---

[3]Experiences from LEP and HERA have also been important but are not included due to space limitations.

any sequence the reader or teacher finds most beneficial. The third part, Chapters 8 and 9, where core experimental findings are confronted with theoretical predictions, again is independent of the second part, although for a better understanding of theoretical subtleties it may be advantageous to be acquainted with certain aspects there.

Finally, a list of updates, clarifications and corrections to this book is maintained at the following website:

```
http://www.ippp.dur.ac.uk/BlackBook
```

2
Hard Scattering Formalism

Before embarking, in this chapter, on a first discussion of the factorization formula and some of its immediate consequences in terms of actual phenomena and calculations, in Section 2.1 an intuitive picture of high-energy reactions involving hadrons in the initial state will be developed. This picture in fact forms the physical background of the factorization formalism, which in turn provides the theoretical foundations of this book.

In the next section, Section 2.2, the ideas formulated in the previous section will be further formalized and condensed into a discussion of the perturbative treatment of high-energy reactions at hadron colliders at fixed order. To illustrate the factorization formalism in action, the case of inclusive W-boson production will be analysed at leading and at next-to-leading order.

In Section 2.3 the perturbative formalism developed so far will be further expanded to include the most important effects to all perturbative orders. This is achieved through the identification and subsequent resummation of the corresponding leading terms. Again, the case of W boson production serves as the main illustrative example.

Finally, in Section 2.4, some general thoughts and issues related to the description of hard processes through perturbation theory will be deepened.

2.1 Physical picture of hadronic interactions

2.1.1 Electromagnetic analogy

To illustrate the physical picture underlying factorization in processes initiated by colliding strongly interacting particles such as protons at high energies, the simpler case of collisions with leptons in the initial state is considered first. In this, an intuitive understanding of the theory of strong interactions, QCD, quantum chromodynamics, will be developed in close analogy with the simpler theory of QED, quantum electrodynamics. In both cases, although ultimately very different, the colliding particles will emit secondary quanta in an intricate radiation pattern, which must be taken into account to gain full theoretical control over the collision dynamics.

Experimentally, for example in collisions of electron–positron pairs, it is of course typically not difficult to require that the energy in the actual collision is close to the

The Black Book of Quantum Chromodynamics. John Campbell, Joey Huston and Frank Krauss, Oxford University Press. © John Campbell, Joey Huston, and Frank Krauss 2018, First published in paperback 2022. DOI: 10.1093/oso/9780192871961.001.0002

centre-of-mass energy of the colliding beams, thus effectively reducing the amount of energy carried away from the leptons through electromagnetic radiation. However, most of the time, especially when their combined initial invariant mass is above the mass of a resonance, such as the Z boson, the leptons will react with actual energies that are reduced with respect to their full available energy. This effect is sometimes called "**radiative return**" The corresponding energy loss is due to the emission of photons from the incident leptons, a process denoted as QED initial-state radiation (ISR). .

While in QED the treatment of ISR potentially is a tedious but essentially straightforward exercise, tractable with perturbation theory, in QCD the problem is much more involved and fundamentally different. This is because in QCD, the colliding particles cannot be interpreted as the fundamental quanta of the theory but rather as bound states, hadrons such as protons, which cannot be quantitatively understood and described through the language of perturbation theory. This conceptual gap necessitates the construction of a framework to provide direct and systematically improvable contact between the proven language of perturbative calculations of corresponding cross-sections and the non-perturbative structure of the hadronic bound states. This section is devoted to developing an intuitive picture of how this **factorization** framework actually works, by dwelling on the limited analogy with the emissions of secondary quanta in QED and the differences between electromagnetic and strong interactions when considering such collisions in greater detail.

2.1.1.1 Equivalent quanta

Classically, the phenomenon of initial-state radiation can be understood with the **equivalent photon picture** [523, 884, 892]. In its own rest frame, the lepton acts as the point-like source of a purely radial electric field, with no magnetic field present. This situation is depicted in the left panel of Fig. 2.1. Boosting the lepton into any frame where it is not at rest and, in particular, into the laboratory frame transforms the static source of an electric field into an electromagnetic current, which in turn produces also a magnetic field. With increasing lepton velocity, $v \to c$, the two fields become increasingly confined to a plane perpendicular to the axis of motion and they also become more and more perpendicular to each other, with the electrical field radial and the magnetic field circular around the axis of motion, *cf.* the middle panel of Fig. 2.1. This orthogonality allows the identification, in a straightforward way, of this classical configuration with quanta of the electromagnetic interaction, the photons, shown in the right panel of Fig. 2.1. The accumulated energy flux – that is, the total energy carried by the fields accompanying the lepton – is obtained by integrating the Poynting vector over the plane orthogonal to the lepton's axis of motion. It is thus parallel to the lepton's motion. Interpreting the continuous energy flux with a flux of equivalent quanta, the photons, the number density of accompanying photons per energy interval and distance from the lepton can be deduced as

$$\mathrm{d}n_\gamma = e_e^2 \, \frac{\alpha}{\pi} \cdot \frac{\mathrm{d}\omega}{\omega} \cdot \frac{\mathrm{d}b_\perp^2}{b_\perp^2} \,. \tag{2.1}$$

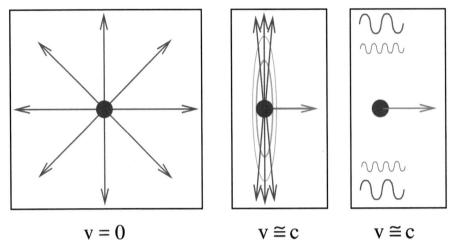

$$v = 0 \qquad\qquad v \cong c \qquad\qquad v \cong c$$

Fig. 2.1 Electrical and magnetic fields in blue and green of a lepton at rest ($v = 0$, left panel) and with a velocity $v \approx c$ (middle panel). The equivalent photons are depicted on the right panel.

Here ω is the energy of the equivalent photon, and the constant of proportionality is obtained by integrating over the transverse plane, parameterized by the impact factor $b_\perp$, and by the electromagnetic coupling constant. The latter gets modified by the relative charge of the electron, e_e, which of course equals -1. There is a maximal energy available for these equivalent photons; naively it is given by $\omega_{\mathrm{max}} = E$, the energy of the lepton.

In a more quantum field-theoretical way of thinking about this, the impact parameter is replaced with the transverse momentum, $b_\perp \longleftrightarrow k_\perp$, through a Fourier transform, and the equivalent photons are considered to be part of the lepton's wave function. Their spectrum then reads

$$\mathrm{d}n_\gamma = \frac{\alpha}{\pi} \cdot \frac{\mathrm{d}\omega}{\omega} \cdot \frac{\mathrm{d}k_\perp^2}{k_\perp^2} . \tag{2.2}$$

In such a picture, the physical lepton is given by a superposition of states with varying photon multiplicity,

$$|e\rangle_{\mathrm{phys}} = |e\rangle + |e\gamma\rangle + |e\gamma\gamma\rangle + \dots , \tag{2.3}$$

where the photons have different energies and transverse momenta. Due to momentum conservation, they are typically off their mass shell. This limits the lifetime of the quantum-fluctuations like $|e\gamma\rangle$ or $|e\gamma\gamma\rangle$.

To see how this works in somewhat more detail, consider the case of the $|e\gamma\rangle$ Fock state. Assuming massless on-shell electrons in the initial and final state, but allowing the photons to go off-shell, the kinematics of a splitting $e(P) \to e(p) + \gamma(k)$ can be written as

$$P = p + k \quad \longrightarrow \quad 0 = 2p \cdot k + k^2 . \tag{2.4}$$

Parameterizing the splitting of the energy, E of P such that $\omega \equiv E_\gamma = xE$ and therefore $E_e = (1-x)E$ for the energy of the outgoing electron, leads to

$$k^2 \approx -k_\perp^2 = \vec{k}_\perp^2 \tag{2.5}$$

in the case where x is small, *i.e.*, for the bulk of the photons. This also implies that in this limit the momentum component of the photon parallel to the electron is about $k_\parallel \approx \omega$, the energy of the photon. The fact that the emitted photons (or the electrons or both) are off-shell is equally true for massive electrons, and it limits the lifetime of the $(e\gamma)$-component of the wave function through the uncertainty principle. With the photon momentum given by $k^\mu \approx (\omega, \vec{k}_\perp, \omega)$, the energy shift necessary to move it on its mass shell is given by $\delta\omega \approx k_\perp^2/2\omega$ yielding

$$\tau_\gamma \sim \frac{1}{\delta\omega} \approx \frac{2\omega}{k_\perp^2} = \frac{2xE}{k_\perp^2} \tag{2.6}$$

as the **lifetime of the fluctuation**. This implies that such fluctuations live longer, as the energy of the photon increases and the relative transverse momentum of the photon and the electron decrease. Note that in this book natural units are being used, so effectively $\hbar = c = 1$.

In addition, of course, the photons can split into a fermion–anti-fermion pair, as yet another quantum fluctuation, which in turn may emit further photons. This complicates the wave-function picture even more. However, altogether, the distance from the original lepton and the lifetime of all these virtual particles, the quantum fluctuations which they manifest, are given by the amount they are off their mass shell and by the amount of energy they carry.

2.1.1.2 Dokshitser–Gribov–Lipatov–Altarelli–Parisi equations

The wave-function idea can be related, at leading order, by probabilities of finding the original lepton, the photons, or secondary leptons at given energies and transverse momenta. Due to the nature of the electromagnetic interaction, these probabilities (or, the lepton wave function), can be calculated from first principles. In the following the probability of finding a lepton or photon at energy fraction x (with respect to the original lepton) and at transverse momentum $k_\perp$ will be denoted as $\ell(x, k_\perp)$ and $\gamma(x, k_\perp)$, respectively. At **leading** order in α, that is without any emission of secondary quanta, they are of course given by

$$\ell(x, k_\perp^2 = 0) = \delta(1-x) \quad \text{and} \quad \gamma(x, k_\perp^2 = 0) = 0. \tag{2.7}$$

The radiation that actually gives rise to the aforementioned secondary particles can now be described, to all orders, by equations that are known as **evolution equations**, since they relate the probability to find quanta with certain kinematics x and $k_\perp^2$ to similar probabilities at other scales, through emission of secondaries. In general, they can be obtained in different approximations related to different kinematical situations, which can be intimately related to different **factorization schemes**. In the **collinear factorization scheme**, which will be employed in most of the remainder of this book, the energy and especially transverse momentum of the secondary particles are considered to be small with respect to the energy of the original lepton, and therefore the kinematical effect merely amounts to a successive reduction of the original lepton energy. In this scheme, the probability densities evolve with the logarithm of the transverse momentum, *cf.* Eq. (2.2), as

$$
\frac{\mathrm{d}\ell(x, k_\perp^2)}{\mathrm{d}\log k_\perp^2} = \frac{\alpha(k_\perp^2)}{2\pi} \int_x^1 \frac{\mathrm{d}\xi}{\xi} \mathcal{P}_{\ell\ell}\left(\frac{x}{\xi}, \alpha(k_\perp^2)\right) \ell(\xi, k_\perp^2)
$$

$$
\frac{\mathrm{d}\gamma(x, k_\perp^2)}{\mathrm{d}\log k_\perp^2} = \frac{\alpha(k_\perp^2)}{2\pi} \int_x^1 \frac{\mathrm{d}\xi}{\xi} \mathcal{P}_{\gamma\ell}\left(\frac{x}{\xi}, \alpha(k_\perp^2)\right) \ell(\xi, k_\perp^2),
$$

(2.8)

where the effect of photons splitting into virtual lepton–anti-lepton pairs has been omitted. The first line of these equations exhibit how the probability density of leptons with smaller energy fraction x is driven by leptons with a larger energy fraction $x/\xi > x$, which can be interpreted as the leptons on the right hand side of the equation losing an energy fraction $1 - \xi$ in the emission of a photon. In the kinematical approximation employed here, terms like $\mathcal{P}_{\ell\ell}(x/\xi, \alpha)$, the splitting kernels, can be understood as reduced matrix elements for the emission of one photon off the lepton. In general, $\mathcal{P}_{ba}$ denotes such kernels for a transition of a particle of type a to a particle of type b, while emitting a particle of type c, which is not been made explicit. These kernels have been taken at leading order, as manifest by the explicit order in α in front, but of course they can also be evaluated to higher orders in a perturbative expansion in powers of the coupling with corresponding coefficient functions.

Closer inspection reveals that this set of equations is nothing but the celebrated **Dokshitser–Gribov–Lipatov–Altarelli–Parisi (DGLAP) equation**, specified for QED. In this equation, the **splitting kernels** at leading order are independent of the coupling constant and read

$$
\mathcal{P}_{\ell\ell}(z) = e_q^2 \left[\frac{1+z^2}{(1-z)_+} + \frac{3}{2}\delta(1-z) \right]
$$

$$
\mathcal{P}_{\gamma\ell}(z) = e_q^2 \left[\frac{1 + (1-z)^2}{z} \right].
$$

(2.9)

Here, the notation of "+"–functions has been employed, which will crop up again at various places, especially in conjunction with splitting kernels such as the ones discussed here. They are defined through their integral together with a test function $g(z)$ such that

$$
\int_0^1 \mathrm{d}z [f(z)]_+ g(z) = \int_0^1 \mathrm{d}z f(z) \left[g(z) - g(1) \right].
$$

(2.10)

For further details, the reader is refered to Appendix A.1.2. To gain a more intuitive understanding, consider this prescription to work in such a way that the pole for $z \to 1$ in $\mathcal{P}_{\ell\ell}$ is excluded in the +-function and reinserted through the second part of the splitting kernel, proportional to the δ-function. It will be seen later, how such terms become necessary to ensure the correct physical behaviour of the splitting functions, such as satisfying momentum conservation.

However, these equations will be revisited and discussed in more detail in later chapters of the book. To follow the reasoning in this more introductory chapter it should suffice to mention that the $1/\omega$ pole present in the naive classical picture has its counterpart in the $1/(1-z)$ or $1/z$ terms appearing in the splitting function.

2.1.1.3 Initial-state radiation

Turning away from the details of how the lepton wave function is evaluated and back to the physical consequences of the existence of the fluctuations, the obvious question is: what happens in a collision? Can the emerging picture be put to stringent tests and can the existence of such fluctuations be quantified by suitable probes?

To answer this, at least qualitatively, and to gain some insight into the physical processes taking place in a collision, consider first the case of no collision at all. There, the fluctuations have a finite lifetime and distance to the original lepton, related to the kinematics of the virtual particles. Four-momentum conservation thus forces these fluctuations to eventually collapse back into the original lepton, guaranteeing the quantum coherence of the lepton – it must remain intact. If, on the other hand, a collision takes place, one or more of the quantum excitations, the particles forming the lepton's Fock state, may, through the exchange of four-momentum with the incident projectile, go on their mass shell. In this way they acquire an infinite lifetime. In such a situation, the corresponding particles will not fall back onto the original lepton and the coherence of the fluctuations is therefore broken. In this way, one or more of the hitherto virtual particles may become real and hence physically observable in the final state of the collision.

To illustrate this, consider the case of e^+e^- collisions, which have been investigated in great detail for example at LEP, and assume that it is the electron and positron that take part in the collision. Then the centre-of-mass energy $\hat{E}_{\text{c.m.}}$ of this actual reaction may be reduced with respect to the energy obtained from the nominal beam energies, $\hat{E}_{\text{c.m.}} < E_{\text{c.m.}} = 2E_{\text{beam}}$.[1] The difference is stored in the energies of the photons accompanying the electron–positron pair; their distribution in transverse momenta and energies is given approximately by Eq. (2.2). In more modern language these photons would be attributed to **initial-state radiation** off the incident "original" pair, rather than to the quantum manifestation of the breakdown of coherence of the Fock states describing the leptons and their accompanying electromagnetic fields. Nevertheless, this radiation would, of course, be described by equations very similar to the approximate one.

It should be stressed, however, that, taken on its own, this initial-state radiation is a manifestation of the breakdown of quantum coherence of complicated Fock states, binding the photons to the original leptons.

---

[1]Here, and throughout the book, kinematical quantities q related to the colliding partons are supplemented with a $\hat{q}$, while those related to the (beam-)particles are left without it as q.

2.1.1.4 Final-state radiation

A similar picture also emerges when charged particles are produced in the final state. As an example, consider the case of, say, muon pair production in electron-positron annihilations, $e^+e^- \to \mu^+\mu^-$. Classically, the production of the muons can be understood as their instantaneous acceleration after emerging from a finite energy density related to the previous annihilation of the electron and positron into electromagnetic fields, the intermediate photon in quantum field theory. In classical electrodynamics such an acceleration of charged particles triggers the radiation of additional photons off the charges, known as **Bremsstrahlung**.[2] Interpreting the muon pair as a current going from a velocity $\vec{v}'$ to $\vec{v}$ at the origin of the coordinate system, the double differential classical radiation spectrum I in the direction $\vec{n}(\Omega)$ with polarization $\vec{\epsilon}$ reads

$$\frac{\mathrm{d}^2 I}{\mathrm{d}\omega\mathrm{d}\Omega} = \frac{e^2}{4\pi^2}\left|\vec{\epsilon}^{\,*}\cdot\left(\frac{\vec{v}}{1-\vec{v}\cdot\vec{n}} - \frac{\vec{v}'}{1-\vec{v}'\cdot\vec{n}}\right)\right|^2 , \tag{2.11}$$

in the dominant region of small energies, where the squared term is known as the **radiation function** W [645]. For massless particles travelling at the speed of light, the radiation function can be rewritten as

$$\mathrm{W}_{\beta',\beta} = \frac{1 - \cos\theta_{vv'}}{(1 - \cos\theta_{nv})(1 - \cos\theta_{nv'})}. \tag{2.12}$$

Cast in its covariant form and interpreted as a photon spectrum and tacitly inserting $\alpha = e^2/(4\pi)$, the radiation spectrum becomes

$$\mathrm{d}N = \frac{\alpha}{\pi}\left|\epsilon_\mu^*\left(\frac{p^\mu}{p\cdot k} - \frac{p'^\mu}{p'\cdot k}\right)\right|^2 \frac{\mathrm{d}^3 k}{(2\pi)^3 2k_0} \tag{2.13}$$

for the number of photons N emitted in the process. Here, k and ϵ denote the photon's four-momentum and polarization four-vector, while p and p' denote the four-momenta of the muons. The form above,

$$\mathcal{W}(p, p'; k, \epsilon) = \epsilon_\mu^*\left(\frac{p^\mu}{p\cdot k} - \frac{p'^\mu}{p'\cdot k}\right) \tag{2.14}$$

is also known as the **eikonal** form, and it is identical with the result of a full-fledged calculation in quantum field theory, as follows.

For the case at hand, consider the photon emission part off the muons, which are assumed to be massless. The muons are produced through a vertex factor called Γ. The leading-order Feynman diagrams are depicted in Fig. 2.2. The corresponding matrix element for $X \to \mu^-(p)\mu^+(-p')\gamma(k)$ is given by

$$\mathcal{M}_{X\to\mu^+\mu^-\gamma} = e\bar{u}_{\mu^-}(p)\left[\gamma^\mu \frac{\not{p}+\not{k}}{(p+k)^2}\Gamma - \Gamma\frac{\not{p}'-\not{k}}{(p'-k)^2}\gamma^\mu\right]u_{\mu^+}(p')\epsilon_\mu^*(k)$$

---

[2]In the quantum picture, equivalently, this translates into the radiation of Bremsstrahlung photons.

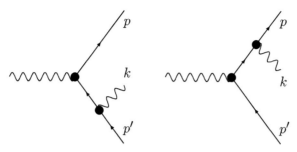

Fig. 2.2 Feynman diagrams for the emission of one photon by a muon pair at the lowest perturbative order.

$$= e\bar{u}_{\mu^-}(p)\left[\frac{2p^\mu + k^\mu - \frac{1}{2}[\gamma^\mu, \not{k}]}{2p\cdot k}\Gamma - \Gamma\frac{2p'^\mu - k^\mu + \frac{1}{2}[\gamma^\mu, \not{k}]}{2p'\cdot k}\right]u_{\mu^+}(p')\epsilon_\mu^*(k)\,,$$

(2.15)

where the equations of motion

$$\not{p}'u(p') = \bar{u}(p)\not{p} = 0 \quad \text{and} \quad p^2 = p'^2 = k^2 = 0 \tag{2.16}$$

for massless particles as well as the anti-commutator relation $\{\gamma^\mu, \gamma^\nu\} = 2g^{\mu\nu}$ have been used. Note that the only effect of the muon spin is the occurrence of emissions through magnetic terms, $\frac{1}{2}[\gamma^\mu, \not{k}]$, while the terms $2p^\mu - k^\mu$ do not bear any memory of the muons having spin. Going to the limit of soft photon emission, terms $\propto k$ in the numerator vanish, resulting in

$$\mathcal{M}_{X\to\mu^+\mu^-\gamma} = e\epsilon_\mu^*(k)\left[\frac{p^\mu}{p\cdot k} - \frac{p'^\mu}{p'\cdot k}\right]\bar{u}_{\mu^-}(p')\Gamma u_{\mu^+}(p)$$

$$= e\mathcal{W}(p, p'; k, \epsilon)\,\mathcal{M}_{X\to\mu^+\mu^-}\,. \tag{2.17}$$

In other words, in the limit of soft photon emission, the emission term completely factorizes from the production of the system emitting the photon and is just given by the **eikonal** term $\mathcal{W}(p, p'; k, \epsilon)$.

It is worth noting that soft photon emission is independent from the spin of the emitting particles and other internal properties and by its very nature it is a classical phenomenon. This is the essence of **Low's theorem** [734]. As a consequence, soft photons can be thought of as essentially not carrying any quantum numbers, and, therefore, the emission of soft photons will not lift any veto – for example, due to C-parity or angular momentum – for a transition to happen. Such transitions, in the case at hand, for instance, from one helicity to another, necessitate the emission of a **hard** photon, described by the terms $\propto k$. In contrast to the soft, classical terms, these essentially quantum-mechanical contributions do not exhibit any soft divergence $\mathrm{d}\omega/\omega$ but rather behave like $\mathrm{d}\omega \cdot \omega$.

The simple classical picture of soft photon emission in principle allows a description of the pattern of photon radiation from the muon pair, by iterating emissions through the eikonal term. This implicitly assumes that the individual emissions are independent from one other. The key point here is the introduction of the notion of photon resolution. The reason for this is the observation that the eikonal factor diverges for $k \to 0$, the soft divergence, or for k parallel to p or p', the collinear divergence. This is nothing but the well-known infrared-catastrophe of QED, a pattern that occurs for every theory with massless spin-1 bosons such as QED or QCD. Essentially it can be explained by the fact that it makes no physical sense to ask how many photons are emitted by a particle, without specifying how the photons are measured. In practice, photons can be too soft for a detector to respond; at the same time, if they are too parallel with the emitting particle they will end up in the same detector cell, which must have a finite size, and thus will not be distinguished. Therefore, the phase space of the photons must be constrained to detectable photons to make any sense. This is achieved by appropriate cuts, for instance by demanding a minimal energy and a minimal angle with respect to the emitting particle.

There are now in principle two ways of how the full photon radiation pattern can be described: in a more direct approach, the integral of the eikonal factor over the constrained photon phase space could be used as the relevant term in a Poissonian distribution of the number of photons being emitted, the independence of the individual emissions guaranteeing the Poissonian character of this distribution. For each photon then, the phase space could be individually fixed. Alternatively, the emissions could be ordered in, say, the energy of the emitted photons, or, as a somewhat preferred choice, the relative transverse momentum with respect to the emitter. This would allow a redefinition of the radiation pattern through a probabilistic picture, driven by the **Sudakov form factor**. In order to see how this works in more detail, *cf.* Section 2.3 and Chapter 5.

In any case, it is worth noting that the eikonal picture here, condensed in Eq. (2.17) can be directly translated to the equivalent quanta picture above and the slightly more sophisticated version encoded in the DGLAP evolution equations, Eq. (2.8).

2.1.2 Bound states, strong interactions and α_s

This simple, qualitative picture will now be extended to the case of incident hadrons - for simplicity they will be assumed to be (anti-)protons. Naively, they consist of three **valence quarks**, which carry the quantum numbers of the proton — electrical charge, spin, and isospin. In a naive picture, where the quarks just stick together with no discernible interactions responsible for it, it would be a fair guess to assume that they are all equally important and thus all carry the same energy fraction of the proton, namely $1/3$.

Switching on naive interactions, acting like rubber bands gluing the valence quarks together, one could assume that the original sharp distribution is washed out - but that the energy fraction distribution of the valence quarks still has a mean value of $1/3$. Of course, in view of the previous discussion of the QED case, and because the QCD coupling constant is much larger than the QED one, this simplistic picture cannot hold true. In fact, taking into account quantum corrections to the strong coupling constant, α_s, which are mediated by loop diagrams depicted in Fig. 2.3, the picture is a bit more complicated.

Fig. 2.3 The leading quantum corrections to the running of the QCD coupling constant α_s: gluon self energy at one loop order (ghost diagrams are ignored).

2.1.2.1 The running of α_s

Including such corrections, and denoting all couplings — QED and QCD — collectively with $\alpha = g^2/(4\pi)$ they vary with the scale μ_R, also known as **(renormalization) scale**, as

$$\mu_R^2 \frac{\partial \alpha(\mu_R^2)}{\partial \mu_R^2} = \beta(\alpha)\,, \tag{2.18}$$

where the β–**function** $\beta(\alpha)$ can be expanded in a perturbative series and reads

$$-\beta(\alpha) = \sum_{n=0}^{\infty} b_n\, \alpha^{2+n} = \frac{\beta_0}{4\pi}\,\alpha_s^2 + \frac{\beta_1}{(4\pi)^2}\alpha_s^3 + \dots\,. \tag{2.19}$$

Here the customary relation $\alpha_s = g_S^2/(4\pi)$, and, similarly, $\alpha = e^2/(4\pi)$ has been assumed.

In $SU(N)$ gauge theory, the first coefficients β_i of the perturbative expansion of the β-function read

$$
\begin{aligned}
\beta_0 &= \frac{11}{3}\,C_A - \frac{4}{3}T_R n_f \\
\beta_1 &= \frac{34}{3}\,C_A^2 - \frac{20}{3}\,C_A T_R n_f - 4\,C_F T_R n_f \\
\beta_2 &= \frac{2857}{54}\,C_A^3 + 2\,C_F^2 T_R n_f - \frac{205}{9}\,C_F C_A T_R n_f - \frac{1415}{27}\,C_A^2 T_R n_f \\
&\quad + \frac{44}{9}\,C_F T_R^2 n_f^2 + \frac{158}{27}\,C_A T_R^2 n_f^2\,,
\end{aligned}
\tag{2.20}
$$

where the number of active fermions is given by n_f, and where the **Casimir operators** of the gauge group in its **fundamental** and **adjoint** representations are C_F and C_A. It is worth stressing here that the first two coefficients, β_0 and β_1 are renormalization scheme–independent, while all further contributions, starting with β_2, depend on the **renormalization scheme**; the result given here for β_2 is the one in the $\overline{\text{MS}}$ scheme. For a very brief review, *cf.* Appendix B.1.

In QCD, they are given by

$$C_F \equiv C_q = \frac{N_c^2 - 1}{2N_c} \quad \text{and} \quad C_A \equiv C_g = N_c \tag{2.21}$$

and represent the **colour charges** of the particles. In addition,

$$T_R = \frac{1}{2} \tag{2.22}$$

has been used.

To first order accuracy, and with the help of a reference scale Q, the solution of Eq. (2.18) can be written as

$$\alpha(\mu_R^2) \equiv \frac{g^2(\mu_R^2)}{4\pi} = \frac{\alpha(Q^2)}{1 + \alpha(Q^2)\frac{\beta_0}{4\pi}\log\frac{\mu_R^2}{Q^2}} . \tag{2.23}$$

For the case of QCD, it has become customary to write the solution as

$$\alpha_{\rm s}(\mu_R^2) \equiv \frac{g_s^2(\mu_R^2)}{4\pi} = \frac{1}{\frac{\beta_0}{4\pi}\log\frac{\mu_R^2}{\Lambda_{\rm QCD}^2}} . \tag{2.24}$$

$\Lambda_{\rm QCD} \approx 250\,{\rm MeV}$ has been introduced as the **QCD scale**. Conversely, $\alpha_{\rm s}$ could be fixed experimentally, for instance through $\alpha_{\rm s}(m_Z^2) \approx 0.118$ [799], taken from the PDG.

The difference with respect to QED, at this level, can be traced back to the difference in the β-function, which for QED to first order reads

$$\beta_0 = -\frac{2n_f}{3} . \tag{2.25}$$

This difference in the overall sign, stemming from the gluon loops which of course are absent in the QED case, implies that, in striking contrast to the electromagnetic coupling, the strong coupling increases with increasing distance or decreasing scale and decreases with decreasing distance or increasing scale, a property known as **asymptotic freedom**. In physical terms this phenomenon can be interpreted as the response of the vacuum to the presence of an electromagnetic or strong charge. In quantum field theory, the vacuum is not really an empty space, but it consists of a superposition of short-lived quantum fluctuations with total quantum numbers 0. In the case of QED, such fluctuations consist, to leading perturbative order, of fermion–anti-fermion pairs. In the presence of a fixed charged particle, say an electron, these pairs will not be completely unpolarized. Instead they will orient themselves in such a way that the positively charged fermions are closer to the electron, and the negatively charged virtual particles are further away. Trying to measure the charge of the electron by probing it with a photon thus introduces a scale dependence: a low-Q^2 photon will only be able to "resolve" a comparably large cloud of charged virtual particles, partially screening the charge of the electron. This can be seen as a close analogy to a charge in a dielectric material in classical electrodynamics, where the internal polarization fields partially counteract the electric field related to the charge. Increasing the virtual mass Q^2 of the photon will allow it to resolve smaller distances and thus probe the electron with diminished influence of the screening cloud. The same mechanism, of course, also accounts for the case of QCD. Taking here a quark as a fixed charge, by far and large, the virtual quark–anti-quark pairs will orient themselves in such a way that the anti-quarks are closer to the quark. Similar to the case of QED, this effectively will lead to a screening of the fixed charge inside the cloud of short-lived virtual charged particles. However, in contrast to QED, also the carriers of the interaction, the gluons, carry colour charges and thus contribute to screening effects. Their contribution comes with a sign opposite to the fermionic contribution, in classical electrodynamics such a be-haviour would be classified with the pathological case of a dielectric constant smaller than unity.

The effect of this **running coupling of the strong interaction** has been con-firmed experimentally, as shown in the pictorial summary of theory and experimental data in Fig. 2.4. For small scales, *i.e.* for $\mu_R \rightarrow \Lambda_{\mathrm{QCD}}$, the running coupling diverges,

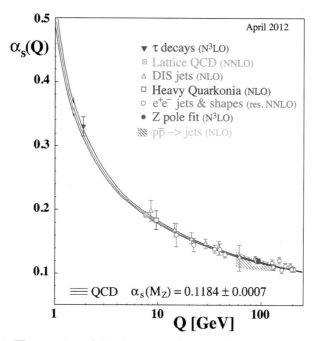

Fig. 2.4 The running of the strong coupling constant: experimental data confront theory, with the theoretical uncertainty shown as the blue band (Reprinted with permission from PDG [229]).

signalling the breakdown of perturbation theory and any ideas of treating quarks as quasi-free particles. This is also known as the **Landau pole** of QCD.

2.1.2.2 Equivalent quanta (QCD)

Applied to the situation encountered in hadrons this means that there is a proliferation of particles, gluons emitted by the quarks and in turn splitting into gluon or quark–anti-quark pairs, at small transverse momenta, *i.e.*, scales of the order of a few times

Λ_{QCD}^2. This radiation of secondary particles is even stronger than in electrodynamics: the $1/k_\perp^2$-term present in both QED and QCD, is further enhanced by multiplying it with the strongly increasing $\alpha_{\mathrm{s}}(k_\perp^2)$ rather than with $\alpha_{\mathrm{QED}}(k_\perp^2)$. In addition, since the gluons carry a colour charge, they also emit secondary gluons, such that the simple picture of **QED Bremsstrahlung** off an electron as introduced in Eq. (2.2) adapted for QCD becomes

$$\mathrm{d}n_g^{q,g} = C_{q,g} \cdot \frac{\alpha_{\mathrm{s}}(k_\perp^2)}{\pi} \cdot \frac{\mathrm{d}\omega}{\omega} \cdot \frac{\mathrm{d}k_\perp^2}{k_\perp^2}, \tag{2.26}$$

where the **colour factors** $C_{q,g} = C_{F,A}$ again are the Casimir operators of the fundamental (C_F) and adjoint (C_A) representation of the gauge group underlying QCD, $SU(3)$. From the relevant equation, Eq. (2.21), it can be seen that in the **large-N_c limit**, $N_c \to \infty$, the colour charge of the quark becomes $N_c/2$, half as large as the gluon colour charge, reflecting the fact that gluons carry two colour indices, while quarks carry only one. They are therefore, in this limit, twice as likely to emit a gluon than the quarks. Note here that it is quite often useful to consider the large-N_c limit, as corrections to this limit are typically suppressed by $1/N_c^2$. Parametrically this is a correction of the order of 10%, but often, and in particular for sufficiently inclusive quantities, the effects are significantly smaller and can thus be neglected.

It should be stressed, though, that due to the Landau pole the perturbative language must fail at transverse momenta around or below the Landau pole, with the result that the bound-state structure of the hadrons is more complicated than three quarks somehow bound together. Coming back to the qualitative picture, where hard interactions with constituents of the complex Fock state describing an incident fermion break its coherence, it is clear that something similar will also occur in the case of QCD. There are, however, a number of differences to this simple picture. First of all, the incident particle will be a hadron, a bound state in itself, with a difficult structure that cannot be calculated from first principles. Ignoring this complication, assume that at scales sufficiently high above the Landau pole, where a perturbative description becomes sensible, the hadron is made of a number of valence partons. Apart from some colour exchange binding them together, these objects will start emitting softer quanta, in analogy to the Fock state picture in QED, which are also known as sea partons. The coherence of the hadrons enforces the recombination of these secondary quanta, and a picture like the one depicted in the left panel of Fig. 2.5 emerges. In the right panel of the same figure the effect of a hard probe, a photon interacting with the partonic ensemble is shown: one of the partons — the one the photon interacted with — is "kicked" out of the hadron. As a consequence the colour field is missing a quantum number, and the remaining partons cannot recombine anymore. The coherence of the hadron breaks down and a more complex final state emerges through a reconfiguration of the coloured partons.

2.1.3   Hadrons in the initial state: The full picture

Starting from a naive constituent picture (three quarks for a nucleon, like *uud* for a proton, *udd* for a neutron, or a quark–anti-quark pair for a meson, for example, $u\bar{d}/\bar{u}d$ for $\pi^\pm$), multiple emissions of secondary **sea** partons off these primary **valence**

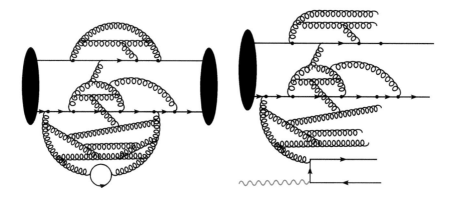

Fig. 2.5 Hard interactions break coherence of QCD initial states, In the left panel, coherence of the stable hadrons enforces the recombination of emitted partons, in the right panel the interaction with a hard photon destroys this coherence by "kicking" out the parton it couples to.

partons will occur, again governed approximately by Eq. (2.26). Similar to the case of QED, also in QCD, the lifetime of the sea partons is proportional to $\omega/k_\perp^2$ and their distance from the emitting valence quark is given by $1/k_\perp$. The coherence of the quantum states, encoded as negative virtual mass, also guarantees, like in QED, that no partons can evade their hadron. In a hard collision, again, this coherence may break down if a parton is "kicked out" of the radiation pattern, and the partons that stay behind may go on-shell, thus becoming physically meaningful objects with observable consequences. Of course, also in QCD, the more modern way to look at this employs the notion of initial-state radiation, described by evolution equations such as the simplified ones encountered already in the naive discussion of QED, see Eq. (2.8).

2.1.3.1 Deep-inelastic scattering

In order to discuss this in more detail, consider the case of **deep-inelastic scattering** (**DIS**), where an electron interacts with a proton by exchanging a highly virtual photon. The process is called "deep" for the absolute value of the virtual mass of the photon, Q, being much larger than the proton mass m_P or its inverse radius, R_p, which is of about the same size: $Q \gg m_P \approx 1/R_p$, and therefore probing its deep internal structure. It is "inelastic" if the electron loses enough energy for the proton to break up. To further analyse the process, typically the **Breit-frame** is employed; in this Lorentz frame the proton moves with large velocity, and the virtual photon has no energy. In other words, their respective momenta P and q are given by

$$P^\mu = (P_0, \vec{0}, P_z) \text{ with } P_z = \sqrt{P_0^2 - m_P^2} \approx P_0$$
$$q^\mu = (0, \vec{0}, -q_z) \text{ with } Q^2 = -q^2 = q_z^2. \tag{2.27}$$

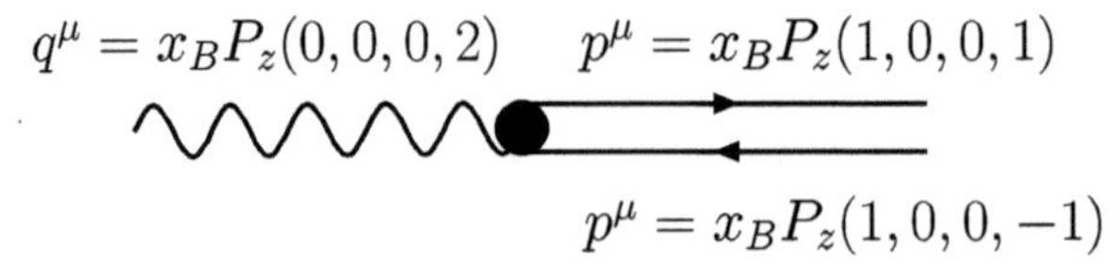

$$q^\mu = x_B P_z(0,0,0,2) \qquad p^\mu = x_B P_z(1,0,0,1)$$

$$p^\mu = x_B P_z(1,0,0,-1)$$

Fig. 2.6 Sketch of the photon–parton interaction in the Breit-frame. The parton only changes its momentum in the collision — its z-component effectively is reflected.

Introducing the **Bjorken variable** x_B as

$$x_B = -\frac{q^2}{2P \cdot q} = \frac{q_z^2}{2P_z q_z} \tag{2.28}$$

implies that $q_z = 2x_B P_z$. A sketch of the Breit-frame in these parameters is depicted in Fig. 2.6. The time during which the interaction between photon and proton takes place is given by the (longitudinal) wavelength of the photon, which of course is the component being hit by the proton. Hence,

$$\tau_{\text{int}} \sim \lambda_z \sim \frac{1}{q_z}. \tag{2.29}$$

In order for the photon to see the partons inside the hadron, they must have wavelengths at least as large as the photon, which is purely longitudinal; at the same time for the partons to see the photon, the photon's wavelength must at least be as large as theirs. Therefore, their respective longitudinal wavelengths must be the same and, consequently, also their momenta must be of the same size: $p_z = q_z$. In the approximation of quasi-collinear partons then the momentum-fraction x of the struck parton with respect to the incident proton must be of about the same order as x_B. The **lifetime** of these partons, as discussed in Eq. (2.6), is given by $\tau_{\text{parton}} \sim p_z/p_\perp^2$, which is larger than the interaction time $\tau_{\text{int}} \sim 1/p_z$, provided that $p_z \gg p_\perp$.

To summarize the discussion up to now: the parton picture of interactions developed so far is a sensible construction. Demanding that the interaction time of the parton with the photon is much smaller than the lifetime of the quantum fluctuation that actually **is** the parton, yields a condition on the parton kinematics: the transverse momentum of the parton must be much smaller than its longitudinal momentum, $p_z^2 \gg p_\perp^2$. In this setup the point-like photons measure the number of partons with similar longitudinal momentum in an area of size $1/Q^2$, or phrased in slightly different terms, they probe partons with momentum fraction $x \approx x_B$ at scale Q^2. The cross-section of this process must then be proportional to the sum of probabilities to find partons with this kinematics, that can interact with the photon, *i.e.*, ignoring higher-order diagrams:

$$\sigma_{ep} \sim \sum_q e_q^2 f_{q/p}(x, Q^2), \tag{2.30}$$

if the photon that is responsible for the interaction has a (negative) virtual mass squared with absolute value Q^2 and a longitudinal momentum given by x.

Furthermore, if this picture holds true — if $p_z^2 \gg p_\perp^2$ — then the photon–parton collision can be treated as the collision of two independent quasi-free particles; the struck parton does not feel the presence of the surrounding partons forming the proton, since the interaction time is smaller than the lifetime of this parton, *i.e.*, the characteristic time after which the strong colour fields force the parton to recombine with the other partons. In such a framework the probabilities to find partons with a given momentum fraction x at a scale Q^2 inside a proton is process-independent, so the DIS setup with exchanged photons can be replaced by other processes with similar kinematics, such as Drell–Yan production, jet production, etc.. Thus, encoding these probabilities of finding a parton p in a hadron h as process-independent **parton distribution functions** (PDFs) $f_{p/h}(x, Q^2)$ is sensible. This is the physical basis of the factorization formula presented in the next section.

The PDFs cannot, at the moment, be calculated from first principles, since they are truly non-perturbative objects. Of course, assuming the validity of the factorization picture, they can be measured in different processes and at different scales. These can be related to each other through the DGLAP equations, *cf.* Eq. (2.8) for the case of QED, which for QCD read,

$$
\frac{\partial}{\partial \log Q^2} \begin{pmatrix} f_{q/h}(x, Q^2) \\ f_{g/h}(x, Q^2) \end{pmatrix}
$$
$$
= \frac{\alpha_{\mathrm{s}}(Q^2)}{2\pi} \int\limits_x^1 \frac{\mathrm{d}z}{z} \begin{pmatrix} \mathcal{P}_{qq}\left(\frac{x}{z}\right) & \mathcal{P}_{qg}\left(\frac{x}{z}\right) \\ \mathcal{P}_{gq}\left(\frac{x}{z}\right) & \mathcal{P}_{gg}\left(\frac{x}{z}\right) \end{pmatrix} \begin{pmatrix} f_{q/h}(z, Q^2) \\ f_{g/h}(z, Q^2) \end{pmatrix}, \tag{2.31}
$$

where the sum over different $2n_f$ quark flavours and their anti-flavours is implicit and will be further detailed in Chapter 6. Schematically, the convolution in Eq. (2.31) could be written as

$$
\frac{\partial}{\partial \log Q^2} \begin{pmatrix} f_{q/h}(Q^2) \\ f_{g/h}(Q^2) \end{pmatrix} = \frac{\alpha_{\mathrm{s}}(Q^2)}{2\pi} \begin{pmatrix} \mathcal{P}_{qq} & \mathcal{P}_{qg} \\ \mathcal{P}_{gq} & \mathcal{P}_{gg} \end{pmatrix} \otimes \begin{pmatrix} f_{q/h}(Q^2) \\ f_{g/h}(Q^2) \end{pmatrix}, \tag{2.32}
$$

The kernels of the evolution equation, the **splitting functions**, are given, again, at leading order,[3] by

---

[3] For further reference, the splitting functions are decomposed here into a splitting kernel P and the anomalous dimensions of quarks or gluons, $\gamma_{q,g}$, where applicable.

$$\mathcal{P}_{qq}^{(1)}(x) = C_F \left[\frac{1+x^2}{(1-x)_+} + \frac{3}{2}\delta(1-x) \right] = \left[P_{qq}^{(1)}(x) \right]_+ + \gamma_q^{(1)}\delta(1-x)$$

$$\mathcal{P}_{qg}^{(1)}(x) = T_R \left[x^2 + (1-x)^2 \right] = P_{qg}^{(1)}(x)$$

$$\mathcal{P}_{gq}^{(1)}(x) = C_F \left[\frac{1+(1-x)^2}{x} \right] = P_{gq}^{(1)}(x) \qquad (2.33)$$

$$\mathcal{P}_{gg}^{(1)}(x) = 2C_A \left[\frac{x}{(1-x)_+} + \frac{1-x}{x} + x(1-x) \right]$$

$$+ \frac{11C_A - 4n_f T_R}{6}\delta(1-x) = \left[P_{gg}^{(1)}(x) \right]_+ + \gamma_g^{(1)}\delta(1-x).$$

Here x is the splitting parameter, essentially governing the light-cone momentum fraction of the offspring with respect to its emitter. It is worth noting that the indices of both $\mathcal{P}_{ba}$ and of P_{ba} are related to a parton splitting process $a \to bc$, where the type c of the third parton is fixed by a and b. The $\mathcal{P}^{(o)}(x)$ are also called **regularized splitting functions** to order o.

A pictorial way to construct the anomalous dimensions is to analyse the splitting functions and to realize that they can be written as the sum of a part that diverges for $z \to 1$ and some finite remainders which are linked to the δ-function from the $+$ prescription. This link is given by sum rules, namely

$$\sum_i \int_0^1 \mathrm{d}z \mathcal{P}_{ij}(z) = 0$$

$$\int_0^1 \mathrm{d}z \mathcal{P}_{qq}(z) = 0$$

$$\int_0^1 \mathrm{d}z z \mathcal{P}_{gg}(z) = 0.$$

Here the first sum rule satisfies the condition that the splitting functions can be interpreted as probability densitiies for a splitting to take place, while the second and third one ensure flavour and momentum conservation.

However, it should be stressed here that, strictly speaking, the picture above, called **collinear factorization**, which factorizes cross-sections of processes involving hadrons in the initial state into process-independent PDFs times the matrix element with quasi-free partons in the initial state, has been proved only for the cases of deep-inelastic scattering and Drell–Yan production [143, 409, 410]. The fact that the same ideas and the same formalism are also employed for other processes is in fact justified only by the success of this description rather than on a strict mathematical proof. In fact, recent work hints at the existence of rather subtle, factorization-breaking effects at higher orders [343], which are beyond the scope of this book.

2.1.4 Hadrons in the final state: The full picture

2.1.4.1 Partons in the final state

Similar to the parton distribution functions $f_{p/h}(x, Q^2)$, which at leading order encode the probability to find at scale Q^2 the parton p in the hadron h with a light-cone momentum fraction x with respect to the hadron, the **fragmentation functions** (FFs) $D_{p/h}(x, Q^2)$ yield, again at leading order, the probabilities of finding the hadron or photon h emerging from parton p, at scale Q^2 and carrying the fraction x of the parton's light-cone momentum. Similar to the PDFs, the FFs typically cannot be calculated but must be measured, and to complete the analogy, they enjoy the same kind of evolution equation, *cf.* Eq. (2.31), as the PDFs.

Starting at some scale with a parton with some momentum, the DGLAP equation therefore describes, how, by successive emissions, momentum is carried away by the secondary partons, until the scale of where the hadron is measured is reached. This is, pictorially speaking, the inverse of the PDF situation, where, starting from a low scale, a parton constituent of a hadron with some momentum fraction evolves to the larger scale of a relatively harder scattering process via multiple emissions of secondary partons.

Practically speaking there is yet another difference: in the case of the PDFs and the DGLAP equation for them, which is related to the evolution of initial-state partons from a known and fixed incident hadron, the final state is in many cases far less defined. In fact, the FFs come into play only, if the process signature is sensitive to finding a specific hadron in the final state – a situation in which the respective cross-sections are often denoted as single-hadron inclusive cross-sections. If, on the other hand, the presence of specific hadron species does not matter, then the FFs are typically not relevant. In such a situation, however, the **DGLAP equations** can still be used in order to describe the evolution of the partonic ensemble down to scales which are low enough for the perturbative description based on partons to cease being useful. The emergence of the then relevant degrees of freedom, the hadrons, however, cannot be treated from first principles and will rely on modelling with various levels of theoretical input.

2.1.4.2 Emissions in QCD

Compared to QED there is an important difference in the radiation pattern in QCD, driven by the fact that the gluons themselves carry colour charges, allowing them to radiate further gluons. Inspecting the form of the splitting function, Eq. (2.33), reveals that the emission of secondary gluons off gluons exhibits divergent structures in the splitting parameter z, namely $1/z$ and $1/(1-z)$. It is in remarkable contrast to the finite behaviour of the splitting function for gluons or photons splitting into fermion pairs. This additional channel thus drives a further acceleration of radiation beyond the effect of the larger coupling in QCD compared to QED. From a theoretical point of view, ignoring the finite splitting of bosons into fermion pairs, the colour of the gluons means that the simple eikonal picture of iterating independent emissions in QED will not directly translate to QCD. While in QED, the eikonal for any photon emission is always spanned by the muons acting as eikonal partners p and p', and

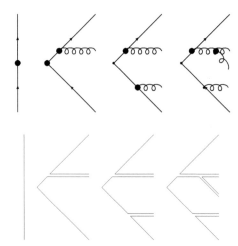

Fig. 2.7 Some Feynman diagrams for the emission of multiple gluons by a quark pair contrasted with corresponding leading colour diagrams.

emissions of photons by photons are absent in the QED case, in QCD the gluons carry colour and thus can also act as gluon emitters. Then, after each gluon emission the structure of the eikonals will change, as the emitted gluons introduce new directions of the colour field. In the limit of infinite colours, also known as the large-N_c limit, this leads to the notion of **colour dipoles** emitting the gluons. The emergent more complicated radiation pattern can be represented by some colour-flow diagrams of the type depicted in Fig. 2.7.

The leading behaviour of the QCD emission process follows a pattern given by

$$
\begin{aligned}
\mathrm{d}w^{q \to qg} &= \frac{\alpha_s(k_\perp^2)}{2\pi} C_F \frac{\mathrm{d}k_\perp^2}{k_\perp^2} \frac{\mathrm{d}\omega}{\omega} \left[1 + \left(1 - \frac{\omega}{E} \right)^2 \right] \\
&= \frac{\alpha_s(k_\perp^2)}{2\pi} C_F \frac{\mathrm{d}k_\perp^2}{k_\perp^2} \, \mathrm{d}z \, \frac{1 + z^2}{1 - z} = \frac{\alpha_s(k_\perp^2)}{2\pi} C_F \frac{\mathrm{d}k_\perp^2}{k_\perp^2} \, \mathrm{d}z \, P_{qg}^{(1)}(z) \,.
\end{aligned}
\tag{2.34}
$$

for gluon emission off a quark with energy E. In the soft limit, for vanishing gluon energies $\omega = E(1 - z)$, this reproduces the dipole form in Eq. (2.26), which qualitatively is defined by a logarithmic distribution in both the gluon energy ω and its transverse momentum $k_\perp$ with respect to the quark direction. They are often denoted as **transverse logarithms**, related to the **collinear divergence** or **mass singularity** for the logarithms related to the $1/k_\perp^2$ term and as **longitudinal logarithms** related to the **soft divergence** or **infrared singularity** stemming from the $1/\omega$ part of Eq. (2.34).[4]

---

[4]This automatically also leads to a qualification of parton emissions: for the production of a "jet", a parton needs to be emitted at relatively large angles and energies, $k_\perp$, $\omega \sim E$, leading to $\mathrm{d}w^{q \to qg} \sim C_F \alpha_s(k_\perp^2)/(2\pi)$. Conversely, emissions not giving rise to additional jets are characterized by $k_\perp$, $\omega \ll E$ and therefore $\mathrm{d}w^{q \to qg} \sim C_F \alpha_s(k_\perp^2)/(2\pi) \log^2 E$.

2.1.4.3 Interplay of final state radiation and hadronization

As can be seen in Eqs. (2.26) and (2.34), the emission of massless final state partons leads to divergences of the form $1/k_\perp^2$ and $1/\omega$ that must be cured. Typically, a cut on the phase space of resolvable emissions is applied, which in the case of photons could for instance be expressed by minimal photon energies and minimal opening angles with respect to its emitters. While in QED these cuts are often dictated by practical considerations, the technology and sensitivity of the detectors employed, the situation is different in QCD. There, a cut-off would typically be provided by the scale, at which the phase transition of the partons to the hadrons occurs. It has become customary to parameterize this scale through a cut on the relative transverse momentum $k_\perp$ of the emitted parton with respect to its emitters, and by the requirement that the minimal transverse momentum is sufficiently far above the Landau pole of QCD. Thus, such cuts are of the order of about $k_\perp^{(\mathrm{min})} \approx 1\,\mathrm{GeV} > \Lambda_{\mathrm{QCD}}$. This relatively rough estimate can be further refined by some simple estimates.

Consider, again, the electromagnetic field accompanying an accelerated charge moving with velocity near the speed of light $v \approx 1$. While at asymptotically large times the electromagnetic fields will be the Lorentz-contracted discs of Fig. 2.1, right panel, it is also evident that this state would not just "jump" into existence, *cf.* [477], which also forms the base of the further reasoning below.

Assuming the existence and acceleration of the charge started at $t = 0$ with $v = 0$, the field spreads out as a sphere with $r' \leq t'$ in the (primed) rest frame of the charge. In the (unprimed) laboratory frame, of course, time dilation by a factor $\gamma = E/m$ the field at distances $r_\perp$ from the axis of movement of the charge will show up at the earliest at $t = \gamma t' = E r_\perp / m$.

For quarks or other colour charges being produced at energy E in a hard interaction the situation is very similar. However, the reasoning there has far-reaching consequences: identifying with m the constituent mass of a quark, typically $\mathcal{O}(\Lambda_{\mathrm{QCD}}) \approx R^{-1}$ for light quarks and m_Q for heavy quarks, and with $r_\perp$ a typical hadronic size, $R = \mathcal{O}(1\,\mathrm{fm})$, immediately yields a hadronization time $t^{(\mathrm{had})}$ proportional to the energy of the quark:

$$t^{(\mathrm{had})} \approx \begin{cases} ER^2 & \text{for light quarks} \\ \dfrac{ER}{m_Q} & \text{for heavy quarks.} \end{cases} \tag{2.35}$$

Qualitatively, this naive picture emerges also in a quantum mechanical treatment. For hadrons of characteristic size R, the confining QCD forces are associated with gluon fields with typical momenta

$$k \approx k_\parallel \approx k_\perp \approx R^{-1} \tag{2.36}$$

in the hadronic rest–frame. They emerge after typical times

$$t \approx R \approx \frac{1}{k_\perp}, \tag{2.37}$$

which, including a Lorentz–boost factor as before leads to hadronization times

$$t^{(\text{had})} \approx \frac{k_\parallel}{k_\perp^2} \approx \frac{ER}{m}, \tag{2.38}$$

where m is a typical hadronic mass scale; the result is the same as in the more classical treatment above.

At the same time it is important to see how long it actually takes for a gluon to form in the emission off, say, a quark. Naively, this formation time can be estimated from the invariant mass of the quark–gluon pair and its energy as

$$t^{(\text{form})} \approx \frac{1}{m_{qg}} \frac{E}{m_{qg}} \approx \frac{E}{kE\theta_{qg}^2} \approx \frac{k}{k_\perp^2} \approx \frac{k_\parallel}{k_\perp^2}. \tag{2.39}$$

Here, the starting point is given by a combination of uncertainty principle and Lorentz time dilation effects, and in some intermediate steps a small opening angle of the resulting pair has been assumed.

Comparing this with the hadronization time and demanding — reasonably — that gluons be formed before they hadronize, yields a constraint on the emission kinematics:

$$\frac{k_\parallel}{k_\perp^2} \approx t^{(\text{form})} \leq t^{(\text{had})} \approx k_\parallel R^2 \longrightarrow k_\perp \geq \frac{1}{R} = \mathcal{O}\left(\text{few } \Lambda_{\text{QCD}}\right). \tag{2.40}$$

Extending this to the case of heavy quarks Q, such as top quarks, it is worth noting that in their case the lifetime given by

$$\tau_Q \sim \left(\frac{m_W}{m_Q}\right)^3 \frac{E}{m_q} \ll \frac{E}{m_q} \sim t^{(\text{had})} \tag{2.41}$$

is smaller than their hadronization time; they behave as truly free quarks.

2.1.4.4   Hadronization

At this low scale, QCD enters a new regime, where the strong interaction becomes so strong that the partons start feeling the effect of being bound together in hadronic states. As in the case of the initial state, the transition between the parton and the hadron description is beyond our current quantitative understanding. This, of course, is where the parameterization provided by the fragmentation functions, obtained from measurements, becomes important. There is, however, a significant difference with respect to the PDFs in the initial state. In the final state there may well be more than one hadron emerging from a parton; consequently the fragmentation functions describe the transition of the parton into the corresponding hadron plus any other hadrons. They are therefore mainly useful for describing the inclusive production of this one given hadron in a collision, ignoring all other hadrons which may eventually emerge as well. To reach a more inclusive picture, other methods are thus necessary, which will be introduced and discussed at later stages of this chapter and in Chapter 7.

Fig. 2.8 Scetch of a hard hadron-hadron collision at a hadron collider experiment, such as the LHC. It depicts the hard interaction as the central, red blob. The initial and final state particles experience initial and final state radiation, in blue and red, respectively, leading to a proliferation of a large number of secondaries. The underlying event — in purple — has other partons scattering and producing further activity through multiple secondary emissions. All emerging partons will at some point arrive at low scales, and hadronize, before the primary hadrons produced there will decay further (in green).

2.1.5 Summary: space-time picture of hadronic collisions

In the following, the different aspects important in hadronic collisions that have been discussed so far will be put into a wider, more coherent context. In order to supplement this discussion, in Fig. 2.8 an event at a hadron collider has been scetched. The following discussion is summarized in Fig. 2.9.

2.1.5.1 Hard interaction

Ordering this picture with the help of typical scales related to the momentum transfers at different stages of an individual event, the starting point is the hard interaction. Because, by definition, this is the hardest part of the event, the scales there are the largest, and this opens the possibility to describe it by fixed-order perturbation theory.

There, the relevant features of the event — the emergence and decay of heavy states such as gauge or Higgs bosons or top quarks or the occurrence of a number of hard **QCD jets** — can systematically be included. This is, at leading order and for small final state multiplicities, typically achieved by the well-known textbook methods of constructing the relevant Feynman diagrams, summing and squaring them by employing completeness relations, convolution with the PDFs, and by finally integrating over the relevant phase space of the final state and over the momentum fractions of the initial-state partons with respect to the incoming hadrons. The last step is done with numerical methods, **Monte Carlo integration**, which is due to the non-analytical structure of the PDFs. For increasing multiplicities of final state particles, still at leading order, the number of Feynman diagrams exhibits a faster than factorial growth, rendering the textbook methods prohibitively time-consuming, and numerical methods must be employed throughout. For a summary of such methods, the reader is referred to Section 3.2.1. In order to improve the accuracy of such fixed order calculations, of course higher perturbative orders must be included, which typically leads to additional problems due to the occurrence of ultraviolet and infrared divergences. For a more in-depth discussion of the overall mechanism of such fixed-order calculations the reader is referred to Section 2.2, and to Chapter 3, while in Chapter 4 specific issues related to different process will be detailed. The role of the PDFs, on the other hand, and how they are obtained from data, will be dwelt on in Chapter 6.

2.1.5.2   Secondary emissions

Having produced a number of particles at large scales, they will undergo some final state radiation, emitting Bremsstrahlung-quanta at lower scales. At the same time, the incident partons will experience initial-state radiation, due to the breakdown of the coherence of the Fock states related to the incident particles induced by the hard interaction. As can be seen, for example in Eq. (2.2) for the case of QED, these extra emissions may lead to logarithmically enhanced contributions, where parametrically large logarithms due to the $k_\perp$ integration compensate, at least partially, the small coupling. This leads to the rough, schematic distinction of two different regimes of parton emission:

- a regime of jet production, where $k_\perp \sim k_\parallel \sim \omega$ and emission probabilities scale like $w \sim \alpha_s(k_\perp) \ll 1$; and
- a regime of jet evolution, where $k_\perp \ll k_\parallel \sim \omega$ and therefore emission probabilities scale like $w \sim \alpha_s(k_\perp) \log^2 k_\perp^2 \gtrsim 1$.

This implies that in this jet evolution regime, typical fixed-order counting of coupling factors alone ceases to be sensible, and it becomes more important to resum terms of the form $\alpha^n \log^{2n}$, $\alpha^n \log^{2n-1}$, and so on. For an introduction into resummation the reader is referred to Section 2.3 and to Chapter 5 for more details. This additional radiation is particularly important for QCD particles, because, first of all, the strong coupling is larger than the electromagnetic one, but also because, in contrast to the QED case, the gluons are coloured and thus emit secondary gluons. Such an enhanced radiation manifests itself in the emergence of jets, which turn out to be the relevant objects to discuss QCD final states. In simulation programs, also known as **event**

generators, the resummation of these logarithms and the resulting jet structure in typical QCD events is achieved numerically invoking **parton showers**, *cf.* the second half of Chapter 5.

Taken together, the logarithmically enhanced radiation pattern may have some sizable impact on the structure of the events, as it may very well change the kinematical distribution of particles in the final state. In addition, it should not be forgotten here that the experimentally observable objects are hadrons rather than the quanta of QCD, quarks and gluons, which introduces yet another layer of complication when discussing the full structure of hard collision events.

2.1.5.3 Hadronization and hadron decays

At the low scales of $\mathcal{O}\left(1\,\mathrm{GeV}^2\right)$, where the perturbative description breaks down, the impact of the transition to hadrons becomes important for a number of observables. There are various ways of how to approach this problem. By now the most often used method is to rely on Monte Carlo-event generators which, aiming at a complete simulation of particle collisions including all facets, typically provide a phenomenological model that translates the partons into hadrons. These models usually contain about a dozen parameters which need to be adjusted (*"tuned"*) to existing data. In the construction of these models, various almost always rather qualitative ideas of how the phase transition proceeds are turned into some algorithmic prescription. Such models include **independent fragmentation** [137, 527, 636], statistical hadronization [207], the **string model** [162, 163, 172, 281], and the **cluster model** [528, 528, 533, 896, 899], based on the idea of **preconfinement** [151].

Alternatively, analytical or semi-analytical descriptions are sometimes employed, which are often based on the analytical continuation of the strong coupling, α_s into the non-perturbative regime and on employing the kinematics of perturbative parton splitting also in this infrared regime. Examples include **power corrections** [481, 483, 691, 887] and have mainly been employed in the description of hadronization effects in electron–positron annihilations into hadrons and deep-inelastic scattering.

In many cases, the hadrons that emerge in the hadronization process are unstable and decay further, leading to an additional proliferation of particles in the final state. This is typically dealt with in event generators, where, typically, for lighter and better-known hadrons such as the η, η', ρ's, ω, or ϕ hadrons, the corresponding branching ratios can be taken from the experimental results published in the PDG, subject to some rescaling to ensure that the branching ratios add up to exactly unity. For heavier or less well-known states this, of course, ceases to be an option, and additional decay channels must be invented and branching ratios must be fixed by considerations involving, among others, phase-space or flavour-symmetry arguments. The kinematical distributions of the final state particles are quite often fixed by assuming isotropic decays — this may be a poor approximation. Therefore, in some cases matrix elements from effective theories such as **(Resonance) Chiral Perturbation Theory (χPT)** [494, 495] or from spin consideration only are invoked. At this point it is worth noting that the hadronic decays of the τ-lepton are treated in a similar fashion, with matrix elements either stemming from RχPT [491, 492, 844] or from the Kühn–Santamaria model [703].

For the decays of hadrons with heavy quarks, quite often only those final states are known, which involve a few final state particles only, which then can be taken, again, from the PDG. There, kinematical distributions typically are taken from suitably chosen matrix elements and often involve additional form factors, both obtained from **Heavy Quark Effective Theory (HQET)** [497, 640, 745, 784, 813, 847, 848]. For high-multiplicity final states (which in the case of, say, the B-mesons amount to about 50% of the total decay rate), often the decay is treated by going back to the parton level. In these cases, the parton-level matrix elements are supplemented with parton showers and hadronization. For a more detailed description, the reader is referred to Chapter 7.

2.1.5.4 Underlying event

An additional complication, when discussing collisions with hadron initial states, arises from the fact that hadrons are extended objects, which are composed from a multitude of partons. In standard textbook formalism, this is reflected by employing PDFs to account for the transition of the incident hadrons into the partons which then experience a hard scattering. However, this formalism does not account for the possibility that more than one parton pair coming from the two hadrons may interact. This effect, also known as **double-parton scattering** or **multi-parton interactions**, in fact is beyond standard factorization; in the absence of a first-principles approach it thus must be modelled. Up to today, even the simplest models, which are based on a simple factorized *ansatz* of merely multiplying the partonic cross-sections and including a trivial symmetry factor, appear to be in agreement with data. This suggests that, possibly, probable correlation effects when going from one- to two-particle PDFs or non-trivial final state interactions are not dominant and can be treated as some kind of higher-order correction to the simple picture.

Apart from such hard double or multiple interactions there are other, softer additional interactions when hadrons collide — some of them will just fall into the category of multiple-parton interactions at lower scales, where no hard objects like jets with some tens of GeV in transverse momentum are produced. In addition, there is another contributor to the overall particle multiplicity, namely the remnants of the incident hadrons. After one or more partons have been extracted from them, the parton ensemble forming them typically cannot combine back into a single colourless object but instead must hadronize into a set of hadrons. Quite often, however, the two beam remnants appear to be connected in colour-space which must be included into the reasoning. In addition, although the partons are by far and large moving in parallel to the beam, *i.e.*, to the hadron they form, there is no reason to assume that they do not have some transverse momentum of the order of $\Lambda_{\rm QCD}$ up to some few GeV. This **intrinsic transverse momentum** of the beam-remnant partons guarantees that the hadrons stemming from their hadronization do not necessarily vanish along the beam pipe, but can instead reach the detector due to their finite transverse momentum.

For a further discussion of current models, the reader is referred to Chapter 7; some comparisons to data are presented in Chapters 8 and 9.

2.1.5.5   Pile-up

Another important effect in hadron collisions, especially at the LHC, is the interaction of multiple pairs of protons in the same bunch crossing, the so-called **pile-up**. This is driven by the instantaneous luminosity of the collider experiment in question, and while at maximal luminosity at the TEVATRON there were about 5 proton–anti-proton interactions per bunch crossing, at the LHC about 25 pairs of protons are expected to interact in each bunch crossing at design luminosity and the centre-of-mass energy $E_{\text{c.m.}} = 14\,\text{TeV}$. For a possible SLHC, this number would go up to about 250 proton–proton collisions per bunch crossing. In addition to that, the size of the detector allows for the possibility that the particles produced in more than one bunch crossing interact with different parts of the detector at the same time, an effect known as **temporal pile-up**. At the LHC, with the anticipated maximal frequency of 40 MHz, there will be about three such generations of particles at the same time in the detector.

2.1.6   Definition of physical objects: leptons, photons, and jets

Before turning to calculating the first example cross-sections in perturbation theory and analysing their behaviour, especially at higher order in QCD, in this section proper definitions of relevant physical objects will be discussed.

2.1.6.1   Leptons

While naively this discussion may seem a bit awkward, it is important to understand why this is a relevant issue. To gain some understanding, consider first an example in electrodynamics, namely the production of Z bosons and their subsequent decay to a lepton pair. Without QED final-state radiation, and in an ideal world with perfect detectors, the invariant mass of the lepton pair would roughly follow the Breit–Wigner form expected from resonance production, of course modulated by PDF effects. Their combined four-momentum, due to energy conservation, would of course be identical to the intermediate boson's state. In turn its transverse momentum and rapidity could be directly reconstructed. This, however, is not how reality presents itself. Instead the lepton will emit photons that follow, at leading order, the eikonal pattern already discussed in Section 2.1.1, resulting in a logarithmic enhancement of soft and collinear emissions. Consequently, the lepton four-momentum is reduced and, even for ideal detectors, their reconstructed kinematics would typically differ from those in the absence of such QED FSR effects. There are various strategies to deal with this apparent problem. Of course it is possible to take such effects into account by including them in the calculation. A popular way of achieving this is employed in some modern Monte Carlo simulations. It is based on the algorithm of Yennie, Frautschi, and Suura [901] and allows for a resummation of leading logarithms in an energy cut-off and an angular cut-off, which can be systematically improved by fixed-order calculations. Since this algorithm also provides a transparent way of guaranteeing four-momentum conservation,[5] it is particularly well-suited to describing the kinematic effects on leptons due to photon emissions in both the initial and the final state. Emissions below the cut-offs

---

[5]For some specific realizations of such infrared-safe phase space mappings between states before and after QED radiation, see for instance [838].

Detector Event

Hadron-hadron collision

Parton-parton collision   Hard

> hardest interaction: the *"signal process"*
> description: fixed-order perturbation theory (Chapters 3 and 4)
> important input: PDFs, factorization theorem (Chapter 6)

Initial- & final-state radiation

> emission of secondary particles
> description: resummation or parton shower (Chapter 5)
> important: matching to fixed order & logarithmic precision
> note: final state radiation is perturbative part of *"fragmentation"*

Underlying event

> multiple parton-parton interactions
> description: 2 → 2 parton-parton scatterings in perturbation theory
>   or other QCD-inspired models (Chapter 7)
> treated only in full simulations, comes with further parton showering

Hadronization

> transition of partons to primordial hadrons
> description: fragmentation functions in calculations
>   non-perturbative models in simulations (Chapter 7)

Hadron decays

> decay cascades of the primordial hadrons
> description: data, effective theories, symmetries, models (Chapter 7)

                                                                           Soft

Pile-up

> multiple hadron–hadron collisions, typically soft.
> description: models, often QCD-inspired (Chapter 7)

Fig. 2.9 The various components of a hadronic collision event as seen in the experiment. Typically, in one collision, a number of hadron-hadron collisions occur, most of which are soft and as *"pile-up"* blur the picture of the one hadronic collision of interest. This usually has a hard component, the partonic *"signal event"*, which is calculable from first principles in perturbation theory. The interacting particles undergo initial and final state radiation, before the partons hadronize. The emerging primordial hadrons then decay further to the particles finally visible in the detector.

are considered to be too soft or too collinear to yield any visible effect; conversely, such photons can practically be taken as recombining with the original lepton. However the singularities related to their emission cancel the corresponding soft and collinear divergences in the virtual contributions, to all orders. This is yet another manifestation of the KLN and BN theorems, *cf.* Section 2.2.5.

On the other hand, the production of leptons, especially in hadronic collisions, may essentially proceed through two mechanisms. Direct production, the case discussed up to now, is where the leptons directly stem from the hard interactions. However, leptons may also emerge in the decay of hadrons and, in particular, in the weak decay of heavy hadrons containing charm or bottom quarks. In fact the leptons there play an important role in the identification of such objects, with the finite lifetime of weakly-decaying heavy particles resulting in **displaced vertices** measurably different from the primary one. However this production channel may also mimic direct production and its characteristics. Typically the production cross-sections of heavy flavours convoluted with corresponding decay branching ratios are orders of magnitude larger than the cross-section for direct lepton production, implying that this is an issue that must be dealt with carefully. A straightforward solution relies on two considerations. First, weak decays of heavy hadrons producing the leptons also yield other, lighter hadrons. In addition, more often than not, the heavy flavour is part of a larger system containing more hadrons — a **jet** — that will be discussed further shortly. Especially for leptons with a transverse momentum larger than the mass of typical heavy hadrons, these two effects mean that the lepton more often than not is part of a roughly collimated bunch of other particles, mostly hadrons. As a consequence, demanding leptons to be isolated in rapidity and transverse angle from any hadronic activity will significantly reduce the impact on an analysis of leptons produced in hadronic decays. This isolation requirement is typically realized experimentally by demanding that the sum of the hadronic energy or the transverse momentum of other charged tracks in a radius R_{crit} around the lepton be smaller than a critical value. Here, the **distance** ΔR_{ij} between two objects i and j is given by

$$\Delta R_{ij} = \sqrt{\Delta \eta_{ij}^2 + \Delta \phi_{ij}^2} = \sqrt{(\eta_j - \eta_i)^2 + (\phi_j - \phi_i)^2} \qquad (2.42)$$

with η and ϕ the pseudorapidity and the azimuthal angle, respectively.

At this point some terminology needs to be defined. The lepton four-vectors provided by a fixed-order parton-level program are termed to be at the **Born** level.[6] A **bare** or **undressed** lepton is one that has undergone QED radiation, it has lost energy due to photon emission. A large fraction of that photon radiation is relatively collinear with the lepton direction (corresponding to about 3% of the lepton energy carried away by radiation off of electrons and 1.5% for radiation off of muons) and a smaller fraction is emitted at wider angles.[7]

This is highlighted in Fig. 2.10, which demonstrates the effect of photon radiation on the lepton in the decay of a W boson. The emission of a single photon depletes

---

[6]Of course the term Born level here refers to Born level with respect to QED corrections, so leptons in higher-order QCD calculations would still be coined as Born level in this sense.

[7]Note that the QED corrections are strongly kinematic and phase-space-dependent.

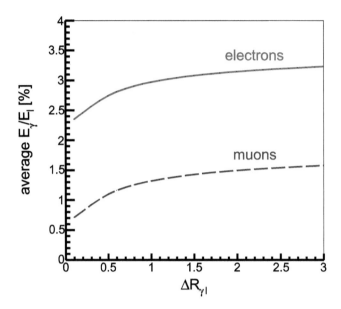

Fig. 2.10 The effect of photon radiation on the average lepton energy in W boson decay, as a function of the angular separation between the photon and the lepton.

the energy of the lepton (measured in the rest frame of the decaying W boson) by a fraction that depends on the angular separation of the photon and the lepton ($\Delta R_{\gamma\ell}$). This fraction is larger for electrons than muons due to the collinear enhancement that goes with the lepton mass (m_ℓ) as $\log(m_\ell/m_W)$. For nearly-collinear photons, in the region $\Delta R_{\gamma\ell} < 0.1$, the fraction of energy radiated away is around 2.5% for electrons and 0.7% for muons. Emission at wider angles is responsible for approximately 1% of additional depletion in both cases.

In an experimental lepton reconstruction algorithm, photon radiation is absorbed inside a cone of small radius around the lepton, typically $\Delta R < 0.1$ using the definition in Eq. (2.42). The resulting lepton four-vector, that has been corrected for collinear photon radiation effects by such an algorithm, is designated as a **dressed** lepton. There is still a residual correction of around 1% (for both electrons and muons) to account for wide angle photon emission. Often, data and theory comparisons are made at the dressed lepton level, if the theoretical calculation derives from a Monte Carlo program, which is capable of including photon radiation effects. For comparison to fixed-order calculations, the data typically is corrected for QED radiation to the Born level.

2.1.6.2 Photons

Similar considerations also hold true when considering the production of photons. Once again, the dominant contributions stem from secondary mechanisms of photon production rather than direct production in the hard interaction. Such **secondary photons** can be emitted as final-state radiation off quarks or, more often, originate from the decay of hadrons. In the latter case the major source of concern is not tied to the production and decay of heavy flavours, which typically gives rise to other, secondary hadrons in addition to charged leptons, but the annihilation of neutral, mostly light mesons in processes such as $\pi^0 \to \gamma\gamma$ or $\eta^{(\prime)} \to \gamma\gamma$. These particles are created in abundance in hadronic final states and, since both they and their decay products are neutral, they can only be seen calorimetrically. Because they are so light, their decay products more often than not end up in the same calorimeter cells when looking for high-$p_\perp$ objects, which makes it fairly tricky to disentangle them from single photons. This is particularly relevant for signals containing **primary photons**, which are directly produced in the hard process and thereby have large backgrounds from such secondary production channels. This adds another level of complication with respect to the lepton case, but the overall solution remains the same.

Again, in order to disentangle **direct** (or **prompt**) photons from those emerging in the fragmentation or decay of strongly interacting particles, isolation criteria are introduced. In past years, it has become customary to use the **Frixione isolation** criterion [540] in theoretical calculations. This criterion is defined by a cone with opening angle δ_0 around the photon in η–ϕ space, a critical exponent n and a scaling factor ε_γ.[8] Photons are considered isolated from the hadronic environment, if the accumulated hadronic transverse energy inside any cone with size $\delta < \delta_0$, weighted by the distance from the photon, is smaller than the photon's transverse energy multiplied with a cone-size-dependent weight:

$$\sum_{i \in \text{hadrons}} \left[E_{\perp,i} \Theta(\delta - \Delta R_{i\gamma}) \right] \leq \varepsilon_\gamma \, E_{\perp,\gamma} \left[\frac{1 - \cos\delta}{1 - \cos\delta_0} \right]^n \quad \forall \, \delta < \delta_0 \, . \tag{2.43}$$

It is worth noting that the modifying factor on the right-hand side of the equation above enjoys the property that

$$\lim_{\delta \to 0} \left[\frac{1 - \cos\delta}{1 - \cos\delta_0} \right]^n = 0 \, . \tag{2.44}$$

This guarantees that in the strictly collinear limit any hadronic emission will lead to the photon being considered as not isolated. At the same time, the scaling with energies on both sides ensures that the case of partons splitting into partons in some final-state radiation process does not hamper the isolation criterion too badly – it will remain sufficiently infrared-safe.

From the theoretical perspective, the use of the Frixione prescription greatly simplifies any calculation. It removes the need for including non-perturbative photon fragmentation contributions that describe the parton to photon transition, *cf.* Section 2.1.4. Since these fragmentation contributions are a purely collinear phenomenon, they are explicitly removed when using the Frixione approach. However, experimen-

---

[8]Customary choices for the parameters of this algorithm are, for instance, $\varepsilon_\gamma = 1$, $n = 1$ and $\delta_0 \approx 0.5$, mimicking the typical cone-size of jets, see the section about jets, 2.1.6.3.

talists have typically not used the Frixione approach for photon isolation either at the TEVATRON or at the LHC, not least because the condition in Eq. (2.43) cannot be directly implemented for a detector with finite granularity. Instead an isolation cone is defined, typically of radius $R=0.4$, and any transverse momentum contained in the isolation cone (not assigned to the photon) is required to be less than either a fixed value, or a percentage of the photon tranverse momentum. The isolation definition is designed to remove most of the jet fragmentation background to photons, while having a high efficiency for retaining true, isolated photons. While the primary purpose of the isolation algorithm is to remove the jet backgrounds, it also has the effect of effectively removing much of the photon production from fragmentation processes.

Another procedure, also known as a **democratic approach**, is based on treating photons as if they were jets and applying a jet clustering algorithm on *all* outgoing particles. In this method a photon is isolated, if it would be forming a jet and if its hadronic content contributes less than a critical value to its overall energy or transverse momentum. In the next section, such jet finding and clustering algorithms will be discussed in more detail.

Ultimately both algorithms can be used in parton-level calculations and in hadron–level simulations or measurements, facilitating a direct comparison between the two. In reality the energy in the isolation region around any photon candidate is dominated by underlying event energy, and at high luminosities, by pile-up. There are techniques to effectively subtract this energy that is either completely (in the case of pile-up), or partially (in the case of the underlying event energy) uncorrelated with the hard scatter, see [586]. However, the stochastic uncertainties in the subtraction tend to over-whelm any fragmentation energy that may be in the cone. Given these uncertainties it is therefore not unreasonable to use Frixione isolation in theoretical calculations for comparison to data on which typical experimental isolation cuts have been applied.

2.1.6.3 Jet algorithms: general considerations

The situation becomes even more complicated when trying to analyse the hadronic components of final states. As already discussed in Section 2.1.4, strongly interacting particles, quarks and gluons, experience significant final state radiation, driven by the large logarithms related to the usual soft and collinear divergences when emitting massless quanta. The radiation is further enhanced by the size of the strong coupling, which is an order of magnitude larger than the electromagnetic one. This last effect is even more pronounced for emissions at relatively small scales, due to the faster running of the strong coupling. The resulting ensemble of QCD quanta fragments into hadrons, many of which are unstable and decay. This leads to a massive proliferation of hadrons, which, however, tend to "clump" into fairly energetic clusters, called **jets**. While this picture is very intuitive qualitatively, a more quantitative way of dealing with this phenomenon is necessary. One obvious requirement is that any reliable solution must allow the direct comparison of perturbative calculations on the parton level, based on first principles, with the experimental reality. Thus quantitative descriptions, called **jet algorithms**, must be applicable at the level of partons, observable hadrons, and even energy deposits in calorimeter cells. In addition, it must be guaranteed that the chosen jet algorithm holds water irrespective of the actual perturbative order. Simply put: considering the same observable at a higher perturbative order typically involves additional parton emissions, which actually may be soft and/or collinear with respect to already present partons. These additional and potentially unresolved emissions *must not* lead to unwanted artifacts such as ambiguities in the actual number of jets observed or their position in phase space. If this requirement is satisfied, the jet algorithm is called **infrared-safe**.

In this section, the focus will be on jet definitions and algorithms applied at the perturbative level, *i.e.*, using partons. Some of the difficulties that result when hadrons are clustered to jets instead will be discussed in Chapter 8.

At leading order, each jet is modelled by a single quark or gluon. As already noted in Section 2.2, in general this leads to infrared singularities corresponding to kinematic configurations in which two partons are collinear, or a gluon is soft. It is only after the application of a jet algorithm, which ensures that all partons are both sufficiently hard and well-separated, that any sensible theoretical prediction can be made.

In general, jet algorithms can be considered as constructed from two ingredients. First, all objects belonging to a jet must be identified, for which there are essentially two categories of algorithm. One is based on purely geometric considerations and proceeds by identifying a jet axis and assigning all objects within a radius R_0 around this axis to the jet. In contrast, sequential algorithms proceed by combining and clustering pairs of particles in turn until only hard quasi-particles identified with jets are left. Second, the momenta of the jet constituents must be combined to yield the overall jet momentum, which in modern algorithms is realized by recombination schemes acting sequentially on four-momenta. These two stages in principle are independent, but in sequential clustering algorithms they are, to a certain degree, intertwined.

It is crucial to describe this construction of jets from their constituents in terms of an appropriate set of kinematic variables. The standard set consists of the transverse momentum $(p_\perp)$, rapidity (y), azimuthal angle (ϕ), and invariant mass (m) of the jet.[9]

2.1.6.4 Cone algorithms

Cone algorithms are the prime example for assigning objects to jets on geometric grounds and typically work as follows. In a first step, a seed is found, defining the direction of the prospective jet centre or centroid (η_c, ϕ_c) in the η-ϕ plane.[10] For instance, such a seed could be any particle in the event — an idea typically used in parton-level calculations — or any calorimeter cell with an energy deposit larger than some critical energy, an idea usually used for hadron- or detector-level analyses. Then a circle with radius R_0 is drawn around the centre and all particles i with radius $R_{i,c} < R_0$ are assigned as members the prospective jet. Its overall momentum can then be defined as the vectorial sum of the four-momenta of all its constituents. Such candidate jets are then accepted if their transverse momentum is larger than some critical value, $p_\perp^{(j)} > p_{\perp,\mathrm{crit}}$. As a by-product, of course, the jet position $(\eta^{(j)}, \phi^{(j)})$, its mass, etc., can readily be evaluated.

---

[9]Early studies of jets at the TEVATRON did not employ this set, for example using E_T rather than p_T, and η rather than y, and largely ignoring the information provided by the jet mass.

[10]In past usage, cone jet algorithms tended to use pseudo-rapidity (η) rather than rapidity (y). Modern jet algorithms all use y.

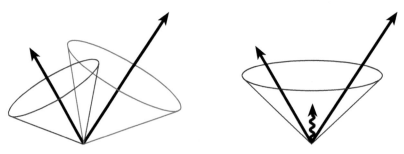

Fig. 2.11 A kinematic configuration demonstrating the lack of a proper all-orders definition of the simplest cone algorithm. The addition of a soft parton acting as a potential jet seed between the two original cones "pulls" the hard partons in, resulting in one rather than two jets. This is a typical example for infrared-safety problems in perturbation theory. Reprinted with permission from Ref. [508].

This is a very simple algorithm but, like many nice ideas, it is rather too simplistic and it cannot be used to describe the full wealth of features predicted in QCD without introducing further complications. A first problem arises when trying to find more than one jet. Depending on the parameters of the collision and the cone algorithm, it is not unlikely that in a multijet environment jets may overlap, leading to particles that may in principle contribute to more than one jet. Such assignments would most certainly lead to unwanted features like non-conservation of momentum and therefore must be avoided. A simple way of doing this is by ensuring that, for instance, the harder jet includes the particles in the overlap zone. So, this problem can to some degree be solved algorithmically.

More complications arise when considering seeded cone algorithms. As an example at the parton level, consider the situation depicted in Fig. 2.11, where two hard partons i and j have a relative distance $R_0 < R_{ij} < 2R_0$. As long as these are the only two partons in the vicinity, they will form *two separate jets* since $R_{ij} > R_0$, however, the addition of a soft gluon may completely change the picture. Using it as a potential seed it is entirely possible that *both* partons have a distance smaller than R_0, thus being sucked into *one combined jet*, which has a larger total energy and momentum and will therefore be accepted as a candidate. So, basically, the presence of additional radiation at higher orders allows *new seed directions* about which jets can be formed. The appearance of such new jet axes, especially if it is due to the presence of arbitrarily soft radiation, will cause real and virtual corrections to fall into bins of different jet multiplicity. This will quite often hamper the mutual cancellation of the associated soft singularities and in such cases will invalidate the perturbative calculation. In general, this kind of breakdown is termed a problem with **infrared-safety**. The consequence of this is that although simple cone algorithms are perfectly feasible at the hadron level they have no relevant theoretical interpretation in terms of perturbation theory. This renders the comparison of theory and experiment, using these algorithms, seriously flawed at best.

A remedy to this problem of the simple cone algorithm above is to introduce additional seed directions throughout; this was the philosophy underlying the introduction of the "Midpoint cone" algorithm. In this algorithm, extra seeds are placed between every pair of stable cones having a separation of less than $2R_0$, twice the size of the clustering cones. However, a careful analysis reveals that such solutions only postpone the infrared-safety problem to some higher order in perturbation theory, which of course means that they are not very satisfying. Considering all possible directions of jet cones would eliminate any infrared-safety problems to all orders; but while this is feasible for a low-multiplicity parton final state, it becomes computationally expensive for true experimental data, or indeed parton shower predictions, tamed only by the granularity of real-world detectors. In summary, this problem of infrared-safety renders "seeded" cone-algorithms in principle problematic for any meaningful comparison between theoretical calculations and experimental data. In practice, however, a careful analysis of the midpoint and SISCone algorithms (see below), as applied to jet physics at the TEVATRON, shows only marginal differences between the two. This ultimately motivates the use of TEVATRON data with jets defined through the midpoint algorithm which was the standard there.

A practical solution maintaining the idea of perfectly cone-sized jets was found by abandoning the idea of seeds and replacing them with innovative geometrical methods to reduce the computational complexity [836] and resulted in the **Seedless Infrared-Safe Cone (SISCone)** algorithm. This algorithm suffers none of the limitations of its forebears but retains a relatively intuitive physical picture. However, in the first years of data-taking at the LHC another class of jet finders, the k_T-algorithms became the standard tool.

2.1.6.5 k_T algorithms

The k_T family of algorithms [304, 345–347, 480, 511, 898] defines jets through a procedure that follows an idea rather different from the one employed in simple cone algorithms. The latter use a predefined, regular shape — the cone — to capture the relevant objects in the event, either partons, hadrons, or calorimeter entries. The k_T algorithm instead uses these basic objects as inputs from which to build jets in a recursive, iterative way. When these k_T-algorithms were originally proposed in [345–347, 511], they were constructed in such a way as to follow the natural pattern of QCD radiation. Such a procedure may of course lead to jets with fairly irregular shapes. With the introduction of a less well motivated clustering algorithm in form of the anti-k_T algorithm [304], this possible obstacle was overcome and the k_T-type of jet algorithms was established.

To cluster the objects in k_T-algorithms, it is necessary to introduce a generalized distance measure in momentum space, namely

$$
\begin{aligned}
d_{iB} &= (p_{\perp i})^{2p} & &\text{for each object } i \\
d_{ij} &= \min\left\{ (p_{\perp i})^{2p}, (p_{\perp j})^{2p} \right\} \frac{R_{ij}}{R_0} & &\text{for each pair of objects } i \text{ and } j .
\end{aligned}
\tag{2.45}
$$

The clustering now proceeds iteratively, where in each step the object(s) i (and j) with the smallest d are clustered, either with the beam, if d_{iB} is the smallest, or with each other, if d_{ij} is the smallest. This is repeated until all distances d are larger than a critical value d_{cut}, which in turn defines the relative transverse momentum two jets have with respect to the beam or to each other. The generalized form of the algorithm discussed here is thus controlled by two parameters, one denoting the cone-like size of the jets (R_0), and one specifying the power to which the transverse momenta entering these expressions are raised (p). The R_0 parameter here introduces a useful discrepancy in how relative transverse momenta impact on the jet definition: for typical values of $R_0 < 1$ this implies in particular that the relative transverse momentum between two jets, which is given by

$$p_\perp^2 = \min\{p_{\perp i},\, p_{\perp j}\}^2\, R_{ij} \qquad (2.46)$$

must be larger than their transverse momenta with respect to the beam, which effectively translates into a different treatment of initial and final state QCD radiation when analysing the structure of the overall radiation pattern, where the former is more susceptible to relative transverse momenta with respect to the beam and the latter is more sensitive to the relative transverse momentum of two final state objects.

The original k_T algorithm [346, 511] corresponds to the choice $p = 1$ in the above criteria, and in this realization jets are clustered together in order of increasing transerse momenta. The Cambridge–Aachen (CA) algorithm [480, 898] uses $p = 0$ and therefore clusters in a sequence of proximity in η-ϕ space.[11] In both specific implementations jets tend to have a fairly irregular shape, which renders any subtraction of hadronic activity due to pile-up or to the underlying event a cumbersome task. In contrast, the variant in which $p = -1$, known as the anti-k_T algorithm [304], results in fairly regular-shaped jets. Because of this feature, by now it has become the preferred choice in experimental analyses in recent years: it combines infrared-safety with fairly conical jets.

A variety of jet sizes are commonly used with the anti-k_T jet algorithm at the LHC, but unfortunately differ between ATLAS (0.4, 0.6) and CMS (0.5, 0.7). The tools now available, though, both for jet calibration and jet reconstruction, should allow for physics analyses to be carried out with multiple jet sizes and multiple jet algorithms.

2.1.6.6 Choice of jet size

Within an NLO calculation, a jet can consist of either one or two partons. The phase space in which two partons are included in the same jet is different for cone and k_T algorithms using the same jet size parameter. Define d as the separation in $\eta - \phi$ space between two partons and z as the ratio of the transverse momentum of the softer parton over the harder parton. The (d, z) phase space is shown in Fig. 2.12. In Region I, the two partons are closer than the jet radius R; both partons will be included in the same jet for both a cone and a k_T algorithm. Partons in Region III

---

[11]Note that in their original form, both the k_T [345] (or Durham) and the Cambridge–Aachen algorithm were defined for e^-e^+ annihilations into hadrons. In this realizationx, of course, R_{ij} is replaced by $\cos\theta_{ij}$, $R_0 = 1$, and the role of the transverse momenta in Eq. (2.45) is played by either the particles' energies or the absolute value of their three-momenta. It is interesting to also note that for e^-e^+ annihilations into hadrons another jet definition based on a clustering according to the $(p_i + p_j)^2$, the Jade-algorithm [204], was a popular choice. However, it turned out that this algorithm was inferior to the other two in its behaviour when mapping out the structure of QCD radiation and therefore providing a link between parton-level calculations and hadron-level data.

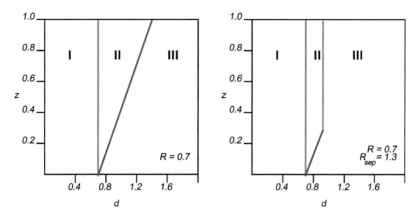

Fig. 2.12 The phase space for the two partons that can form a jet in a NLO calculation, parametrized by $z = p_{T,2}/p_{T,1}$ ($p_{T,1} \geq p_{T,2}$) and $d = \sqrt{(y_1 - y_2)^2 + (\phi_1 - \phi_2)^2}$. Reprinted with permission from Ref. [508].

will not be clustered into the same jet by either algorithm, while partons in Region II would nominally be clustered into the same jet with a cone jet algorithm, but not with a k_T jet algorithm. Thus, a cone jet effectively has a larger catchment area than a k_T jet with the same size parameter. In practice, with real data or Monte Carlo events, Region II is truncated for cone algorithms. This will be developed further in Chapter 8.

Inclusive measurements involving jets at the TEVATRON tended to involve relatively large jet sizes, typically with jet radius $R = 0.7$, such that most of the energy of the jet is included within the jet radius [61, 86, 119, 122]. For complex final states, such as $W + n$ jets, or $t\bar{t}$ production, it has been useful to use smaller jet sizes, in order to resolve the n-jet structure of the final state [59, 60, 81, 85, 88, 96, 99, 112].

Each jet size comes with its own benefits and drawbacks, both theoretically and experimentally. A smaller jet size reduces the impact of pile-up and the underlying event, but fragmentation effects are inversely proportional to the jet size, thus becoming more important as the jet size decreases. In addition, as R decreases, terms proportional to $\log R$ start becoming important, requiring a resummation of such terms (not present in fixed-order calculations) for a precise prediction.

2.1.6.7 Boosted objects and fat jets

With the advent of the LHC, another question related to how to deal with jets has emerged. It refers to situations where heavy particles, such as gauge or Higgs bosons or the top quark, are produced at transverse momenta much larger than their rest mass. Their decay products will, in such circumstances, be fairly collimated. If these heavy objects decay hadronically this will thus result in jet-like structures, also known as **fat jets**. To disentangle such **boosted objects** from ordinary jets, a careful analysis of their structure is mandatory. The first ideas of how to achieve this were presented by Seymour [842] and have received renewed attention more than a decade later. This was mainly driven by the work of Butterworth and collaborators [297], who showed that it may be possible to use such techniques to find a Higgs boson that decays to b-quarks at the LHC in gauge-boson associated production and thereby reviving an important search and measurement channel in Higgs boson physics. One of the key observations is that the angular separation of the decay products depends on the Higgs boson transverse momentum in a simple way.

To see this, one can write the Higgs boson four-momentum as $p = p_i + p_j$, with $p^2 = m_H^2$, and i and j the two decay products. Choosing a frame in which one of the decay products is directed along the x-axis, the momenta p_i and p_j are,

$$
\begin{aligned}
p_i &= p_\perp^i \left(\cosh \eta, \cos \phi, \sin \phi, \sinh \eta \right) , \\
p_j &= p_\perp^j \left(1, 1, 0, 0 \right) ,
\end{aligned}
\tag{2.47}
$$

In this frame, i has rapidity η and azimuthal angle ϕ while j has zero rapidity and azimuthal angle. Therefore the angular separation of the decay products is given by,

$$
\Delta R_{ij} = \sqrt{\eta^2 + \phi^2} , \quad (p_i + p_j)^2 = 2 p_\perp^i p_\perp^j \left(\cosh \eta - \cos \phi \right) .
\tag{2.48}
$$

If both η and ϕ are not too large, *i.e.* i and j are collimated, a series expansion yields

$$
\begin{aligned}
(p_i + p_j)^2 &= p_\perp^i p_\perp^j \left(\eta^2 + \phi^2 + \mathcal{O}(\eta^4, \phi^4) \right) \\
&= p_\perp^i p_\perp^j \left(\Delta R_{ij} \right)^2 + \mathcal{O}(\eta^4, \phi^4).
\end{aligned}
\tag{2.49}
$$

In the limit in which i and j are highly collimated one can replace the four-momenta of the decay products by the Higgs boson four momentum scaled by an appropiate energy fraction, *i.e.* $p_i = z p$ and $p_j = (1 - z)p$. From Eq. (2.49) it is then clear that,

$$
m_H^2 \approx z(1 - z) p_\perp^2 (\Delta R_{ij})^2 ,
\tag{2.50}
$$

and hence that,

$$
\Delta R_{ij} \approx \frac{m_H}{\sqrt{z(1 - z)} \, p_\perp} .
\tag{2.51}
$$

This also shows that there is an intrinsic angular scale to such a configuration since the separation is bounded below, within the approximation, by the value $2 m_H / p_\perp$. This observation paved the way for more detailed studies of boosted objects and jet substructure. Some of the recent ideas and studies that are of particular phenomenological importance are summarized in Refs. [106, 124, 144].

All jet algorithms and a number of jet substructure algorithms discussed in this section are incorporated in the FASTJET plugin [305].

2.2 Developing the formalism: W boson production at fixed order

2.2.1 Factorization formula

In collinear factorization introduced and motivated above, cross-sections for the hadronic production of an n-parton final state in a reaction of the type $h_1 h_2 \to n + X$ with hadrons h_1 and h_2 in the initial state can be written as

$$\sigma_{2\to n} = \sum_{a,b} \int_0^1 \mathrm{d}x_a \mathrm{d}x_b \, f_{a/h_1}(x_a, \mu_F) \, f_{b/h_2}(x_b, \mu_F) \, \hat{\sigma}_{ab\to n}(\mu_F, \mu_R)$$

$$= \sum_{a,b} \int_0^1 \mathrm{d}x_a \mathrm{d}x_b \, f_{a/h_1}(x_a, \mu_F) \, f_{b/h_2}(x_b, \mu_F) \frac{1}{2\hat{s}} \int \mathrm{d}\Phi_n \, |\mathcal{M}_{ab\to n}|^2(\Phi_n; \mu_F, \mu_R)$$

$$= \frac{1}{2s} \sum_{a,b} \int_0^1 \frac{\mathrm{d}x_a}{x_a} \frac{\mathrm{d}x_b}{x_b} \, f_{a/h_1}(x_a, \mu_F) f_{b/h_2}(x_b, \mu_F) \int \mathrm{d}\Phi_n \, |\mathcal{M}_{ab\to n}|^2(\Phi_n; \mu_F, \mu_R) \,.$$

$$(2.52)$$

A number of objects enters this master formula, namely:

- The $f_{a/h}(x, \mu)$ denote the *parton distribution functions* (PDFs), which depend on the light-cone momenta fractions x parton a has with respect to the hadron h, which incorporates it, and on the factorization scale μ_F.[12] The x also connect the centre-of-mass energies squared of the hadronic collision $E_{\text{c.m.}}^2(h_1 h_2)^2 = s$ with its partonic counterpart, $E_{\text{c.m.}}^2(ab) = \hat{s}$ through

$$\hat{s} = x_a x_b s \,. \qquad (2.53)$$

  At leading order, $f_{a/h}(x, \mu)$ can be interpreted as probabilities to find a in h with momentum fraction x at the (space-like)[13] resolution scale μ – an interpretation deeply connected to the case of deep-inelastic scattering, where factorization has been proven and where PDFs have traditionally been defined and measured. With this probabilistic interpretation at hand, the integral over the momentum fractions x is well understood.

- The **parton-level cross-section**, $\mathrm{d}\hat{\sigma}_{ab\to n}(\mu_F, \mu_R)$ which is given by the product of the incoming partonic flux

$$\frac{1}{4\sqrt{(p_a \cdot p_b)^2 - p_a^2 p_b^2}} \xrightarrow{m_{a,b}\to 0} \frac{1}{2\hat{s}} = \frac{1}{2x_a x_b s} \qquad (2.54)$$

  of the massless partons a and b, and the integral of the partonic transition amplitude squared, $|\mathcal{M}_{ab\to n}|^2(\Phi_n; \mu_F, \mu_R)$, over the available n-**parton phase-space element**, $\mathrm{d}\Phi_n$.[14] This phase-space element is given by

$$\mathrm{d}\Phi_n = \prod_{i=1}^n \left[\frac{\mathrm{d}p_i}{(2\pi)^4} (2\pi) \, \delta(p_i^2 - m_i^2) \Theta(p_i^{(0)}) \right] (2\pi)^4 \delta^4 \left(p_a + p_b - \sum_{i=1}^n p_i \right). \qquad (2.55)$$

---

[12]For some reminder of basic scattering kinematics, the reader is referred to Appendix A.3. There, the light-cone decomposition of momenta will be discussed.

[13]Systems with total four-momentum squared $Q^2 < 0$ are called space-like, those with $Q^2 > 0$ are time-like, and those with $Q^2 = 0$ are called light-like.

[14]Here and in the following, quantities at parton level are denoted by a circumflex (for example, $\hat{\sigma}$), while quantities at hadron level are left without it.

Each term in the square brackets is the Lorentz-invariant phase-space element of a particle i with momentum p_i and mass m_i. The additional δ-function at the end implements total four-momentum conservation.

- The **factorization scale** μ_F and the **renormalization scale** μ_R are process-dependent quantities. The former, μ_F signifies the scale at which the parton distribution functions are determined, while the latter, μ_R indicates the scale at which the coupling constants are evaluated, thus taking into account the quantum nature of the underlying theory. Typically, for simple processes which are characterized by one scale only, μ_F and μ_R are assumed identical to this scale. As an example for such a process, consider the case of W-production. On the parton level this is given by $q\bar{q}' \rightarrow W$; since the incoming partons can safely be taken massless, the only meaningful partonic scale in this process is the centre-of-mass energy $E_{\text{c.m.}} = \sqrt{\hat{s}}$ of the partonic system, which for an on-shell W boson is just the W-mass, m_W. However, for more complicated processes, and especially for those with multiple QCD emissions, picking a suitable scale is far from being trivial, *cf.* Section 2.2.6 for a further discussion.

Some general considerations on these objects will be presented in the following sections, employing the classical example of the production of a single vector boson.

2.2.2 Parton Distribution Functions

Having discussed the basic idea behind the factorization formula in Section 2.1 and its formalization in the master formula Eq. (2.52) for the calculation of scattering cross-sections at hadron colliders in Section 2.2.1, it is now time to discuss in a bit more detail the structure of the parton distribution functions (PDFs). They act as the link between the incoming hadrons in such a scattering and the incident elementary quanta, the partons, which are employed to calculate the cross-sections.

2.2.2.1 Valence quarks and simple sum rules

Concentrating on the case of incident protons, one could naively assume that they consist of three valence quarks. If they did not interact, there would be no reason to assume why one of these valence quarks would be preferred over the others. Therefore their respective PDFs would be given by

$$
\begin{aligned}
f_{u/p}(x, \mu^2) &= 2\delta\left(x - \frac{1}{3}\right) \\
f_{d/p}(x, \mu^2) &= \delta\left(x - \frac{1}{3}\right),
\end{aligned}
\tag{2.56}
$$

with all other PDFs being exactly zero. This would guarantee the **flavour sum rule**, given by

$$
\int_0^1 \mathrm{d}x \left[f_{u/p}(x, \mu^2) - f_{\bar{u}/p}(x, \mu^2) \right] = 2
$$

$$\int\limits_{0}^{1} \mathrm{d}x \, \left[f_{d/p}(x, \mu^2) - f_{\bar{d}/p}(x, \mu^2) \right] = 1$$

$$\int\limits_{0}^{1} \mathrm{d}x \, \left[f_{q/p}(x, \mu^2) - f_{\bar{q}/p}(x, \mu^2) \right] = 0 \text{ for } q \in \{s, c, b\}, \tag{2.57}$$

and the **momentum sum rule**,

$$\int\limits_{0}^{1} \mathrm{d}x \, x \sum_{i} f_{i/h}(x, \mu^2) = 1 \;\; \forall \mu^2 \text{ and for all hadrons } h, \tag{2.58}$$

where i runs over all partons, $\{u, \bar{u}, d, \bar{d}, \ldots, g\}$.

Switching on elastic interactions between the quarks, but still neglecting any emission of additional quanta, would not change this picture dramatically. The only effect of such a form of interaction, which can be thought of as some kind of "rubber bands" holding the quarks together, would be to smear out the sharp peak at $x = 1/3$ and replace it with a probability density such that its expectation value would be $1/3$:

$$\left\langle x|_{f_{u,d/p}(x, \mu^2)} \right\rangle = \frac{1}{3}. \tag{2.59}$$

2.2.2.2 QCD effects and scaling violations

A drastic change, however, occurs, when the emission of quanta without their immediate reabsorption is considered, along the lines of what was discussed in Section 2.1. As has been shown, these additional partons, which are summarily denoted as "sea" or "sea-partons", would have a finite lifetime that increases with decreasing momentum fraction x. This leads to these sea partons typically having larger probability distributions at small rather than at large values of x. In fact, it is worth analysing the kinematics of parton emission as described by the QCD analogue of the splitting functions given in Eq. (2.33).

They will lead to gluon emissions off valence quarks typically taking place for values of z close to 1, *i.e.* for small momentum fractions $(1 - z)$ of the gluon with respect to the quark. As sea quarks as well as additional gluons emitted by these gluons inherit their kinematics, the perturbative production of sea partons favours them to have small values of x. Consequently, the combination of lifetime arguments and the form of the splitting functions induces non-zero sea PDFs, which typically steeply increase for decreasing x. Naively, and to a fairly good approximation,

$$f_{\text{sea}/p}(x, \mu^2) \propto x^{-\lambda}, \tag{2.60}$$

where $\lambda \approx 1$ for gluons and sea quarks and $\lambda \approx -1/2$ for valence quarks. This behaviour is contrasted with the cases of no or elastic, "rubber-band"-type interactions in a sketchy way in Fig. 2.13. Apart from the increase for small x due to the sea partons, it is worth noting that also the distribution of valence partons, which for elastic interactions

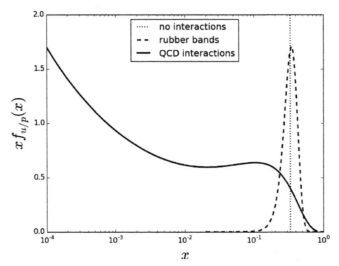

Fig. 2.13 Sketch of the up-quark PDF for different interactions: valence quarks only, without and with elastic interactions and full QCD.

is centred around $x \approx 1/3$, shifts towards smaller x values of about 0.1 and broadens out. This of course is due to them losing energy due to the emission of sea partons. And in fact, while the PDF – from a CT10 NLO set [551] – displayed in Fig. 2.13 for an u-quark is taken at relatively low scales of $\mu_F = 10\,\text{GeV}$, this slight "valence bump" at $x \approx 0.1$ persists also to higher scales.

This picture of secondary parton radiation indeed is very similar to the case of photons emitted by an electron. There, the photons play the role of the sea, allowing the photons to split into electron–positron pairs supplements this sea with additional fermions.

Putting this together yields a picture of the proton in the x-Q^2 plane as sketched in Fig. 2.14. The behaviour of the PDFs is entirely calculable with perturbative methods, and is in fact given by the **Dokshitser–Gribov–Lipatov–Altarelli–Parisi (DGLAP) equation** for QED, *cf.* Eq. (2.8), with the starting condition

$$f_{e/e}(x, 0) = \delta(1 - x) \quad \text{and} \quad f_{\gamma/e}(x, 0) = 0\,. \tag{2.61}$$

In contrast to this fully known and entirely perturbative QED case, the case of QCD is more complicated, owing to the essentially non-perturbative infrared structure of the theory. There, the perturbative regime is bound from below; typically it is assumed that perturbative QCD breaks down for scales of the order of $1 - 2$ GeV and below. This implies that the starting conditions for a DGLAP evolution must be taken from data, a subject that will be the focus of a more detailed discussion in Chapter 6. As already stated in Section 2.1.3, the scale evolution of the PDFs is given by the DGLAP equation, Eq. (2.31), where q denotes all quark flavours q and their anti-quarks.

A consequence of this scaling behaviour is that for increasing scales μ the sea

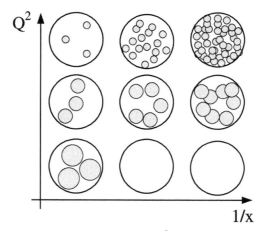

Fig. 2.14 Sketch of the proton in the x-Q^2 plane. At large x ($\approx 1/3$), the number of partons in the proton remains roughly the same, given by the valence partons. In contrast, at smaller x and larger Q^2 significant scaling of the parton number takes place. The "transverse size" of the resolved partons roughly scales with $1/Q$.

contribution is logarithmically increasing, a phenomenon known as **scaling violation**. Pictorially speaking, the probability of finding a parton within a parton increases by probing with larger transverse resolution power, as provided by the scale. In this respect, partons are not elementary objects, *i.e.* point-like particles. Rather, they are objects which carry the quantum numbers of the fundamental fields of QCD, quarks and gluons, but they cannot be interpreted as "naked" particles at all scales. In other words, partons are non-elementary particles, with a complicated internal structure, but they behave like fundamental quanta if probed at a low enough scale. In this respect, the partons are very similar to the colliding electron of the electromagnetic analogy in Section 2.1, which is a superposition of various Fock states involving secondary quanta.

2.2.3 Partonic cross-section at leading order

The production of a lepton-neutrino pair $\ell^-\bar{\nu}_\ell$ or $\ell^+\nu_\ell$ in hadron collisions provides a great starting point for a more in-depth discussion of particle production in such processes. In the Standard Model, it is related to the exchange of a single $W^\pm$ boson, with subsequent decay into a lepton and its neutrino.

These processes are mediated by the weak interaction. Due to the underlying electroweak symmetry, the **weak coupling** g_W can be related at LO to the electromagnetic coupling through

$$g_W = \left[4\sqrt{2} G_F m_W^2 \right]^{\frac{1}{2}} = \frac{e}{\sin\theta_W} \tag{2.62}$$

with the **Weinberg angle** θ_W and the **Fermi constant** G_F. Numerically, the electromagnetic coupling in the Thompson limit and at the Z-pole are roughly

$$\alpha(\mu) = \frac{e^2(\mu)}{4\pi} \approx \begin{cases} \dfrac{1}{137} & \text{for} \quad \mu \to 0 \\[2mm] \dfrac{1}{128} & \text{for} \quad \mu = m_Z, \end{cases} \tag{2.63}$$

the sine of the Weinberg angle is about

$$\sin\theta_W \approx 0.23 \tag{2.64}$$

and the Fermi constant approximately is given by

$$G_F \approx 1.166 \cdot 10^{-5}\,\text{GeV}^{-2}. \tag{2.65}$$

2.2.3.1 Matrix element for on-shell W production

As a first step, consider the matrix element for the production of an on-shell W^+ boson. At leading order, *i.e.*, in the approximation of tree amplitudes only, there is only one diagram, and the corresponding matrix element reads

$$\mathcal{M}_{u\bar{d}\to W^+} = -\frac{iV_{ud}g_W\delta_{ij}}{\sqrt{2}}\,\bar{d}_i(p_2)\gamma^\mu\frac{1-\gamma_5}{2}u_j(p_1)\epsilon_\mu^{(W)}, \tag{2.66}$$

where the spinor arguments and subscripts indicate their momenta and colour indices and V_{ud} is the relevant element of the **Cabibbo–Kobayashi–Maskawa matrix**. This amplitude leads to the summed and squared expression

$$\begin{aligned}\overline{\sum}|\mathcal{M}_{u\bar{d}\to W^+}|^2 &= \frac{3}{9\cdot 4}\frac{|V_{ud}|^2 g_W^2}{2}\text{Tr}\left[\slashed{p}_2\gamma^\mu\slashed{p}_1\gamma^\nu\frac{1-\gamma_5}{2}\right]\left[-g_{\mu\nu}+\frac{Q_\mu Q_\nu}{m_W^2}\right]\\ &= \frac{|V_{ud}|^2 g_W^2}{12}Q^2 = \frac{|V_{ud}|^2 g_W^2}{12}m_W^2,\end{aligned} \tag{2.67}$$

where $Q = p_1 + p_2$ and the invariant mass of the boson, $\hat{s} = (p_1+p_2)^2 = 2(p_1 p_2) = Q^2 = m_W^2$, have been introduced. The factor 3 stems from the sum over three possible quark-line colours, the $1/9$ takes care of taking the average over all possible colour configurations of the quark and the anti-quark, and the factor $1/4$ reflects the average over the incoming quark spins.

2.2.3.2 Matrix elements for $u\bar{d}\to\nu_\ell\bar{\ell}$

The two diagrams displayed in Fig. 2.15 relate to the two different charge states W^+ and W^-. At leading order each of them is the only relevant one for each of the charge channels. The matrix element for the process $u\bar{d}\to\bar{\ell}\nu_\ell$, W^+ production and decay, reads

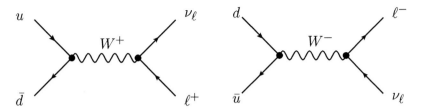

Fig. 2.15 Production of a $W^\pm$ boson and its leptonic decay at leading perturbative order. Here u and d stand for arbitrary up- and down-type quarks, respectively.

$$\mathcal{M}_{u\bar{d}\to\nu_\ell\bar{\ell}} = \left[\bar{v}_{\bar{d}}\left(\frac{-ig_W V_{ud}}{\sqrt{2}}\gamma_{\mu L}\right)u_u\right]\left[\bar{u}_\nu\left(\frac{-ig_W}{\sqrt{2}}\gamma_{\nu L}\right)v_{\bar{\ell}}\right]$$

$$\times \frac{-i}{(p_u+p_{\bar{d}})^2 - m_W^2 + im_W\Gamma_W}\left[g^{\mu\nu} - \frac{(p_u+p_{\bar{d}})^\mu(p_u+p_{\bar{d}})^\nu}{m_W^2}\right].$$

$$(2.68)$$

The terms in the first line correspond to the respective left-handed fermion currents, made manifest by the short-hand notation

$$\gamma_{\mu L} = \gamma_\mu \frac{1-\gamma_5}{2} \tag{2.69}$$

while the second line represents the W propagator connecting them.

A similar expression can be found for the case of W^- production, by suitably permuting the labels of the fermion spinors. Squaring yields

$$\sum|\mathcal{M}_{u\bar{d}\to\ell^+\nu_\ell}|^2 = \frac{3}{9\cdot4}\frac{|V_{ud}|^2 g_W^4}{4}\mathrm{Tr}\left[\not{p}_{\bar{d}}\gamma^\mu\not{p}_u\gamma^\rho\frac{1-\gamma_5}{2}\right]\mathrm{Tr}\left[\not{p}_{\nu_\ell}\gamma^\nu\not{p}_{\bar{\ell}}\gamma^\sigma\frac{1-\gamma_5}{2}\right]$$

$$\times \frac{\left(g_{\mu\nu} - \frac{Q_\mu Q_\nu}{m_W^2}\right)\left(g_{\rho\sigma} - \frac{Q_\rho Q_\sigma}{m_W^2}\right)}{(Q^2 - m_W^2)^2 + m_W^2\Gamma_W^2}$$

$$= \frac{|V_{ud}|^2 g_W^4}{12}\frac{\hat{t}^2}{(Q^2 - m_W^2)^2 + m_W^2\Gamma_W^2},$$

$$(2.70)$$

where the average over the initial quarks' spins and colours and the sum over the lepton spins in the final state is implicit. Here the **Mandelstam variables**

$$\hat{s} = Q^2 = (p_u+p_{\bar{d}})^2 \quad \text{and} \quad \hat{t} = (p_u-p_\ell)^2 \tag{2.71}$$

have been employed.

Rewriting the **phase-space** integral over the outgoing particles as

$$\int d\Phi_n = \int \frac{d^4 p_\ell}{(2\pi)^4} (2\pi)\delta(p_\ell^2) \frac{d^4 p_\nu}{(2\pi)^4} (2\pi)\delta(p_\nu^2)(2\pi)^4 \delta^4(p_u + p_d - p_\ell - p_\nu)$$
$$= \frac{1}{32\pi^2} \int d^2\Omega_\ell^* \,, \tag{2.72}$$

where Ω_ℓ^* is the solid angle of the outgoing lepton with respect to the incident u-type quark in the rest frame of the collision. In the same system, the $\hat{t} = -2p_u \cdot p_\ell$ becomes

$$\hat{t} = -2p_u \cdot p_\ell \xrightarrow{\text{cms}} -\frac{\hat{s}}{2}(1 - \cos\theta^*)\,, \tag{2.73}$$

where θ^* is the polar angle of the lepton with respect to the incoming u-type quark. This allows to rewrite

$$\hat{\sigma}^{(LO)} = \frac{1}{2\hat{s}} \int \frac{d^2\Omega_\ell^*}{32\pi^2} |\mathcal{M}|^2_{u\bar{d}\to\nu_\ell\bar{\ell}} = \frac{g_W^4 |V_{ud}|^2}{12 \cdot 2\hat{s}} \int_{-1}^{1} \frac{2\pi d\cos\theta^*}{4 \cdot 32\pi^2} \frac{\hat{s}^2(1 - \cos\theta^*)^2}{\left[(\hat{s} - m_W^2)^2 + m_W^2\Gamma_W^2\right]}$$

$$= \frac{g_W^4 |V_{ud}|^2}{576\pi} \frac{\hat{s}}{(\hat{s} - m_W^2)^2 + m_W^2\Gamma_W^2}\,, \tag{2.74}$$

and thus

$$\sigma^{(LO)}_{h_1 h_2 \to \nu_\ell \bar{\ell}} = \frac{g_W^4 |V_{ud}|^2}{576\pi} \int dy_W \, d\hat{s} \left[\frac{1}{[\hat{s} - m_W^2]^2 + m_W^2\Gamma_W^2} \right.$$

$$\left. \times \sum_{u,\bar{d}} x_u f_{u/h_1}(x_u, \mu_F) x_{\bar{d}} f_{\bar{d}/h_2}(x_{\bar{d}}, \mu_F) \right]\,, \tag{2.75}$$

where the integral over the energy fractions has been written as

$$dx_u dx_{\bar{d}} = \frac{d\hat{s}}{s} dy_W \,. \tag{2.76}$$

This is because the $x_{u,\bar{d}}$ can be related to the centre-of-mass energy squared and the centre-of-mass rapidity through

$$\hat{s} = x_u x_{\bar{d}} s$$
$$y_{\text{c.m.}} = \frac{1}{2} \log \frac{x_u}{x_{\bar{d}}} \,. \tag{2.77}$$

In other words, the cross-section is written as an integral over the invariant mass squared and rapidity of the produced system.

Going a step further allows to calculate a differential cross-section with respect to the lepton rapidity. The trick here is to relate its **rapidity** $\hat{y}_{\bar{\ell}}$ in the partonic c.m.-frame with its rapidity $y_{\bar{\ell}}$ in the hadronic c.m.-frame. This is fairly straightforward, since rapidities are additive along the same boost axis and therefore

$$y_{\bar{\ell}} = \hat{y}_{\bar{\ell}} + y_W \, . \tag{2.78}$$

For massless particles, such as the lepton in the process here, also rapidities coincide exactly with **pseudorapidities**, therefore, in the partonic c.m.-frame

$$\hat{y}_{\bar{\ell}} = \log \cot \frac{\theta^*}{2} = \frac{1}{2} \log \frac{1 + \cos \theta^*}{1 - \cos \theta^*} \tag{2.79}$$

or

$$\sin \theta^* = \frac{1}{\cosh \hat{y}_{\bar{\ell}}} \, . \tag{2.80}$$

This means that

$$\mathrm{d} \cos \theta^* = \sin^2 \theta^* \mathrm{d} \hat{y}_{\bar{\ell}} = \sin^2 \theta^* \mathrm{d} y_{\bar{\ell}} \, . \tag{2.81}$$

Finally, therefore

$$\frac{\mathrm{d}\sigma_{h_1 h_2 \to \bar{\ell} \nu_{\ell}}}{\mathrm{d}y_{\bar{\ell}}} = \int \mathrm{d}x_u \mathrm{d}x_{\bar{d}} f_{u/h_1}(x_u, \mu_F) f_{\bar{d}/h_2}(x_{\bar{d}}, \mu_F) \frac{\sin^2 \theta^* \mathrm{d}\hat{\sigma}_{u\bar{d} \to \bar{\ell} \nu_{\ell}}}{\mathrm{d} \cos \theta^*} \, . \tag{2.82}$$

2.2.3.3   Narrow-width approximation

The result in Eq. (2.75) can be further simplified by noting that the propagator — a Breit–Wigner form — suppresses values of $\hat{s}$ away from m_W^2. This observation is manifest in a simplification known as the **narrow-width approximation** (NWA). In this approximation, the propagator factor for an internal particle X with mass M_X and width Γ_X is replaced according to

$$\frac{\mathrm{d}\hat{s}}{(\hat{s} - M_X^2)^2 + M_X^2 \Gamma_X^2} \longrightarrow \frac{\pi}{M_X \Gamma_X} \mathrm{d}\hat{s} \, \delta(\hat{s} - M_X^2) \, , \tag{2.83}$$

where the overall factor outside the δ-function ensures that the replacement does not change the value of the integral. Applying such an approximation will result in a sharp mass distribution of the decay products of the propagator particle. Depending on the actual measurement, and, of course, the values of the internal particles mass and width, this then may yield unphysical results. As a rule of thumb it can be argued that this may be the case if a measurement of kinematic observables of the decay products is more accurate than Γ_X / M_X. In NWA,

$$\sigma_{h_1 h_2 \to \nu_{\ell} \bar{\ell}}^{(\mathrm{LO})} = \frac{g_W^4 \, |V_{ud}|^2}{576 s} \frac{m_W}{\Gamma_W} \int\limits_{-y_{\mathrm{max}}}^{y_{\mathrm{max}}} \mathrm{d}y_W \sum_{u, d} f_{u/h_1} \left(\frac{m_W e^{y_W}}{\sqrt{s}}, \mu_F \right) f_{\bar{d}/h_2} \left(\frac{m_W e^{-y_W}}{\sqrt{s}}, \mu_F \right) , \tag{2.84}$$

where the W rapidity y_W is constrained by $x_u x_{\bar{d}} s = m_W^2$ and therefore

$$|y_W| \leq y_{\mathrm{max}} = \frac{1}{2} \log \frac{s}{m_W^2} \, . \tag{2.85}$$

Quite often, the fact that this is an approximation only is further aggravated by also ignoring the Lorentz structures in the numerator of the propagator, thus effectively forfeiting any knowledge of eventual correlations between initial and final state particles. In such a case, the matrix element squared for a typical $2 \to 2$ s-channel process $ab \to X \to cd$ with an intermediate particle X (like the one studied here, W-production and decay) becomes proportional to the respective branching ratios:

$$|\mathcal{M}|^2_{ab \to X \to cd} \propto \mathcal{BR}_{X \to ab}\, \mathcal{BR}_{X \to cd} \,. \tag{2.86}$$

2.2.3.4 W forward-backward asymmetry

Alternatively one could ignore the decay of the intermediate particle and focus on its on-shell production. Employing this approximation for the production of the W boson for the moment, it is possible to discuss the on-shell production of a W boson. Its matrix element squared at leading order is given by

$$|\mathcal{M}|^2_{u\bar{d} \to W^+} = \frac{g_W^2 |V_{ud}|^2 m_W^2}{12}\,, \tag{2.87}$$

resulting in the parton-level cross-section

$$
\begin{aligned}
\hat{\sigma}^{(\mathrm{LO})}_{u\bar{d} \to W^+} &= \frac{1}{2\hat{s}} \int \frac{\mathrm{d}^4 p_W}{(2\pi)^4} (2\pi)^4 \delta^4(p_u + p_{\bar{d}} - p_W)(2\pi)\delta(p_W^2 - m_W^2)|\mathcal{M}|^2_{u\bar{d} \to W^+} \\
&= \frac{\pi\delta(\hat{s} - m_W^2)}{\hat{s}}|\mathcal{M}|^2_{u\bar{d} \to W^+} = \frac{\pi\delta(\hat{s} - m_W^2)}{\hat{s}} \frac{g_W^2 |V_{ud}|^2 m_W^2}{12} \\
&\longrightarrow \frac{4\pi^2 \alpha |V_{ud}|^2}{12 \sin^2\theta_W m_W^2}\,,
\end{aligned}
\tag{2.88}
$$

where in the last line the implicit integration over $\hat{s}$ with has been used together with the δ function and where $g_W = e/\sin\theta_W$ has been employed. The production cross-section in hadronic collisions thus becomes

$$
\begin{aligned}
\sigma^{(\mathrm{LO})}_{h_1 h_2 \to W^+} &= \int_0^1 \mathrm{d}x_u \mathrm{d}x_{\bar{d}} \sum_{u,\bar{d}} f_{u/h_1}(x_u, \mu_F)\, f_{\bar{d}/h_2}(x_{\bar{d}}, \mu_F)\, \hat{\sigma}^{(\mathrm{LO})}_{u\bar{d} \to W^+} \\
&= \frac{\pi g_W^2 |V_{ud}|^2}{12} \int \frac{m_W^2 \mathrm{d}\hat{s}}{\hat{s}^2}\, \delta(\hat{s} - m_W^2) \\
&\qquad\qquad \times \int_{-y_{\max}}^{y_{\max}} \mathrm{d}y_W \sum_{u,\bar{d}} x_u f_{u/h_1}(x_u, \mu_F)\, x_{\bar{d}} f_{\bar{d}/h_2}(x_{\bar{d}}, \mu_F)\Big|_{x_u x_{\bar{d}} s = m_W^2} \\
&= \frac{\pi g_W^2 |V_{ud}|^2}{12s} \int_{-y_{\max}}^{y_{\max}} \mathrm{d}y_W \sum_{u,\bar{d}} f_{u/h_1}(x_u, \mu_F)\, f_{\bar{d}/h_2}(x_{\bar{d}}, \mu_F)\,,
\end{aligned}
$$

$$\tag{2.89}$$

where $E = \sqrt{s}$ is the hadronic centre-of-mass energy and the limits of the rapidity integral $\pm y_{\max} = \pm\frac{1}{2}\log\frac{s}{m_W^2}$ follow from momentum conservation, $\hat{s} = x_u x_{\bar{d}} s = m_W^2$ and the relation of the $x_{1,2}$ with the rapidity of the produced system $y_{\mathrm{c.m.}}$, *cf.* Eqs. (2.76) and (2.77). For future reference it is also useful to introduce

$$\sigma^{(LO)}_{u\bar{d}\to W^+} = \frac{\pi g_W^2 \, |V_{ud}|^2}{12s} \tag{2.90}$$

so that

$$\sigma^{(LO)}_{h_1 h_2 \to W^+} = \sigma^{(LO)}_{u\bar{d}\to W^+}(s) \int\limits_{-y_{\max}}^{y_{\max}} \mathrm{d}y_W \sum_{u,\bar{d}} f_{u/h_1}(x_u, \mu_F)\, f_{\bar{d}/h_2}(x_{\bar{d}}, \mu_F). \tag{2.91}$$

The result above, Eq. (2.91), implies that the rapidity distribution of the W boson is entirely defined by the PDFs, and, at leading order, by the quark PDFs only. While the sea is more or less flavour symmetric, the valence contribution is not. In protons (anti-protons) there are twice as many valence u quarks ($\bar{u}$ anti-quarks) than d quarks ($\bar{d}$ anti-quarks). This has the following implications: in proton–anti-proton collisions, like at the Fermilab TEVATRON, both species of W bosons, W^+ and W^- bosons, have exactly the same production cross-section. However, W^+ bosons tend to fly more likely in the direction of the protons, the forward direction, while the W^- bosons tend to follow more likely the direction of the anti-protons, the backward direction. This is because they are more likely to obtain a strong "kick" into the respective direction by an up-type rather than a down-type valence quark. In proton–proton collisions like at the CERN LHC, this picture does not hold true any longer. There, the cross-sections for W^+ production is larger than the cross-section for W^- production — in fact, if in both cases a valence quark would have to be involved, the would differ by a factor of two. This is not true, since there is a sizable contribution from the other sea quarks and, at higher orders, from incident gluons. Also, their rapidity distributions are not a reflection of each other around central rapidity any longer, but each of them of course is symmetric under reflections around $y = 0$. In addition, the W^+ bosons tend to have a slightly larger rapidity, due to the higher probability to obtain a strong "kick" from an incident valence quark, while the W^- bosons are more central. This behaviour and the comparison of it at the TEVATRON and the LHC at different c.m.-energies is exhibited in Fig. 2.16. It is worth noting that the shape at symmetric proton–proton collisions also changes quite dramatically, as the c.m.-energy of the colliding hadrons increases from 8 to 100 TeV. This is due to the fact that with increasing energies the x the quarks need decreases, leading to the contributions of the sea-quarks becoming larger and larger. The valence quarks, when annihilating anti-quarks from the sea, lead to a pronounced boost of the W boson, and, thus, a depletion in the region of central rapidities. This region is being filled by the more symmetric annihilation of pairs of sea partons. At 8 and 14 TeV this leads to the plateau-shape of the distribution. This plateau of course widens in rapidity with the energy of the colliding hadrons. At 100 TeV, however, the sea contribution takes over shaping a mount at central rapidities.

This behaviour is completely driven by the PDFs and the interplay of valence and

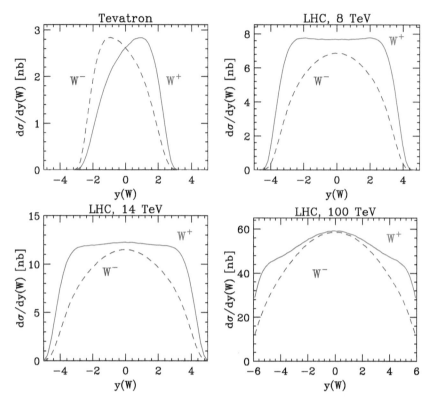

Fig. 2.16 Rapidity distributions of $W^{\pm}$ bosons at the TEVATRON, at the LHC at c.m.-energies of 8 and 14 TeV, and at a future hadron collider with a c.m.-energy of 100 TeV. The calculation has been performed at leading order with the CT10 PDF [713].

sea contributions at high and low values of x, respectively. In fact, measurements of the W^{+}–W^{-} asymmetries are a very useful way of constraining high-x valence quark PDFs [551]. However, the simple picture outlined above is not easy to translate into a measurement, the problem of course being that the W bosons decay into a lepton and an invisible neutrino. The latter makes it very hard to reconstruct the W kinematics, since neutrinos manifest themselves as missing transverse energy at hadron colliders, which leaves the charged lepton only. The corresponding **lepton asymmetry** thus reads

$$\mathcal{A}_{\ell} = \frac{\left(\dfrac{\mathrm{d}\sigma_{\ell^{+}}}{\mathrm{d}y_{\ell^{+}}}\right) - \left(\dfrac{\mathrm{d}\sigma_{\ell^{-}}}{\mathrm{d}y_{\ell^{-}}}\right)}{\left(\dfrac{\mathrm{d}\sigma_{\ell^{+}}}{\mathrm{d}y_{\ell^{+}}}\right) + \left(\dfrac{\mathrm{d}\sigma_{\ell^{-}}}{\mathrm{d}y_{\ell^{-}}}\right)}, \qquad (2.92)$$

where the $y_{\ell^{\pm}}$ are the rapidities of the positively and negatively charged leptons.

This complication gives rise to a subtle effect impacting on the measurement. Analysing the interplay of momentum transfer and parton centre-of-mass energy in

Eq. (2.73), which defines the form of the differential cross-section, it becomes clear that the positively charged leptons prefer to travel anti-parallel to the incident up-type quark. At the TEVATRON, this means that positively charged leptons preferably move in the direction of the anti-proton, and the negatively charged leptons preferably move in the direction of the proton, *i.e.*, in both cases, the leptons tend to move against the direction of the *W* boson they come from, partially compensating their initial boost.

In proton–proton collisions, the picture then looks a bit more confusing. Naively, one would expect to have more W^+ than W^- bosons, and typically they would stretch out to larger rapidities, leading to the asymmetry always being positive and actually increasing with increasing rapidity. However, the *W* bosons have a maximal rapidity, about $y_{max} \approx 4.5$ at a 8 TeV LHC, while the leptons, being effectively massless, can reach rapidities well beyond this point. This implies that at some point the typical relative rapidity the leptons have with respect to the original *W* boson becomes the dominating factor, and, as discussed, this is where the negatively charged leptons will become more abundant than the positively charged ones. This means that the asymmetry will turn negative. The fact that most of the bosons are not at maximal rapidity (which is possible only if one of the x, typically x_u equals 1) and the fact that the positively charged leptons are typically oriented against the direction of the W^+ boson, translate into this point being significantly more central than the maximal rapidity available to the *W* bosons. This behaviour is exhibited in Fig. 2.17, where the lepton asymmetry at leading order is displayed.

2.2.4   *W* + jet production at leading order

In this section, the emission of an additional parton will be interpreted as the production of the vector boson in association with a jet. The structure of this process will be analysed, with special emphasis on the collinear and soft limits of the matrix element.

2.2.4.1   Structure of the matrix elements

Consider **real corrections** to a given process, for example the real corrections to *W* production at hadron colliders. Relevant diagrams are exemplified in Fig. 2.18. If the additional parton is energetic enough, it will leave a trace in the detector by depositing some hadronic energy. If this deposit is well-separated enough, it is typically interpreted as an extra jet. Of course, the cross-section for such a process depends strongly on the separation criterion; here just a minimal transverse momentum of the extra parton will be demanded. By virtue of momentum conservation this will then immediately lead to the *W* boson recoiling against this jet, *i.e.*, acquiring transverse momentum as well.

Looking at them, it is important to stress that the *Feynman diagrams are nothing but pictorial representations of quantum mechanical transition amplitudes, interfering with each other.* At the same time, their quantum nature renders any question like "which of the two incident quarks emitted the outgoing gluon?" completely unphysical and meaningless. Such questions represent a futile and uncomprehending attempt to transport classical concepts to the quantum world. They are fully equivalent to the question of which slit the electron passed through in the double-slit experiment.

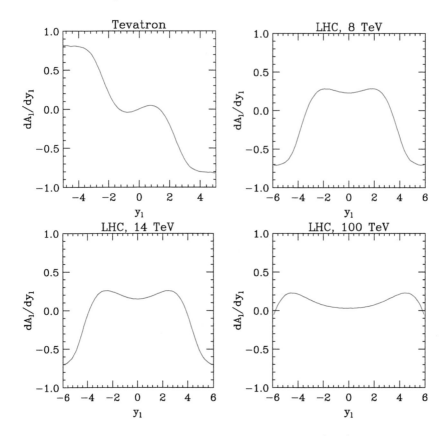

Fig. 2.17 The lepton asymmetry, defined in Eq. (2.92), at the TEVATRON, at the LHC at c.m.-energies of 8 and 14 TeV, and at a future hadron collider with a c.m.-energy of 100 TeV. The calculation has been performed at leading order with the CT10 PDF [713].

With this in mind, the real contributions can be decomposed into three sets of diagrams each with two interfering amplitudes: one set where a gluon is emitted into the final state, *i.e.* the sub-process $u\bar{d} \to gW^+[\to \nu_\ell\bar{\ell}]$, and two sub-processes where the initial gluon gives rise to either a down-type quark or an up-type anti-quark, $ug \to dW^+[\to \nu_\ell\bar{\ell}]$ and $\bar{d}g \to \bar{u}W^+[\to \nu_\ell\bar{\ell}]$, respectively. These three sets of sub-processes do not interfere, since their initial and their final states are composed of particles that, in principle, could possibly be distinguished.

Ignoring, for simplicity, the decay of the W-boson, the resulting amplitudes are given by

$$\mathcal{M}_{u\bar{d}\to gW^+} = \frac{ig_s g_W V_{ud}}{\sqrt{2}} \, \bar{v}_{d,i} \left[\gamma_\nu \, T_{ij}^a \frac{\slashed{p}_{\bar{d}} - \slashed{p}_g}{(p_{\bar{d}} - p_g)^2} \gamma_{\mu L} \right.$$
$$\left. + \gamma_{\mu L} \frac{\slashed{p}_u - \slashed{p}_g}{(p_u - p_g)^2} \gamma_\nu \, T_{ij}^a \right] u_{u,j} \epsilon_W^\mu \epsilon_g^{\nu,a} \,, \tag{2.93}$$

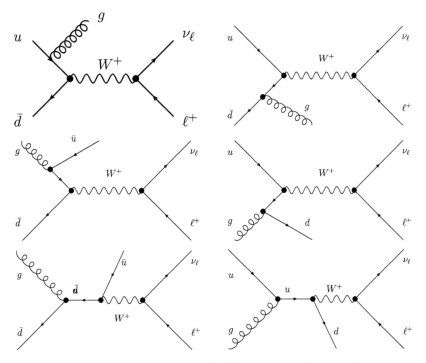

Fig. 2.18 Real contributions to the NLO correction of the production and leptonic decay of a W^+ boson. Here u and d stand for arbitrary up- and down-type quarks, respectively.

$$
\mathcal{M}_{ug \to dW^+} = \frac{ig_s g_W V_{ud}}{\sqrt{2}}\, \bar{u}_{d,i} \left[\gamma_\nu\, T^a_{ij}\, \frac{\not{p}_g - \not{p}_d}{(p_g - p_d)^2}\, \gamma_\mu L \right.
$$
$$
\left. + \gamma_\mu L\, \frac{\not{p}_u + \not{p}_g}{(p_u + p_g)^2}\, \gamma_\nu\, T^a_{ij} \right] u_{u,j}\, \epsilon^\mu_W\, \epsilon^{*\nu,a}_g
$$

(2.94)

and

$$
\mathcal{M}_{\bar{d}g \to \bar{u}W^+} = \frac{ig_s g_W V_{ud}}{\sqrt{2}}\, \bar{v}_{d,i} \left[\gamma_\nu\, T^a_{ij}\, \frac{\not{p}_g + \not{p}_{\bar{d}}}{(p_g + p_{\bar{d}})^2}\, \gamma_\mu L \right.
$$
$$
\left. + \gamma_\mu L\, \frac{\not{p}_g - \not{p}_{\bar{u}}}{(p_g - p_{\bar{u}})^2}\, \gamma_\nu\, T^a_{ij} \right] v_{u,j}\, \epsilon^\mu_W\, \epsilon^{*\nu,a}_g
$$

(2.95)

Here, all colour indices and structures have been made explicit, by adding the fundamental (or triplet) colour indices i and j of the quarks and the adjoint (or octet) colour index a of the gluon as well as the colour matrix T^a_{ij} appearing in the quark–quark–gluon vertex. Looking at these expressions, it becomes apparent that the diagrams contain potentially divergent structures. The divergences emerge in those cases, where the additional parton in the final state becomes **soft** or **collinear**. The propagator of the intermediate parton line reads

$$\frac{1}{(p_q - p_g)^2} = -\frac{1}{2E_q E_g (1 - \cos\theta)}, \tag{2.96}$$

which diverges for the energy of the outgoing parton, E_{out}, or its opening angle θ approaching 0. This type of divergent structure is also known as an **infrared divergence**.

2.2.4.2 Matrix elements squared

After squaring and averaging/summing over initial/final-state polarizations and colours and some colour algebra, the squared matrix elements are given by

$$|\mathcal{M}|^2_{u\bar{d}\to gW^+} = \frac{4\pi\alpha_s C_F \, g_W^2 |V_{ud}|^2}{12} \frac{\hat{t}^2 + \hat{u}^2 + 2m_W^2 \hat{s}}{\hat{t}\hat{u}} \tag{2.97}$$

and

$$|\mathcal{M}|^2_{ug\to dW^+} = |\mathcal{M}|^2_{\bar{d}g\to\bar{u}W^+} = \frac{4\pi\alpha_s T_R \, g_W^2 |V_{ud}|^2}{12} \frac{\hat{s}^2 + \hat{u}^2 + 2m_W^2 \hat{t}}{-\hat{s}\hat{u}}. \tag{2.98}$$

In all cases **Mandelstam variables** have been used, namely

$$\begin{aligned}
\hat{s} &= (p_a + p_b)^2 = (p_1 + p_2)^2, \\
\hat{t} &= (p_a - p_1)^2 = (p_b - p_2)^2, \\
\hat{u} &= (p_a - p_2)^2 = (p_b - p_1)^2,
\end{aligned} \tag{2.99}$$

where the incoming partons are labelled with a and b, and the outgoing particles are labelled with 1 and 2. These variables satisfy

$$\hat{s} + \hat{t} + \hat{u} = m_a^2 + m_b^2 + m_1^2 + m_2^2, \tag{2.100}$$

the sum of the squares of the masses of the external particles.

Closer inspection reveals that the squared matrix elements above can be written as the leading-order matrix element squared times a QCD emission term, which consists of the strong coupling and a colour factor times an expression representing the kinematics of the extra emission:

$$\begin{aligned}
|\mathcal{M}|^2_{u\bar{d}\to gW^+} &= \frac{|\mathcal{M}^{(\text{LO})}|^2_{u\bar{d}\to W^+}}{m_W^2} \cdot (4\pi\alpha_s C_F) \frac{\hat{t}^2 + \hat{u}^2 + 2m_W^2 \hat{s}}{\hat{t}\hat{u}} \\
|\mathcal{M}|^2_{ug\to dW^+} = |\mathcal{M}|^2_{\bar{d}g\to\bar{u}W^+} &= \frac{|\mathcal{M}^{(\text{LO})}|^2_{u\bar{d}\to W^+}}{m_W^2} \cdot (4\pi\alpha_s T_R) \frac{\hat{s}^2 + \hat{u}^2 + 2m_W^2 \hat{t}}{-\hat{s}\hat{u}}.
\end{aligned} \tag{2.101}$$

2.2.4.3 Soft and collinear limits

It becomes apparent that the result of the processes with gluon emission into the final state diverge, if either $\hat{t} \to 0$ or $\hat{u} \to 0$. This is the case, when the gluon is either parallel with one of the incoming particles, the collinear divergence, or if its energy vanishes, the soft divergence, in wich case both $\hat{t}$ and $\hat{u}$ go to zero. The divergent regions of the gluon emission phase space can be avoided by suitable cuts. A convenient choice here is to demand that the gluon has a minimal transverse momentum, which would be interpreted as the minimal transverse momentum of the corresponding jet.

Similar reasoning also applies to the case where the gluon appears in the initial state, which exhibits a divergence with $\hat{u} \to 0$. There the case $\hat{s} \to 0$ is prohibited by the finite and sufficiently large mass of the W boson. This also prohibits the limit of the gluon energy going to 0, with the result that this process is less divergent than the one with the gluon in the final state. However, its remaining, purely collinear divergence, can also be avoided by a cut on the transverse momentum of the extra parton.

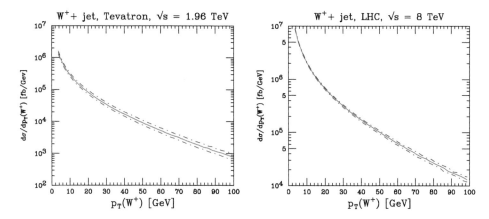

Fig. 2.19 The differential cross-section for $W^+ + j$ production with respect to the transverse momentum of the W^+ boson at the TEVATRON (left) and the LHC (right) at c.m.-energies of 1.96 TeV and 8 TeV, respectively. The error bands are obtained from a variation of the factorization and renormalization scales by a factor of two.

Assuming massless incoming and outgoing partons, the outgoing momenta can be written as

$$
\begin{aligned}
p_W^\mu &= (m_{\perp W} \cosh y_W,\ p_\perp \cos\phi,\ p_\perp \sin\phi,\ m_{\perp W} \sinh y_W) \\
p_{q,g}^\mu &= (p_\perp \cosh y_{q,g},\ -p_\perp \cos\phi,\ -p_\perp \sin\phi,\ p_\perp \sinh y_{q,g})\ ,
\end{aligned}
\tag{2.102}
$$

where $p_\perp$ is the transverse momentum of the jet and, by momentum conservation, of the W boson, and $m_{\perp W}$ is its transverse mass given by

$$
m_{\perp W} = \sqrt{m_W^2 + p_{\perp W}^2} = \sqrt{m_W^2 + p_\perp^2}\ .
\tag{2.103}
$$

The incoming momenta read

$$
p_{1,2}^\mu = x_{1,2} \frac{\sqrt{s}}{2}\, (1, 0, 0, \pm 1)
\tag{2.104}
$$

and therefore, with $x_\perp = p_\perp/\sqrt{s}$

$$
\begin{aligned}
\hat{s} &= x_1 x_2 s \\
\hat{t} &= -2p_1 p_{q,g} = -x_1 x_\perp s\, e^{-y_{q,g}} \\
\hat{u} &= -2p_2 p_{q,g} = -x_2 x_\perp s\, e^{+y_{q,g}}
\end{aligned}
\tag{2.105}
$$

The kinematic part of the gluon-emission matrix element squared thus becomes

$$
\frac{\hat{t}^2 + \hat{u}^2 + 2m_W^2 \hat{s}}{\hat{t}\hat{u}} = \frac{x_\perp^2 \left(x_1^2 e^{-2y_g} + x_2^2 e^{2y_g} \right) + 2x_1 x_2 x_M^2}{x_1 x_2 x_\perp^2} \,,
\tag{2.106}
$$

where $x_M = m_W/\sqrt{s}$. This shows that for $p_\perp^2 \to 0$ the matrix element diverges like $2m_W^2/p_\perp^2$, a logarithmic divergence. This becomes apparent in Fig. 2.19, where the $p_\perp$ distribution of the W^+ bosons in the hadronic production of the bosons in association with a single jet at the TEVATRON and the LHC, calculated at leading order is displayed.

To actually properly calculate differential cross-sections like the one depicted in Fig. 2.19, the phase-space element has to be determined in useful quantities such as the rapidity of the gauge boson and its transverse momentum. To this end, the phase-space element for the outgoing particles can be written as

$$
\begin{aligned}
&\frac{\mathrm{d}^4 p_W}{(2\pi)^4} \frac{\mathrm{d}^4 q}{(2\pi)^4} (2\pi)^4 \delta(p_u + p_{\bar{d}} - p_W - q)\,(2\pi)\delta(p_W^2 - m_W^2)\,(2\pi)\delta(q^2) \\
&= \frac{m_{\perp W}\mathrm{d}m_{\perp W}\mathrm{d}y_W \mathrm{d}^2 Q_\perp}{(2\pi)^2} \delta(m_{\perp W}^2 - Q_\perp^2 - m_W^2)\,\delta(\hat{s} + \hat{t} + \hat{u} - m_W^2) \\
&= \frac{\mathrm{d}y_W \mathrm{d}Q_\perp^2}{4\pi} \delta(\hat{s} + \hat{t} + \hat{u} - m_W^2)\,.
\end{aligned}
\tag{2.107}
$$

In the transformation of the δ functions, the fact has been used that

$$
\delta(q^2) = \delta((p_u + p_{\bar{d}} - p_W)^2) = \delta(\hat{s} + m_W^2 - 2(p_u + p_{\bar{d}})\cdot p_W) = \delta(\hat{s} + \hat{t} + \hat{u} - m_W^2) \tag{2.108}
$$

encodes the Mandelstam identity.

Using this form of the phase space element, the fixed order cross-section for the production of a W^+ boson in conjunction with a gluon in the final state can therefore be written as

$$
\mathrm{d}\sigma_{AB \to Wg} = \int_0^1 \mathrm{d}x_A \mathrm{d}x_B\, \mathcal{L}_{u\bar{d}}(x_A, x_B, \mu_F) \int \frac{\mathrm{d}y_W \mathrm{d}Q_\perp^2}{4\pi} |\mathcal{M}|^2_{u\bar{d} \to gW^+}\, \delta(\hat{s} + \hat{t} + \hat{u} - m_W^2)
\tag{2.109}
$$

with the **parton luminosity** given by

$$
\begin{aligned}
\mathcal{L}_{u\bar{d}}(x_A,\, x_B,\, \mu_F) \; &= \; \frac{1}{2\hat{s}} \left[f_{u/A}(x_A,\, \mu_F) f_{\bar{d}/B}(x_B,\, \mu_F) + \{u \longleftrightarrow \bar{d}\} \right] \\
&= \; \frac{1}{2x_A x_B s} \left[f_{u/A}(x_A,\, \mu_F) f_{\bar{d}/B}(x_B,\, \mu_F) + \{u \longleftrightarrow \bar{d}\} \right].
\end{aligned}
\tag{2.110}
$$

Ignoring all information concerning the additional parton by basically integrating over its phase space, this can now be rewritten as the double differential cross-section for the production of a gauge boson,

$$
\begin{aligned}
\frac{\mathrm{d}\sigma_{AB \to Wg}}{\mathrm{d}Q_\perp^2 \,\mathrm{d}y_W} \; &= \; \frac{1}{4\pi} \int_0^1 \mathrm{d}x_A \mathrm{d}x_B \, \mathcal{L}_{u\bar{d}}(x_A,\, x_B,\, \mu_F) \, |\mathcal{M}|^2_{u\bar{d} \to gW^+} \, \delta(\hat{s} + \hat{t} + \hat{u} - m_W^2) \\
&= \int_{\tilde{x}_A}^1 \frac{\mathrm{d}x_A}{x_A} \int_{\tilde{x}_B}^1 \frac{\mathrm{d}x_B}{x_B} \left[f_{u/A}\,(x_A,\, \mu_F)\, f_{\bar{d}/B}\,(x_B,\, \mu_F)\, \delta(\hat{s} + \hat{t} + \hat{u} - m_W^2) \right. \\
&\qquad\qquad\qquad\qquad\qquad \left. \times\; \sigma^{(LO)}_{u\bar{d} \to W^+}(s) \frac{1}{Q_\perp^2} \frac{\alpha_s C_F}{2\pi} \frac{\hat{t}^2 + \hat{u}^2 + 2m_W^2 \hat{s}}{\hat{s}} \right].
\end{aligned}
\tag{2.111}
$$

Here, the relations

$$
Q_\perp^2 \; = \; \frac{\hat{t}\hat{u}}{\hat{s}} \quad \text{and} \quad \sigma^{(LO)}_{u\bar{d} \to W^+}(s) \; = \; \frac{1}{s} \frac{\pi g_W^2 |V_{ud}|^2}{12},
\tag{2.112}
$$

cf. Eq. (2.90) for the latter, have been used to rewrite the matrix element squared in Eq. (2.97) as

$$
|\mathcal{M}|^2_{u\bar{d} \to gW^+} \; = \; \sigma^{(LO)}_{u\bar{d} \to W^+} \frac{8\pi s}{Q_\perp^2} \frac{\alpha_s C_F}{2\pi} \frac{\hat{t}^2 + \hat{u}^2 + 2m_W^2 \hat{s}}{\hat{s}}.
\tag{2.113}
$$

The lower limits of the integration over the Bjorken parameters x_A and x_B in Eq. (2.111), $\tilde{x}_A$ and $\tilde{x}_B$ are fixed by the dynamics of the W boson:

$$
m_W^2 \; = \; \tilde{x}_A \tilde{x}_B \, s \quad \text{and} \quad \tilde{x}_{A,B} \; = \; \frac{m_W}{\sqrt{s}} \, e^{\pm y}.
\tag{2.114}
$$

Taking a closer look at Eq. (2.111) exhibits the divergent structures related to the emission of a gluon. First of all, there is the term $1/Q_\perp^2$, giving rise to a logarithmic divergence of the form $\mathrm{d}Q_\perp^2/Q_\perp^2$. In addition, and less obvious, there is the implicit dependence on the Bjorken parameters x_A and x_B of all kinematic quantities, which will lead to further divergences, which are partially related to the evolution of the PDFs. They will be treated in the next section, by identifying and suitably absorbing them.

2.2.5 Partonic cross-section at next-to leading order

In order to achieve higher precision for the theoretical calculation, the evaluation of **next-to-leading order (NLO)** or **next-to-next-to-leading order (NNLO)** contributions or even beyond, is mandatory. Here, the focus will be on the discussion of the first perturbative correction, called the NLO correction. Looking at the Feynman diagrams of the example process, Fig. 2.15, higher-order corrections will involve higher orders in either the electroweak or the strong coupling constant. At hadron colliders, however, it it usually the latter, the QCD correction, which turns out to be significantly larger than the electroweak correction.

Typically, such an NLO QCD correction is related to either the emission of an additional leg, *i.e.*, an extra parton, into the final state (**real correction**) or to the emission and reabsorption of a parton, through a loop (**virtual correction**). The former actually has already been discussed to some extent in Section 2.2.4.

2.2.5.1 Divergent structures in the matrix elements

Similar to the divergences already encountered in the case of the real emission diagrams before, divergences also appear in the virtual correction, which is due to Feynman diagrams involving closed loops. Here, however, the divergences correspond to two cases:

- **Infrared divergences** may show up when the momentum squared of one of the propagators approaches zero. In other words, when a particle in the loop goes on its mass shell.

- **Ultraviolet divergences** may emerge in the limit $k \to \infty$, where the momentum running in the loop becomes infinite. In this case, terms like $\mathrm{d}^4 k / k^n$ naively diverge, if $n \le 4$.

In all cases, **dimensional regularization** is the method of choice when such divergent integrals need to be evaluated. In this method, the originally 4-dimensional phase space of a single particle, is replaced with a D-dimensional expression,

$$\frac{\mathrm{d}^4 p}{(2\pi)^4}(2\pi)\delta(p^2 - m^2)\Theta(E) \longrightarrow \frac{\mathrm{d}^D p}{(2\pi)^D}(2\pi)\delta(p^2 - m^2)\Theta(E)\,, \qquad (2.115)$$

and the divergences manifest themselves as poles in $2/(D-4) = 1/\varepsilon$.

Concentrating on the case of infrared divergences, the complication here is that, for a process with N outgoing particles at the Born level, these integrals are over the real correction phase space with $(N+1)$ particles and over the loop momentum in the N-body virtual correction term. For simple processes such as the example of inclusive W-production, these integrals are fairly straightforward to calculate directly, leading to the desired cancellation of infrared divergences as described by the BN and KLN theorems [257, 678, 724]. This direct approach, however, becomes increasingly complicated and forbids itself, if the phase-space integration is too complicated for direct analytical evaluation. In such a case, if the results of the phase space integration can only be obtained numerically, other methods have to be invoked. In the next chapter, this will be highlighted with a toy model, leading to a master formula for the calculation of cross-sections at next-to-leading order. In a first step, however, the structure of such a calculation in its analytic form will be presented.

In contrast, the ultraviolet divergences typically do not cancel each other, and they reside in the loop contributions only. Naively, whenever such a loop contribution is encountered, an ultraviolet divergence may not be far away. In such a case, the programme of regularization and renormalization of the diagrams and the theory, respectively, needs to be invoked, which, however, is a standard topic across textbooks on quantum field theory. In particular, in the case of non-Abelian gauge theories, renormalization becomes a somewhat tedious exercise due to many identities stemming from the gauge symmetry relating the different contributions to all orders.

2.2.5.2 Matrix elements for virtual corrections

The real emission contributions have already been discussed in the previous section, Section 2.2.4, where this part of the next-to-leading order calculation has been interpreted as a new leading order process. Here, the results from Section 2.2.4 will be used in a slightly different way, namely as part of the higher-order correction to the **inclusive vector boson production**. This implies that, at first perturbative order in the strong coupling, the emission of an extra jet is a part of the inclusive cross-section. But while this calculation allows for a next-to-leading order description of the inclusive production properties with correspondingly improved perturbative uncertainties, the **exclusive vector boson plus jet production** part of the calculation still is at the leading-order accuracy.

In the case of W^+ production the Feynman diagram related to the loop correction is depicted in Fig. 2.20.

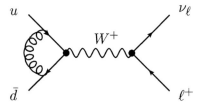

Fig. 2.20 Vertex correction contributing to the NLO correction of the production and leptonic decay of a W^+ boson. Here u and d stand for arbitrary up- and down-type quarks.

For the virtual contribution, the amplitude reads

$$
\mathcal{M}^{(1)}_{u\bar{d}\to W^+} = \frac{g_W}{\sqrt{2}} g_s^2 \mu^{4-D}
$$

$$
\bar{v}_i(\bar{d}) \left\{ \int \frac{\mathrm{d}^D k}{(2\pi)^D} \frac{g^{\nu\rho}\delta^{ab}}{k^2} \left[\gamma_\nu T^a_{ik} \frac{\slashed{p}_d + \slashed{k}}{(p_d+k)^2} \gamma^\mu L \frac{\slashed{p}_u - \slashed{k}}{(p_u-k)^2} \gamma_\rho T^b_{kj} \right] \right\} u_j(u)\, \epsilon_\mu(W^+),
$$

(2.116)

where $D = 4 - 2\varepsilon$.

Closer inspection reveals that the ultraviolet divergences cancel exactly with the renormalization of the external particles, essentially obtained from self-energy diagrams. This ultraviolet cancellation presents testimony to the fact that the $u\bar{d}$-current is conserved. Therefore, the only remaining divergences are infrared, and the result for the virtual matrix element multiplying the Born-level matrix element reads [15]

$$
2\left|\mathcal{M}^{(1*)}_{u\bar{d}\to W^+}\mathcal{M}^{(0)}_{u\bar{d}\to W^+}\right| =
$$

$$
\left|\mathcal{M}^{(0)}_{u\bar{d}\to W^+}\right|^2 \frac{\alpha_s}{2\pi} C_F \left(\frac{\mu^2}{Q^2}\right)^\varepsilon c_\Gamma \left(-\frac{2}{\varepsilon^2} - \frac{3}{\varepsilon} - 8 + \pi^2\right), \tag{2.117}
$$

with $Q^2 = (p_u + p_d)^2 = m_W^2$ and $c_\Gamma = 4\pi^\varepsilon/\Gamma(1-\varepsilon)$. For a detailed discussion of how this result may be obtained, the reader is referred to the following chapter, Section 3.3.1.

2.2.5.3 Next-to-leading order: Real corrections

For the real correction diagrams, the result emerges from a D-dimensional integration over the phase space of the emitted parton, as discussed in the previous section. In the case of the additional gluon in the final state, with its momentum denoted by k, this yields (again, *cf.* Section 3.3.2 for details of this calculation)

$$
\mu^{4-D}\int \frac{\mathrm{d}^D k}{(2\pi)^D}\left|\mathcal{M}^{(0)}_{u\bar{d}\to W^+g}\right|^2 = \left|\mathcal{M}^{(0)}_{u\bar{d}\to W^+}\right|^2 \frac{\alpha_s}{2\pi} C_F \left(\frac{\mu^2}{Q^2}\right)^\varepsilon c_\Gamma
$$

$$
\times \left[\left(\frac{2}{\varepsilon^2} + \frac{3}{\varepsilon} + \frac{\pi^2}{3}\right)\delta(1-z) + \left(\frac{4}{1-x}\log\frac{(1-z)^2}{z}\right)_+ - 2(1+z)\log\frac{(1-z)^2}{z}\right.
$$

$$
\left. -\frac{2}{\varepsilon}\frac{\mathcal{P}^{(1)}_{qq}(z)}{C_F}\right]. \tag{2.118}
$$

The divergences in the first line of the result for this real correction exactly cancel the divergences in the virtual result. The additional divergence in front of the splitting function $\mathcal{P}^{(1)}_{qq}$ is in fact universal, *i.e.*, process-independent. This process-independent term can thus be absorbed into the renormalization of the PDF, see Section 3.3. This, however, is a scheme-dependent procedure. In total then, a finite result is obtained.

The method of directly calculating the phase space of the real emission part in D dimensions followed here of course becomes prohibitively complicated for processes with an increasing number of external particles. For such situations better algorithms have been developed, among them what is by now known as **infrared subtraction algorithms** such as Catani–Seymour dipole subtraction [344, 353] or the method by Frixione, Kunszt, and Signer [539, 542]. The former will be further discussed in Section 3.3.2.

---

[15] More correctly, this is the result obtained in conventional dimensional regularization, a regularization scheme that is defined in Section 3.3.1.

2.2.5.4 Next-to-leading order: Final results

Using the techniques briefly outlined in the previous section, and further worked out in detail in the following chapter, the total partonic cross-section for the $u\bar{d} \to W^+$ production cross-section at NLO reads

$$
\hat{\sigma}^{(\text{NLO})}_{u\bar{d}\to W^+} = \hat{\sigma}^{(\text{LO})}_{u\bar{d}\to W^+} \left\{ 1 + \frac{\alpha_\text{s}(\mu_R)}{2\pi} C_F \left[\left(\frac{4\pi^2}{3} - 8 \right) \delta(1-z) \right. \right. \tag{2.119}
$$
$$
\left. \left. + \left(\frac{4}{1-x} \log \frac{(1-z)^2}{z} \right)_+ - 2(1+z) \log \frac{(1-z)^2}{z} - 2\frac{\mathcal{P}^{(1)}_{qq}(z)}{C_F} \log \frac{\mu_F^2}{Q^2} \right] \right\} .
$$

Note that the result still needs to be convoluted with the PDFs. The last term in the square bracket above, proportional to the splitting function and containing a term of $\log(\mu_F^2/Q^2)$, ensures that there is a compensating factor if the PDFs are evaluated away from their "natural" scale (if $\mu^2 \neq Q^2$); for the process at hand it is the only scale in the process, Q^2. This corresponds to the "running" of the PDF, mediated by the corresponding splitting function and resluting in the logarithm of the scale ratio. If the process at leading order was already containing n factors of α_s, a similar running of the strong coupling, mediated by n first order terms of the β-function would appear as well, this time of course with logarithms of the ratio of μ_R^2 and Q^2.

Another contribution, stemming from the gluon initial states,

$$
\hat{\sigma}^{(\text{NLO})}_{ug\to dW^+} = \hat{\sigma}^{(\text{LO})}_{u\bar{d}\to W^+}
$$
$$
\cdot \frac{\alpha_\text{s}(\mu_R)}{2\pi} T_R \left[\mathcal{P}^{(1)}_{qg}(z) \left(\log \frac{(1-z)^2}{z} - \log \frac{\mu_F^2}{m_W^2} \right) + \frac{1}{2}(1-z)(1+7z) \right], \tag{2.120}
$$

needs to be added, which of course starts at $\mathcal{O}(\alpha_\text{s})$.

2.2.6 Scales

2.2.6.1 Sketching the issue

In order to estimate theoretical uncertainties, it has become customary to vary the **renormalization** and **factorization scales** by a factor up or down. Typically this factor is chosen to be two, and in most cases both scales are changed in parallel, *i.e.*, both are at the same time multiplied with either 2 or 1/2. There is some dispute on whether 2 is a sufficiently large factor to catch all uncertainties and on whether or not the scales are indeed chosen in parallel. There is, however, an additional caveat, which is more related to the actual choice of scale. Usually, the default scale $\mu = \mu_F = \mu_R$ is determined by identifying it as the "*characteristic scale*" of the process under consideration, either given by some intermediate particle's mass or some function of the final-state momenta.

In such an arrangement, and at leading order, the factorization scale is typically interpreted as the scale up to which softer partons and, correspondingly, the emission of additional partons are ignored in the actual matrix element. Such emissions are rather subjected to a more inclusive treatment in the evolution of the parton content of the hadrons from hadronic scales to the actual harder, *i.e.* perturbative, scale of the process. When considering more complicated processes, this simple picture is likely to change drastically. As an example, consider the case of W production in conjunction with additional partons, defined as jets at leading order. Very broadly speaking, one may distinguish two extreme cases of how the kinematics of this process works out. There will be a region, where the W boson is accompanied by softer partons; here, the parton emission can be thought of as QCD correction to an electroweak process. This is a case, where one may still think of using m_W or similar as the relevant scale for μ_F, and, maybe, even for μ_R. On the other hand, there will also be a region in phase space, where the partons are produced at transverse momenta, which are much larger than the W mass. In such a configuration, one may interpret the W emission as a weak correction to jet production, intrinsically a QCD process. Here, m_W, $p_\perp^W$, or similar as the scale characterising the process would not be a great choice. In fact, recent work [225] suggests that choices that interpolate between these two regimes, such as $H_T/2$ or so are much better suited. H_T is defined as the scalar sum of the transverse momenta of all jets, leptons, and the missing transverse energy,

$$H_T = \sum_{j \in \text{jets}} p_{\perp,j} + \sum_{l \in \ell} p_{\perp,l} + \not{E}_\perp \,. \tag{2.121}$$

What is clear, though, is that the scale dependence essentially stems from an inability to calculate cross-sections correctly, *i.e.*, to all orders. Instead, in all calculations, the perturbative series is truncated at some fixed order, which leaves a residual logarithmic dependence on the renormalization and factorization scales. With the perturbative series thought to be asymptotic, there is some assertion that this dependence would diminish with each additional order in the calculation. While this seems to be by far and large correct, cases like $W+3$ jet production discussed in [225] indicate that with apparently bad choices of scale, $p_\perp^W$ in this case, this decreasing dependence is not always realized at next-to-leading order. However, apart from such pathological cases, higher-order calculations indeed always diminish the scale dependence. It is fair to state that, in order to have any reasonable and reliable estimate of the corresponding **scale uncertainty**, the inclusion of at least one additional perturbative order, *i.e.*, a next-to-leading order calculation, is mandatory.

2.2.6.2 A quantitative example

In order to see more quantitatively how higher-order calculations reduce the artificial scale uncertainty due to the truncation of the perturbative series, consider an example worked out in [585], namely the single-jet-inclusive $p_\perp$-distribution at the TEVATRON. At leading order, this distribution can be related to diagrams such as the ones shown in Fig. 2.21. At large transverse momentum — and therefore at large centre-of-mass energies and consequently at large parton x — the dominant contributions stem from quark–anti-quark initial states that can be seen as initial valence quarks. The lowest order differential *hadronic* cross-section is given by

$$\frac{d\sigma^{(\text{LO})}}{dp_\perp} = f_{q/p}(\mu_F) f_{\bar{q}/\bar{p}}(\mu_F) \otimes \alpha_{\text{s}}^2(\mu_R)\, \hat{\sigma}^{(0)} \tag{2.122}$$

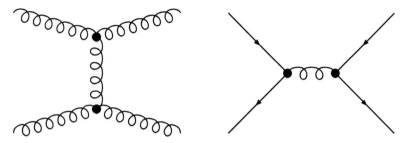

Fig. 2.21 Some leading-order diagrams for inclusive jet production from a quark–anti-quark initial state.

where $\hat{\sigma}^{(0)}$ represents the lowest order *partonic* cross-section. Including the next-to-leading order corrections, this can be written as

$$
\begin{aligned}
\frac{\mathrm{d}\sigma^{(\mathrm{NLO})}}{\mathrm{d}p_{\perp}} &= f_{q/p}(\mu_F) f_{\bar{p}/\bar{q}}(\mu_F) \\
&\cdot \left[\alpha_{\mathrm{s}}^2(\mu_R)\hat{\sigma}^{(0)} + \alpha_{\mathrm{s}}^3(\mu_R) \left(\hat{\sigma}^{(1)} + 2b_0 \log \frac{\mu_R}{p_{\perp}} \hat{\sigma}^{(0)} - 2\mathcal{P}_{qq} \log \frac{\mu_F}{p_{\perp}} \hat{\sigma}^{(0)} \right) \right],
\end{aligned}
\tag{2.123}
$$

where logarithms that explicitly involve the renormalization or factorization scales have been exposed. The remainder of the $\mathcal{O}\left(\alpha_{\mathrm{s}}^3\right)$ corrections are all incorporated in the function $\hat{\sigma}^{(1)}$.

From this expression, the sensitivity of the distribution to the renormalization scale is easily calculated using the expression for the running of the strong coupling,

$$
\frac{\partial \alpha_{\mathrm{s}}(\mu_R)}{\partial(\log \mu_R)} = -b_0 \alpha_{\mathrm{s}}^2(\mu_R) - b_1 \alpha_{\mathrm{s}}^3(\mu_R) + \mathcal{O}\left(\alpha_{\mathrm{s}}^4\right),
\tag{2.124}
$$

cf. Eq. (2.18), where the two leading coefficients in the β-function, b_0 and b_1, are given in Eq. (2.20). Looking at the terms in Eq. (2.123) it becomes apparent that the first term in the square bracket, the leading-order contribution, and the term proportional to $2b_0$ in the round bracket cancel out, such that the remaining dependence on μ_R is contained in terms $\mathcal{O}\left(\alpha_{\mathrm{s}}^4\right)$.

In a similar way, the factorization scale dependence can be calculated using the non-singlet DGLAP equation

$$
\frac{\partial f_{j/h}(\mu_F)}{\partial(\log \mu_F)} = \frac{\alpha_{\mathrm{s}}(\mu_F)}{2\pi} \mathcal{P}_{ji} \otimes f_{i/h}(\mu_F),
\tag{2.125}
$$

This time, the partial derivative of each parton distribution function, multiplied by the first term in Eq. (2.123), cancels with the final term. Thus, once again, the only remaining terms are $\mathcal{O}\left(\alpha_{\mathrm{s}}^4\right)$.

This is a generic feature of a next-to-leading order calculation. Any observable predicted to $\mathcal{O}\left(\alpha_{\mathrm{s}}^n\right)$ is independent of the choice of either renormalization or factorization scale, up to the next higher order in the strong coupling, $\mathcal{O}\left(\alpha_{\mathrm{s}}^{n+1}\right)$. Of course, this is only a formal statement and the numerical importance of such higher-order terms

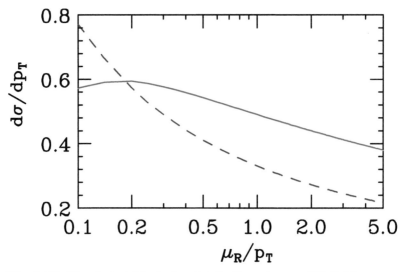

Fig. 2.22 The single-jet-inclusive distribution at $p_\perp = 100$ GeV, appropriate for Run I of the TEVATRON. Theoretical predictions are shown at LO (dashed blue) and at NLO (solid red).

may be large. As a concrete example, consider jet production at the TEVATRON Run I. In that case the numerical values of the partonic cross-sections entering Eq. (2.123) are $\hat{\sigma}^{(0)} = 24.4$ and $\hat{\sigma}^{(1)} = 101.5$. Equipped with these values the LO and NLO scale dependence can be calculated, as shown in Fig. 2.22, adapted from Ref. [585]. In this case the factorization scale has been kept fixed at $\mu_F = p_\perp$ and only the dependence on the renormalization scale is shown. The figure indicates the expected result, which is that the renormalization scale dependence is reduced over a wide range of values for μ_R when going from LO to NLO.

While the reasoning above, culminating in Fig. 2.22, is a fairly accurate representative of the situation found at NLO, the exact details depend upon the kinematics of the process under study and on choices such as the running of α_s and the PDFs used. It is worth noting, though, that due to the actual structure of NLO corrections, as exemplified in Eq. (2.123), there will normally be a peak in the NLO curve, around which the scale dependence is minimized. The scale at which this peak occurs is often favoured as a choice specific for the process, its kinematics, and additional cuts such as the jet definition. For example, for inclusive jet production at the TEVATRON, using a cone size of $R = 0.7$, a central scale of

$$\mu_F = \mu_R = p_\perp^{\text{jet}}/2 \qquad (2.126)$$

is usually chosen. This is near the peak of the NLO cross-section for a large set of different observables, *cf.* the specific case shown in Fig. 2.22. Adjusting the scale choice to be in the region of the peak is often referred to as the *"principle of minimal sensitivity"* [864]. It is worth keeping in mind that such a choice is also usually near the scale at which the LO and NLO curves cross, *i.e.* for the value of the scale where the NLO

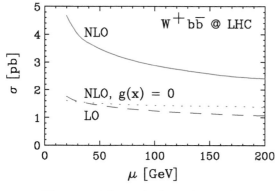
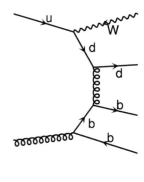

Fig. 2.23 The scale dependence of the cross-section for $W b\bar{b}$ production at the 14 TeV LHC (left). A real radiation diagram leading to the large corrections is also shown (right).

corrections do not significantly change the LO cross-section. Setting the scale by this means assumes *"fastest apparent convergence"* (FAC) of the perturbative series [600].

Finally, a rather different motivation comes from the consideration of a "physical" scale for the process. As examples take $p_\perp^{\text{jet}}$ for the inclusive jet production process or the W mass in the case of inclusive W production. These typical methods for choosing the scale do in general not agree, leading to somewhat different results in dependence on the actual choice and thus a corresponding theoretical uncertainty. If, on the other hand, the scales or the respective results do agree — quite often this seems fairly accidental at first sight — this may be viewed as a sign for the perturbative expansion to be very well-behaved.

A word of caution is due at this point. Although the improved scale dependence sketched out here is typical, it is worth reiterating that this is by no means guaranteed. In addition to effects due to apparently bad scale choices which do not appreciate the full intricacy of the kinematics, like the case hinted at above, there is also another source of potential pitfalls. They are related to cross sections, in particular at the LHC, which at leading order are driven by quark–anti-quark initial states. Since the LHC produces an abundance of gluons, real radiation diagrams containing one or maybe even two gluons in the initial state can give rise to very large NLO corrections. Since they enter for the first time at NLO, strictly speaking they are some kind of additional leading-order contribution appearing at higher orders. As such they typically also give rise to a sizable additional scale dependence. A well-known example is shown in Fig. 2.23 for the case of $W b\bar{b}$ production at the 14 TeV LHC. In the absence of gluons the NLO calculation has the canonical behaviour; in their presence the rate is not well-controlled due to diagrams such as the one shown in the same figure.

2.2.7 Other considerations

2.2.7.1 Perturbative orders

There is a somewhat tricky point in the discussion of perturbative orders. As a simple example, consider the case of the $p_\perp$-distribution of a W boson produced in hadronic collisions. At LO in collinear factorization, there is no distribution and $p_\perp^{(W)} = 0$, since there is no final state parton to compensate the recoil introduced by a finite $p_\perp$. So, strictly speaking, the $p_\perp$–distribution of the W boson at hadron colliders is, at leading order, an observable at $\mathcal{O}(\alpha_\mathrm{s})$. This, however, is already the next-to-leading order level for the total cross-section. Not surprisingly, then, the theoretical uncertainty, *i.e.*, the scale uncertainty, of the total cross-section is smaller than the induced shape uncertainty due to scale variations of the $p_\perp$–distribution of the W. To make things even more confusing, note that, in contrast, the rapidity distribution of the W is $\mathcal{O}\left(\alpha_\mathrm{s}^0 = 1\right)$ at leading order, since it is due to the PDFs which are of course present at leading order. At the same time, the production cross-section for a Higgs boson in gluon, $gg \to H$, is $\mathcal{O}\left(\alpha_\mathrm{s}^2\right)$ already at leading order, because the coupling of the Higgs boson to the gluons proceeds through a quark loop. Even integrating out the heavy quark does not change this picture - the emerging effective vertex is also proportional to α_s. In summary, this means that the correct assignment of the perturbative order as LO, NLO, and so on, is not a fact of merely counting orders of α_s. Instead it is a process- and, even more confusing, observable-dependent characterization.

2.2.7.2 Total cross-sections and K-factors

Another important point here is related to what is called a K-factor, which usually denotes the ratio of a higher-order result for a total cross-section and the leading order one:

$$K_X^{\mathrm{(N)NLO}} = \frac{\sigma_{\mathrm{tot}}^{\mathrm{(N)NLO}}(X)}{\sigma_{\mathrm{tot}}^{\mathrm{LO}}(X)} \tag{2.127}$$

with X denoting a specific final state. Such a K factor is proportional to powers of $\alpha_\mathrm{s}/(2\pi)$ times factors which are usually of the order of 1. A prime example is the hadronic cross-section in electron-positron annihilations, related at leading order to the process $e^+e^- \to q\bar{q}$. Higher-order corrections now introduce either additional loops, without changing the final state, or the emission of further partons, for instance $e^+e^- \to q\bar{q}g$. The corresponding K factor is given by

$$\begin{aligned} K_{e^+e^- \to hadrons}^{NNLO} &= \frac{\sigma_{\mathrm{tot}}^{\mathrm{NNLO}}(e^+e^- \to \mathrm{hadrons})}{\sigma_{\mathrm{tot}}^{\mathrm{LO}}(e^+e^- \to \mathrm{hadrons})} \\ &= 1 + \frac{\alpha_\mathrm{s}}{2\pi}\left(C_F\right) + \left(\frac{\alpha_\mathrm{s}}{2\pi}\right)^2 (\dots) . \end{aligned} \tag{2.128}$$

At hadron colliders the situation is not as straightforward. There, the inclusion of higher-order corrections often allows other partonic channels to open up, which may lead to significant higher-order corrections. As an example in the spirit of $e^+e^- \to$ hadrons, consider the "inverse" process at hadron colliders, namely the production of lepton pairs. At leading order, the parton-level process is $q\bar{q} \to \ell^+\ell^-$. Higher orders now introduce, as before, additional partons in the final state or loops. There is one difference, however, namely the opening of channels with a gluon in the initial state, for instance $qg \to q\ell^+\ell^-$. Such processes may yield a significant contribution, because, dependent on the phase space taken by the lepton pair invariant mass, the gluon

PDF may become significantly larger than the corresponding quark PDFs. In the case discussed here, this leads to NLO K-factors of the order of $K_Z^{NLO} \approx 1.3$ compared to the significantly smaller $K_{e^+e^- \to \text{hadrons}}^{NLO} \approx 1.03$.

This example is just the very innocent tip of an iceberg of a number of other processes, where the opening of additional channels leads to K factors which are drastically different from one.

Nevertheless, there is yet another class of processes which obtain large NLO corrections, even without opening additional channels. The prime example for this is the gluon-induced Higgs production, $gg \to H$, where the NLO K-factor is about a factor of two. The origin of the large K-factor is explained by two effects, that can be seen by consideration of the virtual amplitude for this process. The one-loop amplitude for $gg \to H$, working in the large top-mass effective theory is,

$$\mathcal{M}_{gg \to H}^{(1)} = \mathcal{M}_{gg \to H}^{(0)} \times \frac{\alpha_s}{4\pi} C_A \left(\frac{\mu^2}{Q^2}\right)^\varepsilon c_\Gamma \left[-\frac{2}{\varepsilon^2} + \frac{11}{3} + \pi^2\right], \tag{2.129}$$

where the dimensional reduction scheme has been employed. Due to the analytic continuation that must be performed in order to evaluate the virtual amplitude for time-like momentum transfer Q^2, this formula contains a factor of $-\pi^2/2$ for every factor of $1/\epsilon^2$ present. This expression is to be contrasted with the relevant part of the virtual amplitude for W production in the same scheme given in Eq. (3.64),

$$\mathcal{M}_{u\bar{d} \to W^+}^{(1)} = \mathcal{M}_{u\bar{d} \to W^+}^{(0)} \times \frac{\alpha_s}{4\pi} C_F \left(\frac{\mu^2}{Q^2}\right)^\varepsilon c_\Gamma \left[-\frac{2}{\varepsilon^2} - \frac{3}{\varepsilon} - 7 + \pi^2\right]. \tag{2.130}$$

In the formula above the non-pole term is $(\pi^2 - 7)$, with the π^2 factor resulting from the analytic continuation mostly cancelled by the numerical constant. However in Eq. (2.129) there is no such cancellation and the factor of π^2 remains an important contribution. Moreover, as can be seen from these two formulae, this effect is amplified by the overall colour factor of C_A for the Higgs case compared to C_F for Drell–Yan. Taken together, this explains why the K-factor for Higgs production is much larger than for the Drell-Yan process. Since the π^2 terms are so important numerically, renormalization group techniques have recently been used to resum them to all orders and thus provide improved predictions for the Higgs boson cross-section [130].

2.2.7.3 Differential cross-sections at fixed order and giant K factors

Another caveat arises when considering differential distributions. Often, either due to the limitations of the calculation or because of specific cuts that are applied, some distributions have a kinematic limit at LO. Adding extra radiation at higher orders, and in the context of the discussion here, at NLO will frequently extend hitherto constrained kinematic ranges.

An example of such a situation is shown in Fig. 2.24, which depicts the transverse momentum distribution of a W^+ boson at the 7 TeV LHC. At leading order, it is computed from process $pp \to W^+ + 1$ jet with a cut on the jet p_T at 25 GeV which immediately translates into the same cut on the W^+ boson's p_T. At NLO there is another parton in the final state of the real emission contribution, with an arbitrary transverse momentum. This additional parton, together with the original one, can produce a summed transverse momentum below the cut, and therefore the W^+ boson recoiling against the partonic system will populate the transverse momentum region below the 25 GeV cut. This has two consequences. Since the region below 25 GeV

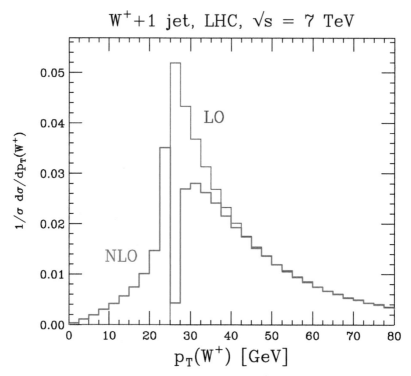

Fig. **2.24** The transverse momentum of a W^+ boson produced at the 7 TeV LHC, computed at LO and NLO in QCD, recoiling against a hard jet with $p_T > 25$ GeV. The discontinuity in the NLO prediction around 25 GeV is caused by the opening of the kinematic region $p_T < 25$ GeV at that order.

originates solely from real radiation events, it should only be trusted as much as a LO calculation. In addition, the region around the kinematic boundary at 25 GeV is not well-described. The values of the histogram bins there are simply artifacts of the calculation and are not reliable. Smoothing out this sort of problem, and providing hadron-level predictions that can be directly compared with experimental results, is the domain of the parton shower and resummation.

Another example for this has been discussed in [830], where the term "giant K-factors" has also been coined. Following previous work, in particular [200, 205, 321, 521] for the case of vector bosons accompanied by jets, but also, for the case of vector boson pairs, [252, 311], it has been observed that there are substantial NLO corrections for observables at large scales of the order of the vector boson masses and beyond, which actually fall far outside the respective bands obtained from a simple scale variation by a factor of two applied on the leading-order result. As an example, Fig. 2.25 shows the NLO correction for the $p_\perp$-distribution of the additional jet produced in $V + j$ at the LHC. In the publications dealing with this problem, this apparently tremendous

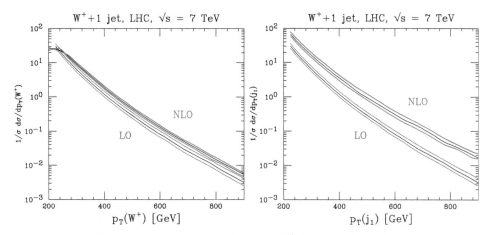

Fig. 2.25 The transverse momenta of a W^+ boson (left) and accompanying leading jet, j_1 (right) produced at the 7 TeV LHC, computed at LO and NLO in QCD. The leading jet is required to have $p_\perp > 200$ GeV, after jets are clustered with the $k_\perp$-algorithm using $R = 0.7$. The central scale choice is $\mu_F = \mu_R = \sqrt{m_W^2 + p_\perp^2}$ and the bands correspond to variation of this scale by a factor of two in each direction.

K factor[16] has been related to new configurations, like the ones exhibited in Fig. 2.26. Essentially, the idea there is that for jets at very large transverse momenta compen-

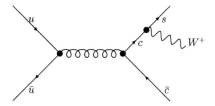

Fig. 2.26 A typical configuration giving rise to "giant K-factors". This configuration cannot be interpreted as real correction to any underlying $W + 1$ parton contribution, but rather as some electroweak correction to a QCD process, in this case a parton-level configuration related to dijet production.

---

[16]In some loose sense, the term K factor introduced for total cross-sections could also be applied to the ratio of higher-order and lower-order results for distributions, with the effect that the K factor then becomes "local".

sating each other, an additional relatively softer vector boson could be interpreted as a real electroweak correction to a process that essentially is a QCD process. One of the strategies for addressing this sort of issue, presented in Ref. [830], will be discussed further in Section 3.4.2.

2.2.7.4 *K*-factors: rules of thumb

The K-factor for a process is a useful concept, but often ill-defined, as it depends on the order of the PDFs (and the $\alpha_s(m_Z)$ value) used in the evaluation of the matrix elements in the numerator and denominator of the K-factor, the scale choices in those matrix elements, and even the chosen jet algorithm and size, for processes involving jets in the final state. See for example, the discussion in Ref. [320] and in Section 4.3. NLO corrections are most-often thought of as positive additions to the leading-order cross-section, but, depending on the parameters listed above, can just as well be negative. It is important to remember that the K-factor is a ratio of cross-sections at two orders, and it is most often the leading-order cross-section that is changing most rapidly, and the NLO cross-section that has a more stable scale dependence.

There are some rules of thumb, however. NLO corrections tend to be large for processes for which there is a great deal of colour annihilation in the interaction. The prime example is the process previously discussed, $gg \to H$, in which two colour octet gluons collide to produce a colour-singlet Higgs boson, with a very large correction from LO to NLO. In addition, NLO corrections tend to decrease as more final state legs are added. For example, the K-factor for $gg \to H$+jet is less than that for $gg \to H$.

A useful, albeit simplistic, rule is given by the equation below: [17]

$$C_{i_1} + C_{i_2} - C_{f,max} \tag{2.131}$$

The relative size of the NLO corrections for a given process depends on the sum of the Casimir colour factors for the initial state minus the Casimir colour factor for the biggest colour representation possible for the final state. Again, this is a rule of thumb and not a rigorous statement. It is also not yet clear whether this argument can be extended to calculations at NNLO.

2.3   Beyond fixed order: *W* boson production to all orders

In this section, the basic ideas underlying resummation techniques are discussed. After introducing some of the technology and terminology through a classic example in QED, the findings will be generalized to the case of hadronic W production.

2.3.1   A QED example

In the following the discussion will follow the classic example of resummation, first described in [796], namely the low $p_\perp$ distribution of lepton pairs produced in e^+e^- collisions, where the contribution of the Z^0 bosons has been ignored.

---

[17]The so-called **Dixon conjecture**.

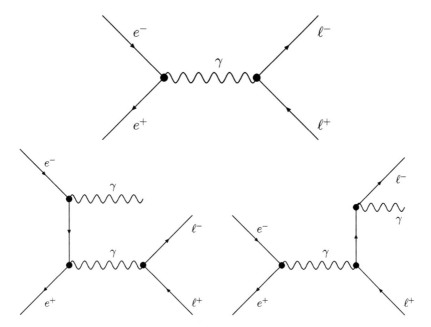

Fig. 2.27 Diagrams for the process $e^-e^+ \to \ell^+\ell^-$ (upper panel) and for $e^-e^+ \to \ell^+\ell^-\gamma$ (lower panel) in QED. The diagrams relating to the Born term in the upper panel and to the photon emission off an initial line, like the one in the left lower panel, are relevant for the following discussion. Diagrams corresponding to photon emission off a final-state leg will be suppressed by fixing the $\ell^+\ell^-$ invariant mass.

2.3.1.1 $p_\perp$-spectrum in the single emission approximation

Typical diagrams related to the emission of additional photons in the process $e^-e^+ \to \gamma^* \to \ell^-\ell^+$ with γ^* signifiying a virtual photon are exhibited in Fig. 2.27. The emission of photons off the final-state leptons, like in the right diagram of Fig. 2.27, will be ignored for simplicity. This is well justified, as in the case of QCD discussed below, they will also be irrelevant, since gluons do not couple to leptons. In this approximation the cross-section for this process can be written as a combination of $e^-e^+ \to \gamma^*\gamma$ with the subsequent decay of the virtual photon to the leptons, $\gamma^* \to \ell^-\ell^+$, which is supplemented in a seond step.

Then the former part can be obtained from a similar process already encountered, namely the emission of an additional gluon in the production of a heavy vector boson with mass M in the annihilation of a quark–anti-quark pair, *cf.* Eq. (2.97). There, of course, the massive boson was a W boson. Adjusting couplings and ignoring colour factors, therefore

$$|\mathcal{M}|^2_{e^-e^+\to\gamma\gamma^*} = 32\pi^2\alpha^2 \, \frac{\hat{t}^2 + \hat{u}^2 + 2Q^2\hat{s}}{\hat{t}\hat{u}} \,, \tag{2.132}$$

where $Q^2 \equiv M^2$ denotes the invariant mass of the virtual photon and $\hat{s}$, $\hat{t}$, and $\hat{u}$ are the usual Mandelstam variables, satisfying

$$\hat{s} + \hat{t} + \hat{u} = M^2 = Q^2 \tag{2.133}$$

or

$$Q^2 - \hat{t} - \hat{u} = \hat{s} \, . \tag{2.134}$$

Using

$$\frac{\mathrm{d}\hat{\sigma}}{\mathrm{d}\hat{t}} = \frac{|\mathcal{M}|^2}{16\pi\hat{s}^2} \tag{2.135}$$

one therefore finds

$$\frac{\mathrm{d}\hat{\sigma}_{e^- e^+ \to \gamma^* \gamma}}{\mathrm{d}\hat{t}} = \frac{2\pi\alpha^2}{\hat{s}^2} \frac{\hat{t}^2 + \hat{u}^2 + 2Q^2\hat{s}}{\hat{t}\hat{u}} \, . \tag{2.136}$$

for the differential cross-section. In the following the limit of $\hat{t}$, $\hat{u} \to 0$ or, equivalently, $\hat{s} \approx Q^2$ will be considered. Factoring out the cross-section for the production of a virtual photon,

$$\hat{\sigma}_{e^- e^+ \to \gamma^*}^{(\mathrm{LO})} = \frac{4\pi^2\alpha}{Q^2} \approx \frac{4\pi^2\alpha}{\hat{s}} \, , \tag{2.137}$$

yields

$$\frac{\mathrm{d}\hat{\sigma}_{e^- e^+ \to \gamma^* \gamma}}{\mathrm{d}\hat{t}} = \hat{\sigma}_{e^- e^+ \to \gamma^*}^{(\mathrm{LO})} \cdot \frac{\alpha}{2\pi\hat{s}} \frac{\hat{t}^2 + \hat{u}^2 + 2Q^2\hat{s}}{\hat{t}\hat{u}} \, . \tag{2.138}$$

This ultimately also allows the replacement of this cross-section for the production of a virtual photon with the cross-section for producing a lepton pair instead,

$$\hat{\sigma}_{e^- e^+ \to \ell^- \ell^+}^{(\mathrm{LO})} = \frac{4\pi\alpha^2}{3Q^2} \, , \tag{2.139}$$

to arrive at the double-differential cross-section

$$\frac{\mathrm{d}\hat{\sigma}_{e^- e^+ \to \ell^- \ell^+ \gamma}}{\mathrm{d}\hat{t}\mathrm{d}Q^2} = \hat{\sigma}_{e^- e^+ \to \ell^- \ell^+}^{(\mathrm{LO})} \cdot \frac{\alpha}{2\pi\hat{s}Q^2} \frac{\hat{t}^2 + \hat{u}^2 + 2Q^2\hat{s}}{\hat{t}\hat{u}} \, . \tag{2.140}$$

The limit of small transverse momenta of the virtual photon, or, equivalently, the lepton pair, $Q_\perp \to 0$, is related to $\hat{t} \to 0$, $\hat{u} \to 0$, or both $\hat{t}$ and $\hat{u}$ approaching zero. Due to the symmetry of the cross-section in Eq. (2.140) under the exchange $\hat{t} \leftrightarrow \hat{u}$ it is sufficient to only consider the case where one of the two Mandelstam variables goes to zero, say $\hat{t} \to 0$, with the other one being less singular, but potentially approaching zero as well. Assuming that $\hat{t}$ is the relevant Mandelstam variable allows the replacements

$$\hat{t} \to -Q_\perp^2 \to 0 \quad \text{and} \quad \hat{u} = Q^2 - \hat{s} - \hat{t} \to Q^2 - \hat{s} \, . \tag{2.141}$$

To capture the potential divergence for $\hat{u} \to 0$, one must integrate over $\hat{u}$. This is equivalent to integrating over Q^2, which in turn exposes the logarithmic divergence for $\hat{u} \to 0$ as the upper limit at the kinematic boundary of $Q^2 \approx \hat{s}$, namely

$$Q^2 \leq \hat{s} - Q_\perp^2 \, . \tag{2.142}$$

Keeping in mind that the region $Q_\perp^2 \ll \hat{s}$ is being analysed therefore results in

$$
\begin{aligned}
\frac{\mathrm{d}\hat{\sigma}_{e^- e^+ \to \ell^- \ell^+ \gamma}}{\mathrm{d}Q_\perp^2} &= \hat{\sigma}_{e^- e^+ \to \ell^- \ell^+}^{(\mathrm{LO})} \frac{\alpha}{2\pi\hat{s}} \cdot \int^{\hat{s}-Q_\perp^2} \frac{\mathrm{d}Q^2}{Q^2} \frac{\hat{s}^2 + Q^4}{\hat{s} - Q^2} \\
&= \hat{\sigma}_0 \frac{\alpha}{\pi} \frac{1}{Q_\perp^2} \left[\log \frac{\hat{s}}{Q_\perp^2} + \mathcal{O}(1) \right] \longrightarrow \mathrm{d}\hat{\sigma}_R \approx \hat{\sigma}_0 \frac{\alpha}{\pi} \frac{\mathrm{d}Q_\perp^2}{Q_\perp^2} \log \frac{\hat{s}}{Q_\perp^2} \, .
\end{aligned} \tag{2.143}
$$

Consider now the $Q_\perp$-integrated scaled cross-section

$$\frac{\alpha}{\pi} \int_0^{p_\perp^2} \frac{\mathrm{d}Q_\perp^2}{Q_\perp^2} \log \frac{\hat{s}}{Q_\perp^2} = \frac{1}{\hat{\sigma}_0} \int_0^{p_\perp^2} \mathrm{d}Q_\perp^2 \frac{\mathrm{d}\hat{\sigma}_R}{\mathrm{d}Q_\perp^2} \, , \tag{2.144}$$

with $p_\perp$ some arbitrary but finite cut-off on the maximally allowed transverse momentum of the lepton pair. This integral will diverge due to the collinear divergence that is manifest in the $1/Q_\perp^2$ term in the real emission term $\hat{\sigma}_R$. So, naively, one would now expect that this cross-section diverges. In fact, however, this is not the case; as discussed in Section 2.2.5 the Block–Nordsieck and Kinoshita–Lee–Nauenberg theorems guarantee that this divergence will be cancelled exactly by another divergence in the corresponding virtual contribution, where again the final state fermions will be ignored. This results in a finite overall cross-section at order α [257, 678, 724]. In other words, adding in $\hat{\sigma}_V \propto \delta(Q_\perp^2)$ and normalizing to $\hat{\sigma}_0$ results in

$$\Sigma^{(1)}(\hat{s}) = \frac{1}{\hat{\sigma}_0} \int_0^{\hat{s}} \mathrm{d}Q_\perp^2 \frac{\mathrm{d}(\hat{\sigma}_R + \hat{\sigma}_V)}{\mathrm{d}Q_\perp^2} = 1 + \mathcal{O}(\alpha) \, , \tag{2.145}$$

with the coefficient of the order-α contribution being free of any potentially **large logarithms**. Decomposing the integral as

$$\frac{1}{\hat{\sigma}_0} \int_0^{\hat{s}} \mathrm{d}Q_\perp^2 \frac{\mathrm{d}(\hat{\sigma}_R + \hat{\sigma}_V)}{\mathrm{d}Q_\perp^2} = \frac{1}{\hat{\sigma}_0} \left[\int_0^{p_\perp^2} \mathrm{d}Q_\perp^2 \frac{\mathrm{d}(\hat{\sigma}_R + \hat{\sigma}_V)}{\mathrm{d}Q_\perp^2} + \int_{p_\perp^2}^{\hat{s}} \mathrm{d}Q_\perp^2 \frac{\mathrm{d}\hat{\sigma}_R}{\mathrm{d}Q_\perp^2} \right] \, , \tag{2.146}$$

where the fact that the virtual contribution $\hat{\sigma}_V$ is concentrated in the region of $Q_\perp^2 = 0$ has been accounted for, allows one to rewrite

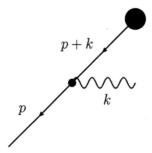

Fig. 2.28 Photon emission off a fermion line. Here, the thick blob represents the rest of the process, while the fermion line corresponds to one of the incoming electrons.

$$\Sigma^{(1)}(p_\perp^2) = \frac{1}{\hat\sigma_0} \int\limits_0^{p_\perp^2} \mathrm{d}Q_\perp^2 \frac{\mathrm{d}(\hat\sigma_R + \hat\sigma_V)}{\mathrm{d}Q_\perp^2} \approx 1 - \frac{1}{\hat\sigma_0} \int\limits_{p_\perp^2}^{\hat s} \mathrm{d}Q_\perp^2 \frac{\mathrm{d}\hat\sigma_R}{\mathrm{d}Q_\perp^2}$$

$$\approx 1 - \frac{\alpha}{2\pi} \log^2 \frac{\hat s}{p_\perp^2} .$$

In the approximations up to now all terms that would yield constant or single logarithmic contributions have been ignored — therefore the result obtained in Eq. (2.147) is correct in the **double leading logarithmic approximation (DLLA)**, also known as the **Dokshitzer–Dyakonov–Troyan approximation (DDT)**, which was first developed in [476].

2.3.1.2   Multiple soft photon emissions

In order to deal with multiple photon emissions it will prove useful to remember the way the **eikonal** expression in Eq. (2.17) has been derived. In a sketchy way, the relevant photon emission contribution exhibited in Fig. 2.28 can be written as

$$e\, \epsilon_\mu(k)\, \bar u(p)\gamma^\mu \frac{\not p + \not k}{(p+k)^2} \ \ldots \ \approx \ e\, \bar u(p)\frac{p\cdot\epsilon}{p\cdot k} \ \ldots \ , \tag{2.147}$$

where terms have been ignored that are finite in the soft limit $k \to 0$. This result emerges from using both the anti-commutator of the Dirac matrices and, subsequently, the equation of motion for the spinor.

In the same way, the two photon contributions can be written as a factor

$$\frac{e^2}{2!} \frac{p\cdot\epsilon_1}{p\cdot k_1} \frac{p\cdot\epsilon_2}{p\cdot k_2} \ \ldots \ , \tag{2.148}$$

where the factor of $1/(2!)$ stems from the proper symmetrization of the two identical particles, the two photons, in the final state. From there, it is straightforward to see the form of the n-photon contribution,

$$\frac{e^n}{n!} \frac{p \cdot \epsilon_1}{p \cdot k_1} \frac{p \cdot \epsilon_2}{p \cdot k_2} \cdots \frac{p \cdot \epsilon_n}{p \cdot k_n} \cdots . \tag{2.149}$$

The corresponding n-photon contribution to the scaled cross-section reads

$$\Sigma^{(n)}(p_\perp^2) = \frac{1}{n!} \left[\frac{\alpha}{\pi} \int_0^{p_\perp^2} \mathrm{d}\log Q_\perp^2 \log \frac{\hat{s}}{Q_\perp^2} \right]^n = \frac{1}{n!} \left[-\frac{\alpha}{2\pi} \log^2 \frac{\hat{s}}{p_\perp^2} \right]^n , \tag{2.150}$$

taking into account the leading logarithms only. This implies that

$$\Sigma(Q_\perp^2) = \sum_{n=0}^{\infty} \frac{1}{n!} \left[-\frac{\alpha}{2\pi} \log^2 \frac{\hat{s}}{Q_\perp^2} \right]^n = \exp \left[-\frac{\alpha}{2\pi} \log^2 \frac{\hat{s}}{Q_\perp^2} \right] \tag{2.151}$$

after summation over all possible number of emissions.

From this the normalized differential cross-section with respect to the transverse momentum squared can be obtained to all orders in the DLLA as

$$\frac{1}{\hat{\sigma}_0} \frac{\mathrm{d}\hat{\sigma}}{\mathrm{d}Q_\perp^2} = -\frac{\mathrm{d}}{\mathrm{d}Q_\perp^2} \Sigma(Q_\perp^2) = \frac{\alpha}{\pi} \frac{1}{Q_\perp^2} \log \frac{\hat{s}}{Q_\perp^2} \exp \left[-\frac{\alpha}{2\pi} \log^2 \frac{\hat{s}}{Q_\perp^2} \right] . \tag{2.152}$$

Inspection reveals that in this expression the resummation of the double leading logarithms has indeed tamed the divergence for $Q_\perp^2 \to 0$, and in fact the cross-section tends to zero in this limit. It can be shown that this very nice suppression in fact is too strong and sub-leading logarithms will ameliorate this situation.

$\Sigma(Q_\perp^2)$ is called the **Sudakov form factor**, and it encodes the probability for not emitting a photon with a transverse momentum larger than $Q_\perp$. This probability tends to zero for $Q_\perp \to 0$; in other words it is impossible not to emit photons with arbitrary small transverse momentum.

2.3.1.3   Impact parameter space

Up to now, the individual photon emissions have explicitly been treated as independent, uncorrelated processes, which is not true. After all, the transverse momentum $Q_\perp$ of the lepton pair is given by the sum of all individual emissions,

$$\vec{Q}_\perp = -\sum_{i=0}^{n} \vec{k}_{\perp,i} . \tag{2.153}$$

This constraint can be cast in the form of a δ-function, which in turn can be expressed through a Fourier transform to **impact parameter space**. The impact parameter is conjugate to the transverse momentum,

$$\delta^2 \left(\vec{Q}_\perp + \sum_{i=0}^{n} \vec{k}_{\perp,i} \right) = \frac{1}{2\pi} \int \mathrm{d}^2 b_\perp \exp \left[i\vec{b}_\perp \cdot \left(\vec{Q}_\perp + \sum_{i=0}^{n} \vec{k}_{\perp,i} \right) \right] . \tag{2.154}$$

Similarly, the individual emission factors

$$\nu(k_{\perp,i}) = \frac{\alpha}{\pi} \frac{1}{k_{\perp,i}} \log \frac{\hat{s}}{k_{\perp,i}^2} \tag{2.155}$$

can be Fourier transformed to yield

$$\nu(b_{\perp,i}) = \frac{1}{2\pi} \int \mathrm{d}^2 k_{\perp,i} \exp\left[-i\vec{b}_{\perp,i} \cdot \vec{k}_{\perp,i}\right] \nu(k_{\perp,i}). \tag{2.156}$$

The contribution of exactly n photons that conspire to yield an overall transverse momentum of $Q_\perp$ of the lepton pair thus reads

$$\frac{1}{\hat{\sigma}_0} \frac{\mathrm{d}\hat{\sigma}^{(n)}}{\mathrm{d}Q_\perp^2} = \frac{1}{2\pi n!} \int \mathrm{d}^2 b_\perp \exp\left[i\vec{b}_\perp \cdot \vec{Q}_\perp\right] \left[\nu(b_\perp)\right]^n \tag{2.157}$$

and summing over all numbers n of photon emissions results in

$$\begin{aligned} \frac{1}{\hat{\sigma}_0} \frac{\mathrm{d}\hat{\sigma}^{(\mathrm{all})}}{\mathrm{d}Q_\perp^2} &= \frac{1}{2\pi} \int \mathrm{d}^2 b_\perp \exp\left[i\vec{b}_\perp \cdot \vec{Q}_\perp\right] \exp\left[\nu(b_\perp)\right] \\ &= \frac{1}{2\pi} \int b_\perp \mathrm{d}b_\perp \, \mathrm{J}_0(b_\perp Q_\perp) \exp\left[\nu(b_\perp)\right], \end{aligned} \tag{2.158}$$

where J_0 is the typical Bessel function stemming from angular integrals.

A few comments are in order here. This method to include transverse momentum conservation seems at first a bit opaque, it is however necessary. It is crucial in a situation where the photons balance each other and lead to a zero net transverse momentum of the lepton pair. In such a situation the vanishing $Q_\perp$ guarantees that the Sudakov form factor goes to zero, indicating that at least one photon must have been emitted. Configurations, where only some of the photons are correlated such that their transverse momenta compensate each other, are of non-leading order in the leading logarithmic expansion for arbitrary, and possibly large, transverse momentum. These configurations become leading when all photons are correlated and mostly soft, because this is the only way to obtain the zero total transverse momentum of the lepton pair. In that respect the Fourier transformation to impact parameter space is a straightforward way to incorporate sub-leading effects due to transverse momentum conservation systematically.

In the example here, small transverse momenta of the lepton pair translate to large impact parameters, and vice versa. When discussing the example with hadrons in the initial state, however, the naive Fourier transform encountered up to now will be further modified by form factors which take into account that the incident partons have a non-zero intrinsic transverse momentum, which is typically non-perturbative in origin. In such a picture, where multiple parton emissions are considered, the factorization scale will decrease from the hard scale of the process down to the low scale, where the perturbative treatment is assumed to break down. This scale is then identified with the inverse of the large corresponding conjugate **impact parameter** $b_\perp$.

2.3.1.4 Adding electron structure functions

There is yet another physics effect that needs to be taken into account. As we have seen in previous sections of this chapter, quantum fluctuations introduce some internal structure for the initial leptons, *cf.* the discussion in Section 2.1.1. This structure can be captured by the introduction of electron structure functions, here denoted by $f_{e/e}(x, \mu_F^2)$, where the factorization scale typically is chosen to be of the order of the transverse momentum of the final-state lepton pair. This feature will become more important in the QCD case; here it suffices to say that an obvious choice for the structure function in b-space will be such that the factorization scale is replaced by $1/b_\perp$.

2.3.2 Standard resummation for $p_{\perp,W}$

2.3.2.1 Summing soft gluons

Applying the technology developed so far in a naive translation from the QED to the analogous QCD case, the $p_\perp$-distribution of the lepton–neutrino pair in $q\bar{q}' \to \nu_\ell \ell^+$, essentially reduces to replacing the QED emission factor of Eq. (2.155) with a suitable QCD factor. This can be done by merely replacing the α with α_s and adding the corresponding quark **Casimir operator** C_F.

There is an important difference, however. While in QED the coupling does not change dramatically with the scale, the QCD coupling does. This implies that the choice of scale in this emission factor is important and will have numerically significant consequences. By merely glancing at the structure of $\nu(k_{\perp,i})$, the only available scale within this factor is the transverse momentum, and thus

$$\nu^{(QCD)}(k_\perp) = \frac{\alpha_s(k_\perp^2)}{\pi} C_F \frac{1}{k_\perp^2} \log \frac{\hat{s}}{k_\perp^2}. \tag{2.159}$$

2.3.2.2 Leading-order leading logarithmic expression (LOLL)

At the accuracy level achieved so far, **leading logarithmic accuracy**, the doubly differential cross-section for the production of a singlet, here a W^+ boson, reads

$$\frac{d\sigma_{AB \to X}}{dy dQ_\perp^2} = \sum_{ij} \pi \hat{\sigma}_{ij \to X}^{(LO)} \left\{ \int \frac{d^2 b_\perp}{(2\pi)^2} \left[\exp(i\vec{b}_\perp \cdot \vec{Q}_\perp) \tilde{W}_{ij}(b; Q, x_A, x_B) \right] \right\}, \tag{2.160}$$

with y and $Q_\perp$ the rapidity and transverse momentum of the singlet. The factor π on the right-hand side of the equation stems from the integration over the azimuthal angle of the produced system X. The leading-order cross-section for the production of the singlet X would of course be identifed with the W-production cross-section:

$$\hat{\sigma}_{ij \to X}^{(LO)} \equiv \hat{\sigma}_{ij \to W^+}^{(LO)} = \frac{4\pi^2 \alpha |V_{ij}|^2}{12 \sin^2 \theta_W m_W^2}. \tag{2.161}$$

Eq. (2.160) already exhibits the choice of scale at which the PDFs are evaluated, namely $1/b_\perp$, as seen in the equation below for $\tilde{W}_{ij}$,

$$\tilde{W}_{ij}(b; Q, x_A, x_B) =$$

$$f_{i/A}\left(x_A, \frac{1}{b_\perp}\right) f_{j/B}\left(x_B, \frac{1}{b_\perp}\right) \exp\left[-\int\limits_{1/b_\perp^2}^{Q^2} \frac{dk_\perp^2}{k_\perp^2}\left(A(k_\perp^2)\log\frac{Q^2}{k_\perp^2}\right)\right].$$

$$(2.162)$$

The sum runs over all relevant parton flavours i and j contributing to the production of X. The energy fractions with respect to the incoming hadrons A and B, x_A and x_B, are fixed by the rapidity of the singlet system,

$$x_{A,B} = \frac{M_X}{\sqrt{s}}\, e^{\pm y},$$

$$(2.163)$$

which is also the solution for the zeroes of the residual δ-function in Eq. (2.107).

Comparing terms with previous equations also shows that the exponential in the resummation part $\tilde{W}_{ij}$ is nothing but the **Sudakov form factor** in $k_\perp$ space. Thus, the term $A(k_\perp^2)$, at the logarithmic order considered up to now, *i.e.* double leading logarithms or DDT approximation,[18] and ignoring the potentially dangerous region of large $b_\perp$, is given by

$$A(k_\perp^2) = C_F \frac{\alpha_s(k_\perp^2)}{\pi}.$$

$$(2.165)$$

2.3.2.3 Improving the accuracy: the master formula

In order to increase the accuracy of Eq. (2.160), higher-order terms must be added, extending the equation by various bits which have hitherto not been included. In particular, this includes emission terms beyond leading logarithms which can be exponentiated as well and thus enter the Sudakov form factor. This will translate into supplementing the term $A\log(Q^2/k_\perp^2)$ with another, non-logarithmic term B in the exponential of Eq. (2.162). In addition, higher-order terms H and C stemming from finite virtual and collinear parts, which come with the same underlying Born kinematics, must be included; they will multiply the resummation bit $\tilde{W}_{ij}$, which is therefore modified. Furthermore, exact real emission matrix elements can be added in such a way that the terms which describe the emissions in the soft leading-logarithmic approximation through the Sudakov form factor are not double-counted. These additional hard emission terms will not be resummed and therefore enter as some finite remainder $Y_{ij \to X}$, effectively the difference of the exact real emission matrix element and the correspondimng $\mathcal{O}(\alpha_s)$ expansion of the emission pattern encoded in $\tilde{W}_{ij}$. With

---

[18]In fact, in the original DDT result, the Sudakov form factor was given in a form also including some universal sub-leading logarithms (the term 3/2):

$$\Sigma_{\text{DDT}}(q_T, Q) = \exp\left[-\int\limits_{q_T^2}^{Q^2} \frac{dk_\perp^2}{k_\perp^2} \frac{\alpha_s(k_\perp^2)}{\pi} C_F \left(\log\frac{Q^2}{k_\perp^2} - \frac{3}{2}\right)\right],$$

$$(2.164)$$

this in mind, the **master equation for $Q_\perp$ resummation in the CSS formalism** discussed up to now reads

$$\frac{\mathrm{d}\sigma_{AB\to X}}{\mathrm{d}y\mathrm{d}Q_\perp^2} = \sum_{ij} \pi\hat{\sigma}_{ij\to X}^{(LO)} \left\{ \int \frac{\mathrm{d}^2 b_\perp}{(2\pi)^2} \left[\exp(i\vec{b}_\perp \cdot \vec{Q}_\perp) \, \tilde{W}_{ij}(b; Q, x_A, x_B) \right] \right.$$

$$\left. + Y_{ij\to X}(Q_\perp; Q, x_A, x_B) \right\} , \qquad (2.166)$$

where $\tilde{W}_{ij}$ and $Y_{ij\to X}$ will be given in Eqs. (2.167) and (2.170), respectively.

For the resummation bit, increasing the precision amounts to extending the function $\tilde{W}_{ij}$ such that

$$\tilde{W}_{ij}(b; Q, x_A, x_B) = \sum_{ab} \left\{ \int_{x_A}^1 \frac{\mathrm{d}\xi_A}{\xi_A} \int_{x_B}^1 \frac{\mathrm{d}\xi_B}{\xi_B} \left[f_{a/A}\left(\xi_A, \frac{1}{b_\perp}\right) f_{b/B}\left(\xi_B, \frac{1}{b_\perp}\right) \right. \right.$$

$$\times C_{ia}\left(\frac{x_A}{\xi_A}, b_\perp; \mu\right) C_{jb}\left(\frac{x_B}{\xi_B}, b_\perp; \mu\right) H_{ab}\left(\frac{x_A}{\xi_A}, \frac{x_B}{\xi_B}; \mu\right)$$

$$\left. \left. \times \exp\left[-\int_{1/b_\perp^2}^{Q^2} \frac{\mathrm{d}k_\perp^2}{k_\perp^2} \left(A(k_\perp^2) \log\frac{Q^2}{k_\perp^2} + B(k_\perp^2) \right) \right] \right] \right\} . \qquad (2.167)$$

All terms A, B, C, and H can be expanded in a perturbative series, which will define the formal accuracy of the result in terms of the logarithmic order (LL, NLL, ...) and the fixed-order (LO, NLO, ...) accuracy. Note that the origin of the terms A and B can also be traced to the splitting functions, see also Sections 2.3.3 and 5.2. This expansion yields, for example,

$$A(\mu) = \sum_N \left(\frac{\alpha_s(\mu)}{2\pi}\right)^N A^{(N)} \qquad (2.168)$$

and similar for the other terms. For instance, by direct comparison with the previous result $A^{(1)} = 2C_F$. In contrast to $A^{(1)}$, however, the result for $B^{(1)}$ depends on the lower limit of the $k_\perp^2$ integration. Choosing $(2e^{-\gamma_E}/b_\perp)^2 = (b_0/b_\perp)^2$, leads to $B^{(1)} = -3C_F/2$, while in [410] it was given by $B^{(1)} = 2C_F \log \frac{e^{\gamma_E - 3/4}}{2}$.

The expansion of the C and H terms starts at $N = 0$, with

$$C_{ia}^{(0)}(z) = \delta_{ia}\delta(1-z)$$
$$H_{ab}^{(0)}(z_A, z_B; \mu) = \delta_{ia}\delta_{jb}\delta(1-z_A)\delta(1-z_B), \qquad (2.169)$$

which allows one to recover Eq. (2.160).

It is worth stressing that, to the first few orders, the terms A and B depend on the incoming flavours only. For the case of W production these are quarks, which is reflected in the corresponding colour factor, C_F. For other processes, such as, e.g., the production of a Higgs boson in gluon fusion, $gg \to H$, these terms would be proportional to C_A, a factor of two larger, apart from sub-leading colour corrections. This reflects of course the fact that, trivially, a gluon has two colour degrees of freedom, while a quark has only one such degree of freedom. A more detailed discussion of the technical steps summarized here, and an application of this formalism to a variety of processes, will be presented in Chapter 5. There, other different resummation techniques will also be briefly discussed.

The **finite remainders** Y can be expanded in a similar series,

$$
Y_{ij \to X}(Q_\perp; Q, x_A, x_B) =
$$

$$
\int_{x_A}^1 \frac{d\xi_A}{\xi_A} \int_{x_A}^1 \frac{d\xi_B}{\xi_B} \left\{ f_{i/A}(\xi_A, \mu) f_{j/B}(\xi_B, \mu) \sum_N \left[\left(\frac{\alpha_s}{2\pi}\right)^N R^{(N)}_{ij \to X}\left(Q, \frac{x_A}{\xi_A}, \frac{x_B}{\xi_B}\right) \right] \right\}
$$

$$
(2.170)
$$

with the first non-trivial terms $R^{(1)}_{ij \to X}$ different from zero listed in relevant sections in Chapter 5.

In principle, scales in the hard remainder could be chosen differently from the choices made in the resummation term; this has been made explicit by introducing separate factorization and renormalization scales μ_F and μ_R. In addition, the hard scale Q can be identified through $Q = M_X \equiv m_W$ and the Mandelstam variables which will emerge in the functions $R_{ij \to X}$ can be expressed through the other kinematic parameters as

$$
\hat{s} = \frac{1}{\xi_A \xi_B} Q^2 \quad \text{and} \quad \hat{t}, \hat{u} = \left(1 - \frac{\sqrt{1 + \frac{Q_\perp^2}{Q^2}}}{\xi_{B,A}} \right) Q^2 . \tag{2.171}
$$

The connection of resummed and fixed-order cross-sections is exemplified for the case of W^+ production at an 8 TeV LHC in Fig. 2.29. As the transverse momentum of the W^+ boson approaches zero, the resummation bit, the exponential term in $\tilde{W}_{ij}$ takes over and guarantees that the triple differential cross-section goes to zero, although the fixed (first) order result diverges. On the other hand, the finite remainder terms, the differences between resummed and fixed-order cross-section are visible only in the hard region, where $Q_\perp \to Q = m_W$. Note that the soft region, where $b \to \infty$, has been regularized with the Collins-Soper scheme, which will be discussed in more detail in Chapter 5. There, in addition, more results for this and other processes will be given and the anatomy of different contributions will be discussed in more detail.

2.3.2.4 Dealing with the soft region

This looks very good, but it has an additional problem, not present in QED. Integrating over all values of $k_\perp$ down to zero results in the need to evaluate the strong coupling at scales around and below the Landau pole, where it diverges.

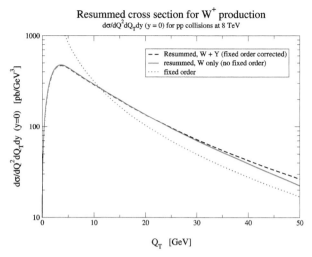

Fig. 2.29 The triply differential production cross-section for a W^+ boson at an 8 TeV LHC, from [300, 716]. As customary, the cross-section is evaluated for $Q^2 = m_W^2$ and at central rapidity $y = 0$. The soft region has been regularized according to the Collins–Soper scheme.

This is typically cured by modifying the Sudakov form factor in impact parameter space. A naive way of achieving this is by adding a **soft form factor** through a function ρ, multiplying the Sudakov form factor for all values of $b_\perp$. A typical parameterization of this function ρ would look like

$$\rho(b_\perp) = \exp\left(-\frac{b_\perp^2}{4A}\right),$$
(2.172)

where A usually is identified with the average intrinsic transverse momentum of the incident partons due to Fermi motion inside the nucleon or similar effects,

$$A = \langle p_{\perp,\text{intrinsic}}^2 \rangle^{-1}.$$
(2.173)

The effect of such a modification is negligible for small $b_\perp$, but effectively amounts to a dampening for large $b_\perp$ or small $k_\perp$:

$$\Sigma^{(QCD)}(b_\perp) \longrightarrow \Sigma^{(QCD)}(b_\perp)\,\rho(b_\perp).$$
(2.174)

Of course, there are more methods to tame the divergent behaviour of QCD around the Landau pole, which will not be further discussed here.

2.3.3 Aside: resumming jet rates in $e^+e^- \to$ hadrons

2.3.3.1 Origin of the resummation terms A and B

The formalism developed for the case of the $p_\perp$ spectrum of the W boson in the previous section can also be applied to a different case, namely the formation of jets in the annihilation of electron–positron pairs into hadrons. This additional example will further highlight the versatility of the resummation approach, and it will also motivate why parton shower Monte Carlo event generators do a fairly decent job in describing the bulk of existing data.

As a first step the **Sudakov form factor**, encoding the resummation of multiple emissions, needs to be re-examined. This will lead to an alternative explanation for the specfic form of the terms $A^{(1)}$ and $B^{(1)}$. To see how this works, consider gluon emission off a quark in the approximation, where the gluon is collinear with the quark. Then, the respective matrix element is well approximated by the **splitting function**, $\mathcal{P}_{qq}(z)$. This splitting function would merge when keeping also terms of order $k^\mu/(p\cdot k)$ in Eq. (2.147). Correspondingly the emission factor $\nu^{(\mathrm{QCD})}$ in Eq. (2.159) becomes

$$\nu^{(\mathrm{QCD})}(k_\perp) = \frac{\alpha_\mathrm{s}}{\pi} C_F \frac{1}{k_\perp} \log \frac{Q^2}{k_\perp^2} \longrightarrow \nu^{(\mathrm{QCD})}_{q \to qg}(k_\perp; z) = \frac{\alpha_\mathrm{s}}{\pi} C_F \frac{1}{k_\perp} P_{qq}(z), \quad (2.175)$$

with

$$P_{qq}(z) = \frac{1 + z^2}{1 - z}. \quad (2.176)$$

The integration over $k_\perp$ must also be supplemented with one over z. This integration needs furthere manipulation, because of the divergent structures in splitting functions for $z \to 1$ or $z \to 0$, which must be regularized. While this is typically achieved through the "+"-prescription, *cf.* Eq. (2.33), another way to guarantee finite integrals will be pursued here.

In the application of the resummation formalism to the description of jet production, the emitted partons, the gluons, must be resolved, for example by demanding that they have a minimal transverse momentum Q_0 with respect to the emitting quark, $k_\perp > Q_0$. Momentum conservation ensures that the momentum fraction they carry away from the quark must be non-zero, and the residual momentum fraction of the quark therefore must be smaller than 1 by an amount of $\varepsilon = k_\perp^2/Q^2$, if Q is the scale of the quark momentum. This allows to drop the "+"-prescription in the splitting function, and, correspondingly, the δ-function compensating for it. This form of modification, from now on, will be made obvious by replacing $\mathcal{P}_{ij} \longrightarrow P_{ij}$.

Thus the Sudakov form factor can be rewritten for this specific case as

$$S(Q, Q_0) = \exp\left[-\int_{Q_0^2}^{Q^2} \frac{\mathrm{d}k_\perp^2}{k_\perp^2} \left(\frac{\alpha_\mathrm{s}(k_\perp^2)}{2\pi} \int_0^{1-\varepsilon} \mathrm{d}z\, P_{qq}(z) \right) \right] \quad (2.177)$$

Inspection of the z-integral reveals that it can be approximated by

$$\int_0^{1-\varepsilon} \mathrm{d}z\, P_{qq}(z) = C_F \int_0^{1-\varepsilon} \mathrm{d}z\, \frac{1+z^2}{1-z} \approx C_F \left[\int_0^{1-\varepsilon} \mathrm{d}z\, \frac{2}{1-z} - \int_0^1 \mathrm{d}z\, (1+z) \right] \quad (2.178)$$

$$= 2C_F \left[\log \frac{Q^2}{k_\perp^2} - \frac{3}{4} \right] \equiv \Gamma_q(Q^2, k_\perp^2).$$

Here, for the sake of the following discussion, the **integrated splitting function** Γ has been introduced. From these results, the coefficients $A^{(1)} = C_F$ and $B^{(1)} = -\frac{3}{2}C_F$ are readily read off. Of course, they are identical to the coefficients in the standard resummation procedure outlined in Section 2.3.2.

2.3.3.2 Sudakov form factor and its interpretation

This allows one to write the Sudakov form factor in this case as

$$
S(Q^2, Q_0^2) \equiv \Delta(Q^2, Q_0^2) = \exp\left[-\int_{Q_0^2}^{Q^2} \frac{dk_\perp^2}{k_\perp^2} \frac{\alpha_s(k_\perp^2)}{\pi} \Gamma(Q^2, k_\perp^2) \right]
\tag{2.179}
$$

How can this be interpreted? In Section 5.3 it will be argued that the Sudakov form factor is nothing but a **probability for a given particle not to radiate a secondary particle** between the two scales Q^2 and Q_0^2. This can be motivated by realizing that

(a) its kernel is related to the emission of a particle;

(b) its form as an exponential of a negative definite argument guarantees that

$$
S(Q^2, Q_0^2) \in [0, 1]
\tag{2.180}
$$

as expected from a properly defined probability;

(c) in the limits $Q^2 \gg Q_0^2$ it approaches 0, which is what one would expect: it becomes increasingly unlikely that particles do not radiate in an increasing interval of scales.

This reasoning will be further extended at the beginning of Section 5.3, by elaborating on a correspondence with the decay of a radioactive isotope.

2.3.3.3 Sudakov form factor for fixed and running α_s

In general, the integrated splitting kernels are given by the usual coefficents

$$
\Gamma_{q,g}(Q^2, q^2) = A_{q,g}^{(1)} \log \frac{Q^2}{q^2} + B_{q,g}^{(1)}.
\tag{2.181}
$$

Note that in the case of gluons the corresponding integrated splitting function consists of two parts, the $g \to gg$ and $g \to q\bar{q}$ splittings. This leads to Sudakov form factors given by

$$
\Delta_{q,g}(Q^2, q^2) = \exp\left[-\frac{\alpha_s}{2\pi} \left(A_{q,g}^{(1)} \log^2 \frac{Q^2}{q^2} + B_{q,g}^{(1)} \log \frac{Q^2}{q^2} \right) \right]
\tag{2.182}
$$

for a fixed α_s or

$$
\Delta_{q,g}(Q^2, q^2) = \exp\left[-\frac{\alpha_s}{2\pi} \left(A_{q,g}^{(1)} \log^2 \frac{\alpha_s(Q^2)}{\alpha_s(q^2)} + B_{q,g}^{(1)} \log \frac{\alpha_s(Q^2)}{\alpha_s(q^2)} \right) \right]
\tag{2.183}
$$

for $\alpha_{\rm s}$ running at one-loop.

2.3.3.4 Emissions and jet rates

The two-jet rate $\mathfrak{R}_2$ in $e^-e^+ \to$ hadrons is just given by the combined probability that neither the quark nor the anti-quark in the final state of the underlying Born-level diagram radiate a gluon above a jet resolution scale given by $p_\perp^{(j)} > Q_{\rm cut}$. The probabilistic interpretation of the Sudakov form factor yields

$$\mathfrak{R}_2(Q_{\rm cut}) = \left[\Delta_q(Q^2, Q_{\rm cut}^2)\right]^2 . \tag{2.184}$$

In order to obtain the three-jet rate, it is important to realize that the probability density $\mathrm{d}\mathcal{P}_{\rm rad}(q_\perp^2)/\mathrm{d}q_\perp^2$ to have a radiation off a quark at $q_\perp$ is given by

$$
\begin{aligned}
\frac{\mathrm{d}\mathcal{P}_{\rm rad}(q_\perp^2)}{\mathrm{d}q_\perp^2} &= -\frac{\mathrm{d}\Delta_q(Q^2, Q_{\rm cut}^2)}{\mathrm{d}q_\perp^2} \\
&= \frac{\alpha_{\rm s}(q_\perp^2)}{\pi} C_F \frac{1}{q_\perp^2} \Gamma_q(Q^2, q_\perp^2) \Delta_q(Q^2, Q_{\rm cut}^2) \Theta(Q^2 - q_\perp^2)\Theta(q_\perp^2 - Q_{\rm cut}^2) .
\end{aligned}
\tag{2.185}
$$

Here, the Θ-functions guarantee that $q_\perp^2 \in [Q_{\rm cut}^2, Q^2]$.

This allows one to write the three-jet rate as

$$
\begin{aligned}
\mathfrak{R}_3(Q_{\rm cut}) = 2\Delta_q(Q^2, Q_{\rm cut}^2) \int\limits_{Q_{\rm cut}^2}^{Q^2} \frac{\mathrm{d}q_\perp^2}{q_\perp^2} &\left[\left(\frac{\alpha_{\rm s}(q_\perp^2)}{\pi} C_F \Gamma_q(Q^2, q_\perp^2) \frac{\Delta_q(Q^2, Q_{\rm cut}^2)}{\Delta_q(Q^2, q_\perp^2)}\right) \right. \\
&\left. \times \Delta_q(q_\perp^2, Q_{\rm cut}^2)\Delta_g(q_\perp^2, Q_{\rm cut}^2)\right] ,
\end{aligned}
\tag{2.186}
$$

where the term in the round brackets accounts for the emission of the gluon off one of the two quark lines, and the two additional Sudakov form factors ensure that the two offsprings of this splitting do not radiate further. The ratio of Sudakov form factors can be interpreted as the probability for the intermediate quark line not to experience any radiation resolvable above $Q_{\rm cut}$, between Q^2 and $q_\perp^2$.

Similar expressions can also be constructed for higher jet multiplicities. A convenient way to do this is by the introduction of generating functionals, *cf.* [345] for more details.

2.3.3.5 Scaling patterns in jet production

Generating functional technqiues can also be used to study the scaling behaviour of exclusive jet multiplicities, *i.e.*, the cross-sections σ_n for the production of exactly n jets, possibly in association with other particles. There are two extreme scenarios or scaling patterns, namely

1. **staircase scaling**, which is characterized by the ratios

$$R_{(n+1)/n} = \frac{\sigma_{n+1}}{\sigma_n} \equiv R \quad \text{or} \quad R_{(n+1)/n} = \frac{\mathfrak{R}_{n+1}}{\mathfrak{R}_n} = R \quad \text{with} \quad \mathfrak{R}_n = \frac{\sigma_n}{\sigma_{\text{tot}}}$$
(2.187)

being constant, and where R depends on the core process and the requirements on the jets only, but not on the multplicity of jets. This pattern is also known as **Berends scaling** [223], and it has been observed for example by UA2 [140], at the TEVATRON [107], and the LHC [20, 364].

2. **Poisson scaling**, where the exclusive jet rates P_N behave like

$$\mathfrak{R}_n = \frac{\bar{n}^n e^{-\bar{n}}}{n!} \quad \text{or} \quad R_{(n+1)/n} = \frac{\bar{n}}{n+1}.$$
(2.188)

Such patterns emerge when individual events, in the case at hand jet formation through hard parton radiation, repeat themselves and are independent from each other.

In order to see how this pattern emerges, consider the subsequent emission of two gluons off a primary quark line, given by

$$\sigma^{(1)}_{q \to qgg} \propto \frac{1}{2} \left[\int_{Q_0^2}^{Q^2} dt \Gamma_{q \to qg}(Q^2, t) \Delta_g(Q^2, t) \right] \left[\int_{Q_0^2}^{Q^2} dt' \Gamma_{q \to qg}(Q^2, t') \Delta_g(Q^2, t') \right],$$
(2.189)

where Q_0 represents the jet resolution scale and Q is the hard scale of the process, and the factor of $1/2$ accounts for the ordering of emissions. The Sudakov form factors guarantee that the gluons do not experience any further splitting, and they form jets. Similarly, one could have a second contribution, where the second gluon emerges from a secondary gluon splitting,

$$\sigma^{(2)}_{q \to qgg} \propto \left[\int_{Q_0^2}^{Q^2} dt \Gamma_{q \to qg}(Q^2, t) \Delta_g(Q^2, t) \right] \left[\int_{Q_0^2}^{t} dt' \Gamma_{g \to gg}(t, t') \Delta_g(t, t') \right].$$
(2.190)

There are two important limits to consider, namely

1. $\frac{\alpha_s}{\pi} \log^2 \frac{Q}{Q_0} \gg 1$. Expanding the results for $\sigma^{(1)}$ and $\sigma^{(2)}$ around $Q_0/Q \to 0$, the leading contributions are given by

$$\sigma^{(2)}_{q \to qgg} \propto \frac{1}{4} \left[\frac{\alpha_s}{C_A} \log^2 \frac{Q}{Q_0} - \sqrt{4\alpha_s} C_A^3 \log \frac{Q}{Q_0} + \mathcal{O}\left(\frac{Q_0^2}{Q^2} \right) \right]$$

$$\sigma^{(2)}_{q \to qgg} \propto \frac{1}{4} \left[(\sqrt{2} - 1) \sqrt{\alpha_s} C_A^3 \log \frac{Q}{Q_0} + \mathcal{O}\left(\frac{Q_0^2}{Q^2} \right) \right],$$
(2.191)

indicating that the pattern of subsequent, primary emissions off the quarks is enhanced with respect to the secondary gluon splittings. This is the limit of independent emissions, therefore exhibiting Poisson scaling.

2. $\frac{\alpha_s}{\pi} \log^2 \frac{Q}{Q_0} \ll 1$. In this limit

$$\sigma^{(1)}_{q \to qgg} \propto \frac{\alpha_s^2}{4(2\pi)^2} \log^4 \frac{Q}{Q_0} + \mathcal{O}\left(\alpha_s^3 \log^6 \frac{Q}{Q_0}\right) \propto \sigma^{(2)}_{q \to qgg}, \qquad (2.192)$$

and primary and secondary emissions off the quark or gluon, respectively, contribute in roughly equal, democratic measure, with relatively small emission probability. This modifies the Poisson scaling. In QED such a non-Abelian contribution from secondary emissions is absent, and as a consequence Poisson scaling is not modified in QED.

At this point it should be clear that in order to observe Poisson scaling, the individual emissions must be decoupled, to be treated as independent. This can be enforced by selecting events with a strong hierarchy, for example by demanding that the hardest jet has very large transverse momentum, while all other jets can be much softer. In such configurations, the large scale ratio of the hard jet to the softer ones guarantees large logarithms. Increasing the number of jets leads to more scale ratios with vanishing logarithms, and this also introduces a jet emission phase space that is increasingly constrained through momentum conservation. As an overall effect, the absence of logarithmic enhancement means that the emissions cannot be treated as independent any more, negating the condition for Poisson scaling, and staircase scaling patterns emerge.

2.4 Summary

In this chapter the technology underlying every discussion of high-energy phenomena at hadron colliders based on first principles, perturbative methods, has been introduced. In order to calculate cross-sections, the idea of factorization must be invoked. This idea is at the heart of perturbative QCD at hadron colliders. First of all, factorization guarantees that the partons, the constituents of the protons, can be treated as quasi-free particles. This is the case if the characteristic time-scales related to the process probing them are sufficiently smaller than the typical response time of the strong field. Only then, when the partons can be treated as quasi-free, can they be quantized like any other field in quantum field theory. This allows a perturbative expansion, typically represented by Feynman diagrams, which is a systematic treatment based on the Lagrangian of QCD. At the same time, the validity of factorization allows a determination of the parton distribution functions (PDFs) in a process-independent way which can be used to evaluate cross-sections for all other processes. At leading order the PDF $f_{a/h}(x, \mu_F)$ describes the probability to find a parton a in hadron h with a momentum fraction x at the factorization scale μ_F. Since they can be related to the bound state stucture of the respective hadron they fulfil a number of sum rules, which in turn act as theoretical constraints. While the PDFs are non-perturbative objects, their evolution with the factorization scale is governed by the perturbative DGLAP equations. This evolution in turn can be employed to understand multiple softer emissions. In particular, initial-state radiation can be interpreted as the breakdown of the coherence of the multi-particle Fock state of the incident particles, where the quantum fluctuations populating the Fock state are governed by the evolution equation.

At the same time, final-state radiation leads to a proliferation of particles in the final state by repeated emissions of secondary quanta. Again, the behaviour of the radiation pattern is governed by an evolution equation. Due to the presence of massless quanta, such as photons or gluons, however, some care is necessary to obtain physically meaningful results when analysing this additional radiation. This is typically reflected by certain phase-space conditions which lead to subsuming emitted quanta into observable particles. In the case of QED this effectively leads to adding sufficiently soft and/or collinear photons to the leptons emitting them; in contrast, in QCD the picture is further blurred by the fact that the quanta of the strong interaction, quarks and gluons, only occur in bound states, the hadrons. In order to arrive at meaningfully defined physical objects, the combination procedures must therefore be valid on the parton and the hadron level. They must not destroy the perturbative precision in the desciption of how these multi-particle states emerged from radiation in the first place. This then leads to the technical definition of jets through suitable jet algorithms.

The development of the perturbative formalism has been exemplified by W production. This process has been one of the "standard candles" in the experimental programme at the TEVATRON and the first run of the LHC and it will continue to play a similarly central role also for the physics programme in Run II and beyond.

Already at leading order, a rapidity asymmetry in the production of the two charge eigenstates $W^{\pm}$ of the W bosons emerges, which translates into a more convoluted measurable asymmetry of the charged leptons coming from the W decay. This observable is being used to further constrain the PDFs.

Of course, going to higher perturbative orders increases the precision of the theoretical results, but this enhanced quality of the prediction comes with a price. First of all, the corresponding matrix elements tend to diverge, necessitating regularization and renormalization of the ultraviolet divergences present in the virtual contributions, and requiring the treatment of infrared divergences occuring in the real and virtual contributions. Since the latter cancel exactly for all physically sensible observables, it is mandatory to regularize them in each contribution. Due to the different dimensions of the respective phase spaces and the different sources of the divergent structures, loop vs. phase space integrals, this presents a considerable nuisance in higher order calculations. To highlight this point, note that this problem, in fact, has been fully and robustly solved for next-to-leading order (NLO) calculations only. It should also be noted that only the inclusion of higher-order corrections allows the quantification of theoretical uncertainties in a meaningful way, by varying the renormalization and factorization scales appearing in the calculation because of the truncation of the perturbative series. While in principle this seems to be a fairly straightforward exercise, there is some subtlety involved, especially when choosing a meaningful central scale.

At the same time, the NLO calculations contain a contribution, the real emission part, that can interpreted as a new process. Instead of the inclusive production of a system, like the W boson discussed here, they describe the production of the system plus an additional parton, which can be interpreted as an additional jet. This indeed signals the start of a tower of ever increasing jet multiplicities described by corresponding tree-level Feynman diagrams, which contain more and more partons in the final state. With respect to higher-order calculations they form, on the other hand, towers of multiple real emission corrections. For the measurement of the W mass, the knowledge of the $p_{\perp}$ spectrum of the produced W bosons is indispensable, and the process with one extra parton together with the W in the final state merely yields the leading-order expression for this observable. In order to increase the precision one could, again, invoke higher-order corrections. However the validity of a normal NLO calculation would be limited to non-negligible values of the W transverse momentum. This is because, already at leading order, the differential cross-section with respect to the transverse momentum of the W diverges for small transverse momenta. This divergence appears again at any fixed order of perturbation theory. However, a careful analysis of the situation indicates that a finite result can be obtained for small values of the W transverse momentum by resumming the soft and/or collinear limits of extra parton emission to all orders.

3

QCD at Fixed Order: Technology

In this chapter, the perturbative description at fixed order of processes at the LHC and other hadron collider experiments, outlined in Chapter 2, will be further discussed. In particular, theoretical issues related to calculations for more complex final states will be addressed. First, the terminology used in this book to denote the accuracy of a given calculation is described in Section 3.1.

Section 3.2 is then devoted to a discussion of the technology used in fixed-order calculations, in particular for the calculation of multi-particle final states at leading order (LO).

This is followed by a discussion of techniques employed in next-to-leading-order calculations in Section 3.3, containing a presentation of subtraction methods for the treatment of the infrared divergences which facilate their mutual cancellation between virtual and real corrections and advanced methods to evaluate the former in processes for multi-particle final states.

In the final part of this chapter, Section 3.4, some ideas are presented on how even higher orders in perturbation theory can be treated.

3.1 Orders in perturbation theory

To discuss in more detail technicalities and results of fixed-order calculations in QCD, it is mandatory to establish and define a suitable language first. While most people appear to have a clear idea what they mean when talking about calculations at "leading order" (LO), at "next-to-leading order" (NLO) or even at "next-to-next-to-leading order" (NNLO), these ideas do not always coincide. In the framework of this book, these notions will be used as follows: the order, at which an observable is calculated, is denoted as "leading order", if the result of the calculation yields the first non-trivial contribution to the perturbative expansion *to this very observable*. Consequently, "next-to-leading order" denotes the order at which the first perturbative correction to the same observable is being evaluated.

To see how this works in more detail, consider the by-now notorious example of W production, invoked in the previous chapter. Calculating the process, or to be more precise its cross-section, at leading order introduces the Feynman diagram in the left panel of Fig. 3.1. Due to the dependence on the value of x probed in the PDFs, this

The Black Book of Quantum Chromodynamics. John Campbell, Joey Huston and Frank Krauss, Oxford University Press. © John Campbell, Joey Huston, and Frank Krauss 2018, First published in paperback 2022. DOI: 10.1093/oso/9780192871961.001.0003

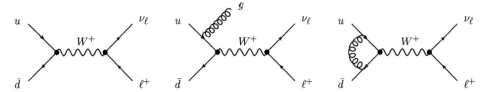

Fig. 3.1 Feynman diagrams for W production at hadron colliders. The left panel is related to the calculation of the cross-section, the rapidity distribution of the W boson, or the transverse momentum and rapidity distributions of the individual decay products at leading order. In order to have a leading-order expression for the boson's transverse momentum, real-emission diagrams such as the one in the middle panel must be considered. Together with the virtual corrections. such as the loop diagram in the right panel, they also form part of the next-to-leading order correction to the total cross-section, etc.

Table 3.1 The order at which various observables related to W production are computed, as a function of the overall power of the strong coupling in the theoretical calculation.

strong coupling order	σ_{tot}, $d\sigma/dy$	$d\sigma/dp_T(W)$	$d\sigma/d\phi_{jj}$
α_s^0	LO	-	-
α_s^1	NLO	LO	-
α_s^2	NNLO	NLO	LO
α_s^3	N^{3}LO	NNLO	NLO

diagram also predicts a non-trivial rapidity distribution. On the other hand, since the partons are collinear with the incoming protons, this diagram will not produce any transverse momentum distribution of the W boson. Instead it will generate such distributions only for its decay products. In order to have a leading-order expression for the $p_\perp$ distribution of the W boson, real-emission diagrams such as the one in the middle panel must be evaluated. There the boson can recoil against the additional parton. As seen in the previous chapter, for $p_\perp^{(W)} \to 0$ the expression related to this diagram diverges, a problem that actually persists for each fixed-order calculation and which can only be resolved through the use of resummation. However, this diagram not only yields *a leading-order result for* $p_\perp^{(W)}$ but it is also part of the *next-to-leading-order calculation of the total cross-section or of observables such as the boson rapidity or the kinematical distributions of the W boson's decay products.* This situation is summarized in Table 3.1.

3.2   Technology of leading-order calculations

Going back to the master formula for the evaluation of cross-sections for the production of an n body final state in hadronic collisions at leading order, Eq. (2.52),

$$\sigma_n^{(\mathrm{LO})} \equiv \sigma_n^{(\mathrm{Born})} = \int \mathrm{d}\,\Phi_{\mathcal{B}}\,\mathcal{B}(\Phi_{\mathcal{B}})$$

$$= \sum_{a,b} \int_0^1 \mathrm{d}x_a \mathrm{d}x_b\, f_{a/h_1}(x_a, \mu_F) f_{b/h_2}(x_b, \mu_F) \int \mathrm{d}\hat{\sigma}_{ab \to n}^{(\mathrm{LO})}(\mu_F, \mu_R)$$

$$= \sum_{a,b} \int_0^1 \mathrm{d}x_a \mathrm{d}x_b \int \mathrm{d}\Phi_n\, f_{a/h_1}(x_a, \mu_F) f_{b/h_2}(x_b, \mu_F) \frac{1}{2\hat{s}} |\mathcal{M}_{ab \to n}|^2 (\Phi_n; \mu_F, \mu_R) \quad (3.1)$$

it is fairly obvious that already for this simplest result two major obstacles must be faced.

First of all, the squared matrix element $|\mathcal{M}_{ab \to n}|^2$ must be evaluated. Since the number of diagrams increases very quickly, typically faster than factorial, the traditional methods encountered so far become impractical. Squaring the amplitudes, using the completeness relations to arrive at traces which may be evaluated analytically is just too complex an operation. As a consequence, in modern techniques the focus is on evaluating individual amplitudes as functions of their internal and external degrees of freedom, which means that every amplitude becomes just a complex number and in turn renders their summation and squaring a straightforward exercise. Modern algorithms to achieve this are introduced in Section 3.2.1.

Second, the complicated structure of the phase space resulting in an integral in $(3n - 2)$-dimensions, possibly supplemented with complicated cuts, renders this high-dimensional integration impossible to be evaluated analytically, even if the PDFs were tractable in this way. As this is not the case, mainly numerical methods must take over, both in the evaluation of the matrix elements and in the phase-space integration. For the latter, essentially only Monte Carlo integration techniques are viable, since for them the error estimate scales like $1/\sqrt{N}$ for the number N of integrand evaluations, independent of the number of dimensions.[1] A general discussion of Monte Carlo techniques can be found in Section 3.2.2.

Sampling in such a way over the phase-space degrees of freedom to conveniently obtain a numerical estimate for the cross-section — the actual result — begs the question, how far sampling can also be extended to other degrees of freedom such as particle spins, colours, or similar. In other words, for the matrix element evaluation, a choice between summation and sampling over quantum degrees of freedom must be made. Since the computation times for summation and sampling naively differ by the sth power of the number of possible states, usually $\approx 2^s$ for the possible helicity assignments and $\approx 3^s \ldots 8^s$ for the possible colour assignments, dependent on whether

---

[1]For traditional integration methods such as trapezoid quadratures or similar, the number of dimensions enters such that usually the error scales like $N^{-k/n}$ where n is the number of integral dimensions and $k > 0$ depends on the method of choice.

the particles in question are quarks or gluons, this may have a significant impact on the overall evaluation time, especially if there are strong correlations also with the phase space, which would thus favour a simultaneous sampling over external (phase space) and internal (spins, colours, etc.) degrees of freedom. By suitably eliminating common sub-expressions in the matrix elements, however, these naive factors can often be reduced quite considerably, which renders this choice strongly dependent on the process and the particle multiplicity in question.

3.2.1 Evaluating matrix elements at leading order

3.2.1.1 Helicity amplitudes from Feynman diagrams

In this section, modern methods for the fast numerical calculation of cross-sections like the one in Eq. (3.1) will be introduced, focusing on leading-order or tree-level evaluations. One of the first apparent problems in evaluating scattering amplitudes, already at leading order, is due to summing over external polarizations through completeness relations, resulting in the need to evaluate about $N^2/2$ terms for a process with N Feynman diagrams. If, instead, this summation is not being performed, every single one of the N individual contributions would just be a complex number, dependent solely on external momenta, colours, and spins. Naively, for n massless external particles with two helicity states each this would thus lead to 2^n complex numbers to be evaluated. Considering this behaviour, the actual gain in computational effort through the latter method is based on the realization that the number of Feynman diagrams, N usually scales faster than factorially with n and therefore the number of contributions to be evaluated when using the traditional method becomes more like $N^2/2 \leq (n!)^2$, an exploding number of terms.

This reasoning is the basic idea behind a set of methods which could collectively be denoted as the method of **helicity amplitudes**. In this method, any Feynman amplitude (represented by propagators and vertices for the internal lines and spinors and polarization vectors for the external particles) is translated into a complex number, dependent on external helicities and momenta. This method will ultimately rely on smart ways to represent the building blocks of the amplitudes to allow for a reasonably fast and efficient numerical evaluation. However, before even entering a quick discussion of how this could be achieved there are some tricks to accelerate the computation time.

First, the amplitude may vanish for certain helicity combinations. For instance, for massless fermions, a **helicity-flip** in the interaction with a spin-1 particle is disallowed due to the **conservation of angular momentum**,[2] translating into the conservation of helicity along a fermion line and thus the vanishing of half the helicity combinations in the full process.

As a second trick, it is of course possible to recycle certain sub-amplitudes. As an example consider the two Feynman diagrams contributing to the process $ug \to dg\ell^+\nu_\ell$ displayed in Fig. 3.2. The parts in the boxes are identical and thus only need to be

---

[2] As an analogy, one may think about allowed and suppressed electromagnetic transitions of excited atoms, which are identified with $\vec{E}$ and $\vec{B}$ transitions. In the **Gordon representation**, the latter scale with a term proportional to the particle mass.

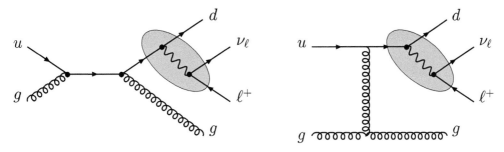

Fig. 3.2 Two Feynman diagrams contributing to the process $ug \to dg\ell^+\nu_\ell$ at leading order. The parts of the diagrams in the coloured blobs are identical; when constructing the helicity amplitudes, they can be factored out and recycled to speed up the evaluation time for the total amplitude.

evaluated once for both diagrams depicted and for similar ones. Using such recycling recursively will reduce the number of complex multiplications like the ones in the spinor products and thus increase the efficiency of the calculation, especially when such common sub-amplitudes are identified in the construction of the helicity amplitudes and factored out from the beginning.

In order to identify and isolate such sub-amplitudes, it is important to realize that tensor structures in Dirac space in the numerators of propagators can be re-expressed through spinors as follows:

$$\not{p} + m = \frac{1}{2} \sum_\lambda \left[\left(1 + \frac{m}{\sqrt{p^2}} \right) u(p, \lambda)\bar{u}(p, \lambda) + \left(1 - \frac{m}{\sqrt{p^2}} \right) v(p, \lambda)\bar{v}(p, \lambda) \right]. \quad (3.2)$$

A similar relation also holds true for the propagator numerators for vector particles, with the added complication of gauge choices for gauge bosons. Typically to fix the gauge, a light-like gauge vector q^μ is introduced, essentially fixing the axis of an axial gauge and resulting in

$$-g_{\mu\nu} + \frac{q_\mu p_\nu + q_\nu p_\mu}{pq} = \sum_{\lambda=\pm} \epsilon_\mu(p, \lambda)\epsilon_\nu^*(p, \lambda). \quad (3.3)$$

As further discussed in Appendix A.2, this allows us to rewrite the polarization vectors as spinor products such that the full amplitude can be recast in the form of spinor products.

3.2.1.2   Dealing with spinors and polarizations

Of course, the reasoning above necessitates the representation of the spinors and polarization vectors of the external states in a suitable way. While this is perfectly possible, it will most likely also be very cumbersome if treated in such a blunt way due to the occurrence of massive matrix multiplications, stemming for instance from the notorious Dirac γ matrices. A better way is to try to decompose the amplitude into scalar products of spinors and of Lorentz structures, where the latter ones are fairly straightforward to implement. There are different ways to represent spinors and their products, as discussed in more detail in Appendix A.2.

One frequently used **spinor representation** is based on **two-component spinors**, also known as **Weyl spinors**, which indeed form a basic representation of the Lorentz group. To see, in short, how this works, it should be realized that the Lorentz group is generated by the usual six infinitesimal boost and rotation transformations along and around the x, y, and z axes, respectively. While individually the commutators of these generators form a somewhat nonintuitive algebra, they can be rearranged into a set of six linearly independent linear combinations which decompose into two $SU(2)$ algebras. In other words, the Lorentz group is a (locally isomorphic) product of two independent $SU(2)$ groups. These groups, in turn have their first non-trivial representation through two-component objects, or Weyl spinors, associated with **left-handed** and **right-handed spinors**. The components of these two kinds of Weyl spinors are indicated by **dotted** and **undotted indices** α and $\dot{\alpha}$, each of which ranges only from 1 to 2, $a, \dot{a} \in \{1, 2\}$.

For instance, for a massless four-vector k,

$$\zeta_a(k) = \begin{pmatrix} \sqrt{k^+} \\ \sqrt{k^-} e^{i\phi_k} \end{pmatrix}, \tag{3.4}$$

where the **light-cone momenta** $k^{\pm} = k^0 \pm k^3$ and ϕ_k is the angle of the transverse momentum $\vec{k}_\perp$. By using the complex plane, $k_\perp = k_1 + ik^2$ and $\phi_k = \arg(k_\perp)$. There is a fair amount of freedom in this definition of the spinors, like, for instance, the definition of the axis defining the light-cone momentum directions or the orientation used in the definition of the angle ϕ_k. This freedom can then be used for checks of the actual calculation, which must be independent of these choices.

However, with these definitions, the scalar products of such spinors $\zeta_a(k) = |k\rangle$ and $\zeta_{\dot{a}}(k) = |k]$ can be written as

$$\eta_a(k)\zeta^a(q) = \eta_a(k)\epsilon^{ab}\zeta_b(q) = \langle kq \rangle$$
$$\eta_{\dot{a}}(k)\zeta^{\dot{a}}(q) = [kq] = \langle kq \rangle^*, \tag{3.5}$$

where the latter relation holds true because trivially one could identify $\eta_{\dot{a}} = \eta_a^*$ and where the fact that the individual spinor components must be Grassmann numbers, anti–commuting, is encapsulated in the "spinor" metric ϵ^{ab} given by

$$\epsilon_{ab} = \epsilon^{ab} = \epsilon_{\dot{a}\dot{b}} = \epsilon^{\dot{a}\dot{b}} = \begin{pmatrix} 0 & 1 \\ -1 & 0 \end{pmatrix}. \tag{3.6}$$

Note that it has been customary to use the momentum label of the spinors in the two scalar products and to identify whether it is over the dotted or undotted spinors only by the shape of the bracket, a convenient short-hand notation. Using them, regular massive Dirac fermion spinors known from the usual textbook methods can be written as two such Weyl spinors, arranged in a bi-spinor; this, however, introduces a "gauge"-like degree of freedom into their definition.

Another important building block of scattering amplitudes is the polarization vector for external or internal gauge bosons. In general, four-vectors k^μ can also be represented by Weyl spinors, by realizing that

$$k^\mu = \sigma^\mu_{\dot{a}b}\, \zeta^{\dot{a}}(k)\zeta^b(k), \tag{3.7}$$

where $\sigma^0 = 1$ is the two-dimensional unit matrix and the σ^i ($i \in \{1, 2, 3\}$) are the Pauli matrices. By introducing a light-like gauge vector q, polarization vectors for external massles particles with four-momentum p^μ can be written as

$$\epsilon^\mu_\pm(p,\, q) = \pm\frac{1}{\sqrt{2}}\frac{\langle q^\mp|\sigma^\mu|p^\mp\rangle}{\langle q^\mp p^\pm\rangle}. \tag{3.8}$$

For polarization vectors for massive external gauge bosons, *cf.* Appendix A.2.

It is also interesting to note that regular scalar products of four-vectors can be rewritten in the form of spinor products, for instance

$$2k \cdot q = \langle kq\rangle\,[kq]\,. \tag{3.9}$$

It is relations such as the one above which render this spinor representation a versatile and powerful tool to rewrite Feynman amplitudes in a form that lends itself to the automation of scattering amplitude calculations through numerical methods. For more details, *cf.* Appendix A.2.

3.2.1.3   Dealing with colour

An interesting problem when constructing the amplitudes presents itself with the inclusion of colour. Quite often, relevant theoretical result are formulated in the **large-N_c** limit, and sub-leading terms usually suppressed by $1/N_c^2$ or similar are omitted. It is worth remembering that quarks, being in the **fundamental representation of** $SU(N_c)$ carry colour indices $i \in [1, N_c]$, while gluons, residing in the **adjoint representation**, carry an adjoint index $a \in [1, (N_c^2 - 1)]$. The fundamental interaction between both is mediated by terms proportional to the **generators T^a_{ij} of the fundamental representation** of $SU(N_c)$, normalized through

$$T^a_{i\bar{j}}\, T^b_{\bar{j}i} = \delta^{ab}. \tag{3.10}$$

The self-coupling of gluons, mediated by the **generators f^{abs} in the group's adjoint representation**, can be rewritten through the T^a_{ij} as

$$if^{abc}\, T^c_{i\bar{j}} = [T^a,\, T^b]_{i\bar{j}}. \tag{3.11}$$

Ultimately, the colour algebra allows the decomposition of every n-gluon amplitude, which includes the most complicated colour structures, according to

$$\mathcal{A}(1,\, 2,\, \ldots,\, n) = \sum_{\sigma \in S_{n-1}} \mathrm{Tr}\,[T^{a_1}\, T^{a_2}\, \ldots\, T^{a_n}]\, \mathrm{A}(1,\, \sigma_2,\, \ldots,\, \sigma_n), \tag{3.12}$$

where σ denotes all $(n-1)!$ permutations of the indices $2 \ldots n$, and the colour-stripped or **colour-ordered** amplitude is denoted by $A(1, \sigma_2, \ldots, \sigma_n)$, which only depends on the momenta and helicities of the external particles. Of course, similar decompositions also emerge when some of the particles are quarks or even colour-neutral objects, which typically leads to a simplification of the overall result in terms of colour. As a welcome by-product, the colour-ordered amplitudes $A(1, \sigma_2, \ldots, \sigma_n)$ correspond to planar graphs only, when it comes to their QCD parts, which is due only a small set of Feynman amplitudes contributing to each of them and which renders them simpler to calculate. Some of their most remarkable properties are defined by the **Kleiss–Kuijf relations** [681] yielding linear relations among such amplitudes. One of the consequences of these relations is that the maximal number of colour-independent amplitudes is $(n-2)!$, showing that colour ordering (CO) does not yield a minimal set of such amplitudes. Such a set can be achieved by using the adjoint representation instead, *cf.* [452, 453] for details.

Another decomposition to coloured amplitudes, known as **colour dressing**, has been proposed in [651, 739]. Based on actual colour flows it treats the $SU(N_c)$ gauge field, the gluons, as $N_c \times N_c$ matrix, thus making the matrix character of the gluon field $G^a_{\mu, i\bar{j}} = G_\mu T^a_{i\bar{j}}$ in the fundamental representation more explicit than by denoting it with an index a in the adjoint representation. Considering a term $T^a_{ij} T^a_{k\bar{l}}$, corresponding to a gluon exchange motivates why this may be helpful for numerical implementations:

$$
T^a_{ij} T^a_{k\bar{l}} = \delta_{i\bar{l}} \delta_{k\bar{j}} - \frac{1}{N_c} \delta_{ij} \delta_{k\bar{l}} \longleftrightarrow \qquad - \frac{1}{N_c} \qquad \tag{3.13}
$$

This sketch also explains why this is based on colour flows — both terms correspond to connecting indices of fundamental $SU(N_c)$ objects, with a sum over independent colours that yields exactly the $(N_c^2 - 1)$ degrees of freedom present in the gluon field. This also means that *every* QCD vertex can be written as a sum of δ functions in colour space, connecting the quark and gluon colour attached to it in all allowed combinations. This allows for a fairly straightforward implementation in terms of a computer code, and replacing the potentially cumbersome colour algebra with factors of one or zero further accelerates the evaluation of the amplitudes.

3.2.1.4   Recursion relations

This idea of recycling identical parts of the amplitude is brought to perfection by using recursion relations from the beginning. The basic idea here is to create **one-particle off-shell parts** of the amplitude recursively in the spirit of the **Dyson-Schwinger equations** [493, 840] known from text-books on quantum field theory, where they are used to construct one-particle off-shell Greens functions. This very idea has been put into effect in different realizations, directly in the HELAC code [308, 486], or in some variations such as the ALPHA algorithm [326, 327], on which ALPGEN [743] and O'MEGA [770] are based, or as **Berends–Giele relations** [219–222, 681], implemented in COMIX [582].

The idea in all of them is to construct generalized currents $\mathcal{J}_\alpha(\pi)$ for a set π of external particles on their mass shell plus one internal one, which then of course may be off-shell. In this construction, α denotes the combined quantum numbers of this internal particle, including spin, colour, etc.. As a somewhat special realization, the external particles may also be considered as currents; then the quantum numbers α are directly identified with the ones of this particle. From these special cases, recursively more and more complex currents are built, by applying

$$\mathcal{J}_\alpha(\pi) = P_\alpha(\pi) \left\{ \sum_{\mathcal{P}_2(\pi)} \sum_{\mathcal{V}_\alpha^{\alpha_1 \alpha_2}} [S(\pi_1, \pi_2) \mathcal{V}_\alpha^{\alpha_1 \alpha_2} \mathcal{J}_{\alpha_1}(\pi_1) \mathcal{J}_{\alpha_2}(\pi_2)] \right.$$

$$\left. + \sum_{\mathcal{P}_3(\pi)} \sum_{\mathcal{V}_\alpha^{\alpha_1 \alpha_2 \alpha_3}} [S(\pi_1, \pi_2, \pi_3) \mathcal{V}_\alpha^{\alpha_1 \alpha_2 \alpha_3} \mathcal{J}_{\alpha_1}(\pi_1) \mathcal{J}_{\alpha_2}(\pi_2) \mathcal{J}_{\alpha_3}(\pi_3)] \right\}. \tag{3.14}$$

In this equation, $P_\alpha(\pi)$ denotes the propagator denominator, which of course depends on the properties α of the propagating particle and on its momentum given by the momenta in π. The terms $S(\pi_1, \pi_2)$ and $S(\pi_1, \pi_2, \pi_3)$ take care of symmetry factors, which emerge in the partitions $\mathcal{P}_2(\pi) : \pi \to \pi_1 \oplus \pi_2$ and $\mathcal{P}_2(\pi) : \pi \to \pi_1 \oplus \pi_2 \oplus \pi_3$ of the original set. Finally, the $\mathcal{V}_\alpha$ signify the three- and four-leg vertices connecting the particles α_i of the sub-currents with the emerging new particle α. This allows us to write the total amplitude $\mathcal{A}(\pi)$ for one specific configuration π as

$$\mathcal{A}(\pi) = \mathcal{J}_{\alpha_\rho}(\rho) \frac{1}{P_{\bar{\alpha}_\rho}(\pi|\rho)} \bar{\alpha}_\backslash(\pi|\rho), \tag{3.15}$$

where $\pi|\rho$ denotes the residual subset of π after the subset ρ has been taken off. Overall conservation of quantum numbers also guarantees that the combined quantum number $\alpha_{\pi|\rho}$ of this conjugate subset is the conjugate of the set ρ, $\bar{\alpha}_\rho$. It is worth noting, however, that it is computationally advantageous to map the four-vertices onto vertices with three legs only.

For some specific cases it is possible to solve the recursion equation in Eq. (3.14) in closed form [220, 692]. For instance, a current with n external colour-ordered like-sign gluons is given by

$$J^\mu(1^+, 2^+, \ldots, n^+) = g_s^{n-2} \frac{\langle q^- | \gamma^\mu \not{P}_{1,n} | q^+ \rangle}{\sqrt{2} \langle q1 \rangle \langle 12 \rangle \langle 23 \rangle \ldots \langle (n-1)n \rangle \langle nq \rangle}, \tag{3.16}$$

where the four-momentum $P_{i,j}^\mu$ is constructed from the outgoing momenta through

$$P_{i,j}^\mu = \sum_{l=1}^{j-1} p_l^\mu, \tag{3.17}$$

and where q^μ is the gauge vector. This current can be used, to prove the form of the **maximally helicity violating (MHV)** or **Parke–Taylor amplitudes**, which correspond to all gluonic, colour-ordered amplitudes with all equal signs apart from two, labelled by i and j in the equation below. Their form was conjectured for the first time in [797] and proven in [220] to be

$$
\begin{aligned}
\mathcal{A}(1^+, 2^+, \ldots, i^-, \ldots, j^-, \ldots\ldots, n^+) &= ig_s^{n-2} \frac{\langle ij \rangle^4}{\langle 12 \rangle \langle 23 \rangle \ldots \langle (n-1)n \rangle \langle n1 \rangle} \\
\mathcal{A}(1^-, 2^-, \ldots, i^+, \ldots, j^+, \ldots\ldots, n^-) &= ig_s^{n-2} \frac{[ij]^4}{[12][23] \ldots [(n-1)n][n1]}.
\end{aligned}
\tag{3.18}
$$

Note that the all-sign identical amplitudes vanish due to the conservation of angular momentum.

3.2.1.5 On-shell methods

The study of more general properties of scattering amplitudes, especially in the context of the super-renormalizable $\mathcal{N} = 4$ Super-Yang-Mills theory, has experienced a somewhat surprising renaissance in the early 2000s, leading to a remarkable progress in the understanding of perturbative QCD from twistor-inspired methods [897]. Building on a correspondence with some well-understood type of string theory, it could be shown that colour-ordered tree-level amplitudes can be related to curves in twistor space, giving rise to the **Cachazo–Svrček–Witten (CSW) vertex rules** [307]. Even loop amplitudes were shown to follow the same principles and rules [306]. These rules state that arbitrary colour-ordered scattering amplitudes can be constructed from MHV amplitudes [306, 307, 897]. They serve as generalized MHV vertices and are connected by scalar propagators, resulting in a full n-gluon amplitude being built from $(n - l)$ same-sign helicity gluons (and l gluons with helicity of the opposite sign) arranged in $(l - 1)$ of these vertices.

As an example, consider Fig. 3.3, displaying the construction of the colour-ordered six-gluon amplitude $\mathcal{A}(1^-, 2^-, 3^-, 4^+, 5^+, 6^+)$ from MHV vertices. As a consequence of this construction, for any n-gluon amplitude MHV vertices for up to n particles may contribute, implying that the number of such vertices that are needed for a cross-section calculation grows steadily with the number of external legs. This problem has been addressed in [211], reformulating the CSW rules in a fully recursive fashion.

A further refinement of the CSW rules has been worked out in by **Britto, Cachazo, Feng, and Witten** in [286, 287], stating that any colour-ordered tree-level amplitude can be constructed from two on-shell amplitudes with a scalar off-shell propagator in between them. This yields the BCF recursion relations, which can be summarized as

$$
\mathcal{A}_n(1, 2, \ldots, n) = \sum_{k=2}^{n-2} \mathcal{A}_{k+1}(\hat{1}, 2, \ldots, k, -\hat{p}_{1,k}^{-h}) \frac{1}{p_{1,k}^2} \mathcal{A}_{n-k+1}(\hat{p}_{1,k}^h, k+1, \ldots, \hat{n}), \tag{3.19}
$$

where the momentum of the propagator is given by the sum of external momenta,

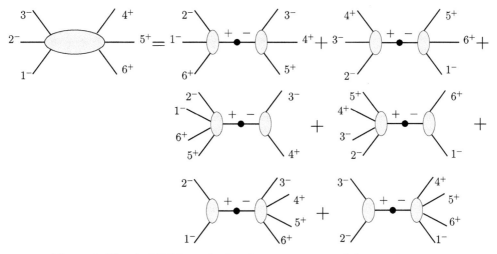

Fig. 3.3 The six MHV graphs for the construction of the six-gluon amplitude $\mathcal{A}(1^-, 2^-, 3^-, 4^+, 5^+, 6^+)$ in the CSW formalism.

$$p_{1,k}^{\mu} = \sum_{i=1}^{k} p_i^{\mu}, \tag{3.20}$$

and where the $\hat{\ }$ denotes momenta that are shifted by an amount z,

$$\begin{aligned}
\hat{p}_{1,k} &= p_{1,k} + z\lambda_n\tilde{\lambda}_1 \\
\hat{p}_1 &= p_1 + z\lambda_n\tilde{\lambda}_1 \\
\hat{p}_n &= p_n + z\lambda_n\tilde{\lambda}_1.
\end{aligned} \tag{3.21}$$

Here the λ and $\tilde{\lambda}$ denote the co- and contra-variant components of the spinors for light-like momenta,

$$p_i^{a\dot{b}} = \lambda_i^a\tilde{\lambda}_i^{\dot{b}}, \tag{3.22}$$

and the shift parameter is given by

$$z = \frac{p_{1,k}^2}{\langle n|p_{i,k}|1]}. \tag{3.23}$$

In contrast to the CSW rules this implies that the sub-amplitudes contain on-shell particles only, which makes them easier to calculate. They are, however, not gauge-invariant objects due to the need of a gauge vector to define the opposite gluon helicities, which enter through the shift parameter z. It is interesting to note in this context that the shifted momenta $\hat{p}_1$ and $\hat{p}_n$ are complex valued but still on-shell and light-like, adding an interesting twist to the calculation. Finally, it is important to stress that of course these formalisms have also been extended to include quarks or the case of QED interactions.

3.2.2 Phase-space integration

Turning now to the issue of phase-space integration through numerical methods, already simple considerations show that traditional quadrature techniques like the trapezoidal or Simpson method will not converge fast enough for large numbers n of final-state particles.[3] Instead, the method of choice turns out to be Monte Carlo integration, which due to its statistical nature is guaranteed to have an uncertainty estimate that scales like $1/\sqrt{N}$ with N the number of phase-space points. The basic idea of Monte Carlo integration is to replace the integration of a function with sampling it:

$$
I = \int_V \mathrm{d}^D x f(\vec{x}) \implies \langle I(f) \rangle_x = \frac{V}{N} \sum_{i=1}^N f(\vec{x}_i) = \langle f \rangle_x \,.
\tag{3.24}
$$

Here $\vec{x}$ denotes a point in the D-dimensional finite phase space with volume V, and the $\vec{x}_i$ are uniformly randomly distributed in it. In Monte Carlo integration the exact value of the integral I is estimated as an average over N test points in the volume. The central limit theorem guarantees that with infinitely many calls $N \to \infty$ the estimator $\langle I \rangle \to I$. The name of the game in Monte Carlo integration is to control this convergence, through an error estimate $\langle E \rangle$ that will scale like $1/\sqrt{N}$. A convenient error estimate is the variance, written in useful form as

$$
\langle E(f) \rangle_x = \sqrt{\frac{1}{N} \sum_{i=1}^N \left(f^2(\vec{x}_i) \right) - \left(\frac{1}{N} \sum_{i=1}^N f(\vec{x}_i) \right)^2} = \sqrt{\langle f^2 \rangle_x - \langle f \rangle_x^2}.
\tag{3.25}
$$

Various methods have been discussed in the literature to accelerate error reduction. The most prominent ones are known as **importance sampling** and **stratified sampling**.

3.2.2.1 Importance sampling

The underlying idea of importance sampling is to use a mapping of random numbers to function arguments $\vec{x}$, the four-momenta in the case of particle physics, which captures as far as possible the distribution given by the function, the squared matrix element for the process in question. In this method a probability density $g(\vec{x})$ for a vector of random numbers is constructed in such a way that $f(\vec{x})/g(\vec{x})$ is better behaved than $f(\vec{x})$ alone. Then the D-dimensional Monte Carlo integration of $f(\vec{x})$ over D uniformly distributed random numbers x_i becomes

$$
\begin{aligned}
\langle I(f) \rangle_x &= \int \mathrm{d}^D x\, f(\vec{x}) = \int \mathrm{d}^D x\, g(\vec{x}) \frac{f(\vec{x})}{g(\vec{x})} \\
&= \int \mathrm{d}^D \rho\, \frac{f(\vec{\rho})}{g(\vec{\rho})} = \langle I(f/g) \rangle_\rho = \left\langle \frac{f}{g} \right\rangle_g \,,
\end{aligned}
\tag{3.26}
$$

---

[3]For a summary of Monte Carlo integration methods, see for example the review [648].

where the ρ_i are distributed according to the phase-space density $g(\vec{x})$. If f and g are sufficiently similar, their ratio f/g will fluctuate less than f alone, leading to a smaller variance and therefore an accelerated convergence. The limiting factor in this is obvious: a function g must be found, which has to be inverted in order to generate the points ρ. Furthermore, to keep things simple, the $g(\vec{x})$ must be non-negative in V and integrate to unity

$$\int_V \mathrm{d}^D x\, g(\vec{x}) = 1, \tag{3.27}$$

which can be trivially achieved by normalizing them to their integral over V.

3.2.2.2 Stratified sampling

Simply put, stratified sampling aims at improving the integration performance by binning the integration region V and by distributing the phase-space points with a different density in the bins. The overall integral and the square of the overall variance are the sum and the quadratic sum of the single integrals and squares of variances in the bins b:

$$\langle I(f)\rangle_x = \sum_b \langle I(f)\rangle_{x\in b}$$
$$\langle E(f)\rangle_x^2 = \sum_b \langle E(f)\rangle_{x\in b}^2. \tag{3.28}$$

The overall variance will be minimized by making the individual variances in each bin of equal size. This can be achieved by introducing a different a priori probability a_b for picking a bin b to generate a phase-space point in it, and by updating them regularly after a sufficiently large number of function calls. Bins with larger variance after such an optimization step, with more fluctuations, will obtain a larger a_b and therefore an increased number of points sampling the phase space in this bin; conversely bins with smaller variance, less fluctuations of $f(\vec{x})$, will have a smaller a_b and fewer sampling points. Ultimately, the best theoretical solution is to have a priori weights a_b which behave like the variance in each bin. Therefore updating them by multiplying them with the variance or similar will accelerate the convergence.

3.2.2.3 Isotropic phase space

Ignoring for the moment issues related to the sampling over the initial-state parameters, x_1 and x_2 or, complementarily y_{cms} and $\hat{s}$, a logical first attempt at Monte Carlo phase-space integration for the final state consists of an isotropic phase-space sampling. In such a sampling, for a given centre-of-mass energy squared $E_{\mathrm{cms}}^2 = P^2 = \hat{s}$, the n momenta p_i^μ populate the phase space homogenously, but still respect overall four-momentum conservation and mass-shell conditions for the individual four-momenta:

$$0 = P^\mu - \sum_{i=1}^{n} p_i^\mu = \left(E, \vec{0}\right) - \sum_{i=1}^{n} p_i^\mu$$
$$0 = p_i^2 - m_i^2 = s_i - m_i^2 \quad \forall i \in \{1, \dots, N\}. \tag{3.29}$$

These requirements are incorporated through δ functions in

$$\mathrm{d}\Phi_N = \prod_{i=1}^{N} \left[\frac{\mathrm{d}^4 p_i}{(2\pi)^4} (2\pi)\delta(p_i^2 - m_i^2) \, \Theta(E_i) \right] (2\pi)^4 \delta^4 \left(P^\mu - \sum_{i=1}^{N} p_i^\mu \right), \qquad (3.30)$$

where the $\Theta(E_i)$ account for the projection on physical, positive energy solutions for the outgoing particles.

For massless particles, it is fairly straightforward to calculate the volume of the final state phase space. As all ingredients in the N-particle phase space of Eq. (3.30) are formulated in Lorentz-invariant quatities, in the following $P^\mu = (E_{\mathrm{cms}}, \vec{0})$ can be safely assumed. Following [685], first an "unconstrained" phase-space volume $\tilde{\Phi}_N$ is introduced, based on unconstrained momenta q_i not fulfilling overall momentum conservation like the constrained p_i:

$$\int \mathrm{d}\tilde{\Phi}_N = \int \prod_{i=1}^{N} \left[\frac{\mathrm{d}^4 q_i}{(2\pi)^4} (2\pi)\delta(q_i^2) \, \Theta(q_i^0) \, f(q_i^0) \right] = \left[\frac{1}{(2\pi)^2} \int_0^\infty \mathrm{d}x \, x f(x) \right]^N \qquad (3.31)$$

$$\xrightarrow{f(x) \to e^{-x}} \frac{1}{(2\pi)^{2N}}.$$

Here, $f(q_i^0)$ denotes an arbitrary regulator function that keeps the overall volume finite — for the choice made here the result is displayed.[4] In a second step a transformation between unconstrained momenta q_i^μ and constrained momenta p_i^μ is found as a combiniation of a scaling operation parameterized by x and a Lorentz boost given by $\vec{b}$,

$$p_i^\mu = x H_{\vec{b}}^\mu(q_i) = x \begin{pmatrix} \gamma q_i^0 + \vec{b} \cdot \vec{q}_i \\ \vec{q}_i + \vec{b} q_i^0 + \dfrac{(\vec{b} \cdot \vec{q}_i)\vec{b}}{1 + \gamma} \end{pmatrix}. \qquad (3.32)$$

Consequently, the inverse transformation is given by

$$q_i^\mu = \frac{1}{x} H_{-\vec{b}}^\mu(p_i) \qquad (3.33)$$

With $M = \sqrt{Q^2}$ the invariant mass of the unconstrained system, the boost parameters $\vec{b}$ and γ and the scaling parameter x read

$$\vec{b} = -\frac{\vec{Q}}{M}, \quad \gamma = \frac{Q^0}{M}, \quad \text{and} \quad x = \frac{E}{M}. \qquad (3.34)$$

---

[4]Integrating over the spatial components of the massless momenta is trivial: the δ function guarantees that $|\vec{q}_i| = q_i^0$ and therefore

$$\int \frac{\mathrm{d}^3 q_i}{(2\pi)^4} (2\pi)\delta(q_i^2) = \int \frac{(q_i^0)^2 \mathrm{d}^2 \Omega}{16\pi^3 q_i^0} = \frac{q_i^0}{4\pi^2}.$$

Inserting these transformations into the unconstrained phase space and after a few manipulations detailed in [685], the two phase-space volumes are related with each other through

$$\int \mathrm{d}\tilde{\Phi}_N = \int \frac{\mathrm{d}x\,\mathrm{d}^3\vec{b}}{(2\pi)^4}\frac{E^4}{x^{2N+1}\gamma}\left\{(2\pi)^4\delta\left(E-\sum_{i=1}^{N}E_i\right)\delta^3\left(\sum_{i=1}^{N}\vec{p}_i\right)\right.$$

$$\left.\prod_{i=1}^{N}\left[\frac{\mathrm{d}^4p_i(2\pi)\delta(p_i^2)\Theta(E_i)}{(2\pi)^4}f\left(\frac{1}{x}H^0_{-\vec{b}}(p_i)\right)\right]\right\} \qquad (3.35)$$

$$= \int \frac{\mathrm{d}x\,\mathrm{d}^3\vec{b}}{(2\pi)^4}\left\{\frac{E^4}{x^{2N+1}\gamma}\prod_{i=1}^{N}\left[f\left(\frac{1}{x}H^0_{-\vec{b}}(p_i)\right)\right]\mathrm{d}\Phi_N\right\}.$$

With the choice made for the limiting function $f(x)$,

$$\prod_{i=1}^{N}f\left(\frac{1}{x}H^0_{-\vec{b}}(p_i)\right) = \exp\left[-\frac{\gamma E}{x}\right], \qquad (3.36)$$

such that only the integration over $\vec{b}$ and x remains, which results in

$$S_N = \int \frac{\mathrm{d}^3\vec{b}}{(2\pi)^3}\int_0^\infty \frac{\mathrm{d}x}{(2\pi)}\frac{E^4\exp\left(-\frac{\gamma E}{x}\right)}{\gamma x^{2N+1}} = \frac{E^{4-2N}}{(2\pi)^3}\frac{\Gamma\left(\frac{3}{2}\right)\Gamma\left(n-1\right)\Gamma\left(2n\right)}{\Gamma\left(n+\frac{1}{2}\right)}. \qquad (3.37)$$

This fixes the unconstrained phase-space volume to

$$\int \mathrm{d}\Phi_N = \frac{E^{2N-4}}{2(4\pi)^{2N-3}\Gamma(N)\Gamma(N-1)}. \qquad (3.38)$$

To turn this into a Monte Carlo algorithm, it is sufficient to generate N *unconstrained momenta* q_i^μ, with isotropic angular distribution and an energy density given by $q_i^0\exp(-q_i^0)$. Denoting with # a random number uniformly distributed in the interval $[0, 1]$, the q_i are obtained from

$$\phi_i = 2\pi \cdot \#, \quad \cos\theta_i = 1 - 2\cdot\#, \quad q_i^0 = -\log(\#\cdot\#). \qquad (3.39)$$

They are transformed into the constrained momenta p_i^μ through the transformations detailed here. This is the RAMBO algorithm invented in [685], where also the generalization to massive momenta has been worked out.

3.2.2.4 Democratic phase-space mappings for pure QCD processes

The isotropic phase-space generation introduced above is a good example of a **democratic approach**, where all particles are treated on the same footing. It should be intuitively clear that a uniform phase space, while simple, elegant, and transparent, will not deliver good results in cases where the actual distribution of final-state momenta is not uniform, which typically is the case in all relevant applications. There is, however, a class of processes where another democratic approach to phase-space integration has been worked out and indeed delivers results superior to the ones provided by the RAMBO algorithm above. The processes in question are those with QCD only final states, where the leading amplitudes assume a particularly simple form. For instance, for purely gluonic processes the most singular amplitudes are the MHV and $\overline{\text{MHV}}$ ones, which for a colour-ordered n-gluon process behave like

$$\mathcal{A}_n^{\text{MHV},\overline{\text{MHV}}} \propto \prod_{i=1}^{n} \frac{1}{\hat{s}_{i(i+1)}} \quad \text{with} \quad \hat{s}_{n(n+1)} = \hat{s}_{n1}, \tag{3.40}$$

cf. Eq. (3.18). This particularly simple form is also recovered in other processes; the inclusion of quarks typically just leads to the absence of some of the $1/\hat{s}$ factors, therefore rendering the amplitudes somewhat less complicated to integrate.

A first method that has been optimized for the integration of such amplitudes has been introduced in [487] and goes by the name of SARGE. It builds on sequentially filling the phase space through emissions that follow the basic antenna density

$$\mathrm{d}\mathcal{A}_{ij}^k = \frac{1}{\pi}\mathrm{d}^4 p_k \delta p_k^2 \Theta(p_k^0) \frac{(p_i p_j)}{(p_i p_k)(p_j p_k)} \, g\left(\xi_{ij}^{ik}\right) \, g\left(\xi_{ij}^{jk}\right), \tag{3.41}$$

where the

$$\xi_{ij}^{jk} = \frac{(p_j p_k)}{(p_i p_j)} \tag{3.42}$$

are the arguments of the regulator functions

$$g(\xi) = \frac{1}{2 \log \xi_m} \Theta(\xi - \xi_m^{-1})\Theta(\xi_m - \xi), \tag{3.43}$$

which ensure that the divergences for $(p_i p_k) \to 0$ and $(p_j p_k) \to 0$ are avoided and are chosen such that $\mathrm{d}\mathcal{A}$ integrates to unity. The cut-off value ξ_m depends on the actual cuts being applied on the outgoing momenta. Demanding that for each pair of momenta i and j, $(p_i + p_j)^2 \geq s_0$, the expression for ξ_m is,

$$\xi_m = \frac{\hat{s}}{s_0} - \frac{(n+1)(n+2)}{2}, \tag{3.44}$$

where $\hat{s}$ is the energy squared of the partonic system in the centre-of-mass frame.

To generate a momentum p_k according to the basic antenna structure, Eq. (3.41), SARGE proceeds along the following steps. The initial momenta p_i and p_j are boosted into their centre-of-mass frame. Then, two numbers ξ_{ij}^{ik} and ξ_{ij}^{jk} are generated with a probability density of $g(\xi)/\xi$. They allow the calculation of the energy of particle k, p_k^0, and its polar angle θ with respect to p_i. The azimuthal angle is chosen uniformly in $[0, 2\pi]$, which then enables the construction of p_k in the rest frame of i and j. Boosting p_k back into their lab frame finishes the generation of one emitted momentum. It is worth noting that this algorithm does not respect four-momentum conservation, since recoil effects on the emitters i and j are not captured here.

This algorithm is now iterated to yield all outgoing momenta along the following lines. Generating two outgoing momenta q_1 and q_n in the centre-of-mass frame of the partonic collision triggers the consecutive generation of the $(n-2)$ other momenta through the basic antennae $\mathrm{d}\mathcal{A}^2_{1n}\mathrm{d}\mathcal{A}^3_{2n}\ldots\mathrm{d}\mathcal{A}^{n-1}_{(n-2)n}$ already discussed. Using the same boost and scaling transformations introduced above for the RAMBO algorithm [685] yields the final momenta p_i, which indeed, by construction, satisfy four-momentum conservation.

This democratic algorithm can be further improved by intelligent symmetrization over colour orderings (outgoing momenta) and better maps, *cf.* the orginal publication [487]. Furthermore, going from a symmetric way of antenna generation to a more hierarchical way, as implemented in the HAAG algorithm introduced in [880], leads to a further refinement and, consequently, an accelerated convergence of the integration. These more specific algorithms are good examples of how improvements in Monte Carlo integration can be achieved by employing knowledge about the characteristics of the underlying process. In the following this intuitive picture will be substantiated a bit more through a more detailed discussion of optimization procedures in the literature and the introduction of the by now widely employed method for Monte Carlo phase space integration in particle physics.

3.2.2.5 Hierarchical mappings

In contrast to democratic approaches with fairly symmetric final states there are of course also processes for which the opposite is true: the final-state particles emerge from a very specific set of Feynman diagrams which could be interpreted as a sequence of production processes of usually resonant particles followed by their decay into lighter particles. A good example of this would be the production of a $t\bar{t}$ pair and its subsequent decay into two b quarks and two W bosons, which in turn decay further. In such a case democratic and, in particular, isotropic mappings will not be very efficient, since the particles are just not distributed independently in phase space but will enjoy strong correlations due to the intermediate resonances. Knowing this and understanding the underlying phase-space structure lends itself to the textbook Monte Carlo method of importance sampling for efficient phase-space generation.

In the example of $t\bar{t}$ production and decay this could be achieved for instance along the following steps. Assuming first a useful distribution of the $t\bar{t}$ pair in terms of its centre-of-mass energy and rapidity, which for an e^-e^+ collider could naively be fixed, while for hadron colliders PDF effects would have to be taken into account. Then, in the centre-of-mass frame of the pair, the top and the anti-top would be isotropically distributed, with back-to-back kinematics, and with their invariant masses individually given by a Breit–Wigner distribution. Each of the top decays, again, could be treated in their respective rest frame, with the invariant masses of the W's again given through a Breit–Wigner form. Of course, this process would be finally repeated also for the Ws.

This implies a hierarchical structure of Breit-Wigner distributed invariant masses of the intermediate resonant particles (or a similar distribution for the overall centre-of-mass energy of the total system) followed by their binary decays, which in the respective rest frames reduces to choosing one pair (θ, ϕ) fixing the orientation in the solid angle of the back-to-back binary decay kinematics.

3.2.2.6   Multi-channel method

The case of the example above can be further extended also for processes where a number of such hierarchies compete with each other. Up to now the best solution for such cases can be constructed by combining different mappings in a dynamic form. This is known as the **multi-channel method**, which has been presented for the first time in the framework of particle physics in Ref. [682]. Expressed in technical terms, this method presents itself as an interesting amalgamation of the traditional Monte Carlo integration methods of stratified and importance sampling, where the binning of the stratified sampling is not in actual integration parameters or phase-space but rather in independent functional forms of individual mappings. In other words, multi-channeling performs a stratified sampling over various importance samplings.

Mathematically speaking, a mapping function $g(\vec{x})$ is constructed as a sum of individual mappings $g_j(\vec{x})$, each of them with an a priori weight a_j:

$$g(\vec{x}) = \frac{1}{\sum\limits_j a_j} \sum_j a_j g_j(\vec{x}). \tag{3.45}$$

As before, the estimate for the integral is given by

$$\langle I \rangle_x = \langle f \rangle_x = \langle f/g \rangle_g \tag{3.46}$$

and the error estimate is

$$\langle E(a) \rangle = \sqrt{\left\langle \left(\frac{f}{g}\right)^2 \right\rangle_g - \left(\left\langle \frac{f}{g} \right\rangle_g\right)^2}. \tag{3.47}$$

While in principle the integral does not depend on the a_j, its estimator and, more importantly, the error estimator does. This can easily be understood by writing the expressions above again as integral. In I the mapping function g cancels, but this is not the case in the first term of E:

$$I = \int_V d^D x\, g(\vec{x})\, \frac{f(\vec{x})}{g(\vec{x})} = \int_V d^D x\, f(\vec{x})$$

$$E = \int_V d^D x\, g(\vec{x}) \left(\frac{f(\vec{x})}{g(\vec{x})}\right)^2 - I^2 \int_V d^D x\, \frac{f^2(\vec{x})}{g(\vec{x})} - I^2. \tag{3.48}$$

From here it would be trivial to rediscover stratified sampling by using mappings $g_j(x)$ which are unity inside the bin j and zero outside.

This, however, is not how multi-channel sampling is being used for the phase space integration in scattering processes in particle physics. There, the form of the transition matrix element is known. Using, for instance, Feynman diagrams to construct it, one can obtain individual phase-space mappings g_j respecting the kinematics of an individual diagram, which is mainly driven by its propagator structure. In principle, they can be represented by intuitive mappings: a simple pole of the form $1/\hat{s}$ for a massless particle, a Breit–Wigner form, defined by the mass M and width Γ of the propagating particle, $1/[(\hat{s} - M^2)^2 + M^2\Gamma^2]$, and a t-channel propagator, $1/(\hat{t} - m^2)$. Similar structures also exist for decays, for example isotropic two-body decays of a massive particle, or anisotropic decays to capture, for example, the emission of a gluon or photon in the final state. The individual mappings are then composed by a sequence of decays and propagators, basically a hierarchical mapping for each diagram. This is fairly straightforward. What is typically not clear, though, is which specific phase-space configurations dominate the behaviour of the cross-section, especially in the presence of non-trivial cuts on some of the outgoing momenta. In such a setup, when a suitable linear combination of the most important mappings is hard to find, importance sampling loses its power. Then, the stratified sampling part of the integration does the work and automatically finds a well-suited integration setup. A variant of this technology, coined single-diagram enhanced (SDE) multi-channel phase-space integration has been introduced in [740].

It is worth noting that this technology can be further refined, for example by using stratified sampling in the phase-space mappings first to find the most efficient combination of them, and by then using, for instance, VEGAS [725] to increase the efficiency of the "best" channels.

3.2.3 Automated leading-order tools

The methods presented in Sections 3.2.1 and 3.2.2 form the core of a number of fully automated tools which allow the generation of events at the parton level. Such tools are sometimes also called "matrix element generators". Their performance depends mainly on two construction choices, namely

- on the methods used for the evaluation of the matrix elements squared: the construction of Feynman amplitudes, traditional squaring, and using completeness relations, or translating the Feynman amplitudes into helicity amplitudes, or abandoning the language of Feynman diagrams altogether and using on-shell or off-shell recursion relations instead; and

- on the methods used for phase-space integration and, possibly, the treatment of colour degrees of freedom.

In Table 3.2 publicly available leading-order tools are listed and roughly categorized.

3.3 Technology of next-to-leading-order calculations

When promoting the calculations at leading order discussed in some detail in the previous section to next-to-leading order accuracy, the master formula for the evaluation of production cross-sections of an n-body final state in hadronic collisions, *cf.* Eq. (2.52), will be altered. This is because the partonic cross-section, $\hat{\sigma}$ now receives contributions not only from Born-level matrix elements but also from virtual and real corrections. Therefore, at NLO

Table 3.2 Publicly available tools for leading–order cross-section calculations at hadron colliders. For the matrix element part of the evaluation different methods are being used: off-shell recursion relations in different algorithms, denoted by "off-shell", Feynman diagram based helicity amplitudes, denoted by "hel. amps", and the traditional method based on squaring the amplitudes and using completeness relations for external particles and traces over the Dirac algebra, denoted by "traces". Related to this, different methods for the treatment of coloured particles are also listed, in particular the direct evaluation of the colour algebra ("explicit") and the method of colour dressing (CD) or colour connections (CC) supplemented with a sampling over colour and helicities. For the phase-space integration, various versions of automated multi-channel sampling methods are being used ("multi"), in addition to some more process-specific solutions.

	ME: $\|\mathcal{M}_{ab \to n}\|^2$ colours & helicities	PS: $d\Phi_N$
ALPGEN [743]	off-shell [326, 327] explicit	process-specific
AMEGIC++ [696]	hel. amps [197, 684, 741] explicit	automated multi [682]
COMIX [582]	off-shell [220] CD [651, 739] & sampling	recursive multi [299]
COMPHEP [266]/ CALCHEP [210]	traces explicit	specific single-channel
HELAC/ PHEGAS [308]	off-shell [651] CC [869] & sampling	automated multi [795]
MADGRAPH [150]	hel. amps [773] explicit	automated SDE multi [740]
O'MEGA/ WHIZARD [677, 770]	off-shell [327, 328] explicit	specific [789]

$$\sigma^{(\text{NLO})} = \sum_{a,b} \int_0^1 dx_a dx_b \, f_{a/h_1}(x_a, \mu_F) f_{b/h_2}(x_b, \mu_F) \int d\hat{\sigma}^{(\text{NLO})}_{ab \to n}(\mu_F, \mu_R)$$

$$= \int d\Phi_{\mathcal{B}} \left[\mathcal{B}_n(\Phi_{\mathcal{B}}; \mu_F, \mu_R) + \mathcal{V}_n(\Phi_{\mathcal{B}}; \mu_F, \mu_R) \right] + \int d\Phi_{\mathcal{R}} \, \mathcal{R}_n(\Phi_{\mathcal{R}}; \mu_F, \mu_R),$$

$$(3.49)$$

where the various contributions — Born term $\mathcal{B}$, virtual correction term $\mathcal{V}$, and real correction term $\mathcal{R}$ — are given by suitably helicity-summed or averaged matrix elements. Denoting the perturbative order of the Born-level contribution with b, and indicating the orders of the matrix element $\mathcal{M}^{(b)}$ accordingly, therefore

$$\mathcal{B}_n(\Phi_\mathcal{B}; \mu_F, \mu_R) = \sum_h^{\overline{}} \left| \mathcal{M}_n^{(b)}(\Phi_\mathcal{B}, h; \mu_F, \mu_R) \right|^2$$

$$\mathcal{V}_n(\Phi_\mathcal{B}; \mu_F, \mu_R) = 2 \sum_h^{\overline{}} \mathfrak{Re} \left[\mathcal{M}_n^{(b)}(\Phi_\mathcal{B}, h; \mu_F, \mu_R) \mathcal{M}_n^{*(b+1)}(\Phi_\mathcal{B}, h; \mu_F, \mu_R) \right] \quad (3.50)$$

$$\mathcal{R}_n(\Phi_\mathcal{R}; \mu_F, \mu_R) = 2 \sum_h^{\overline{}} \left| \mathcal{M}_{n+1}^{(b+1)}(\Phi_\mathcal{R}, h; \mu_F, \mu_R) \right|^2 .$$

At this point it is important to stress that the Born-level and virtual matrix elements are multiplied to yield the overall virtual correction. Thereby, the latter matrix element emerges from the former by adding a closed loop without changing the external particles of the process $ab \to n$. Both matrix elements share the same phase space, essentially an n-body final-state phase space, $\Phi_\mathcal{B}$, see below. In contrast the real correction emerges as the square of matrix elements with one additional outgoing particle; this may lead to replacing an incoming particle type a or b with a' or b'. For instance, an incoming quark a may be replaced by a gluon, which splits into an outgoing anti-quark and a quark a such that for the process in question $ab \to n$ is replaced by $gb \to n + \bar{a}$. With this change also the PDFs must change accordingly.

Turning to the phase-space elements, the expressions for the Born-level and real correction phase space are given by

$$\begin{aligned} \mathrm{d}\Phi_\mathcal{B} &= \mathrm{d}x_a \mathrm{d}x_b \, f_{a/h_1}(x_a, \mu_F) f_{b/h_2}(x_b, \mu_F) \frac{1}{2\hat{s}_{ab}} \, \mathrm{d}\Phi_n \\ \mathrm{d}\Phi_\mathcal{R} &= \mathrm{d}x_{a'} \mathrm{d}x_{b'} \, f_{a'/h_1}(x_{a'}, \mu_F) f_{b'/h_2}(x_{b'}, \mu_F) \frac{1}{2\hat{s}_{a'b'}} \, \mathrm{d}\Phi_{n+1} \end{aligned} \quad (3.51)$$

with

$$\mathrm{d}\Phi_n = \prod_{i=1}^{n} \frac{\mathrm{d}^4 p_i}{(2\pi)^4} (2\pi)\delta(p_i^2 - m_i^2) (2\pi)^4 \delta^4 \left(p_a + p_b - \sum_i p_i \right) \Theta(E_i), \quad (3.52)$$

cf. Eq. (2.55).

As already discussed in Section 2.2.5, there typically are additional complications when evaluating cross-sections at higher orders beyond merely having to calculate more and possibly more complicated diagrams. First of all, the loops in the virtual contribution introduce a new integration variable, the four-momentum in the loop l, which is not constrained by overall four-momentum conservation or on-shell conditions encoded in the δ functions, but instead can take any value in any of its components allowed by four-momentum conservation in the interaction vertices. In particular, the four-momentum in the loop can become infinitely large; for expressions of the general form $\mathrm{d}^4 l / l^m$ this will potentially yield infinite results if the power of the loop momentum $m \leq 4$. Such **ultraviolet divergences** have to be **regularized**, which usually is achieved by employing **dimensional regularization**, corresponding to integrating in $D = 4 - 2\varepsilon$ rather than in four dimensions. The benefit of this method is that it can be formulated in such a way that Lorentz and gauge invariance are guaranteed which is not the case for other, traditional methods such as a regularization through a cut-off or through the **Pauli–Villars method**. In dimensional regularization the ultraviolet divergences manifest themselves as simple poles $1/\varepsilon$. Having thus "quantified" the degree of divergence the theory is **renormalized** through a suitable, scheme-dependent redefinition of the Lagrangian, adding suitable **counterterms**. Reviews of this formalism may be found in most of the many excellent textbooks on quantum field theory. However, some generic methods to evaluate the new structures introduced by the loop integration are discussed in Section 3.3.1.

However, in addition to ultraviolet divergences infrared divergences appear, both in the virtual and the real correction. In both cases they are related to emissions, either in the loop in the virtual correction or in the additional particle radiated in the real correction, where the additional particle has zero energy or is parallel to another particle. This has already been discussed in Section 2.2.5, where it was also pointed out that due to the **Bloch–Nordsieck (BN)** and **Kinoshita–Lee–Nauenberg (KLN)** theorems [257, 678, 724] these divergences need to cancel each other in physically meaningful observables.

Again, in order to deal with the infrared divergences, they have to be regularized, with dimensional regularization as the method of choice, as in the case of ultraviolet divergences and for the same reasons. This time, however, the divergences manifest themselves as double or single poles, $1/\varepsilon^2$ or $1/\varepsilon$, respectively. As already implied, these poles do not need to be renormalized since they will cancel in the total result.

There is a practical problem in the cancellation, though. Inspecting Eq. (3.49), these poles show up in two different parts of the calculation, and in two in principle independent integrations: for the infrared divergences in the virtual contribution, they appear in the integral over the n-body phase space associated with Born configurations, while for their counterpart in the real contribution, they of course appear in the $(n+1)$-body final state. Even if these integrations could be performed analytically, they are cumbersome to trace in D dimensions. This will be exemplified in Section 3.3.2, where NLO corrections to W production are discussed. In the more general case, as already noted the phase-space integration has to be performed numerically, with Monte Carlo methods. In this case it is impossible to integrate in D dimensions, and other methods have to be found to isolate the divergences before the integration can be successfully achieved. Typically this is by now achieved through subtraction methods, introduced in Section 3.3.2. A general subtraction method, **Catani–Seymour** or **dipole subtraction** [344, 353], is discussed in Section 3.3.3.

3.3.1   Evaluating virtual corrections at one-loop

3.3.1.1   The simplest one-loop amplitude: The Drell–Yan process

In order to understand some of the subtleties of one-loop calculations, it is instructive to return to the simplest case, Drell–Yan production, and go through the calculation in some detail.

The formula for the virtual amplitude is, *cf.* Eq. (2.116),

$$
\mathcal{M}^{(1)}_{u\bar{d}\to W^+} = g_W g_s^2 C_F \mu^{2\varepsilon} \int \frac{\mathrm{d}^D k}{(2\pi)^D} \left\{ \frac{g^{\nu\rho}}{k^2} \bar{v}(p_d) \left[\gamma_\nu \frac{\not{p}_d + \not{k}}{(p_d+k)^2} \gamma^{\mu L} \frac{\not{p}_u - \not{k}}{(p_u-k)^2} \gamma_\rho \right. \right. \tag{3.53}
$$
$$
\left. \left. + \gamma_\nu \frac{\not{p}_d + \not{k}}{(p_d+k)^2} \gamma_\rho \frac{\not{p}_d}{p_d^2} \gamma^{\mu L} + \gamma^{\mu L} \frac{\not{p}_u}{p_u^2} \gamma_\nu \frac{\not{p}_u - \not{k}}{(p_u-k)^2} \gamma_\rho \right] u(p_u) \, \epsilon_\mu(W^+) \right\}.
$$

The contribution of the vertex correction, represented by the first term in square brackets, is in fact the only lengthy calculation that needs to be performed. This term can be written as

$$
\mathcal{M}^{(\mathrm{vertex})}_{u\bar{d}\to W^+} = g_W g_s^2 C_F \mu^{4-D} \int \frac{\mathrm{d}^D k}{(2\pi)^D} \frac{V^\mu \, \epsilon_\mu(W^+)}{k^2 (p_d+k)^2 (p_u-k)^2}, \tag{3.54}
$$

where the current V^μ is

$$
V^\mu = \bar{v}(p_d) \, \gamma_\nu \, (\not{p}_d + \not{k}) \, \gamma^{\mu L} \, (\not{p}_u - \not{k}) \, \gamma^\nu u(p_u). \tag{3.55}
$$

This expression is readily simplified with a little γ-matrix algebra to the form

$$
V^\mu = \bar{v}(p_d) \left[-2 \, (\not{p}_u - \not{k}) \, \gamma^{\mu R} \, (\not{p}_d + \not{k}) + 2\varepsilon a^{\mathrm{CDR}} \, (\not{p}_d + \not{k}) \, \gamma^{\mu R} \, (\not{p}_u - \not{k}) \right] u(p_u). \tag{3.56}
$$

This equation introduces a constant, a^{CDR}, that specifies the variant of the **regularization scheme** to be used in the calculation. In **conventional dimensional regularization** all quantities are continued into D dimensions such that $\gamma_\nu \gamma^\nu = D = 4-2\varepsilon$. This corresponds to the choice $a^{\mathrm{CDR}} = 1$. In **dimensional reduction**, only the loop momentum is continued into D dimensions, and then $\gamma_\nu \gamma^\nu = 4$ implying that $a^{\mathrm{CDR}} = 0$.

In order to evaluate the loop integral, it is simplest to combine the loop propagators in Eq. (3.54) by introducing **Feynman parameters**, here x, y, and z.[5] Applying the identity Eq. (A.27) to the case at hand yields,

$$
\frac{1}{(p_d+k)^2 (p_u-k)^2 k^2} = \int_0^1 \mathrm{d}x \, \mathrm{d}y \, \mathrm{d}z \, \frac{2\delta(1-x-y-z)}{[x(p_d+k)^2 + y(p_u-k)^2 + zk^2]^3}
$$
$$
= \int_0^1 \mathrm{d}x \int_0^{1-x} \mathrm{d}y \, \frac{2}{[k^2 + 2k \cdot (xp_d - yp_u)]^3}
$$

where one of the Feynman parameters has been eliminated immediately using the δ function. In a second step, the denominator has been simplified with the on-shell conditions for p_u and p_d. It is now useful to shift the variable of integration from k to ℓ with the relation, $k = \ell - xp_d + yp_u$. After this shift the denominator takes the simple form $(\ell^2 + Q^2 xy)^3$, where $Q^2 = 2p_d \cdot p_u$ as usual.

---

[5] Feynman parameter identities are discussed in Appendix A.1.3

At first glance, performing the same shift on the current V^μ leads to a proliferation of terms. However, since the denominator is an even function of ℓ, any odd powers of ℓ may be dropped in the numerator since they will vanish upon integration. Moreover, the integral of the term $\ell^\alpha \ell^\beta$ must be proportional to $g^{\alpha\beta}$ and contracting with $g_{\alpha\beta}$ fixes the overall constant. It is therefore sufficient to replace

$$\ell^\alpha \ell^\beta \to g^{\alpha\beta}\, \ell^2 / D, \tag{3.57}$$

under the integral. Finally, the equations of motion for the spinors, $\bar{v}(p_d)\slashed{p}_d = \slashed{p}_u u(p_u) = 0$ greatly simplify many of the resulting expressions.

After this simplification the integral takes the form

$$\int \frac{\mathrm{d}^D k}{(2\pi)^D} \frac{V^\mu}{k^2(p_d + k)^2(p_u - k)^2} = \int_0^1 \mathrm{d}x \int_0^{1-x} \mathrm{d}y \int \frac{\mathrm{d}^D \ell}{(2\pi)^D} \frac{4N\, \bar{v}(p_d)\, \gamma^{\mu L}\, u(p_u)}{(\ell^2 + Q^2 xy)^3} \cdot \tag{3.58}$$

where,

$$N = Q^2 \left[(1-x)(1-y) - \varepsilon a^{\mathrm{CDR}} xy\right] - \ell^2(1 - a^{\mathrm{CDR}}\varepsilon)(1 - 2/D). \tag{3.59}$$

As a consequence of the spinor structure the vertex correction is proportional to the leading order amplitude. The integrals over the loop momentum are now easily performed by using a **Wick rotation**, for which the general result is given in Eq. (A.33). The integrals appearing here are

$$\int \frac{\mathrm{d}^D \ell}{(2\pi)^D} \frac{\ell^2}{(\ell^2 + Q^2 xy)^3} = \frac{i}{(4\pi)^{D/2}} \left(\frac{D}{4}\right) \Gamma(\varepsilon) \left[-Q^2 xy\right]^{-\varepsilon}$$
$$\int \frac{\mathrm{d}^D \ell}{(2\pi)^D} \frac{1}{(\ell^2 + Q^2 xy)^3} = -\frac{i}{(4\pi)^{D/2}} \left(\frac{1}{2}\right) \Gamma(1+\varepsilon) \left[-Q^2 xy\right]^{-1-\varepsilon}. \tag{3.60}$$

Hence,

$$\int \frac{\mathrm{d}^D \ell}{(2\pi)^D} \frac{4N}{(\ell^2 + Q^2 xy)^3} = \frac{i\Gamma(1+\varepsilon)}{(4\pi)^{D/2}}(-Q^2)^{-\varepsilon}$$
$$\times \left\{2\left[(1-x)(1-y) - \varepsilon a^{\mathrm{CDR}} xy\right] x^{-1-\varepsilon} y^{-1-\varepsilon} - 2(1 - a^{\mathrm{CDR}}\varepsilon)(1-\varepsilon)\frac{1}{\varepsilon}x^{-\varepsilon}y^{-\varepsilon}\right\}$$

The integral over y can now be performed in a straightforward manner. The remaining x integral is immediately in the form of a beta function that can in turn be expressed in terms of gamma functions, *cf.* Eq. (A.6). After manipulating these and dropping terms of order ε the final result for the integral is

$$\int \frac{\mathrm{d}^D k}{(2\pi)^D} \frac{V^\mu}{k^2(p_d+k)^2(p_u-k)^2}$$

$$= \bar{v}(p_d)\,\gamma^{\mu L}\,u(p_u)\,\frac{i(-Q^2)^{-\varepsilon}}{(4\pi)^{2-\varepsilon}}\,\frac{\Gamma(1+\varepsilon)\Gamma^2(1-\varepsilon)}{\Gamma(1-2\varepsilon)}\left[\frac{2}{\varepsilon^2}+\frac{3}{\varepsilon}+7+a^{\mathrm{CDR}}\right]$$

$$= -i\,\bar{v}(p_d)\,\gamma^{\mu L}\,u(p_u)\left(-\frac{4\pi}{Q^2}\right)^\varepsilon\frac{1}{16\pi^2}\frac{1}{\Gamma(1-\varepsilon)}\left[-\frac{2}{\varepsilon^2}-\frac{3}{\varepsilon}-7-a^{\mathrm{CDR}}\right].$$

In order to arrive at the last line it is necessary to use the relation in Eq. (A.4) to simplify the combination of gamma functions that naturally occurs.

Restoring the overall factors present in Eq. (3.54) and extracting the leading order amplitude, the result for the vertex correction contribution to the amplitude is

$$\mathcal{M}_{u\bar{d}\to W^+}^{(\mathrm{vertex})} = \mathcal{M}_{u\bar{d}\to W^+}^{(0)}\frac{\alpha_s}{4\pi}C_F\left(-\frac{4\pi\mu^2}{Q^2}\right)^\varepsilon\frac{1}{\Gamma(1-\varepsilon)}\left[-\frac{2}{\varepsilon^2}-\frac{3}{\varepsilon}-7-a^{\mathrm{CDR}}\right]. \quad (3.61)$$

It is common to identify the factor

$$c_\Gamma = (4\pi)^\varepsilon/\Gamma(1-\varepsilon) = 1+\varepsilon\left(1+\log(4\pi)-\gamma_E\right)+\mathcal{O}(\varepsilon^2), \quad (3.62)$$

which is a universal factor in one-loop calculations. In addition, for the Drell–Yan process, the kinematics require $Q^2>0$ so that it is also necessary to expand the factor

$$(-1)^\varepsilon = 1+i\pi\varepsilon-\pi^2\varepsilon^2/2+\mathcal{O}(\varepsilon^3). \quad (3.63)$$

Putting this together yields the final expression

$$\mathcal{M}_{u\bar{d}\to W^+}^{(\mathrm{vertex})} = \mathcal{M}_{u\bar{d}\to W^+}^{(0)}\frac{\alpha_s}{4\pi}C_F\left(\frac{\mu^2}{Q^2}\right)^\varepsilon c_\Gamma$$

$$\times\left[-\frac{2}{\varepsilon^2}-\frac{3}{\varepsilon}-7-a^{\mathrm{CDR}}+\pi^2-i\pi\left(\frac{2}{\varepsilon}+3\right)\right]. \quad (3.64)$$

This amplitude enters the cross-section in the virtual correction term $\mathcal{V}$ as

$$\mathcal{V} = 2\,\mathfrak{Re}\left[\mathcal{M}_{u\bar{d}\to W^+}^{(1)}\mathcal{M}_{u\bar{d}\to W^+}^{*(0)}\right], \quad (3.65)$$

cf. Eq. (3.50).

In conventional dimensional regularization, with $a^{\mathrm{CDR}}=1$, this thus leads to the result for the full virtual contribution already given in Eq. (2.117) in Section 2.2.5. To complete the derivation it is therefore only necessary to demonstrate that the contribution of the self-energy corrections is zero. Repeating the same steps as above for the self-energy contributions, the result is proportional to the value of the integral,

$$\int \frac{\mathrm{d}^D\ell}{(2\pi)^D}\frac{1}{(\ell^2)^2}. \quad (3.66)$$

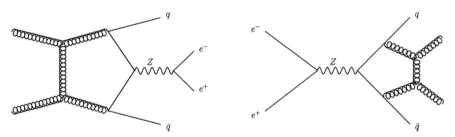

Fig. 3.4 A pentagon diagram entering the loop amplitude for $Z + 2$ jet production (left) and the corresponding $e^+e^- \to 4$ jets diagram from which it can be obtained by crossing (right).

However, since this integral has dimension $D - 4$ but contains no dimensionful quantities with which to express the result, it must vanish. As a result, self-energy corrections on massless external lines are zero within dimensional regularization.

3.3.1.2   General approach to one-loop amplitudes

Although the previous example illustrates the basic concepts of loop-integral calculations, this technique for evaluating the amplitude does not extend to more complicated cases. The calculation is particularly simple for a number of reasons. The one-loop amplitude is represented by a single diagram that contains three external legs, all of the propagators are massless and as a result the calculation is relatively simple. For processes involving more particles in the final state the number of diagrams, and their complexity, grows substantially. For instance, the calculation of $V + 2$ jet production at NLO requires the evaluation of 5-point diagrams such as the one shown in Fig. 3.4 (left). In general, loop diagrams often are referred to as **n-point diagrams**, where n is the number of loop propagators. Alternatively, a nomenclature referring to them as **tadpole** (1-point), **bubble** (2-point), **triangle** (3-point), **box** (4-point), **pentagon** (5-point), or **hexagon** (6-point) diagrams, is often used. The example depicted in the figure would therefore constitute a pentagon diagram.

One might expect the computation of diagrams involving such loops of virtual particles to be the limiting factor that determines the complexity of calculations that can be performed at NLO. Indeed for a long time this was the case, and the efficient treatment of loop diagrams continued to be the bottleneck of NLO calculations for multi-particle processes. The nature of such calculations is exemplified by the computation of a single one-loop diagram that contributes to the process of V+jet production at hadron colliders, depicted in Fig. 3.5. In this example all momenta are labelled as outgoing, which is a particularly convenient notation since momentum conservation is represented by a simple sum, $p_1 + p_2 + p_3 + p_4 = 0$. This of course implies that $p_4 = -p_{123} = -(p_1 + p_2 + p_3)$ in obvious nomenclature. With this very symmetric notation it is also easy to manipulate the matrix element to obtain the result for any **crossed process** with some other external particles in the initial state. Basically, incoming and outgoing particles are related to one another by conjugation which effectively amounts to the inclusion of complex phases ("barring"). This means that computing an amplitude with an outgoing particle of a given chirality also provides

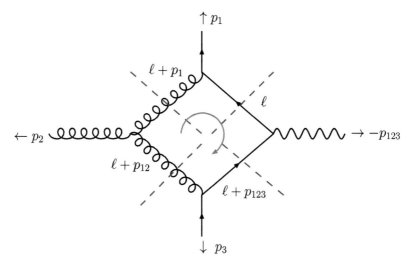

Fig. 3.5 Example one-loop diagram contributing to V+jet production. All external momenta are outgoing and the loop momenta labelled on the internal line flow clockwise around the diagram. The arrow on the fermion line indicates the direction of particle flow, so that p_1 represents an outgoing quark. The dashed blue lines indicate propagators that may be put on-shell when computing integral coefficients using the OPP procedure of Section 3.3.1.3.

the corresponding amplitude for an incoming anti-particle of the opposite chirality. For some of the more technical details of this procedure the reader is referred to Section 3.2.1.

Hence, as written, the diagram in Fig. 3.5 corresponds to the amplitude

$$0 \to q(p_1) + g(p_2) + \bar{q}(p_3) + V(-p_{123}), \tag{3.67}$$

but may equivalently also describe

$$V(p_{123}) \to q(p_1) + g(p_2) + \bar{q}(p_3), \tag{3.68}$$

$$\bar{q}(-p_1) + q(-p_3) \to g(p_2) + V(-p_{123}), \tag{3.69}$$

or

$$\bar{q}(-p_1) + g(-p_2) \to q(p_3) + V(-p_{123}). \tag{3.70}$$

The behaviour of this crossing operation, when applied to the leading-order amplitudes, can be verified in Eqs. (2.97) and (2.98). Note that, similarly, the amplitudes for $p\bar{p} \to V + 2$ jet production, *cf.* Fig. 3.4 (left), can be related to those for four-jet production in e^+e- collisions, *cf.* Fig. 3.4 (right).

The matrix element corresponding to the diagram in Fig. 3.5 reads

$$\int \frac{\mathrm{d}^D \ell}{(2\pi)^D} \, \bar{u}(p_1)\gamma_\mu \frac{\ell}{\ell^2} \gamma_\sigma \frac{\ell + \slashed{p}_{123}}{(\ell + p_{123})^2} \gamma_\nu u(p_3) \frac{1}{(\ell + p_1)^2 (\ell + p_{12})^2} \tag{3.71}$$
$$\times \, V^{\mu\nu\rho}(\ell + p_1, -\ell - p_{12}, p_2) \, \epsilon_\rho(p_2) \, J^\sigma,$$

where J^σ represents the external vector boson current and $V^{\mu\nu\rho}$ is the triple gluon vertex factor,

$$V^{\mu\nu\rho}(\ell + p_1, -\ell - p_{12}, p_2) = (2\ell^\rho + 2p_1^\rho + p_2^\rho)g^{\mu\nu} - (\ell^\nu + 2p_2^\nu + p_1^\nu)g^{\nu\rho}$$
$$+ (p_2^\nu - p_1^\nu - \ell^\nu)g^{\mu\rho}. \tag{3.72}$$

Inspecting this expression, the amplitude can be described in terms of a basis set of loop integrals, categorized according to the number of loop-momentum factors that appear in the numerator. This number is referred to as the **rank** of the **tensor integral** and integrals without any additional numerator factors are usually called **scalar integrals**. One way of evaluating this contribution, for a long time the standard method, is to separately tabulate integrals of each rank. Contracting the Lorentz structures in the numerator with external momenta and polarizations produces the matrix element in Eq. (3.71) Inspection of Eq. (3.71) reveals that, for the case at hand, tensor integrals of up to rank 3 are required,

$$\int \frac{d^D \ell}{(2\pi)^D} \frac{\{1, \ell^\alpha, \ell^\alpha \ell^\beta, \ell^\alpha \ell^\beta \ell^\gamma\}}{\ell^2 (\ell + p_1)^2 (\ell + p_{12})^2 (\ell + p_{123})^2}. \tag{3.73}$$

For the simplest — the scalar — integrals, numerical libraries for their evaluation are now widely available for up to 4-point functions [499, 879, 881]. In fact these libraries are sufficient for all one-loop calculations since the more complicated scalar integrals, pentagons and beyond, can be written in terms of linear combinations of scalar box integrals. However the integrals with additional tensor structure are more complicated to evaluate. A number of systematic methods for reducing them to the form of scalar integrals do exist, the most widely used of which is the **Passarino and Veltman reduction method** [798]. The basic method can be illustrated by considering the integral

$$I^\mu = \int \frac{d^D \ell}{(2\pi)^D} \frac{\ell^\mu}{\ell^2 (\ell + p_1)^2 (\ell + p_{12})^2 (\ell + p_{123})^2}. \tag{3.74}$$

Since the integral I^μ can only depend on the momenta that appear in the denominator, the Lorentz structure can be decomposed as

$$I^\mu = p_1^\mu D_1 + p_2^\mu D_2 + p_3^\mu D_3. \tag{3.75}$$

The problem is then immediately reduced to determining the coefficients D_1, D_2 and D_3. Contracting I^μ with each of the external momenta in turn gives a system of three independent equations that can be used to solve for the three unknowns,

$$\begin{pmatrix} I \cdot p_1 \\ I \cdot p_2 \\ I \cdot p_3 \end{pmatrix} = \begin{pmatrix} 0 & p_1 \cdot p_2 & p_1 \cdot p_3 \\ p_1 \cdot p_2 & 0 & p_2 \cdot p_3 \\ p_1 \cdot p_3 & p_2 \cdot p_3 & 0 \end{pmatrix} \begin{pmatrix} D_1 \\ D_2 \\ D_3 \end{pmatrix}. \tag{3.76}$$

Now, for example, contracting each side of Eq. (3.74) with $p_{3\,\mu}$ results in

$$
\begin{aligned}
I \cdot p_3 &= \frac{1}{2} \int \frac{d^D \ell}{(2\pi)^D} \frac{[(\ell + p_{123})^2 - \ell^2 - p_{123}^2] - [(\ell + p_{12})^2 - \ell^2 - p_{12}^2]}{\ell^2 (\ell + p_1)^2 (\ell + p_{12})^2 (\ell + p_{123})^2} \\
&= \frac{1}{2} \int \frac{d^D \ell}{(2\pi)^D} \left\{ \frac{1}{\ell^2 (\ell + p_1)^2 (\ell + p_{12})^2} - \frac{1}{\ell^2 (\ell + p_1)^2 (\ell + p_{123})^2} \right. \\
&\qquad\qquad \left. - \frac{2 p_{12} \cdot p_3}{\ell^2 (\ell + p_1)^2 (\ell + p_{12})^2 (\ell + p_{123})^2} \right\}.
\end{aligned}
\tag{3.77}
$$

The quantity $I \cdot p_3$ is thus expressible in terms of a combination of scalar integrals, as are $I \cdot p_1$ and $I \cdot p_2$. The complete integral I^μ, and in a similar manner $I^{\mu\nu}$, $I^{\mu\nu\rho}$, ... can all be written in terms of scalar integrals in this way. For rank 2 and beyond, the Lorentz structure must also include the metric tensor. For instance, the most general rank 2 box integral can be decomposed as

$$
I^{\mu\nu} = \sum_i p_i^\mu p_i^\nu D_{ii} + \sum_{i \neq j} \left(p_i^\mu p_j^\nu + p_j^\mu p_i^\nu \right) D_{ij} + g^{\mu\nu} D_{00}.
\tag{3.78}
$$

The complication in this approach stems from the inversion of the matrix relation in Eq. (3.76). The determinant of this matrix that is introduced in solving for the D_i, the **Gram determinant**, is an artefact of the computation introduced by the expansion in terms of scalar integrals. The matrix element itself contains no singularity in the limit in which these determinants vanish, yet one inverse power of a determinant is introduced for each loop momentum factor present in the numerator of the original integral. This redundancy leads to expressions for individual Feynman diagrams that can not only be very lengthy but also be affected by significant numerical cancellations between terms.

It is possible to reformulate the reduction in a number of ways in order to alleviate problems such as Gram determinant singularities. To see how one such solution works, consider a triangle integral with $p_1^2 \neq 0$ and $p_2^2 \neq 0$. The basic scalar integral is,

$$
C_0(p_1, p_2) = \int \frac{d^D \ell}{(2\pi)^D} \frac{\ell^\mu}{\ell^2 (\ell + p_1)^2 (\ell + p_{12})^2}
\tag{3.79}
$$

and the rank 1 tensor integral C^μ can be decomposed in similar fashion to the box example above,

$$
C^\mu = \int \frac{d^D \ell}{(2\pi)^D} \frac{\ell^\mu}{\ell^2 (\ell + p_1)^2 (\ell + p_{12})^2} = C_1 \, p_1^\mu + C_2 \, p_2^\mu.
\tag{3.80}
$$

The matrix equation for the reduction reads, *cf.* Eq. (3.76),

$$
\begin{pmatrix} C \cdot p_1 \\ C \cdot p_2 \end{pmatrix} = \begin{pmatrix} p_1^2 & p_1 \cdot p_2 \\ p_1 \cdot p_2 & p_2^2 \end{pmatrix} \begin{pmatrix} C_1 \\ C_2 \end{pmatrix},
\tag{3.81}
$$

and the explicit expressions for the quantities that appear on the left-hand side are,

$$C \cdot p_1 = \frac{1}{2} \left[B_0(p_{12}) - B_0(p_2) - p_1^2 C_0(p_1, p_2) \right], \tag{3.82}$$

$$C \cdot p_2 = \frac{1}{2} \left[B_0(p_1) - B_0(p_{12}) + (p_1^2 - p_{12}^2) C_0(p_1, p_2) \right], \tag{3.83}$$

in terms of the scalar bubble integral $B_0(q)$ which is defined by

$$B_0(q) = \int \frac{d^D \ell}{(2\pi)^D} \frac{1}{\ell^2 (\ell + q)^2}. \tag{3.84}$$

The direct solution of Eq. (3.81) is,

$$\begin{pmatrix} C_1 \\ C_2 \end{pmatrix} = \frac{1}{\Delta} \begin{pmatrix} p_2^2 & -p_1 \cdot p_2 \\ -p_1 \cdot p_2 & p_1^2 \end{pmatrix} \begin{pmatrix} C \cdot p_1 \\ C \cdot p_2 \end{pmatrix}, \tag{3.85}$$

where $\Delta = p_1^2 p_2^2 - (p_1 \cdot p_2)^2$ is the Gram determinant. Rather than following this path to the solution, note that multiplying the top and bottom rows of Eq. (3.81) by p_2^2 and $(-p_1 \cdot p_2)$ respectively, and adding, yields the equation,

$$(C \cdot p_1) \, p_2^2 - (C \cdot p_2) \, p_1 \cdot p_2 = \left[p_1^2 p_2^2 - (p_1 \cdot p_2)^2 \right] C_1 = \Delta \, C_1. \tag{3.86}$$

From the explicit solutions for $C \cdot p_1$ and $C \cdot p_2$ given in Eq. (3.82) this equation becomes,

$$\Delta \, C_1 = -\frac{1}{2} \left[p_1^2 p_2^2 + p_1 \cdot p_2 (p_1^2 - p_{12}^2) \right] C_0(p_1, p_2) + \{\text{scalar bubbles}\}, \tag{3.87}$$

where, for brevity, the exact form of the linear combination of scalar bubble integrals has been suppressed. After rearranging this yields an expression for the scalar triangle integral,

$$C_0(p_1, p_2) = \frac{2}{p_1 \cdot p_2 (p_{12}^2 - p_1^2) - p_1^2 p_2^2} \left[\Delta \, C_1 + \{\text{scalar bubbles}\} \right]. \tag{3.88}$$

This equation indicates that, in the limit that the Gram determinant vanishes, the scalar triangle integral can be written as a sum of bubble integrals. Note that the same conclusion could have been reached more directly by noting that, in this limit, p_1 and p_2 are collinear and that the correct expansion of C^μ should therefore be,

$$C^\mu = C_1^\star p_1^\mu. \tag{3.89}$$

However, the advantage of Eq. (3.88) is that it provides the $\mathcal{O}(\Delta)$ correction to this relation, if the quantity C_1 is known. The same pattern is reproduced at higher-tensor ranks so that, for instance,

$$C_1 = \frac{2}{p_1 \cdot p_2(p_{12}^2 - p_1^2) - p_1^2 p_2^2} \left[\Delta \sum_{i,j} \alpha_{ij} C_{ij} + \{\text{up to rank 1 bubbles}, C_0\} \right], \quad (3.90)$$

where C_{ij} are rank 2 reduction coefficients and so on.[6] This means that if the rank r bubble integral is already determined, the rank r triangle integral can be determined up to corrections of order Δ. This suggests an iterative approach to the problem, that begins by using Eq. (3.88) with $\Delta = 0$ to determine the first approximation to C_0. With this in hand, Eq. (3.90) can be used with $\Delta = 0$ to determine C_1 for the first time. At this point Eq. (3.88) can be used once more to refine the value of C_0, which is now accurate to order Δ. This procedure involves denominators such as the one shown in Eq. (3.88), which in general do not vanish at the same time as the Gram determinant. However, it may be that the coefficients appearing in this iterative scheme are such that the procedure does not converge. Alternative reduction methods to handle such exceptional cases, together with explicit algorithms for reductions of up to six-point integrals, are given in Ref. [456].

Despite the existence of libraries implementing the sort of rescue procedures described above, this approach still becomes more cumbersome as the number of particles in the final state increases, simply due to the rising number of Feynman diagrams that must be computed. Nevertheless, such methods have been pushed to their limits in the build-up to the LHC era, providing NLO predictions for final states such as $b\bar{b}b\bar{b}$ [253], $W^+W^-b\bar{b}$ [459] and $t\bar{t}b\bar{b}$ [242, 284].

3.3.1.3 Reduction at the integrand level: OPP method

A very powerful method for obtaining one-loop amplitudes relies on a different reduction technique that works at the integrand level and naturally takes advantage of the powerful leading-order amplitude techniques discussed previously. To see how this works, consider again the V+jet process represented by Eq. (3.67).

This process depends on three independent four-vectors, p_1, p_2, and p_3. In what follows it will be convenient to parameterize a general four-vector — the one that defines the momentum running round the loop — in terms of the vectors that lie in the *physical space* spanned by p_1, p_2, and p_3, and the *transverse space* that can be spanned by a single vector, n. For the physical space, rather than using the momenta of the particles themselves, it is more convenient to use the **van Neerven–Vermaseren basis**, v_1, v_2, v_3, defined by

$$v_1^\mu = \frac{\delta_{k_1 k_2 k_3}^{\mu k_2 k_3}}{\Delta}, \quad v_2^\mu = \frac{\delta_{k_1 k_2 k_3}^{k_1 \mu k_3}}{\Delta}, \quad v_3^\mu = \frac{\delta_{k_1 k_2 k_3}^{k_1 k_2 \mu}}{\Delta}, \quad (3.92)$$

where $k_1 = p_1$, $k_2 = p_1 + p_2$ and $k_3 = p_1 + p_2 + p_3$ are the momenta that naturally appear in propagator factors of loop diagrams. The Kronecker delta appearing here is

---

[6]To show this identity requires some further work to demonstrate that the coefficient C_{00} can be written in terms of bubble integrals and a scalar triangle, plus rank 2 triangle coefficients of order Δ. This is indeed the case and, for instance, one can deduce the relation,

$$4\Delta C_{22} + 4(D-1)p_1^2 C_{00} + p_1^4 C_0(p_1, p_2) = \{\text{bubbles}\}. \quad (3.91)$$

a shorthand notation with the following definition,

$$\delta^{\mu\nu\rho}_{k_1 k_2 k_3} = \begin{vmatrix} k_1^\mu & k_2^\mu & k_3^\mu \\ k_1^\nu & k_2^\nu & k_3^\nu \\ k_1^\rho & k_2^\rho & k_3^\rho \end{vmatrix}, \tag{3.93}$$

and the factor Δ is the corresponding Gram determinant,

$$\Delta = \delta^{\mu\nu\rho}_{k_1 k_2 k_3} k_{1\mu} k_{2\nu} k_{3\rho} \equiv \delta^{k_1 k_2 k_3}_{k_1 k_2 k_3}. \tag{3.94}$$

These momenta satisfy the orthogonality condition

$$v_i \cdot k_j = \delta_{ij}. \tag{3.95}$$

The vector n in the transverse space can be defined by

$$n^\mu = \frac{\epsilon^{\mu k_1 k_2 k_3}}{\sqrt{\Delta}}, \tag{3.96}$$

where the epsilon tensor ensures orthogonality

$$n_i \cdot k_j = n_i \cdot v_j = 0, \tag{3.97}$$

and the normalization is such that $n^2 = 1$. The properties of these vectors make it straightforward to see that a four-dimensional loop momentum can be expanded as

$$\ell^\mu = \sum_{i=1}^{3} (\ell \cdot k_i) v_i^\mu + (\ell \cdot n) n^\mu. \tag{3.98}$$

   This is a particularly useful form for the following reason: the diagram shown in Fig. 3.5 contains four propagators $d_0, \ldots, d_3$, given by (assuming massless quarks)

$$d_0 = \ell^2, \quad d_i = (\ell + k_i)^2, \text{ for } i = 1, 2, 3. \tag{3.99}$$

The dot products that appear in the loop momentum decomposition can thus be re-expressed as

$$\ell \cdot k_i = \frac{1}{2}(d_i - d_0) - \frac{1}{2} k_i^2. \tag{3.100}$$

The most complicated term in the evaluation of this diagram contains three powers of the loop momentum in the numerator and can be treated by realizing that

$$\ell^{\mu}\ell^{\nu}\ell^{\rho} = \frac{1}{2}\ell^{\mu}\ell^{\nu}\left(\sum_{i=1}^{3}(d_i - d_0 - k_i^2)v_i^{\rho} + (\ell \cdot n)n^{\rho}\right)$$

$$= \frac{1}{4}\ell^{\mu}\left(\sum_{j=1}^{3}(d_j - d_0 - k_j^2)v_j^{\nu} + (\ell \cdot n)n^{\nu}\right)$$

$$\times \left(-\sum_{i=1}^{3}k_i^2 v_i^{\rho} + (\ell \cdot n)n^{\rho}\right) + \text{triangles}$$

$$= \frac{1}{8}\left(-\sum_{m=1}^{3}k_m^2 v_m^{\mu} + (\ell \cdot n)n^{\mu}\right)\left(-\sum_{j=1}^{3}k_j^2 v_j^{\nu} + (\ell \cdot n)n^{\nu}\right)$$

$$\times \left(-\sum_{i=1}^{3}k_i^2 v_i^{\rho} + (\ell \cdot n)n^{\rho}\right) + \text{triangles, bubbles}$$

$$\equiv \delta_0^{\mu\nu\rho} + \delta_1^{\mu\nu\rho}(\ell \cdot n) + \delta_2^{\mu\nu\rho}(\ell \cdot n)^2 + \delta_3^{\mu\nu\rho}(\ell \cdot n)^3 + \text{lower points.}$$

In other words, the integral can be manipulated into a very particular form. It can be represented by coefficients of a scalar box integral (δ_0) and tensor integrals that are still of rank 3, but where the loop momentum is contracted with the transverse vector, n. The remaining terms are all integrals in which at least one propagator has been cancelled.

One crucial simplification remains. Contracting the loop momentum in Eq. (3.98) with itself yields the relation

$$\ell^2 = \sum_{i,j=1}^{3}(\ell \cdot k_i)(\ell \cdot k_j)v_i \cdot v_j + (\ell \cdot n)^2$$

$$\Rightarrow (\ell \cdot n)^2 = d_0 - \frac{1}{4}\sum_{i,j=1}^{3}(d_i - d_0 - k_i^2)(d_j - d_0 - k_j^2)v_i \cdot v_j$$

(3.101)

This means that a factor of $(\ell \cdot n)^2$ in Eq. (3.101) may be replaced by a constant, up to terms in which a denominator has been cancelled and that the results thus represent lower-point integrals. Redefining δ_0 and δ_1 suitably to absorb the additional constant terms, the rank 3 integral can thus be parameterized most simply by

$$\ell^{\mu}\ell^{\nu}\ell^{\rho} \longrightarrow \delta_0^{\mu\nu\rho} + \delta_1^{\mu\nu\rho}(\ell \cdot n) + \text{lower points}$$

(3.102)

After integration, the ℓ-dependent term cannot contribute to the result since the integral will only produce contributions of the form $k_i \cdot n = 0$, thus vanishing by definition. Therefore the rank 3 box integral has been reduced to a scalar box integral, together with lower-point integrals of rank at most 2.

A similar line of reasoning can now be followed for the lower point triangle and bubble, integrals. In these cases the parameterization of the loop momentum becomes more complicated since additional transverse directions are required in order to span the four-dimensional space of loop momenta. However, the basic line of reasoning follows and, in similar fashion, these integrals can also be reduced to their scalar counterparts. This allows the *entire diagram* to be reduced to a simple sum of coefficients multiplied by basic scalar integrals *at the integrand level*

$$\mathcal{D} = \frac{\{\ell^\mu \ell^\nu \ell^\rho A_{\mu\nu\rho} + \ldots\}}{d_0 d_1 d_2 d_3} = c_4 I_4 + \sum_i c_3^{(i)} I_3^{(i)} + \sum_{i,j} c_2^{(i,j)} I_2^{(i,j)}. \tag{3.103}$$

In this equation the n-point scalar integral is denoted by $I_n^{(\ldots)}$, where the box propagators that have been cancelled are labelled in the superscript. This decomposition is at the heart of the approach to computing one-loop amplitudes that was originally formulated by Ossola, Pittau, and Papadopoulos [792], now commonly referred to as the **OPP method**.

It is convenient to rewrite this decomposition one more time, with the cancelled propagators in the lower-point integrals explicit,

$$\mathcal{D} = \frac{1}{d_0 d_1 d_2 d_3} \left[c_4 + \sum_i c_3^{(i)} d_i + \sum_{i,j} c_2^{(i,j)} d_i d_j \right]. \tag{3.104}$$

If a particular loop momentum could be chosen such that $d_i = 0$ for all i, then all the terms on the right-hand side of Eq. (3.104) would vanish except for c_4. Therefore the coefficient c_4 can be determined by evaluating the diagram for this special value of the loop momentum. The form of the loop momentum that satisfies these additional constraints is already determined by Eq. (3.98), Eq. (3.100), and Eq. (3.101). Setting all the propagators to zero in these expressions, the loop momentum reads

$$\ell^\mu = -\frac{1}{2} \sum_{i=1}^3 k_i^2 \, v_i^\mu + \frac{1}{2} \left(\sum_{i,j=1}^3 k_i^2 k_j^2 \, v_i \cdot v_j \right)^{\frac{1}{2}} n^\mu. \tag{3.105}$$

Evaluating the expression for the diagram, $\mathcal{D}$, at this particular value of ℓ^μ yields the box integral coefficient c_4. With this coefficient determined, Eq. (3.104) can be rewritten as

$$\mathcal{D} - \frac{c_4}{d_0 d_1 d_2 d_3} = \frac{1}{d_0 d_1 d_2 d_3} \left[\sum_i c_3^{(i)} d_i + \sum_{i,j} c_2^{(i,j)} d_i d_j \right] \tag{3.106}$$

to clarify the strategy for determining the triangle coefficients. Another parameterization of the loop momentum is required, which differs in important respects from Eq. (3.98). Importantly, for the triangle case, there are only two independent momenta so that spanning the physical space requires *two* transverse directions n_1 and n_2,

$$\ell^\mu = \sum_{i=1}^2 (\ell \cdot k_i) v_i^\mu + (\ell \cdot n_1) n_1^\mu + (\ell \cdot n_2) n_2^\mu, \tag{3.107}$$

where v_1 and v_2 are now redefined appropriately. Despite this difference, the essential strategy remains the same: use an appropriate parameterization of the loop momentum that allows the reduction of tensor integrals to the scalar case, then demand that it also satisfy $d_i = 0$ for the three propagators that appear in the integral. With this choice, choosing for instance $d_1 = d_2 = d_3 = 0$ in Eq. (3.106) singles out the coefficient $c_3^{(0)}$. Repeating this procedure, at the level of bubble and tadpole coefficients, where appropriate, eventually leads to values for all of the integral coefficients appearing in Eq. (3.104). For further details, the interested reader is referred to Refs. [501, 792].

At this point two important points must be stressed. The first is that by setting various combinations of propagators to zero, the integrands actually reduce to combinations of on-shell tree-level amplitudes. Combining this approach with the efficient methods for computing such amplitudes discussed in the previous section has yielded powerful new tools for computing NLO corrections in an automated fashion, far beyond the limits of purely analytic calculations. These techniques are equally adept at handling massive particles propagating in the loop and the total particle multiplicity is limited purely by computing power. Examples of such numerical codes are listed below.

The second point is that the discussion presented above neglects an important complication. The discussion was rooted in four dimensions, which is sufficient to determine all of the contributions proportional to scalar integrals. However, when working consistently in $d = 4 - 2\varepsilon$ dimensions, the algorithm presented so far must be extended. Explicitly, the triangle loop momentum decomposition is now described not only by the transverse vectors n_1 and n_2 but also by the unit vector in (-2ε) dimensions, n_ε,

$$\ell^\mu = \sum_{i=1}^{2} (\ell \cdot k_i) v_i^\mu + (\ell \cdot n_1) n_1^\mu + (\ell \cdot n_2) n_2^\mu + (\ell \cdot n_\varepsilon) n_\varepsilon^\mu. \tag{3.108}$$

While at first glance this seems like a rather small change, it results in an expression for the decomposition of a general tensor integral that is more complicated than the form given previously. In particular, the relation analogous to Eq. (3.101) is,

$$(\ell \cdot n_1)^2 + (\ell \cdot n_2)^2 + (\ell \cdot n_\varepsilon)^2 = d_0 - \frac{1}{4} \sum_{i,j=1}^{3} (d_i - d_0 - k_i^2)(d_j - d_0 - k_j^2) v_i \cdot v_j \tag{3.109}$$

so that the reduction of a triangle integral of rank at least 2 leads to integrals with $(\ell \cdot n_\varepsilon)^2$ numerators. The contribution of such integrals is straightforward to evaluate by Passarino–Veltman reduction,

$$\int d^D \ell \frac{(\ell \cdot n_\varepsilon)^2}{d_0 d_1 d_2} = n_\varepsilon^\mu n_\varepsilon^\nu \left[g_{\mu\nu} C_{00} \right] = (-2\varepsilon) C_{00} \tag{3.110}$$

where most of the components of the integral vanish by orthogonality, but the term proportional to the metric tensor, C_{00}, survives (*cf.* the equation defining the box integral counterpart of C_{00}, Eq. (3.78)). Since C_{00} contains an ultraviolet $1/\varepsilon$ singularity, this integral thus gives a non-zero contribution. These additional terms are called **rational parts** and additional work is required to determine their contributions. Although this is an important aspect of the procedure, it is not conceptually different from the strategy outlined here and does not require much more computational effort.

3.3.1.4   Reduction at the integrand level: OPENLOOPS

The idea underlying the OPENLOOPS algorithm [338] is to combine tensor integral reduction and OPP methods to generate one-loop amplitudes based on a recursive construction of Feynman diagrams. To see how this works in more detail, consider the colour-stripped n-point one-loop diagram

$$\mathcal{A}^{(D)} = \int \frac{\mathrm{d}^D q \, \mathcal{N}(\mathcal{I}_n; q)}{D_0 D_1 \ldots D_{n-1}}, \tag{3.111}$$

which connects the n-ordered subtrees

$$\mathcal{I}_n = \{i_1, i_2, \ldots, i_n\} \tag{3.112}$$

with n loop propagators

$$D_i = (q + p_i)^2 - m^2 + i\epsilon. \tag{3.113}$$

All other propagators, the ones in the subtrees, as well as all numerators of the propagators and all structures from vertices and external particles are absorbed in the overall "numerator" $\mathcal{N}$, which effectively is a polynomial in the loop momentum q. In gauge theories, its rank R usually cannot be larger than n: $R \leq n$, which allows to write

$$\mathcal{N}(\mathcal{I}_n; q) = \sum_{r=0}^{R} \mathcal{N}^{(r)}_{\mu_1 \mu_2 \ldots \mu_r}(\mathcal{I}_n) q^{\mu_1} q^{\mu_2} \ldots q^{\mu_n}. \tag{3.114}$$

This leaves the following generic one-loop tensor integrals to evaluate,

$$\mathcal{T}^{\mu_1 \mu_2 \ldots \mu_r}_{n, r} = \int \frac{\mathrm{d}^D q \, q^{\mu_1} q^{\mu_2} \ldots q^{\mu_n}}{D_0 D_1 \ldots D_{n-1}}. \tag{3.115}$$

As already discussed, in traditional tensor-reduction techniques this is achieved by reducing the tensor integrals to scalar ones. Technically this reduction results from analytically cancelling the momenta q in the numerator with the denominators D. Alternatively, the OPP method numerically expresses the numerator as a polynomial of the denominators, where the residual scalar integrals and their coefficients are obtained by solving systems of equations stemming from simultaneous multicut relations $D_i = D_j = \ldots = 0$ for up to four different propagators at the same time.

Setting $p_0 \equiv 0$ eliminates all momentum shift ambiguities and singles out D_0. Cutting the propagator D_0 and removing the other loop denominators D_i leaves the numerator term $\mathcal{N}^\beta_\alpha(\mathcal{I}_n; q)$. Pictorially, it can be reconstructed through the recursion

$$\mathcal{N}_\alpha^\beta(\mathcal{I}_n; q) = \quad$$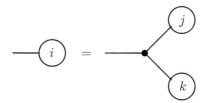

or, written out, as

$$\mathcal{N}_\alpha^\beta(\mathcal{I}_n; q) = \mathcal{N}_\alpha^\gamma(\mathcal{I}_{n-1}; q)\, \mathcal{X}_{\gamma\delta}^\beta(\mathcal{I}_n, i_n, \mathcal{I}_{n-1})\, w^\delta(i_n). \tag{3.116}$$

Here, $\mathcal{X}_{\gamma\delta}^\beta$ and w^δ are the vertices and subtrees that enter the tree algorithm. The former are similar or identical to the vertices $\mathcal{V}_\alpha$ used in the off-shell recursion relations, *cf.* Eq. (3.14), while the latter are different. In contrast to the currents in the off-shell recursion relations that capture all possible combinations of given subsets of external particles, the w^δ are recursively constructed from Feynman (sub-)amplitudes. Pictorially,

or

$$w^\beta(i) = \frac{\mathcal{X}_{\gamma\delta}^\beta(i, j, k)\, w^\gamma(j)\, w^\delta(k)}{p_i^2 - m_i^2 + i\epsilon}, \tag{3.117}$$

with polarization vectors or spinors being the wave functions representating external vector bosons or fermions. Expanding[7] the vertex functions in orders of q^μ,

$$\mathcal{X}_{\gamma\delta}^\beta = \mathcal{Y}_{\gamma\delta}^\beta + q^\nu\, \mathcal{Z}_{\nu;\,\gamma\delta}^\beta, \tag{3.118}$$

similar to Eq. (3.114), allows to decompose the recursion for the numerator terms in Eq. (3.116) to be written as

$$N_{\mu_1\mu_2\ldots\mu_r;\alpha}^\beta(\mathcal{I}_n) = \left[\mathcal{Y}_{\gamma\delta}^\beta N_{\mu_1\mu_2\ldots\mu_r;\alpha}^\gamma(\mathcal{I}_{n-1}) + \mathcal{Z}_{\nu;\,\gamma\delta}^\beta N_{\mu_2\ldots\mu_r;\alpha}^\gamma(\mathcal{I}_{n-1}) + \right] w^\delta(i_n). \tag{3.119}$$

The number of coeffients for this kind of decomposition grows polynomially, with a degree determined by the tensorial rank r of the original numerator term. Recycling subtrees, symmetrization over open-loop tensorial indices $\mu_1, \mu_1, \ldots, \mu_r$, "pinching"

---

[7]In the equation below, only Feynman rules from gauge theory have been assumed, limiting the expansion to first order in the four-momentum; including also effective theories such as the Higgs Effective Theory with its more complicated vertex structure renders the inclusion of second-order terms necessary.

propagators, and thereby factoring out common factors $\mathcal{I}_n$ further enhances the efficiency of this algorithm. Once the coefficients of the decomposition are known, the polynomial in Eq. (3.114) can be evaluated multiple times at very low CPU cost, thereby yielding a massive acceleration of the overall calculation. As a further benefit, the OPENLOOPS algorithm can be used in conjunction with both OPP and traditional tensor reduction methods or any admixture, which enhances it even more. As a consequence, after its initial implementation in the OPENLOOPS package with a first non-trivial application in [337], the algorithm has also been adopted by the MADGRAPH collaboration [148].

3.3.2 Dealing with infrared divergences: General ideas

In Section 3.3.1, methods to calculate the virtual correction $\mathcal{V}$ in the expression for cross-sections at next-to-leading order, *cf.* Eq. (3.49), have been discussed. As already stated, this term exhibits two kinds of divergences: ultraviolet and infrared ones. While the former are confined inside the virtual correction and can be dealt with there through regularization and renormalization, the latter are a bit more tricky. This is due to the fact that these divergences, the infrared ones, must cancel between virtual and real contributions, due to the **Bloch–Nordsieck (BN)** and **Kinoshita–Lee–Nauenberg (KLN)** theorems [257, 678, 724]. This cancellation has already been observed in the previous chapter, for the example of inclusive W production at hadron colliders, *cf.* Section 2.2.5. Direct calculation of the virtual contribution in this example resulted in the terms exhibited in Eq. (2.117), while the direct evaluation of the real correction term, including an integration over the two-body final-state phase space resulted in Eq. (2.118). It is important to stress here that the exact cancellation of the infrared divergences in this case was only recovered upon integration over the respective phase space of the final-state particles.

With more complicated processes, this way of directly calculating the matrix elements and performing phase-space integrals to cancel the divergences very quickly exhausts its applicability. This can be traced back to the fact that, in general, final-state phase-space integrals for final states with more than three particles cannot analytically be evaluated, neither in four nor in the even more complicated case of D dimensions. For such cases numerical methods — Monte Carlo integration techniques — must be invoked, which by construction only work in an integer number of dimensions, another approach to deal with infrared divergences and their mutual cancellation is mandatory. By now the method of choice is **infrared subtraction**. The underlying idea is to isolate the divergent structures through suitable terms such that Eq. (3.49) becomes

$$\sigma^{(\mathrm{NLO})} = \sum_{a,b} \int_0^1 \mathrm{d}x_a \mathrm{d}x_b \, f_{a/h_1}(x_a, \mu_F) f_{b/h_2}(x_b, \mu_F) \int \mathrm{d}\hat{\sigma}_{ab \to n}^{(\mathrm{NLO})}(\mu_F, \mu_R)$$

$$= \int \mathrm{d}\Phi_{\mathcal{B}} \left[\mathcal{B}_n(\Phi_{\mathcal{B}}; \mu_F, \mu_R) + \mathcal{V}_n(\Phi_{\mathcal{B}}; \mu_F, \mu_R) \right] + \int \mathrm{d}\Phi_{\mathcal{R}} \, \mathcal{R}_n(\Phi_{\mathcal{R}}; \mu_F, \mu_R)$$

$$= \int \mathrm{d}\Phi_{\mathcal{B}} \left[\mathcal{B}_n(\Phi_{\mathcal{B}}; \mu_F, \mu_R) + \mathcal{V}_n(\Phi_{\mathcal{B}}; \mu_F, \mu_R) + \mathcal{I}_n^{(\mathcal{S})}(\Phi_{\mathcal{B}}; \mu_F, \mu_R) \right]$$

$$+ \int \mathrm{d}\Phi_{\mathcal{R}} \left[\mathcal{R}_n(\Phi_{\mathcal{R}}; \mu_F, \mu_R) - \mathcal{S}_n(\Phi_{\mathcal{R}}; \mu_F, \mu_R) \right],$$

$$(3.120)$$

where the **real subtraction term** $\mathcal{S}_n$, which lives in the $(n+1)$-particle phase space of the real correction, and the **integrated subtraction term** $\mathcal{I}_n^{(\mathcal{S})}$, which lives in the (n)-particle phase space of the Born and virtual contributions, cancel each other in the integrals

$$0 \equiv \int \mathrm{d}\Phi_{\mathcal{B}} \, \mathcal{I}_n^{(\mathcal{S})}(\Phi_{\mathcal{B}}; \mu_F, \mu_R) - \int \mathrm{d}\Phi_{\mathcal{R}} \, \mathcal{S}_n(\Phi_{\mathcal{R}}; \mu_F, \mu_R). \qquad (3.121)$$

The catch in this method is that due to the **universal structure of infrared divergences in gauge theories** it is possible to construct subtraction terms $\mathcal{S}$ in a process-independent way, guaranteeing that the difference $(\mathcal{R} - Subtraction)$ is finite in every single phase-space point in $\Phi_{\mathcal{R}}$. Even more, these terms can be constructed as the product of some Born-level configurations times some individual terms, accounting for the emission of an additional particle. There terms can be integrated over the phase space of the additional particle in D dimensions, yielding infrared divergences that manifest as terms proportional to $1/\varepsilon^2$ and $1/\varepsilon$. Before, however, turning to the explicit construction of these terms in a specific formalism, **Catani–Seymour subtraction** [353], the underlying idea will be elucidated in a toy model and applied to the by-now familiar example of W production at hadron colliders.

3.3.2.1 Infrared divergences: A toy model

In the toy model, assume that Born, virtual, and real emission matrix elements squared to $\mathcal{O}(\alpha)$ read

$$\mathcal{B}_n = \sum |\mathcal{M}_n^{(\mathcal{B})}|^2,$$

$$\mathcal{V}_n = \frac{V_n}{\varepsilon} = \sum |\mathcal{M}_n^{(\mathcal{B})} \mathcal{M}_n^{*(\mathcal{V})}|, \qquad (3.122)$$

$$\mathcal{R}_n(x) = \frac{R_n)(x)}{x} = \sum |\mathcal{M}_{n+1}^{(\mathcal{R})}(x)|^2,$$

where both the Born and the virtual contributions are constant in the toy model and the latter has already been regularized in D dimensions and its ultraviolet divergences have been regularized and renormalized. The remaining infrared divergences then lead to the pole in $1/\epsilon$, which has been made explicit in the virtual contribution — here, in the toy model, V_n indeed is infrared finite and ultraviolet renormalized. The real-emission part in the toy model depends on a one-dimensional phase-space parameter $x \in [0, 1]$ and it diverges for $x \to 0$, producing a single pole $1/\epsilon$, as again made explicit, while the function $R_n(x)$ is regular in the full interval.

Written in these quantities the NLO cross-section is given by

$$\sigma^{(\text{NLO})} = [\mathcal{B}_n + \mathcal{V}_n] \, F_n^J + \int\limits_0^1 \mathrm{d}x \, \mathcal{R}_n(x) F_{n+1}^J(x)$$

$$= \left[\mathcal{B}_n + \frac{V_n}{\varepsilon}\right] F_n^J + \int\limits_0^1 \frac{\mathrm{d}x}{x} R_n(x) F_{n+1}^J(x),$$

(3.123)

where F^J is a (jet) criterion applied in order to ensure that the Born-level cross-section and the corresponding kinematics are free of singularities. The concept of jets has already been introduced in more detail in Section 2.1.6. Here, it suffices to remember that in the language of QCD the application of a jet criterion implies that the m outgoing particles are all sufficiently energetic and well separated from each other such that the Born cross-section is free of infrared singularities. In fact, this jet criterion could be replaced with any observable that ensures that the Born-level part is well behaved.

Typically, when adding virtual corrections, no new phase-space regions become available because the incoming and outgoing particles are still the same as at the Born level. Therefore, the function F can be applied without much ado also to the virtual part. This is not true when adding additional real radiation. In fact, infrared divergences appear; in the toy model they are represented by the $1/x$-poles. In order to deal with them the observable F^J must be defined in such a way that soft or collinear emissions — those with $x \to 0$ in the toy model — do not affect it. This catches the essence of the intuitive definition of an observable being infrared-safe: it must be defined in such a way that further soft and/or collinear emissions do not affect it. Mathematically speaking this means that

$$\lim_{x \to 0} F_{n+1}^J(x) = F_{n+1}^J(0) = F_n^J.$$

(3.124)

It is important to stress, though, that without infrared safe observables the full concept of higher-order calculations becomes entirely meaningless.

The Bloch–Nordsieck and Kinoshita–Lee–Nauenberg theorems [257, 678, 724] now state that for infrared-safe quantities the infrared divergences in the real and virtual contributions cancel, which implies that

$$\lim_{x \to 0} R_n(x) = R_n(0) = V.$$

(3.125)

In the framework of NLO calculations two systematically different methods have been developed to isolate the pole in the real-emission contributions, namely phase-space slicing [576] and subtraction [352, 353, 542]. In both cases the real-emission contribution will be written in D dimensions — in the toy model here this amounts to replacing the x^{-1} pole by $x^{-1-\epsilon}$. Then the ideas underlying both methods are as follows:

1. **Phase-space slicing**
   The idea [573] is to introduce an arbitrary cut-off δ and to rewrite the NLO part of the cross-section as

$$
\begin{aligned}
\sigma^{(1)} &= \frac{V_n}{\varepsilon} F_n^J + \int_0^1 \frac{\mathrm{d}x}{x^{1+\epsilon}} R_n(x) F_{n+1}^J(x) \\
&= \frac{V_n}{\varepsilon} F_n^J + \int_0^\delta \frac{\mathrm{d}x}{x^{1+\epsilon}} R_n(x) F_{n+1}^J(x) + \int_\delta^1 \frac{\mathrm{d}x}{x^{1+\epsilon}} R_n(x) F_{n+1}^J(x).
\end{aligned}
\tag{3.126}
$$

This can be approximated up to first order in ϵ as

$$
\begin{aligned}
\sigma^{(1)} &= \frac{V_n}{\varepsilon} F_n^J + R_n(0) F_n^J \int_0^\delta \frac{\mathrm{d}x}{x^{1+\epsilon}} + \int_\delta^1 \frac{\mathrm{d}x}{x^{1+\epsilon}} R_n(x) F_{n+1}^J(x) + \mathcal{O}(\varepsilon) \\
&= \left[1 - \delta^{-\epsilon}\right] \frac{V_n}{\varepsilon} F_n^J + \int_\delta^1 \frac{\mathrm{d}x}{x^{1+\epsilon}} R_n(x) F_{n+1}^J(x) + \mathcal{O}(\varepsilon) \\
&= \log \delta \cdot V_n F_n^J + \int_\delta^1 \frac{\mathrm{d}x}{x^{1+\epsilon}} R_n(x) F_{n+1}^J(x) + \mathcal{O}(\varepsilon),
\end{aligned}
\tag{3.127}
$$

where $\delta^{-\epsilon} = 1 - \epsilon \log \delta + \mathcal{O}(\varepsilon^2)$ has been used. This procedure has a nice additional feature, that the answer should be independent of δ. However, there is a tension between retaining a good singular approximation (which is obtained by choosing a very small δ) and still avoiding corresponding large logarithmic cancellations, which leads to a preference of larger δ. An illustration of this check is shown in Fig. 3.6, taken from a calculation of $Wb\bar{b}$ production at NLO [520]. This ambiguity in choosing an optimal value of δ has made this method in practice going somewhat out of fashion, since manual interference and careful monitoring of the numerical stability of the final results are mandatory.

2. **Subtraction methods**
   This problem is alleviated by subtraction methods, where a zero is introduced into the result by adding and subtracting a term

$$
R_n(0) F_n^J \int_0^1 \frac{\mathrm{d}x}{x^{1+\epsilon}}
\tag{3.128}
$$

such that the first-order contribution reads

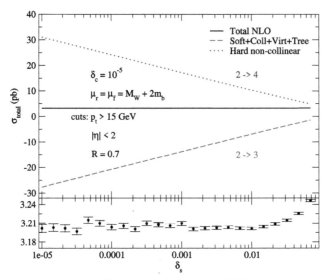

Fig. 3.6 An illustration of the dependence of a full NLO calculation on the phase-space slicing parameter. Reprinted with permission from Ref. [520].

$$
\begin{aligned}
\sigma^{(1)} &= \frac{V_n}{\varepsilon} F_n^J + \int_0^1 \frac{\mathrm{d}x}{x^{1+\epsilon}} R_n(x) F_{n+1}^J(x) \\
&= \frac{V_n}{\varepsilon} F_n^J + R_n(0) F_n^J \int_0^1 \frac{\mathrm{d}x}{x^{1+\epsilon}} \\
&\qquad - R_n(0) F_n^J \int_0^1 \frac{\mathrm{d}x}{x^{1+\epsilon}} + \int_0^1 \frac{\mathrm{d}x}{x^{1+\epsilon}} R_n(x) F_{n+1}^J(x) \\
&= \frac{V_n}{\varepsilon} F_n^J [1 - 1] + \int_0^1 \frac{\mathrm{d}x}{x^{1+\epsilon}} \left[R_n(x) F_n^J(x) - V_n F_n^J \right].
\end{aligned}
\tag{3.129}
$$

This subtraction is possible because the infrared structure of the toy model is entirely known and fixed by Eq. (3.125). In real applications the case is not quite as trivial, although the infrared structure still is fixed by the fact that in the soft and collinear limits of parton emission the respective real-emission contributions factorize into a process-dependent Born part and a process-independent parton splitting part, where the spin dependence is given by the Altarelli–Parisi splitting kernels already encountered for the case of QED in Eq. (2.9).

3.3.2.2 Toy model vs. cross-section in the subtraction formalism

Translating the toy model of the previous section to the language employed in the rest of the book, in general a cross-section at leading and at next-to-leading order is written as

$$
\sigma^{(\mathrm{LO})} = \int \mathrm{d}\Phi_{\mathcal{B}}\, \mathcal{B}_n(\Phi_{\mathcal{B}}; \mu_F, \mu_R)
$$

$$
\sigma^{(\mathrm{NLO})} = \int \mathrm{d}\Phi_{\mathcal{B}} \left[\mathcal{B}_n(\Phi_{\mathcal{B}}; \mu_F, \mu_R) + \mathcal{V}_n(\Phi_{\mathcal{B}}; \mu_F, \mu_R) + \mathcal{I}_n^{(\mathcal{S})}(\Phi_{\mathcal{B}}; \mu_F, \mu_R) \right]
$$

$$
+ \int \mathrm{d}\Phi_{\mathcal{R}} \left[\mathcal{R}_n(\Phi_{\mathcal{R}}; \mu_F, \mu_R) - \mathcal{S}_n(\Phi_{\mathcal{R}}; \mu_F, \mu_R) \right],
\tag{3.130}
$$

cf. Eq. (3.120). As before, $\Phi_{\mathcal{B}}$ and $\Phi_{\mathcal{R}}$ denote the Born and real-emission phase space, respectively, and $\mathcal{B}_n$, $\mathcal{V}_n$, $\mathcal{I}_n^{(\mathcal{S})}$, $\mathcal{R}_n$, and $\mathcal{S}_n$, are the matrix elements for the Born, renormalized virtual, integrated subtraction, real-emission, and real subtraction contributions.

The integrated subtraction and the real subtraction contributions actually correspond to the terms

$$
\pm R_m(0)\, F_m^J \int_0^1 \frac{\mathrm{d}x}{x^{1+\varepsilon}},
\tag{3.131}
$$

which are combined with the virtual correction $V_m F_m^J/\varepsilon$ and with the real correction term in the toy model. By now, there are different, well-established methods to construct such subtraction terms in a process-independent way. Note that for hadronic initial states $\hat{V}$ has been defined such as to include the collinear mass-factorization counter-terms related to divergences that are absorbed into the definition of the PDFs for the incoming partons, essentially the terms in the last line of Eq. (2.118). All integrands include parton luminosity, symmetry, and flux factors.

3.3.2.3 Example: W production at NLO with simple subtraction

To see how the subtraction method works in more detail, it is instructive to consider the real radiation contributions that enter the NLO corrections to W production.

The starting point is the real matrix element squared for the process $u\bar{d} \to gW^+$ given in Eq. (2.101). As already observed, it is divergent in the limits $\hat{t} \to 0$ and $\hat{u} \to 0$. The key to a general approach for handling these infrared singularities lies in the ability to deal with the collinear singularities associated with each of these limits separately. This can immediately be seen by analysing the kinematic dependence, isolating the divergent terms and then using partial fractioning,

$$
\frac{\hat{t}^2 + \hat{u}^2 + 2m_W^2 \hat{s}}{\hat{t}\hat{u}} = \frac{\left(\hat{t} + \hat{u}\right)^2 + 2m_W^2 \hat{s}}{\hat{t}\hat{u}} - 2
\tag{3.132}
$$

$$= \left(\frac{1}{\hat{t}} + \frac{1}{\hat{u}}\right) \frac{(\hat{t} + \hat{u})^2 + 2m_W^2 \hat{s}}{\hat{t} + \hat{u}} - 2 = \left(\frac{1}{\hat{t}} + \frac{1}{\hat{u}}\right) \left[(m_W^2 - \hat{s}) + \frac{2m_W^2 \hat{s}}{m_W^2 - \hat{s}}\right] - 2.$$

In the last line, overall momentum conservation has been employed, embodied by the identity $\hat{s} + \hat{t} + \hat{u} = m_W^2$. Furthermore, the terms in square brackets in the final line can be written in terms of the single dimensionless quantity

$$x = m_W^2/\hat{s}, \tag{3.133}$$

leading to

$$\left|\mathcal{M}_{u\bar{d}\to gW^+}^{(\mathrm{LO})}\right|^2 = \frac{2\pi C_F \alpha_\mathrm{s}(\mu_R)}{m_W^2} \left|\mathcal{M}_{u\bar{d}\to W^+}^{(\mathrm{LO})}\right|^2 \cdot \left[\frac{\hat{t}^2 + \hat{u}^2 + 2m_W^2 \hat{s}}{\hat{t}\hat{u}}\right] \tag{3.134}$$

$$= \frac{2\pi C_F \alpha_\mathrm{s}(\mu_R)}{x} \left|\mathcal{M}_{u\bar{d}\to W^+}^{(\mathrm{LO})}\right|^2 \cdot \left[\left(\frac{1}{\hat{t}} + \frac{1}{\hat{u}}\right)\left(-\frac{2}{1-x} + x + 1\right) - \frac{2x}{m_W^2}\right]$$

for the amplitude squared of the real-emission contribution. Isolating the divergent terms $\propto 1/\hat{t}$ or $1/\hat{u}$, the equation above can be cast into the form

$$\left|\mathcal{M}_{u\bar{d}\to gW^+}^{(\mathrm{LO})}\right|^2 = \frac{1}{x} \left|\mathcal{M}_{u\bar{d}\to W^+}^{(\mathrm{LO})}\right|^2 \cdot \left[\mathcal{D}(\hat{t}, x) + \mathcal{D}(\hat{u}, x) + \mathcal{R}(x)\right]. \tag{3.135}$$

The infrared-singular terms are expressed in terms of the function $\mathcal{D}(\hat{t}, x)$, where

$$\mathcal{D}(\hat{t}, x) = 8\pi\alpha_\mathrm{s} C_F \left[-\frac{1}{\hat{t}}\left(\frac{2}{1-x} - 1 - x\right)\right], \tag{3.136}$$

and the non-singular remainder reads

$$\mathcal{R}(x) = 8\pi\alpha_\mathrm{s} C_F \left[-\frac{2x}{m_W^2}\right]. \tag{3.137}$$

The sum of the two singular terms therefore forms the subtraction term,

$$\mathcal{S}(\Phi_\mathcal{R}) = \frac{1}{x} \left|\mathcal{M}_{u\bar{d}\to W^+}^{(\mathrm{LO})}\right|^2 \left[\mathcal{D}(\hat{t}, x) + \mathcal{D}(\hat{u}, x)\right]. \tag{3.138}$$

It is worth noting that they are identical to the **dipole subtraction** terms used in the seminal Catani–Seymour paper [353], but they have been simply derived from the original singular matrix element. The general form of the subtraction terms in **Catani–Seymour subtraction** will be discussed in the next section. This dipole term is also closely related to the standard Altarelli–Parisi splitting function for a quark into a quark and a gluon. There is one dipole for each of the $\hat{t}$ and $\hat{u}$ singularities, but due to the simplicity of the process and its symmetry with respect to the initial states their forms are exactly identical.

To compute the corresponding **integrated subtraction term**, $\mathcal{I}^{(\mathcal{S})}$ in Eq. (3.130), first the phase space must be appropriately factorized to allow an analytic integration over the **one-particle phase space** of the emitted gluon. The phase space is given by

$$
\mathrm{d}\Phi_{Wg} = \frac{\mathrm{d}^D p_W}{(2\pi)^D}(2\pi)\delta(p_W^2 - m_W^2)\frac{\mathrm{d}^D p_g}{(2\pi)^D}(2\pi)\delta(p_g^2)(2\pi)^D\delta^D(p_a + p_b - p_W - p_g)
$$

$$
= (2\pi)^{2-D}\frac{\mathrm{d}^{D-1}p_g}{2E}\,\delta\left((p_a + p_b - p_g)^2 - m_W^2\right) \tag{3.139}
$$

where, in the last line, E represents the energy of the gluon. It is straightforward to evaluate this in the c.m. frame. The result reads

$$
\mathrm{d}\Phi_{Wg} = \frac{(2\pi)^{2\varepsilon-2}}{2\sqrt{\hat{s}}}\left(\frac{\hat{t}\hat{u}}{\hat{s}}\right)^{-\varepsilon}\mathrm{d}\left(-\frac{\hat{t} + \hat{u}}{2\sqrt{\hat{s}}}\right)\mathrm{d}\hat{t}\,\mathrm{d}\Omega^{1-2\varepsilon}\delta\left(\hat{s} + \hat{t} + \hat{u} - m_W^2\right), \tag{3.140}
$$

where $\mathrm{d}\Omega^{D-3}$ is the solid angle element in $D - 3$ dimensions,

$$
\int \mathrm{d}\Omega^{1-2\varepsilon} = \frac{2\pi}{\pi^\varepsilon\,\Gamma(1-\varepsilon)}. \tag{3.141}
$$

It is convenient to use the two dimensionless variables defined by

$$
x = \frac{\hat{s} + \hat{t} + \hat{u}}{\hat{s}} = \frac{m_W^2}{\hat{s}}, \qquad v = -\frac{\hat{t}}{\hat{s}}, \tag{3.142}
$$

where the former has been simplified using the Mandelstam relation $\hat{s} + \hat{t} + \hat{u} = m_W^2$ from Eq. (2.100). The phase-space integral can be expressed through them as

$$
\mathrm{d}\Phi_{Wg} = \frac{\hat{s}^{1-\varepsilon}}{16\pi^2}\frac{\mathrm{d}\Omega^{1-2\varepsilon}}{(2\pi)^{1-2\varepsilon}}\,\mathrm{d}x\,\mathrm{d}v\,v^{-\varepsilon}(1 - x - v)^{-\varepsilon}\left[2\pi\,\delta\left(x\hat{s} - m_W^2\right)\right]. \tag{3.143}
$$

The factors in square brackets can now be recognized as the final-state phase space for W production, at the reduced c.o.m. energy squared, $x\hat{s}$. The phase space is thus an integral over x of the convolution of this reduced phase space with the **dipole phase space** defined by

$$
\mathrm{d}\phi(x, v, \hat{s}) = \frac{\hat{s}^{1-\varepsilon}}{16\pi^2}\frac{\mathrm{d}\Omega^{1-2\varepsilon}}{(2\pi)^{1-2\varepsilon}}\,\mathrm{d}v\,v^{-\varepsilon}(1 - x - v)^{-\varepsilon}, \tag{3.144}
$$

where the integral over v ranges from 0 to $(1 - x)$. In this case, since the only partonic content in the leading order process is in the initial state, this corresponds to an initial emitter and initial spectator, anticipating the language of the Catani–Seymour approach that is used in the next section.

   The dipole phase space derived above can be used to integrate the single dipole term of Eq. (3.136). Restoring the correct overall dimensions with a factor of $\mu^{2\varepsilon}$,

$$
\mu^{2\varepsilon}\int \mathcal{D}(\hat{t}, x)\,\mathrm{d}\phi(x, v, \hat{s})
$$

$$= \frac{\alpha_s C_F}{2\pi} \left(\frac{\mu^2}{m_W^2} \right)^\varepsilon c_\Gamma x^\varepsilon \int_0^{1-x} dv \, v^{-\varepsilon} (1 - x - v)^{-\varepsilon} \frac{1}{v} \left[\frac{2}{1-x} - 1 - x \right]. \quad (3.145)$$

Here, the convenient constant c_Γ of Eq. (3.62) has been used. The v integration can be evaluated by substituting the variable $v \to (1 - x)y$ and then using the result in Eq. (A.6):

$$\int_0^{1-x} dv \, v^{-\varepsilon} (1 - x - v)^{-\varepsilon} \frac{1}{v} = -\frac{1}{\varepsilon} \frac{\Gamma^2(1 - \varepsilon)}{\Gamma(1 - 2\varepsilon)} (1 - x)^{-2\varepsilon}. \quad (3.146)$$

The result is a function of x that is well defined as $\varepsilon \to 0$, except at the point $x = 1$ where a non-zero value of ε regulates the divergence. A convenient way of extracting this singularity is to employ the same + distributions that were already used for the splitting functions, *cf.* Appendix A.1.2. Applied to the case at hand, this allows the following replacement to be made,

$$\frac{x^\varepsilon (1 - x)^{-2\varepsilon}}{1 - x} \to \left[\frac{x^\varepsilon (1 - x)^{-2\varepsilon}}{1 - x} \right]_+ - \frac{1}{2\varepsilon} \left(1 + \frac{\varepsilon^2 \pi^2}{3} \right) \delta(1 - x) \quad (3.147)$$

$$= \left[\frac{1 + \varepsilon \log x - 2\varepsilon \log(1 - x)}{1 - x} \right]_+ - \frac{1}{2\varepsilon} \left(1 + \frac{\varepsilon^2 \pi^2}{3} \right) \delta(1 - x)$$

where terms of order ε^2 or higher have been dropped. The remaining terms are non-singular in the limit that $x \to 1$ and so do not require the introduction of further +-distributions. Thus the integrand becomes

$$-\frac{1}{\varepsilon} x^\varepsilon (1 - x)^{-2\varepsilon} \left(\frac{2}{1 - x} - 1 - x \right)$$

$$= -\frac{1}{\varepsilon} \left\{ \frac{\mathcal{P}_{qq}^{(1)} x}{C_F} - \frac{3}{2\varepsilon} \delta(1 - x) - \varepsilon \left[\frac{2}{1 - x} \log \frac{(1 - x)^2}{x} \right]_+ \right.$$

$$\left. - \frac{1}{\varepsilon} \left(1 + \frac{\varepsilon^2 \pi^2}{3} \right) \delta(1 - x) + \varepsilon(1 + x) \log \frac{(1 - x)^2}{x} \right\}$$

$$= \left(\frac{1}{\varepsilon^2} + \frac{3}{2\varepsilon} + \frac{\pi^2}{3} \right) \delta(1 - x) - \frac{1}{\varepsilon} \frac{\mathcal{P}_{qq}^{(1)}(x)}{C_F}$$

$$+ \left[\frac{2}{1 - x} \log \frac{(1 - x)^2}{x} \right]_+ - (1 + x) \log \frac{(1 - x)^2}{x}. \quad (3.148)$$

In this equation the **regularized splitting function**, to first order in α_s, $\mathcal{P}_{qq}^{(1)}(x)$, appears. It has been given in Eq. (2.33).

Using the relation between gamma functions given in Eq. (A.5) ultimately yields

$$\mu^{2\varepsilon} \int \mathcal{D}(\hat{t}, x) \, \mathrm{d}\phi(x, v, \hat{s}) = \frac{\alpha_{\mathrm{s}} C_F}{2\pi} \left(\frac{\mu^2}{m_W^2}\right)^{\varepsilon} c_{\Gamma} \left[\left(\frac{1}{\varepsilon^2} + \frac{3}{2\varepsilon} + \frac{\pi^2}{6}\right) \delta(1 - x)\right.$$

$$+ \left[\frac{2}{1 - x} \log \frac{(1 - x)^2}{x}\right]_+ - (1 + x) \log \frac{(1 - x)^2}{x} \bigg]$$

$$- \frac{\alpha_{\mathrm{s}}}{2\pi} \left(\frac{\mu^2}{m_W^2}\right)^{\varepsilon} c_{\Gamma} \frac{1}{\varepsilon} \mathcal{P}_{qq}^{(1)}(x).$$

$$(3.149)$$

There are two such contributions that take identical forms originating from the $\mathcal{D}(\hat{t}, x)$ and $\mathcal{D}(\hat{u}, x)$ terms in Eq. (3.135). The sum of the two is the form of the **integrated dipole term** represented by $\mathcal{I}^{(S)}$ in Eq. (3.130). Note that, since the starting expression in Eq. (3.136) was defined in four dimensions, this calculation has implicitly been performed in the dimensional reduction scheme. Working in conventional dimensional regularization would have introduced an extra term $-\varepsilon(1 - x)$ in that equation that would eventually manifest itself as an additional $+(1 - x)$ inside the square brackets in Eq. (3.149)

With the explicit expression for this contribution at hand it is possible to see exactly how the factorization of singularities into the parton distribution functions occurs. The relevant singular contribution in Eq. (3.149) is given by

$$\frac{\alpha_{\mathrm{s}}}{2\pi} \left(\frac{\mu_F^2}{m_W^2}\right)^{\varepsilon} c_{\Gamma} \left\{-\frac{1}{\varepsilon} \mathcal{P}_{qq}^{(1)}(x)\right\}, \qquad (3.150)$$

since the remaining singularities proportional to $\delta(1 - x)$ cancel with those obtained from the virtual diagrams. The NLO quark PDF $f_{q/h}(y)$ can then be defined in terms of the bare PDF $f_{q/h}^{(B)}(y)$, which does not exhibit any scaling violations, as follows

$$f_{q/h}(y) = f_{q/h}^{(B)}(y) + \frac{\alpha_{\mathrm{s}}}{2\pi} \left(-\frac{c_{\Gamma}}{\varepsilon}\right) \int_y^1 \frac{\mathrm{d}x}{x} \mathcal{P}_{qq}^{(1)}(x) f_{q/h}^{(B)}\left(\frac{y}{x}\right), \qquad (3.151)$$

in order to absorb this singularity. The factor of $1/x$ present in this equation can be traced back to Eq. (3.135) while the dependence on $f_{q/h}^{(B)}(y/x)$ is a consequence of the fact that $x\hat{s} = m_W^2$. In addition to the pure $1/\varepsilon$ singularity, this definition also includes constant terms that are obtained by expanding out the universal factor c_{Γ}.

Note that it is only the pole term in Eq. (3.151) that must be absorbed in order to arrive at a finite result while the constant terms are a matter of choice. The choice of additional, finite constant terms that are absorbed defines the factorization scheme that is used for the PDFs. The definition in Eq. (3.151) corresponds to the **modified minimal subtraction**, or the $\overline{\mathrm{MS}}$-scheme, which is the preferred definition for all modern PDF sets. This scheme dependence must also account for the particular regularization scheme used in the calculation. The redefinition in Eq. (3.151) corresponds to conventional dimensional regularization. In the dimensional reduction scheme one must make the replacement,

$$\frac{1}{\varepsilon}\mathcal{P}_{qq}^{(1)}(x) \rightarrow \frac{1}{\varepsilon}\mathcal{P}_{qq}^{(1)}(x) + C_F(1-x) - \frac{C_F}{2}\delta(1-x). \tag{3.152}$$

Hence, after this factorization has been accounted for, the contribution of a single dipole term in Eq. (3.149) can be written as,

$$\mu^{2\varepsilon}\int \mathcal{D}(\hat{t},x)\,\mathrm{d}\phi(x,v,\hat{s}) = \frac{\alpha_s C_F}{2\pi}\left(\frac{\mu^2}{m_W^2}\right)^{\varepsilon} c_\Gamma\left[\left(\frac{1}{\varepsilon^2} + \frac{3}{2\varepsilon} + \frac{\pi^2}{6} - \frac{1-a^{\mathrm{CDR}}}{2}\right)\delta(1-x)\right.$$
$$+1-x+\left[\frac{2}{1-x}\log\frac{(1-x)^2}{x}\right]_+ - (1+x)\log\frac{(1-x)^2}{x}\right]$$
$$-\frac{\alpha_s}{2\pi}\log\left(\frac{\mu^2}{m_W^2}\right)\mathcal{P}_{qq}^{(1)}(x). \tag{3.153}$$

In this equation the regularization scheme dependence has been captured by the constant a^{CDR}, previously introduced in Section 3.3.1. Notice that, as must be the case, once both dipole contributions are accounted for the dependence on a^{CDR} in this equation exactly cancels that from the virtual corrections, given in Eq. (3.64).

In fact this result is also very close to the one for the full real corrections previously quoted in Eq. (2.118). The two differ by a single term which is given by the integral of the non-singular remainder in Eq. (3.137),

$$\mu^{2\varepsilon}\int \mathcal{R}(x)\,\mathrm{d}\phi(x,v,\hat{s}) = \frac{\alpha_s C_F}{2\pi}\left(\frac{\mu^2}{m_W^2}\right)^{\varepsilon} c_\Gamma\left[-2(1-x)\right]. \tag{3.154}$$

3.3.3   Catani–Seymour dipole subtraction

3.3.3.1   General idea

In general, subtraction methods like the one above, based on the form of the actual real-emission matrix elements, are not very useful when considering more complicated topologies or even the automation of the subtraction methods in form of computer code. In such cases, process-independent methods such as the **FKS method** [539, 542] certainly are advantageous.

Maybe the most frequently used of such algorithms is known as **Catani–Seymour dipole subtraction** [344, 353]. The idea underlying this method is to treat the emission of additional particles in phase space in such a way that the soft and/or collinear limits of the individual emission factorizes from underlying Born-type configurations. This is possible, in a process-independent way, by realizing that in the soft limit the emission cross-section of an additional particle — typically a gluon or photon — can be written as the product of an eikonal, *cf.* Eq. (2.14), and a corresponding Born level cross-section. Similarly, in the collinear limit, emissions can be written as the product of a Born-level cross-section with a splitting kernel of the form introduced in Eq. (2.33). In Catani–Seymour subtraction these two limits, soft and collinear, of the emission are analysed by introducing a spectator parton k. This spectator parton also allows the construction of phase-space mappings $i + j + k \rightarrow \{ij\} + k$ for the combination of partons i and j into an emitter $\{ij\}$ with momentum $\tilde{p}_{ij}$, while parton k accounts for the recoil and changes its momentum from p_k to $\tilde{p}_k$. This not only allows all particles to be kept on their mass-shell all the time, $p_k^2 = \tilde{p}_k^2 = m_k^2$ and $\tilde{p}_{ij}^2 = m_{ij}^2$, but it also directly leads to a factorization not only of the matrix elements, but of phase space. The latter fact is important to allow a calculation of Born-level matrix elements with one less particle. The catch in this construction now is that these extra emission bits, $\{ij\} + k \to i + j + k$ can be integrated analytically in D dimensions, thereby allowing them to be subtracted in differential form from the real-emission matrix element and added back to the virtual contribution in integrated form.

In order to put both types of contributions together, it is sensible to decompose the eikonal cross-section term $\mathrm{W}(p_1, p_2; k)$ with its symmetry between both emitting partons, with light-like momenta p_1 and p_2, into two individual terms with a definite emitter and spectator assignment. Including colour factors, they schematically read

$$
\begin{aligned}
\mathrm{W}(p_1, p_2; k) &= -\frac{\mathbf{T}_1 \cdot \mathbf{T}_2}{2} \left(\frac{p_1^\mu}{p_1 k} - \frac{p_2^\mu}{p_2 k} \right)^2 = \mathbf{T}_1 \cdot \mathbf{T}_2 \, \frac{p_1 p_2}{(p_1 k)(p_2 k)} \\
&= \mathbf{T}_1 \cdot \mathbf{T}_2 \left[\frac{p_1 p_2}{(p_1 k)(p_1 k + p_2 k)} + \frac{p_1 p_2}{(p_2 k)(p_1 k + p_2 k)} \right] \\
&= \tilde{\mathcal{D}}_{1k;2} + \tilde{\mathcal{D}}_{2k;1},
\end{aligned}
\tag{3.155}
$$

where the two dipole terms $\tilde{\mathcal{D}}_{ik;j}$ have been introduced, with i denoting the **emitter** or **splitter**, k being the emitted particle, and j denoting the **spectator**. Note that the colour generators $\mathbf{T}_{i,j}$ related to the particles i and j act as matrices on the full colour structure of the Born-level cross-section the eikonal is being attached to in order to facilitate the extra emission.

Analysing these terms shows that each of them diverges in both the soft limit of the energy of the emitted parton approaching zero, $\omega_k \to 0$ and the collinear limit of its momentum becoming parallel to the momentum of the splitter, $k^\mu \parallel p_i^\mu$. At the same time, the individual terms do not diverge for the emitted momentum k going parallel to the spectator momentum. This helps to disentangle soft and collinear divergences. It also helps to analyse singluarities related to the eikonal — essentially the soft and soft-collinear ones — in colour space, including leading and sub-leading colour contributions and to add the hard collinear divergences, which are encoded in splitting functions and are a leading colour effect. This means that the eikonal-based dipoles above have been generalized in such a way that they also contain the collinear bits encoded in suitable splitting kernels. Full dipole subtraction terms $\mathcal{D}$ therefore are introduced that emerge from the combination of Born-level matrix elements squared, including all symmetry and PDF factors, with terms similar to the $\tilde{\mathcal{D}}$ from Eq. (3.155). The catch here is that this factorization of the full subtraction matrix element into Born-level parts and individual emission terms also includes a factorization of the phase space such that for these terms

$$
\Phi_{\mathcal{R}} = \Phi_{\mathcal{B}} \otimes \Phi_1.
\tag{3.156}
$$

Considering in the following a $2 \rightarrow n$–process at Born-level with momenta $p_a p_b \rightarrow p_1 p_2 \ldots p_n$, the subtraction procedure uses two terms: a differential term, which can be written in the form of a sum of differential dipole cross-sections with an additionally emitted parton, $\mathcal{D}(p_a, p_b; p_1, p_2, \ldots, p_{n+1})$ to be subtracted from the real-emission contribution,

$$
\begin{aligned}
\mathcal{S}(\Phi_\mathcal{R}) &= \sum_{\text{dipoles}} \mathcal{D}(p_a, p_b; p_1, p_2, \ldots, p_{n+1}) \\
&= \sum_{ij,k} \mathcal{B}_{ij;k}(\Phi_\mathcal{B}) \otimes \tilde{\mathcal{D}}_{ij;k}(\Phi_1) \longrightarrow \mathcal{B}(\Phi_\mathcal{B}) \otimes \mathbf{D}(\Phi_1).
\end{aligned}
\tag{3.157}
$$

Here, to simplify the notation, the sum over dipole terms $\tilde{\mathcal{D}}_{ij;k}(\Phi_1)$ has been replaced with the product of a dipole operator, $\mathbf{D}(\Phi_1)$, with the summation implicit.

This structure is reflected by a sum of the corresponding integrated terms, to be added back to the virtual contribution,

$$
\begin{aligned}
\mathcal{I}^{(\mathcal{S})}(\Phi_\mathcal{B}, \varepsilon) &= \sum_{\text{dipoles}} \mathcal{I}^{(\mathcal{D})}(p_a, p_b; p_1, p_2, \ldots, p_{i-1}, p_{i+1}, \ldots, p_{n+1}) \\
&= \sum_{ij,k} \mathcal{B}_{ij;k}(\Phi_\mathcal{B}) \otimes \mathcal{I}^{(\mathcal{D})}_{ij;k}(\Phi_\mathcal{B}) \longrightarrow \mathcal{B}(\Phi_\mathcal{B}) \otimes \mathbf{D}(\Phi_\mathcal{B}).
\end{aligned}
\tag{3.158}
$$

Here, one final-state particle has been integrated out. This stresses that there are only n particles in the final state of this integrated terms, which therefore have Born-level kinematics.

The following discussion will proceed closely along the lines of the seminal paper by Catani and Seymour (CS) [353]; a closer relation with their paper will be worked out in Appendix C.2, where individual equations in CS will be directly linked to the expressions below.

3.3.3.2 Catani–Seymour dipole subtraction for final-state partons only

For example, for the case of a dipole with both splitter ij and spectator k being final-state particles, these differential dipole subtraction terms, denoted as $\mathcal{D}_{ij;k}$, schematically read

$$
\begin{aligned}
&\mathcal{D}_{ij;k}(p_a, p_b; p_1, p_2, \ldots, p_n) \\
&= \mathcal{B}(p_a, p_b; p_1, p_2, \ldots, \tilde{p}_{ij}, \ldots, \tilde{p}_k, \ldots, p_n) \otimes \tilde{\mathcal{D}}_{ij;k}(p_i, p_j, p_k), \quad (3.159)
\end{aligned}
$$

where the $\tilde{\mathcal{D}}_{ij;k}$ are the actual dipoles.

The momenta $\tilde{p}_{ij}$ and $\tilde{p}_k$ in the Born term emerge from the combination of the momenta p_i (the splitter), p_j (the emitted parton), and p_k (the spectator). In the case considered here, massless final-state particles,

$$\tilde{p}_{ij} = p_i + p_j - \frac{y_{ij,k}}{1 - y_{ij,k}} p_k$$

$$\tilde{p}_k = \frac{1}{1 - y_{ij,k}} p_k,$$

(3.160)

so that the spectator parton keeps its direction but its momentum is stretched by a factor $1/(1 - y_{ij,k})$, where the dimensionless quantity $y_{ij,k}$ is given by

$$y_{ij,k} = \frac{p_i p_j}{p_i p_j + p_j p_k + p_k p_i}.$$

(3.161)

The splitting functions depend on both $y_{ij,k}$ and the splitting parameter $\tilde{z}_i$,

$$\tilde{z}_i = \frac{p_i p_k}{(p_i + p_j) p_k} = \frac{p_i \tilde{p}_k}{\tilde{p}_{ij} \tilde{p}_k} \quad \text{and} \quad \tilde{z}_j = 1 - \tilde{z}_i.$$

(3.162)

With these parameters, the typical form of the divergences can be written as

$$\frac{p_i p_j + p_j p_k + p_k p_i}{(p_i + p_j) p_k} = \frac{1}{1 - \tilde{z}_i (1 - y_{ij,k})}.$$

(3.163)

In the collinear limit, $p_i \parallel p_j$, or, more formally,

$$p_i^\mu = z p^\mu + k_\perp^\mu - \frac{k_\perp^2}{z} \frac{n^\mu}{2pn} \quad \text{and} \quad p_j^\mu = (1 - z) p^\mu - k_\perp^\mu - \frac{k_\perp^2}{1 - z} \frac{n^\mu}{2pn},$$

(3.164)

where p^μ denotes the light-like collinear direction, and the transverse momentum $k_\perp$ is perpendicular to it and to the auxiliary light-like vector n^μ. Here, z is a splitting parameter, which can readily identified with $\tilde{z}_i$ from above. Then, the collinear limit is given by

$$2 p_i p_j = -\frac{k_\perp^2}{z(1 - z)} \quad \text{with} \quad k_\perp \to 0.$$

(3.165)

In this limit,

$$y_{ij,k} \to -\frac{k_\perp^2}{2\tilde{z}_i (1 - \tilde{z}_i) \tilde{p}_{ij} \tilde{p}_k}$$

(3.166)

and, of course, $\tilde{p}_k \to p_k$ and $\tilde{p}_{ij} \to p_i + p_j = p$. Similarly, in the soft limit, $y_{ij,k} \to 0$, $\tilde{z}_i \to 1$ and, again, $\tilde{p}_k \to p_k$, while $\tilde{p}_{ij} \to p_i$.

For all three particles, splitter, emitted parton, and spectator in the final state, the dipole reads

$$\tilde{\mathcal{D}}_{ij,k} = -\frac{1}{2\tilde{p}_{ij} \tilde{p}_k} \frac{\mathbf{T}_{ij} \cdot \mathbf{T}_k}{\mathbf{T}_{ij}^2} \langle s | V_{ij;k} | s' \rangle.$$

(3.167)

The colour factors T must be inserted at the right place into the Born matrix element and in addition the spin or polarization states labelled by $|s\rangle$ must be accounted for in the product of Born-level matrix element and dipole. The actual information concerning the flavour of the partons and their impact on the kinematics is encoded in the kernel $V_{ij,k}$, which for the case of a quark emitting a gluon is given by

$$\langle s|V_{q_i g_j;k}|s'\rangle = 8\pi\mu_R^{2\varepsilon}C_F\alpha_s(\mu_R)\left[\frac{2}{1-\tilde{z}_i(1-y_{ij,k})} - (1+\tilde{z}_i) - \varepsilon(1-\tilde{z}_i)\right]\delta_{ss'}.$$

$$(3.168)$$

The corresponding integrated dipole term $\mathcal{I}^{(\mathcal{D})}(\varepsilon)$ is given by the integral of the dipole over the emission phase space of parton j, namely

$$\mathcal{I}_{ij;k}^{(\mathcal{D})}(\varepsilon) = -\frac{\alpha_s(\mu_R^2)}{2\pi\Gamma(1-\varepsilon)}\left(\frac{4\pi\mu_R^2}{2\tilde{p}_{ij}\tilde{p}_k}\right)^\varepsilon \frac{\mathbf{T}_{ij}\cdot\mathbf{T}_k}{\mathbf{T}_{ij}^2}\mathcal{V}_{ij}(\varepsilon), \qquad (3.169)$$

where the $\mathcal{V}_{ij}(\varepsilon)$ can be constructed from the following expressions:

$$\mathcal{V}_{ij}(\varepsilon) = \mathbf{T}_i^2\left(\frac{1}{\varepsilon^2} - \frac{\pi^2}{3}\right) + \gamma_i\left(\frac{1}{\varepsilon}+1\right) + K_i$$

$$K_{q,g} = \begin{cases} C_F\left(\dfrac{7}{2} - \dfrac{\pi^2}{6}\right) & \text{for } i = q \\ C_A\left(\dfrac{67}{18} - \dfrac{\pi^2}{6}\right) - \dfrac{10}{9}T_R n_f & \text{for } i = g \end{cases} \qquad (3.170)$$

$$\gamma_q = \begin{cases} \dfrac{3}{2}C_F & \text{for } i = q \\ \dfrac{11}{6}C_A - \dfrac{2}{3}n_f T_R & \text{for } i = g. \end{cases}$$

The anomalous dimensions $\gamma_{q,g}^{(1)}$ have first been encountered as the terms proportional to $\delta(1-z)$ in the construction of the DGLAP splitting kernels, Eq. (2.33). The term K_g actually will return in another context, namely in resummation, where it will be denoted as K and relate to a generic, flavour-independent higher-order correction to the emission of a soft gluon, *cf.* Eq. (5.65) in Section 5.2.1. These terms effectively also parameterize the contribution of collinear logarithms in $Q_\perp$ resummation, see also Eq. (5.66).

As before, for the differential dipole terms, an integrated dipole operator $\mathbf{I}(\Phi_\mathcal{B};\varepsilon)$ is being constructed such that, again,

$$\mathcal{I}^{(S)}(\Phi_\mathcal{B}) = \sum_{ij;k}\mathcal{B}_{ij;k}(\Phi_\mathcal{B}) \otimes \mathcal{I}_{ij;k}^{(\mathcal{D})}(\varepsilon) \longrightarrow \mathcal{B}(\Phi_\mathcal{B}) \otimes \mathbf{I}(\Phi_\mathcal{B};\varepsilon). \qquad (3.171)$$

3.3.3.3   Example: Catani–Seymour dipole subtraction for $e^-e^+ \to q\bar{q}$

To see how this works in practice, consider the case of $e^-e^+ \to q\bar{q}$ at next-to-leading order, closely following the example in Appendix D of [353]. To keep things simple, the exchange of a Z boson will be ignored and only the case of a virtual photon in the s channel will be considered. In this case, the dimensionless parameters $y_{ij,k}$ reduce to the scaled invariant mass of the pair,

$$y_{ij,k} = \frac{2p_i p_j}{Q^2}, \tag{3.172}$$

where Q is the centre-of-mass energy of the hadronic system, $\hat{s} = Q^2$. With e_q the partial charge of the quark, and the Born phase space $\Phi_{\mathcal{B}}$ being a function of the quark and anti-quark momenta p_1 and p_2, $\hat{s} = (p_1 + p_2)^2$, the leading-order cross-section can be written as

$$\sigma^{(\mathrm{LO})} = \int \mathrm{d}\Phi_{\mathcal{B}}\, \mathcal{B}(\Phi_{\mathcal{B}}) = \frac{4\pi\alpha^2 e_q^2}{3Q^2}. \tag{3.173}$$

Note that in comparison to the original work in [353], the **jet function** $F_J^{(n)}(\{p_i\})$ has been ignored by setting it to unity; this corresponds to a calculation of the total cross-section. Other choices would translate into cuts on the n–particle final state: two for contributions with Born-level kinematics and three for real correction terms. This implies that in the limit of soft or collinear emissions $F_J^{(n+1)}$ must reduce to $F_J^{(n)}$,

$$F_J^{(n+1)} \xrightarrow{\text{soft, collinear}} F_J^{(n)}. \tag{3.174}$$

This is a fairly trivial manifestation of the requirement of **infrared safety** for cuts on kinematic configurations in physical processes.

The three-parton matrix element for $q\bar{q}g$ production of course corresponds to the real correction at NLO to the process and it is given by

$$\mathcal{R}(p_1, p_2, p_3) = \frac{8\pi C_F \alpha_{\mathrm{s}}(\mu_R)}{Q^2} \frac{x_1^2 + x_2^2}{(1 - x_1)(1 - x_2)} \mathcal{B}(\Phi_{\mathcal{B}}) \tag{3.175}$$

with $x_i = 2p_i Q/Q^2$. The respective real-emission phase-space element, conventionally expressed through the fractions x_i reads

$$\mathrm{d}\Phi_{\mathcal{R}} = \mathrm{d}\Phi_{\mathcal{B}} \frac{Q^2}{16\pi^2} \mathrm{d}x_1 \mathrm{d}x_2\, \Theta(1 - x_1)\Theta(1 - x_x)\Theta(x_1 + x_2 - 2). \tag{3.176}$$

The dipole subtraction terms are easily constructed. The splitting function is of course the qg splitting function from Eq. (3.168). Inserting it into the Born-level matrix element is fairly straightforward due to the Kronecker δ in the spins, and the only somewhat tricky bit is the colour factor coming with it. Taking a closer look, the relevant term actually reads

$$-\frac{1}{2p_i p_j} \otimes \frac{\mathbf{T}_{ij} \cdot \mathbf{T}_k}{\mathbf{T}_{ij}^2} V_{ij,k}, \tag{3.177}$$

where the $\mathbf{T}_{ij}$ and $\mathbf{T}_k$ are the colour matrices related to the splitter and spectator that are inserted into the matrix element. In the dipole term here, $\mathbf{T}_{ij} = \mathbf{T}_q = \mathbf{T}_{13}$ and $\mathbf{T}_k = \mathbf{T}_{\bar{q}} = \mathbf{T}_2$. By virtue of the fact that the overall amplitude must be a colour singlet, for each splitter ij the sum over all spectators k of the term $\mathbf{T}_k$ will result in $-\mathbf{T}_{ij}$. In the case here, this implies that

$$-\frac{1}{2p_i p_j} \otimes \frac{\mathbf{T}_{ij} \cdot \mathbf{T}_k}{\mathbf{T}_{ij}^2} = \frac{1}{2p_i p_j}. \tag{3.178}$$

The dipole term relating to the gluon p_3 being emitted from quark p_1 therefore is given by

$$\mathcal{D}_{13;2}(p_1, p_2, p_3)^{(\varepsilon=0)} = \mathcal{B}(\tilde{p}_{13}, \tilde{p}_2) \otimes \left[-\frac{1}{2p_1 p_3} V_{q_1 g_3, \bar{q}_2}^{(\varepsilon=0)} \frac{\mathbf{T}_{13} \cdot \mathbf{T}_2}{\mathbf{T}_{13}^2} \right]$$

$$= \mathcal{B}(\tilde{p}_{13}, \tilde{p}_2) \otimes \frac{8\pi C_F \alpha_s(\mu_R)}{2p_1 p_3} \left[\frac{2}{1 - \tilde{z}_1(1 - y_{13,2})} - (1 + \tilde{z}_1) \right]. \tag{3.179}$$

The Born-level momenta are connected to the real-emission phase space through

$$\tilde{p}_2^\mu = \frac{1}{x_2} p_2^\mu \quad \text{and} \quad \tilde{p}_{13}^\mu = Q^\mu - \tilde{p}_2^\mu = Q^\mu - \frac{1}{x_2} p_2^\mu, \tag{3.180}$$

where the quantities $y_{13,2}$ and $\tilde{z}_1$ in the general expression for the mapping in Eq. (3.160) can be expressed through the x_i in the specific case here as

$$y_{13,2} = 1 - x_2 \quad \text{and} \quad \tilde{z}_1 = \frac{1 - x_3}{x_2}. \tag{3.181}$$

Inserting this and using that $x_i = 2 - x_j - x_k$,

$$\mathcal{D}_{13;2}(p_1, p_2, p_3)^{(\varepsilon=0)}$$

$$= \mathcal{B}(\tilde{p}_{13}, \tilde{p}_2) \cdot \frac{8\pi C_F \alpha_s(\mu_R)}{(1 - x_2)Q^2} \left[\frac{2}{2 - x_1 - x_2} - 1 + \frac{1 - x_1 - x_2}{x_2} \right]$$

$$= \mathcal{B}(\tilde{p}_{13}, \tilde{p}_2) \cdot \frac{8\pi C_F \alpha_s(\mu_R)}{Q^2} \left[\frac{1}{1 - x_2} \left(\frac{2}{2 - x_1 - x_2} - 1 - x_1 \right) + \frac{1 - x_1}{x_2} \right], \tag{3.182}$$

and a similar term for $\mathcal{D}_{23,1}$ with the replacement $1 \leftrightarrow 2$. Together, these two terms constitute the subtraction term,

$$\mathcal{S}(\Phi_\mathcal{R}) = \mathcal{S}(p_1, p_2, p_3) = \mathcal{D}_{13;2}(p_1, p_2, p_3)^{(\varepsilon=0)} + \mathcal{D}_{23;1}(p_1, p_2, p_3)^{(\varepsilon=0)}. \tag{3.183}$$

The subtracted real-emission term therefore reads

$$
\begin{aligned}
\mathrm{d}\sigma^{(R-S)} &= \mathrm{d}\Phi_{\mathcal{R}}\left[\mathcal{R}(\Phi_{\mathcal{R}}) - \mathcal{S}(\Phi_{\mathcal{R}})\right] \\
&= \frac{C_F\alpha_\mathrm{s}(\mu_R)}{2\pi}\int\limits_0^1 \mathrm{d}x_1\mathrm{d}x_2\left\{\mathrm{d}\Phi_{\mathcal{B}}\,\mathcal{B}\,\frac{x_1^2 + x_2^2}{(1-x_1)(1-x_2)}\right. \\
&\quad - \mathrm{d}\Phi_{\mathcal{B}}\,\mathcal{B}(\tilde{p}_{13},\tilde{p}_2)\cdot\left[\frac{1}{1-x_2}\left(\frac{2}{2-x_1-x_2}-1-x_1\right)+\frac{1-x_1}{x_2}\right] \\
&\quad \left. -\mathrm{d}\Phi_{\mathcal{B}}\,\mathcal{B}(\tilde{p}_{23},\tilde{p}_1)\cdot\left[\frac{1}{1-x_1}\left(\frac{2}{2-x_1-x_2}-1-x_2\right)+\frac{1-x_2}{x_1}\right]\right\}.
\end{aligned}
$$
(3.184)

Casting the three-parton matrix element into a form similar to the subtraction terms

$$
\frac{x_1^2 + x_2^2}{(1-x_1)(1-x_2)} = \frac{1}{1-x_2}\left(\frac{2}{2-x_1-x_2}-1-x_1\right) + \{x_1 \leftrightarrow x_2\}
$$
(3.185)

makes the cancellation of divergences explicit. The subtracted real-emission contribution therefore is given by

$$
\mathrm{d}\sigma^{(R-S)} = -\frac{C_F\alpha_\mathrm{s}(\mu_R)}{4\pi}\mathrm{d}\Phi_{\mathcal{B}}\,\mathcal{B}.
$$
(3.186)

The subtracted one-loop correction consists of the genuine virtual contribution and the integrated subtraction term, in this case

$$
\begin{aligned}
\mathrm{d}\sigma^{(V+I)} &= \mathrm{d}\Phi_{\mathcal{B}}\left[\mathcal{V}(\Phi_{\mathcal{B}}) + \mathcal{I}^{(S)}(\Phi_{\mathcal{B}};\varepsilon)\right] \\
&= \mathrm{d}\Phi_{\mathcal{B}}\left[\mathcal{V}(\Phi_{\mathcal{B}}) + \mathcal{B}(\Phi_{\mathcal{B}}) \otimes \mathbf{I}(\Phi_{\mathcal{B}};\varepsilon)\right].
\end{aligned}
$$
(3.187)

The individual integrated dipole terms $\mathcal{I}_{qg;\bar{q}}^{(D)}$ and its barred counterpart, which constitute $\mathbf{I}(\Phi_{\mathcal{B}};\varepsilon)$, are given by

$$
\begin{aligned}
\mathcal{I}_{qg;\bar{q}}^{(D)}(\Phi_{\mathcal{B}},\varepsilon) &= -\frac{\mathbf{T}_q\cdot\mathbf{T}_{\bar{q}}}{\mathbf{T}_q^2}\frac{\alpha_\mathrm{s}(\mu_R)}{2\pi\Gamma(1-\varepsilon)}\left(\frac{4\pi\mu_R^2}{\hat{s}}\right)^\varepsilon \mathcal{V}_{qg}(\varepsilon) \\
&= \frac{C_F\,\alpha_\mathrm{s}(\mu_R)}{2\pi\,\Gamma(1-\varepsilon)}\left(\frac{4\pi\mu_R^2}{\hat{s}}\right)^\varepsilon\left[\frac{1}{\varepsilon^2}+\frac{3}{2\varepsilon}+5-\frac{\pi^2}{2}\right].
\end{aligned}
$$
(3.188)

Again the fact has been used that the overall amplitude must be a colour singlet, implying that, as before, the colour matrix $\mathbf{T}_{\bar{q}} = -\mathbf{T}_q$ and therefore $\mathbf{T}_q\cdot\mathbf{T}_{\bar{q}} = -C_F$. Together these yield the overall result for the subtracted virtual contribution to the cross-section, namely

$$d\sigma^{(V+I)} = d\Phi_\mathcal{B}\,\mathcal{B}(\Phi_\mathcal{B}) \left[\frac{C_F \alpha_s(\mu_R)}{2\pi\Gamma(1-\varepsilon)} \left(\frac{4\pi\mu_R^2}{Q^2}\right)^\varepsilon \left(-\frac{2}{\varepsilon^2} - \frac{3}{\varepsilon} - 8 + \pi^2 + \mathcal{O}(\varepsilon)\right) \right.$$
$$\left. + \frac{C_F \alpha_s(\mu_R)}{2\pi\Gamma(1-\varepsilon)} \left(\frac{4\pi\mu_R^2}{Q^2}\right)^\varepsilon \left(\frac{2}{\varepsilon^2} + \frac{3}{\varepsilon} + 10 - \pi^2 + \mathcal{O}(\varepsilon)\right) \right]$$
$$= d\Phi_\mathcal{B}\,\mathcal{B}(\Phi_\mathcal{B}) \frac{C_F \alpha_s(\mu_R)}{\pi}.$$

$$(3.189)$$

As anticipated, the overall result for the subtracted virtual contribution also is finite, rendering both individual parts of the next-to-leading order correction separately finite and therefore allowing their safe numerical integration.

However, in combination with the real-emission part above this yields the well-known result

$$\sigma^{(\mathrm{NLO})}_{e^+e^- \to q\bar{q}} = \sigma^{(\mathrm{LO})}_{e^+e^- \to q\bar{q}} \left(1 + \frac{3}{4} \frac{C_F \alpha_s(\mu_R)}{\pi}\right).$$

$$(3.190)$$

3.3.3.4 Master equations for Catani–Seymour dipole subtraction

As a consequence of this decomposition into splitters and spectators, there are four types of dipole structures in Catani–Seymour subtraction, namely all combinations of initial and final splitters with initial and final spectators. In the original CS paper, initial state particles are indicated by moving their label from Eq. (3.155) from subscript — reserved for final state particles — to superscript. This is also exhibited in Fig. 3.7.

The full differential subtraction cross-section, including emissions from splitter–spectator pairs in all combinations of initial and final state particles, written in the notation of this book, is given by

$$d\sigma^{(S)}(p_a,\, p_b;\, p_1,\, p_2,\, \ldots,\, p_{n+1})$$

$$= d\Phi_\mathcal{R} \left[\sum_{\substack{y\{ij\} \\ k \neq i,j}} \mathcal{D}_{ij;k}(p_a,\, p_b;\, p_1,\, p_2,\, \ldots,\, p_{n+1}) + \sum_{\substack{\{ij\} \\ a}} \mathcal{D}_{ij}^a(p_a,\, p_b;\, p_1,\, p_2,\, \ldots,\, p_{n+1}) \right.$$

$$\left. + \sum_{\substack{\{aj\} \\ k \neq j}} \mathcal{D}_k^{aj}(p_a,\, p_b;\, p_1,\, p_2,\, \ldots,\, p_{n+1}) + \sum_{\substack{\{aj\} \\ b \neq a}} \mathcal{D}^{aj;b}(p_a,\, p_b;\, p_1,\, p_2,\, \ldots,\, p_{n+1}) \right].$$

$$(3.191)$$

In this equation, the first line in the square bracket refers to emitters in the final state with the first term referring to a spectator in the final state, whereas the second term relates to a spectator in the initial state. The pattern repeats itself in the last line, with the only difference being the splitters in the initial state. Of course, in all cases kinematic maps similar to the one in Eq. (3.160) will be invoked to construct

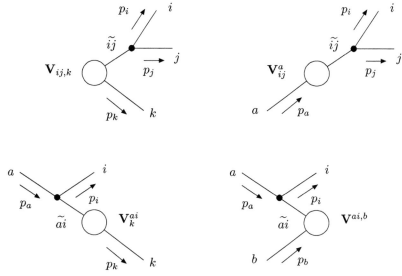

Fig. 3.7 Pictorial representation of the different Catani–Seymour dipole types: final-state splitter with final-state spectator (FF) (top left) and with initial-state spectator (FI) (top right) and initial-state splitter with final-state spectator (IF) (bottom left) and with initial-state spectator (FI) (bottom right). The blob denotes the m-parton matrix element, incoming lines enter from left, and outgoing lines leave the blob to the right.

the corresponding Born-level kinematics. For the details of this construction in the various cases, *cf.* Appendix C.2

Turning to the actual construction of the subtraction terms, the terms relating to final-state emitters in the first line have very similar structures, namely

$$\mathcal{D}_{ij;k}(p_a, p_b; \{p_i\}) = \mathcal{B}(p_a, p_b; \{\ldots, \tilde{p}_{ij}, \ldots, \tilde{p}_k, \ldots\}) \otimes \left[-\frac{1}{2p_ip_j} V_{ij,k} \frac{\mathbf{T}_{ij} \cdot \mathbf{T}_k}{T_{ij}^2} \right]$$

$$(3.192)$$

for the final–state splitter final–state spectator (FF) case, *cf.* Eq. (3.179). For the FI case,

$$\mathcal{D}_{ij}^a(p_a, p_b; \{p_i\}) = \mathcal{B}(\tilde{p}_a, p_b; \{\ldots, \tilde{p}_{ij}, \ldots\}) \otimes \left[-\frac{1}{2p_ip_j} \frac{1}{x_{ij,a}} V_{ij}^a \frac{\mathbf{T}_{ij} \cdot \mathbf{T}_a}{T_{ij}^2} \right],$$

$$(3.193)$$

indicating that the momentum of the spectator parton a in the initial state will be altered. In both cases, FF and FI, only the relevant momenta have been shown as arguments of the Born-part $\mathcal{B}$. It is also understood that due to the implicit cancellation of IR divergences in the difference of real-emission and subtraction terms, the dipoles V ultimately will be evaluated in $D = 4$ dimensions, with $\varepsilon = 0$. The kinematic maps for the two cases, FF and FI case, including in particular the extra term $x_{ij,a}$

in the propagator-like structure before the dipole, is detailed in Appendices C.1.1 and C.1.2. It is worth noting that, due to the initial-state parton compensating for the recoil of the emission process in the FI case, the Born-level matrix element, including the PDFs must be evaluated with different initial state kinematics, as indicated by $\tilde{p}_{aj}$ in its arguments, and of course also flux and symmetry factors must be adjusted correspondingly. This, however, is a fairly trivial manipulation.

This simple picture somewhat changes when initial-state singularities emerge due to initial-state particle being the emitters. These cases are treated by the subtraction terms in the last line of Eq. (3.191). The reason for this added complication is that, first of all, the factorization of the phase space in the soft and especially in the collinear limit must be checked, which has been done in great detail and instructively in the original CS paper [353]. In addition, now the altered initial-state parton momenta entering the Born term translate not only into the need to evaluate the PDF for this parton at a different x, with a potentially changed flavour, as above, but the emission off the initial-state parton also introduced new divergences which must be absorbed into the definition of the PDF. Essentially, this is because the particles in the initial state are fixed to move along a preferred direction, the beam axis. This additional constraint, which is not present for final-state emitters, necessitates special treatment and the emergence of additional terms. Different ways of how this is being done and which finite terms are absorbed together with a collinear divergence give rise to different factorization schemes, all of which of course are variants of the collinear factorization. This implies that scheme-dependent terms may arise, which must be properly accounted for. With this in mind,

$$\mathcal{D}_k^{aj}(p_a,\, p_b;\, \{p_i\}) \;=\; \mathcal{B}(\tilde{p}_{aj},\, p_b;\, \{\dots,\, \tilde{p}_k,\, \dots\}) \;\otimes\; \left[-\frac{1}{2p_a p_j} \frac{1}{x_{jk,a}} V_k^{aj} \frac{\mathbf{T}_{aj} \cdot \mathbf{T}_k}{\mathbf{T}_{aj}^2} \right]$$

(3.194)

for the IF case, and

$$\mathcal{D}^{aj;b}(p_a,\, p_b;\, \{p_i\}) \;=\; \mathcal{B}(\tilde{p}_{aj},\, \tilde{p}_b;\, \{\dots,\, \tilde{p}_k,\, \dots\}) \;\otimes\; \left[-\frac{1}{2p_a p_j} \frac{1}{x_{j,ab}} V^{aj,b} \frac{\mathbf{T}_{aj} \cdot \mathbf{T}_b}{\mathbf{T}_{aj}^2} \right],$$

(3.195)

for the II case. Details of the phase space mappings and the appropriate splitting functions V_k^{aj} and $V^{aj,b}$ can be found in Appendices C.1.3 and C.1.4, respectively. It is worth stressing here that by far and large the kinematic mappings are organized such that spectator partons keep their direction. This of course is not feasible any more when the splitter parton is in the initial state: in the case of IF splittings then the final state spectator must compensate the transverse momentum transfer; in the case of II splittings this is of course not possible. In this case, the transverse momentum compensation is achieved by moving the complete final state.

Turning to the integrated subtraction terms, to be added back and combined with the virtual parts of the NLO calculation, one has to remember that they are essentially given by the integral over the real-emission phase space of the differential ones. In principle, this is fairly straightforward and results in

$$d\sigma^{(I)} = d\Phi_\mathcal{B}\,\mathcal{B}(\Phi_\mathcal{B}) \otimes \mathbf{I}(\Phi_\mathcal{B}; \varepsilon), \tag{3.196}$$

where the integrated dipole term is constructed from terms with final- or initial-state splitters and spectators in intuitive notation as

$$\mathbf{I}(\Phi_\mathcal{B}; \varepsilon) = \mathbf{I}_{FF}(\Phi_\mathcal{B}; \varepsilon) + \mathbf{I}_{FI}(\Phi_\mathcal{B}; \varepsilon) + \mathbf{I}_{IF}(\Phi_\mathcal{B}; \varepsilon) + \mathbf{I}_{II}(\Phi_\mathcal{B}; \varepsilon). \tag{3.197}$$

The integrated dipole for the final-state–final-state case is given, as before, by

$$\mathbf{I}_{FF}(\Phi_\mathcal{B}; \varepsilon) = -\frac{\alpha_\mathrm{s}(\mu_R)}{2\pi\Gamma(1-\varepsilon)} \sum_{\{ij\}} \sum_{k \neq \{ij\}} \left[\left(\frac{4\pi\mu_R^2}{2p_{\{ij\}}p_k} \right)^\varepsilon \frac{\mathbf{T}_{\{ij\}} \cdot \mathbf{T}_k}{\mathbf{T}_{\{ij\}}^2} \mathcal{V}_{\{ij\}}(\varepsilon) \right], \tag{3.198}$$

where the sum is over all pairs of particles $\{ij\}$ and k. Note that here the "splitter" parton has been denoted by $\{ij\}$ in order to make the connection to the splitter–spectator notation in the differential dipole terms more manifest. The terms $\mathcal{V}_{\{ij\}}(\varepsilon)$ for different flavour are given in Eq. (3.170). Naively, of course, terms with an identical form, apart from trivial replacements in momenta and colours, also emerge for the other integrated dipoles.

This, however, is a bit too simple to be entirely correct. In cases with initial state partons a complication emerges, which originates from the fact that the kinematics mappings imply a change of the incoming particle's momentum, irrespective of whether it is a splitter or spectator parton. Schematically, for the Born-level n-particle phase space and the matrix element this will be accounted for by

$$\begin{aligned} d\Phi_\mathcal{B} &\xrightarrow{\mathrm{FF}\to\mathrm{FI,IF,II}} d\xi\,d\Phi_\mathcal{B}(\xi) \\ \mathcal{B}(\Phi_\mathcal{B}) &\xrightarrow{\mathrm{FF}\to\mathrm{FI,IF,II}} \mathcal{B}(\Phi_\mathcal{B}, \xi). \end{aligned} \tag{3.199}$$

For the integrated dipole this translates into an additional integration over the recoil parameter ξ, in the interval $[0, 1]$, which will ultimately lead to the emergence of various functions of ξ to be folded into the resulting expression. In principle, ξ parameterizes how much the momentum of the incident parton acting as splitter or spectator changes, and it will therefore also change the four-momentum balance accordingly, encoded in a δ function in the phase space element $d\Phi_\mathcal{B}$. In addition, if the parton in question actually is treated as the splitter then this change in momentum will also impact on the argument of the PDF, replacing the corresponding x with x/ξ. As already indicated in Eq. (3.149) in Section 3.3.2, there are also additional terms that must be considered when emissions off initial state partons are present. These terms are related to further collinear divergences originating from the fact that the matrix elements must be convoluted with PDFs, which themselves exhibit divergent structures in their scale evolution. A fixed-order part of this evolution actually emerges when considering initial-state singularities alone in the matrix elements and therefore must be treated through a suitable collinear subtraction. It is no surprise that these terms conversely exhibit a dependence on the details of the factorization scheme, manifesting themselves in some finite terms. In the $\overline{\mathrm{MS}}$ scheme used throughout this book these terms reduce to zero, but in other schemes such as the DIS scheme this is not the case anymore.

These collinear divergent terms lead to single poles already encountered in the previous section, *cf.* Eq. (3.150), and are of the form $\mu_F^{2\varepsilon}/\varepsilon\mathcal{P}_{ij}(z)$. For every initial-state parton a, this collinear subtraction term reads

$$
\mathrm{d}\sigma_a^{(C)} \;=\; \sum_{a'} \int_0^1 \mathrm{d}\xi\,\mathrm{d}\Phi_\mathcal{B}(\xi)\,\mathcal{B}(\Phi_\mathcal{B},\,\xi)\left[-\frac{1}{\varepsilon}\left(\frac{4\pi\mu_R^2}{\mu_F^2}\right)^\varepsilon \mathcal{P}_{a'a}^{(1)}(\xi) + K_{(\mathrm{F.S.})}^{aa'}(\xi)\right]. \tag{3.200}
$$

Here, $P_{a'a}^{(1)}(\xi)$ are the Altarelli–Parisi splitting kernels to first order; their four-dimensional version is given in *cf.* Eq. (2.33). The finite terms $K_{(\mathrm{F.S.})}^{aa'}(\xi)$ are related to the choice of factorization scheme. As already indicated, for the $\overline{\mathrm{MS}}$ scheme, they are given by

$$
K_{(\mathrm{F.S.})}^{aa'}(\xi) \overset{\overline{\mathrm{MS}}}{=} 0. \tag{3.201}
$$

Therefore, after absorbing the pole in the PDF, residual terms are left stemming from their expansion in ε.

The form of this subtraction term in fact could have been anticipated without any calculation, they merely connect the PDFs from their scale μ_F with the parton density at the scale μ where the rest of the process and in particular the singularities are evaluated. Not surprisingly the implicit change of parton density must be accounted for, in a form similar to the DGLAP equation, which is why the respective splitting kernels emerge. If the process has one initial-state parton only, merely one of these terms has to be considered; in the presence of two initial-state partons, two of these terms with suitable choices of flavours, etc. , have to be added to the virtual part.

In addition, and as already indicated, there are also terms coming from the initial-state parton in the integral over the emission phase space in the dipole terms, like the ones in the second line of the integrated dipole in Eq. (3.149). It has become customary to construct the integrated dipole terms for initial-state partons through specific collinear subtraction terms, extending the one in Eq. (3.200). These terms are added to the simple dipole functions constituting the integrated subtraction terms,[8]

$$
\mathrm{d}\sigma^{(I+C)} = \mathrm{d}\Phi_\mathcal{B}\mathcal{B}_{ab}(p_a,\,p_b)\otimes\mathbf{I}(\varepsilon)
$$

$$
+\sum_{a'}\int_0^1 \mathrm{d}\xi_a\,\mathrm{d}\Phi_\mathcal{B}(\xi_a)\,\mathcal{B}_{a'b}(\xi_a p_a,\,p_b)\otimes\left[\mathbf{K}^{aa'}(\xi_a)+\mathbf{P}^{aa'}(\xi_a p_a,\,\xi_a;\,\mu_F^2)\right]
$$

$$
+\sum_{b'}\int_0^1 \mathrm{d}\xi_b\,\mathrm{d}\Phi_\mathcal{B}(\xi_b)\,\mathcal{B}_{ab'}(p_a,\,\xi_b p_b)\otimes\left[\mathbf{K}^{bb'}(\xi_b)+\mathbf{P}^{bb'}(\xi_b p_b,\,\xi_b;\,\mu_F^2)\right]
$$

$$
\tag{3.202}
$$

---

[8]It is worth noting that similar reasoning in fact also applies for processes involving specified hadrons in the final state and the corresponding fragmentation functions: also in this case collinear subtraction terms must be added which stem from the evolution of the fragmentation functions through secondary emissions.

where the dependence of the Born cross-sections on the incoming flavours and momenta has been made explicit. The integrated dipole terms are given by Eq. (3.198) and similar. They can be combined to yield

$$
\mathbf{I}(\varepsilon) \equiv \mathbf{I}(p_a, p_b; p_1, \ldots, p_m; \varepsilon)
$$

$$
= -\frac{\alpha_s(\mu_R)}{2\pi\Gamma(1-\varepsilon} \left\{ \sum_i \frac{\mathcal{V}_i(\varepsilon)}{\mathbf{T}_i^2} \left[\sum_{k\neq i} \mathbf{T}_i \cdot \mathbf{T}_k \left(\frac{4\pi\mu_R^2}{2p_i p_k}\right)^\varepsilon + \sum_{c\in\{a,b\}} \mathbf{T}_i \cdot \mathbf{T}_c \left(\frac{4\pi\mu_R^2}{2p_i p_c}\right)^\varepsilon \right] \right.
$$

$$
\left. + \sum_{c\in\{a,b\}} \frac{\mathcal{V}_c(\varepsilon)}{\mathbf{T}_c^2} \left[\sum_i \mathbf{T}_c \cdot \mathbf{T}_i \left(\frac{4\pi\mu_R^2}{2p_c p_i}\right)^\varepsilon + \mathbf{T}_c \cdot \mathbf{T}_d \left(\frac{4\pi\mu_R^2}{2p_c p_d}\right)^\varepsilon \Bigg|_{d\neq c} \right] \right\}.
$$

$$(3.203)$$

The terms in the first square bracket relate to dipoles with final-state emitters i — the first sum is over final-state spectators k and the second sum is over the two initial state spectators c — while the second square bracket relates to dipoles with initial-state emitters c, where, again, the first sum is over final-state spectators i and the second term is for the other initial-state particle $d \neq c$ being the spectator. In all cases, the $\mathcal{V}$ are given by Eq. (3.170).

The operators in the collinear terms, for instance $\mathbf{K}^{aa'}(\xi_a)$ and $\mathbf{P}^{aa'}(\xi_a p_a, \xi_a; \mu_F^2)$, emerge from the considerations above and are given by

$$
\mathbf{K}^{aa'}(\xi_a) = \frac{\alpha_s(\mu_R)}{2\pi} \left\{ \overline{K}^{aa'}(\xi_a) + \delta^{aa'} \sum_{\{ij\}} \gamma_{\{ij\}}^{(1)} \frac{\mathbf{T}_{\{ij\}} \cdot \mathbf{T}_{a'}}{\mathbf{T}_{\{ij\}}^2} \left[\left(\frac{1}{1-\xi_a}\right)_+ + \delta(1-\xi_a) \right] \right.
$$

$$
\left. - \frac{\mathbf{T}_b \cdot \mathbf{T}_{a'}}{\mathbf{T}_{a'}^2} \tilde{K}^{aa'}(\xi_a) - K_{F.S.}^{aa'}(\xi_a) \right\}
$$

$$(3.204)$$

and

$$
\mathbf{P}^{aa'}(\xi_a p_a, \xi_a;, \mu_F^2)
$$

$$
= \frac{\alpha_s(\mu_R)}{2\pi} P_{aa'}^{(1)}(\xi_a) \left[\sum_{\{ij\}} \frac{\mathbf{T}_{\{ij\}} \cdot \mathbf{T}_{a'}}{\mathbf{T}_{a'}^2} \log \frac{\mu_F^2}{2\xi_a p_a p_{\{ij\}}} + \frac{\mathbf{T}_b \cdot \mathbf{T}_{a'}}{\mathbf{T}_{a'}^2} \log \frac{\mu_F^2}{2\xi_a p_a p_b} \right],
$$

$$(3.205)$$

cf. Eq. (C.45). The $\gamma_{\{ij\}}^{(1)}$ in the first line of Eq. (3.204) are the anomalous dimensions to first order of the DGLAP splitting functions introduced in Eq. (2.33). The functions $\overline{K}$ and $\tilde{K}$ can be found in Appendix C.2. As already stated, the term $K_{F.S.}$ related to the choice of factorization scheme vanishes for the choice of the minimal subtraction scheme, the one made in this book, $K_{\overline{MS}}(\xi) = 0$. The first-order splitting kernel $P_{aa'}^{(1)}$ occurring in the $\mathbf{P}$–operator has been given in Eq. (2.33).

Note the occurrence of the sums over final-state particle $\{ij\}$ in both the $\mathbf{K}$ and the $\mathbf{P}$ operator. While in the former case these sums emerge from dipoles where the initial-state particle a acts as spectator and the final-state particle $\{ij\}$ is the splitter, as indicated by the colour factors, these roles are reversed in the latter case. There, in the **P** operators in addition a term emerges which is due to the initial-state particle a being the splitter and the other initial-state particle b being the spectator. Of course, if the corresponding dipoles do not emerge, due to the lack of coloured particles in either initial or final state, the respective terms are just dropped.

3.3.3.5 Example: W production at NLO with Catani–Seymour subtraction

In the following, only the real-emission part with a gluon in the final state will be considered, and of course, only the corresponding integrated terms need to be taken into account then. Starting with the real-emission contribution $\mathcal{R}(\Phi_\mathcal{R})$ and the corresponding subtraction term in four dimensions $\mathcal{S}(\Phi_\mathcal{R})^{(\varepsilon=0)}$, the subtracted real-emission contribution reads

$$d\sigma^{(R-S)} = d\Phi_\mathcal{R} \left[\mathcal{R}(\Phi_\mathcal{R}) - \mathcal{S}(\Phi_\mathcal{R})^{\varepsilon=0} \right]. \tag{3.206}$$

As before, *cf.* Eq. (3.134), the real contribution can be written as

$$d\sigma^{(R)} = d\Phi_\mathcal{R} \frac{2\pi C_F \alpha_s(\mu_R)}{x} \left| \mathcal{M}^{(\mathrm{LO})}_{u\bar{d}\to W^+} \right|^2 \left[\left(\frac{1}{\hat{t}} + \frac{1}{\hat{u}} \right) \left(-\frac{2}{1-x} + x + 1 \right) - \frac{2x}{m_W^2} \right], \tag{3.207}$$

where, again, the dimensionless quantity $x = m_W^2/\hat{s}$ has been employed and where the real phase-space element including PDFs and flux factors is given by

$$d\Phi_\mathcal{R} = \frac{1}{2\hat{s}} \, dx_u dx_{\bar{d}} \, f_{u/h_1}(x_u, \mu_F) f_{\bar{d}/h_2}(x_{\bar{d}}, \mu_F) \, d\Phi_{Wg} \tag{3.208}$$

with the final-state phase-space element for the $\{Wg\}$ system, $d\Phi_{Wg}$, given in Eq. (3.139).

At the same time, including averaging over incoming spins, the subtraction cross-section is constructed as

$$d\sigma^{(S)} = d\Phi_\mathcal{R} \frac{1}{4} \left| \mathcal{M}^{(\mathrm{LO})}_{u\bar{d}\to W^+} \right|^2$$
$$\times \left[-\frac{1}{2p_u p_g} \frac{1}{x_{g,u\bar{d}}} V^{ug,\bar{d}} \frac{\mathbf{T}_{ug} \cdot \mathbf{T}_{\bar{d}}}{\mathbf{T}_{ug}^2} - \frac{1}{2p_{\bar{d}} p_g} \frac{1}{x_{g,\bar{d}u}} V^{\bar{d}g,u} \frac{\mathbf{T}_{\bar{d}g} \cdot \mathbf{T}_u}{\mathbf{T}_{\bar{d}g}^2} \right]. \tag{3.209}$$

Using the Mandelstam identity Eq. (2.100) for this process, the recoil parameter $x_{g,u\bar{d}}$ in the dipole of Eq. (3.195) is given by

$$x_{g,u\bar{d}} = \frac{p_u p_{\bar{d}} - p_g(p_u + p_{\bar{d}})}{p_u p_{\bar{d}}} = \frac{\hat{s} + \hat{t} + \hat{u}}{\hat{s}} = \frac{m_W^2}{\hat{s}} = x_{g,\bar{d}u} \overset{!}{=} x. \tag{3.210}$$

This is identical to the x of Eq. (3.133), in the first discussion of subtraction with dipole terms from a suitable ad-hoc definition. In Catani–Seymour subtraction the splitting kernel $V^{qg,q}$ in four dimensions, *i.e.* $\varepsilon = 0$, reads

$$V^{ug,\bar{d}} = 8\pi C_F \alpha_s(\mu_R) \left[\frac{2}{1-x} - (1+x) \right]. \tag{3.211}$$

Therefore,

$$d\sigma^{(S)} = d\Phi_\mathcal{R} \left[\frac{8\pi C_F \alpha_s(\mu_R)}{x} \frac{1}{4} |\mathcal{M}^{(\mathrm{LO})}_{u\bar{d}\to W^+}|^2 \left(\frac{1}{\hat{t}} + \frac{1}{\hat{u}} \right) \left(-\frac{2}{1-x} + (1+x) \right) \right]. \tag{3.212}$$

This result is exactly the sum of the terms $\mathcal{D}(\hat{t}, x)$ and $\mathcal{D}(\hat{u}, x)$ from Eq. (3.136). Thus, for the subtracted real correction contribution one finally arrives at

$$
\begin{aligned}
d\sigma^{(R-S)} &= d\Phi_\mathcal{R} \left| \mathcal{M}^{(\mathrm{LO})}_{u\bar{d}\to W^+} \right|^2 \frac{2\pi C_F \alpha_s(\mu_R)}{x} \left\{ \left[\frac{1}{\hat{t}} + \frac{1}{\hat{u}} \right] \left[-\frac{2}{1-x} + (1+x) \right] - \frac{2x}{m_W^2} \right. \\
&\qquad\qquad \left. - \left[\frac{1}{\hat{t}} + \frac{1}{\hat{u}} \right] \left[-\frac{2}{1-x} + (1+x) \right] \right\} \\
&= - d\Phi_\mathcal{R} \frac{4\pi C_F \alpha_s(\mu_R)}{m_W^2} \left| \mathcal{M}^{(\mathrm{LO})}_{u\bar{d}\to W^+} \right|^2.
\end{aligned}
\tag{3.213}
$$

This constitutes a perfectly finite result, as anticipated.

Turning now to the virtual contribution for the $u\bar{d} \to W$ process and invoking Eq. (3.203) for the integrated dipole terms shows that the only relevant contributions to $\mathbf{I}(\varepsilon)$ are those with both splitter and spectator in the initial state. Therefore

$$
\begin{aligned}
\mathbf{I}_{u\bar{d}\to W}(\varepsilon) &= \mathbf{I}_{u\bar{d}\to W}(p_u, p_{\bar{d}}; \varepsilon) \\
&= -\frac{\alpha_s(\mu_R)}{2\pi\Gamma(1-\varepsilon)} \left(\frac{4\pi\mu_R^2}{2p_u p_{\bar{d}}} \right)^\varepsilon \left\{ \frac{\mathbf{T}_u \cdot \mathbf{T}_{\bar{d}}}{\mathbf{T}_u^2} \mathcal{V}_u(\varepsilon) + \frac{\mathbf{T}_{\bar{d}} \cdot \mathbf{T}_u}{\mathbf{T}_{\bar{d}}^2} \mathcal{V}_{\bar{d}}(\varepsilon) \right\} \\
&= \frac{C_F \alpha_s(\mu_R)}{\pi} c_\Gamma \left(\frac{\mu_R^2}{m_W^2} \right)^\varepsilon \left[\frac{1}{\varepsilon^2} + \frac{3}{2\varepsilon} + 5 - \frac{1 - a^{\mathrm{CDR}}}{2} - \frac{\pi^2}{2} \right], \tag{3.214}
\end{aligned}
$$

where it has been realized that, at Born-level, the centre-of-mass energy of the annihilating quark pair equals the W mass, consequently allowing the identification $\hat{s} \equiv m_W^2$.

Finally, the collinear subtraction terms from Eq. (3.202) have to be added. Specifying to the $u\bar{d} \to Wg$ case only, so ignoring gluon-initiated processes, they are given by

$$
\begin{aligned}
d\sigma^{(C)} &= \int_0^1 d\xi_u \, d\Phi_\mathcal{B}(\xi_u) \mathcal{B}_{u\bar{d}\to W}(\xi_u p_u, p_{\bar{d}}) \otimes \left[\mathbf{K}^{qq}(\xi_u) + \mathbf{P}^{qq}(\xi_u p_u, \xi_u;, \mu_F^2) \right] \\
&\quad + \int_0^1 d\xi_{\bar{d}} \, d\Phi_\mathcal{B}(\xi_{\bar{d}}) \mathcal{B}_{u\bar{d}\to W}(p_u, \xi_{\bar{d}} p_{\bar{d}}) \otimes \left[\mathbf{K}^{qq}(\xi_{\bar{d}}) + \mathbf{P}^{qq}(\xi_{\bar{d}} p_{\bar{d}}, \xi_{\bar{d}};, \mu_F^2) \right].
\end{aligned}
$$

The **K** and **P** operators can be obtained from Eqs. (3.204) and (3.205) with the contributing functions for the case at hand given by

$$
\overline{K}^{qq}(\xi) = C_F \left[\left(\frac{2}{1-\xi} \log \frac{1-\xi}{\xi} \right)_+ - (1+\xi) \log \frac{1-\xi}{\xi} + (1-\xi) - \delta(1-\xi) \left(5 - \pi^2 \right) \right]
$$

$$
\tilde{K}^{qq}(\xi) = C_F \left[\left(\frac{2}{1-\xi} \log(1-\xi) \right)_+ - (1+\xi) \log(1-\xi) - \frac{\pi^2}{3} \delta(1-\xi) \right].
$$

$$(3.215)$$

Not surprisingly, the sum over all particles contained inside these terms collapses to the other initial-state particle only. Concentrating on the first term for the time being, and making the dependence on the PDFs explicit, therefore,

$$
\begin{aligned}
\mathrm{d}\sigma_u^{(C)} &= \int\limits_0^1 \mathrm{d}x_u \mathrm{d}x_{\bar{d}} \int\limits_{x_u}^1 \mathrm{d}\xi_u \, f_{u/h_1}\!\left(\frac{x_u}{\xi_u}, \mu_F^2\right) f_{\bar{d}/h_1}\!\left(x_{\bar{d}}, \mu_F^2\right) \frac{\pi\delta(x_u x_{\bar{d}} s - m_W^2)}{m_W^2} \\
&\quad \cdot \left|\mathcal{M}_{u\bar{d}\to W^+}^{(\mathrm{LO})}\right|^2 \frac{\alpha_s(\mu_R)}{2\pi} \left\{ \overline{K}^{qq}(\xi_u) + \left[\tilde{K}^{qq}(\xi_u) - P_{qq}^{(1)}(\xi_u) \log \frac{\mu_F^2}{m_W^2} \right] \right\} \\
&= \int\limits_0^1 \mathrm{d}x_u \mathrm{d}x_{\bar{d}} \int\limits_{x_u}^1 \mathrm{d}\xi_u \, f_{u/h_1}\!\left(\frac{x_u}{\xi_u}, \mu_F^2\right) f_{\bar{d}/h_1}\!\left(x_{\bar{d}}, \mu_F^2\right) \frac{\pi\delta(x_u x_{\bar{d}} s - m_W^2)}{m_W^2} \\
&\quad \cdot \left|\mathcal{M}_{u\bar{d}\to W^+}^{(\mathrm{LO})}\right|^2 \frac{\alpha_s(\mu_R) C_F}{2\pi} \left\{ \left[\frac{2}{1-\xi_u} \left(\log \frac{1-\xi_u}{\xi_u} + \log(1-\xi_u) - \log \frac{\mu_F^2}{m_W^2} \right) \right]_+ \right. \\
&\quad - (1+\xi_u) \left(\log \frac{1-\xi_u}{\xi_u} + \log(1-\xi_u) - \log \frac{\mu_F^2}{m_W^2} \right) \\
&\quad \left. + (1-\xi_u) - \delta(1-\xi_u) \left(5 - \frac{2\pi^2}{3} + \frac{3}{2} \log \frac{\mu_F^2}{m_W^2} \right) \right\}.
\end{aligned}
$$

$$(3.216)$$

Here, as before, the colour insertion term $\mathbf{T}_u \cdot \mathbf{T}_{\bar{d}}/\mathbf{T}_u^2 = -1$ has already been evaluated and replaced.

Together with the virtual contribution and the subtraction term $I_{u\bar{d}\to W}(\varepsilon)$ this results in the following contribution to the overall cross-section:

$$
\begin{aligned}
\mathrm{d}\sigma_{u\bar{d}\to W^+}^{(V+I+C)}(p_u, p_{\bar{d}}) &= \frac{C_F \alpha_s(\mu_R)}{2\pi} \left|\mathcal{M}_{u\bar{d}\to W^+}^{(\mathrm{LO})}\right|^2 \int\limits_0^1 \mathrm{d}x_u \mathrm{d}x_{\bar{d}} \, \frac{\delta(x_u x_{\bar{d}} s - m_W^2)}{m_W^2} \\
&\quad \int\limits_{x_u}^1 \mathrm{d}\xi_u \int\limits_{x_d}^1 \mathrm{d}\xi_d \, f_{u/h_1}\!\left(\frac{x_u}{\xi_u}, \mu_F^2\right) f_{\bar{d}/h_1}\!\left(\frac{x_{\bar{d}}}{\xi_{\bar{d}}}, \mu_F^2\right)
\end{aligned}
$$

$$\times \left\{ \left[c_\Gamma \left(\frac{\mu_R^2}{m_W^2} \right)^\varepsilon \left(-\frac{2}{\varepsilon^2} - \frac{3}{\varepsilon} - 7 - a^{\text{CDR}} + \pi^2 \right) \right] \delta(1 - \xi_u) \, \delta(1 - \xi_{\bar{d}}) \right.$$

$$+ \left[c_\Gamma \left(\frac{\mu_R^2}{m_W^2} \right)^\varepsilon \left(+\frac{2}{\varepsilon^2} + \frac{3}{\varepsilon} + 9 + a^{\text{CDR}} - \pi^2 \right) \right.$$

$$\left. + 2 \left(\frac{2\pi^2}{3} - 5 - \frac{3}{2} \log \frac{\mu_F^2}{m_W^2} \right) \right] \delta(1 - \xi_u) \, \delta(1 - \xi_{\bar{d}})$$

$$+ \left[\left(\frac{2}{1 - \xi_u} \left(\log \frac{(1 - \xi_u)^2}{\xi_u} - \log \frac{\mu_F^2}{m_W^2} \right) \right)_+ \right.$$

$$\left. - (1 + \xi_u) \left(\log \frac{(1 - \xi_u)^2}{\xi_u} - \log \frac{\mu_F^2}{m_W^2} \right) + (1 - \xi_u) \right] \delta(1 - \xi_{\bar{d}})$$

$$+ \left[\left(\frac{2}{1 - \xi_{\bar{d}}} \left(\log \frac{(1 - \xi_{\bar{d}})^2}{\xi_{\bar{d}}} - \log \frac{\mu_F^2}{m_W^2} \right) \right)_+ \right.$$

$$\left. - (1 + \xi_{\bar{d}}) \left(\log \frac{(1 - \xi_{\bar{d}})^2}{\xi_{\bar{d}}} - \log \frac{\mu_F^2}{m_W^2} \right) + (1 - \xi_{\bar{d}}) \right] \delta(1 - \xi_u) \right\}.$$

$$(3.217)$$

The double and single poles $1/\varepsilon^2$ and $1/\varepsilon$ cancel, which allows the prefactor to be replaced by unity,

$$c_\Gamma \left(\frac{\mu_R^2}{m_W^2} \right)^\varepsilon \longrightarrow 1. \qquad (3.218)$$

Therefore the total contribution to the cross-section from these terms is,

$$d\sigma_{u\bar{d}\to W^+}^{(V+I+C)}(p_u, p_{\bar{d}}) = \frac{C_F \alpha_s(\mu_R)}{2\pi} |\mathcal{M}_{u\bar{d}\to W^+}^{(\text{LO})}|^2 \int_0^1 dx_u dx_{\bar{d}} \frac{\delta(x_u x_{\bar{d}} s - m_W^2)}{m_W^2}$$

$$\int_{x_u}^1 d\xi_u \int_{x_{\bar{d}}}^1 d\xi_{\bar{d}} \, f_{u/h_1} \left(\frac{x_u}{\xi_u}, \mu_F^2 \right) f_{\bar{d}/h_1} \left(\frac{x_{\bar{d}}}{\xi_{\bar{d}}}, \mu_F^2 \right)$$

$$\times \left\{ \left[\frac{4\pi^2}{3} - 8 + 3 \log \frac{m_W^2}{\mu_F^2} \right] \delta(1 - \xi_u) \, \delta(1 - \xi_{\bar{d}}) \right.$$

$$+ \left[\left(\frac{2}{1 - \xi_u} \log \frac{(1 - \xi_u)^2 m_W^2}{\xi_u \mu_F^2} \right)_+ \right.$$

$$- (1 + \xi_u) \left(\log \frac{(1 - \xi_u)^2 m_W^2}{\xi_u \mu_F^2} \right) + (1 - \xi_u) \right] \delta(1 - \xi_{\bar{d}})$$

$$+ \left[\left(\frac{2}{1 - \xi_{\bar{d}}} \log \frac{(1 - \xi_{\bar{d}})^2 m_W^2}{\xi_{\bar{d}} \mu_F^2} \right)_+ \right.$$

$$\left. - (1 + \xi_{\bar{d}}) \left(\log \frac{(1 - \xi_{\bar{d}})^2 m_W^2}{\xi_{\bar{d}} \mu_F^2} \right) + (1 - \xi_{\bar{d}}) \right] \delta(1 - \xi_u) \Big\} .$$

$$(3.219)$$

Not surprisingly, this agrees with results that have previously been constructed in Sections 2.2.5 and 3.3.2. Here, however, the full result is quoted, including the convolution with the PDFs.

3.3.3.6   Practical aspects of the evaluation

In the previous sections, the Catani–Seymour dipole subtraction has been introduced as a specific implementation of the more general idea of infrared subtraction. This method adds and subtracts terms to the virtual and real corrections in such a way that the resulting subtracted contributions are individually infrared-finite. Ultimately, this allows their phase-space integration with Monte Carlo techniques, which is necessary due to the complex structure of the high-dimensional integration region in such calculations.

In other words, in

$$\sigma^{(\mathrm{NLO})} = \int \mathrm{d}\Phi_{\mathcal{B}} \left[\mathcal{B}_n(\Phi_{\mathcal{B}}; \mu_F, \mu_R) + \mathcal{V}_n(\Phi_{\mathcal{B}}; \mu_F, \mu_R) + \mathcal{I}_n^{(\mathcal{S})}(\Phi_{\mathcal{B}}; \mu_F, \mu_R) \right]$$
$$+ \int \mathrm{d}\Phi_{\mathcal{R}} \left[\mathcal{R}_n(\Phi_{\mathcal{R}}; \mu_F, \mu_R) - \mathcal{S}_n(\Phi_{\mathcal{R}}; \mu_F, \mu_R) \right],$$

$$(3.220)$$

the two integrals over $\mathrm{d}\Phi_{\mathcal{B}}$ and $\mathrm{d}\Phi_{\mathcal{R}}$ are individually finite and can therefore be integrated with the fairly evolved and efficient methods discussed in Section 3.2.2.

There are two *caveats*, which have a purely technical reason. First of all, the collinear terms encountered in the $\mathrm{d}\sigma^{(C)}$ contribution of Eq. (3.202) feature "+" functions. They are introduced in order to regulate possible divergences in the integration in the limit where the variable approaches 1. Looking at their definition in Eq. (2.10),

$$\int\limits_0^1 \mathrm{d}z [f(z)]_+ g(z) = \int\limits_0^1 \mathrm{d}z f(z) \left[g(z) - g(1) \right], \qquad (3.221)$$

where $g(z)$ is a regular function, this necessitates the inclusion of such terms — typically the respective Born-level cross-sections — in the integration. This, however, is not difficult per se but merely a slight annoyance in actual implementations. The only tricky issue there consists in the fact that for $z \to 1$ the divergence in $f(z)$ is countered by the simultaneous vanishing of $[g(z) - g(1)]$, which numerically is not always perfect due to the limited accuracy of numerical methods. One way to account for this is to set a numerically very small limit for the regulator $[g(z) - g(1)]$ below which it is replaced by exactly zero, thereby setting the full integrand to zero.[9]

A second problem emerges in the subtracted real contribution, for similar reasons. By construction $[\mathcal{R}(\Phi_\mathcal{R}) - \mathcal{S}(\Phi_\mathcal{R})]$ approaches zero in singular regions of the phase space $\Phi_\mathcal{R}$. This cancellation, however, is a cancellation of the type $\infty - \infty$, which notoriously is tricky to achieve numerically. It has therefore become customary to set the difference above to exactly zero, when two four-momenta in $\Phi_\mathcal{R}$ become very collinear or one of the momenta becomes very soft. In reality this means that scalar products of the four-momenta are being checked, and if they fall below some very small cut-off value α, $[\mathcal{R}(\Phi_\mathcal{R}) - \mathcal{S}(\Phi_\mathcal{R})]$ will be replaced by zero. The name of the game then is to optimize the choice of α in such a way that the overall result is numerically stable.

Both these issues have been discussed and tested quite exhaustively in a number of automated implementations [536, 537, 584, 617].

3.3.3.7 Taming the growing number of dipole terms

The two most popular subtraction methods, the algorithm of Frixione, Kunszt, and Signer (FKS) [539, 542] and the formalism of Catani and Seymour (CS) [344, 353], exhibit a different behaviour in the scaling of the number of subtraction terms N_sub with the number of external particles n. While, naively, the former method demands three subtraction terms per external particle, $N_\mathrm{sub} \propto 3n$, the latter method requires two dipoles per pair of particles. Thus, in this case, $N_\mathrm{sub} \propto 3n(n+1)/2$. While this does not play any significant role for processes with only few external particles, $n \leq 6$ or so, it becomes of course more pronounced with the number of external particles increasing.

One way to handle this is to realize that the FKS method bases the construction of subtraction terms on a decomposition of the additional particle's emission phase space in soft, collinear, and soft-collinear regions, with one term per region. Conversely, in non-singular regions of the phase space, no subtraction term is being constructed. In contrast, the dipoles of the CS method are always constructed, and the subtraction is performed over all phase space. This of course can be changed in such a way, that phase-space criteria are defined which allow to identify potentially dangerous, singular, regions of phase space based on the dipole kinematics. Then, in uncritical regions, no subtraction is being performed. One particular way is to define the criterion through

---

[9]A typical value such a cut-off would be $\mathcal{O}\left(10^{-\delta}\right)$, where δ is the number of significant digits up to which the numerics of the program are stable; for double precision this is typically $\delta \geq 10$.

some α_{dip}, which parameterizes the phase space. For instance, for FF dipoles, the "constrained" dipoles are given by [781]

$$\mathcal{D}'_{ij;k} \;=\; \mathcal{D}_{ij;k}\,\Theta(\alpha - y_{ij;k}), \tag{3.222}$$

where $y_{ij;k}$ has been defined in Eq. (3.161) and parameterizes the soft and collinear limits of the splitting, as can be seen in Eq. (3.166). This idea has been extended to other dipole configurations, namely to II, IF, and FI dipoles in [777], see also Appendix C.2. Of course, the integrated contributions, the **I**, **K**, and **P** operators, inherit this dependence, leading to some α dependent additional terms.

This parameter-dependent identification of singular regions and the ensuing decomposition into subtracted and unsubtracted phase-space regions lead to a potentially large saving in the number of terms. In addition, it also allows a very convenient check, as the overall results for cross-sections must be parameter-independent.

3.3.4   Next-to-leading order tools

The methods presented in Section 3.3.1, to calculate one-loop amplitudes have been used to construct libraries of NLO calculations or even tools for their automated calculation. These tools typically need to be interfaced with other tools, which take care of the real correction contribution, the treatment of infrared singularities, phase-space integration, and the leading order part of the calculation. The latter two of these tasks — and quite often also the first two — are conveniently handled by tools presented in Table 3.2 in Section 3.2.3. The interface between these two classes of tools has been standardized in the **Binoth–Les Houches accord** [138, 250].

In Table 3.3, publicly available tools for the evaluation of one-loop amplitudes are listed and roughly categorized. This comprises

- their scope: evaluating integrals ("integral evaluator"), reducing the integrands to integrals ("integral reduction"), acting as a library of NLO or one-loop calculations ("library"), or automated creation and calculation of one-loop amplitudes ("one-loop generator")
- the way the amplitudes are being reduced and evaluated: Passarino–Veltman ("PV"), Ossola–Pittau–Papadopoulos ("OPP"), or OPENLOOPS ("OL") reduction.

These tools are often supplemented with other, automated programs, that take care of the infrared subtraction of the real-emission corrections. These are listed in Table 3.4. Most of them are also capable of providing the real-emission matrix elements, taking care of the phase-space integration, etc.

3.3.5   Next-to-leading order practicalities

Once a calculation at NLO, or even higher order, has been completed, it has its greatest utility if it is available in a form in which a non-author can produce results for a specific kinematic configuration, or choice of scales, choice of PDFs, etc. Often the calculation is available in a program publicly available; there are also collections of such calculations, as for example are available in MCFM [311]. In some cases, especially for complex final states, the program may not be available per se but the information needed to provide

Table 3.3 Tools for calculating virtual corrections: there are different types of relevant tools, including those that list scalar (and other) basic integrals ("integrals"), reduce integrands ("reduction"), provide libraries of virtual corrections or the full cross-sections ("libraries"), or generate one-loop amplitudes automatically ("generators"). They use different reduction technologies, such as Passarino–Veltman ("PV"), Ossola–Pittau–Papadopoulos ("OPP"), or OPENLOOPS ("OL") reduction. Some of them are set up to directly produce full calculations ("full") including all contributions.

	type	technology dependencies on other codes
LOOPTOOLS [605]	integrals	
ONELOOP [879]	integrals	
QCDLOOP [499]	integrals	
COLLIER [457]	reduction	
CUTTOOLS [793]	reduction	OPP
FORMCALC [605]	reduction	PV
NINJA [802]	reduction	Laurent expansion
SAMURAI [756]	reduction	
BLACKHAT [226]	library (amplitudes)	OPP (unitarity)
MCFM [309]	library (full calculation)	PV & OPP
NJET [181]	library (amplitudes)	OPP
GOSAM [420]	generator (amplitudes)	OPP SAMURAI +NINJA +
MADLOOP [625]	generator (full calculation)	OL+OPP CUTTOOLS +
OPENLOOPS [338]	generator (amplitudes)	OL+OPP COLLIER +CUTTOOLS +
HELAC–NLO [241]	generator (full calculation)	OPP CUTTOOLS +

a specific prediction is available in ROOT-ntuple form, where each event stores a phase-space point, along with the matrix elements and other information, such as the four-momenta of all final-state particles. At NLO, events are generated separately for each of four types of contributions: Born, virtual, integrated-subtraction, and real emission.[10] A positive-definite physical cross-section is only formed by the addition of many events of all four types. The ntuples can be used with an analysis script to allow for the construction of cross-sections with the appropriate kinematic cuts corresponding to a given experimental analysis.

For a process containing n partons at Born level, the emission of a real particle results in a $(n+1)$ parton phase space. The subtracted real-emission events correspond to regions of this phase space where two massless partons become collinear or soft. These divergent regions have been regularized by a subtraction procedure, often the

---

[10]At higher multiplicities, a further sub-division may be optimal.

Table 3.4 Tools for automated infrared subtraction: there are different types of relevant tools, those including the full underlying Born-level amplitudes and tools for phase-space integration in the construction of the subtraction terms and those tools which only provide the differential and integrated subtraction terms to be added on top of matrix elements provided from outside.

	type	technology dependencies on other codes
AMEGIC++ [584]	full: ME+PS	CS dipoles SHERPA
AUTODIPOLE [617]	subtraction only	CS dipoles MADGRAPH
COMIX [582]	full: ME+PS	CS dipoles SHERPA
MADDIPOLE [537, 538]	full: ME+PS	CS dipoles MADGRAPH
MADFKS [536]	full: ME+PS	FKS subtraction MADGRAPH
MATCHBOX [808]	subtraction only	CS subtraction

Catani–Seymour subtraction scheme discussed in Section 3.3.3. For this type of contribution, one event corresponds to the original emission, while the rest correspond to the subtraction terms. The dipole terms, evaluated in n-parton phase space, are then added back in the integrated subtraction events. For complex final states, there can be many Catani–Seymour subtraction terms and thus the need for many subtracted real-emission events in the ROOT ntuple. In fact, the subtracted real events take the majority of the disk space required for storing the ROOT-ntuples. The statistical uncertainty for events from the Born, virtual, and integrated subtraction ntuples can be calculated in the standard Monte Carlo way. As the real-emission and the corresponding subtraction configurations for a given event are strongly anti-corrrelated, the anti-correlation must be taken into account in order not to over-estimate the statistical error. It is possible to use a number of different jet algorithms and sizes, as long as the appropriate subtracted real counter-events are present. Typically, a wide range of jet sizes, $R \in [0.2, 1]$, can be allowed with only a small overhead (number of additional subtracted real events needed).

As mentioned above, the partons in each event can be re-clustered on the fly to form cross-sections for different jet algorithms, for example using FASTJET. The appropriate weight information is stored for each event, allowing the matrix element for that event to be reweighted for different PDFs (and the appropriate value of $\alpha_s(m_Z)$ for that PDF), and for different values of the renormalization and factorization scales. Thus, the PDF, $\alpha_s(m_Z)$ and scale uncertainties can be automatically calculated in one run over the ntuples. For the Born and subtracted real events, this reweighting is relatively straightforward. The virtual events have an additional dependence on the renormalization scale resulting from the one-loop amplitudes. The integrated subtraction events also depend on the factorization scale and the PDFs as a result of off-diagonal splittings.

There is a standard format for storing information from NLO calculations in ROOT ntuples that was originally developed by the BLACKHAT +SHERPA collaboration [233], but has now been adopted for use in a number of NLO calculations from other groups as well [182, 420]. Thus, an analysis code developed for the evaluation of ROOT ntuple events for one process, can easily be adapted for use with any other NLO process.

Consider the production of Higgs $(+ \geq 3)$ jets through gg fusion at NLO as an example. This calculation has been performed by the GOSAM collaboration [420] and the output has been made available in the BLACKHAT +SHERPA ROOT ntuple format.[11] This is one of the most difficult NLO calculations carried out to-date. For cross-section predictions for reasonable statistical uncertainties at 13 TeV, approximately 70 GB of disk storage is needed for the Born ntuples, 2.5 GB for the virtual ntuples, 130 GB for the integrated subtraction ntuples and 2.2 TB for the subtracted real ntuples. As discussed earlier, the subtracted real events by far require the most storage space. Although the total disk space required is large, each ntuple file is restricted to a size of a few GB, allowing for the option of parallel processing.

In Fig. 3.8 (left), the cross-section for the production of a third jet for $gg \rightarrow H+ \geq 3$ jets at 13 TeV is shown as a function of the third jet transverse momentum. Jets are clustered using the anti-k_T $(D = 0.4)$ jet algorithm and must have a minimum transverse momentum of 30 GeV and a maximum absolute rapidity of 4.4. The prediction in the ntuple uses the CT10 NLO PDFs and the calculation is carried out at a central scale of $\mu_R = \mu_F = H_T/2$. The reweighting information in the ntuples was used to produce the PDF uncertainty band (using the CT10 Hessian error set) and the scale uncertainty band (varying the renormalization and factorization scales independently up and down by a factor of two, while keeping the difference between the two scales within a factor of two). The scale uncertainty dominates except at high p_T where the PDF uncertainty becomes comparable.

Fig. 3.8 (right) shows the jet mass distribution for the third jet, calculated from the same ntuples. The scale dependence is significantly larger than for the jet p_T distribution since the jet mass, in this context, is a leading-order quantity, non-zero only when an additional gluon is emitted. While the high jet mass behaviour shown in the figure is reasonable, the low mass region lacks the Sudakov suppression that would be present in a resummation calculation or parton shower Monte Carlo. Similar distributions can be calculated for the jet mass distributions for the leading and second leading jets, with the same caveat.

---

[11]One of the authors (JH) is grateful to the GOSAM collaboration for providing and assisting in the use of these ntuples.

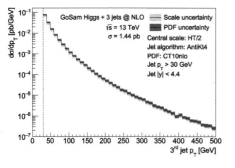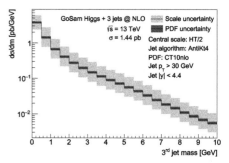

Fig. 3.8 The third jet transverse momentum distribution (left) and third jet mass distribution (right), for Higgs + ≥ 3 jets, calculated at NLO using the GoSam ROOT ntuples. The predictions are shown with scale and PDF uncertainties.

3.4   Beyond next-to-leading order in QCD

3.4.1   Next-to-next-to-leading order

Before discussing some of the technical aspects of calculations at NNLO, it is useful to review the ingredients that must enter such a calculation. Consider a computation of the cross-section for the the production of a Z boson and a jet, which has already been discussed at some length at NLO. At NLO this calculation involves one-loop diagrams and real radiation diagrams, with some mechanism for cancelling the divergences that are present in each contribution. What is the corresponding situation at NNLO?

One way to understand the different contributions is to consider all possible distinct cuts of the $\mathcal{O}(\alpha_s^3)$ four-loop diagram shown in Fig. 3.9. In total there are four types of contribution, corresponding to cuts (a)–(d) in the diagram. They can be understood as follows.

(a) The first contribution corresponds to an interference involving a two-loop diagram on one side of the cut and can be written schematically as

$$\mathfrak{Re}\left[\mathcal{A}^{2-\mathrm{loop}}(Zq\bar{q}g) \times \mathcal{A}^{\mathrm{tree}}(Zq\bar{q}g)^*\right]. \tag{3.223}$$

In the past this type of contribution had been the focus of the most attention since the evaluation of two-loop amplitudes is highly non-trivial.

(b) This contribution corresponds to the square of the one-loop three-parton matrix elements,

$$\left|\mathcal{A}^{1-\mathrm{loop}}(Zq\bar{q}g)\right|^2. \tag{3.224}$$

Note that these are the same amplitudes that appear in the NLO calculation, although in that case they enter as an interference with the tree-level amplitude.

(c) The third contribution also contains one-loop matrix elements, this time with four partons, and enters interfered with the corresponding tree-level amplitude,

$$\mathfrak{Re}\left[\mathcal{A}^{1-\mathrm{loop}}(Zq\bar{q}gg) \times \mathcal{A}^{\mathrm{tree}}(Zq\bar{q}gg)^*\right]. \tag{3.225}$$

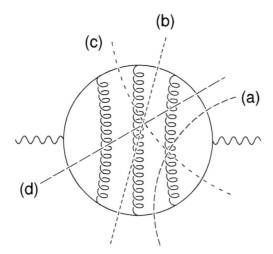

Fig. 3.9 A four-loop diagram representing NNLO contributions to the Z+jet cross-section (the Z boson is shown as an external wavy line). The diagram should be cut in all possible ways, shown by the dashed lines (a)–(d), with the cut contributions described in the text.

Notice that exactly this contribution would be present in the NLO calculation of $Z + 2$ jet production. The difference in this case is that one of the partons may be unresolved, leading to additional soft and collinear singularities.

(d) The final contribution involves only tree-level matrix elements that contain five partons,

$$\left| \mathcal{A}^{\text{tree}}(Zq\bar{q}ggg) \right|^2 . \tag{3.226}$$

In this case two partons may be unresolved, giving rise to singularities of a new form than encountered at NLO. Isolating these has provided the toughest challenge to completing such NNLO calculations.

Any NNLO calculation will therefore be substantially more complicated than its NLO counterpart and involve the introduction of new techniques. Unlike at NLO, at present there is no automatic procedure for generating predictions at this level and calculations are currently being performed on a case-by-case basis. It is therefore important to understand the benefits of having a NNLO calculation at hand in each case. Of course, by extending the perturbative calculation by an additional order, one expects that the quality of the prediction should improve. Certainly more effects can be accounted for at this order in perturbation theory. For instance, contribution (d) allows a single jet to be composed of three partons, a situation that is impossible at NLO. Such configurations are more sensitive to details of the jet algorithm that may be reflected in real data. In addition, as already discussed in Section 2.2.6, the scale

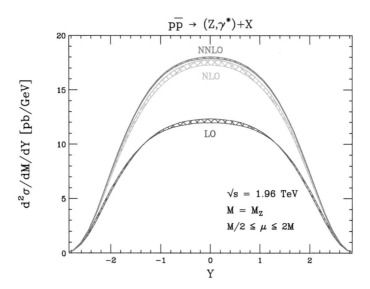

Fig. 3.10 Predictions for the rapidity distribution of an on-shell Z boson in Run II at the TEVATRON. The bands indicate the scale uncertainty of the LO, NLO and NNLO predictions, when renormalization and factorization scales are varied within the range $M_Z/2$ to $2M_Z$.

uncertainty is expected to be further reduced at NNLO compared to NLO. Finally, in cases where NLO corrections are large, it is a chance to check the convergence of the perturbative expansion and, hopefully, regain good theoretical control of the prediction. Of all of the motivations, the issue of paramount importance is obtaining a theoretical prediction with as small an uncertainty as possible in order to extract information about fundamental parameters of the Standard Model.

Most of these factors are exhibited by the calculation of the Z rapidity spectrum at the TEVATRON up to NNLO [154]. As shown in Fig. 3.10 the normalization and the shape of this distribution changes significantly between LO and NLO. However the inclusion of NNLO corrections does little to further alter the shape and the overall size of the correction amounts to only a few per cent. Finally, the scale uncertainty which is represented by the width of the bands in the figure, is greatly reduced with each successive order. Of course, the improved theoretical prediction for the Drell–Yan process is essential for calibrating many measurements at the TEVATRON and the LHC, *cf.* Chapters 8 and 9

Returning to the elements of a NNLO calculation, one of the most complicated aspects is the evaluation of the two-loop diagrams. One reason for this is that the integrals themselves have a much richer analytic structure and, importantly, a higher degree of divergence. At one-loop the singularities occur when the momentum flowing through the loop puts one of the internal particles on its mass shell and, in dimensional regularization, this results in poles as deep as $1/\varepsilon^2$. For two-loop amplitudes there are two such unconstrained momenta in the loop, so that the leading pole can be as high as $1/\varepsilon^4$. Furthermore, at one loop every amplitude can be expressed in terms of a basis of scalar integrals containing up to four propagators, *cf.* Eq. (3.103). At the two-loop level no such basis is known. Instead the usual procedure for performing such a calculation is as follows. All of the integrals appearing in the calculation of the amplitude are collected and reduced to a set of **master integrals**, to which all the others can be straightforwardly related. This reduction usually proceeds through the **Laporta algorithm** [721] which takes advantage of integration-by-parts [393] and Lorentz-invariance [559] identities and symmetry relations. A number of tools for performing this reduction, through variants of this algorithm, are commonly available. At that point the remaining work lies in evaluating the master integrals. For the most complicated cases, the current method of choice is to solve a system of differential equations and match to known solutions in particular limits. At present most of the important amplitudes for $2 \to 1$ and $2 \to 2$ processes are known, with the most complicated being those relevant for pair production of massive gauge bosons. However, relatively little is known about amplitudes for $2 \to 3$ processes and beyond.

With the two-loop contribution in hand, the remaining difficulty — which is very significant indeed — is to isolate and cancel the various singularities that enter all of the contributions. For contributions of type (c), corresponding to one-loop amplitudes in which an external particle may become unresolved, the form of the singularities and methods for handling them have been known for some time (see for instance Ref. [184] and references therein). However, a general algorithm, along the lines of the Catani–Seymour dipole subtraction method developed for NLO, has not yet been formulated. The elements of the calculation that are hardest to handle correspond to contribution (d), when two of the partons are unresolved. The factorization of both the amplitudes and the phase space in these limits is more complicated than in the single-unresolved case. For instance, except in special cases the factorization of amplitudes does not involve a product of two Altarelli–Parisi splitting functions and instead new functions must be introduced to describe all limits [319, 349].

So far, a number of techniques have been successfully applied in order to isolate and cancel the singularities appearing in NNLO calculations. The method of **sector decomposition** has been used in the calculation of corrections to the Drell–Yan process [159] and to Higgs production [160]. In this approach a special factorization of the phase space is used to ensure that all singularities can be extracted analytically by decomposing in terms of simple plus distributions, as in Eq. (A.7). A similar approach, termed **sector-improved residue subtraction** [425], has been used to compute NNLO corrections to top pair production [427] and the all-gluonic Higgs+jet channel [270]. An alternative approach, based on well-known properties of transverse momentum distributions in simple processes, is called q_T **subtraction** [350]. It has been applied to Higgs production [594], vector boson production [341], associated WH and ZH production [524], photon pair production [399] and top pair production [265]. A closely related method is **jettiness subtraction** [274, 555] which uses ideas from **soft-collinear effective theory (SCET)** to isolate and cancel infrared singularities. It has been used to compute a number of $2 \to 2$ processes at NNLO such as Higgs+jet [271], W+jet [274], and Z+jet production [269]. Last, the method of **antenna subtraction** [563] has been used to compute NNLO corrections to dijet production through gluon fusion [566] and Higgs+jet [392] and Z+jet production [564]. This approach is most similar to the subtraction methods discussed at NLO above, with counter-terms constructed that can render individual contributions finite but that can also be analytically integrated to explicitly collect singularities. For example, returning to the example above, the schematic form of the NNLO contribution to the cross-section for the parton-level process $i + j \to Z$+parton would be

$$d\hat{\sigma}_{ij}^{\mathrm{NNLO}} = \int \Phi_1 \left[d\hat{\sigma}_{ij}^{(a)} + \hat{\sigma}_{ij}^{(b)} - \hat{\sigma}_{ij}^{C_1} \right] + \int \Phi_2 \left[d\hat{\sigma}_{ij}^{(c)} - \hat{\sigma}_{ij}^{C_2} \right] + \int \Phi_3 \left[d\hat{\sigma}_{ij}^{(d)} - \hat{\sigma}_{ij}^{C_3} \right].$$

(3.227)

In these equations the superscripts label the different contributions indicated in Fig. 3.9. Each contribution is integrated over the appropriate phase space for n final-state particles, Φ_n. The counter-terms for each contribution are indicated by $\hat{\sigma}_{ij}^{C_k}$. In this formulation the counter-term C_3, for instance, removes all singularities resulting from single- and double-unresolved limits of the contribution from the matrix elements (d).

Of course, not all calculations at NNLO require the full machinery discussed here. In particular, for the simplest $2 \to 1$ and $2 \to 2$ processes, particularly if one is only interested in total cross-sections and not exclusive properties of the final-state particles, the calculations can be performed more simply. In fact, for the most important such cases NNLO results have been available for some time. The total inclusive cross-section for the Drell–Yan process, production of a lepton pair by a W or Z in a hadronic collision, has long been known to NNLO accuracy [607]. Similarly, the inclusive Higgs boson cross-section, which is a one-scale problem in the limit of large top mass, was also first computed at NNLO some time ago [158, 615].

To conclude, the frontier of NNLO calculations is currently evolving very rapidly. There are many competing methods for performing the calculations and undoubtedly these techniques will continue to be honed. At present, $2 \to 2$ reactions can be comfortably handled, even with coloured and massive objects in the final state. These pioneering calculations will lead to an even greater availability of NNLO predictions in the near future. An important final note is that, for a consistent description of the entire hard process at NNLO, all of the calculations discussed above rely on the availability of parton densities evolved at the same order. Such PDF sets are indeed available, as will be discussed in detail in Chapter 6, thanks to the calculation of the QCD three-loop splitting functions [767, 883].

3.4.2   Approximate NNLO: LoopSim

Rather than computing exact NNLO corrections it may instead be useful to consider an approximation of the full result that captures some of the most important effects that enter at that order. One such approximation is provided by the **LoopSim** method [830], that was developed to handle observables that receive significant higher order corrections resulting from new channels or topologies (*cf.* Section 2.2.7).

In this method, contributions (b), (c), and (d) are treated exactly. Only the two-loop contribution (a) is approximated, although its singular behaviour is of course known and can thus be included in its exact form. Given a set of input momenta

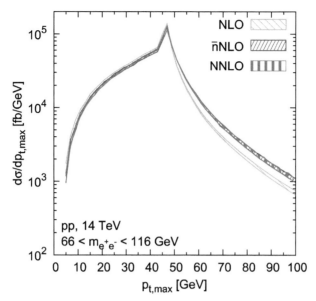

Fig. 3.11 Comparison between the approximate NNLO result from Loop-Sim (blue) and the exact result from DYNNLO (red). The corresponding NLO result is shown in green. Reprinted with permission from Ref. [830].

that correspond to the real radiation contributions entering at NNLO, the method finds a corresponding Born configuration and sequence of subsequent emissions. This step can be performed using a sequential-recombination jet algorithm, for which the Cambridge–Aachen algorithm is preferred. By considering all possible ways in which emitted particles can be combined with emitters, LoopSim determines exactly the singular (or logarithmic) contributions of loop diagrams, which unitarize the corresponding singular terms in the real radiation diagrams.[12] This approximate NNLO prediction is then finite and differs from the full NNLO calculation only by constant terms. Indeed, the expected precision of the method for an observable A is,

$$\frac{\mathrm{d}\sigma_{LoopSim}^{NNLO}}{\mathrm{d}A} = \frac{\mathrm{d}\sigma^{NNLO}}{\mathrm{d}A} \left[1 + \mathcal{O}\left(\frac{\alpha_s^2}{K^{NNLO}(A)} \right) \right] \qquad (3.228)$$

where $\mathrm{d}\sigma^{NNLO}/\mathrm{d}A = K^{NNLO}(A)\,\mathrm{d}\sigma_{LO}/\mathrm{d}A$ defines the "local" NNLO K factor as a function of A. As $K^{NNLO}(A)$ increases, the quality of the approximation is expected to dramatically improve. In cases where this can be explicitly tested, such as the Drell–Yan process, this expectation is confirmed, as shown in Fig. 3.11. Comparisons of the predictions of this method with LHC data representing more complicated final states will be discussed in Chapter 9.

---

[12]As such, this method bears an interesting similarity to multijet merging methods for matrix elements and parton showers that will be discussed in Chapter 5. At present this connection has not been fully explored.

3.4.3   Beyond NNLO

Since NNLO predictions for hadron collider observables are only just becoming more widespread, the prospects for extending these to even higher orders in perturbation theory are hard to assess. However, given the complexity of any such calculation, there must be a very strong need for an improved theoretical prediction. The case of the Higgs cross-section is a strong candidate for such a calculation since this knowledge could be put to use immediately in trying to unravel the properties, such as coupling strengths, of the Higgs boson. This calculation is also among the most simple that could be conceived at a hadron collider, since it is a $2 \to 1$ calculation involving a scalar particle.

For this reason there has already been a remarkable amount of progress on computing the Higgs boson cross-section at N^{3}LO. The first partial results for this calculation were presented in Ref. [155], where the calculation is performed as an expansion about the threshold region. Specifically, the partonic cross-section is expanded as,

$$\hat{\sigma}_{ij}(m_H^2, \hat{s}) \propto \sum_{k=0}^{\infty} \left(\frac{\alpha_s}{\pi}\right)^k \eta_{ij}^{(k)}(z), \tag{3.229}$$

where $z = m_H^2/\hat{s}$ so that $(1-z)$ is the variable that parameterizes the distance from threshold. The results for the first term in the expansion of $\eta_{ij}^{(3)}(z)$ are presented in Ref. [155], neglecting terms of order $(1-z)$. The same technique is of course also applicable to the Drell–Yan process, where a similar level of approximation has also already been applied [129]. The phenomenological impact of these results is ambiguous, due to the fact that there can be substantial differences resulting from equivalent parameterizations of the neglected terms. Nevertheless, the same methodology has very recently been extended to compute the sub-leading terms to an arbitrary level [157]. This effectively results in a full N^{3}LO calculation, paving the way for a similar level of precision for a range of inclusive cross-sections.

3.4.4   Detour: Electroweak corrections

When making theoretical predictions for a hadron collider it is natural to consider a perturbative expansion in the strong coupling, as described so far. The hard scattering must involve strongly interacting particles and the expansion parameter α_s is an order of magnitude larger than the corresponding electroweak coupling, α_w. Nevertheless, one may still be interested in the effect of electroweak corrections in a number of circumstances.

First of all, simply estimating the size of the corrections from the numerical values of the couplings, one might expect that, since $\alpha_w \sim \alpha_s^2$, the effect of NLO EW corrections should be considered at the same time as NNLO QCD, when the precision of the theoretical precision is paramount. A further motivation, that has received considerable attention in recent years, is due to the expected nature of electroweak effects at high energies. As in QCD, the emission of gauge bosons in the electroweak theory can be described by an eikonal factor. The crucial difference is that, when integrated over phase space, the mass of the W and Z bosons provides a cut-off for the integral so that it does not diverge. As a result the eikonal approximation results in the EW Sudakov factor,

$$\hat{\sigma}_{\text{EW real}} \sim \frac{\alpha_{\text{w}}}{4\pi} \log^2 \left(\frac{s}{m_W^2} \right) \hat{\sigma}_0. \tag{3.230}$$

where s is the hard scale at which the process is being probed, *cf.* the QCD equivalent in Chapter 2. There is a corresponding virtual contribution, generated by one-loop diagrams in which a W or Z boson is exchanged internally, and just as in the QCD case this enters with the opposite sign,

$$\hat{\sigma}_{\text{EW virtual}} \sim -\frac{\alpha_{\text{w}}}{4\pi} \log^2 \left(\frac{s}{m_W^2} \right) \hat{\sigma}_0. \tag{3.231}$$

However, there is a crucial difference between the QCD and electroweak cases. The Sudakov factor is associated with a combination of isospin generators and, for a fixed initial state, there is a mismatch when the forms appearing in Eq. (3.230) and Eq. (3.231) are promoted to exact relations [866]. This violation of the Bloch–Nordsieck theorem is due to electroweak symmetry breaking and the fact that the initial states present in a given process are not averaged over, but weighted by different PDFs. In addition, and perhaps even more importantly for the interpretation of collider data, it is of course straightforward to isolate the effects of EW radiation in data — unlike the case of QCD radiation. The lack of infrared divergences means that it is customary that events that are identified as containing additional W or Z bosons form a separate data sample. Of course, there are regions where W and Z radiation escapes detection and some effect from these contributions can partially cancel the virtual corrections, depending on the particular process and the experimental setup [206]. However, the net effect of real EW radiation is typically rather small and therefore the virtual corrections have been the subject of the most intense theoretical scrutiny.

Note that the form of the Sudakov correction in Eq. (3.231), that is enhanced by two powers of the logarithm, multiplies the leading-order amplitude. This means that it is easy to estimate the size of the leading relative EW corrections for the simplest processes,

$$\delta^{\text{EW}} = \frac{\hat{\sigma}^{\text{EW virtual}}}{\hat{\sigma}_0} = -\,(\text{constant})\, \frac{\alpha_{\text{w}}}{4\pi} \log^2 \left(\frac{s}{m_W^2} \right). \tag{3.232}$$

This approximation is rather crude since it assumes that all relevant kinematic scales are large and can be approximated by a single value s, for instance for a $2 \to 2$ process $s \approx |t| \approx |u|$. The constant that appears in Eq. (3.232) is a combination of isospin Casimirs and depends on the identities of the particles participating in the reaction. It can be written in terms of electroweak parameters such as the weak mixing angle and is usually of order unity. To illustrate the size of the correction that can be anticipated, Fig. 3.12 shows the value of δ^{EW} obtained from Eq. (3.232) as a function of $\sqrt{s}$, with the constant set to 1. In reality this constant varies for each sub-process and the large negative contribution is partially mitigated by collinear (single) logarithms, but Fig. 3.12 gives a good guide to the size of the corrections that can be expected. With corrections of order 10% or larger for scales of 1 TeV and beyond, in regions of large

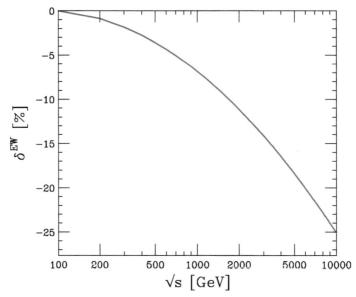

Fig. 3.12 An estimate of the importance of leading electroweak Sudakov logarithms, obtained from Eq. (3.232) with the constant set equal to 1, as a function of $\sqrt{s}$.

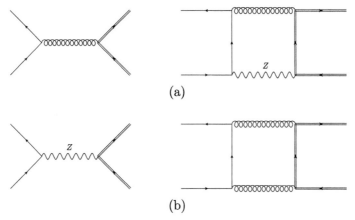

Fig. 3.13 Sample box diagrams entering the NLO EW corrections to the process $q\bar{q} \to t\bar{t}$.

invariant mass or transverse momentum it is possible that the EW corrections can be as large as their QCD counterparts.

A complication is that the consistent inclusion of EW effects often means the calculation of a wider set of contributions that must be handled with care. As a concrete example, consider the reaction $q\bar{q} \to t\bar{t}$ that proceeds at $\mathcal{O}(\alpha_s)$ through a single diagram with an s-channel gluon. The virtual electroweak corrections to this process include a number of self-energy, vertex, and box diagram contributions. One of the contributing box diagram contributions is shown in panel (a) of Fig. 3.13. It consists of the tree-level $\mathcal{O}(\alpha_{\mathrm{s}})$ process interfered with the $\mathcal{O}(\alpha_{\mathrm{s}}\alpha_{\mathrm{w}})$ one-loop correction involving an internal Z-boson in the loop. However, at the same order one must also consider the interference of the diagrams shown in panel (b): the tree-level $\mathcal{O}(\alpha_{\mathrm{w}})$ process and the one-loop $\mathcal{O}(\alpha_{\mathrm{s}}^2)$ box. In this way the weak and strong amplitudes become entangled when computing EW corrections if a given final state may be obtained at tree-level through both strong and weak interactions. In fact the inclusion of all the diagrams in Fig. 3.13 results in an infrared-divergent contribution that must be cancelled by the radiation of real gluons. These diagrams correspond to dressing the tree-level ones in Fig. 3.13 with a gluon, to obtain $\mathcal{O}(\alpha_{\mathrm{s}}\sqrt{\alpha_{\mathrm{s}}})$ and $\mathcal{O}(\alpha_{\mathrm{w}}\sqrt{\alpha_{\mathrm{s}}})$ amplitudes, and performing the interference. Singularities can be extracted and cancelled against the virtual contributions using the same methods, such as dipole subtraction, that are applied in the pure QCD case [704].

Although the motivation for including EW corrections has been presented in terms of Sudakov logarithms, for the simplest processes it is possible to compute the corrections exactly, to include all sub-leading effects. In this way EW corrections have been obtained and investigated for most $2 \to 2$ processes, including the production of dijets, vector bosons and a jet, top pairs, and dibosons. A recent review [765] summarizes the most important results and contains references to the original calculations. Alternatively, one can use the known factorization properties of the Sudakov logarithms to obtain an approximate form of the EW corrections [461], a strategy that can be used to provide the corrections for more complex final states [395].

3.5 Summary

This chapter has discussed in some detail the essential elements involved in making perturbative predictions for hadron colliders. There has been rapid progress in this area in the years leading up to data-taking at the LHC, as the theoretical predictions have had to evolve to match the expected breadth and precision afforded by such a machine. The latest tools are able to provide NLO corrections for configurations involving many jets and NNLO predictions and beyond for the most important processes. In many cases even the effect of electroweak corrections can be included.

Poised at the brink of yet another substantial jump in machine capability — not only an increase in energy, but also an unprecedented amount of data resulting from the increased luminosity — it is imperative that perturbative predictions continue to improve at a similar rate. The preparation of the "Les Houches wishlist" [294] has provided a forum for discussing the most useful calculations that could be performed, in order to ensure that such progress is achieved. By now, all of the NLO perturbative QCD calculations contained in the original list have been completed, primarily due to the emergence of the unitarity techniques discussed in Section 3.3.1. The latest iteration of the list [296] reflects the high-precision calculations that are expected to be required during the expected lifetime of the LHC snd typically demands the inclusion of both NNLO QCD and NLO EW effects. As an example, Table 3.5 shows the Higgs-related calculations for which a strong need is anticipated in the future. For each calculation presented in the table, Ref. [296] discusses the motivation and degree of need, which is particularly important given the challenging nature of many of the demands. Note that such discussions are certainly not limited to the case of Higgs boson cross-sections and Ref. [296] contains wishlists for other Standard Model processes.

As an example of the motivation, consider the Higgs + 2 jets final state. This channel is crucial in order to understand the Higgs boson coupling to vector bosons, through the vector boson fusion (VBF) channel. For the VBF channel the NNLO QCD corrections are known in a fully differential form, in the double-DIS approximation, and EW effects are known to NLO. However, the search for this production mode suffers from a background consisting of Higgs production through gluon fusion, when two additional hard jets are also radiated. Currently this channel is known only to NLO QCD in the infinite top mass approximation, and only to LO QCD when the full dependence on the top quark mass is retained. If both the VBF and gluon fusion Higgs + 2 jets cross-section are known to NNLO QCD, and NLO EW, accuracy then with 300fb^{-1} of data it may be possible to measure the HWW coupling strength to the order of 5%.

Having discussed the frontier of fixed-order, parton-level treatments, the following chapter will describe in more detail the application of these predictions to a wide range of hadron collider processes. A variety of alternative approaches that go beyond the ones presented so far will be discussed in Chapter 5. These represent all-orders treatments that either address regions where the calculations presented here break down, or which in addition allow predictions to be made at the hadron level for a direct comparison with data.

Table 3.5 The "Les Houches wishlist" for processes involving the Higgs boson, taken from Ref. [296].

Process	Comments
H	State of the Art
	$d\sigma$ @ NNLO QCD (expansion in $1/m_t$)
	full m_t/m_b dependence @ NLO QCD and @ NLO EW
	NNLO+PS, in the $m_t \to \infty$ limit
	Desired
	$d\sigma$ @ NNNLO QCD (infinite-m_t limit)
	full m_t/m_b dependence @ NNLO QCD and @ NNLO QCD+EW
	NNLO+PS with finite top quark mass effects
$H + j$	State of the Art
	$d\sigma$ @ NNLO QCD (g only)
	and finite-quark-mass effects @ LO QCD and LO EW
	Desired
	$d\sigma$ @ NNLO QCD (infinite-m_t limit)
	and finite-quark-mass effects @ NLO QCD and NLO EW
$H + 2j$	State of the Art
	σ_{tot}(VBF) @ NNLO(DIS) QCD and $d\sigma$(VBF) @ NLO EW
	$d\sigma(gg)$ @ NLO QCD (infinite-m_t limit)
	and finite-quark-mass effects @ LO QCD
	Desired
	$d\sigma$(VBF) @ NNLO QCD + NLO EW
	$d\sigma(gg)$ @ NNLO QCD (infinite-m_t limit)
	and finite-quark-mass effects @ NLO QCD and NLO EW
$H + V$	State of the Art
	$d\sigma$ @ NNLO QCD and $d\sigma$ @ NLO EW
	$\sigma_{tot}(gg)$ @ NLO QCD (infinite-m_t limit)
	Desired
	with $H \to b\bar{b}$ @ same accuracy
	$d\sigma(gg)$ @ NLO QCD with full m_t/m_b dependence
tH and $\bar{t}H$	State of the Art
	$d\sigma$(stable top) @ LO QCD
	Desired
	$d\sigma$(top decays) @ NLO QCD and NLO EW
$t\bar{t}H$	State of the Art
	$d\sigma$(stable tops) @ NLO QCD
	Desired
	$d\sigma$(top decays) @ NLO QCD and NLO EW
$gg \to HH$	State of the Art
	$d\sigma$ @ NLO QCD (leading m_t dependence)
	$d\sigma$ @ NNLO QCD (infinite-m_t limit)
	Desired
	$d\sigma$ @ NLO QCD with full m_t/m_b dependence

4

QCD at Fixed Order: Processes

In Chapter 2 the description of hadron collider processes at fixed order in pertu-
bation theory was introduced and in Chapter 3 the technology for performing such
calculations was reviewed. This chapter will discuss the application of these ideas
to describing important Standard Model processes that can be probed at the LHC
and in other hadron collider experiments. In particular, theoretical issues related to
calculations for more complex final states will be addressed. The discussion of these
processes begins with jet production in Section 4.1. This is followed by a discussion of
processes that are, in a theoretical sense, closely related: the production of photons in
association with jets (Section 4.2). The extension to the production of gauge bosons
plus, potentially, some jets is the focus of Section 4.3. Other processes considered include the production of pairs of electroweak bosons (Section 4.4), top pair production
(Section 4.5) and single-top processes (Section 4.6). A selection of processes that are
more rare than these are briefly discussed in Section 4.7. The chapter concludes in
Section 4.8 with an overview of the main channels by which a Higgs boson is produced and observed in the LHC experiments. These processes will remain central to
the continuing LHC program for the foreseeable future.

A more in-depth comparison of these fixed order results with data will be presented
in later chapters, *cf.* Chapters 8 and 9.

4.1 Production of jets

In the previous chapter, an outline of the perturbative approach was introduced by
considering the production of a lepton and neutrino through the weak interaction. A
far more likely outcome of a collision is the production of a final state through the
strong interaction, which in the partonic picture is represented by quarks and gluons.
In contrast to the simplest examples discussed previously, in this case even at LO there
are many contributions corresponding to various combinations of partons in both the
initial and final states. Therefore the two-jet cross-section is given by (*cf.* Eq. (2.52)),

$$\sigma_{2-\text{jet}} = \frac{1}{2s} \sum_{a,b,c,d} \int_0^1 \frac{\mathrm{d}x_a}{x_a} \frac{\mathrm{d}x_b}{x_b} \, f_{a/h_1}(x_a, \mu_F) f_{b/h_2}(x_b, \mu_F) \int \mathrm{d}\Phi_n \, |\mathcal{M}_{ab \to cd}|^2, \quad (4.1)$$

The Black Book of Quantum Chromodynamics. John Campbell, Joey Huston and Frank Krauss, Oxford University Press. © John Campbell, Joey Huston, and Frank Krauss 2018, First published in paperback 2022. DOI: 10.1093/oso/9780192871961.001.0004

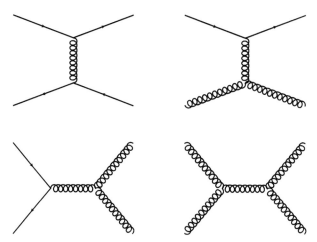

Fig. 4.1 Representative Feynman diagrams for four categories of LO reactions entering the calculation of two-jet production at hadron colliders. These are: $q\bar{q}' \to q\bar{q}'$ (top left), $qg \to qg$ (top right), $q\bar{q} \to gg$ (bottom left), and $gg \to gg$ (bottom right).

where the sum runs over all permissible combinations of quarks, anti-quarks, and gluons. Four such parton-level processes that must be included at leading order in the strong coupling, which shall be discussed in more detail shortly, are shown in Fig. 4.1. These interactions should produce strongly interacting particles in opposite hemispheres, with zero net transverse momentum. The rate for this process is extremely high at the LHC, with cross-sections for typical jet cuts, $p_T(\text{jet}) > 50$ GeV, in the tens of microbarn range, *cf.* Fig. 4.2. The cross-section for two jets with transverse momenta of 1 TeV or higher is still at the level of a nanobarn. Such energetic jets are of course much easier to produce at a possible future hadron collider: 1 TeV jets at a 100 TeV collider are just as common as 100 GeV jets at a 14 TeV LHC, with a cross-section of a few microbarns.

This process represents an important probe of QCD in a number of ways. For instance, by measuring cross-sections as a function of jet transverse momenta it is possible to assess the running of the strong coupling. The large cross-section enables measurements of this process to be made for a wide range of transverse momenta and rapidities that, due to the very simple kinematics involved, can easily be translated into information about the PDFs (as will be discussed further in Chapter 6). Besides these areas where the two-jet process itself is the process of interest, jet production can easily be a source of background events for many analyses, for instance when one of the jets is misidentified as a lepton or a photon, or when mis-measurement of one of the jets results in substantial missing transverse momentum. Since the cross-section is so large, even a very small fake rate can lead to significant background rates. It is therefore essential to understand this process in some detail.

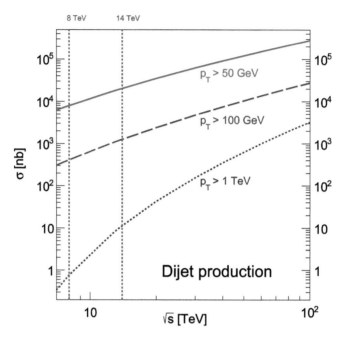

Fig. 4.2 Dijet production cross-sections at proton–proton colliders as a function of centre-of-mass operating energy, $\sqrt{s}$. Three values of the minimum jet transverse momentum are considered: 50 GeV (solid), 100 GeV (dashed), and 1 TeV (dotted).

4.1.1   Dijet production

In the simple picture outlined so far the theoretical definition of the dijet process is clear, a $2 \to 2$ scattering involving only quarks and gluons. For the sake of illustration consider the four sub-processes, the reactions $qq' \to qq'$, $qg \to qg$, $q\bar{q} \to gg$, and $gg \to gg$ that are depicted in Fig. 4.1. The matrix elements for these processes, summed over final-state colour and spins and averaged over the colour and spins in the initial state, are,

$$|\mathcal{M}_{qq' \to qq'}|^2 = \frac{1}{4N^2}\, m^{(0)}_{q\bar{q}' \to q\bar{q}'} = \frac{V}{2N^2}\left(\frac{\hat{s}^2 + \hat{u}^2}{\hat{t}^2}\right), \tag{4.2}$$

$$|\mathcal{M}_{qg \to qg}|^2 = \frac{1}{4NV}\, m^{(0)}_{qg \to qg} = \frac{-1}{2N^2}\left(\frac{V}{\hat{u}\hat{s}} - \frac{2N^2}{\hat{t}^2}\right)(\hat{s}^2 + \hat{u}^2), \tag{4.3}$$

$$|\mathcal{M}_{q\bar{q} \to gg}|^2 = \frac{1}{4N^2}\, m^{(0)}_{q\bar{q} \to gg} = \frac{V}{2N^3}\left(\frac{V}{\hat{u}\hat{t}} - \frac{2N^2}{\hat{s}^2}\right)(\hat{t}^2 + \hat{u}^2), \tag{4.4}$$

$$|\mathcal{M}_{gg \to gg}|^2 = \frac{1}{4V^2}\, m^{(0)}_{gg \to gg} = \frac{2N^2}{V}\left(3 - \frac{\hat{u}\hat{t}}{\hat{s}^2} - \frac{\hat{s}\hat{u}}{\hat{t}^2} - \frac{\hat{s}\hat{t}}{\hat{u}^2}\right), \tag{4.5}$$

which also implicitly defines the non-averaged leading-order matrix elements $m^{(0)}$. The matrix elements are written in terms of the usual invariant quantities $\hat{s} = (p_1 + p_2)^2$, $\hat{t} = (p_1 - p_3)^2$ and $\hat{u} = (p_2 - p_3)^2$. Note that, as is clear from Fig. 4.1, the diagrams entering the calculations of the reactions $qg \rightarrow qg$ and $q\bar{q} \rightarrow gg$ are identical. The processes differ only by the identities of the partons that are present in the initial state. Therefore the matrix elements are expected to be related by a **crossing symmetry**, in this case the exchange of p_2 and $-p_3$, where the minus sign accounts for the fact that the particles are exchanged between the initial and final states. By noting that this corresponds to the exchange $\hat{s} \leftrightarrow \hat{t}$, the crossing relation can be straightforwardly verified by inspecting Eq. (4.5), up to an overall factor related to colour averaging in the initial state and a minus sign from crossing.

To turn these matrix elements into a cross-section they must be combined with the appropriate two-particle phase space. As shown explicitly in Appendix A.3, the phase space for two massless particles can be written directly in terms of the transverse momentum and rapidity of one of them as

$$\mathrm{d}\Phi_2 = \frac{p_\perp \mathrm{d}p_\perp \mathrm{d}\eta \mathrm{d}\phi}{2(2\pi)^3} (2\pi) \delta\left((p_1 + p_2 - p_3)^2\right) \tag{4.6}$$

where it is convenient to work in the lab frame, in which the four-momentum of the first jet can be written as,

$$p_3 = p_\perp (\cosh \eta, \sin \phi, \cos \phi, \sinh \eta) \tag{4.7}$$

(see Appendix A.3). After performing the trivial integration over ϕ and using the $p_\perp$ integration to remove the δ function, the phase space takes the simple form,

$$\mathrm{d}\Phi_2 = \frac{1}{4\pi} \frac{p_\perp^2}{\hat{s}} \mathrm{d}\eta. \tag{4.8}$$

This is the phase-space element for a jet of a given transverse momentum ($p_\perp$), with the remaining degree of freedom parameterized by its rapidity (η). An even more useful variable is one that reflects more directly the kinematic configuration of both jets. To that end it is convenient to write the momentum of the other jet as

$$p_4 = p_\perp (\cosh \eta', -\sin \phi, -\cos \phi, \sinh \eta') \tag{4.9}$$

so that momentum conservation ($p_1 \cdot p_2 = p_3 \cdot p_4$) then implies the relation

$$\sqrt{\hat{s}} = 2p_\perp \cosh\left(\frac{\eta_3 - \eta_4}{2}\right) = p_\perp \left(\frac{\chi + 1}{\sqrt{\chi}}\right), \tag{4.10}$$

where $\chi = \exp(\eta_3 - \eta_4)$. The variable χ is clearly a natural one to describe the problem and, since it only depends on the difference of two rapidities, it is invariant under longitudinal boosts. This property means that it has a very simple definition in terms of the angle θ between the jet and the beam in the centre-of-mass frame. To see this relation it is useful to write expressions for the invariants $\hat{t}$ and $\hat{u}$ in the two frames,

$$\hat{t} = -\frac{\hat{s}}{2}(1 - \cos\theta) = -\frac{\hat{s}}{\chi + 1},$$

$$\hat{u} = -\frac{\hat{s}}{2}(1 + \cos\theta) = -\frac{\hat{s}\chi}{\chi + 1}. \tag{4.11}$$

From these it is clear that χ and θ are related by

$$\chi = \frac{1 + \cos\theta}{1 - \cos\theta}. \tag{4.12}$$

Performing the change of variables $\mathrm{d}\eta \to \mathrm{d}\chi/\chi$ in Eq. (4.8) yields a particularly simple parameterization of the phase space,

$$\mathrm{d}\Phi_2 = \frac{1}{4\pi}\frac{\mathrm{d}\chi}{(\chi + 1)^2}. \tag{4.13}$$

Finally, combining the final-state phase space in Eq. (4.13) with the matrix elements of Eq. (4.5) recast in terms of χ, yields the relevant quantities for each partonic channel,

$$\mathrm{d}\Phi_2 \left|\mathcal{M}_{q\bar{q}' \to q\bar{q}'}\right|^2 = \frac{1}{4\pi}\frac{V}{2N^2}\mathrm{d}\chi\left[1 + \left(\frac{\chi}{\chi + 1}\right)^2\right], \tag{4.14}$$

$$\mathrm{d}\Phi_2 \left|\mathcal{M}_{qg \to qg}\right|^2 = \frac{1}{4\pi}\frac{1}{2N^2}\mathrm{d}\chi \tag{4.15}$$

$$\times \left(V\left[\frac{1}{\chi(\chi + 1)} + \frac{\chi}{(\chi + 1)^3}\right] + 2N^2\left[1 + \left(\frac{\chi}{\chi + 1}\right)^2\right]\right),$$

$$\mathrm{d}\Phi_2 \left|\mathcal{M}_{q\bar{q} \to gg}\right|^2 = \frac{1}{4\pi}\frac{V}{2N^3}\frac{\mathrm{d}\chi}{(\chi + 1)^2}\left[V\left(\chi + \frac{1}{\chi}\right) - 2N^2\frac{1 + \chi^2}{(\chi + 1)^2}\right], \tag{4.16}$$

$$\mathrm{d}\Phi_2 \left|\mathcal{M}_{gg \to gg}\right|^2 = \frac{1}{4\pi}\frac{2N^2}{V}\mathrm{d}\chi\frac{(1 + \chi + \chi^2)^3}{\chi^2(\chi + 1)^4}. \tag{4.17}$$

With the exception of the $gg \to gg$ reaction, these expressions depend rather weakly on χ in the physical region, $\chi \geq 1$. As a result, a measurement of $\mathrm{d}\sigma/\mathrm{d}\chi$ for jet production is quite insensitive to the details of the parton distribution functions that have yet to be folded in. Indeed, this weak dependence on χ is none other than the statement that these processes are dominated by the t-channel exchange of a spin-1 gluon, analogous to Rutherford scattering.

Indeed, this observable can be used to search for evidence of **contact interactions** that would indicate quark substructure. For instance, in the presence of an additional 4-quark contact term [394, 718],

$$\mathcal{L}_{\text{contact}} = \frac{2\pi}{\Lambda^2}(\bar{\psi}_L\gamma^\mu\psi_L)(\bar{\psi}_L\gamma_\mu\psi_L), \tag{4.18}$$

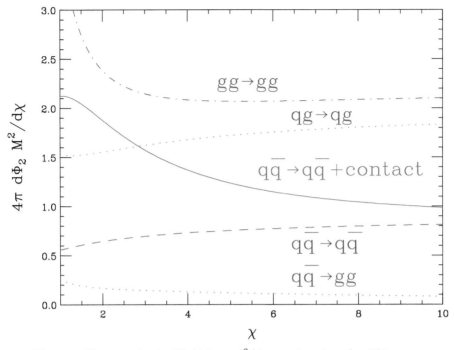

Fig. 4.3 The quantity $4\pi\,\mathrm{d}\Phi_2|\mathcal{M}_{ab\to cd}|^2/\mathrm{d}\chi$ as a function of χ. This quantity is plotted for $qg \to qg$ (upper dotted), $q\bar{q} \to q\bar{q}$ (dashed), $q\bar{q} \to gg$ (lower dotted), $gg \to gg$ (dot-dashed), and $q\bar{q} \to q\bar{q}$ in the presence of a contact interaction with $\hat{s} = 0.5$ TeV, $\Lambda = 1$ TeV, and $\alpha_s = 0.1$ (solid).

the expression for the quark–anti-quark contribution to the cross-section is modified to,

$$\mathrm{d}\Phi_2\,|\mathcal{M}_{q\bar{q}'\to q\bar{q}'}|^2 = \frac{\mathrm{d}\chi}{4\pi}\left\{\frac{V}{2N^2}\left[1+\left(\frac{\chi}{\chi+1}\right)^2\right]+\left(\frac{\hat{s}^2}{\alpha_s^2\Lambda^4}\right)\frac{\chi^2}{(\chi+1)^4}\right\}. \qquad (4.19)$$

An illustration of the behaviour of these various contributions is shown in Fig. 4.3. As observed previously, the curves for the quark–gluon and quark–anti-quark initial states depend only very mildly on χ. The reaction $gg \to gg$ shows a rapid rise as $\chi \to 1$, but the curve is rather flat for $\chi > 2$. In contrast, in the presence of a contact interaction with $\hat{s} = 0.5$ TeV, $\Lambda = 1$ TeV, and $\alpha_s = 0.1$ there is a marked rise for moderate χ, $2 \lesssim \chi \lesssim 6$. In reality this simple picture is distorted somewhat once the convolution with the parton distribution functions has been performed and all partonic channels summed over, but the essential difference remains. The issue of contact interactions will be revisited, in the context of experimental searches at the TEVATRON and the LHC, in Chapters 8 and 9.

A further simple remark concerning the kinematics of the dijet process is useful in order to demonstrate its importance in determinations of the PDFs. For a collision involving two partons with momentum fractions x_1 and x_2 of the parent hadrons, momentum conservation yields the relations,

$$\frac{\sqrt{s}}{2}(x_1 + x_2) = p_\perp(\cosh\eta + \cosh\eta'), \tag{4.20}$$

$$\frac{\sqrt{s}}{2}(x_1 - x_2) = p_\perp(\sinh\eta + \sinh\eta'). \tag{4.21}$$

These equations are easily solved for x_1 and x_2,

$$x_1 = \frac{p_\perp}{\sqrt{s}}\left(e^\eta + e^{\eta'}\right), \qquad x_2 = \frac{p_\perp}{\sqrt{s}}\left(e^{-\eta} + e^{-\eta'}\right). \tag{4.22}$$

Hence different ranges of transverse momenta and rapidities represent probes of particular regions of x_1 and x_2. In order to probe large momentum fractions one can investigate the behaviour of jets with large p_T or η. Since a jet has a certain minimum p_T, in order to access small momentum fractions it is most useful to examine jets of moderate transverse momenta but that lie at large rapidities. These features are exploited in global fits of PDFs to hadron collider jet data, as will be discussed further in Chapter 6.

4.1.2   Dijets at next-to-leading order

The next-to-leading-order corrections to dijet production have been known for a long time [509, 574]. The virtual corrections are relatively simple due to the fact that it is a $2 \to 2$ process involving only massless particles, so that the scalar integrals can be written in terms of only logarithms and constants.

The perturbative expansion for the partonic process $a + b \to c + d$, without the average over initial state colours and spins, reads,

$$m_{ab \to cd} = m^{(0)}_{ab \to cd} + \left(\frac{\alpha_S}{2\pi}\right) m^{(v)}_{ab \to cd} + \dots, \tag{4.23}$$

where the virtual matrix elements $m^{(v)}_{ab \to cd}$ represent the interference of tree-level and loop diagrams. For the process $qq' \to qq'$ the virtual contribution is given by [503],

$$
\begin{aligned}
m^{(v)}_{qq' \to qq'} = c_\Gamma \bigg\{ & \left[C_F\left(-\frac{4}{\varepsilon^2} - \frac{1}{\varepsilon}\left(6 + 8l(s) - 8l(u) - 4l(t)\right)\right) \right. \\
& + \frac{N_c}{\varepsilon}\left(4l(s) - 2l(u) - 2l(t)\right) + n_f T_R\left(\frac{4}{3}\left(l(t) - l(-\mu^2)\right) - \frac{20}{9}\right) \\
& + \left(C_F\left(-16 - 2l^2(t) + l(t)\left(6 + 8l(s) - 8l(u)\right)\right)\right. \\
& \left. + N_c\left(\frac{85}{9} + \pi^2 + 2l(t)\left(l(t) + l(u) - 2l(s)\right) + \frac{11}{3}\left(l(-\mu^2) - l(t)\right)\right)\right] m^{(0)}_{qq' \to qq'} \\
& - 4VC_F\frac{s^2 - u^2}{t^2}\left(2\pi^2 + 2l^2(t) + l^2(s) + l^2(u) - 2l(s)l(t) - 2l(t)l(u)\right)
\end{aligned}
$$

$$+ N_c V \left(\frac{s^2 - u^2}{t^2} \left(3\pi^2 + 3l^2(t) + 2l^2(s) + l^2(u) - 4l(s)l(t) - 2l(t)l(u) \right) \right)$$
$$+ 4V \left(C_F \left(\left(l(u) - l(s) \right) - \frac{u - s}{t} \left(2l(t) - l(s) - l(u) \right) \right) \right)$$
$$+ N_c \left(-\frac{s}{2t} \left(l(t) - l(u) \right) + \frac{u}{t} \left(l(t) - l(s) \right) \right) \right) \Bigg\}. \tag{4.24}$$

This equation is written in terms of the colour factor $V = N_c^2 - 1$ and the logarithmic function,

$$l(x) = \log \left(-\frac{x}{Q^2} \right), \tag{4.25}$$

where $Q^2 > 0$ is an arbitrary momentum scale. Note that this means the function develops an imaginary part for $x > 0$. Since only the real part should be kept in Eq. (4.24), this only results in contributions from terms of the form,

$$l^2(t) = \log^2 \left(-\frac{t}{Q^2} \right) \to \log^2 \left(\frac{t}{Q^2} \right) - \pi^2 \qquad \text{if } t > 0. \tag{4.26}$$

The result in Eq. (4.24) also demonstrates that, unlike the simplest Drell–Yan case considered in Chapter 3, in general the virtual matrix element exhibits a rich kinematic structure. Only the singular terms are proportional to the lowest-order matrix element, while the finite remainder depends on s, t, and u in a more complicated way.

The other partonic contributions can be written in a similar fashion. The virtual contribution to the process $q\bar{q} \to gg$ is,

$$
\begin{aligned}
m^{(v)}_{q\bar{q} \to gg} = c_\Gamma \Bigg\{ & \left[C_F \left(-\frac{2}{\varepsilon^2} - \frac{3}{\varepsilon} - 7 \right) + N_c \left(-\frac{2}{\varepsilon^2} - \frac{11}{3\varepsilon} + \frac{11}{3} l(-\mu^2) \right) \right. \\
& \left. + n_f T_R \left(\frac{4}{3\varepsilon} - \frac{4}{3} l(-\mu^2) \right) \right] m^{(0)}_{q\bar{q} \to gg} \\
& + \frac{l(s)}{\varepsilon} \left[\left(2N_c^2 V + \frac{2V}{N_c^2} \right) \frac{t^2 + u^2}{ut} - 4V^2 \frac{t^2 + u^2}{s^2} \right] \\
& + \frac{4N_c^2 V}{\varepsilon} \left[l(t) \left(\frac{u}{t} - \frac{2u^2}{s^2} \right) + l(u) \left(\frac{t}{u} - \frac{2t^2}{s^2} \right) \right] \\
& - \frac{4V}{\varepsilon} \left(\frac{u}{t} + \frac{t}{u} \right) (l(t) + l(u)) \Bigg\} + f_1(s, t, u) + f_1(s, u, t), \tag{4.27}
\end{aligned}
$$

which makes explicit the form of the singular terms. The auxiliary function f_1 contains only finite contributions and is defined by

$$f_1(s,t,u) = 4N_c V \Bigg\{$$

$$\frac{l(t)l(u)}{N_c}\frac{t^2+u^2}{2tu} + l^2(s)\left[\frac{1}{4N^3}\frac{s^2}{tu} + \frac{1}{4N_c}\left(\frac{1}{2} + \frac{t^2+u^2}{tu} - \frac{t^2+u^2}{s^2}\right) - \frac{N_c}{4}\frac{t^2+u^2}{s^2}\right]$$

$$+l(s)\left[\left(\frac{5}{8}\frac{V}{N_c} - \frac{1}{2N_c} - \frac{1}{N_c^3}\right) - \left(N_c + \frac{1}{N_c^3}\right)\frac{t^2+u^2}{2tu} - \frac{V}{4N_c}\frac{t^2+u^2}{s^2}\right]$$

$$+\pi^2\left[\frac{1}{8N_c} + \frac{1}{N_c^3}\left(\frac{3(t^2+u^2)}{8tu} + \frac{1}{2}\right) + N_c\left(\frac{t^2+u^2}{8tu} - \frac{t^2+u^2}{2s^2}\right)\right]$$

$$+\left(N_c + \frac{1}{N_c}\right)\left(\frac{1}{8} - \frac{t^2+u^2}{4s^2}\right)$$

$$+l^2(t)\left[N_c\left(\frac{s}{4t} - \frac{u}{s} - \frac{1}{4}\right) + \frac{1}{N_c}\left(\frac{t}{2u} - \frac{u}{4s}\right) + \frac{1}{N_c^3}\left(\frac{u}{4t} - \frac{s}{2u}\right)\right]$$

$$+l(t)\left[N_c\left(\frac{t^2+u^2}{s^2} + \frac{3t}{4s} - \frac{5u}{4t} - \frac{1}{4}\right) - \frac{1}{N_c}\left(\frac{u}{4s} + \frac{2s}{u} + \frac{s}{2t}\right) - \frac{1}{N_c^3}\left(\frac{3s}{4t} + \frac{1}{4}\right)\right]$$

$$+l(s)l(t)\left[N_c\left(\frac{t^2+u^2}{s^2} - \frac{u}{2t}\right) + \frac{1}{N_c}\left(\frac{u}{2s} - \frac{t}{u}\right) + \frac{1}{N_c^3}\left(\frac{s}{u} - \frac{u}{2t}\right)\right]\Bigg\}. \qquad (4.28)$$

In this case it is clear that the singular contributions are not all proportional to the leading-order matrix element, $m_{q\bar{q}\to gg}^{(0)}$. Only the soft divergence, represented by the $1/\varepsilon^2$ term, multiplies this factor. The collinear divergence, given by terms proportional to $1/\varepsilon$, is more complicated due to the fact that the collinear factorization of the matrix elements is modified by colour connections between the gluons. This is a general feature of calculations for more complex final states.

Finally, the result for the all-gluon process $gg \to gg$ is

$$m_{gg\to gg}^{(v)} = c_\Gamma\Bigg\{\left[-\frac{4N_c}{\varepsilon^2} - \frac{22N_c}{3\varepsilon} + \frac{8n_f T_R}{3\varepsilon} - \frac{67N_c}{9} + \frac{20n_f T_R}{9}\right.$$

$$\left.+N_c\pi^2 + \frac{11N_c}{3}l(-\mu^2) - \frac{4n_f T_R}{3}l(-\mu^2)\right]m_{gg\to gg}^{(0)}$$

$$+\frac{16VN_c^3}{\varepsilon}\left[l(s)\left(3 - \frac{2tu}{s^2} + \frac{t^4+u^4}{t^2u^2}\right) + l(t)\left(3 - \frac{2us}{t^2} + \frac{u^4+s^4}{u^2s^2}\right)\right.$$

$$\left.+l(u)\left(3 - \frac{2st}{u^2} + \frac{s^4+t^4}{s^2t^2}\right)\right]\Bigg\}$$

$$+4VN_c^2\left[f_2(s,t,u) + f_2(t,u,s) + f_2(u,s,t)\right] \qquad (4.29)$$

where f_2 is defined by

$$f_2(s,t,u) = N_c \left\{ \left(\frac{2(t^2+u^2)}{tu} \right) l^2(s) + \left(\frac{4s(t^3+u^3)}{t^2u^2} - 6 \right) l(t)l(u) \right.$$

$$+ \left[\frac{4}{3}\frac{tu}{s^2} - \frac{14}{3}\frac{t^2+u^2}{tu} - 14 - 8\left(\frac{t^2}{u^2} + \frac{u^2}{t^2} \right) \right] l(s) - 1 - \pi^2 \right\}$$

$$+ n_f T_R \left\{ \left(\frac{10}{3}\frac{t^2+u^2}{tu} + \frac{16}{3}\frac{tu}{s^2} - 2 \right) l(s) - \frac{s^2+tu}{tu} l^2(s) \right.$$

$$\left. - \frac{2(t^2+u^2)}{tu} l(t)l(u) + 2 - \pi^2 \right\}. \tag{4.30}$$

In all of these virtual matrix elements one can also see the emergence of the logarithms of the renormalization scale, μ, that were previously highlighted in Eq. (2.123). Inspecting Eqs. (4.24), (4.27), and (4.30) one can see that they all contain a term proportional to the appropriate leading-order matrix element multiplied by the factor

$$\left(\frac{11 N_c}{3} - \frac{4}{3} n_f T_R \right) l(-\mu^2) = \beta_0 \, l(-\mu^2). \tag{4.31}$$

There is a term of precisely this form, proportional to the beta-function coefficient β_0, in Eq. (2.123).

4.1.3   Scale uncertainties in inclusive jet production

The dijet process is sufficiently simple that it provides a good laboratory in which to perform a more thorough examination of the dependence of the theoretical prediction on the renormalization and factorization scales, μ_R and μ_F. To do so it is useful to consider **inclusive jet production**, where the notion of an inclusive cross-section refers to the idea that it describes the observed properties of any one of the jets in a given event. This means that, for a n-jet event, there are n contributions to an inclusive jet observable. A natural choice for the hard scale in the theoretical calculation of inclusive jet quantities is the transverse momentum of the jet under consideration, p_T^j. Therefore a scale choice $\mu = p_T^j$ means that an event containing n jets requires an evaluation of the corresponding matrix elements and PDFs n times, each time with a scale proportional to the p_T of the jet under consideration. An alternative scale choice is given by the transverse momentum of the lead jet in the event ($p_T^{j_1}$), although this introduces a new scale that is not a natural one for an inclusive observable. At the Born level the two choice are identical but there are non-trivial differences at higher orders which will be discussed later.

Rather than choosing a common value for μ_R and μ_F, as in Section 2.2.6, it is instead instructive to allow them to vary independently so that the theoretical predictions can be pictured in a two-dimensional surface spanned by μ_R and μ_F. An example of such a surface is shown in Fig. 4.4, for the case of the NLO prediction for the inclusive jet cross-section at $\sqrt{s} = 7$ TeV.[1] Since the theoretical prediction con-

---

[1] We thank Pavel Starovoitov for providing the ROOT ntuples.

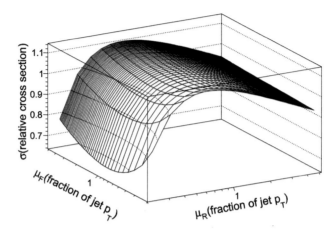

Fig. 4.4 The scale dependence for the inclusive jet cross-section using the anti-$k_\perp$ jet algorithm with $R = 0.4$ at $\sqrt{s} = 7$ TeV. Jets satisfy $60 < p_T < 80$ GeV and lie in the rapidity range $0 < y < 0.3$. Cross-sections have been computed with NLOJET++ [777] interfaced with Applgrid [329] and are normalized to the prediction at scales of $\mu_R = \mu_F = 2.5 p_T$.

tains logarithms of the form $\log(\mu_R/\mu_F)$ it is prudent to vary them in a similar range, so that the surface extends as far as $p_T^j/5 < \mu_R, \mu_F < 5 p_T^j$. The two-dimensional equivalent of the NLO scale dependence observed, for instance in Fig. 2.22, is a saddle shape. In this particular example the peak cross-section is at the saddle point, which corresponds to a scale of approximately $\mu_R = \mu_F = p_T^j$.

A simpler way to see the saddle point, and the region of mild dependence on the scales around it, is to examine a contour plot such as the one shown in Fig. 4.5 (left). From this plot it is clear that the cross-section depends much less on the factorization scale than on the renormalization scale. For a much higher jet transverse momentum range the corresponding plot is qualitatively different, *cf.* Fig. 4.5 (right). The plot appears to have been rotated by an angle of -45° with respect to the vertical, such that the saddle point is at somewhat smaller scales and a scale choice of p_T^j no longer corresponds to the peak cross-section.

This behaviour can be understood as follows. This process is dominated by jets produced in the central rapidity region, thus probing partonic momentum fractions $x \sim 2p_T^j/(7000 \text{ GeV})$, *cf.* Eq. (4.22). Hence, for low jet transverse momenta $x \sim 0.02$ and the dominant sub-process is $gg \to gg$. At this x value, there is very little μ_F-dependence in the gluon distribution, as will be discussed in Chapter 6. This is the behaviour observed in Fig. 4.5 (left), where the jet cross-section is relatively independent of the factorization scale. In contrast, the higher p_T region shown in Fig. 4.5 (right) probes much larger parton x values where sub-processes such as $gq \to gq$ and $qq \to qq$ are also important. In this region both the quark and gluon distributions depend more strongly on μ_F, leading to the "rotation" noted above.

The same analysis can be performed in different kinematic ranges, for instance other

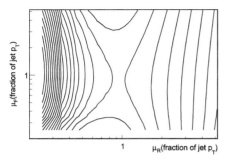
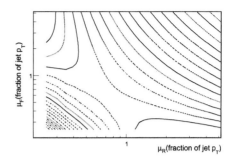

Fig. 4.5 Left: scale-dependence contours for the NLO inclusive jet cross–section, as defined in Fig. 4.4. Right: the same contours for higher jet transverse momenta, in the range $1200 < p_T < 1500$ GeV.

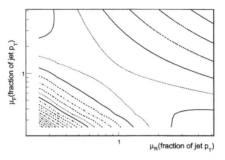
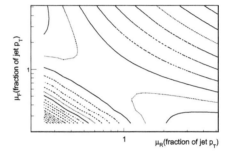

Fig. 4.6 Scale-dependence contours for the NLO inclusive jet cross-section using the anti-$k_\perp$ jet algorithm with $R = 0.4$ (left) and $R = 0.6$ (right). Jets satisfy $1200 < p_T < 1500$ GeV and lie in the rapidity range $1.2 < y < 2.1$. Cross-sections have been normalized to the prediction at $\mu_R = \mu_F = 2.5 p_T$.

regions of jet rapidity and jet separation. Fig. 4.6 (left) shows the contours of scale dependence in the same high transverse momentum range but now also corresponding to much larger rapidities. Again, the rotation has taken place but the saddle point is once more at scales near $\mu_R = \mu_F = p_T$. Fig. 4.6 (right), depicts the scale-dependence contours when the jet size is increased from 0.4 to 0.6. The peak cross-section, the saddle point, corresponds to smaller values for both scales as the jet radius increases.

Note that the standard scale uncertainty analysis corresponds to a one-dimensional projection of the contour plot along the diagonal $\mu_R = \mu_F$. In all these cases this would result in a curve with a maximum at or near the saddle point. However, as demonstrated above, the exact form of the scale dependence clearly depends on the kinematic region being studied and such a similarity between the one- and two-dimensional analyses is not guaranteed. Such considerations need to be kept in mind, not only for inclusive jet production, but more generally for all theoretical predictions for cross-sections at the LHC.

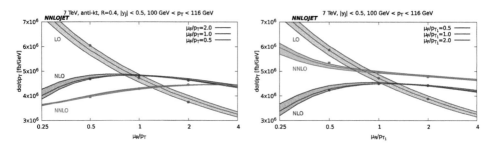

Fig. 4.7 Scale dependence of the inclusive jet cross-section at the 7 TeV LHC, computed at LO, NLO, and NNLO, for the central scale choices $\mu_R = \mu_F = p_T^j$ (left) and $\mu_R = \mu_F = p_T^{j_1}$ (right). The calculation uses the anti-k_T algorithm with $R = 0.4$ and jets are defined by $|y| < 0.5$ and $100 < p_T < 116$ GeV [422]. Reprinted with permission from the authors.

This discussion should of course be extended to even higher orders, when such theoretical predictions are available. Very recently, pioneering calculations have extended the accuracy of inclusive jet production cross-sections to NNLO [335, 422, 566], as shown in Fig. 4.7. The inclusive jet cross-section is shown for a particular slice of jet transverse momenta at the 7 TeV LHC, with the central scale given by either p_T^j or the p_T of the leading jet, $p_T^{j_1}$. The behaviour when moving from LO to NLO to NNLO clearly depends quite strongly on the choice of central scale. For instance, the improvement in the scale dependence at each order that is observed for $\mu = p_T^{j_1}$ is not observed for $\mu = p_T^j$; for a detailed discussion of this and other related subtleties, see Ref. [422].

4.1.4 Jet shape

At NLO there can be two partons in a jet and a description of the **jet shape** — the distribution of energy inside the jet — is possible. Note that by definition, a NLO calculation of the inclusive jet cross-section is a LO calculation of the jet shape. Soon after the inclusive jet cross-section was first calculated to NLO, comparisons of the NLO predictions and the experimental data from the TEVATRON were carried out [510]. Perhaps surprisingly, the description of the jet shape using only one parton agreed well with the experimental data, as shown in Fig. 4.8. Two caveats apply though: first, the non-perturbative effects were ignored in this comparison (expected to be small for a jet transverse energy of 100 GeV), and second, the jet shape at large distances from the jet centre did not agree with the fixed-order prediction unless an ad hoc parameter (R_{sep}) was introduced. Two partons would normally be combined within the same cone jet if they were separated by a distance of less than $2R_{cone}$. Agreement with the data only occurred if the two partons in the NLO calculation were required to be within a distance $2R_{cone} \times R_{sep}$, where R_{sep} had to have a value around 1.3. This was later explained (see for example Ref. [508]) by the stochastic instabilities of jet clustering using the multiple hadrons comprising a jet, as compared to the original two (fixed order) partons. These are complications due to the use of cone-jet algorithms

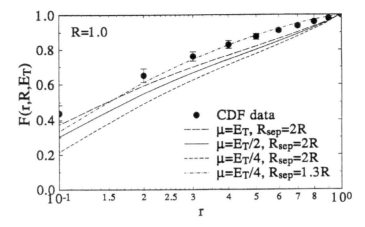

Fig. 4.8 The fraction of energy for a jet of radius $R = 1.0$ that is inside a radius r. Data from the CDF experiment for jets of 100 GeV is compared to fixed order theoretical predictions at NLO. The curves correspond to different values of the renormalization and factorization scales and/or values of the parameter R_{sep}. Reprinted with permission from Ref. [510].

and are absent for example for the anti-k_T jet algorithm. The main point is that jet shapes can be reasonably well described using only the one extra gluon present in a NLO calculation.

Jet shapes will be revisited in the context of experimental data from the TEVATRON in Section 8.3. With the increasing precision of parton shower Monte Carlos, it has become much more common to compare experimental jet shape measurements with the predictions of those Monte Carlo programs (where the jet shape is described by the effects of multiple gluon emissions, as well as non-perturbative effects).

4.1.5 Multijets

Final states containing more than two jets are also of considerable interest at hadron colliders. Foremost, their study constitutes an essential test of the theory of strong interactions and our ability to describe jets of hadrons using the partonic description. In addition, the production of multijets is a considerable background in many searches for new physics, typically when one or more of the jets fakes either a lepton or a photon.

From a theoretical point of view the description of multijet states becomes significantly more complicated than the dijet case. The number of Feynman diagrams contributing to a given n-jet calculation increases more than factorially with n. In addition, as n grows so does the complexity of the colour configurations that must be included in the calculation of the scattering amplitudes. The use of recursion relations, as described in Section 3.2.1, has been instrumental in taming the factorial growth and enabling the calculation of leading-order predictions for $n > 4$ [220]. These recursive techniques, coupled with the D-dimensional numerical unitarity methods mentioned earlier, have led to algorithms capable of computing one-loop virtual corrections for

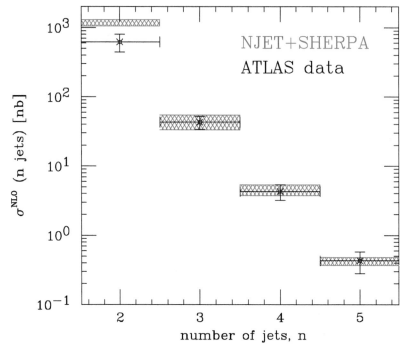

Fig. 4.9 NLO predictions for multijet cross-sections at the 7 TeV LHC [183], compared to ATLAS data from Ref. [4].

essentially arbitrary multijet processes [181, 575]. In view of the factorial growth in the number of diagrams it is especially interesting to observe the scaling behaviour of the computational time using such techniques. The NJET code [181] scales only as a power of n, the number of external partons in the scattering amplitude. Configurations that involve only gluons are most time-consuming to evaluate and scale as approximately n^5. As pairs of gluons are replaced by a quark–anti-quark pair the computation becomes less complicated and thus faster, although the scaling behaviour is harsher. Ref. [181] indicates that a single such computation for $n = 13$, corresponding to 11-jet production, may be completed in less than a second. This is a testament to the power of modern numerical unitarity approaches.

The combination of these ingredients into full NLO calculations of multijet production is an arduous task, due to the high number of partonic channels and their associated infrared singularities. The three-jet case was first available in the NLOJET code [777] in 1997, while the four-jet calculation has only more recently been completed by two independent groups [180, 231]. Results for up to five jets have been presented in Ref. [183], allowing for a comprehensive study of multijet production over a large kinematic range. Fig. 4.9 shows the NLO predictions for the n-jet cross-sections for $n = 2, 3, 4, 5$ at the $\sqrt{s} = 7$ TeV LHC, computed using NJET virtual matrix elements and SHERPA for the assembly into a full NLO prediction. The predictions use a central scale of $H_T/2$, which is found to be a scale at which the NLO corrections are small and the scale dependence flat. For $n = 3, 4, 5$ jets the agreement between the theoretical prediction and the data is excellent, with differences at the level of 30%. Given the complexity of the QCD interactions being described, this agreement is remarkable. Moreover, as can be seen in the figure, the difficulty of experimentally measuring such cross-sections means that the theoretical uncertainties for $n > 3$ are smaller than the experimental ones. The situation is different in the two-jet bin, where the perturbative expansion is not under such good control. There the effect of the NLO corrections is very large under this set of cuts and the NNLO corrections should be important. Note that the cross-section falls by almost the same factor, about ten, when another jet is included in the final state. This is an example of **Berends scaling**, that was originally observed in vector boson+jet production [224]. This observation motivates the study of the quantity,

$$R_n = \frac{\sigma(n+1 \text{ jet})}{\sigma(n \text{ jet})} \qquad (4.32)$$

which is especially interesting since various sources of theoretical and experimental uncertainty may cancel in such a ratio. Since R_n is proportional to α_s at first order, this allows determinations of the strong coupling from measurements of jet rates at hadron colliders, in a similar manner to previous determinations at LEP [174].

Although no results have been presented for more than five jets, it is clear from the discussion above that the virtual matrix elements are readily available in the NJET code and could be utilized in the future. However, due to the time taken to evaluate both the virtual matrix elements and the real corrections, and also because of its limited phenomenological interest, NLO predictions for more than five jets may not be available for some time.

4.2 Production of photons and jets

The production of photons, either on their own or accompanied by additional jets, plays an important role at hadron colliders. Since the photon–quark and gluon–quark interactions are very similar the Feynman diagrams representing these processes are cousins, despite the lack of self-interactions for photons. Of course, photons are produced far less copiously than jets due to the difference in strength between the electromagnetic and strong forces. However, since a photon is typically a well-measured object, these processes can provide useful constraints on hadronic quantities. These processes also provide significant backgrounds to the detection of the Higgs boson through its decay $H \to \gamma\gamma$.

4.2.1 Theoretical considerations

At the most basic theoretical level the description of final states containing photons and jets is no more complicated than that of jets alone. To see how to make a direct correspondence, consider the process involving a quark-antiquark pair and two gluons,

$$0 \to \bar{q}^+(p_1) + q^-(p_2) + g^-(p_3) + g^+(p_4). \qquad (4.33)$$

where each particle is in a definite (postive or negative) helicity state as indicated by the superscript. The leading order amplitude for this process can be written as,

$$\mathcal{M}(\bar{q}_1^+,q_2^-,g_3^-,g_4^+) =$$
$$ig^2 \left[(T^{a_3}T^{a_4})_{i_1 i_2} M(\bar{q}_1^+,q_2^-,g_3^-,g_4^+) + (T^{a_4}T^{a_3})_{i_1 i_2} M(\bar{q}_1^+,q_2^-,g_4^+,g_3^-) \right], \quad (4.34)$$

where the amplitude has been expressed in terms of subamplitudes that are proportional to products of colour matrices in two different orders. This decomposition is made possible by the relationship between the QCD structure constants that appear in the diagram containing the triple-gluon vertex and the anti-commutator of two colour matrices,

$$[T^{a_3},T^{a_4}]_{i_1 i_2} = i f^{a_3 a_4 b} T^b_{i_1 i_2} \qquad (4.35)$$

The subamplitudes that appear in Eq. (4.34) are simple examples of **colour-ordered amplitudes**.

The helicity choice under consideration is a **maximal helicity violating (MHV)** one, with two partons of a given helicity and the remainder of the opposite helicity. As a result the corresponding colour-ordered helicity amplitudes are given by simple expressions,

$$M(\bar{q}_1^+,q_2^-,g_3^-,g_4^+) = \frac{\langle 1\,3\rangle \langle 2\,3\rangle^3}{\langle 1\,2\rangle \langle 2\,3\rangle \langle 3\,4\rangle \langle 4\,1\rangle}, \qquad (4.36)$$

$$M(\bar{q}_1^+,q_2^-,g_4^+,g_3^-) = -\frac{\langle 1\,3\rangle \langle 2\,3\rangle^3}{\langle 1\,2\rangle \langle 2\,4\rangle \langle 3\,4\rangle \langle 3\,1\rangle}. \qquad (4.37)$$

In these two amplitudes the colour-ordering of the gluons is apparent from the denominators: $\langle 2\,3\rangle \langle 4\,1\rangle$ and $\langle 2\,4\rangle \langle 3\,1\rangle$ respectively. In addition, the denominator factors $\langle 1\,2\rangle$ and $\langle 3\,4\rangle$ are signatures of the presence of the diagram containing a triple-gluon vertex.

The corresponding amplitude for the process in which one of the gluons is replaced by a photon,

$$0 \to \bar{q}^+(p_1) + q^-(p_2) + \gamma^-(p_3) + g^+(p_4). \qquad (4.38)$$

receives contributions from similar diagrams, and apart from the overall coupling factor that is different, the coupling of a photon to quarks does not introduce a colour matrix. Therefore the factor T^{a_3} in Eq. (4.34) can be replaced by an identity matrix in colour space and the overall coupling changed to yield,

$$\mathcal{M}(\bar{q}_1^+,q_2^-,\gamma_3^-,g_4^+) = ieQ_q g \, (T^{a_4})_{i_1 i_2} \left[M(\bar{q}_1^+,q_2^-,g_3^-,g_4^+) + M(\bar{q}_1^+,q_2^-,g_4^+,g_3^-) \right]$$
$$\equiv ieQ_q g \, (T^{a_4})_{i_1 i_2} M(\bar{q}_1^+,q_2^-,\gamma_3^-,g_4^+), \qquad (4.39)$$

where Q_q is the electric charge of the quark, in units of e. The photon amplitude is thus obtained by a sum over both colour orderings of the gluon amplitude. One might worry about the presence of the triple-gluon diagram in the original result, which should not be present for the photon. However, since this diagram enters each colour subamplitude with opposite sign, *cf.* Eq. (4.35), the sum in Eq. (4.39) ensures that this diagram does not contribute to the photon amplitude. Explicitly, the result is,

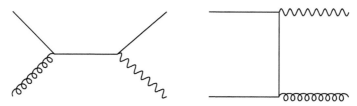

Fig. 4.10 Example Feynman diagrams for direct photon production, $qg \to q\gamma$ (left) and $q\bar{q} \to g\gamma$ (right). The complete set of Born diagrams is obtained by considering also the diagrams with the opposite ordering of the gluon and photon on the quark line.

$$M(\bar{q}_1^+, q_2^-, \gamma_3^-, g_4^+) = \frac{\langle 1\,3 \rangle \langle 2\,3 \rangle^3}{\langle 1\,2 \rangle \langle 2\,3 \rangle \langle 2\,4 \rangle \langle 3\,4 \rangle \langle 3\,1 \rangle \langle 4\,1 \rangle} \left(\langle 2\,4 \rangle \langle 3\,1 \rangle - \langle 2\,3 \rangle \langle 4\,1 \rangle \right)$$

$$= \frac{\langle 1\,3 \rangle \langle 2\,3 \rangle^3}{\langle 2\,3 \rangle \langle 2\,4 \rangle \langle 3\,1 \rangle \langle 4\,1 \rangle}, \tag{4.40}$$

where in the final line the denominators $\langle 1\,2 \rangle$ and $\langle 3\,4 \rangle$ cancel after using the Schouten identity (*cf.* Section 3.2.1) to simplify the numerator, as expected. It is also clear from Eq. (4.39) that the two-photon amplitude with the replacement $g^+(p_4) \to \gamma^+(p_4)$, must be identical up to an overall coupling and colour factor.

With such rules it is straightforward to compute amplitudes for processes involving photons from results obtained in pure QCD [455]. An amplitude for a process containing a photon can be simply obtained from a colour-ordered gluon amplitude by appropriate symmetrization. Thus, once care has been taken in order to define the photon appropriately according to the requirements in Section 2.1.6, the calculation of an m photon plus n-jet final state is straightforward once the $(m+n)$-jet calculation is at hand.

4.2.2 Direct photon production

The simplest process to consider at a hadron collider is direct photon production. It is represented at leading order by Feynman diagrams such as the ones shown in Fig. 4.10, for which one of the helicity amplitudes has already been given in Eq. (4.40). At leading order the kinematics of this process are rather simple, with the parton transverse momentum equal to that of the photon. This property means that a measurement of final states consisting of a photon and a single jet is of particular significance in assessing detector performance. The well-measured photon can be used to calibrate the response of the hadronic calorimeters, providing a measurement of the **jet energy scale** and its uncertainty.

Since the hadronic and photonic activity lie in different hemispheres at LO, the issue of photon isolation does not enter the perturbative calculation at this order. At NLO this is no longer the case, with real radiation contributions allowing partons to populate areas of phase space close to the photon. Moreover, diagrams such as the one shown in Fig. 4.11 (left) can result in a strong dependence on the manner in which the photon is isolated since it contains a singularity when the quark and photon are

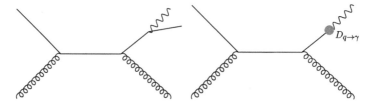

Fig. 4.11 A Feynman diagram entering the NLO calculation of direct photon production (left) and one representative of the fragmentation contribution (right). The diagram on the left is drawn to emphasize the singularity when the photon and quark are collinear. On the right, the function $D_q \to \gamma$ represents the fragmentation of a quark into a photon.

collinear. This singularity can be absorbed into the definition of the bare function representing a quark fragmenting into a photon. The remaining finite **fragmentation function** $D_{i \to \gamma}$ is thus dependent on the physical scale at which this separation is performed, the **fragmentation scale** M_F. The inclusion of such fragmentation contributions, dependent on $D_{q \to \gamma}(M_F)$, is indicated in Fig. 4.11 (right). Just like the parton distribution functions, the fragmentation functions are non-perturbative quantities that must be experimentally determined but whose evolution is governed by perturbative QCD.

Schematically, a differential cross-section for photon production can thus be written as the sum of two components,

$$\mathrm{d}\sigma = \mathrm{d}\sigma_{\gamma + X}(M_F) + \sum_i \mathrm{d}\sigma_{i+X} \otimes D_{i \to \gamma}(M_F). \tag{4.41}$$

The direct, or prompt component, is represented by the first term and the fragmentation contribution by the second. The sum runs over all partons, with σ_{i+X} the inclusive differential cross-section for the production of parton i. As is clear from the equation, this separation is well-defined only at a given value of the fragmentation scale M_F (and is also scheme dependent). Note that in the case of the Frixione isolation criterion discussed in Section 2.1.6, the isolation constraint removes the collinear singularity present in Fig. 4.11 (left). As a result there is no need to introduce the concept of a fragmentation function in this approach. From a theoretical standpoint this is very attractive since the calculation becomes more straightforward and amenable to the usual techniques of pure QCD.

The direct photon process can also be used as an effective probe of the parton distribution functions, in particular of the gluon distribution. Indeed, until 1999 direct photon data were routinely used in the global extraction of PDFs. Since DIS measurements are not directly sensitive to the gluon PDF this was a useful complement to the wealth of available HERA data. However, the inability of NLO predictions of the time to accommodate data from some fixed target and collider experiments led to the abandonment of this approach. TEVATRON collider photon data at moderate to high p_T do have the potential to provide information on the PDFs, but the dominant subprocess at a $p\bar{p}$ collider of this energy is $q\bar{q} \to \gamma g$. Since the quark distributions are well

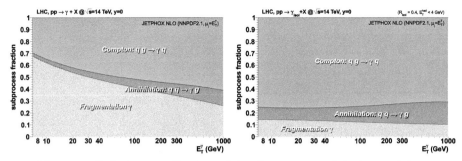

Fig. 4.12 Relative contributions of the Compton, annihilation, and fragmentation contributions at the 14 TeV LHC, for photons at $y = 0$ and as a function of their transverse momentum. The figures are produced with the NLO program JETPHOX. Reprinted with permission from Ref. [462].

constrained in this region already, little useful additional information is provided. This is not true at the LHC, where the availability of a wealth of data has led to renewed interest in constraining the gluon PDF with photon measurements [462]. Fig. 4.12, taken from this reference, shows the size of various contributions to the direct photon cross-section at the LHC, at central rapidity ($y = 0$) and as a function of the photon transverse momentum. As expected, the "annihilation" contribution — resulting from diagrams such as Fig. 4.10 (right) and its corresponding NLO corrections — is much smaller than the "Compton" process (Fig. 4.10 (left)). For inclusive direct photon production, Fig. 4.12 (left), the fragmentation component is large, even for very high E_T photons. However after application of typical experimental isolation cuts the less well-known fragmentation contribution is much reduced, as shown in Fig. 4.12 (right). As a result the direct photon process can be used to probe the gluon PDF at the LHC, particularly now that the theoretical prediction is known to NNLO [316].

4.2.3   Photon pair production

The production of photon pairs arises through the leading-order process,

$$q + \bar{q} \to \gamma\gamma. \tag{4.42}$$

As an example of its importance, it provides the principal background to Higgs boson production in the decay mode $H \to \gamma\gamma$. It should therefore be understood in some detail. Experimentally, this final-state signature also receives a significant contribution from both the direct photon and dijet processes, where one or more of the associated jets is identified as a photon. This section will focus on aspects of the perturbative calculation of the process in Eq. (4.42) and its higher-order corrections.

To be clear about the role of different diphoton amplitudes in the higher-order corrections it is useful to make it explicit in the following way, which mirrors the discussion in Section 3.1. Consider the quantity $\mathcal{M}_n$ that represents the amplitude for the process $q\bar{q} \to \gamma\gamma + n$ gluons. It may be expanded in perturbation theory as,

Table 4.1 The contribution of various diphoton amplitudes to relevant observables and the order at which the obervable is then predicted.

Amplitude contribution	Coupling	$\sigma_{\rm tot}$	$p_\perp^{\gamma\gamma}$ distribution	$\sigma_{\rm jet\ gap}$
$\left\| M_0^{(0)} \right\|^2$	1	LO	-	-
$\left\| M_1^{(0)} \right\|^2$	g^2	NLO	LO	-
$2\,\Re\left(M_0^{(0)} M_0^{(1)\,*} \right)$	g^2	NLO	-	-
$\left\| M_2^{(0)} \right\|^2$	g^4	NNLO	NLO	LO
$2\,\Re\left(M_1^{(0)} M_1^{(1)\,*} \right)$	g^4	NNLO	NLO	-
$2\,\Re\left(M_0^{(2)} M_0^{(0)\,*} \right)$	g^4	NNLO	-	-
$\left\| M_0^{(1)} \right\|^2$	g^4	NNLO	-	-

$$\mathcal{M}_n = g^n \left[M_n^{(0)} + g^2 M_n^{(1)} + g^4 M_n^{(2)} + \ldots \right], \tag{4.43}$$

where $M_n^{(l)}$ represents the corresponding l-loop amplitude. Thus $M_n^{(0)}$ simply represents a tree-level calculation, with $M_0^{(0)}$ just the diphoton production process in Eq. (4.42). It is then a straightforward exercise to list all the contributions that enter the calculation of the diphoton cross-section at a given order, remembering that real-emission contributions (i.e. $n > 0$ in Eq. (4.43)) must also contribute. An explicit list, up to NNLO, is given in the first column of Table 4.1. The corresponding ingredients for two other observables, the transverse momentum of the diphoton system ($p_\perp^{\gamma\gamma}$) and the cross-section for diphoton production in association with two jets widely separated in rapidity ($\sigma_{\rm jet-gap}$), are also shown. The table makes clear the order at which each observable can be calculated, in the sense of Section 3.1. For instance, a NNLO calculation of the total cross-section also contains a NLO calculation of $p_T^{\gamma\gamma}$ and a LO one of $\sigma_{\rm jet-gap}$. However the converse is not true — a NLO calculation of $p_T^{\gamma\gamma}$ is not equivalent to a NNLO calculation of $\sigma_{\rm tot}$ since, for example, it does not contain genuine two-loop corrections originating from $M_0^{(2)}$.

The NLO QCD corrections to the total cross-section are included in the DiPhox Monte Carlo [254]. This calculation includes fragmentation contributions and thus allows for a traditional implementation of isolation. The gluon PDF enhancement at small momentum fraction x leads to a large flux of gluons at high energy, in particular at current LHC energies. Therefore, although diphoton production does not involve gluons at leading order, it can receive significantly corrections from such contributions at higher orders. Although formally suppressed by additional powers of the strong coupling, such contributions are numerically important because of the size of the gluon PDF. One class of diagrams that contributes to $\sigma_{\rm tot}$ at NNLO, those involving loops of quarks as shown in Fig. 4.13, can give a significant additional contribution [152, 463, 775]. As indicated in Table 4.1, such diagrams enter the calculation as $\left| M_0^{(1)} \right|^2$. However, they are particularly interesting since their contribution is

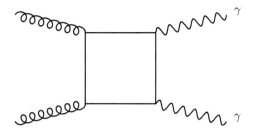

Fig. 4.13 Gluon-induced loop contributions to photon pair production.

separately finite. This is guaranteed by the fact that there is no tree-level $gg \to \gamma\gamma$ amplitude. In fact this contribution has been computed to the next order in the strong coupling, including some of the N^3LO corrections to the total diphoton rate [230, 235]. Despite the apparent high order of the correction terms, since the $\mathcal{O}(\alpha_s^2)\, gg \to \gamma\gamma$ process is finite the calculation of the strong corrections to it only requires the technical machinery used in a normal NLO computation.

By now a number of full NNLO calculations of σ_{tot} have been performed, albeit without accounting for the effects of photon fragmentation [313, 340]. As expected, the NNLO calculation becomes sensitive to kinematic regions that are inaccessible or poorly described by the NLO prediction. One such example is shown in Fig. 4.14, taken from Ref. [313], which illustrates the predictions for the azimuthal angle between the two photons obtained in the NLO and NNLO calculations of σ_{tot}. At leading order in the total cross-section the photons are produced back to back, so the prediction for this distribution is a δ function at $\Delta\phi_{\gamma\gamma} = \pi$. At NLO this requirement is relaxed, but as a result the distribution for $\Delta\phi_{\gamma\gamma} < \pi$ is predicted only at leading order. At NNLO in the total cross-section $\Delta\phi_{\gamma\gamma}$ may only truly be predicted at NLO. In this way it is very similar to the diphoton transverse momentum $(p_\perp^{\gamma\gamma})$ which is also only predicted at NLO in this calculation, *cf.* Table 4.1. It is also clear from Fig. 4.14 that the effect of the NNLO contribution results in a much-improved agreement with the data collected by CMS [378]. However, the theoretical prediction is far from perfect, with differences of up to 40%, as might have been expected from a prediction that is effectively only at the NLO level. Similar agreement has also been obtained in ATLAS, as discussed in Chapter 9.

A limitation of the fixed-order approach can be made clear by considering the effect of typical experimental photon cuts. It is common in diphoton analyses to use staggered cuts,

$$p_T^{\gamma_1} > 40 \text{ GeV}, \qquad p_T^{\gamma_2} > 40 + \delta \text{ GeV}, \qquad (4.44)$$

where $p_T^{\gamma_1}$ and $p_T^{\gamma_2}$ are the transverse momenta of the hardest and softest photon respectively and δ is of order 10 GeV. Such cuts can be useful for reasons of purity of the signal, or rejection of fake backgrounds. In the perturbative calculation it raises a problem due to the fact that, at LO, the two photons have equal transverse momenta. At NLO this is no longer the case and indeed the NLO cross-section is quite sensitive to the value of δ, as shown in Fig. 4.15. For negative δ the cross-section grows rapidly and, with a LO cross-section of 4.8 pb with these cuts, the NLO corrections become

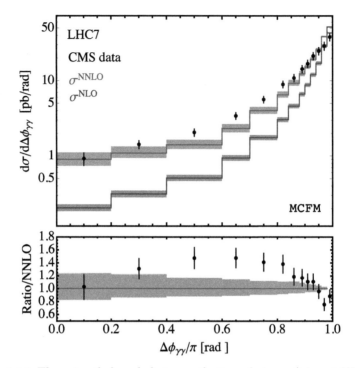

Fig. 4.14 The azimuthal angle between the two photons, $\Delta\phi_{\gamma\gamma}$ at NLO (lower histogram) and NNLO (upper histogram) [313]. The data points are as observed by the CMS collaboration [378]. The height of the histogram bins indicates the scale uncertainty. The lower panel shows the ratio of the data to the NNLO prediction. Reprinted with permission from Ref. [378].

very large. Moreover, the prediction exhibits an interesting cusp in the region of small δ. This is indicative of the presence of terms proportional to $\delta \log \delta$ resulting from the emission of soft gluons [545]. In order to provide a well-controlled prediction for the diphoton cross-section in this case, when the staggered cuts are very close together, some form of resummation of these logarithms should be performed, *cf.* Chapter 5. However, it is worth noting that away from the threshold region, for instance in the region of the Higgs signal where $m_{\gamma\gamma} \sim 125$ GeV, the reliability of the calculation is not spoiled by such logarithms.

By now, NLO calculations of the diphoton process in association with up to three jets are available [185], using the Frixione isolation criterion. On the one hand these predictions can be used to provide further tests of QCD and the understanding of photon production. For instance, just as in the case of pure-jet production (*cf.* Section 4.1), one can now construct NLO predictions for cross-section ratios that are sensitive to the strong coupling. In addition, such processes are important backgrounds for Higgs production, particularly in the weak boson fusion channel, when the Higgs boson subsequently decays to photons.

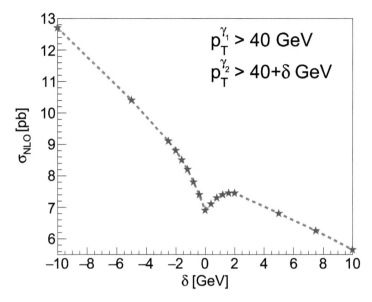

Fig. 4.15 The effect of staggered transverse momentum cuts in a NLO calculation of diphoton production.

4.3   Production of V+jets

The production of vector bosons in association with jets is a benchmark process in the SM for a number of reasons. As already discussed, it lends itself to a good description in perturbation theory because of the intrinsic hard scale given by the vector boson mass. Moreover it probes a wide range of parton luminosities and final state kinematics and it is thus able to test the perturbative description of collider processes in a very broad way. In addition, calculations of vector boson plus jets production have been the test-bed for providing systematic improvements to parton shower predictions. Although beyond the scope of this chapter, Chapter 5 will discuss in some detail various schemes for matching the approximate treatment of parton branching in shower Monte Carlos to exact results for one or more emissions at LO or NLO. Although more widely applicable, most of these methods were developed in the framework of V+jet production. The fact that both multijet parton-level predictions and experimental data are widely available has been crucial to the development and testing of such ideas.

On the experimental side the motivation for studying such processes is also clear. The decays of the W and Z bosons can lead to final states containing charged leptons and/or missing transverse energy, which are then observed in association with jets. Understanding such final states is crucial to many searches for new physics. For example, the production of a Z boson that decays to neutrinos, in association with jets, provides the dominant background to missing E_T (MET) plus jets searches for dark matter and supersymmetry. In addition, such final states represent considerable backgrounds to

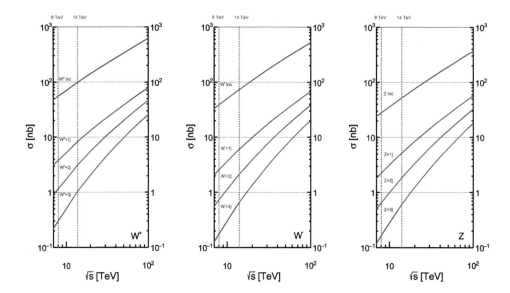

Fig. 4.16 LO cross-sections for W^+, W^-, and Z production at proton-proton colliders as a function of centre-of-mass operating energy, $\sqrt{s}$. For production of a vector boson in association with jets, the jets satisfy $p_T > 40$ GeV, $|y| < 5$ and are found using the k_T algorithm with $D = 0.4$.

known SM processes; for example, the production of top quarks and the investigation of the newly discovered Higgs boson. Probing the properties of these known particles requires a good understanding of these high-rate backgrounds. The fact that the rates for the production of a vector boson in association with many jets are often significant in many LHC analyses can be appreciated by inspecting Fig. 4.16. At current LHC operating energies the cross-section for producing a vector boson in association with a single 40 GeV jet is a few picobarns. For each additional jet the cross-section falls by a factor of about three or four, a further example of the approximate Berends scaling discussed in Section 4.1.

4.3.1 Tree-level and one-loop amplitudes for $V + 1$ jet

For the production of W and Z bosons, with couplings to leptons and quarks that depend on their helicities, it is most useful to express partonic matrix elements in terms of helicity amplitudes. The helicity amplitudes can be easily dressed with the appropriate couplings as necessary.

For $V+1$ jet production this is a particularly compact way of writing the matrix elements since there is only one independent helicity configuration for the leading-order and virtual amplitudes. These amplitudes were first computed in Ref. [576] and, using the notation of this book, in Ref. [234]. To establish the notation, the process under consideration is

$$0 \to q^+(p_1) + g^+(p_2) + \bar{q}^-(p_3) + \bar{\ell}^-(p_4) + \ell^+(p_5), \qquad (4.45)$$

where all particles are outgoing and their helicities are indicated by the superscripts. Following Ref. [234], the leading-order amplitude for this helicity configuration is given by

$$A^{\mathrm{LO}} = 2e^2 g T^{a_2}_{i_1 i_3} A^{\mathrm{tree}} \qquad (4.46)$$

where a_2, i_1, and i_3 are the colour indices of the gluon, quark, and anti-quark respectively, and the basic function is

$$A^{\mathrm{tree}} = -i \frac{\langle 3\,4\rangle^2}{\langle 1\,2\rangle \langle 2\,3\rangle \langle 4\,5\rangle}. \qquad (4.47)$$

The one-loop amplitude contains an additional decomposition according to the colour prefactor that appears. The decomposition is into leading-colour (lc) and sub-leading-colour (slc) amplitudes, where the designation corresponds to an expansion in (the inverse power of) the number of colours (N_c),

$$A^{1-\mathrm{loop}} = 2e^2 g \left(\frac{\alpha_s N_c}{4\pi}\right) T^{a_2}_{i_1 i_3} \left(A^{\mathrm{lc}} + \frac{1}{N_c^2} A^{\mathrm{slc}}\right). \qquad (4.48)$$

The leading-colour amplitude is given by

$$A^{\mathrm{lc}} = c_\Gamma A^{\mathrm{tree}} \left\{ -\frac{1}{\varepsilon^2}\left(\frac{\mu^2}{-s_{12}}\right)^\varepsilon - \frac{1}{\varepsilon^2}\left(\frac{\mu^2}{-s_{23}}\right)^\varepsilon - \frac{3}{2\varepsilon}\left(\frac{\mu^2}{-s_{23}}\right)^\varepsilon - 3 \right\}$$
$$+ i \left\{ \frac{\langle 3\,4\rangle^2}{\langle 1\,2\rangle \langle 2\,3\rangle \langle 4\,5\rangle} \mathrm{Ls}_{-1}\left(\frac{-s_{12}}{-s_{45}}, \frac{-s_{23}}{-s_{45}}\right) \right.$$
$$\left. - \frac{\langle 3\,4\rangle \langle 1\,3\rangle [1\,5]}{\langle 1\,2\rangle \langle 2\,3\rangle} \frac{\mathrm{L}_0\left(\frac{-s_{23}}{-s_{45}}\right)}{s_{45}} + \frac{1}{2} \frac{\langle 1\,3\rangle^2 [1\,5]^2 \langle 4\,5\rangle}{\langle 1\,2\rangle \langle 2\,3\rangle} \frac{\mathrm{L}_1\left(\frac{-s_{23}}{-s_{45}}\right)}{s_{45}^2} \right\} \qquad (4.49)$$

which is written in terms of poles that are proportional to the tree-level amplitude and a finite remainder. The one-loop factor c_Γ has previously been defined in Eq. (3.62). The remainder introduces a dependence on additional functions that are defined by

$$\mathrm{L}_0(x) = \frac{\ln(x)}{1-x}, \qquad \mathrm{L}_1(x) = \frac{\mathrm{L}_0(x)+1}{1-x},$$
$$\mathrm{Ls}_{-1}(x,y) = \mathrm{Li}_2(1-x) + \mathrm{Li}_2(1-y) + \ln x \ln y - \frac{\pi^2}{6}, \qquad (4.50)$$

where the dilogarithm function $\mathrm{Li}_2(x)$ is defined in Eq. (A.9). The dilogarithm is ubiquitous in general one-loop amplitudes since it naturally appears in the analytic expression for scalar box integrals. Indeed, the function Ls_{-1} is simply related to a scalar box integral evaluated in six space-time dimensions, which has neither an infrared nor an ultraviolet divergence. Note that the functions $\mathrm{L}_0(x)$ and $\mathrm{L}_1(x)$ have the property that they are finite in the limit $x \to 1$, which reflects a useful reorganization of the amplitude into functions that are particularly numerically stable.

The sub-leading-colour contribution can be expressed in terms of the same functions as,

$$
A^{\text{slc}} = -c_\Gamma A^{\text{tree}} \left\{ -\frac{1}{\varepsilon^2} \left(\frac{\mu^2}{-s_{13}} \right)^{\varepsilon} - \frac{3}{2\varepsilon} \left(\frac{\mu^2}{-s_{45}} \right)^{\varepsilon} - \frac{7}{2} \right\}
$$

$$
+ i \left\{ -\frac{\langle 3\,4 \rangle^2}{\langle 3\,2 \rangle \langle 2\,1 \rangle \langle 4\,5 \rangle} \text{Ls}_{-1} \left(\frac{-s_{13}}{-s_{45}}, \frac{-s_{12}}{-s_{45}} \right) \right.
$$

$$
+ \frac{\langle 3\,4 \rangle \left(\langle 1\,3 \rangle \langle 2\,4 \rangle - \langle 1\,4 \rangle \langle 3\,2 \rangle \right)}{\langle 3\,2 \rangle \langle 1\,2 \rangle^2 \langle 4\,5 \rangle} \text{Ls}_{-1} \left(\frac{-s_{13}}{-s_{45}}, \frac{-s_{23}}{-s_{45}} \right) + 2\, \frac{[1\,2] \langle 1\,4 \rangle \langle 3\,4 \rangle}{\langle 1\,2 \rangle \langle 4\,5 \rangle} \frac{\text{L}_0 \left(\frac{-s_{23}}{-s_{45}} \right)}{s_{45}}
$$

$$
+ \frac{\langle 1\,4 \rangle^2 \langle 3\,2 \rangle}{\langle 1\,2 \rangle^3 \langle 4\,5 \rangle} \text{Ls}_{-1} \left(\frac{-s_{13}}{-s_{45}}, \frac{-s_{23}}{-s_{45}} \right) - \frac{1}{2} \frac{\langle 1\,4 \rangle^2 [1\,2]^2 \langle 3\,2 \rangle}{\langle 1\,2 \rangle \langle 4\,5 \rangle} \frac{\text{L}_1 \left(\frac{-s_{45}}{-s_{23}} \right)}{s_{23}^2}
$$

$$
+ \frac{\langle 1\,4 \rangle^2 \langle 2\,3 \rangle [1\,2]}{\langle 1\,2 \rangle^2 \langle 4\,5 \rangle} \frac{\text{L}_0 \left(\frac{-s_{45}}{-s_{23}} \right)}{s_{23}} - \frac{\langle 3\,1 \rangle [1\,2] \langle 4\,2 \rangle [2\,5]}{\langle 1\,2 \rangle} \frac{\text{L}_1 \left(\frac{-s_{45}}{-s_{13}} \right)}{s_{13}^2}
$$

$$
\left. - \frac{\langle 3\,1 \rangle [1\,2] \langle 2\,4 \rangle \langle 1\,4 \rangle}{\langle 1\,2 \rangle^2 \langle 4\,5 \rangle} \frac{\text{L}_0 \left(\frac{-s_{45}}{-s_{13}} \right)}{s_{13}} - \frac{1}{2} \frac{[2\,5] \left([1\,2] [3\,5] + [3\,2] [1\,5] \right)}{[1\,3] [3\,2] \langle 1\,2 \rangle [4\,5]} \right\}.
\tag{4.51}
$$

After summing over all helicity combinations and crossing to account for each partonic configuration, these formulae are sufficient to construct the complete virtual matrix elements for a generic $V + 1$ jet process.

4.3.2   Next-to-leading-order calculations

Using numerical implementations of the types of on-shell unitarity methods discussed in Section 3.3.1, NLO predictions for V+jet production have been computed for up to five additional jets [232]. At present these calculations are now being limited not by the complexity of the one-loop calculations, but by the demands of the real radiation component. The sheer number of infrared-singular phase-space regions presents a serious challenge to the numerical stability and running-time of the code. With these calculations in hand it is instructive to examine the effect of these NLO corrections on the simplest possible observable, the $V + n$ jet cross-sections with a given jet definition. As can be seen in Fig. 4.17, the NLO corrections reduce the scale dependence of the predictions considerably. For $n \geq 2$ the NLO calculation falls within the standard LO uncertainty bands, but for $n = 1$ this is not the case. In this case the NLO prediction is also a factor of two larger than at LO. This is explained by the fact that, as noted earlier, the $V + 1$ jet cross-section is atypical in that it does not depend on either the gg or qq parton luminosities, both of which are large at the LHC. These contributions only enter at NLO, in the form of real radiation corrections corresponding to diagrams such as the one in Fig. 2.26, and cause an unusually large NLO enhancement.

The situation becomes more interesting when turning to less-inclusive observables. The jet transverse momentum distribution predicted for V+jet events can exhibit

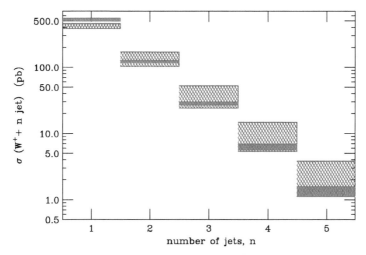

Fig. 4.17 LO and NLO cross-sections for W^++ jet production at the 7 TeV LHC, for anti-k_T jets with $R = 0.5$, adapted from Table I of Ref. [232]. The uncertainties are shown as blue (hatched) and red (solid) bands, for LO and NLO respectively.

the type of "giant K factors", corresponding to very large NLO corrections, already discussed in Section 2.2.7. Similar issues have also been observed for higher jet multiplicities. In these cases the behaviour of the NLO calculation is perfectly normal in the regions that dominate the cross-section, that is the production of an on-shell vector boson with jets that are central and fairly close to the transverse momentum threshold. It is away from these regions, in more extreme kinematic configurations, that the bad behaviour is observed. When considering these regions it is clear that one should not expect the use of a fixed scale, for instance the mass of the vector boson, to produce reliable results. In the tail of a distribution, for instance at very high jet transverse momentum, that very scale provides an alternative physical choice that might be expected to provide a more sensible prediction.

The choice of a scale that changes event by event rather than one that is fixed, for instance at the vector boson mass, must still be performed with care so as to capture the correct behaviour. Fig. 4.18, taken from Ref. [225], shows LO and NLO predictions for the transverse momentum of the second jet in $W+3$ jet events, computed at the LHC for two different event-by-event scale choices. When using the scale choice $\mu = E_T^W$, the transverse energy of the W boson, the NLO prediction differs substantially from the LO one and, for sufficiently high transverse momentum, becomes negative and therefore unphysical. This is clearly a breakdown in the perturbative expansion due to a poor choice of scale. For jets at large transverse momenta E_T^W is not the relevant physical scale. The dominant kinematic configuration is two jets that are produced approximately back to back, with the W boson and the third jet relatively soft. A sketch of such a configuration is shown in Fig. 4.19. In that case a scale such as $\mu = \hat{H}_T$, the scalar sum of all partonic transverse energies, is more appropriate. Using such a scale restores the good behaviour of the NLO prediction, which now remains

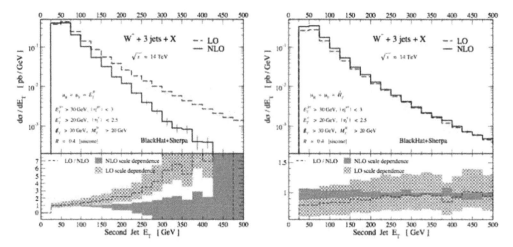

Fig. 4.18 The E_T distribution of the second jet in $W^- + 3$ jet events at the 14 TeV LHC, predicted at LO and NLO [225]. Predictions are shown for two central scale choices, $\mu = E_T^W$ (left) and $\mu = \hat{H}_T$ (right). Reprinted with permission from Ref. [225].

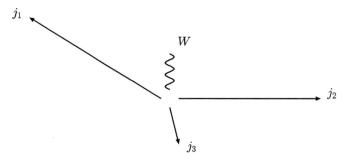

Fig. 4.19 A $W + 3$ jet configuration for which E_T^W would not be a good scale choice. As indicated by the length of the arrows, $E_T^{j_1} \approx E_T^{j_2} \gg E_T^{j_3}, E_T^W$.

physical and rather close to the leading-order distribution, as can be seen in Fig. 4.18 (right).

As already discussed in Section 4.1, the approximate scaling of the V+jets cross-section with the number of jets was first observed in the original leading-order calculations of up to $V + 4$ jets [224]. At that time such calculations were important for assessing the leading backgrounds to top pair production in the semi-leptonic decay mode. With the availability of NLO calculations for such quantities these observations can now be re-examined at higher order. The NLO predictions for the ratio

$$R_n = \frac{\sigma(W + n \text{ jets})}{\sigma(W + (n-1) \text{ jets})} \tag{4.52}$$

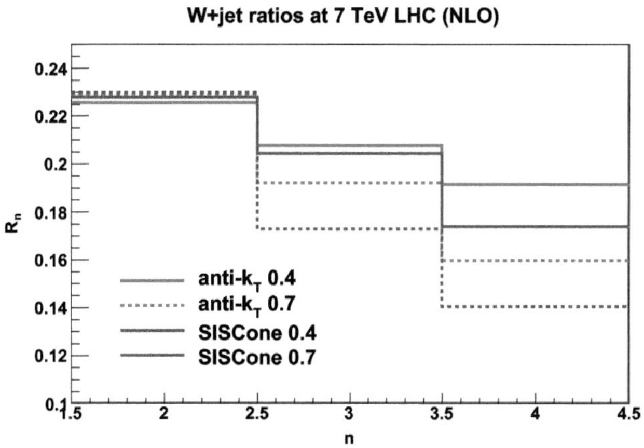

Fig. 4.20 The dependence of the ratio R_n, as defined in Eq. (4.52), on the jet multiplicity n, for four different jet algorithms. The theoretical prediction is computed at NLO using the BLACKHAT code and is shown for jets with $p_T > 30$ GeV at the 7 TeV LHC.

are shown in Fig. 4.20, for $n \leq 4$. The ratio is shown for two different jet algorithms, anti-k_T and SISCone, and two choices of the jet separation, $\Delta R = 0.4$ and 0.7. Thus, for a variety of typical jet algorithms, the cross-section falls by approximately a factor of five for each additional jet present in the final state. The exact value of this ratio, and its approximation by a constant, is clearly dependent on the details of the cuts and algorithm. In particular it fails for larger jet separations where the phase space for multijet production becomes significantly constrained. Nevertheless, the idea that cross-sections for high jet multiplicities — beyond the scope of any present calculation — could be approximated in this way is very attractive. Such ideas have been the focus of renewed interest in recent years [570, 571].

Calculations to NNLO are now available for the W+jet [274] and Z+jet [269, 564] processes. In both cases the NNLO calculations indicate only a small correction at this order, which is especially reassuring given the large NLO K factor observed in Fig. 4.17. Most importantly, the theoretical scale uncertainty is reduced to the level of about 5%. These developments pave the way for precision studies using these final states in upcoming LHC runs.

4.3.3 Jet algorithms and scale dependence

These processes provide an ideal arena for investigating the effects of different jet algorithms and their interplay with, for instance, the issue of scale dependence. The dependence of multi-parton fixed-order jet cross-sections, at both LO and NLO, depend on these details in ways that are not always obvious. For instance, at leading order each parton is associated with a single jet and the imposition of a jet algorithm with size parameter R simply requires that all pairs of partons are separated by the distance R in (η, ϕ) space. The larger the distance requirement, R, the smaller the cross-section — with a rather strong dependence due to the collinear singularities associated with parton splittings in the underlying matrix elements. At NLO there can be two partons in a jet, so that for the first time, jets can have internal structure. Although the collinear singularities have been cancelled against the virtual corrections to yield a finite result, one remnant of this cancellation is the fact that the theoretical prediction inherits a logarithmic dependence on the jet size. Controlling these logarithms can become important if the value of R is small. For this reason it is possible, at NLO, to increase the cross-section by increasing the value of R. The larger the number of jets in the final state, the greater the potential difference between the behaviour of the LO and NLO predictions for the dependence on the jet size.

It has already been argued that scale choices based on the variable H_T are well motivated for the $W + 3$ jet process. As the number of jets present in the theoretical calculation increases this becomes ever more true: the number of possible relevant scale choices grows rapidly, while the definition of H_T tries to capture some of the essential hardness of the scattering process in a generic fashion. The results presented in this section use a central scale choice of $\mu = H_T/2$, where the prefactor is chosen for a mixture of theoretical and pragmatic reasons. This scale yields well-behaved cross-section predictions at both LO and NLO, with K factors near unity [225]. In addition, as will be shown in Chapter 9, the use of such a scale results in good agreement with LHC data.

The results of a study of the dependence of the cross-section on the jet algorithm is shown in Fig. 4.21. Predictions from the BLACKHAT +SHERPA collaboration [225] are shown for both the anti-k_T and SISCone jet algorithms, as a function of the jet size parameter R. The leading-order cross-section for $W + 1$ jet is independent of the jet size and algorithm because the jet consists of only one parton. At NLO the SISCone cross-section is slightly larger than the anti-k_T cross-section for the same jet size, as expected, since the phase space for two partons to be in the same jet is greater for the SISCone algorithm than for the anti-k_T algorithm. The NLO cross-section grows with jet size since it is more likely to find two partons in the same jet as the jet size increases.

For $W + 2$ jet production, the LO cross-sections decrease with increasing jet size since the two partons must be separated by a distance ΔR. The effective separation is larger for SISCone than for anti-k_T ($\Delta R = R_{jet}$ for anti-k_T), as discussed earlier, leading to smaller cross-sections for SISCone. For $W + 2$ jets at NLO there can be either two or three partons in the final state, and either one or two partons in a jet. Now there are two competing effects with regards to the dependence of the jet cross-section on jet size: the ΔR requirement and the larger phase space for two of the three partons to be in the same jet. The second effect wins, with the net result that the jet cross-sections increase with increasing jet size, with SISCone being slightly larger than anti-k_T.

For three and four jets in the final state the cross-sections decrease with increasing jet size at both LO and NLO, with the slope becoming steep at LO. The first (ΔR) effect becomes dominant over the second (phase-space). The anti-k_T cross-sections are larger than the SISCone ones at both LO and NLO. It is interesting to note that while the scale uncertainties increase dramatically with increasing number of jets at LO (since each extra jet requires an extra power of α_s), the uncertainties at NLO are

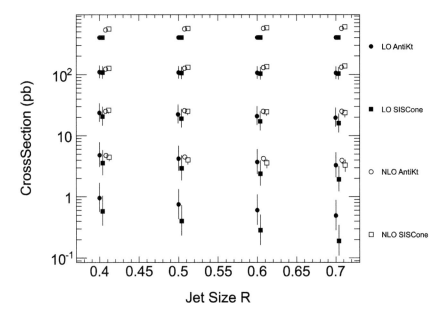

Fig. 4.21 Cross-sections for $W + n$ jet production at the 7 TeV LHC. Predictions are computed at NLO and LO, with jets defined by $p_T > 30$ GeV, and are shown (from top to bottom) for n=1,2,3, and 4 jets (5 jets at LO only). The predictions were generated using BLACKHAT +SHERPA ROOT ntuples, using CTEQ6.6 PDFs, a central scale of $H_T/2$, and uncertainties obtained by varying this scale by a factor of two in each direction.

relatively stable.

The scale dependence for the $W + 4$ jet cross-section at the 7 TeV LHC is shown in Fig. 4.22, for LO and NLO predictions, using the anti-k_T jet algorithm. At LO the scale dependence is fairly trivial. The cross-sections decrease monotonically with increasing scale. The size of the cross-section decreases with increasing jet size since, again, the larger the jet size the further the partons are required to be separated at LO. The LO cross-sections are smaller for the SISCone algorithm than for the anti-k_T algorithm, for the same jet size parameter, given the larger effective separation required by the SISCone algorithm.

At NLO, the cross-section behaviour is non-monotonic, with the anti-k_T cross-sections having a peak cross-section around a scale of $H_T/2$ (the peak cross-section for the SISCone jet algorithm occurs at lower scales, at or less than $H_T/4$). At this scale, for the anti-k_T jet algorithm, the K factor is almost exactly 1; this is not strictly true for other jet sizes or for the SISCone algorithm, although the K factors do not differ greatly from unity for these other choices. Since the scale $H_T/2$ is at or near the peak of the anti-k_T cross-section, any scale variations (for example in the range $H_T/4$ to H_T) will be one-sided. At smaller scales, the NLO cross-sections decrease; in fact the cross-section for a scale of $H_T/8$ for the anti-k_T jet algorithm is negative. The

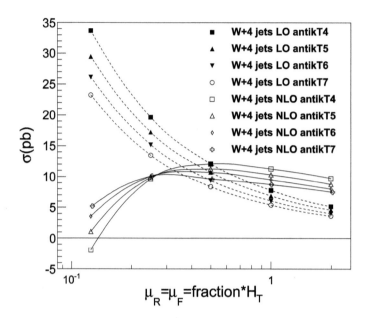

Fig. 4.22 The cross-sections for $W+4$ jet production as a function of the renormalization and factorization scale for pp collisions at 7 TeV. The predictions were generated using BLACKHAT +SHERPA ROOT-ntuples, with CTEQ6.6 PDFs.

decrease in the cross-sections at low scales becomes milder as the jet size increases (or if SISCone is used instead of anti-k_T).

If the NLO and LO cross-sections were evaluated at a scale of about $H_T/5$ (corresponding to 80 GeV, or approximately m_W), the K factor would be much larger than one, with the K factor increasing as the jet size decreases. This latter behaviour is partially due to the divergent behaviour of the LO cross-section and the fact that the NLO cross-section decreases more rapidly with low scales for smaller jet sizes.

An understanding of the dependence of the cross-section on jet algorithm and jet size can be crucial when comparing data to theory. For example, in Ref. [227], the $Z+3$ jet cross-section was calculated at the TEVATRON, using both the SISCone and anti-k_T algorithms (with $R = 0.7$ for both). Using a central scale of $H_T/2$ the calculations give

$$\sigma^{\text{NLO}}_{\text{anti-}k_T} = 48.7^{+3.8}_{-7.9} \text{ fb}, \qquad \sigma^{\text{NLO}}_{\text{SISCone}} = 40.3^{+8.6}_{-8.5} \text{ fb.} \qquad (4.53)$$

At first glance the anti-k_T cross-section is noticeably higher than the SISCone one and has a smaller scale dependence. However, if the peak cross-section is used for both jet algorithms (i.e. $H_T/2$ for anti-k_T and $H_T/4$ for SISCone), the cross-sections and uncertainties become very similar.

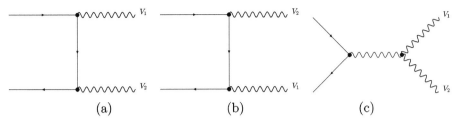

Fig. 4.23 Born-level diagrams representing the production of vector bosons V_1 and V_2 at hadron colliders. Depending on the identity of V_1 and V_2, only a subset of these diagrams contributes to a given amplitude.

4.4 Diboson production

The production of pairs of vector bosons at a hadron collider is represented by the diagrams shown in Fig. 4.23, where V_1 and V_2 represent either W or Z bosons or photons. Depending on the final state, some or all of these diagrams may contribute. For example, the production of a $Z\gamma$ pair only includes diagrams (a) and (b), since the Z boson is neutral and therefore does not couple directly to the photon. However, all of these processes produce final states that are important to measure experimentally, for a number of reasons.

In the first instance these processes provide a wide range of benchmark cross sections with which to test the SM. This is illustrated in Fig. 4.24, which shows the proton–proton cross-section for each final state as a function of the collider energy. Measuring these cross-sections, first at the TEVATRON and then at the LHC, provides a demonstration of the ability of an experiment to measure cross-sections that are comparable to new physics effects that are being sought. For instance, the cross-section for the Higgs-mediated process, $gg \to H \to WW$ at 8 TeV, for a SM Higgs boson with mass $m_H = 125$ GeV, is approximately 5 pb. This is at the same level as the smallest diboson cross-sections shown in Fig. 4.24. Moreover, any analysis of the Higgs boson decays to W and Z pairs requires an accurate description of these diboson processes in order to disentangle them from a potential Higgs boson signal.

Apart from their use as calibration measurements and their relation to the search for the Higgs boson, these processes also provide stringent tests of the non-Abelian gauge structure of the theory. This structure manifests itself in the form of the triple-boson couplings shown in Fig. 4.23 (c). The possibility of anomalous triple gauge-boson couplings (aTGCs), beyond those predicted in the SM, can be investigated by making precision measurements of the different diboson cross-sections, as will be discussed in Chapter 9.

4.4.1 Tree-level and one-loop amplitudes for diboson processes

Once the decays of the vector bosons are taken into account, the matrix elements for diboson production are most easily expressed in terms of helicity amplitudes. As shown in Ref. [471], the tree-level and one-loop amplitudes for all diboson processes can be described in terms of a set of primitive amplitudes that are dressed according to the particular process at hand. For the sake of illustration, consider the WW process,

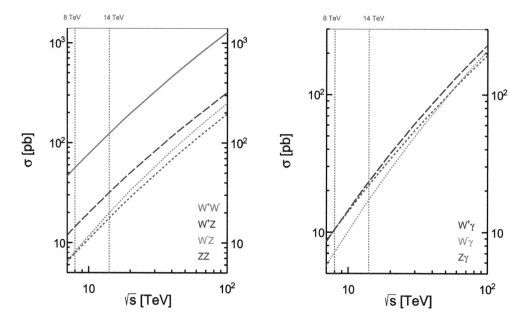

Fig. 4.24 NLO cross-sections for diboson production at proton–proton colliders as a function of centre-of-mass operating energy, $\sqrt{s}$. Photons satisfy $p_T > 40$ GeV and $|y| < 5$.

$$0 \to u^-(p_1) + \bar{u}^+(p_2) + \ell^-(p_3) + \bar{\nu}^+(p_4) + \bar{\ell}'^+(p_5) + \nu'^-(p_6) \qquad (4.54)$$

that is, with all momenta outgoing and the superscripts labelling the particle helicities, "−" and "+" representing left- and right-handed helicities respectively. The tree-level amplitude for this process can be written as

$$A^{\text{tree}} = \left(\frac{e^2}{\sin^2\theta_W}\right)^2 \delta_{i_1 i_2} P_W(s_{34}) P_W(s_{56}) \left[A^{\text{tree},a} + C_{L,u} A^{\text{tree},b}\right] \qquad (4.55)$$

where the propagator and coupling factors are denoted by

$$P_W(s) = \frac{s}{s - m_W^2 + i\Gamma_W m_W},$$
$$C_{L,u} = 2Q_u \sin^2\theta_W + \frac{s_{12}(1 - 2Q_u \sin^2\theta_W)}{s_{12} - m_Z^2}. \qquad (4.56)$$

Note that the factor $C_{L,u}$ contains two terms, representing the coupling of a left-handed up quark to an intermediate photon and Z boson. $A^{\text{tree},a}$ and $A^{\text{tree},b}$ are gauge invariant primitive amplitudes: the coupling structure of $A^{\text{tree},a}$ corresponds to Feynman diagrams containing electroweak bosons directly attached to the quark line, while $A^{\text{tree},b}$ reflects a triple gauge-boson vertex. These same amplitudes can be recycled for other diboson processes; clearly there is no contribution from $A^{\text{tree},b}$ in the case of ZZ production. The specific forms of the tree-level amplitudes are,

$$A^{\text{tree},a} = i\,\frac{\langle 1\,3\rangle\,[2\,5]\,\langle 6|(2+5)|4\rangle}{s_{34}\,s_{56}\,t_{134}},$$

$$A^{\text{tree},b} = \frac{i}{s_{12}\,s_{34}\,s_{56}}\Big[\langle 1\,3\rangle\,[2\,5]\,\langle 6|(2+5)|4\rangle + [2\,4]\,\langle 1\,6\rangle\,\langle 3|(1+6)|5\rangle\Big], \quad (4.57)$$

where $t_{ijm} \equiv (k_i + k_j + k_m)^2$.

At one-loop the amplitude has the same basic structure,

$$A^{1-\text{loop}} = \left(\frac{\alpha_s}{4\pi}\right)\left(\frac{N_c^2-1}{N_c}\right)\left(\frac{e^2}{\sin^2\theta_W}\right)^2 \delta_{i_1 i_2} P_W(s_{34}) P_W(s_{56})\left[A^a + C_{L,u}A^b\right]\;(4.58)$$

where the one-loop primitive amplitudes A^a and A^b correspond to all possible dressings of the corresponding tree-level primitives. It is convenient to decompose the amplitudes further according to

$$A^a = c_\Gamma\left[A^{\text{tree},a}V + i\,F^a\right], \qquad (4.59)$$

and similarly for A^b. In this decomposition V contains a divergent contribution that is common to both primitives,

$$V = -\frac{1}{\varepsilon^2}\left(\frac{\mu^2}{-s_{12}}\right)^\varepsilon - \frac{3}{2\varepsilon}\left(\frac{\mu^2}{-s_{12}}\right)^\varepsilon - \frac{7}{2}, \qquad (4.60)$$

and F is the finite remainder. Since A^b corresponds to a single diagram, a vertex correction, the result for this piece is very simply written in terms of V alone and there is no remainder,

$$F^b = 0. \qquad (4.61)$$

The primitive A^a is more complicated since it corresponds to box diagrams, in particular ones that contain two external momenta that are not light-like. It is given by

$$F^a = \left[\frac{\langle 1\,3\rangle^2\,[2\,5]^2}{\langle 3\,4\rangle\,[5\,6]\,t_{134}\,\langle 1|(5+6)|2\rangle} - \frac{\langle 2|(5+6)|4\rangle^2\langle 6|(2+5)|1\rangle^2}{[3\,4]\,\langle 5\,6\rangle\,t_{134}\,\langle 2|(5+6)|1\rangle^3}\right]\widetilde{\text{Ls}}_{-1}^{2mh}(s_{12},t_{134};s_{34},s_{56})$$

$$+\left[\frac{1}{2}\frac{\langle 6|1|4\rangle^2\,t_{134}}{[3\,4]\,\langle 5\,6\rangle\,\langle 2|(5+6)|1\rangle}\frac{\text{L}_1\!\left(\frac{-s_{34}}{-t_{134}}\right)}{t_{134}^2} + 2\,\frac{\langle 6|1|4\rangle\langle 6|(2+5)|4\rangle}{[3\,4]\,\langle 5\,6\rangle\,\langle 2|(5+6)|1\rangle}\frac{\text{L}_0\!\left(\frac{-t_{134}}{-s_{34}}\right)}{s_{34}}\right.$$

$$-\frac{\langle 1\,6\rangle\,\langle 2\,6\rangle\,[1\,4]^2\,t_{134}}{[3\,4]\,\langle 5\,6\rangle\,\langle 2|(5+6)|1\rangle^2}\frac{\text{L}_0\!\left(\frac{-t_{134}}{-s_{34}}\right)}{s_{34}} - \frac{1}{2}\frac{\langle 2\,6\rangle\,[1\,4]\,\langle 6|(2+5)|4\rangle}{[3\,4]\,\langle 5\,6\rangle\,\langle 2|(5+6)|1\rangle^2}\log\!\left(\frac{(-t_{134})(-s_{12})}{(-s_{34})^2}\right)$$

$$\left.-\frac{3}{4}\frac{\langle 6|(2+5)|4\rangle^2}{[3\,4]\,\langle 5\,6\rangle\,t_{134}\,\langle 2|(5+6)|1\rangle}\log\!\left(\frac{(-t_{134})(-s_{12})}{(-s_{34})^2}\right) + L_{34/12}\log\!\left(\frac{-s_{34}}{-s_{12}}\right) - \text{flip}\right]$$

$$+ T I_3^{3m}(s_{12}, s_{34}, s_{56}) + \frac{1}{2} \frac{(t_{234}\delta_{12} + 2s_{34}s_{56})}{\langle 2|(5+6)|1\rangle \Delta_3} \left(\frac{[45]^2}{[34][56]} + \frac{\langle 36\rangle^2}{\langle 34\rangle\langle 56\rangle} \right)$$

$$+ \frac{\langle 36\rangle [45] (t_{134} - t_{234})}{\langle 2|(5+6)|1\rangle \Delta_3} - \frac{1}{2} \frac{\langle 6|(2+5)|4\rangle^2}{[34]\langle 56\rangle t_{134}\langle 2|(5+6)|1\rangle}, \quad (4.62)$$

where the symmetry operation 'flip', defined by

$$\text{flip}: \qquad 1 \leftrightarrow 2, \quad 3 \leftrightarrow 5, \quad 4 \leftrightarrow 6, \quad \langle a\, b\rangle \leftrightarrow [a\, b]. \quad (4.63)$$

should only be applied to the terms inside the brackets ([ ]) in which it appears. The amplitude is written in terms of

$$\delta_{12} \equiv s_{12} - s_{34} - s_{56}, \qquad \delta_{34} \equiv s_{34} - s_{12} - s_{56}, \qquad \delta_{56} \equiv s_{56} - s_{12} - s_{34}, \quad (4.64)$$

as well as the quantity Δ_3, which is an example of the appearance of Gram determinants in one-loop amplitudes that was discussed in Section 3.3.1. It is given by

$$\Delta_3 \equiv -4 \begin{vmatrix} s_{12} & p_{12} \cdot p_{34} \\ p_{12} \cdot p_{34} & s_{34} \end{vmatrix} = s_{12}^2 + s_{34}^2 + s_{56}^2 - 2s_{12}s_{34} - 2s_{12}s_{56} - 2s_{34}s_{56}. \quad (4.65)$$

The amplitude is written in terms of a basic set of integral functions, some of which have already been introduced in Eq. (4.50). The remaining functions correspond to a new box configuration that is defined by

$$\widetilde{\text{Ls}}_{-1}^{2mh}(s, t, m_1^2, m_2^2) \equiv -\text{Li}_2\left(1 - \frac{m_1^2}{t}\right) - \text{Li}_2\left(1 - \frac{m_2^2}{t}\right)$$

$$- \frac{1}{2}\log^2\left(\frac{-s}{-t}\right) + \frac{1}{2}\log\left(\frac{-s}{-m_1^2}\right)\log\left(\frac{-s}{-m_2^2}\right). \quad (4.66)$$

and a triangle integral with no light-like external legs,

$$I_3^{3m}(s_{12}, s_{34}, s_{56}) = -\frac{1}{\sqrt{\Delta_3}}\text{Re}\left[2(\text{Li}_2(-\rho x) + \text{Li}_2(-\rho y)) + \log(\rho x)\log(\rho y) \right. \quad (4.67)$$

$$\left. + \log\left(\frac{y}{x}\right)\log\left(\frac{1+\rho y}{1+\rho x}\right) + \frac{\pi^2}{3}\right],$$

where

$$x = \frac{s_{12}}{s_{56}}, \qquad y = \frac{s_{34}}{s_{56}}, \qquad \rho = \frac{2s_{56}}{\delta_{56} + \sqrt{\Delta_3}}. \quad (4.68)$$

The coefficient of the logarithm reads

$$L_{34/12} = \frac{3}{2}\frac{\delta_{56}(t_{134} - t_{234})\langle 3|(1+2)|4\rangle\langle 6|(1+2)|5\rangle}{\langle 2|(5+6)|1\rangle \Delta_3^2} + \frac{3}{2}\frac{\langle 36\rangle [4|(1+2)(3+4)|5]}{\langle 2|(5+6)|1\rangle \Delta_3}$$

$$+ \frac{1}{2}\frac{\langle 3|4|5\rangle [4|(5+6)(1+2)|5]}{[56]\langle 2|(5+6)|1\rangle \Delta_3} + \frac{[14]\langle 26\rangle t_{134}(\langle 36\rangle \delta_{12} - 2\langle 3|45|6\rangle)}{\langle 56\rangle\langle 2|(5+6)|1\rangle^2 \Delta_3}$$

$$+\frac{1}{2}\frac{t_{134}}{\langle 2|(5+6)|1\rangle\Delta_3}\left(\frac{\langle 3\,4\rangle\,[4\,5]^2}{[5\,6]}+\frac{[3\,4]\,\langle 3\,6\rangle^2}{\langle 5\,6\rangle}-2\,\langle 3\,6\rangle\,[4\,5]\right)$$

$$+\left(\frac{\langle 3|(1+4)|5\rangle}{[5\,6]}-\frac{\langle 3\,4\rangle\,[1\,4]\,\langle 2\,6\rangle}{\langle 2|(5+6)|1\rangle}\right)\frac{[4\,5]\,\delta_{12}-2\,[4|36|5]}{\langle 2|(5+6)|1\rangle\Delta_3}$$

$$+4\,\frac{\langle 3|4|5\rangle\langle 6|(1+3)|4\rangle+\langle 6|3|4\rangle\langle 3|(2+4)|5\rangle}{\langle 2|(5+6)|1\rangle\Delta_3}$$

$$+2\,\frac{\delta_{12}}{\langle 2|(5+6)|1\rangle\Delta_3}\left(\frac{[4\,5]\,\langle 3|(2+4)|5\rangle}{[5\,6]}-\frac{\langle 3\,6\rangle\,\langle 6|(1+3)|4\rangle}{\langle 5\,6\rangle}\right),\qquad(4.69)$$

and the three-mass triangle coefficient T is given by

$$T=\frac{3}{2}\frac{s_{12}\,\delta_{12}\,(t_{134}-t_{234})\langle 6|(1+2)|5\rangle\langle 3|(1+2)|4\rangle}{\langle 2|(5+6)|1\rangle\Delta_3^2}\qquad(4.70)$$

$$-\frac{1}{2}\frac{(3\,s_{12}+2\,t_{134})\langle 6|(1+2)|5\rangle\langle 3|(1+2)|4\rangle}{\langle 2|(5+6)|1\rangle\Delta_3}$$

$$+\frac{t_{134}}{\langle 2|(5+6)|1\rangle^2\Delta_3}\Big[[1\,4]\,\langle 2\,6\rangle\left(\langle 3|6|5\rangle\,\delta_{56}-\langle 3|4|5\rangle\,\delta_{34}\right)$$

$$-[1\,5]\,\langle 2\,3\rangle\left(\langle 6|5|4\rangle\,\delta_{56}-\langle 6|3|4\rangle\,\delta_{34}\right)\Big]$$

$$+\frac{\langle 3\,6\rangle\,[4\,5]\,s_{12}\,t_{134}}{\langle 2|(5+6)|1\rangle\Delta_3}-\frac{\langle 3\,4\rangle\,[5\,6]\,\langle 6|(1+2)|4\rangle^2}{\langle 2|(5+6)|1\rangle\Delta_3}+2\,\frac{\langle 1\,6\rangle\,[2\,4]\left(\langle 6|5|4\rangle\,\delta_{56}-\langle 6|3|4\rangle\,\delta_{34}\right)}{[3\,4]\,\langle 5\,6\rangle\,\Delta_3}$$

$$+2\,\frac{\langle 6|(2+5)|4\rangle}{\langle 2|(5+6)|1\rangle\Delta_3}\Big[\frac{\langle 6|5|2\rangle\langle 2|1|4\rangle\,\delta_{56}-\langle 6|2|1\rangle\langle 1|3|4\rangle\,\delta_{34}+\langle 6|(2+5)|4\rangle\,s_{12}\,\delta_{12}}{[3\,4]\,\langle 5\,6\rangle}$$

$$+2\,\langle 3|(2+6)|5\rangle\,s_{12}\Big]$$

$$-\frac{[1\,4]\,\langle 2\,6\rangle\,\langle 3|(2+6)|5\rangle}{\langle 2|(5+6)|1\rangle^2}+2\,\frac{[1\,5]\,\langle 2\,3\rangle\,\langle 6|(2+5)|4\rangle}{\langle 2|(5+6)|1\rangle^2}-\frac{[1\,4]\,\langle 2\,6\rangle\,\langle 6|(2+5)|4\rangle\,\delta_{12}}{[3\,4]\,\langle 5\,6\rangle\,\langle 2|(5+6)|1\rangle^2}$$

$$+\frac{1}{2}\frac{1}{\langle 2|(5+6)|1\rangle}\Big[3\,\frac{\langle 6|2|4\rangle\langle 6|1|4\rangle}{[3\,4]\,\langle 5\,6\rangle}+\frac{\langle 3|2|5\rangle\langle 3|1|5\rangle}{\langle 3\,4\rangle\,[5\,6]}+\frac{[1\,4]\,\langle 1\,6\rangle\,[4\,5]}{[3\,4]}-\frac{[2\,4]\,\langle 2\,6\rangle\,\langle 3\,6\rangle}{\langle 5\,6\rangle}$$

$$+\frac{\langle 2\,3\rangle\,[2\,5]\,\langle 3\,6\rangle}{\langle 3\,4\rangle}-\frac{\langle 1\,3\rangle\,[1\,5]\,[4\,5]}{[5\,6]}+4\,\langle 3\,6\rangle\,[4\,5]\Big]$$

$$+\frac{1}{2}\frac{1}{\langle 1|(5+6)|2\rangle}\left[\frac{\langle 1\,6\rangle^2\,[2\,4]^2}{[3\,4]\,\langle 5\,6\rangle}-\frac{\langle 1\,3\rangle^2\,[2\,5]^2}{\langle 3\,4\rangle\,[5\,6]}\right]-\frac{1}{2}\frac{[1\,4]^2\,\langle 2\,6\rangle^2\,(t_{134}\,\delta_{12}+2\,s_{34}\,s_{56})}{[3\,4]\,\langle 5\,6\rangle\,\langle 2|(5+6)|1\rangle^3}.$$

The primitive amplitudes given here are sufficient to describe all helicity amplitudes for the WW process and, after suitable permutations of momenta, all diboson processes. The interested reader is referred to the original paper for further details [471].

4.4.2 Basic properties of diboson processes

Using results for the amplitudes presented above, all of the diboson processes have been computed to next-to-leading order [449, 471, 472]. The most recent treatments

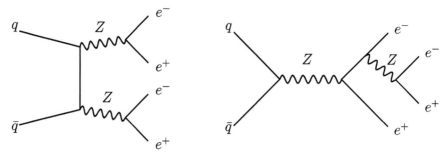

Fig. 4.25 Born-level diagrams entering the calculation of the process $q\bar{q} \rightarrow e^- e^+ e^- e^+$. Shown on the left is a double-resonant contribution, representing the diboson production process $q\bar{q} \rightarrow ZZ$ with subsequent decays, $Z \rightarrow e^- e^+$. The right-hand diagram is a single-resonant contribution where only one of the $e^- e^+$ pairs is produced directly from a Z decay.

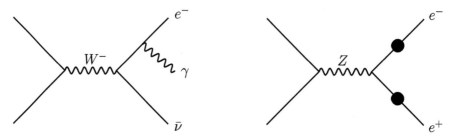

Fig. 4.26 Leading-order diagrams representing photon radiation in the decay of the W (left) and the Z (right). For $e^- e^+ \gamma$ production, the two shaded circles indicate that the photon may be radiated from either of the electrons in the Z decay.

have included the effect of single-resonant diagrams [311, 314, 760]. Examples of such diagrams, that are required for electroweak gauge invariance when the decays of the vector bosons are included, are shown in Fig. 4.25. Such single-resonant diagrams are not important in the calculation of the inclusive cross-section for $e^- e^+ e^- e^+$ production; however, they sculpt other distributions, most notably the invariant mass of the 4-lepton system. In fact, the presence of just such a contribution was useful in cross-checking the first observation of the putative Higgs boson in the decay channel $H \rightarrow ZZ \rightarrow 4$ leptons [369]. Finally, a further important contribution to many of the diboson production processes originates from gluon–gluon initial states. These are the counterparts of the diphoton loop diagram depicted in Fig. 4.13.

An important issue for the $V\gamma$ processes is the treatment of photon radiation. In particular, the photon is radiated copiously from any leptons produced in the decay of the vector boson V, for instance via the leading-order diagrams indicated in Fig. 4.26. The propensity of the leptons to radiate in this way can be assessed by

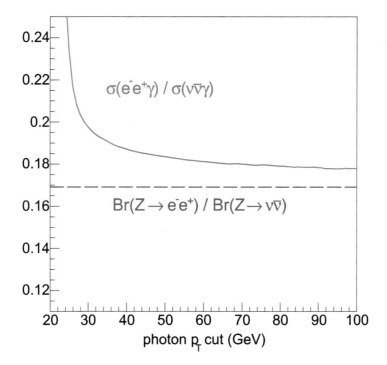

Fig. 4.27 The ratio of LO cross-sections for the $e^-e^+\gamma$ and $\nu\bar{\nu}\gamma$ final states, as a function of the minimum photon p_T.

comparing the cross-sections for $e^-e^+\gamma$ and $\nu\bar{\nu}\gamma$ as a function of the photon transverse momentum. Such a comparison, shown in Fig. 4.27, demonstrates that for sufficiently high photon transverse momentum, radiation in the decay is effectively removed. As a result the ratio $\sigma(e^-e^+\gamma)/\sigma(\nu\bar{\nu}\gamma)$ tends to the ratio of branching fractions, $\mathcal{BR}(Z \to e^-e^+)/\mathcal{BR}(Z \to \nu\bar{\nu}) \approx 1/6$. However, for typical photon p_T cuts used at the LHC, the ratio exceeds this by up to a factor of two. Therefore the proper inclusion of this radiation is essential in order to provide a good theoretical description of the whole event sample. Conversely, in probes of the diboson *production* mechanism, such as in searches for anomalous couplings, it is advantageous to suppress the radiation in the decay by using a higher photon p_T cut.

The dependence of the WW cross-section, for the LHC operating at 8 TeV, on the choice of the renormalization and factorization scale is shown in Fig. 4.28. Since this is an electroweak processes the scale dependence of the LO result is very mild, originating solely from the factorization scale inherent in the definition of the PDF. Of course, this dependence is not indicative of the theoretical uncertainty of the prediction, as can be seen from the NLO curves. These lie outside the range of the LO scale uncertainties due to the $\mathcal{O}(\alpha_s)$ real radiation corrections that are sensitive to the gluon PDF. The NLO corrections are quite large, increasing the theoretical prediction

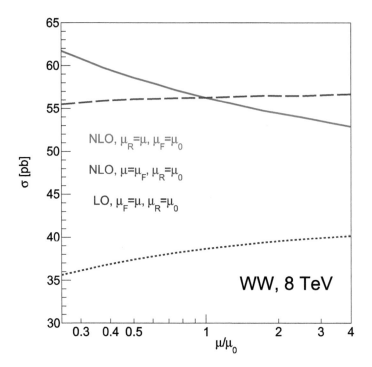

Fig. 4.28 The scale dependence of the WW cross-section at the 8 TeV LHC. The choice of scale is shown relative to the value $\mu_0 = m_W$. At LO the cross-section only depends on the factorization scale, μ_F, while at NLO it also develops a dependence on the renormalization scale, μ_R.

by approximately 40% and include a 3% contribution from gluon–gluon box diagrams that are the counterparts of those shown for the diphoton process in Fig. 4.13. The newly introduced renormalization scale dependence is lessened in a full NNLO calculation of these processes, for which results are now available [336, 557, 595]. Since they must describe the scattering of two massive particles, the two-loop amplitudes that enter the computation of these processes [560] represent the current frontier of such calculations.

One of the other ingredients in the NNLO diboson cross-section is the NLO calculation of the diboson+jet final state. These processes are interesting in their own right, in particular as backgrounds — for instance in jet-binned Higgs boson searches. The NLO results are known in all cases [187]. The subtleties discussed for inclusive diboson production above again apply with, for instance, gluon–gluon initiated contributions once more representing an important higher-order component.

4.4.3 Radiation amplitude zero

Consider the tree-level amplitude for the process,

$$0 \rightarrow u^-(p_1) + \bar{d}^+(p_2) + \ell^-(p_3) + \bar{\nu}^+(p_4) + \gamma^+(p_5) \tag{4.71}$$

which is relevant for $W^-\gamma$ production, with subsequent leptonic W decay. The amplitude is [471],

$$A^{\text{tree}} = i\sqrt{2} \left(\frac{e^3}{\sin^2\theta_W} \right) V_{ud}\,\delta_{i_1 i_2} \frac{P_W(s_{34})}{(s_{12} - s_{34})} \frac{\langle 1\,3\rangle^2}{\langle 3\,4\rangle \langle 1\,5\rangle \langle 2\,5\rangle} (Q_u\, s_{25} + Q_d\, s_{15}) \tag{4.72}$$

using the same notation as before, *cf.* Eq. (4.56), and where $Q_u = 2/3$ and $Q_d = -1/3$ are the charges of the up- and down-quarks. Note that the amplitude can be written in this way despite the presence of the diagram in which the photon is radiated from the W boson thanks to the relation $Q_{W^-} = Q_d - Q_u$. Stripping away overall coupling and spinor factors, the amplitude is thus proportional to a simple factor,

$$Q_u\, p_2 \cdot p_5 + Q_d\, p_1 \cdot p_5, \tag{4.73}$$

that can in turn be readily evaluated in the partonic centre-of-mass. Introducing the angle $\theta^\star$ between the photon and the up-quark (positive z) direction in this frame, the amplitude is thus proportional to the combination,

$$Q_u(1 + \cos\theta^\star) + Q_d(1 - \cos\theta^\star). \tag{4.74}$$

Quick inspection of this equation reveals that the amplitude exactly vanishes for the scattering angle $\cos\theta^\star = (Q_u + Q_d)/(Q_d - Q_u) = -1/3$. This vanishing of the amplitude for a specific scattering angle is characteristic of this process and a feature of all the contributing helicity amplitudes. It has been termed a **radiation amplitude zero** and its features well-known for some time [764].

At a hadron collider it is more useful to translate this critical scattering angle into the corresponding rapidity, $y_\gamma^\star$. This is easily done, with the result,

$$y_\gamma^\star = \frac{1}{2} \log \left(\frac{1 + \cos\theta^\star}{1 - \cos\theta^\star} \right) \approx -0.35. \tag{4.75}$$

In order to construct a boost invariant quantity, and thus obtain an observable that can be measured in the laboratory frame without recourse to reconstructing the partonic centre-of-mass frame, it is usual to consider the rapidity difference, $\Delta y^\star = y_\gamma^\star - y_W^\star$. For typical experimental analyses, where most events are observed with p_T^γ significantly smaller than m_W, the mass of the W boson means that for this $\theta^\star$ the rapidity of the W boson is positive but significantly closer to zero. Explicitly, for small photon transverse momentum p_T^γ (relative to m_W),

$$y_W^\star \approx \frac{1}{2} \log \left(\frac{m_W - p_T^\gamma \cos\theta^\star}{m_W + p_T^\gamma \cos\theta^\star} \right), \tag{4.76}$$

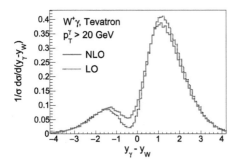 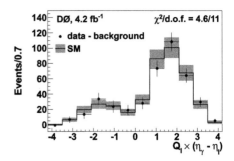

Fig. 4.29 Left: the distribution of $\Delta y = y_\gamma - y_W$ expected in $W^+\gamma$ events at the TEVATRON, at LO (dashed) and NLO (solid, red) in QCD; the radiation zero is manifest as a dip in the distribution around $\Delta y^\star \approx -0.45$. Right: the charge-signed (lepton, photon) rapidity difference in $W\gamma$ events observed by DØ at the TEVATRON [97]. Note the qualitative similarity between the two plots. Reprinted with permission from [97].

so that, plugging in the numerical values,

$$ y_W^\star \approx \frac{p_T^{\gamma,\min}}{3m_W}. \tag{4.77} $$

Therefore for typical experimental cuts, for instance $p_T^\gamma > 20$ GeV, the expected zero in the distribution of the W-photon rapidity difference is for $\Delta y^\star \approx -0.45$ for the subprocess in Eq. (4.71). At the TEVATRON the quark and anti-quark directions correspond reasonably well with those of the protons and anti-protons, leading to a prediction for the radiation zero in $\Delta y^\star$ that is only slightly diluted by PDF effects. At the LHC this is no longer the case; instead the radiation zero is expected to be located at $\Delta y^\star = 0$. The effect of the radiation zero in $W^+\gamma$ production at the TEVATRON is shown in Fig. 4.29 (left), for $p_T^{\gamma,\min} = 20$ GeV. As expected, the zero is replaced by a dip in the distribution that does, however, occur at the expected value of $y^\star$. Also shown is the impact of the NLO corrections, which tend to diminish the effect of the radiation zero. Experimentally one cannot reconstruct the W and, instead, it is simpler to just use the lepton rapidity which retains much of the angular information from the W. However, it does serve to make the radiation zero less pronounced. Despite this, and further contamination from effects such as radiation from leptons in the W decay, evidence of the radiation zero has been observed at the TEVATRON [84]. This is illustrated in Fig. 4.29 (right), which compares DØ data [97] with the theoretical prediction for $Q_\ell(\eta_\gamma - \eta_\ell)$. As expected, the shape of this distribution is very similar to the NLO one shown in the left-panel of the same figure and the excellent agreement between the SM prediction and the measurement confirms the presence of the radiation zero.

4.5   Top-pair production

The large mass of the top quark compared to all the other fermions, which in particular indicates a special role in electroweak symmetry breaking, means that study of top

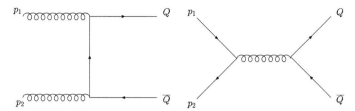

Fig. 4.30 Representative Feynman diagrams for the production of a pair of heavy quarks at hadron colliders, via gg (left) and $q\bar{q}$ (right) initial states.

quark production is of particular significance. The discovery of the top quark was the main triumph of the TEVATRON, as discussed in Chapter 8. From a theoretical point of view, unlike the calculations performed so far, it is necessary to include the non-zero quark mass in order to arrive at sensible predictions. However, it is not only the mass of the top quark that is special but also its lifetime. Assuming that the CKM matrix has $|V_{tb}| = 1$, the width of the top quark can be computed by considering only the decay $t \to Wb$. At leading order it is given by

$$\Gamma_t = \frac{G_F m_t^3}{8\sqrt{2}\pi} \left[(1 - \beta^2)^2 + \omega^2(1 + \beta^2) - 2\omega^4 \right] \sqrt{1 + \omega^4 + \beta^4 - 2(\omega^2 + \beta^2 + \omega^2\beta^2)},$$

(4.78)

where the dimensionless quantities ω and β are the masses of the decay products rescaled by the top quark mass, $\omega = m_W/m_t$, $\beta = m_b/m_t$. Inserting the known masses, this formula gives $\Gamma_t \approx 1.5$ GeV, rather small compared to the top quark mass. The short lifetime of the top quark therefore sets it apart from the other light quarks — unlike them, it is able to decay before hadronizing. As a result there are no bound states of top quarks ("toponium"), but the decay products of the top quark do allow unique studies of its nature (*cf.* Section 2.1.4).

4.5.1 Theoretical considerations

The production of top quark pairs at hadron colliders proceeds via Feynman diagrams such as the ones shown in Fig. 4.30, so that the cross-section is sensitive to the gluon content of the incoming hadrons. This helps to explain why the rate for top quark pair production is significantly larger at the LHC than at the TEVATRON, to the extent that the LHC is often referred to as a "top factory". The extent to which this factory, the LHC, will be able to investigate the properties of the top quark is clear from Fig. 4.31, which shows the total number of top quark pairs produced from the early days of the TEVATRON through projections for future LHC data-taking.

The leading-order matrix elements for the two basic partonic processes,

$$q(p_1) + \bar{q}(p_2) \to t(p_3) + \bar{t}(p_4),$$

(4.79)

$$g(p_1) + g(p_2) \to t(p_3) + \bar{t}(p_4),$$

(4.80)

are, after summing and averaging over colours and spins,

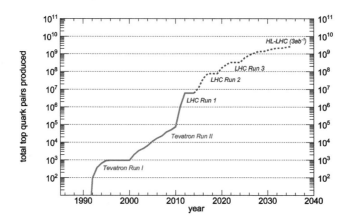

Fig. 4.31 The total number of top quark pairs produced at the Tevatron and LHC colliders, as a function of year. The dotted (blue) line indicates future projections, including estimated shutdown periods, as of 2015.

$$|\mathcal{M}_{q\bar{q}\to t\bar{t}}|^2 = \frac{V}{2N^2}\left(\frac{\hat{t}^2 + \hat{u}^2 + 2m_t^2\hat{s}}{\hat{s}^2}\right),\tag{4.81}$$

$$|\mathcal{M}_{gg\to t\bar{t}}|^2 = \frac{1}{2VN}\left(\frac{V}{\hat{t}\hat{u}} - \frac{2N^2}{\hat{s}^2}\right)\left(\hat{t}^2 + \hat{u}^2 + 4m_t^2\hat{s} - \frac{4m_t^4\hat{s}^2}{\hat{t}\hat{u}}\right),\tag{4.82}$$

where the Mandelstam variables are defined by $\hat{s} = (p_1 + p_2)^2$, $\hat{t} = (p_3 - p_1)^2 - m_t^2$, and $\hat{u} = (p_3 - p_2)^2 - m_t^2$.

In the centre-of-mass frame, the four-momentum of the outgoing top quark can be written as,

$$p_t^\mu = (m_T \cosh y_3, \vec{p}_T, m_T \sinh y_3)\,,\tag{4.83}$$

where $\vec{p}_T$ is the 2-component transverse momentum and $m_T = \sqrt{m_t^2 + p_T^2}$. In this frame the top quark propagator that appears in the left-hand diagram of Fig. 4.30 can easily be evaluated. It is given by

$$(p_3 - p_1)^2 - m_t^2 = \hat{t} = -\sqrt{s}\,x_1 m_T(\cosh y_3 - \sinh y_3),\tag{4.84}$$

where the partonic momentum p_1 corresponds to a probe of the incoming hadron at momentum fraction x_1. This momentum fraction can be related to final-state kinematic quantities through momentum conservation by

$$x_1 = \frac{m_T}{\sqrt{s}}\left(e^{y_3} + e^{y_4}\right),\tag{4.85}$$

so that the propagator can be simplified to,

$$(p_3 - p_1)^2 - m_t^2 = -m_T^2\left(1 + e^{y_4 - y_3}\right).\tag{4.86}$$

Thus the propagator always remains off-shell, since $m_T^2 \geq m_t^2$. The same reasoning applies to all the propagators that appear in the diagrams for top pair production. The addition of the mass scale m_t sets a lower bound for the propagators — that does not occur when considering the production of massless (or light) quarks, where the appropriate cut-off would be the scale Λ_{QCD}. In contrast, as long as the quark is sufficiently heavy, $m_Q \gg \Lambda_{\text{QCD}}$ (as is certainly the case for top, and even bottom, quarks), the mass sets a scale at which perturbation theory is expected to hold. As a result it is possible, for instance, to provide a theoretical prediction for the inclusive top quark pair production cross-section.

Although the large top quark mass therefore provides an easier framework in which to perform theoretical calculations, the actual calculations themselves can become considerably more complex. At the most basic level, retaining non-zero masses for external fermions leads to more complicated expressions for the amplitudes. This can be seen directly by comparing the squared tree-level matrix elements for $gg \to t\bar{t}$ in Eq. (4.82) with those for the process $q\bar{q} \to gg$ in Eq. (4.5). In one-loop calculations of virtual corrections, even the scalar integrals are more complex than their massless counterparts. In addition, obtaining analytic expressions for the loop amplitudes themselves is more complicated since the spinor helicity formalism for massless particles does not readily extend to the massive case. This means that, for instance, the *analytic* on-shell unitarity methods discussed earlier are not immediately applicable to processes containing heavy quarks. In fact, although massive fermions cannot be defined in terms of states of definite helicity, it has been possible to adapt the spinor methods appropriately [683] to obtain amplitudes for some processes of interest, including top quark pair production [186]. Note that the numerical unitarity-based methods discussed in earlier sections do not suffer from most of these problems and are therefore well-suited to calculations involving heavy quarks.

Theoretical predictions for top pair production cross-sections are shown in Fig. 4.32. In addition to the total inclusive cross-section, predictions are also shown for production of an additional jet. Even at 7 TeV there is copious production of additional jets with transverse momenta of 40 GeV or more, with such events accounting for approximately 50% of the total cross-section. At higher energy collisions this fraction grows significantly, with about 80% of all top pair events accompanied by a 40 GeV jet at $\sqrt{s} = 100$ TeV. Such a large proportion of events containing an additional jet calls into question the pertubative description of these cross-sections. Note though that, at a theoretical level, the situation can be easily improved by choosing a higher jet p_T threshold, for instance 100 GeV. For associated production with two jets, results have been obtained using a numerical unitarity approach, dubbed HELAC–NLO [243, 244]. A closely related calculation, for the case where the two associated jets originate from b quarks, is important as an irreducible background to Higgs boson production in the $t\bar{t}H$ channel, with $H \to b\bar{b}$. NLO corrections for this case have been computed by several groups [242, 284, 285].

One can move beyond the NLO calculation for the total top cross-section in a number of ways. In the preceding discussion the top quarks are considered to be stable particles in the theoretical calculation. In reality the top quark is only short-lived and decays to a bottom quark and a W boson that itself subsequently decays. The decays

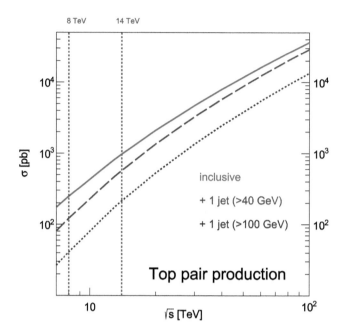

Fig. 4.32 Top pair production cross-section at proton–proton colliders as a function of centre-of-mass operating energy, $\sqrt{s}$ (solid). Also shown are the cross-sections for production in association with a jet satisfying $p_T > 40$ GeV (dashed) and $p_T > 100$ GeV (dotted).

of a pair of top quarks into the different possible combinations of leptons and jets form the overall branching fractions shown in Fig. 4.33. Since $\mathcal{BR}_{t \to Wb} \approx 1$ due to other modes being CKM-suppressed, these branching fractions correspond to the product of two W branching fractions to very good approximation. The dilepton decay mode can be most accurately measured but only captures a relatively small fraction of top quark decays. The converse is true for the "all-jets" channel, so that a good compromise is often found by concentrating on lepton+jets final states. This will be explored further in Chapter 8 and Chapter 9.

One way of accounting for the decay is by using a spin-density approach as in Ref. [236, 237], which has been used to provide predictions at NLO. Alternatively one can directly compute amplitudes including the decay, taking advantage of the considerable simplification that occurs due to the left-handed nature of the W interaction [312, 683]. With this type of approach it is possible to include NLO effects in both the production and decay of the top quarks [312, 762]. Example diagrams that illustrate the division between production and decay stages are shown in Fig. 4.34, for the case of the real radiation contribution. The division of the process into production and decay stages can be performed only in the limit of top quarks that are produced exactly on-shell. The quality of the approximation is then reliant on the fact that the

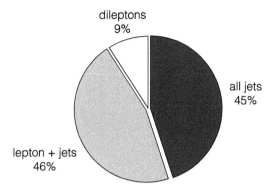

Fig. 4.33 Branching fractions of a top quark pair into leptons and jets.

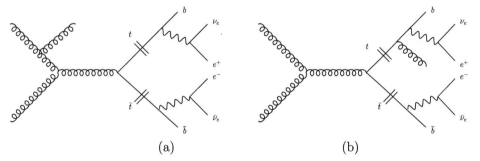

Fig. 4.34 Diagrams entering the real radiation calculation of top pair production in the (a) production and (b) decay stages. The double-barred line indicates that the top quark propagator is considered on-shell at that point in the diagram.

top quark width is quite small, with corrections expected of order Γ_t/m_t. In general, including NLO corrections in the decay of the top quark has only a very small effect on observable quantities. One distribution that is subject to non-trivial corrections is the invariant mass of the lepton and b jet produced in the top quark decay, $m_{\ell b}$ [762]. This observable, and closely related ones — such as the invariant mass of the lepton and b meson observed in the top decay — are interesting since they could be used to measure the top quark mass [256]. This is possible since the distribution shows a distinct kinematic edge, as shown in Fig. 4.35, whose position and shape is sensitive to m_t. However, since the position of the edge is dictated by the kinematics of the decay process, it can be modified by the inclusion of gluon radiation, as is clear from the figure. Hence an extraction of the top quark mass from such a study is best performed with the inclusion of NLO effects in the decay.

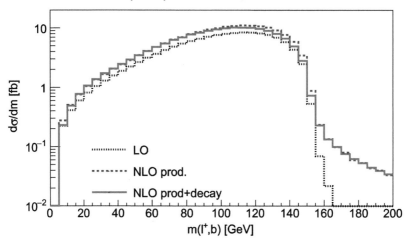

Fig. 4.35 The invariant mass distribution $m_{\ell b}$ expected at the 7 TeV LHC, for typical experimental selection cuts. Predictions are shown at LO and at NLO, with the latter computed both with and without including NLO effects in the top quark decay.

Although the calculations that consider top quark production and decay in a factorized manner can be extended to higher jet multiplicities, the accuracy of the approximation — dropping non-resonant and non-factorizable contributions — can only be judged in the light of a more complete approach. Calculations of the NLO corrections away from the resonance regions are available for the final states $W^+W^-b\bar{b}$ [459] and $\nu_e e^+ b e^- \bar{\nu}_e \bar{b}$ [245]. Fig. 4.36 shows the three classes of diagrams that must be considered in these calculations, where either two, one, or no top quark propagators may be resonant. A detailed study [134] of the two approaches in Refs. [459, 762] showed that, as expected, the doubly resonant approximation of Ref. [762] is adequate for many distributions, resulting in differences of the order of a few percent compared to the full calculation in Ref. [459]. However, some observables, such as the transverse momentum of the $b\bar{b}$ pair, differ by as much as 10–20% for $p_T > 250$ GeV. The improved predictions, throughout a wide kinematic range, provided by these calculations enable a better assessment of these backgrounds in new physics searches.

Another avenue is the calculation of terms in the perturbative expansion beyond parton-level NLO. The total cross-section for $t\bar{t}$ production is known to NNLO [427] and first results have been presented for differential distrbutions [265, 428]. The calculation of the total cross-section to this order represented a tremendous breakthrough in the field of NNLO computations since it was the first calculation involving massive quarks. Fig. 4.37 demonstrates that the resulting theoretical prediction is under excellent control, with a residual 10% uncertainty that accounts for both scale and PDF uncertainties. Note that the accuracy can be further improved in this case by resumming large logarithms that appear as a result of producing a top pair close to

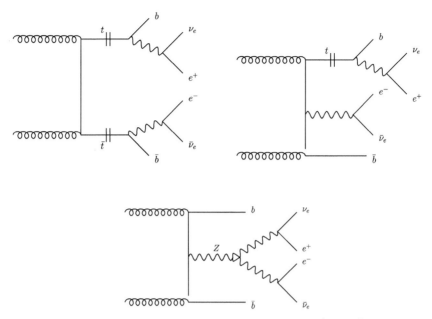

Fig. 4.36 Example diagrams for the process $gg \to \nu_e e^+ b e^- \bar{\nu}_e \bar{b}$, illustrating three categories that may enter the calculation. Clockwise from top left: double-resonant contributions, with two top quark propagators that may be on-shell; single-resonant, with only one such propagator; non-resonant, containing no top propagator that could be resonant.

threshold. This type of resummed calculation will be discussed further in Chapter 5. Finally, the top quark production process was one of the first for which parton shower predictions were available including the effects of NLO corrections [544]. This type of prediction will be discussed at length in Chapter 5.

With the large samples of top quark events that have been collected at the TEVATRON, that has already been increased by two orders of magnitude in the 7 and 8 TeV runs of the LHC, precision studies of the properties of the top quark can be performed. Two particularly interesting examples are observables that are sensitive to the top quark mass and those that probe top quark charge asymmetries. These topics are considered in more detail in the next sections.

4.5.2   Top quark mass

The mass of the top quark has a particularly important role in the SM. Through radiative corrections it affects the mass of the W boson and, indeed, precision measurements of the latter provide an indirect determination of m_t. Due to its large mass compared to the other fermions, the Higgs boson also couples relatively strongly to the top quark. As a result, simultaneous measurements of m_W, m_t, and m_H can provide a stringent test for the presence of BSM physics.

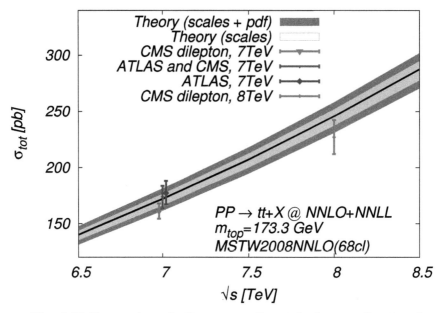

Fig. 4.37 Top quark production cross-section at the LHC as a function of the operating energy. Reprinted with permission from Ref.[427].

Given the importance of the top quark mass, an accurate determination of its value directly from experiment is highly desirable. However, the situation is complicated by the fact that the top mass parameter that appears in the SM Lagrangian, the fundamental parameter of the theory, is subject to renormalization at each order of perturbation theory. As a result, a perturbative calculation at a given order replaces this fundamental parameter, normally referred to as the pole mass m_t, by a renormalization scheme-dependent running mass, $m_t(\mu_R)$. The relationship between the two quantities can be computed in perturbation theory and is currently known up to four loops [752]. For instance, in the $\overline{\text{MS}}$ scheme the pole and running masses are related by

$$m_t = m_t^{\overline{\text{MS}}}(\mu_R) \left[1 + c_1 \frac{\alpha_S}{\pi} + c_2 \left(\frac{\alpha_S}{\pi}\right)^2 + \ldots \right], \tag{4.87}$$

where the coefficients c_1, c_2 are known. At NLO only the one-loop coefficient is required, $c_1 = 4/3 + \log[\mu_R^2/m_t(\mu_R)^2]$ and, evaluating the running mass at the scale $\mu_R = m_t$ gives,

$$m_t = m_t^{\overline{\text{MS}},NLO}(m_t) \left(1 + \frac{4\alpha_S}{3\pi}\right). \tag{4.88}$$

Hence the pole mass is approximately 5%, or 8 GeV, larger than the equivalent NLO $\overline{\text{MS}}$ pole mass. The distinction between the two quantities is therefore of great importance numerically, not just for theoretical consistency. Conventional extractions of the top quark mass, for instance in most of the original TEVATRON analyses, implicitly as-

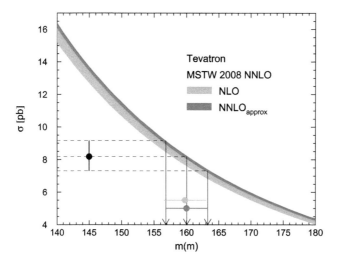

Fig. 4.38 Dependence of the NLO and approximate NNLO top pair cross–section on the running top quark mass, $m(m)$. The data point with vertical error bars represents the experimental measurement of the cross–section given in Ref. [87] and the horizontal error bars the corresponding uncertainties on the extraction of $m(m)$ at the two orders of perturbation theory. Reprinted with permission from Ref. [720].

sume the extraction of the pole mass through kinematic fits of distributions. However, due to the fact that the top quark is a coloured object, it suffers from residual theoretical uncertainties in its definition of the order of 1 GeV [860]. A more well-defined extraction of the top quark mass may be performed by exploiting the dependence of the top pair production cross-section on the mass. The running mass can be extracted order by order in perturbation theory by comparing the cross-section prediction, computed at the same order, with the experimentally measured value. This procedure, which was first used in Ref. [720], is illustrated in Fig. 4.38. The resulting NLO and approximate NNLO running masses are very consistent and, upon converting to the equivalent pole mass, also agree well with other determinations. A number of alternative determinations have been either proposed or implemented, typically based on either kinematic endpoints or clean measurements of leptonic top decays; for a detailed discussion of methods and projections for the LHC the reader is referred to a recent review of this topic [650].

4.5.3 Top quark charge asymmetries

An interesting feature of top pair production is realized for the first time in the next-to-leading-order calculation of the inclusive cross-section. In contrast to the leading-order prediction, at NLO top quarks are not produced with the same distribution in rapidity as anti-top quarks [700]. A sketch of the expected rapidity distributions of top and

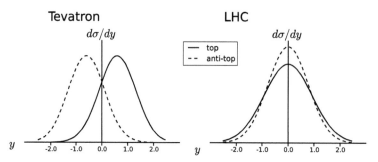

Fig. 4.39 Sketch of the rapidity distributions of top and anti-top quarks
expected at the TEVATRON (left) and LHC (right).

anti-top quarks is shown in Fig. 4.39, from which the asymmetries at the TEVATRON
and LHC can immediately be seen. This asymmetry arises from corrections to the
$q\bar{q}$-initiated process, as indicated in Fig. 4.40. At the TEVATRON the asymmetry is
clearest: top quarks are not produced equally in the forward and backward regions,
where "forward" is defined with respect to one of the beam directions. At the LHC an
asymmetry does not arise in the same fashion since the strong interaction is parity-
invariant and the pp initial state is an eigenstate of parity. However, as seen in Fig. 4.39
(right), there is a difference in the rapidity distributions of the top and anti-top quarks,
which is due to the sub-leading $q\bar{q}$ and qg initial states. A charge asymmetry can be
formed by considering the difference between the distributions in the forward and
central regions, but it is very small and thus hard to measure, *cf.* Chapter 9.

In contrast, the asymmetry at the TEVATRON is a relatively prominent effect that
will be discussed in more detail. At NLO the asymmetry is induced primarily from
the interference between the Born amplitudes for $q\bar{q} \to t\bar{t}$ [2], and the part of the
one-loop amplitude that is anti-symmetric under exchange of the quark and anti-
quark [701, 702]. This contribution arises from box diagrams such as the one shown in
Fig. 4.40 (right). Note that this is an example of a loop-induced effect that does not
change the magnitude of the $t\bar{t}$ cross-section but does affect the kinematic distributions
of the final state. This type of asymmetry is not unique to the $t\bar{t}$ system and, in fact,
it was first discovered in the calculation of QED corrections to the process $e^+e^- \to
\mu^+\mu^-$ [218].

A second type of contribution to the asymmetry arises from diagrams in which
a hard parton is radiated, as shown in Fig. 4.40 (left). The interference of diagrams
with initial and with final-state gluon radiation leads to a smaller asymmetry in the
opposite direction, with a size that depends on the transverse momentum of the top
quark pair. At large $p_T(t\bar{t})$ the pair recoils against a radiated hard gluon. In the leading
colour approximation that is, neglecting contributions of order $1/N_c^2$, the gluon is
colour-connected to either the t–q pair, or the $\bar{t}$–$\bar{q}$ pair (where q, $\bar{q}$ are the initial state
quark and anti-quark). Since gluon radiation indicates an accelerating colour charge,
if the gluon is radiated from the t–q pair then the top quark is more likely to have

---

[2]There is no asymmetry from the gg initial state.

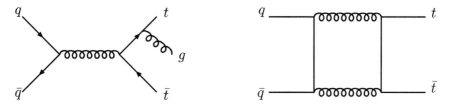

Fig. 4.40 Representative next-to-leading order diagrams responsible for the top quark charge asymmetry. A real radiation diagram is shown on the left and a loop diagram on the right.

accelerated in order to change direction and be produced in the direction opposite to the incoming quark. Hence, assuming that the quark direction is from the negative rapidity hemisphere to the positive one, one can make a qualitative prediction for the lab frame asymmetry $A^{t\bar{t}}_{\text{lab}}$ that is defined by

$$A^{t\bar{t}}_{\text{lab}} = \frac{\sigma(y_t > 0) - \sigma(y_t < 0)}{\sigma(y_t > 0) + \sigma(y_t < 0)}. \tag{4.89}$$

The asymmetry is expected to be negative for sufficiently hard emission and to increase in absolute value with $p_T(t\bar{t})$.[3] This expectation is predicated on the assumption that the beam from which the light quark in the initial state was produced can be determined. As such, it is only a useful observable at the TEVATRON where the asymmetric collisions between protons and anti-protons allow one to use the proton beam as a proxy for the quark direction. The expectations for the asymmetry arising from the real radiation contribution is borne out by the NLO prediction for $A^{t\bar{t}}_{\text{lab}}$ shown in Fig. 4.41. As seen in the figure, the virtual contribution — that because of the $2 \to 2$ kinematics contributes only at exactly $p_T(t\bar{t}) = 0$ — has the opposite sign.

To obtain a more useful prediction at low p_T, one can consider the results obtained using a parton shower. Such a prediction is shown schematically in Fig. 4.41. The parton shower prediction interpolates smoothly between large positive values at small $p_T(t\bar{t})$ and negative values at large top quark pair transverse momentum. The inclusive asymmetry, obtained by integrating out the dependence on $p_T(t\bar{t})$, is small and positive. One expects that its value should not be altered by the parton shower, although further studies have indicated that in fact this is not necessarily the case, due to colour connection effects that may be accounted for in the shower treatment [859]. Finally, one must remember that, although the asymmetry arises at NLO in the cross-section, the non-zero value of the asymmetry is strictly a leading-order prediction. As such it suffers from the usual scale uncertainties, exacerbated by the $\mathcal{O}(\alpha_s^3)$ nature of the observable, resulting in a theoretical uncertainty due to unknown higher orders of around 40%. The calculation of the NNLO corrections to the inclusive asymme-

---

[3]This intuitive understanding must be modified somewhat when considering the NLO corrections to $t\bar{t}$+jet production [763]. The presence of a second scale in the process (p_T^{jet} in addition to m_t) leads to large logarithms in the ratio of the two scales that reduce the predicted asymmetry to a value near zero. This is another lesson in the importance of large logarithms when in the presence of two disparate scales.

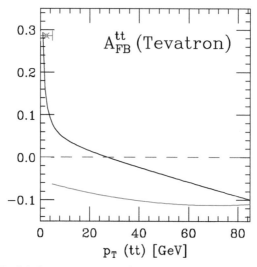

Fig. 4.41 The lab frame top quark forward–backward asymmetry, as defined in Eq. (4.89), at the TEVATRON, The asymmetry is shown as a function of the top quark pair transverse momentum, $p_T(t\bar{t})$. The NLO prediction is shown in red, with a single bin at zero from the virtual corrections, and a NLO+parton shower prediction is shown schematically in black.

try [428] indicate that they are important, at the level of 15–30%. This is crucial in helping to reconcile the theoretical prediction with the TEVATRON data, as will be discussed further in Section 8.6.

4.6   Single-top production

The term single top production encompasses a number of parton level processes that all produce a single top quark via the flavour-changing electroweak vertex coupling a W boson to a top and bottom quark. Although these processes are suppressed relative to QCD production of top quark pairs, since they proceed through an electroweak interaction, they are favoured kinematically since a smaller partonic $\hat{s}$ is required in order to produce only a single heavy top quark. The Feynman diagrams depicting the three such processes that can be explored at hadron colliders are shown in Fig. 4.42.

These processes are important to measure experimentally for a number of reasons. Foremost, they provide a prototype search for the types of final state that are expected in many new physics scenarios. Specifically, they produce bottom quarks, leptons and missing transverse energy with rates that are similarly challenging. Moreover, since the production cross-section for these processes is proportional to $|V_{tb}|^2$. they provide the possibility of directly measuring the CKM matrix element V_{tb}.

The relative importance of the three single-top production processes is illustrated in Fig. 4.43. At all energies the t-channel exchange of a W-boson is the dominant mechanism for producing a single top quark. At the TEVATRON the proton–anti-proton colliding beams mean that s-channel production through a virtual W boson proceeds

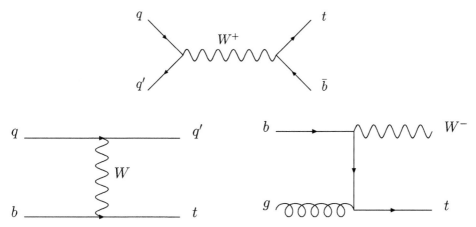

Fig. 4.42 Representative leading-order diagrams for the three single-top channels, *s* channel (top), *t* channel (bottom, left), and associated *Wt* production (bottom, right).

at a significant rate, about half that of the *t*-channel process. Conversely, the associated production of a top quark with a *W* boson is practically inaccessible at the operating energy of 1.96 TeV. At the LHC the situation for the sub-dominant modes is reversed. This is due to the fact that it is far more likely to find a gluon than an anti-quark in the high-energy proton beams of the LHC. The discovery of single top production at the TEVATRON will be discussed in Chapter 8 and the more detailed measurements carried out at the LHC in Chapter 9.

4.6.1   Theoretical issues

The various single top production channels provide a number of different theoretical challenges. The need to retain the mass of the top quark means that the calculations are more complex than similar $2 \to 2$ processes such as jet production. A further difficulty is the fact that the separation of the channels is strictly only possible at the first few orders of perturbation theory. To see how this works, consider the partonic process

$$g(p_1) + q(p_2) \to t(p_3) + \bar{b}(p_4) + q'(p_5) \tag{4.90}$$

that enters at one order higher in the strong coupling than the processes depicted in Fig. 4.43. This amplitude receives contributions from Feynman diagrams such as the ones shown in Fig. 4.44, which are part of a single gauge-invariant set and so must be considered together. From the form of the diagrams it is clear that they could enter the calculation of next-to-leading-order corrections to either the *s*- or *t*-channel processes, but given that they interfere it is not immediately clear how the interference should be apportioned. However, the issue can be quickly resolved by inspection of the colour structure of the different contributions. Using the labelling in Eq. (4.90), the contributions of the two diagrams shown in Fig. 4.44 are

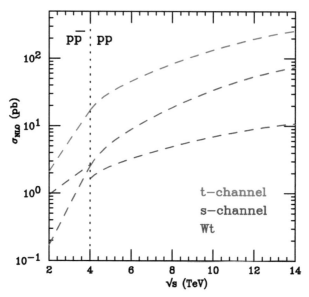

Fig. 4.43 Single top cross-section as a function of the hadron collider operating energy $\sqrt{s}$. For $\sqrt{s} < 4$ TeV the initial state is proton–antiproton, i.e. appropriate for the TEVATRON, while in the rest of the range it is proton–proton, i.e. relevant for the LHC. This causes a noticeable discontinuity in the s-channel cross-section due to the dependence on the anti-quark PDFs.

$$D^{(a)} = T^{a_1}_{i_3 i_4} \delta_{i_5 i_2} K^{(a)}, \ D^{(b)} = T^{a_1}_{i_5 i_2} \delta_{i_3 i_4} K^{(b)}, \tag{4.91}$$

where $K^{(a)}$ and $K^{(b)}$ contain all the kinematic information such as spinors and gamma matrices. When these diagrams are squared the result is

$$\left| D^{(a)} + D^{(b)} \right|^2 = \left| D^{(a)} \right|^2 + \left| D^{(b)} \right|^2 = N_c^2 C_F \left(\left| K^{(a)} \right|^2 + \left| K^{(b)} \right|^2 \right) \tag{4.92}$$

where the interference term vanishes since the colour matrices are traceless, $T^{a_1}_{i_3 i_3} = \text{Tr}\left[T^{a_1} \right] = 0$. Therefore it is possible to simply attribute contributions of the form of Fig. 4.44 (left) to the NLO calculation of the t-channel process and Fig. 4.44 (right) to the effects of NLO in the s channel. However, this argument clearly relies on the fact that one of the fermion lines remains free of any coloured interactions, which is reflected in the δ factors in Eq. (4.91). In the presence of further radiation these can be replaced by colour matrices, so that analogous interference effects do not vanish. Therefore the clean separation of channels breaks down at NNLO and beyond. Nevertheless, the NNLO computation can be performed in the approximation that such contributions are neglected and, in this fashion, fully differential results for the t-channel process have already been presented [288].

Both the t-channel and associated production modes rely on diagrams that contain

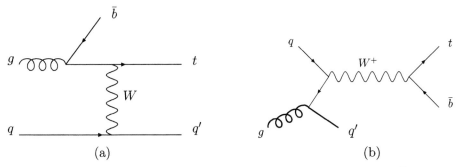

Fig. 4.44 Sample Feynman diagrams for the process $q + g \to t + \bar{b} + q'$. These diagrams may be attributed to the NLO calculation of t- and s-channel single top production (left and right, respectively).

a bottom quark in the initial state. Usually one does not consider any intrinsic bottom quark content in the proton; instead it is generated perturbatively from the gluon and light quark distributions through evolution equations above the bottom quark threshold. Explicitly, at the first order,

$$f_b(x, \mu^2) = \frac{\alpha_s}{2\pi} \log\left(\frac{\mu^2}{m_b^2}\right) \int_x^1 \frac{\mathrm{d}z}{z}\, \mathcal{P}_{qg}(z) f_g\left(\frac{x}{z}, \mu^2\right) + \mathcal{O}(\alpha_s^2) \qquad (4.93)$$

which indicates that the leading contribution to the b-quark PDF is from gluon splitting, $g \to b\bar{b}$, in the proton. Rather than accounting for this effect in the PDF, it is often useful to instead directly compute the t-channel and Wt processes with the gluon splitting present in the matrix elements. Specifically, an equivalent description of the diagrams shown in Fig. 4.42 (b) and (c) is provided by the diagrams shown in Fig. 4.45. One motivation for proceeding in this way is that experimental analyses of these final states often attempt to separate the single top processes from backgrounds by making use of the presence of the additional b quark that is produced in the gluon splitting. However, such a calculation can be considerably more complicated. For instance, the inclusion of the b-quark mass is necessary in order to render a finite result for the diagrams in Fig. 4.45. Without the b-quark mass there would be a collinear divergence when the b quark is not explicitly observed, meaning that an inclusive cross-section could not be defined.

Although explicit information about the anti-bottom quark has been lost in the original approach, logarithms associated with the splitting have been included to all orders in the PDF evolution through the $\mathcal{O}(\alpha_s^2)$ terms that are indicated in Eq. (4.93). At the first order in the evolution the diagrams with the explicit gluon splitting are recovered, albeit in the collinear approximation. If the logarithmic terms, of order $\alpha_s \log\left(\mu^2/m_b^2\right)$ — with μ typically chosen as a physical scale related to the process such as $p_T(b)$ — are important then this approach may be superior. Of course, as calculations in the two schemes are performed at successively higher orders, the results of the calculations should agree to a better degree. The often-used nomenclature for the two calculational schemes contrasts the number of quark flavours included in the

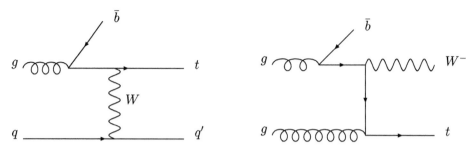

Fig. 4.45 An alternative description of the processes depicted in Fig. 4.42 (b) and (c). The initial bottom quark has been replaced by a gluon that splits into a bottom–anti-bottom quark pair.

initial state: 4-flavour (4F) for the diagrams of Fig. 4.45 and 5-flavour (5F) when computing instead the diagrams of Fig. 4.42 at LO. A comparison of the predictions of the two schemes for t-channel single top cross-sections at the LHC is shown in Fig. 4.46, both at LO and at NLO. At NLO, once the uncertainty from the choice of scale and the PDF sets is included, the two calculations are only in marginal agreement. More importantly, the two calculations give access to kinematic quantities at different levels of precision. Despite the fact that the 5F calculation is performed to NLO, predictions for any properties of the spectator anti-bottom quark only enter the calculation in the real corrections. They are therefore predicted only at leading order. In contrast, the 4F calculation by definition is already sensitive at LO. Indeed, the predictions for such properties are identical in the 5F NLO and 4F LO calculations. However, the 4F NLO calculation raises the precision of the observable to the NLO level, where significant deviations from the LO expectation may be observed.

A further subtlety arises in the consideration of the associated top production process. A careful inspection of the diagrams involved in the NLO calculation of the 5F process reveals, in addition to the LO 4F diagrams such as the one in Fig. 4.45 (right), contributions from diagrams such as the one shown in Fig. 4.47. This is none other than a diagram for $t\bar{t}$ production, with the decay $\bar{t} \to W^-\bar{b}$ included. Since such diagrams, proceeding through resonant top pair production, provide the dominant contribution to the cross-section, a method for effectively excluding them must be devised in order to obtain a useful prediction for the single top tW^- final state. Since gauge invariance requires that all contributing diagrams be included, it is not possible to remove them from the calculation. A number of methods have been proposed for achieving this goal, for instance applying a cut on the mass of the W^-b system in order to remove the resonant top mass region, or insisting on a b-jet veto at moderate p_T. Note that this issue would be present even in the LO calculation of the 4-flavour scheme, with the diagram shown in Fig. 4.47 part of the leading-order contribution. A more sophisticated treatment is to include all diagrams that contribute to this sort of final state, namely $pp \to e^+\nu_e b\mu^-\bar{\nu}_\mu\bar{b}$, and to then see how this compares with the approximate separation into various channels. This sort of calculation is now possible and the NLO results that have been presented indicate that it is possible to make a reasonable approximation in this way [534].

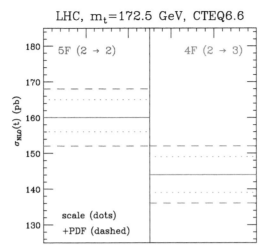

Fig. 4.46 Single top t-channel cross-section at the 14 TeV LHC, computed in the 4- and 5-flavour schemes. Scale uncertainty is indicated by the dotted lines and the total, considering also the uncertainty from the PDFs, is shown as a dashed line. Figure based on the results of Ref. [317].

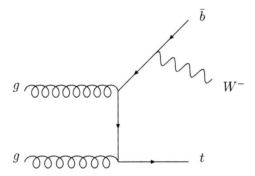

Fig. 4.47 A diagram entering the NLO calculation of Wt single top associated production. It is related by gauge invariance to the one shown in Fig. 4.45 (right) and represents the process $gg \to t\bar{t}$ followed by the decay $\bar{t} \to W^-\bar{b}$.

4.7 Rare processes

This chapter has concentrated on the hadron collider processes with the largest cross-sections, or ones that lead to formidable SM backgrounds. This section discusses a variety of other notable processes that will serve as new benchmarks in the future high-luminosity LHC era. The cross-sections for the processes that will be discussed here, for the LHC operating at $\sqrt{s} = 13$ TeV, are shown in Fig. 4.48. Note that most of the cross-sections are at the level of a few hundred femtobarns or are smaller still, before any branching ratios into observable final states are taken into account. Once

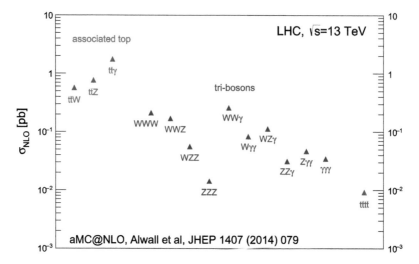

Fig. 4.48 Cross-sections, in picobarns, for a selection of rare processes at hadron colliders: the production of a top quark in association with a vector boson, tri-boson production, and four-top production. The NLO cross-sections are taken from Ref. [148]. When photons are present in a process, the cross-section corresponds to the basic cuts $p_\perp^\gamma > 20$ GeV and $|\eta^\gamma| < 2$.

these are taken into consideration, the expected event rates for many of these processes are so small that their observation will require high-luminosity datasets from future LHC runs.

For the first category of processes in the figure, top pair production in association with a vector boson, the outlook is not so gloomy. Indeed, first evidence for production of these final states at the SM level has already been established in Run I of the LHC [46, 175, 403, 406]. These processes are important for a number of reasons. All three represent significant backgrounds to multi-lepton signals that, for instance, are mainstays of SUSY searches. In particular, the $t\bar{t}W^\pm$ and $t\bar{t}Z$ processes constitute a non-trivial source of same-sign dileptons within the SM. The $t\bar{t}W^\pm$ final state also exhibits a charge asymmetry analogous to the one already discussed for top pair production; in this case the emission of a W boson effectively polarizes the top quarks, leading to a significant $\mathcal{O}(15\%)$ charge asymmetry that should be observable [738]. In contrast to the case of $t\bar{t}W^\pm$ production, where the W boson has only an indirect effect on the top quarks, $t\bar{t}Z$ production directly probes the coupling of the Z boson to top quarks. Direct constraints on the nature of this coupling should be possible with future LHC data [829], where a precise knowledge of the SM cross-section that takes into account both NLO QCD and electroweak effects [541] will be invaluable.

The second category of processes comprises final states containing three vector bosons: "tri-boson" production. Once again, they constitute important backgrounds for not only BSM searches, but for ongoing probes of the Higgs boson. For example, some of these processes represent irreducible backgrounds to the associated production modes of the Higgs boson, $pp \to WH$ and $pp \to ZH$ when the Higgs boson decays to vector boson pairs. Besides this important role, these processes have intrinsic interest due to their ability to probe anomalous gauge boson couplings. Although triple gauge couplings are probed — in general, much more effectively — by diboson processes, quartic couplings are probed for the first time in tri-boson processes. The production rates for these processes, before any branching ratios, are all rather similar and in the 10–100 fb range. Branching ratios into clean final states that could be identified experimentally reduce these to the femtobarn level or smaller. At that level it is clear that, unless anomalous interactions are very large, detailed investigation of these final states will not be possible until hundreds of inverse femtobarns of data have been analysed.

The final process shown in Fig. 4.48, with a cross-section of only about 10 fb, is four-top production. Although this cross-section is tiny, even before top quark decays are accounted for, this process would have a spectacular signature in the detectors. The heavy nature of the top quark means that many models of new physics beyond the SM involve additional interactions of the top quark with new degrees of freedom. As a result such models usually predict a much-enhanced cross-section for four-top production. This means that, despite the small rate expected in the SM, this final state has already attracted considerable interest in the context of BSM searches. Despite the complexity of this final state — containing four coloured, heavy particles — NLO predictions for this cross-section in the SM have been known for some time [246].

Another class of rare processes that will be extensively probed with future data from the LHC corresponds to vector boson scattering. These will be discussed in Section 4.8.8 below.

4.8   Higgs bosons at hadron colliders

4.8.1   Overview

The discovery of a new particle in 2012, consistent with the predictions of a SM Higgs boson, represented the culmination of many years of careful experimental and theoretical study. One reason for this is simple — the SM does not provide a prediction for the mass of the Higgs boson. This means that dedicated searches had to be performed over a wide range of potential masses. On the other hand, once a mass is assumed the SM is very predictive in the sense that its interactions are completely prescribed. This means that, at hadron colliders, the Higgs boson may be produced, and subsequently decay, via many mechanisms. The theoretical study of the various final states in which a Higgs boson may be detected, and the development of suitable experimental techniques with which to extract a viable signal, has been a driving force in hadron collider studies for many years. Due to the expansive nature of the topic it is not possible to discuss all the intricacies of the search for a SM Higgs boson here. Instead, this section will focus on the main themes that have emerged from these studies and concentrate on the issues of particular relevance in the era following the LHC discovery of a Higgs boson with a mass of approximately 125 GeV. Select experimental results will be discussed in more detail in Section 9.6.

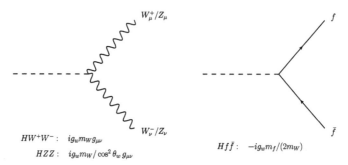

$HW^+W^-: \quad ig_w m_W g_{\mu\nu}$

$HZZ: \quad ig_w m_W / \cos^2\theta_w \, g_{\mu\nu}$

$Hf\bar{f}: \quad -ig_w m_f / (2m_W)$

Fig. 4.49 Feynman rules for the coupling of the SM Higgs boson to W and Z bosons (left) and to fermions (right).

To begin, it is useful to review the ways in which the Higgs boson, in its usual incarnation, may couple to other SM particles. Since the particle was originally introduced as the agent of electroweak symmetry breaking, responsible for giving the W and Z bosons non-zero masses, the Higgs boson has tree-level couplings to those particles. The interactions and the corresponding Feynman rules are shown in Fig. 4.49 (left). The nature of the Higgs boson means that all the couplings are proportional to the masses of the particles with which it interacts (recall that, at tree level, the W and Z masses are related by $m_W = m_Z \cos\theta_w$). If these were the only interactions of the Higgs boson then opportunities for its observation at a hadron collider would be limited to channels with particularly small cross-sections, since production would proceed through Feynman diagrams with multiple powers of the weak coupling. The two such mechanisms that are of primary importance are shown in Fig. 4.50. **Associated production** refers to the modes in which the Higgs boson is produced together with an additional W or Z boson. The last production mode, referred to as **weak boson fusion** or **vector boson fusion**, is especially interesting since it has a very clear experimental signature. The quarks only receive a moderate transverse kick (typically of order $m_V/2$) in their direction when radiating the W or Z bosons, so they can be detected as jets very forward and backward at large absolute rapidities. At the same time, since no coloured particles are exchanged between the quark lines, very little hadronic radiation is expected in the central region of the detector. Therefore the type of event that is expected from this mechanism is often characterized by a "rapidity gap" in the hadronic calorimeters of the experiment.

Although not strictly part of the original formulation of the Higgs boson interactions, it is usual to consider the SM Higgs boson as one that also provides a mechanism by which fermions acquire masses. The Higgs boson then has Yukawa interactions with all the fermions, with a strength proportional to the fermion mass, m_f. This interaction is shown in Fig. 4.49 (right) and, in the SM, it is the coupling to the top quark that is especially relevant due to the large top quark mass. In particular it leads to the channel in which the Higgs boson is produced in association with a pair of top quarks, through leading-order diagrams such as the ones shown in Fig. 4.51.

Finally, but most importantly, the tree-level interactions described above lead to couplings of the Higgs boson to light particles that are mediated by loop diagrams.

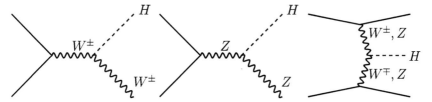

Fig. 4.50 Feynman diagrams for the production of a Higgs boson through electroweak interactions alone. The three production modes are associated production of $W^{\pm}H$ (left) and of ZH (centre) and weak boson fusion (right).

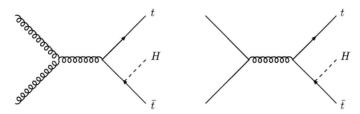

Fig. 4.51 Representative Feynman diagrams for the production of a Higgs boson in association with a top quark pair.

For hadron colliders the key coupling is that of the Higgs boson to two gluons, through loops of heavy quarks, as shown in Fig. 4.52. Since there is no tree-level coupling of the Higgs boson to two gluons it is clear that there is no way to renormalize any possible divergence that this loop diagram could contain. Therefore its contribution must be finite. Explicitly, the colour- and spin-averaged matrix element squared is given by,

$$\overline{|\mathcal{M}|^2} = \frac{1}{8V}\left(\frac{\alpha_s}{3\pi v}\right)^2 p_H^4 \left|\frac{3}{4}I_q\left(m_t^2/m_H^2\right)\right|^2, \tag{4.94}$$

where $v^2 = (\sqrt{2}\,G_F)^{-1} \approx (246\ \text{GeV})^2$ is the squared vacuum expectation value of the Higgs field. The function $I_q(x)$ is defined by,

$$I_q(x) = 4x\left[2 + (4x - 1)F(x)\right], \tag{4.95}$$

where the loop function $F(x)$ is sensitive to whether the top quark in the loop is above or below the pair-production threshold,

$$F(x) = \begin{cases} \frac{1}{2}\left[\log\left((1 + \sqrt{1 - 4x})/(1 - \sqrt{1 - 4x})\right) - i\pi\right]^2 & x < \frac{1}{4} \\ -2\left[\sin^{-1}(1/2\sqrt{x})\right]^2 & x \geq \frac{1}{4}. \end{cases} \tag{4.96}$$

Although the matrix element is formally suppressed by two powers of α_s and a loop factor, in the calculation of the cross-section the gluon parton distribution function enters twice. Thus, despite the loop suppression, this coupling actually results in the

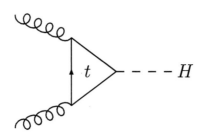

Fig. 4.52 The one-loop diagram representing Higgs production via gluon fusion at hadron colliders. The dominant contribution is from a top quark circulating in the loop, as illustrated.

largest cross-section at the LHC. This mode is usually referred to as Higgs production by **gluon fusion**.

Before going on to discuss each of the production modes in turn, it is useful to consider the size of their cross-sections. Fig. 4.53 shows the cross-sections for each of the Higgs production processes at a pp collider, as a function of the c.o.m. energy. The gluon fusion mode is larger than all other modes combined by an order of magnitude across the range shown, with a cross-section in the range of a few tens of picobarns at LHC energies. In contrast, associated production with top quarks is the mode with the smallest cross-section, a few hundred femtobarns. However, the relative importance of this mode grows far more rapidly with $\sqrt{s}$ than the VBF and associated production modes. This is mainly due to the fact that this channel benefits from the gluon flux that is increasingly important as the typical Feynman-x value that is probed decreases as $\sqrt{s}$ rises. Despite this fact, the VBF mode remains the second-largest cross-section at all foreseeable future hadron collider energies. Also shown, for comparison, is the Higgs boson pair production cross-section. This cross-section is so small that, even with the full 3000 fb^{-1} of integrated luminosity anticipated at the LHC in the future, one could only expect to produce about 10,000 Higgs boson pairs in total. After accounting for Higgs boson branching fractions and experimental efficiencies, this leaves very few events to analyse. Although this cross-section also rises sharply with $\sqrt{s}$, based on cross-sections alone one might expect it as difficult to observe Higgs boson pair production at a 100 TeV collider as to definitively establish the associated production modes of the Higgs boson at the LHC.

4.8.2   Higgs boson decays

An experimental observation of the Higgs boson can of course only be inferred from a measurement of the particles into which it decays. The Feynman rules in Fig. 4.49 also represent the tree-level processes by which the Higgs boson decays into a pair of W or Z bosons, or a fermion–anti-fermion pair. Although the mass of the Higgs boson that has been observed at the LHC is below the threshold for two on-shell vector bosons, the partial width is still significant. For $m_H \lesssim 155$ GeV, the decay can be well described by considering only one of the vector bosons off-shell, the decay $H \to VV^\star$. The partial width for such a decay, including a factor of two to account for either one of the bosons to be off-shell, is given at tree level by,

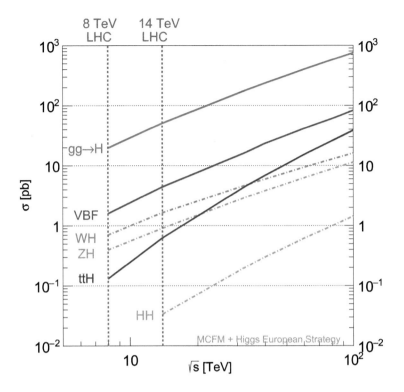

Fig. 4.53 Cross-sections for the production of a SM Higgs boson of mass 125 GeV at a *pp* collider, as a function of the operating energy, $\sqrt{s}$.

$$\Gamma(H \to VV^\star) = \frac{1}{S_V} \frac{g_W^2 m_H^3}{64\pi m_W^2} \left(\frac{m_V \Gamma_V}{\pi} \right) \int_0^{(m_H - m_V)^2} \mathrm{d}p^2 \frac{\sqrt{\lambda(p^2)} \left(\lambda(p^2) + \frac{12 m_V^2 p^2}{m_H^4} \right)}{(p^2 - m_V^2)^2 + m_V^2 \Gamma_V^2}$$

$$(4.97)$$

where the function $\lambda(p^2)$ is given by

$$\lambda(p^2) = \left(1 - \frac{p^2}{m_H^2} - \frac{m_V^2}{m_H^2} \right)^2 - \frac{4 m_V^2 p^2}{m_H^4} \tag{4.98}$$

and the symmetry factor accounts for two identical Z bosons, $S_Z = 2$ and $S_W = 1$. For a heavier Higgs boson above the diboson threshold this reduces to the simpler and more well-known result (*cf.* the NWA given in Eq. (2.83))

$$\Gamma(H \to VV) = \frac{1}{S_V} \frac{g_W^2 m_H^3}{128\pi m_W^2} \sqrt{1 - \frac{4 m_V^2}{m_H^2}} \left(1 - \frac{4 m_V^2}{m_H^2} + \frac{12 m_V^4}{m_H^4} \right), \tag{4.99}$$

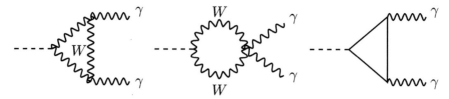

Fig. 4.54 Representative Feynman diagrams for the loop-induced coupling of a Higgs boson to two photons.

after the factor of two that is no longer appropriate is removed.

The partial width into a fermion pair is generated at tree level and is straightforwardly given by,

$$\Gamma(H \to f\bar{f}) = d_f \frac{g_W^2 m_H m_f^2}{32\pi m_W^2} \left(1 - \frac{4m_f^2}{m_H^2}\right)^{3/2}, \tag{4.100}$$

where d_f is a factor representing the additional possible colour degrees of freedom for the fermion. For quarks $d_f = 3$, the number of colours, while for leptons $d_f = 1$. Importantly, the branching ratio in Eq. (4.100) is proportional to the mass-squared of the fermion, so that in the SM the largest fermionic branching ratio is to a pair of the heaviest kinematically possible quarks, bottom quarks for $m_H = 125$ GeV. The partial width into tau pairs is also significant since it provides a non-negligible coupling to leptons. The widths into pairs of charm quarks or muons are very small but could possibly be probed with very large datasets in the future. Decays into lighter quarks and electrons are too rare to have any chance of observation at the LHC.

The gluon-fusion production diagram, illustrated in Fig. 4.52, can also be read as the one-loop decay of the Higgs boson into a pair of gluons. This partial width is not negligible, accounting for about 10% of all Higgs boson decays for $m_H = 125$ GeV. Of course, the resulting events are rather difficult, if not impossible, to observe directly at a hadron collider due to the fact that jet backgrounds are overwhelming.

On the other hand, a very important loop-induced Higgs boson decay is the channel $H \to \gamma\gamma$, which receives contributions from diagrams such as the ones shown in Fig. 4.54. Although the rate for this decay is suppressed by two powers of the electromagnetic coupling and a loop factor, it is very important because the two photons can be reconstructed very cleanly in the detector. The squared matrix element for this decay is

$$|\mathcal{M}|^2 = \frac{e^4}{16\pi^2} \frac{G_F m_H^4}{8\sqrt{2}\pi^2} \left| N_c \left(Q_t^2 I_q(m_t^2/m_H^2) + Q_b^2 I_q(m_b^2/m_H^2)\right) + I_W(m_W^2/m_H^2) \right|^2 \tag{4.101}$$

In this equation the contribution of the top, bottom, and W-boson loops is shown separately, in terms of the function $I_q(x)$ already introduced in Eq. (4.95) and

$$I_W(x) = -2\left[(6x + 1) + 6x(2x - 1)F(x)\right]. \tag{4.102}$$

The function $F(x)$ is defined in Eq. (4.96). It is instructive to evaluate these contributions at their physical values, in this case using $m_b = 4.5$ Gev, $m_t = 175$ GeV, $m_W = 80.4$ GeV and a Higgs boson mass of 125 GeV. Writing the terms in the same order as in Eq. (4.101) the result is

$$|\mathcal{M}|^2 = \frac{e^4}{16\pi^2} \frac{G_F m_H^4}{8\sqrt{2}\pi^2} |(1.838) + (-0.016 + 0.019i) + (-8.323)|^2. \qquad (4.103)$$

The W-boson and top quark contributions enter with opposite signs and therefore interfere destructively. The contribution of the bottom quark loop is complex (*cf.* the behaviour of the function $F(x)$ in Eq. (4.96)) and, as expected, is very small. However, it affects the cross-section at the 0.5% level due to interference with the rest of the amplitude. Beyond its importance to establishing a clear experimental signature of the Higgs boson, the $H \to \gamma\gamma$ decay is also particularly interesting from a theory point of view. Since the partial width is very small, it is rather sensitive to new particles that could couple to the Higgs boson and circulate in the virtual loop. Moreover, the threshold behaviour of the resulting contribution, *cf.* Eq. (4.96), could in principle be observed experimentally. Note that the closely related process, which proceeds through almost identical loop diagrams and thus is qualitatively rather similar, is the decay $H \to Z\gamma$. However, this mode results in an even smaller branching ratio once the decay of the Z boson is also folded in.

Although the mass of the Higgs boson has now been determined at the LHC, it is useful to review the pattern of branching ratios that was expected prior to its discovery. As shown in Fig. 4.55, which depicts the branching ratios for a SM Higgs boson as a function of its mass, the most important decay channels are strongly dependent on the Higgs boson mass. This basic fact necessitated a very broad range of search strategies at the LHC. Note that these results include various higher-order QCD and electroweak corrections, although the basic pattern is very similar to the one that would be obtained using the lowest-order formulae above. As is clear from the figure, the branching ratios to the different decay products are very sensitive to kinematic thresholds, unlike the very smooth mass dependence observed in the Higgs boson production rates. Over the mass range shown, the decay of the Higgs boson to bottom quarks dominates for $m_H < 135$ GeV and, above that value, it is the WW branching ratio that is largest. Above 200 GeV the decays into WW and ZZ remain the most important channels, with a small contribution from $H \to t\bar{t}$ that is as large as 0.2 for $m_H \approx 450$ GeV. It is interesting to note that the dominance of the WW branching ratio is true even below threshold, with at least one of the W bosons produced far from mass-shell. The presence of all of these features results in a very rich phenomenology and a difficult experimental task, with search analyses fine-tuned for different putative Higgs masses. The $ZZ \to 4$ leptons and $\gamma\gamma$ decay modes benefit greatly from the excellent identification and resolution of the detected particles. For the decay $H \to WW \to 2\ell2\nu$, the analysis is hampered by the missing transverse momentum which means that the candidate Higgs boson mass cannot be directly constructed. Although the transverse mass may be used as a substitute, the resulting resolution is not as good as in the fully reconstructed cases. Any of the decay modes involving jets of hadrons suffer from similar resolution issues but are complicated foremost by the fact

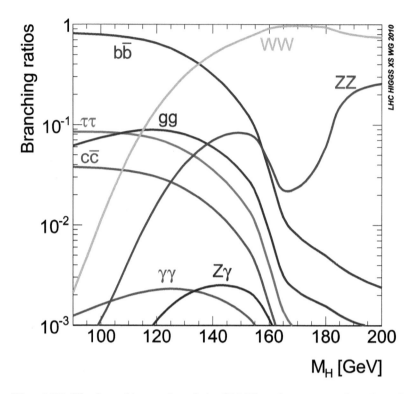

Fig. 4.55 The branching ratios of the SM Higgs boson, as a function of its mass, taken from Ref. [468].

that they must compete with large QCD backgrounds. This is especially true for the decay $H \to b\bar{b}$, with b jets produced prolifically in both pure QCD processes and in top quark decays.

Following the discovery of a Higgs-like boson at the LHC in 2012, it is useful to focus specifically on the case $m_H = 125$ GeV. The branching ratios for this mass are shown in Table 4.2 which is taken from Ref. [460]. For each decay the table also includes the uncertainties due to variation of input parameters (α_s and heavy quark masses) and due to uncalculated higher orders. The order in perturbation theory at which the branching ratio is known is also included, for completeness. One additional branching ratio is shown in this table that did not appear earlier: the one for the decay $H \to \mu^+\mu^-$. The branching ratio for this mode is an order of magnitude smaller than that for $H \to \gamma\gamma$. Since both final states are well-reconstructed and also have smooth, understood backgrounds, one might expect that the luminosity required to observe $H \to \mu^+\mu^-$ should be a factor of 100 larger than that needed for $H \to \gamma\gamma$. A more detailed analysis supports this rough estimate, with hopes that this rare decay mode could indeed be observed with several inverse attobarns of LHC data [401].

Table 4.2 Branching ratios of a SM Higgs boson of mass $m_H = 125$ GeV. The corresponding width is $\Gamma_H = 4.07$ MeV.

Decay mode	Branching ratio	Order of calculation
$b\bar{b}$	$0.577^{+3.2\%}_{-3.3\%}$	N^4LO QCD + NLO EW
WW	$0.215^{+4.3\%}_{-4.2\%}$	NLO QCD + NLO EW
gg	$0.0857^{+10.2\%}_{-10.0\%}$	N^3LO QCD + NLO EW
$\tau\tau$	$0.0632^{+5.7\%}_{-5.7\%}$	NLO EW
$c\bar{c}$	$0.0291^{+12.2\%}_{-12.2\%}$	N^4LO QCD + NLO EW
ZZ	$0.0264^{+4.3\%}_{-4.2\%}$	NLO QCD + NLO EW
$\gamma\gamma$	$0.00228^{+5.0\%}_{-4.9\%}$	NLO QCD + NLO EW
$Z\gamma$	$0.00154^{+9.0\%}_{-8.8\%}$	LO
$\mu\mu$	$0.00022^{+6.0\%}_{-5.9\%}$	NLO EW

4.8.3 Width of the Higgs boson

The predicted total width of the Higgs boson is given by summing over all contributing partial widths so that, for instance, the simple expressions given previously can be used to obtain the leading approximation for the total width. For the boson discovered at the LHC, with a mass around 125 GeV, the total width is dominated by the decay into bottom quark pairs, *cf.* Table 4.2. The total Higgs boson width is thus well approximated, up to a $\mathcal{O}(1)$ factor corresponding to $\mathcal{BR}_{H\to b\bar{b}}$, by Eq. (4.100) with $m_f \to m_b$. Compared to the widths of its partner electroweak bosons the W and the Z, the width of the Higgs boson is thus suppressed by a factor m_b^2/m_W^2. This results in a SM prediction of $\Gamma_H \approx 4$ MeV for a Higgs boson of mass 125 GeV.

This width is much smaller than the typical mass resolution of the LHC experiments, even in the best-measured channels such as $H \to \gamma\gamma$ and $H \to ZZ$. Therefore a typical scan of the threshold region does not yield very much information about the intrinsic width. This type of direct scan yields only a rather weak bound at present, $\Gamma_H \lesssim 1600 \times \Gamma_H^{\text{SM}}$ [402], while an estimate of the eventual sensitivity using this method is at the level of $\Gamma_H \lesssim 50 \times \Gamma_H^{\text{SM}}$ [439]. A number of other methods to constrain the width directly have been proposed, relying on either interference effects that alter the shape of the mass distribution in diphoton decays [473] or on a comparison of Higgs-related cross-sections at the resonance and in the high-mass region beyond it [322].

The latter method rests on the observation that there is a significant fraction of Higgs boson events in the ZZ final state in the region $m(ZZ) > m_H$. This feature can clearly be seen in Fig. 4.56, which shows the cross-section as a function of the 4-lepton invariant mass for $ZZ \to 4\ell$ decays, for typical LHC cuts at 13 TeV. The large off-shell contribution is the result of two effects. First, the branching ratio $\mathcal{BR}_{H\to ZZ}$ grows significantly as the virtuality of the Higgs boson approaches the threshold for producing two real Z bosons. Second, there is an additional enhancement of the spectrum in the region $m_{4\ell} \sim 2m_t$ when there is sufficient energy to resolve the internal top threshold in the loop. The result is that, for typical lepton cuts, the predicted number of $H \to ZZ \to 4$-lepton events in the off-shell region defined by $m_{4\ell} > 130$ GeV can be 15% or more [315, 322, 654]. However, this is not the full story, due to the fact that

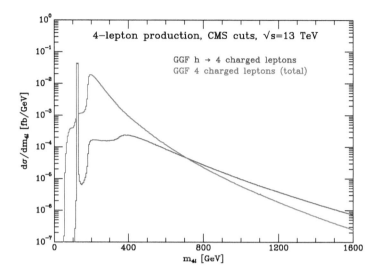

Fig. 4.56 The 4-lepton invariant mass distribution expected from $gg \to ZZ \to 4\ell$ for typical CMS cuts at 13 TeV. The contribution from the Higgs diagram alone is shown in blue, while the total contribution of all gg-initiated diagrams is shown in red.

there is another class of diagrams that contributes to this final state at the same order in perturbation theory. These are indicated in Fig. 4.57 (right) and correspond to box diagrams in which a quark circulates in the loop. In contrast to the Higgs diagram on the left, the contribution of light quarks in the loop is non-negligible. The inclusion of the box diagrams has an important consequence in the off-shell region due to the fact that the two sets of diagrams interfere destructively at high energies. This behaviour is a consequence of the unitarizing effect of the Higgs boson on the production of longitudinally polarized Z bosons [723]. The effect of the destructive interference is clearly seen in Fig. 4.56, where the contribution of all diagrams is *smaller* than the contribution of Higgs diagrams alone in the region $m_{4\ell} > 700$ GeV. Information on the width of the Higgs boson may be obtained by noting that the peak cross-section is related to both the couplings and the width of the Higgs boson, while the off-shell cross-section does not depend strongly on the width. Such methods provide much more stringent limits than direct approaches, $\Gamma_H < (5-9) \times \Gamma_H^{SM}$ [37, 661]. The experimental situation is discussed further in Section 9.6.4.

In addition to these direct constraints, the total width of the Higgs boson can be determined indirectly by comparing measurements of its couplings in different channels. Converting these measurements into constraints on the maximum width requires additional theoretical assumptions, for instance bounds on the couplings of the Higgs to W and Z bosons that are allowed in broad classes of SM extensions [475]. With this caveat, the indirect bounds can be rather strong: $0.3 < \Gamma_H / \Gamma_H^{SM} < 3.56$.

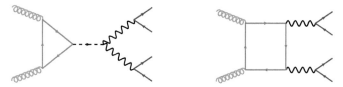

Fig. 4.57 Representative diagrams entering the calculation of $gg \rightarrow ZZ \rightarrow 4\ell$ production, corresponding to Higgs contributions (left) and non-Higgs continuum box diagrams (right).

4.8.4 Higgs boson production in gluon fusion

As already discussed, the gluon fusion process representing the production of a Higgs boson, $gg \rightarrow H$, does not occur at tree level but instead appears for the first time at one-loop. When one starts to consider higher orders in perturbation theory or the radiation of additional hard jets this makes the relevant calculations correspondingly more difficult and, in some cases, impossible to perform at present.

For this reason it is convenient to formulate the diagram in Fig. 4.52 as an effective coupling of the Higgs boson to two gluons in the limit that the top quark is infinitely massive. This corresponds to the effective Lagrangian,

$$\mathcal{L}^{\text{eff}} = \frac{\alpha_{\text{s}}}{12\pi v} \left(1 + \frac{11}{4} \frac{\alpha_{\text{s}}}{\pi} \right) H \operatorname{tr} G_{\mu\nu} G^{\mu\nu} + \mathcal{O}(\alpha_{\text{s}}^3), \qquad (4.104)$$

where the trace is over the colour degrees of freedom. Using the resulting ggH coupling, it is straightforward to compute the LO matrix element squared,

$$\left| \mathcal{M}_{gg \rightarrow H}^{(\text{LO})} \right|^2 = \frac{1}{8V} \left(\frac{\alpha_{\text{s}}}{3\pi v} \right)^2 m_H^4. \qquad (4.105)$$

Alternatively, the same expression could have been derived by taking the limit $m_t/m_H \rightarrow \infty$ in the corresponding expression for the same matrix elements, Eq. (4.94).[4] By taking the ratio of these two expressions one finds that the matrix element squared, and hence the cross-section, in the full theory is related to the one in the effective theory by the factor

$$R \equiv \frac{\sigma_{\text{full}}}{\sigma_{\text{effective}}} = \left| \frac{3}{4} I_q \left(m_t^2/m_H^2 \right) \right|^2. \qquad (4.106)$$

The value of this ratio is shown in Fig. 4.58, for Higgs boson masses between 100 and 300 GeV. Although formally one would expect that this approximation is valid only when all other scales in the problem are much smaller than m_t, in fact one finds that $m_H < m_t$ is sufficient to render the effective theory valid to within 10%. For a Higgs boson of mass 125 GeV, R corresponds to a correction factor of about 6.5%. In more complicated applications of the effective theory, for instance in the presence of jets, it has been found that the requirement $p_T(\text{jet}) < m_t$ is necessary for an accurate approximation [454].

---

[4]In this limit it is easy to show that the function $I_q(m_t^2/m_H^2) \rightarrow 4/3$.

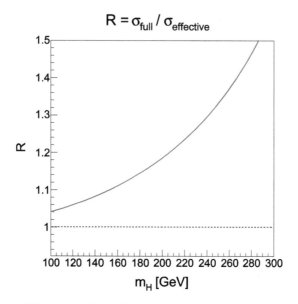

Fig. 4.58 The ratio, R of the LO cross-section for Higgs production through gluon fusion calculated in the full theory to the result in the effective theory. The ratio is given by the expression in Eq. (4.106).

The clear benefit of the effective theory is the ease with which higher-order corrections can be computed. The NLO corrections were first computed long before the discovery of the Higgs boson [438], with the calculation performed in a very similar style to the Drell–Yan case discussed earlier in Sections 3.3.2 and 3.3.3. The contribution of the loop diagrams is, as in the Drell–Yan case, proportional to the leading-order matrix element. Explicitly, using the dimensional reduction scheme, the result is

$$2\mathcal{Re}\left[\mathcal{M}_{gg\to H}^{(1-\text{loop})} \times \mathcal{M}_{gg\to H}^{*(\text{LO})}\right] = \frac{\alpha_s N_c}{2\pi}\left(\frac{\mu^2}{m_H^2}\right)^\varepsilon c_\Gamma\left(-\frac{2}{\varepsilon^2}+\pi^2\right)\left|\mathcal{M}_{gg\to H}^{(\text{LO})}\right|^2 \quad (4.107)$$

Since the LO amplitude contains the strong coupling, it must be renormalized at this order. This is achieved by adding a term,

$$-2\frac{b_0}{\varepsilon}c_\Gamma\frac{\alpha_s}{2\pi}\left|\mathcal{M}_{gg\to H}^{(\text{LO})}\right|^2, \quad (4.108)$$

where the overall factor of two reflects the $\mathcal{O}(\alpha_s^2)$ nature of the LO matrix elements. In addition there is a finite renormalization of the strong coupling associated with the use of dimensional reduction, to bring its definition into the standard $\overline{\text{MS}}$ scheme,

$$+2\frac{N_c}{6}\frac{\alpha_s}{2\pi}\left|\mathcal{M}_{gg\to H}^{(\text{LO})}\right|^2. \quad (4.109)$$

Finally, the full calculation should also include the $\mathcal{O}(\alpha_s)$ correction to the effective Lagrangian shown in Eq. (4.104), which it is natural to include here. Accounting for all of these contributions, the renormalized virtual result is

$$\frac{\alpha_s N_c}{2\pi} c_\Gamma \left[-\frac{2}{\varepsilon^2} \left(\frac{\mu^2}{m_H^2} \right)^\varepsilon - \frac{2}{\varepsilon} \frac{b_0}{N_c} + 4 + \pi^2 \right] \left| \mathcal{M}_{gg \to H}^{(LO)} \right|^2 . \tag{4.110}$$

The real corrections from the process $gg \to Hg$ are easily computed using either an explicit calculation or making use of the Catani–Seymour dipoles that were introduced in Section 3.3.2. The derivation of these contributions is very similar to the calculation of the closely related Drell–Yan process, that was explicitly worked out there. The full result is obtained by adding the real radiation corrections and the appropriate PDF collinear subtractions to the result in Eq. (4.110). In this way one arrives at the expression for the NLO corrections (*cf.* Eq. (3.219) for the Drell–Yan case),

$$d\sigma_{gg \to H}^{NLO}(p_{g_1}, p_{g_2}) = \frac{\alpha_s N_c}{2\pi} |\mathcal{M}_{gg \to H}^{(LO)}|^2 \int_0^1 dx_1 dx_2 \frac{\delta(x_1 x_2 s - m_H^2)}{m_H^2}$$

$$\int_{x_1}^1 d\xi_1 \int_{x_2}^1 d\xi_2 \, f_{g/h_1} \left(\frac{x_1}{\xi_1}, \mu_F^2 \right) f_{g/h_1} \left(\frac{x_2}{\xi_2}, \mu_F^2 \right) \left\{ \left[\frac{11}{3} + \frac{4\pi^2}{3} \right] \delta(1 - \xi_1) \delta(1 - \xi_2) \right.$$

$$+ \left[\left(\frac{2}{1 - \xi_1} \log \frac{(1 - \xi_1)^2 m_H^2}{\xi_1 \mu_F^2} \right)_+ + 2 \left(\frac{1}{\xi_1} - 2 + \xi_1 - \xi_1^2 \right) \left(\log \frac{(1 - \xi_1)^2 m_H^2}{\xi_1 \mu_F^2} \right) \right.$$

$$\left. + \frac{11}{6} \frac{(1 - \xi_1)^3}{\xi_1} \right] \delta(1 - \xi_2)$$

$$+ \left[\left(\frac{2}{1 - \xi_2} \log \frac{(1 - \xi_2)^2 m_H^2}{\xi_2 \mu_F^2} \right)_+ + 2 \left(\frac{1}{\xi_2} - 2 + \xi_2 - \xi_2^2 \right) \left(\log \frac{(1 - \xi_2)^2 m_H^2}{\xi_2 \mu_F^2} \right) \right.$$

$$\left. \left. + \frac{11}{6} \frac{(1 - \xi_2)^3}{\xi_2} \right] \delta(1 - \xi_1) \right\} . \tag{4.111}$$

To compute a complete set of corrections at this order one must also include contributions from the real radiation diagrams, $gq \to Hq$, and all crossings. These contain no virtual contributions and their calculation is more straightforward. The net effect of the NLO corrections is large, which can already be seen from the coefficient of the LO-like term, proportional to $\delta(1 - \xi_1) \delta(1 - \xi_2)$, in Eq. (4.111).

The size of the corrections motivated the calculation of NNLO corrections to the Higgs boson cross-section using the same effective theory approach [158, 615], as already discussed in Section 3.4.1. The inclusion of the NNLO terms provided only a relatively small further correction, thus stabilizing the perturbative expansion of the cross-section. However, the residual scale uncertainty remained at the 10% level until the recent completion of the full N^{3}LO calculation [157]. The results of that calculation are illustrated in Fig. 4.59, which shows the cross-sections and uncertainties at each

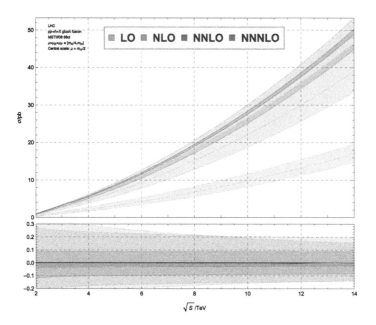

Fig. 4.59 The scale uncertainty of the Higgs gluon fusion cross-section, computed in the range $[m_H/4, m_H]$, versus the collider energy, $\sqrt{s}$. The cross-section is computed at LO, NLO, NNLO, and N^3LO in the strong coupling. Reprinted with permission from Ref. [157].

order of perturbation theory. At N^3LO both the size of the correction, and the total scale uncertainty, are at the level of a few per cent. This has an immediate impact on the precision Higgs programme of the LHC, for example by significantly reducing the theoretical uncertainty associated with the extraction of the Higgs boson coupling strengths. Other improvements in the accuracy of the theoretical prediction can also be taken into account. For instance, the predictions shown in Fig. 4.53 also include contributions from electroweak corrections [123]. These represent a few per cent effect but introduce a small additional uncertainty since one must choose a scheme for combining the separate QCD and electroweak corrections.

An important issue in detecting events in which a Higgs boson is produced in this channel is the presence of additional jet activity. There are many significant backgrounds to Higgs production that naturally contain jets so that it is natural to try to use the jet information to better separate the signal and background processes. One example of this is top pair production, which has a considerable cross-section and leads to final states containing two W bosons and two jets and is thus a background to searches looking for the decay $H \rightarrow WW$. Calculations of the signal rate in association with additional jets have been performed at NLO for up to three jets [420] and at NNLO for $H + 1$ jet [270, 271, 392]. As a further example of the stabilizing effect of higher orders in perturbation theory, the cross-section for Higgs+jet at LO, NLO, and NNLO is shown in Fig. 4.60, as a function of the transverse momentum required to

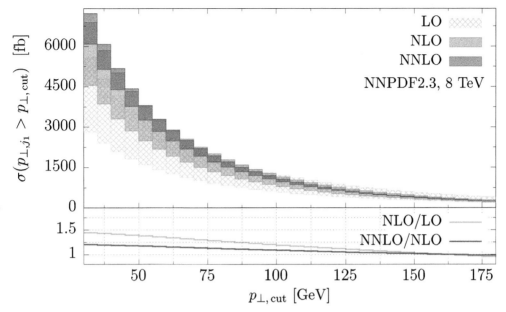

Fig. 4.60 The NNLO cross-section for Higgs production in association with a jet, as a function of the minimum jet $p_\perp$. Reprinted with permission from Ref. [271].

define the jet. However, the real issue is that all these higher-order calculations yield predictions for cross-sections that include *at least* the number of jets present in the leading-order process. That is, the NLO calculations explicitly include up to one additional jet and the NNLO ones up to two. However, the signal discrimination requires predictions for an exact number of jets — and no more than that.[5] In the simplest case, one is thus interested not in the inclusive Higgs cross-section but the cross-section for Higgs production with zero jets, that is, in the presence of a **jet-veto**. However, this type of veto on additional radiation can render the fixed-order perturbative calculations more unreliable. Particularly if the jet-veto scale is much smaller than the Higgs boson mass, as is usually the case, the perturbative expansion develops large logarithms of the form $\log(p_\perp^{\mathrm{veto}}/m_H)$. In order to recover accurate predictions in this case it is necessary to go beyond the fixed-order approach and perform a resummation of these types of logarithm. Such calculations will be discussed further in Chapter 5.

4.8.5 Amplitudes for H+jet production

Although first results for the H+jet process are now available at NNLO, these have all been computed in the effective theory. As already noted, these calculations are expected to break down for values of the jet transverse momentum larger than about m_H. For this reason, it is still useful to consider lower-order perturbative predictions

---

[5]This is not the case for other decay modes such as $H \to \gamma\gamma$ and $H \to ZZ^*$ where inclusive measurements can be carried out. See Section 9.6.

in the full theory. Of course, in the full theory even the lowest-order prediction already contains loop diagrams and so the resulting matrix elements contain the usual one-loop functions. These matrix elements were first computed in Ref. [500] and the results of that calculation are summarized here.

There are two basic parton-level processes that must be considered:

$$(a): \quad 0 \rightarrow g(p_1) + \bar{q}(p_2) + q(p_3) + H, \tag{4.112}$$

$$(b): \quad 0 \rightarrow g(p_1) + g(p_2) + g(p_3) + H. \tag{4.113}$$

The amplitude for process (a) is relatively simple. It can be written as

$$M_{gq \rightarrow qH} = -i \frac{g_s^2}{16\pi^2} \frac{g_W}{4m_W} \frac{1}{2} (t^A)_{i_3 i_2} g_s \tag{4.114}$$

$$\frac{1}{s_{23}} \bar{u}(p_3) \gamma_\mu u(p_2) \left(g^{\alpha\mu} - \frac{p_1^\mu (p_2^\alpha + p_3^\alpha)}{p_1 \cdot (p_2 + p_3)} \right) \epsilon_\alpha(p_1) F(s_{23}, s_H)$$

where $\epsilon_\alpha(p_1)$ is the polarization vector of the gluon, $s_H = (p_1 + p_2 + p_3)^2$ and the loop function $F(s_{23}, s_H)$ for a single quark of mass m_q is given by

$$F(s_{23}, s_H) = -8m_q^2 \Big[2 - (s_H - s_{23} - 4m_q^2) C_0(p_1, p_{23}; m_q, m_q, m_q)$$

$$+ \frac{2s_{23}}{s_H - s_{23}} \Big(B_0(p_{123}; m_q, m_q) - B_0(p_{23}; m_q, m_q) \Big) \Big]. \tag{4.115}$$

This function depends on the bubble and triangle scalar integrals defined by

$$B_0(p_1; m_1, m_2) = \frac{\mu^{4-D}}{i\pi^{\frac{D}{2}}} \Gamma(1 - \varepsilon) \int d^D l \, \frac{1}{(l^2 - m_1^2 + i\epsilon)((l + p_1)^2 - m_2^2 + i\epsilon)},$$

$$C_0(p_1, p_2; m_1, m_2, m_3) = \frac{1}{i\pi^2} \tag{4.116}$$

$$\times \int d^4 l \, \frac{1}{(l^2 - m_1^2 + i\epsilon)((l + p_1)^2 - m_2^2 + i\epsilon)((l + p_1 + p_2)^2 - m_3^2 + i\epsilon)}.$$

The corresponding matrix element squared is

$$|M_{gq \rightarrow qH}|^2 = \left(\frac{g_s^3}{64\pi^2} \frac{g_W}{m_W} \right)^2 N_c C_F \frac{s_{12}^2 + s_{13}^2}{s_{23}(s_H - s_{23})^2} |F(s_{23}, s_H)|^2. \tag{4.117}$$

For process (b) the amplitude is

$$M_{gg \rightarrow gH} = -\frac{g_W}{m_W} \frac{g_s^3}{32\pi^2} s_H^2 f_{ABC} \epsilon_\alpha(p_1) \epsilon_\beta(p_2) \epsilon_\gamma(p_3)$$

$$\Big[F_2^{\alpha\beta\gamma}(p_1, p_2, p_3) A_3(p_1, p_2, p_3) + F_1^{\alpha\beta\gamma}(p_1, p_2, p_3) A_2(p_1, p_2, p_3)$$

$$+F_1^{\beta\gamma\alpha}(p_2,p_3,p_1)A_2(p_2,p_3,p_1)+F_1^{\gamma\alpha\beta}(p_3,p_1,p_2)A_2(p_3,p_1,p_2)\Big], \quad (4.118)$$

where the colour labels of the gluons are denoted by A, B, and C. This equation introduces the projectors F_1 and F_2 that are defined by

$$F_1^{\alpha\beta\gamma}(p_1,p_2,p_3)=\left(\frac{g^{\alpha\beta}}{p_1\cdot p_2}-\frac{p_1^\beta p_2^\alpha}{p_1\cdot p_2^2}\right)\left(\frac{p_2^\gamma}{p_2\cdot p_3}-\frac{p_1^\gamma}{p_1\cdot p_3}\right)$$

$$F_2^{\alpha\beta\gamma}(p_1,p_2,p_3)=\frac{p_3^\alpha p_1^\beta p_2^\gamma-p_2^\alpha p_3^\beta p_1^\gamma}{p_1\cdot p_2\,p_1\cdot p_3\,p_2\cdot p_3}+\frac{g^{\alpha\beta}}{p_1\cdot p_2}\left(\frac{p_1^\gamma}{p_3\cdot p_1}-\frac{p_2^\gamma}{p_3\cdot p_2}\right) \quad (4.119)$$

$$+\frac{g^{\beta\gamma}}{p_2\cdot p_3}\left(\frac{p_2^\alpha}{p_1\cdot p_2}-\frac{p_3^\alpha}{p_1\cdot p_3}\right)+\frac{g^{\alpha\gamma}}{p_1\cdot p_3}\left(\frac{p_3^\beta}{p_2\cdot p_3}-\frac{p_1^\beta}{p_2\cdot p_1}\right).$$

The functions A_2 and A_3 contain the loop integral functions and it is convenient to rewrite A_3 in terms of a further function, A_4, as follows

$$A_3(p_1,p_2,p_3)=\frac{1}{2}\left[A_2(p_1,p_2,p_3)+A_2(p_2,p_3,p_1)+A_2(p_3,p_1,p_2)-A_4(p_1,p_2,p_3)\right]. \quad (4.120)$$

The functions A_2 and A_4 are then defined by

$$A_2(p_1,p_2,p_3)=b_2(s_{12},s_{13},s_{23})+b_2(s_{12},s_{23},s_{13}),$$
$$A_4(p_1,p_2,p_3)=b_4(s_{12},s_{13},s_{23})+b_4(s_{13},s_{23},s_{12})+b_2(s_{23},s_{12},s_{13}) \quad (4.121)$$

where for a single quark loop the functions b_2 and b_4 read

$$b_4(s,t,u)=\frac{m_q^2}{s_H}\left[-\frac{2}{3}+\left(\frac{m_q^2}{s_H}-\frac{1}{4}\right)(W_2(s)-W_2(s_H)+W_3(s,t,u,s_H))\right]$$

$$b_2(s,t,u)=\frac{m_q^2}{s_H^2}\left[\frac{s(u-s)}{s+u}+\frac{2ut(u+2s)}{(s+u)^2}(W_1(t)-W_1(s_H))\right.$$

$$+\left(m_q^2-\frac{s}{4}\right)\left(\frac{1}{2}W_2(s)+\frac{1}{2}W_2(s_H)-W_2(t)+W_3(s,t,u,s_H)\right)$$

$$+s^2\left(\frac{2m_q^2}{(s+u)^2}-\frac{1}{2(s+u)}\right)(W_2(t)-W_2(s_H))+\frac{ut}{2s}(W_2(s_H)-2W_2(t))$$

$$+\frac{1}{8}\left(s-12m_q^2-\frac{4ut}{s}\right)W_3(t,s,u,s_H)\Big]. \quad (4.122)$$

The remaining functions, W_1, W_2 and W_3 are remnants of the scalar integrals that enter the calculation. Their definitions are as follows:

$$W_1(s)=2+\int_0^1 dx\log\left(1-\frac{s}{m_q^2}x(1-x)-i\epsilon\right)$$

$$W_2(s) = 2 \int_0^1 \frac{dx}{x} \log\left(1 - \frac{s}{m_q^2} x(1-x) - i\epsilon\right) \tag{4.123}$$

$$W_3(s,t,u,v) = I_3(s,t,u,v) - I_3(s,t,u,s) - I_3(s,t,u,u)$$

$$I_3(s,t,u,v) = \int_0^1 dx \left(\frac{m_q^2 t}{us} + x(1-x)\right)^{-1} \log\left(1 - \frac{v}{m_q^2} x(1-x) - i\epsilon\right)$$

and explicit results in all kinematic regions of interest are given in the appendix of Ref. [500]. The corresponding matrix element squared is

$$|M_{gg \to gH}|^2 = \frac{g_W^2 g_s^6}{256\pi^4} \frac{8N_c^2 C_F}{s_{12} s_{13} s_{23}} \left[|A_2(p_1, p_2, p_3)|^2 + |A_2(p_2, p_3, p_1)|^2 \right.$$
$$\left. + |A_2(p_3, p_1, p_2)|^2 + |A_4(p_1, p_2, p_3)|^2 \right]. \tag{4.124}$$

These matrix elements can be used to make a comparison of the Higgs+jet cross-section in the full theory with the effective theory approximation. Such a comparison is shown in Fig. 4.61, where the cross-sections have been computed at $\sqrt{s} = 13$ TeV and for $m_H = 125$ GeV. The ratio R, defined analogously to Eq. (4.106), demonstrates the anticipated behaviour. The cross-section at low jet $p_\perp$ is larger by the same factor as for the inclusive Higgs cross-section, *cf.* Fig. 4.61. The difference between the calculations actually decreases as $p_\perp$ increases, until the result of the full calculation is smaller than in the effective theory. Nevertheless, the approximation remains reasonable for $p_\perp(\text{jet}) \lesssim 200$ GeV. At higher jet momenta the approximation quickly breaks down, resulting in a very poor description in the effective theory.

The rate for producing a Higgs boson in association with an additional jet is significant at the LHC, due to the nature of the gluon-fusion process. It is more important than, for instance, the similar Drell–Yan process due to the fact that the gluons naturally radiate more copiously than quarks. Indeed, the cross-section for producing more than one jet is still rather high. This is indicated in Fig. 4.62, which shows cross-sections for the production of a Higgs boson in association with up to three jets as a function of the machine operating energy. At LHC energies, with typical jet cuts, the cross-section for producing H+jet is about half the inclusive Higgs production rate, and the rate for two or more jets is only smaller by about a further factor of three. As $\sqrt{s}$ increases, the jet cross-sections grow relatively more rapidly if the jet definition remains the same. Thus the study of Higgs boson processes that contain additional jets becomes even more of a concern for any future hadron collider.

The first measurements of differential jet production in association with a Higgs boson at the LHC will be discussed in Section 9.6.5.

4.8.6   Higgs boson production in weak boson fusion

As already mentioned, weak boson fusion is an especially valuable production mode for hadron colliders. Although the cross-section is an order of magnitude smaller than for gluon fusion, it provides a valuable probe of the coupling of the Higgs boson to W and Z bosons, independent of the manner in which the Higgs boson decays [902].

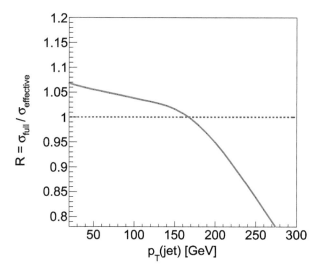

Fig. 4.61 The ratio, R of the LO cross-section for Higgs+jet production in the full theory to the result in the effective theory, as a function of the jet transverse momentum, for $m_H = 125$ GeV.

On the theoretical side, the weak nature of the process means that the cross-section is under good control. The NLO corrections are small and the scale uncertainty at this order is around 10% [171, 228]. To perform a computation beyond NLO it is useful to work in a structure function approach, where the process is described by the independent production of two W or Z bosons by each initial parton. The W or Z bosons then subsequently interact to produce the Higgs boson, as indicated schematically in Fig. 4.63.

   This double deep-inelastic scattering approach is an excellent approximation because of the fact that interference effects between the quark lines are very small. Using this framework it has been possible to compute the NNLO corrections to the WBF process [263, 264], including also the calculation of fully differential observables [301]. Most recently the calculation has been extended to N³LO for the total cross-section [488]. Fig. 4.64 shows the scale dependence of the cross-section through N³LO, computed using the calculation presented in Ref. [488], as a function of the pp collider energy $(\sqrt{s})$. This indicates that the scale uncertainty in the N³LO prediction for the weak boson fusion cross-section at the LHC is at the level of a few per mille. The electroweak corrections to this process have also been computed at NLO and included in the HAWK code [398]. The corrections are of the same size as the NLO QCD contributions and must therefore be taken into account.

   Isolating a clean sample of events in which the Higgs boson is produced through weak boson fusion is complicated not only by the presence of the usual SM backgrounds but also by production of the same final state through other Higgs processes. For instance, the Higgs boson may be accompanied by two jets through associated

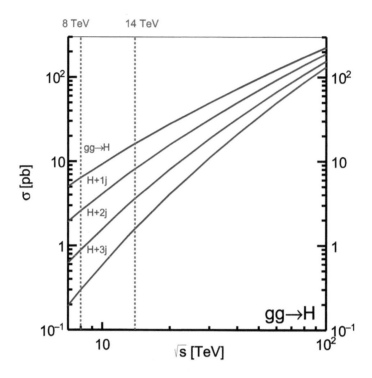

Fig. 4.62 Cross-sections for Higgs production ($m_H = 125$ GeV) through gluon fusion at proton–proton colliders as a function of centre-of-mass operating energy, $\sqrt{s}$. Cross-sections for production of a Higgs boson in association with jets are computed for jets satisfying $p_T > 40$ GeV, $|y| < 5$ and k_T-clustering with $D = 0.4$.

production in which the W or Z boson decays hadronically. However, the biggest such contamination occurs from the gluon fusion process, with the radiation of additional jets from the initial state. Enhancing the efficiency of selecting weak boson fusion events by applying a large rapidity separation between two of the jets in the event still leaves a substantial contribution from gluon fusion events, as shown in Fig. 4.65. This presents an additional complication when trying to make a precision measurement of the couplings of the Higgs boson in this channel.

An interesting aspect of the weak boson fusion process is its ability to probe the tensor structure of HW^+W^- and HZZ couplings [811]. The most general possible structure of the HVV vertex (where $V = W^{\pm}$ or $V = Z$) consistent with gauge invariance can be written as,

$$C_{HVV}^{\mu\nu}(p_1, p_2) = a_1(p_1, p_2)g^{\mu\nu} + a_2(p_1, p_2)\left(p_1 \cdot p_2\, g^{\mu\nu} - p_1^{\nu}p_2^{\mu}\right)$$
$$+ a_3(p_1, p_2)\epsilon^{\mu\nu\rho\sigma}p_{1,\rho}p_{2,\sigma}. \tag{4.125}$$

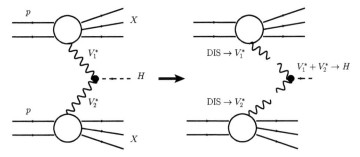

Fig. 4.63 Schematic picture of the double deep-inelastic scattering approach to WBF.

The SM is realized by a constant value for a_1, independent of the vector boson momenta p_1 and p_2 (*cf.* the Feynman rules for the SM Higgs interactions in Fig. 4.49). A non-constant value for a_1, or the new tensor couplings represented by a_2 and a_3 in Eq. (4.125), can be realized by loop-induced couplings within a particular model of new physics. a_2 represents an additional CP-even coupling, while a_3 is CP-odd due to the presence of the epsilon tensor in the interaction. The gold-plated observable for differentiating between these three types of coupling is the azimuthal angle between the two tagging jets, $\Delta\phi_{jj}$. Fig. 4.66 illustrates the expected distribution of this angle, for the SM case and for pure CP-even or CP-odd contributions represented by anomalous couplings a_2 or a_3. The SM expectation is for a relatively flat distribution, while the anomalous CP-even and CP-odd couplings lead to pronounced dips at $\Delta\phi_{jj} = 90°$ and $\Delta\phi_{jj} = 0, 180°$ respectively. This behaviour can be understood from the structure of the interactions and resulting matrix elements [811]. For example, the presence of the epsilon tensor in the CP-odd interaction means that it vanishes when there are fewer than four independent momenta in the process, that is, when the tagging jets are collinear. The picture is more complicated in the presence of an admixture of SM and anomalous CP-odd and CP-even effects, but the observable $\Delta\phi_{jj}$ remains an excellent probe.

Note that a similar analysis can be made for Higgs boson events produced through gluon fusion, where the SM Hgg effective coupling takes exactly the form of the a_2 term in Eq. (4.125). As a result, the SM expectation for the $\Delta\phi_{jj}$ distribution in gluon fusion events has a shape similar to the CP-even (a_2) curve in Fig. 4.66.

4.8.7 Higgs boson production in association with heavy particles

The associated production channels comprise two distinct cases. In the first, the Higgs boson is produced through its coupling to W and Z bosons, through the first two diagrams shown in Fig. 4.50. Although the cross-sections for these production modes are smaller than for weak boson fusion, they do offer the possibility of providing additional information on the couplings to W and Z bosons separately. From the form of these diagrams it is clear that, theoretically, these processes are closely related to the Drell–Yan process: for the most part, they can be described by the off-shell reaction, $pp \to V^*$, followed by a subsequent decay, $V^* \to VH$. This similarity has

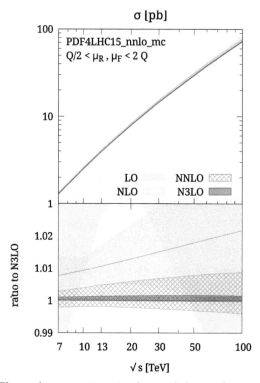

Fig. 4.64 The scale uncertainty in the weak boson fusion cross-section, at each order through N³LO. The cross-sections are shown at each order as a function of the collider energy and are normalized to the N³LO result. Reprinted with permission from Ref. [488].

enabled theoretical predictions for the cross-sections to be made at NNLO in the strong coupling [524]. Moreover, the NLO electroweak corrections are also known [458].

This channel is particularly interesting as a way to get a handle on the bottom quark decay mode, $H \to b\bar{b}$. A standard analysis of this mode would be plagued by large backgrounds, in particular from top pair production that yields events with similar kinematic properties. However, in Ref. [297] it was shown that sensitivity could be recovered in these channels by looking in the *boosted regime* where the vector bosons are produced back to back and at large transverse momenta. Although this significantly reduces the signal cross-section, the dominant backgrounds are impacted even more strongly. However, the key to utilizing the boosted kinematics fully comes from the properties of the jets that should be produced in the signal process: the $b\bar{b}$ pair should be reconstructed in a fat jet, as discussed in Section 2.1.6. In order to provide the best discrimination against background processes it is also necessary to enforce a veto against any additional jet activity. The application of such a requirement greatly affects the size of the QCD corrections in this case. For the inclusive VH cross-section the effect of higher-order corrections is rather mild, but after applying a jet veto this is no longer true. The situation in the presence of the jet veto is indicated in Fig. 4.67, which

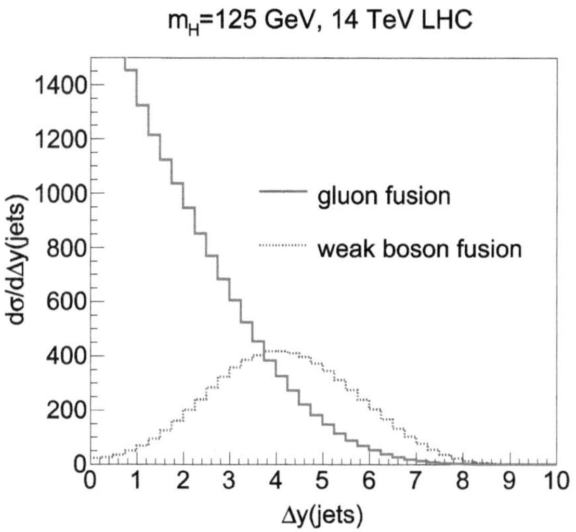

Fig. 4.65 The cross-section for Higgs production through gluon fusion and weak boson fusion, as a function of the rapidity separation between the two jets. Jets are defined using the anti-k_T algorithm with $D = 0.4$ and satisfy $p_T > 40$ GeV.

shows the theoretical prediction for the $p_\perp$ distribution of the fat jet in WH events at LO, NLO, and NNLO. There is a significant negative correction at NLO and a smaller further reduction at NNLO. The fact that the first-order correction is so large indicates that further work is required to obtain a reliable prediction for the cross-section in this region. Since, at this point, an even higher-order calculation is infeasible, a better avenue for improving the prediction is through the use of resummation techniques of the type that will be discussed in Chapter 5.

The second associated production channel results in a final state containing a Higgs boson together with top quarks. A top quark can be produced through one of the normal strong processes, with the Higgs boson coupling to it through the Yukawa interaction. The largest such cross-section is $t\bar{t}H$ production (see Fig. 4.51), taking advantage of the large top quark mass to enhance the Higgs coupling. Even so, at LHC operating energies, this cross-section is the smallest of the SM production modes shown in Fig. 4.53. As a result the LHC has only limited sensitivity in this channel until the accumulation of bigger datasets in Run II and beyond. Nevertheless, such data offer the chance of providing direct evidence of the coupling of the Higgs boson to the top quark. It could also yield information on the coupling to bottom quarks through a striking signature containing four identified b jets, two originating from the top quark decays and the remainder from a $H \to b\bar{b}$ decay. Since the lowest-order process is much more complicated than in the other channels, predictions for the $t\bar{t}H$ cross-section and related observables are only available at the NLO in QCD [209, 440]. The closely related process, where the Higgs boson is produced in association with bottom quarks, is too small to be probed at the LHC in the SM.

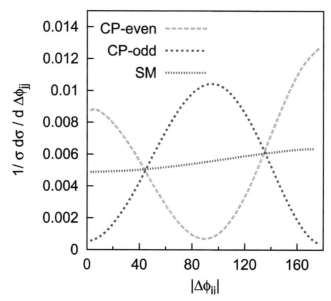

Fig. 4.66 The azimuthal angle between WBF jets for SM Higgs production (magenta), and anomalous production through pure CP-even (green) and CP-odd (blue) couplings. Reprinted with permission from Ref. [613].

It may also be possible to achieve additional sensitivity through the t-channel single top process shown in Fig. 4.68. However, this channel suffers from larger experimental backgrounds and the sensitivity to the $Ht\bar{t}$ coupling is smaller because its effect is washed out by the coupling of the Higgs boson to the t-channel W boson in this process.

4.8.8   Vector boson scattering

A class of processes that will constitute a key part of the future LHC programme is termed **vector boson scattering**. A vector boson scattering (VBS) process is so-called because, at its core, it should probe an amplitude,

$$V_1 + V_2 \rightarrow V_3 + V_4, \tag{4.126}$$

where $V_1, \ldots, V_4$ are a suitable combination of Z and $W^{\pm}$ bosons. Soon after the development of the Higgs theory, such processes were identified as keys to demonstrating the validity of the known SM. The reason is that, at high energies, the behaviour of vector boson scattering amplitudes is very sensitive to the gauge structure of the electroweak sector [723].

Consider the case of W boson scattering, $V_1 = V_3 = W^+$, $V_2 = V_4 = W^-$ in Eq. (4.126). There are three prototype diagrams for this process, as shown in Fig. 4.69, corresponding to s- and t-channel exchange of an intermediate boson and a single diagram involving the quartic coupling. In the high-energy limit of this scattering

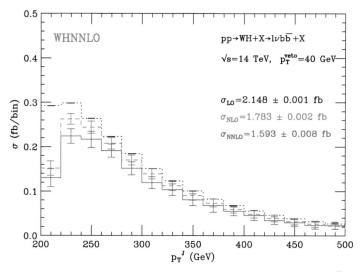

Fig. 4.67 The transverse momentum distribution of the fat $b\bar{b}$ jet in $WH(\to b\bar{b})$ events at LO, NLO, and NNLO, from the calculation of Ref. [524]. Reproduced with permission from Ref. [310].

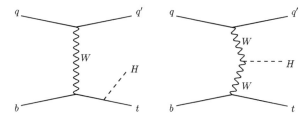

Fig. 4.68 Lowest-order Feynman diagrams for the production of a Higgs boson in association with a single top quark.

process, the amplitude represented by these diagrams is dominated by the longitudinal polarizations of the vector bosons that grow with the energy, E. Naively one thus expects the amplitude to grow as E^4. However, the gauge structure of the SM ensures that the leading behaviour of the quartic-coupling diagram is exactly cancelled by contributions from the exchange of intermediate Z bosons and photons. As a result any anomalous quartic coupling, resulting from simply rescaling the SM quartic coupling strength, would greatly affect the size of this scattering cross-section.

The final component of the puzzle, that prevents the amplitude scaling at the subleading E^2 level, is provided by the Higgs boson. Inclusion of the s- and t-channel Higgs exchange diagrams cancels the remaining E^2 dependence, rendering the SM amplitude well behaved at high energies. Even though the amplitude does not diverge, the perturbative expansion may still not respect unitarity. A partial wave analysis of all vector boson scattering channels [723] leads to a powerful constraint on the Higgs boson mass:

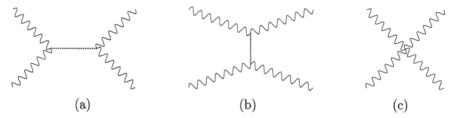

Fig. 4.69 Prototype Feynman diagrams representing the scattering process $W^+W^- \to W^+W^-$. The exchanged particle in diagrams (a) and (b) may be a photon, Z or H boson.

$$m_H < \left(\frac{8\sqrt{2}\pi}{3G_F} \right)^{\frac{1}{2}} \approx 1 \text{ TeV}. \tag{4.127}$$

Indeed, this bound was a powerful argument that a Higgs boson below this mass, or something playing that role, should be observed at the LHC. Even with the discovery of a light Higgs boson, it is still possible that this particle alone is not wholly responsible for unitarizing the high-energy behaviour of the amplitudes. Probing vector boson scattering is therefore an essential closure test of the SM.

At the LHC it is possible to probe vector boson scattering through the $\mathcal{O}(\alpha^4)$ electroweak processes $pp \to V_1V_2jj$. Representative diagrams are obtained by attaching quark lines to two of the vector bosons in Fig. 4.69. However, such diagrams only represent a small fraction of all the possible diagrams one may draw at this order in perturbation theory. The remaining diagrams represent, for instance, simple emission of vector bosons from quark lines. All of the vector boson scattering processes have been computed at NLO in QCD and implemented in the VBFNLO program [187].

A search for VBS is challenging for a number of reasons. First, the weak nature of these processes means that cross-sections are naturally small. Second, the VBS component must be isolated from the relatively uninteresting other production mechanisms. Finally, due to the inherent cancellation mechanism at high energies, the sought-after signal is even smaller than might otherwise be expected. Cross-sections for various VBS processes at 13 and 100 TeV are shown in Table 4.3. Since the cross-sections are so small, and the important aspect to probe is the contribution at high-energies, the study of these processes is much more fruitful at 100 TeV. Note that the t-channel exchange diagrams mean that there is a significant rate for same-sign W-boson production. This is particularly interesting since the same-sign final states have a considerable benefit in that they suffer from far fewer backgrounds than the other channels.

4.9   Summary

This chapter has discussed the application of fixed-order QCD perturbation theory to a variety of hadron collider processes. While the technical achievements in this arena are impressive, from NLO computations of $2 \to 6$ processes to calculations at NNLO and beyond, this survey has also revealed a number of shortcomings of this approach. A central theme of the breakdown of the fixed-order description is its application to

Table 4.3 Cross-sections for vector boson scattering processes at the LHC and a 100 TeV proton-proton collider. Jets are defined by the anti-k_T ($D = 0.4$) algorithm and satisfy $p_T > 20$ GeV, $|\eta| < 5$ and $m_{jj} > 100$ GeV.

Final state	Nominal process	$\sigma(13 \text{ TeV})$ [fb]	$\sigma(100 \text{ TeV})$ [fb]
$e^- \bar{\nu}_e \nu_\mu \mu^+ jj$	$W^- W^+ jj$	11.2	316
$\nu_e e^+ \nu_\mu \mu^+ jj$	$W^+ W^+ jj$	3.35	114
$e^- \bar{\nu}_e \mu^- \bar{\nu}_\mu jj$	$W^- W^- jj$	1.21	69.6
$\nu_e e^+ \mu^- \mu^+ jj$	$W^+ Z jj$	1.44	39.3
$e^- \bar{\nu}_e e^+ \mu^- \mu^+ jj$	$W^- Z jj$	0.89	31.8
$e^- e^+ \mu^- \mu^+ jj$	$Z Z jj$	0.29	10.5

final states where the kinematics are particularly restricted. Examples highlighted were threshold production of photon pairs (Section 4.2.3) and top quarks (Section 4.5), as well as the effects of a jet veto in Higgs production by gluon fusion (Section 4.8.4) and in the associated VH mode (Section 4.8.7).

In these cases the fixed-order approach results in perturbative predictions that do not sufficiently converge at sucessive orders of calculation, or that exhibit unphysical behaviour. In each case the origin of these symptoms is a large logarithm, present at each order of the calculation, which spoils the expected perturbative behaviour. As explained in detail in the following chapter, these logarithms can be systematically identified and analytically resummed to restore the predictive power of QCD in these situations. Moreover, Chapter 5 will also show how similar ideas may be used to perform resummations numerically in order to provide parton shower predictions that offer an extremely wide range of applicability. Ameliorating the fixed-order description discussed in this chapter with such methods will be crucial to achieving the level of understanding necessary to confront the theory of QCD with experimental data. Such comparisons will be presented in some detail in Chapters 8 and 9.

5

QCD to All Orders

As already discussed in Section 2.3, often kinematic situations occur which are characterized by very different scales μ_i. Then logarithms of the type $\log(\mu_0/\mu_1)$ may become so large that they overcome the smallness of the couplings, $\alpha \log(\mu_0/\mu_1) \approx 1$. In such cases, truncating the perturbative expansion at any fixed order will not yield correct results, and rather than attempting to include, order by order, perturbative corrections, it is often more important to resum such dangerous logarithmically enhanced terms to all orders. In this way, the programme of resummation of large logarithms complements the fixed-order efforts described in previous chapters. In fact, there are, broadly speaking, two ways to try to resum large logarithms, namely either through analytic methods, which were introduced in Section 2.3, or numerically, by means of a parton shower. In both cases the name of the game is to further push the accuracy by combining the knowledge of both fixed-order results and of the logarithmic structures in one best theoretical prediction.

In order to develop an intuitive understanding of a number of QCD phenomena further, the chapter starts with revisiting the QCD radiation pattern in Section 5.1. There, some care is taken to better quantify some ideas concerning the interface of perturbative QCD and emissions described by it and of the non-perturbative phase governed by hadrons and hadronization. Some first phenomena, such as angular ordering or the QCD hump-backed plateau, will be elucidated based on these rather quantitative considerations.

In Section 5.2 the discussion of analytic resummation with the example of the $p_\perp$ spectrum of the W boson already sketched in Section 2.3 will be extended to higher accuracy and, in addition, also other processes will be considered. To round this section off, other analytic resummation methods will be introduced, which either aim at different kinematical situations or are based on a different formalism.

Section 5.3 introduces the numerical implementation of resummation in the probabilistic parton shower picture. It will connect the parton shower to the analytic resummation discussed before. In addition, various methods to improve the formal accuracy of event simulation through the parton shower by including higher-order exact matrix elements will be summarized in Sections 5.4 and 5.5. They have been at the centre of formal developments in the framework of simulation tools and continue to play a

The Black Book of Quantum Chromodynamics. John Campbell, Joey Huston and Frank Krauss, Oxford University Press. © John Campbell, Joey Huston, and Frank Krauss 2018, First published in paperback 2022. DOI: 10.1093/oso/9780192871961.001.0005

central role in the quest for an ever-improved precision in the analysis of LHC data.

5.1 The QCD radiation pattern and some implications

This section examines some general features of the QCD radiation pattern. It follows the insightful discussion in various parts of the excellent book by Dokshitzer *et al.* [477].

5.1.1 The QCD radiation pattern revisited

5.1.1.1 Reminder: Characteristic scales

Coming back to Eq. (2.34), the differential probability for the emission of a gluon with energy ω and transverse momentum $k_\perp$ off a quark with energy E is given by

$$dw^{q \to qg} = \frac{\alpha_{\mathrm{s}}(k_\perp^2)}{2\pi} C_F \frac{dk_\perp^2}{k_\perp^2} \frac{d\omega}{\omega} \left[1 + \left(1 - \frac{\omega}{E} \right) \right]. \tag{5.1}$$

This yields a relatively broad spectrum both in transverse momentum and energy of emitted gluons, peaking at small transverse momenta and small energies. As already discussed in Section 2.1, the spectrum is cut off at small transverse momenta and energies by the onset of hadronization. This process takes place at distances of typical hadron radii of $R \approx 1\,\mathrm{fm}$ or at masses of the order of at most a few Λ_{QCD}. Broadly speaking, in a hard scattering process characterized by the hard scale Q, two classes of secondary emissions present themselves. First, there are emissions, where

$$\frac{1}{R} \ll k_\perp \sim \omega \sim Q \longrightarrow w^{q \to qg} \sim \alpha_{\mathrm{s}}(k_\perp^2) \ll 1, \tag{5.2}$$

signalling the production of jets. Second, there are emissions, where

$$\frac{1}{R} \leq k_\perp \leq \omega \ll Q \longrightarrow w^{q \to qg} \sim \alpha_{\mathrm{s}}(k_\perp^2) \log k_\perp^2 \sim 1, \tag{5.3}$$

associated with inner- and intra-jet radiation.

Taking a closer look, different orderings in these emission patterns can be further distinguished, which lead to different physical phenomena, namely

- double-logarithmic enhanced emissions,

$$\frac{1}{R} \leq k_\perp \ll \omega \ll Q, \tag{5.4}$$

  constituting the bulk of the emissions and producing the Bremsstrahlung pattern in inner-jet emissions;

- hard collinear emissions,

$$\frac{1}{R} \leq k_\perp \ll \omega \sim Q, \tag{5.5}$$

  which are responsible for scaling violations in DIS and similar;

- soft, typically wide-angle, emissions,

$$\frac{1}{R} \leq k_\perp \sim \omega \ll Q, \tag{5.6}$$

which are responsible for emissions in the phase space between jets, and which lead to the observable drag effect (see Section 5.1.2.2).

In the following, these different kinematic regions will be investigated in more detail, pointing out their visible consequences.

5.1.1.2   Time scales and the link to hadronization

In Section 2.1.4 typical time scales defining the hadronization process have been discussed, in particular the formation time t^{form} of a parton and its hadronization time, t^{had}. Following Eq. (2.39) and Eq. (2.38) they are given by

$$t^{\text{form}} = \frac{k_\parallel}{k_\perp^2} \quad \text{and} \quad t^{\text{had}} = k_\parallel R^2. \tag{5.7}$$

Demanding that partons are formed before they hadronize (or, in more martial terms, that they are born before they die) automatically implies $k_\perp > 1/R$. Extending this reasoning to study the dynamics of the phase transition from partons to hadrons in more detail, consider partons living at the edge, *i.e.*, those partons for which $k_\perp \approx 1/R$, aptly dubbed "**gluers**" by the authors of [477]. Such gluers are first formed at times R, with momenta given by $k_\parallel \sim k_\perp \sim \omega$. As time increases, more and more of such gluer-partons are being formed.

Assuming that the spectrum of hadrons closely follows that of the partons, a concept known as **local parton-hadron duality (LPHD)** [177], and assuming the absence of parton emissions below the cut-off R allows the identification of final-state hadron energies ϵ with the energies of gluers ω. Keeping only the dominant logarithmic enhanced terms when integrating over the soft emissions encoded in Eq. (5.1) yields the approximate hadron energy spectrum, namely

$$\mathrm{d}N_{(\text{hadrons})} \sim \int\limits_{k_\perp > 1/R}^{Q} \frac{\mathrm{d}k_\perp^2}{k_\perp^2} \frac{C_F \, \alpha_\mathrm{s}(k_\perp^2)}{2\pi} \left[1 + \left(1 - \frac{\omega}{E}\right)\right] \frac{\mathrm{d}\omega}{\omega} \tag{5.8}$$

$$\sim \frac{C_F \alpha_\mathrm{s}(1/R^2)}{\pi} \log(Q^2 R^2) \frac{\mathrm{d}\omega}{\omega} = \frac{C_F \alpha_\mathrm{s}(1/R^2)}{\pi} \log(Q^2 R^2) \, \mathrm{d}\log\omega \tag{5.9}$$

Replacing $\omega \longrightarrow \epsilon$ means that the distribution of hadron energies approximately follows the form

$$\mathrm{d}N_{(\text{hadrons})}/\mathrm{d}\log\epsilon = \text{const.}, \tag{5.10}$$

a *plateau in the logarithm of their energy*. Therefore, the energy distribution of the hadrons peaks at an energy $\epsilon_{\min}$ that can be related to $\omega_{\min}^{(\text{gluer})} \sim 1/R \sim m_{\text{had}}$, the typical hadron length and mass scales.

It is interesting to study how additional hard radiation changes this naive picture. To gain some insight, consider the case of a single secondary parton emitted under an angle θ and introduce the separation time $t^{(\text{sep})}$, after which it reaches a distance

R from the original parton. For such secondary partons, a hierarchy of time scales emerges, namely

$$t^{\text{form}} \sim \frac{k_\parallel}{k_\perp^2}$$

$$t^{\text{sep}} \sim R\theta \sim t^{\text{form}} (Rk_\perp)$$

$$t^{\text{had}} \sim k_\parallel R^2 \sim t^{\text{form}} (Rk_\perp)^2. \tag{5.11}$$

For gluers, as $Rk_\perp \sim 1$, these scales are all identical, but they differ for "proper" gluon emissions.

The natural question now is, how such hard secondary partons enter the hadronization business? The quick answer is fairly straightforward: further gluers are formed, following the secondary parton. This can however be further quantified.

From $\theta^{(\text{gluer})} \sim \theta$ and $k_\perp^{(\text{gluer})} \sim 1/R$ one finds $\omega^{(\text{gluer})} \sim 1/(R\theta)$ and therefore characteristic times R/θ. In other words, at a time $t^{(\text{sep})}$ the secondary parton starts decoupling from the other colour sources through the emission of gluers. They in turn also become hadrons with characteristic energies of about $\omega^{(\text{gluer})}$, which are a factor $1/\theta$ larger than those stemming from the original, primary parton. The secondary parton therefore starts looking like a jet, produced at its own hardness scale $Q \approx k_\perp$, but with energies boosted by $1/\theta$. It therefore does not contribute to the yield of the softest hadrons. The explanation for this is that for hadron energies in the interval

$$1/R \lesssim \omega_{(\text{hadron})} \lesssim 1/(R\theta) \tag{5.12}$$

the primary and secondary parton are not yet separated enough to start emitting gluers independently — the soft colour field manifesting itself in the gluers just "sees" the combined colour charge of both. This is the most striking manifestations of **colour coherence** in the QCD emission pattern.

5.1.1.3 Quantum coherence in the emission pattern

For further insight into this quantum mechanical effect consider its analogue in QED, where it is known as **Chudakov effect** [397]. Recently [882] this effect has been experimentally confirmed. It occurs in the emission of a secondary photon by an electron–positron pair produced in the induced splitting of a primary photon, sketched in Fig. 5.1.

Assume a Lorentz-frame in which the electron and positron carry about the same amount, about half, of the primary photons energy $p_\parallel$ and where the opening angle θ_{ee} of the pair is small. In such a frame, the relative transverse momentum of the e^-e^+-pair, $p_\perp$, satisfies

$$\sin\theta_{ee} \approx \theta_{ee} \approx \frac{p_\perp}{p_\parallel} \tag{5.13}$$

and the transverse component of the pair's wave vector is given by $\lambda_\perp = 1/p_\perp$. The secondary photon must be formed in a time given by the off-shellness of its emitter, here the positron. With a splitting variable z such that the longitudinal component of the photon momentum is $k_\parallel = zp_\parallel$ the positrons virtual mass is about

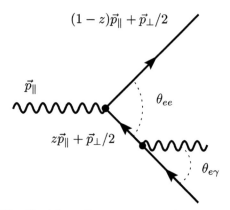

Fig. 5.1 The Chudakov effect: a primary photon with momentum $\vec{p}_\parallel$ splits into an electron–positron pair with momenta $p^{(-)} + p^{(+)}$, which in turn emits a secondary photon with momentum k. The fermion momenta before photon emission are given by $\vec{p}^{(+)} = z\vec{p}_\parallel + \vec{p}_\perp/2$ and $\vec{p}^{(-)} = (1-z)\vec{p}_\parallel + \vec{p}_\perp/2$, respectively, and $p_\perp$ denotes their relative transverse momentum.

$$k_\perp^2 \approx (p^{(+)})^2 \sin^2 \theta_{e\gamma} \approx (zp_\parallel)^2 \theta_{e\gamma}^2 \qquad (5.14)$$

leading to an energy imbalance of

$$\Delta E \approx \frac{k_\perp^2}{zp_\parallel} \approx zp_\parallel \theta_{e\gamma}^2 \qquad (5.15)$$

and therefore a formation time of the secondary photon of

$$\Delta t \approx \frac{1}{\Delta E} \approx \frac{1}{zp_\parallel \theta_{e\gamma}^2} \, . \qquad (5.16)$$

During any time Δt the pair separates in space by Δb

$$\Delta b = \theta_{ee} \, \Delta t = \frac{p_\perp}{p_\parallel} \Delta t \, , \qquad (5.17)$$

applied to the formation of the secondary photon the e^+e^--pair separation is

$$\Delta b = \frac{\theta_{ee}}{zp_\parallel \theta_{e\gamma}^2} \, . \qquad (5.18)$$

In order for the photon to resolve the positron as individual point-like charge rather than just being emitted by the dipole with its zero net charge, the transverse wavelength of the photon

$$\lambda_\perp^\gamma \approx \frac{1}{k_\perp} \approx \frac{1}{zp_\parallel \theta_{e\gamma}} \qquad (5.19)$$

must be smaller than the separation of the pair,

$$\Delta b \approx \frac{\theta_{ee}}{z p_\parallel \theta_{e\gamma}^2} > \frac{1}{z p_\parallel \theta_{e\gamma}} \approx \lambda_\perp^\gamma , \tag{5.20}$$

which is satisfied only if

$$\theta_{ee} > \theta_{\gamma e} . \tag{5.21}$$

This qualitative picture of quantum coherence in the emission pattern above, leading to angular ordering, can be supplemented with a proper calculation, which captures all of its essential features. The relevant part of the matrix element for the process $\gamma^* \to e^+ e^- \gamma$ is given by the eikonal,

$$\mathcal{W}(p, p'; k, \epsilon) = \epsilon_\mu^* \left(\frac{p'^\mu}{p' \cdot k} - \frac{p^\mu}{p \cdot k} \right) ,$$

cf. Eq. (2.14). Squaring it yields the radiation function. For massless particles the velocity equals the speed of light, *i.e.*, both $|\vec{v}|$, $|\vec{v}'| \to 1$. The classical analogy condensed in Eq. (2.12) is recovered, and expressed in angles the radiation function in *both* the quantum mechanical and classical case reads

$$W_{e^+ e^-} = \frac{2(1 - \vec{n}_+ \vec{n}_-)}{(1 - \vec{n}\vec{n}_+)(1 - \vec{n}\vec{n}_-)} = \frac{2(1 - \cos\theta_{e^+ e^-})}{(1 - \cos\theta_{\gamma e^+})(\cos\theta_{\gamma e^-})} . \tag{5.22}$$

Here $\vec{n}$ is the direction of the photon, and $\vec{n}_\pm$ are the direction of the two leptons.

Following [504], this expression can be decomposed into two parts, related to the emission of the photon by either the electron or the positron,

$$W_{e^+ e^-} = W_{e^+ e^-}^{(+)} + W_{e^+ e^-}^{(-)} , \tag{5.23}$$

where

$$W_{e^+ e^-}^{(\pm)} = W_{e^+ e^-} + \frac{1}{1 - \cos\theta_{\gamma e^\pm}} - \frac{1}{1 - \cos\theta_{\gamma e^\mp}} . \tag{5.24}$$

This decomposition will be encountered again, in later parts of this section. The individual emission functions need to be integrated over the full angular region of photon emission. Concentrating on $W_{e^+ e^-}^{(+)}$, the angular integral of the photon direction with respect to the direction of the positron is given by

$$\mathrm{d}^2 \Omega_\gamma = \mathrm{d}\cos\theta_{\gamma e^+} \mathrm{d}\phi_{\gamma e^+} . \tag{5.25}$$

To solve this integral, in a first step, the term $(1 - \cos\theta_{\gamma e^-})$ is expressed as a function of $\phi_{\gamma e^+}$ such that

$$\begin{aligned} 1 - \cos\theta_{\gamma e^-} &= \left(1 - \cos\theta_{e^+ e^-} \cos\theta_{\gamma e^+}\right) - \left(\sin\theta_{e^+ e^-} \sin\theta_{\gamma e^+}\right) \cos\phi_{\gamma e^+} \\ &= a - b \cos\phi_{\gamma e^+} . \end{aligned} \tag{5.26}$$

The $\phi \equiv \phi_{\gamma e^+}$-integration is simplified by introducing the complex variable $z = e^{i\phi}$ and by restricting the integration over z to the unit circle, $|z| = 1$. Therefore,

$$d\phi = i\frac{dz}{z} \tag{5.27}$$

and

$$\cos\phi = \frac{z + z^*}{2}. \tag{5.28}$$

Putting it all together yields

$$
\begin{aligned}
I &= \int_0^{2\pi} \frac{d\phi_{\gamma e^+}}{2\pi} \frac{1}{1 - \cos\theta_{\gamma e^-}} = \frac{i}{2\pi} \oint_{|z|=1} \frac{dz}{z\left(a - b\frac{z+z^*}{2}\right)} \\
&= \frac{1}{i\pi} \oint_{|z|=1} \frac{dz}{bz^2 - 2az + b} = \frac{1}{i\pi b} \oint_{|z|-1} \frac{dz}{(z - z_+)(z - z_-)},
\end{aligned}
\tag{5.29}
$$

where the poles are given by

$$z_\pm = \frac{a}{b} \pm \sqrt{\frac{a^2}{b^2} - 1}. \tag{5.30}$$

For relatively small angles $\theta_{\gamma e^+} < \theta_{e^+ e^-}$ only the pole at z_- resides inside the unit circle, and

$$I = \frac{1}{\sqrt{a^2 - b^2}} = \frac{1}{|\cos\theta_{\gamma e^+} - \cos\theta_{e^+ e^-}|}. \tag{5.31}$$

Combining the $1/(1 - \cos\theta_{\gamma e^-})$ term with the $W_{e^+e^-}$ in Eq. (5.24) leads to

$$
\begin{aligned}
\int_0^{2\pi} \frac{d\phi_{\gamma e^+}}{2\pi} W_{e^+e^-}^{(+)} &= \frac{1}{1 - \cos\theta_{\gamma e^+}} \left[1 + \frac{\cos\theta_{\gamma e^+} - \cos\theta_{e^+e^-}}{|\cos\theta_{\gamma e^+} - \cos\theta_{e^+e^-}|}\right] \\
&= \begin{cases} 0 & \text{if } \theta_{\gamma e^+} > \theta_{e^+e^-} \\ \dfrac{2}{1 - \cos\theta_{\gamma e^+}} & \text{else.} \end{cases}
\end{aligned}
\tag{5.32}
$$

This proves the angular ordering property for each of the two individual radiation functions for the QED case discussed here, and thus also for the combined overall QED radiation pattern.

In QCD a similar effect can be observed. Consider a colour charge, like a quark, emitting a first gluon under an angle Θ. Then, a second gluon emitted under an angle θ would resolve the individual colour charges of the quark and the primary gluon, only if $\theta < \Theta$. If, conversely, $\theta > \Theta$, then this secondary gluon would only feel the combined colour charge of the quark and the primary gluon, *i.e.*, the colour charge of the quark only. Colour coherence therefore results in an **angular ordering** in the emission pattern in **final state radiation**.

5.1.1.4   First consequences: Rapidity distributions in hadron collisions

The same reasoning, so far only employed for final-state radiation, of course also applies to initial state radiation: consider the scattering $ab \to cd$ of two partons a and b in the initial state resulting in two partons c and d in the final state. There are two interesting cases, namely the scattering process being driven by the exchange of either colourless or of colourful particles in the t-channel.

Consider first the case of a colour-singlet particle being exchanged in the t-channel, for example by a photon, such that the two colour flows connecting particles a and c and particles b and d, respectively, decouple. The scattering angle θ_{ac} in the centre-of-mass system of the colliding partons will act as a discriminator for soft emissions. It turns out that emissions of a soft parton k off either the incident parton a or the outgoing parton c under angles $\theta_{ak}, \theta_{ck} > \theta_{ac}$ will be highly suppressed.

To understand this remember that the formation time for a gluon with transverse wavelength $\lambda_\perp$ emitted under an angle θ from either parton a or parton c can be estimated as

$$t^{(\mathrm{form})} \sim \lambda_\perp/\theta\,. \tag{5.33}$$

The transverse displacement of c with respect to the original direction of a during this time is given by

$$\rho_\perp = t^{(\mathrm{form})}\theta_{ac}\,. \tag{5.34}$$

Again, it must be larger than the transverse wavelength of the emitted gluon in order for it to resolve the individual colour charges. For longer wavelengths any potential emission would be off a colour line passing through the process without any deflection, and therefore there is no associated Bremsstrahlung. This ultimately yields the angular ordering constraint

$$\theta_{ak}\,,\ \theta_{ck} \leq \theta_{ac}\,. \tag{5.35}$$

It implies that soft radiation off particles a and c — and similarly for emissions of partons b and d — is confined to a single cone of about twice the size of θ_{ac}. The scale characterizing the interaction is given by the Mandelstam variable $\hat{t}$. Correspondingly, the transverse momentum of the outgoing partons c and d can be estimated as

$$p_\perp \approx \sqrt{-\hat{t}}\,. \tag{5.36}$$

Then, the momenta of the Bremsstrahlung gluons emitted in the process are constrained by $k_\perp < p_\perp$. The amount of the resulting soft radiation is, roughly speaking, given by the sum of the colour charges emitting into the respective combined cones, i.e., the sum of the colour charges $C_a + C_c = 2C_a$ for the forward and $C_b + C_d = 2C_b$ for the backward cone.

The way angular ordering dominates the emission pattern of soft partons in such singlet-exchange processes can be summarized as follows Taking into account their respective orientation, incoming and outgoing particles form coloured dipoles defined by corresponding opening angles θ. Soft emissions off the individual particles forming

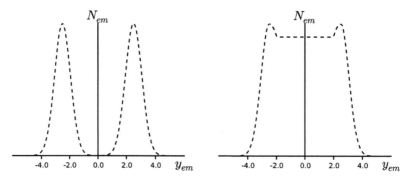

Fig. 5.2 Sketch of the rapidity distributions of soft particles emitted in parton-parton scattering processes, mediated by the exchange of a colour-singlet or a colourful particle in the t-channel. In the latter case, in the high-energy limit, this particle typically is a gluon. The distributions are in the c.m.-frame of the incoming partons, and the outgoing partons in each case are fixed at rapidities of $y - \pm 2.5$.

these dipoles are confined to cones with an opening angle of θ around the particle direction.

Consider now the other case of the reaction $ab \to cd$ being transmitted by the t-channel exchange of a coloured particle. In this case the reasoning above does not apply, since the colour flows are not decoupled any more. Instead, now the region of emission angles θ_{ak}, $\theta_{ck} \geq \theta_{ac}$ will be filled by emissions off the t-channel particle and is therefore susceptible to its colour charge. Since, typically, at large energies and relatively small scattering angles, i.e., in the region $\hat{s}/\hat{t} \gg 1$ gluon t–channel exchange dominates QCD scattering processes, the colour charge usually is C_A. This is a universal property of QCD scattering at large energies and small angles. The additional soft radiation into central rapidity regions is independent of the scattering partons. It can also be shown that the soft particle distributions in the region between the two forward cones are fairly independent of the rapidity and more or less constant. This is quantified in Section 5.2.5.

The findings are summarized in Fig. 5.2, exhibiting sketches of the rapidity distribution of particles being emitted in the two cases discussed. The picture of such processes of course changes when further, hard emissions populate the central rapidity region. As this is a relatively rare process, suppressed by one factor of the strong coupling without any sizeable logarithmic enhancement, however, the overall picture for the bulk of the events is relatively well described by the reasoning above.

A striking and highly relevant example for the impact of this aspect of the QCD radiation pattern in hadron collisions is the production of some potentially heavy system, such as, e.g., a Higgs boson, in the fusion of weakly interacting bosons (WBF). In this case the two outgoing partons c and d form forward tagging jets. These jets typically will have sizeable energies and relatively small scattering angles, as their characteristic transverse momentum scale will be given by the mass of the produced boson. A crucial part of the signal for such a process, allowing a very effective suppression of QCD backgrounds, relies on the fact that soft emissions under large angles are effectively absent due to angular ordering. As a result, in WBF events, QCD radiation in the central region between the two forward, opposite hemisphere tagging jets, is depleted [478, 482]. Therefore, vetoing events with central jets will typically result in a very effective suppression of the regular QCD background while the signal remains nearly completely unaffected [810, 820, 821].

5.1.2   Non-trivial consequences

5.1.2.1   The hump-backed plateau

The "**hump-backed plateau**" of the inclusive particle energy spectrum inside a jet [178] is one of the non-trivial consequences of the QCD radiation pattern. It is due to two competing and opposite effects: on the one hand, the restriction $k_\perp > 1/R$ forces subsequent emission angles to increase with decreasing energies ω,

$$\theta \sim k_\perp/k_\parallel \sim k_\perp/\omega > 1/(\omega R)\,. \tag{5.37}$$

On the other hand, angular ordering leads to shrinking allowed emission angles such that after a few emissions there is no viable phase space left for further softer emissions.

To cast this more quantitatively, consider a jet emerging from a massless quark travelling with energy E. The gluers emitted by the quark translate into hadrons with transverse momentum $p_\perp$ relative to the jet axis and with energy ϵ. These quantities are related to the emission angle θ with respect to the incident quark through

$$p_\perp \sim \epsilon\theta \sim 1/R\,. \tag{5.38}$$

The plateau in the distribution of hadrons with respect to the logarithm of their energy, $\mathrm{d}N/\mathrm{d}\log\epsilon = \mathrm{const.}$, *cf.* Eq. (5.10), can be translated into a spectrum with respect to the angle,

$$N \propto \int^E \frac{\mathrm{d}\epsilon}{\epsilon} \int^1 \frac{\mathrm{d}\theta}{\theta}\, \delta(\epsilon\theta - 1/R)\,, \tag{5.39}$$

where the δ-function encodes the phase space condition defining gluers.

Consider a case where a hard gluon of energy ω is emitted by the quark under an angle θ_0. Assuming that the radiation off the quark and the gluon can naively be added, *i.e.*, assuming independent emission of secondaries without any quantum coherence, the gluon would contribute gluers with energies ϵ given by

$$1/R \leq \epsilon \leq \omega\,, \tag{5.40}$$

and their transverse momenta would be given by

$$p_\perp \sim \epsilon\theta > 1/R\,. \tag{5.41}$$

If this picture was correct, the number of gluers and therefore hadrons would read

$$
\begin{aligned}
N \propto \; & \int\limits_{E}^{} \frac{\mathrm{d}\epsilon}{\epsilon} \int\limits_{}^{1} \frac{\mathrm{d}\theta}{\theta} \, \delta(\epsilon\theta - 1/R) \\
& + \alpha_{\mathrm{s}} \int\limits_{E}^{} \frac{\mathrm{d}\omega}{\omega} \int\limits_{}^{1} \frac{\mathrm{d}\theta_0}{\theta_0} \, \Theta(\omega\theta_0 - 1/R) \int\limits_{}^{\omega} \frac{\mathrm{d}\epsilon}{\epsilon} \int\limits_{}^{\theta_{\mathrm{max}}} \frac{\mathrm{d}\theta}{\theta} \, \delta(\epsilon\theta - 1/R) \,.
\end{aligned}
\tag{5.42}
$$

The condition on the gluon being hard, or, more precisely, harder than a gluer, is encoded in the Heavyside Θ-function. The maximal emission angle of gluers is given by θ_{max}. In the incoherent case outlined above, this angle essentially is unconstrained, and for simplicity one would set it to $\theta_{\mathrm{max}} = 1$.

Following previous discussions it is clear that this picture of incoherent addition of the gluer spectra is overly simplistic. Encoding angular ordering in the upper limit for the angle of the emission of the gluers translates into setting $\theta_{\mathrm{max}} = \theta_0$. Combined with the integral over the angle of the hard gluon emission the effect of including this upper limit or not, *i.e.*, between adding gluers from the gluon coherently or not, corresponds to about a factor of one half:

$$
\int\limits_{}^{1} \frac{\mathrm{d}\theta_0}{\theta_0} \int\limits_{}^{\theta_0} \frac{\mathrm{d}\theta}{\theta} \approx \frac{1}{2} \int\limits_{}^{1} \frac{\mathrm{d}\theta_0}{\theta_0} \int\limits_{}^{1} \frac{\mathrm{d}\theta}{\theta} \,.
\tag{5.43}
$$

This factor exponentiates with additional gluon emissions, thereby leading to a drastic reduction in the overall number of gluers emitted in the parton cascade. Ultimately it leads to a reduction of the soft part of the hadron spectrum produced in such a cascade, a testable result of angular ordering.

Transforming the integral expression for the total number of hadrons into a spectrum with respect to $\log \epsilon$ yields

$$
\frac{\mathrm{d}N}{\mathrm{d}\log \epsilon} =
\begin{cases}
1 + \dfrac{\alpha_{\mathrm{s}}}{2} \left[\log^2(ER) - \log^2(\epsilon R) \right] & \text{for incoherent sum, } \theta_{\mathrm{max}} = 1 \\[2ex]
1 + \alpha_{\mathrm{s}} \log \dfrac{E}{\epsilon} \log \epsilon R & \text{for coherent sum, } \theta_{\mathrm{max}} = \theta_0
\end{cases}
\tag{5.44}
$$

Instead of the incoherent energy spectrum peaking at energies given by hadron masses,

$$
\langle \epsilon \rangle = \langle E_{\mathrm{had}} \rangle = \frac{1}{R} \sim m_{\mathrm{had}} \,,
\tag{5.45}
$$

the depletion of the soft part of the coherent spectrum results in a peak of hadron energies roughly scaling like $\langle E_{\mathrm{had}} \rangle \sim \sqrt{E}$, the energy of the parton giving rise to the jet. In fact, the overall energy spectrum of the particles assumes an approximately Gaussian shape in the variables $\xi = -\log x_p \approx -\log x_E$. For convenience here the scaled momentum or energy of the particles $x_{p,E} = |\vec{p}|/E, \epsilon/E$ is used. In Fig. 5.3 the hadron spectrum following from this discussion is sketched, and it is compared to data taken at $e^- e^+$ colliders at various centre-of-mass energies. The figure shows that, qualitatively, the actual hadron spectra follow the form given in our relatively rough discussion.

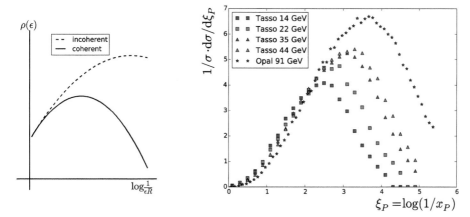

Fig. 5.3 The hump-backed plateau in hadron spectra: in the left panel the impact of coherence is sketched. As discussed in the text, its dominant effect is the depletion of hadrons in the soft regime, *i.e.*, for large values of $\log(1/E)$. This effect is visible also in the right panel, where data taken in e^-e^+ annihilation from TASSO [283] and OPAL [132] at various centre-of-mass energies are displayed.

5.1.2.2 The drag effect

As a second non-trivial effect of quantum coherence in soft gluon emission consider the radiation off a $q\bar{q}$ dipole. As in the QED case, it is given by the eikonal term

$$
\mathrm{d}w^{q\bar{q}} \; = \; C_F \frac{\alpha_{\rm s}}{2\pi} \frac{\mathrm{d}\omega}{\omega} \frac{\mathrm{d}^2\Omega_{\vec{n}}}{4\pi} \frac{2p_i p_j}{(p_i k)(p_j k)} \; = \; C_F \frac{\alpha_{\rm s}}{2\pi} \frac{\mathrm{d}\omega}{\omega} \frac{\mathrm{d}^2\Omega_{\vec{n}}}{4\pi} W_{q\bar{q}}(\vec{n}) \tag{5.46}
$$

cf. Eq. (2.14) and Eq. (5.22), where

$$
\begin{aligned}
W_{q\bar{q}}(\vec{n}) \; &= \; \frac{2(1-\vec{n}_q \vec{n}_{\bar{q}})}{(1-\vec{n}\vec{n}_q)(1-\vec{n}\vec{n}_{\bar{q}})} \\
&= \; \left[\frac{1}{1-\vec{n}\vec{n}_q} + \frac{\vec{n}_q(\vec{n}_{\bar{q}}-\vec{n})}{(1-\vec{n}\vec{n}_q)(1-\vec{n}\vec{n}_{\bar{q}})} \right] + \left[\frac{1}{1-\vec{n}\vec{n}_{\bar{q}}} + \frac{\vec{n}_{\bar{q}}(\vec{n}_q-\vec{n})}{(1-\vec{n}\vec{n}_q)(1-\vec{n}\vec{n}_{\bar{q}})} \right] \\
&= \; W_q(\vec{n};\vec{n}_{\bar{q}}) \; + \; W_{\bar{q}}(\vec{n};\vec{n}_q)
\end{aligned} \tag{5.47}
$$

depends on the directions $\vec{n}_q$ of the quark, $\vec{n}_{\bar{q}}$ of the anti-quark, and $\vec{n}$ of the emitted gluon. The two terms W_q and $W_{\bar{q}}$ are of course nothing but the radiation functions of the individual particles constructed in Eq. (5.24). Each of them decomposes into two parts: the first terms in the square brackets, $1/(1-\vec{n}\vec{n}_{q,\bar{q}})$ represent the incoherent part of the radiation pattern, while the second terms in the square brackets account for interference effects and thus reinstall coherence. As has been shown above, after

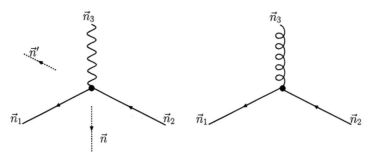

Fig. 5.4 "Mercedes-star" topologies in $q\bar{q}\gamma$ (left) and $q\bar{q}g$ final states (right). They are characterized by relative angles of around 120° between the three particles. Due to momentum conservation their directions $\vec{n}_{1,2,3}$ must lie in a plane in the c.m.-system of the collision. It is interesting to compare the radiation pattern of secondary gluons emitted in the $\vec{n}$ direction midway between the quarks with the counterpart along $\vec{n}'$, between the quark and the boson.

integration over the azimuth angle, this interference part is roughly equal in size to the incoherent part outside the cone with opening angle $\theta_{q\bar{q}}$ such that

$$\langle W_q(\vec{n};\vec{n}_{\bar{q}}) \rangle = \int \frac{\mathrm{d}\phi}{2\pi} W_q(\vec{n};\vec{n}_{\bar{q}}) \approx \frac{2}{1-\cos\theta qg} \Theta(\theta_{q\bar{q}}-\theta qg). \qquad (5.48)$$

The drag (or string) effect in the QCD radiation pattern is best studied in the emission of an additional (second) gluon with direction $\vec{n}$ from "Mercedes-star" $q\bar{q}\gamma$ and $q\bar{q}g$ states. These are configurations formed in electron–positron annihilation, $e^-e^+ \to q\bar{q} + \{\gamma, g\}$, where the energies of the quark, the anti-quark, and the photon or gluon, E_q, $E_{\bar{q}}$, and $E_{\gamma,g}$, and their relative angles are of roughly the same magnitude, *i.e.*,

$$E_q \approx E_{\bar{q}} \approx E_{\gamma,g} \approx \frac{E}{3}$$
$$\theta_{qg} \approx \theta_{\bar{q}g} \approx \theta_{q\bar{q}} \approx \frac{2\pi}{3}, \qquad (5.49)$$

see also Fig. 5.4.

As an extreme example, consider first the radiation of soft quanta in the direction $\vec{n}$ opposite to the boson (the photon or gluon), but in the plane spanned by the three objects — quark, anti-quark, and boson. In case of radiation off the $q\bar{q}\gamma$ final state, the only relevant effect of the photon emission is the corresponding reduced phase space of the quark–anti-quark dipole. The radiation pattern in this case therefore is given by

$$\mathrm{d}w^{q\bar{q}\gamma} = C_F \frac{\alpha_s}{2\pi} \frac{\mathrm{d}\omega}{\omega} \frac{\mathrm{d}^2\Omega_{\vec{n}}}{4\pi} W_{q\bar{q}}(\vec{n}), \qquad (5.50)$$

essentially the radiation of a quark–anti-quark pair at a c.m.-energy reduced by the amount carried away through the photon emission, and boosted back into the lab system. This boost thus induces a cone size of $\theta_{q\bar{q}}^{(\text{lab})} \approx 2\pi/3$ less than $\theta_{q\bar{q}}^{(\text{cms})} \approx \pi$ in the case of the pair in its own c .m.-system, which is compensated for by the suitably increased energies in the lab system.

For the case of a gluon emitted instead off a photon, consider first a "QED" analogue, essentially replacing the gluon with a positron pair and the quark and antiquark with an electron. This mimics the fact that a gluon carries about twice the colour charge of a quark, which in the limit of large N_c is exactly correct, as

$$C_F = \frac{N_c^2 - 1}{2N_c} \overset{N_c \to \infty}{\longrightarrow} \frac{N_c}{2}. \tag{5.51}$$

In this toy model, the radiation pattern is given by

$$
\begin{aligned}
dw^{\text{``}q\bar{q}g\text{''}} &= \frac{\alpha_{\text{s}}}{2\pi} \frac{d^3 k}{(2\pi^3)2\omega} \left(\frac{p_1^\mu}{p_1 k} + \frac{p_2^\mu}{p_2 k} - 2\frac{p_3^\mu}{p_3 k} \right)^2 \\
&= \frac{\alpha_{\text{s}}}{2\pi} \frac{d\omega}{\omega} \frac{d^2 \Omega_{\vec{n}}}{4\pi} \left[W_{qg}(\vec{n}) + W_{g\bar{q}}(\vec{n}) - \frac{1}{2} W_{q\bar{q}}(\vec{n}) \right],
\end{aligned}
\tag{5.52}
$$

A simple calculation based on the individual W shows the complete absence of soft radiation in the direction of $\vec{n}$:

$$
\begin{aligned}
&\left[W_{qg}(\vec{n}) + W_{g\bar{q}}(\vec{n}) - \frac{1}{2} W_{q\bar{q}}(\vec{n}) \right] \\
&= \left[2 \cdot \frac{2\left(1 - \cos\frac{4\pi}{3}\right)}{\left(1 - \cos\frac{2\pi}{3}\right)\left(1 - \cos\pi\right)} - \frac{1}{2} \cdot \frac{2\left(1 - \cos\frac{4\pi}{3}\right)}{\left(1 - \cos\frac{2\pi}{3}\right)^2} \right] \\
&= \left(1 - \cos\frac{4\pi}{3}\right) \left[2 \cdot 2 - \frac{1}{2} \cdot 8 \right] = 0.
\end{aligned}
\tag{5.53}
$$

In a similar way, in the direction $\vec{n}'$, opposite to one of the quarks, soft radiation is given by

$$\left[W_{qg}(\vec{n}') + W_{g\bar{q}}(\vec{n}') - \frac{1}{2} W_{q\bar{q}}(\vec{n}') \right] = \frac{3}{2} \cdot 9. \tag{5.54}$$

This is to be compared with the case of the $q\bar{q}\gamma$ final states, where

$$
\begin{aligned}
W_{q\bar{q}}(\vec{n}) &= \frac{2\left(1 - \cos\frac{4\pi}{3}\right)}{\left(1 - \cos\frac{2\pi}{3}\right)^2} = \frac{3}{2} \cdot 8 \\
W_{q\bar{q}}(\vec{n}') &= \frac{2\left(1 - \cos\frac{4\pi}{3}\right)}{\left(1 - \cos\frac{2\pi}{3}\right)\left(1 - \cos\pi\right)} = \frac{3}{2} \cdot 2.
\end{aligned}
\tag{5.55}
$$

In proper QCD these quantities are of course given by fully including the correct $SU(3)$ colour factors [176]. Denoting with $\vec{n}_\perp$ yet another direction, orthogonal to the event plane, this leads to the following ratios for the emission of a soft secondary gluon in the QED and QCD cases

$$\frac{\mathrm{d}w^{q\bar{q}\gamma}(\vec{n}')}{\mathrm{d}w^{q\bar{q}\gamma}(\vec{n})} = \frac{1}{4}\,, \quad \frac{\mathrm{d}w^{q\bar{q}g}(\vec{n}')}{\mathrm{d}w^{q\bar{q}g}(\vec{n})} = \frac{5N_c^2 - 1}{2N_c^2 - 4} = \frac{22}{7}$$

$$\frac{\mathrm{d}w^{q\bar{q}\gamma}(\vec{n}_\perp)}{\mathrm{d}w^{q\bar{q}\gamma}(\vec{n})} = \frac{1}{4}\,, \quad \frac{\mathrm{d}w^{q\bar{q}g}(\vec{n}_\perp)}{\mathrm{d}w^{q\bar{q}g}(\vec{n})} = \frac{N_c + 2C_F}{2(4C_F - N_c)} = \frac{17}{14}$$

(5.56)

and to the following ratios between the QED and QCD cases in the direction $\vec{n}$ opposite the boson:

$$\frac{\mathrm{d}w^{q\bar{q}g}(\vec{n})}{\mathrm{d}w^{q\bar{q}\gamma}(\vec{n})} = \frac{N_c^2 - 2}{2(N_c^2 - 1)} = \frac{7}{16}\,. \tag{5.57}$$

A number of these findings are remarkable. First of all, in the QCD case the soft radiation in the direction opposite to the gluon is massively suppressed by a factor of about three ($1/N_c$) with respect to the favoured region between the quarks and the gluon, around $\vec{n}'$. Possibly even more noticeable, the destructive interference between the three dipoles renders this direction, $\vec{n}$, more unfavourable than the direction $\vec{n}_\perp$ orthogonal to the event plane, which naively is susceptible mostly to the combined colour charge of the system, namely 0. This depletion is relatively easy to understand with the help of the toy model. There, the "gluon" was made of two equal charges, two positrons, and the quark and anti-quark were identified with two electrons. In the symmetric Mercedes-star configuration the two equal charges compensate each other in the direction of $\vec{n}$. A similar effect also is responsible in the case of QCD where the coloured gluon leaves the quark and the anti-quark with *nearly* opposite charges. Secondly, the radiation midway between the gluon and quark, in direction $\vec{n}'$ is well enhanced, due to constructive interference effects. This becomes most visible when comparing the soft radiation there with the soft radiation in the favoured direction $\vec{n}$ in the $q\bar{q}\gamma$ final state, the QED events. The ratio of the soft radiation in both regimes is about $11/8$, an enhancement by about 30% in the QCD case. Finally, it is worth noting that in the QED case, soft radiation in the directions $\vec{n}'$ and $\vec{n}_\perp$ are similarly disfavoured, by a factor of four with respect to the "best" direction. By far and large, these findings have been confirmed by the JADE collaboration in [203].

5.2 Analytic resummation techniques

5.2.1 Basics of Q_T-resummation in QCD: $b_\perp$-space

In this section, essential ingredients for QCD resummation are recapitulated and further generalized. This will extend the introductory discussion in Section 2.3.2, where the structure of a general expression resumming leading logarithms to all orders has been exemplified for the $p_\perp$ spectrum of a W boson in hadronic collisions.

5.2.1.1 Logarithmic accuracy

In the kinematical situation where the transverse momentum is much smaller than the invariant mass of the boson, $Q_{\perp,X} \ll Q_X$, the production cross-section has contributions of the form

$$\frac{1}{Q_{\perp,X}^2}\, \alpha_s^n \, \log^m \frac{Q_{\perp,X}^2}{Q_X^2} \quad \text{where } m \le 2n - 1\,. \tag{5.58}$$

Although these terms seem to be highly singular for $Q_{\perp,X} \to 0$, when properly re-summed they reorganize into the finite Sudakov suppression factor already encountered in Section 2.3.2. The transverse momentum $Q_{\perp,X}$ of the boson is of course nothing but its recoil against (soft) gluons that have been emitted by the incident partons, lending it the name of "**soft-gluon resummation**" or "**transverse-momentum resummation**".

The formal accuracy of resummation methods is classified according to the order m of the logarithms above that are being taken into account in the resummation of all orders in the strong coupling, n. If all terms $m = 2n - 1$ are included, the resummation is said to be of **leading logarithmic** (LL) accuracy. Of course, integrating the LL expression over all transverse momenta above some minimal value $Q_{\perp,X}$ will result in terms of the form $\alpha_s^n \log^{2n}(Q_{\perp,X}^2/Q_X^2)$.

For higher logarithmic orders, there is some dispute in the literature concerning nomenclature. For the purpose of this book, however, a convention is chosen, where including terms with $m \geq 2n - 2$ refers to **next-to-leading logarithmic** (NLL) accuracy. Including one more logarithmic order, *i.e.*, all terms with $m \geq 2n - 3$, will be dubbed NNLL-accurate and so on.

5.2.1.2 Q_T resummation in the Collins–Soper–Sterman approach

In Section 2.3.2, the CSS expression for the resummed double-differential cross-section has been given as

$$
\frac{\mathrm{d}\sigma_{AB \to X}}{\mathrm{dyd}Q_\perp^2} = \sum_{ij} \pi \hat{\sigma}_{ij \to X}^{(LO)} \left\{ \int \frac{\mathrm{d}^2 b_\perp}{(2\pi)^2} \left[\exp(i\vec{b}_\perp \cdot \vec{Q}_\perp) \tilde{W}_{ij}(b_\perp; Q, x_A, x_B) \right] \right.
$$
$$
\left. + Y_{ij \to X}(Q_\perp; Q, x_A, x_B) \right\},
$$
(5.59)

cf. Eq. (2.166). The two Bjorken-parameters x_A and x_B in this equation are entirely fixed by the invariant mass of the singlet system, Q^2, and its rapidity y. As already discussed, the resummation part $\tilde{W}_{ij}$ and the hard remainder Y_{ij} can be written in terms of various coefficient functions. However, before discussing their form, it is worth noting that it has become customary to evaluate the PDFs in both the resummation part $\tilde{W}_{ij}$ and the hard remainder Y_{ij} at a common scale μ_F. While at LO/LL accuracy this presents an irrelevant shift beyond the actual perturbative accuracy this particular choice has to be taken into account at higher orders. In the resummation part $\tilde{W}_{ij}$ the common scale μ_F is corrected to the "natural" factorization scale $1/b_\perp$; this is achieved in the collinear terms C_{ia} and C_{jb}, which are a part of $\tilde{W}_{ij}$, see Eq. (5.60).

Therefore, employing this choice, and suppressing the scales as arguments of the two parts in the expression for the resummed cross section,

$$
\tilde{W}_{ij}(b_\perp; Q, x_A, x_B)
$$
$$
= \sum_{ab} \int_{x_A}^1 \frac{\mathrm{d}\xi_A}{\xi_A} \int_{x_B}^1 \frac{\mathrm{d}\xi_B}{\xi_B} \left\{ f_{a/A}\left(\xi_A, \mu_F^2\right) f_{b/B}\left(\xi_B, \mu_F^2\right) \right.
$$

$$\times C_{ia}\left(\frac{x_A}{\xi_A}, \mu_R^2, \frac{1}{b_\perp}, \mu_F^2\right) C_{jb}\left(\frac{x_B}{\xi_B}, \mu_R^2, \frac{1}{b_\perp}, \mu_F^2\right) H_{ab}\left(\mu_R^2\right)$$

$$\times \exp\left[-\int_{b_0^2/b_\perp^2}^{Q^2} \frac{\mathrm{d}k_\perp^2}{k_\perp^2}\left(A(k_\perp^2)\log\frac{Q^2}{k_\perp^2} + B(k_\perp^2)\right)\right]\Bigg\} \qquad (5.60)$$

and

$$Y_{ij}(Q_\perp; Q, x_A, x_B) = \int_{x_A}^{1} \frac{\mathrm{d}\xi_A}{\xi_A} \int_{x_A}^{1} \frac{\mathrm{d}\xi_B}{\xi_B}\left[f_{i/A}(\xi_A, \mu_F^2)f_{j/B}(\xi_B, \mu_F^2)\right.$$

$$\left. \times R_{ij\to X}\left(Q_\perp; Q, \frac{x_A}{\xi_A}, \frac{x_B}{\xi_B}\right)\right], \qquad (5.61)$$

see Eqs. (2.167) and (2.170), respectively. The term b_0^2 appearing in the lower limit of the $k_\perp$-integration in the Sudakov form factor typically is chosen as

$$b_0 = 2e^{-\gamma_E} \qquad (5.62)$$

with the Euler–Mascheroni number γ_E (*cf.* Appendix A.1.1).

As already discussed in Section 2.3.2, the coefficient functions in the master equation above are expanded as perturbative series in α_s, and read

$$A(\mu_R^2) = \sum_{N=1} \left(\frac{\alpha_s(\mu_R^2)}{2\pi}\right)^N A^{(N)}$$

$$B(\mu_R^2) = \sum_{N=1} \left(\frac{\alpha_s(\mu_R^2)}{2\pi}\right)^N B^{(N)}$$

$$C_{ia}\left(\frac{x_A}{\xi_A}, \mu_R^2, \frac{1}{b_\perp}, \mu_F^2\right) = \delta_{ia}\delta\left(1 - \frac{x_A}{\xi_A}\right)$$

$$+ \sum_{N=1} \left(\frac{\alpha_s(\mu_R^2)}{2\pi}\right)^N C_{ia}^{(N)}\left(\frac{x_A}{\xi_A}, \frac{1}{b_\perp}, \mu_F^2\right)$$

$$H_{ab\to X}\left(\mu_R^2\right) = 1 + \sum_{N=1} \left(\frac{\alpha_s(\mu_R^2)}{2\pi}\right)^N H_{ab\to X}^{(N)}$$

$$R_{ij\to X}\left(\frac{x_A}{\xi_A}, \frac{x_B}{\xi_B}, Q, \mu_R^2\right) = \sum_{N=1} \left(\frac{\alpha_s(\mu_R^2)}{2\pi}\right)^N R_{ij\to X}^{(N)}\left(Q, \frac{x_A}{\xi_A}, \frac{x_B}{\xi_B}\right). \qquad (5.63)$$

Typically the first terms of the functions A, B, and C depend on the incoming particles only but not on the specifics of the produced singlet system X, while the hard terms H as well as the functions R are process-dependent.

It is interesting to note here that there is some freedom in how different contributions are accounted for. This is in particular true for the hard higher-order (*i.e.*, loop) corrections encoded in $H_{ab \to X}$ and the collinear coefficients C_{ia}. In the original formulation by Collins, Soper, and Sterman[410], implicitly $H_{ab \to X} \equiv 1$ was chosen and the loop corrections were encoded within the C_{ia}. In contrast, in more recent work by Catani, de Florian and Grazzini [342], the term $H_{ab \to X}$ was explicitly introduced and chosen different from unity.

5.2.1.3 Process-independent terms

The first terms in the expansion of A and B do not depend on the process in question but on the incoming partons only, see also Section 2.3.2. Focusing on $q\bar{q}$ and gg annihilation processes into singlets, this is reflected in the respective charge factors

$$C_q = C_F \quad \text{and} \quad C_g = C_A. \tag{5.64}$$

The higher terms $A^{(n)}$ are given by the coefficients of the soft part of the DGLAP splitting function $\mathcal{P}_{aa}$ with $a = q, g$ to the nth order in α_s: the $A^{(1)}$ is the coefficient of the soft terms in the leading-order splitting functions $\mathcal{P}_{aa}^{(1)}(z)$, *cf.* Eq. (2.33), *i.e.*, by the term that comes with $1/(1-z)_+$ with the numerator taken in the soft limit $z \to 1$. The $A^{(2)}$ and $A^{(3)}$ merely modify these terms with universal factors encoding the soft one- and two-loop corrections to these kernels. Therefore [502, 689, 767, 883],

$$A_{q,g}^{(1)} = 2C_{q,g}$$

$$A_{q,g}^{(2)} = 2C_{q,g}\, K = 2C_{q,g} \left[C_A \left(\frac{67}{18} - \frac{\pi^2}{6} \right) - \frac{10}{9} T_R n_f \right]$$

$$A_{q,g}^{(3)} = 2C_{q,g}\, K'$$

$$= 2C_{q,g} \left\{ C_A^2 \left[\frac{245}{24} - \frac{67}{9}\frac{\pi^2}{6} + \frac{11}{6}\zeta(3) + \frac{11}{5}\left(\frac{\pi^2}{6} \right)^2 \right] + C_F n_f \left[-\frac{55}{24} + 2\zeta(3) \right] \right.$$

$$\left. + C_A n_f \left[-\frac{209}{108} + \frac{10}{9}\frac{\pi^2}{6} - \frac{7}{3}\zeta(3) \right] + n_f^2 \left[-\frac{1}{27} \right] \right\}. \tag{5.65}$$

Here, $\zeta(3) \approx 1.2021$ is a special valiue of Riemann's ζ-function (*cf.* Appendix A.1.4). At first sight it may seem a bit of a coincidence that, apart from a global prefactor $2C_{q,g}$, which depends on the particle in question, the higher-order terms in $A_{q,g}^{(2,3)}$ are independent of the particle. This, however, is fairly straightforward to explain: the soft terms $1/(1-z)_+$ which give rise to $A(k_\perp^2)$ in Eq. (5.60) stem from an eikonal expression for soft gluon emission. As already seen, such an eikonal, being quasi-classical does not show any dependence on the details of the particles emitting the gluon beyond charge factors — it is sufficient that there are sources for gluon emission, their exact characteristics such as spin are beyond the resolution power of the long wavelength QCD fields.

The $B^{(1)}$ terms essentially can be identified with the factors in front of the $\delta(1-z)-$ terms in the one-loop splitting functions $P_{aa}^{(1)}(z)$, *cf.* Eq. (2.33). They are also known as the anomalous dimensions $\gamma_a^{(1)}$ of the splitting functions,

$$B_a^{(1)} = -2\gamma_a^{(1)} \,. \tag{5.66}$$

Therefore [502],

$$
\begin{aligned}
B_q^{(1)} &= -3C_F \\
B_g^{(1)} &= -2\beta_0 = -\left(\frac{11}{3}C_A - \frac{4}{3}T_R n_f\right) \,.
\end{aligned}
\tag{5.67}
$$

In contrast to the $A^{(n)}$ terms this way to extract the coefficients through a simple procedure does not trivially extend to higher orders: in fact, for all $n \geq 2$ the $B^{(n)}$ terms also contain parts that depend on the hard process in question.[1]

Furthermore, the first terms in the expansion of the collinear subtraction factor are given by

$$
\begin{aligned}
C_{ia}^{(0)}\left(z, \frac{1}{b_\perp}, \mu_F^2\right) &= \delta_{ia}\, \delta(1-z) \\
C_{ia}^{(1)}\left(z, \frac{1}{b_\perp}, \mu_F^2\right) &= P_{ia}^{(1)} \log \frac{b_0^2}{b_\perp^2 \mu_F^2} - P_{ia}^\epsilon(z) + \delta_{ia}\delta(1-z)C_a\frac{\pi^2}{6} \,.
\end{aligned}
\tag{5.68}
$$

The terms proportional to the first-order splitting kernels $P_{ia}^{(1)}$, *cf.* Eq. (2.33), account for the PDF evolution from μ_F to the correct scale $b_0/b_\perp$, which would be the natural choice in the resummation (to which the C_{ia} contribute). They thus reflect the scale choice already hinted at. Furthermore, the $P_{ia}^\epsilon(z)$ denote the $\mathcal{O}(\epsilon)$ terms of the splitting kernels at next-to-leading order and originate from the way collinear divergences are treated in the $\overline{MS}$-scheme. They read

$$
\begin{aligned}
P_{qq}^\epsilon(z) &= -C_F\,(1-z) \\
P_{gq}^\epsilon(z) &= -C_F\,z \\
P_{qg}^\epsilon(z) &= -2T_R z(1-z) \\
P_{gg}^\epsilon(z) &= 0 \,.
\end{aligned}
\tag{5.69}
$$

These terms as well as the ones proportional to $\pi^2/6$ should not come as a surprise. They are nothing but the finite collinear terms that stem from the ϵ expansion in the subtraction, where $1/\epsilon$ poles conspire with terms proportional to ϵ in the splitting functions to yield finite contributions to be absorbed into the PDFs. These terms, the P^ϵ, are thus readily identified with, for example, the corresponding terms in fixed-order calculations.

---

[1]The identifcation of the $B^{(1)}$ terms with the anomalous dimensions is the reason why parton showers can be shown to provide an approximation to the $p_\perp$ spectrum of colour-singlet objects in hadron collisions, which is accurate up to the next-to-leading logarithmic level in the Sudakov terms.

5.2.1.4   Process-dependent terms: $B^{(2)}$

The $B^{(2)}$ terms have been worked out for example for the cases of Drell–Yan production in $q\bar{q}$–annihilation in [436] and for Higgs boson production in gluon fusion in [443, 444]. Analysing its structure in more detail, the authors of [444] were able to relate it to the anomalous dimensions, $\gamma_a^{(2)}$, in the two-loop splitting functions $P_{aa}^{(2)}(z)$, i.e., their coefficients of the $\delta(1-z)$–terms, plus a term proportional to the one-loop correction to the leading-order amplitude, $\mathcal{A}_{ij\to X}^{(1-\text{loop})}$,

$$B_a^{(2)} = -2\gamma_a^{(2)} + \beta_0 \left(\frac{2\pi^2}{3} C_a + \mathcal{A}_a^{(\text{loop})} \right),\tag{5.70}$$

where

$$\gamma_q^{(2)} = C_F^2 \left[\frac{3}{8} - \frac{\pi^2}{2} + 6\zeta(3) \right] + C_F C_A \left[\frac{17}{24} + \frac{11\pi^2}{18} - 3\zeta(3) \right] - C_F T_R n_f \left[\frac{1}{6} - \frac{2\pi^2}{9} \right]$$

$$\gamma_g^{(2)} = C_A^2 \left[\frac{8}{3} + 3\zeta(3) \right] - C_F T_R n_f - \frac{4}{3} C_A T_R n_f.\tag{5.71}$$

This ultimately results in the expressions known from the literature,[2] namely

$$B_{q\bar{q}\to Z}^{(2)} = C_F^2 \left[\pi^2 - \frac{3}{4} - 12\zeta(3) \right] + C_F C_A \left[\frac{11\pi^2}{9} - \frac{193}{12} + 6\zeta(3) \right]$$

$$+ C_F T_R n_f \left[\frac{17}{3} - \frac{4\pi^2}{9} \right]$$

$$B_{gg\to H}^{(2)H} = C_A^2 \left[\frac{23}{6} + \frac{22\pi^2}{9} - 6\zeta(3) \right] + 4C_F T_R n_f - C_A T_R n_f \left[\frac{2}{3} + \frac{8\pi^2}{9} \right] - \frac{11}{2} C_F C_A.\tag{5.72}$$

5.2.1.5   Process-dependent terms: the finite loop correction

For the phenomenologically relevant cases of Drell–Yan, W-boson, and Higgs-boson production, the first non-trivial term in the expansion of $H_{ab\to X}$ reads

$$H_{ab\to X}^{(1)} = \mathcal{A}_{ab\to X}^{(1-\text{loop})},\tag{5.73}$$

where the finite loop contributions $\mathcal{A}_{ab\to X}^{1-\text{loop}}$ read [142, 438, 474]

$$\mathcal{A}_{q\bar{q}\to Z, q\bar{q}'\to W^\pm}^{(1-\text{loop})} = C_F \left(-8 + \frac{2\pi^2}{3} \right)$$

$$\mathcal{A}_{gg\to H}^{(1-\text{loop})} = C_A \left(5 + \frac{2\pi^2}{3} \right) - 3C_F.\tag{5.74}$$

---

[2]When comparing with literature, care has to be taken, since at this order differences between the by now customary $\overline{MS}$-scheme and the DIS-scheme start showing up, leading to some shifts that are cumbersome to trace.

Note that here the loop correction to Higgs boson production has been evaluated in the $m_t \to \infty$ limit.

5.2.1.6   Process-dependent terms: The hard remainders $R_{ab \to X}$

In order to calculate the hard remainders, it is important to understand the principle underlying their construction. This will be worked out with the example of the term $R_{q\bar{q}' \to W}^{(1)}$, the hard remainder to $\mathcal{O}(\alpha_s)$ for the production of a W boson in the quark–anti-quark annihilation channel. The starting point is provided by the real-emission cross-section for $q\bar{q}' \to W^+ g$ in Eq. (2.111), where they have already been provided in a form suitable for the further discussion here:

$$
\frac{d\sigma_{AB \to Wg}}{dQ_\perp^2 \, dy_W} = \int\limits_{\tilde{x}_A}^{1} \frac{dx_A}{x_A} \int\limits_{\tilde{x}_B}^{1} \frac{dx_B}{x_B} \left[f_{u/A}\left(\xi_A, \mu_F\right) f_{\bar{d}/B}\left(\xi_B, \mu_F\right) \delta(\hat{s} + \hat{t} + \hat{u} - m_W^2) \right.
$$

$$
\left. \times \; \sigma_{u\bar{d} \to W^+}^{(LO)}(s) \, \frac{1}{Q_\perp^2} \frac{\alpha_s C_F}{2\pi} \frac{\hat{t}^2 + \hat{u}^2 + 2m_W^2 \hat{s}}{\hat{s}} \right].
$$

$$\tag{5.75}$$

From this expression all terms at $\mathcal{O}(\alpha_s)$ must be subtracted, which are already present in the resummation part $\tilde{W}_{ij}$, but which do not originate from a genuine higher-order correction. At NLO, this does not include the generic higher-order corrections encoded in the terms $C^{(1)}$ and $H^{(1)}$. While the former originate from the treatment of collinear divergences at NLO, which are absorbed into the PDFs, the latter are genuine loop corrections. Therefore both can be ignored for the $R^{(1)}$. This reasoning results in the following contributions, to be subtracted:

- terms, where one of the incoming partons remains at its light-cone momentum fraction, $z = 1$, and where the PDF evolution of the other parton through the DGLAP splitting function and the corresponding shift in its momentum is taken into account. This will yield terms of the form

$$
\frac{1}{Q_\perp^2} \delta(1 - z_A) P(z_B). \tag{5.76}
$$

- terms which have their origin in the expansion of the Sudakov form factor to the first order in α_s. With them multiplying a Born-like phase space, they are proportional to $\delta(1 - z_A)\delta(1 - z_B)$. Therefore, overall, they amount to a term of the form

$$
\frac{\delta(1 - z_A)\delta(1 - z_B)}{Q_\perp^2} \left[A^{(1)} \log \frac{Q_\perp^2}{Q^2} + B^{(1)} \right]. \tag{5.77}
$$

For the example at hand, therefore

$$
R_{q\bar{q}' \to W}^{(1)} = \frac{C_F}{\pi Q_\perp^2} \left\{ \frac{\hat{t}^2 + \hat{u}^2 + 2m_W^2 \hat{s}}{\hat{s}} \delta(\hat{s} + \hat{t} + \hat{u} - Q^2) \right.
$$

$$-\delta(1-z_A)\delta(1-z_B)\left(2\log\frac{Q^2}{Q_\perp^2}-3\right)$$

$$\left.-\delta(1-z_A)\left(\frac{1+z_B^2}{1-z_B}\right)_+ - \delta(1-z_B)\left(\frac{1+z_A^2}{1-z_A}\right)_+\right\}. \quad (5.78)$$

Contact with the literature, and in particular with [410], where the $R^{(1)}$ term has been worked out for the first time in the context of Q_T resummation, is established by identifying

$$\hat{t}^2 + \hat{u}^2 + 2m_W^2\,\hat{s} = (Q^2 - \hat{t})^2 + (Q^2 - \hat{u})^2\,, \quad (5.79)$$

using the Mandelstam identity

$$m_W^2 \equiv Q^2 = \hat{s} + \hat{t} + \hat{u}\,. \quad (5.80)$$

The hard remainders for the $q\bar{q}$-initiated parton-level process are given by

$$
\begin{aligned}
R^{(1)}_{qg\to W} = R^{(1)}_{\bar{q}g\to W} &= \frac{1}{4\pi}\left\{-\frac{(\hat{s}+\hat{t})^2 + (\hat{t}+\hat{u})^2}{\hat{s}\hat{u}}\delta(\hat{s}+\hat{t}+\hat{u}-Q^2)\right.\\
&\qquad\left.-\delta(1-z_A)\frac{z_B^2 + (1-z_B)^2}{Q_\perp^2}\right\}\\
R^{(1)}_{gq\to W} = R^{(1)}_{g\bar{q}\to W} &= \frac{1}{4\pi}\left\{-\frac{(\hat{s}+\hat{u})^2 + (\hat{t}+\hat{u})^2}{\hat{s}\hat{t}}\delta(\hat{s}+\hat{t}+\hat{u}-Q^2)\right.\\
&\qquad\left.-\delta(1-z_B)\frac{z_A^2 + (1-z_A)^2}{Q_\perp^2}\right\}.
\end{aligned}
\quad (5.81)
$$

To achieve the subtraction in a form which lends itself to the implementation in a computer program or similar, the kinematical quantities in the resummation part, and in particular the $\xi_{A,B}$ and the transverse momentum $Q_\perp$, need to be mapped onto the Mandelstam variables used in the finite remainder. This is achieved through the relations

$$\hat{s} = \frac{1}{\xi_A\xi_B}Q^2\,, \quad \hat{t},\hat{u} = \left(1 - \frac{\sqrt{1+\frac{Q_\perp^2}{Q^2}}}{\xi_{B,A}}\right)Q^2\,, \quad Q_\perp^2 = \frac{\hat{t}\hat{u}}{\hat{s}}\,. \quad (5.82)$$

In addition, there will of course be those terms that originate from real-emission processes initiated by other partons. In the example of W production, such terms are related to processes with a gluon in the inital state, like $qg \to Wq'$. Such terms do not need special care in subtracting out contributions already accounted for in the Sudakov form factor, since they are simply not present, and only those terms involving the PDFs must be subtracted.

When going to higher orders $n \geq 2$, of course, the picture becomes a bit more involved. First of all, of course, the finite multiple-emission matrix elements assume an increasingly complicated structure, eventually mixing loop corrections with real-emission terms, which will make it harder to identify terms in a straightforward way.

This, however, is of course merely a nuisance, since they could in principle be taken from numerical routines. However, at the same time, in the expansion of the resummation part, the terms $A^{(n-k)}$ and $B^{(n-k)}$ from the Sudakov exponents will combine with terms of order k in the finite parts, the $C^{(k)}$ and $H^{(k)}$, which will become process-dependent.

5.2.1.7 Soft regime: $b_\perp \to \infty$

Another problem, up to now more or less swept under the carpet, needs to be addressed. It is related to the fact that the Fourier transform to impact parameter space necessitates an integral over all $b_\perp$, from zero to infinity. While the short distance part, $b_\perp \to 0$ is no problem at all, the long-distance part with $b_\perp \to \infty$ is, and for two reasons. First of all, integrals with infinite integration limits are notoriously unpleasant to deal with and guaranteeing their convergence to the true value is sometimes not trivial, especially when aiming for high numerical precision. While this is a merely technical problem, the physical problem in this region is much harder to solve in a satisfying fashion. It is related to the fact that in all integrations, the PDFs are evaluated at scales $\mu_F \propto 1/b_\perp$, which therefore may be taken into unphysical infrared regions where $Q^2 \to 0$. Similarly, in the Sudakov form factor, the strong coupling is evaluated at scales $k_\perp^2$, but, again, the integration region of $k_\perp^2$ extends down to values of $1/b_\perp^2$ thus also probing the strong coupling in the infrared regime and eventually hitting the Landau pole when $b_\perp \to \infty$.

   As already indicated in Section 2.3.2, the dangerous region of large $b_\perp$ can be dealt with by multiplying the resummation part $\tilde{W}_{ij}$ with a non-perturbative modification factor. This is often supplemented by suppressing large values of $b_\perp$ by replacing it in $\tilde{W}_{ij}$ with a modified b_* defined through

$$\tilde{W}_{ij}(b_\perp, \dots) \longrightarrow \tilde{W}^{(\text{NP})}(b_\perp, \dots) \, \tilde{W}_{ij}(b_*, \dots) \,, \tag{5.83}$$

where

$$b_* = \frac{b_\perp}{\sqrt{1 + (b_\perp/b_{\max})^2}} \,. \tag{5.84}$$

There are different parameterizations for this non-perturbative suppression factor beyond the simple Gaussian of Eq. (2.3.2), namely

$$\tilde{W}_{ij}^{(\text{CSS})}(b_\perp^2) = \exp\left[-F_1(b_\perp) \log\left(\frac{Q^2}{Q_0^2}\right) - F_{i/h_1}(x_1,\, b_\perp) - F_{j/h_1}(x_2,\, b_\perp)\right]$$

$$\tilde{W}_{ij}^{(\text{DWS})}(b_\perp^2) = \exp\left[-g_1 b_\perp^2 - g_2 b_\perp^2 \log\left(\frac{Q}{2Q_0}\right)\right]$$

$$\tilde{W}_{ij}^{(\text{LY})}(b_\perp^2) = \exp\left[-g_1 b_\perp^2 - g_2 b_\perp^2 \log\left(\frac{Q}{2Q_0}\right) - g_1 g_3 b_\perp \log(100 x_1 x_2)\right]$$

$$\tilde{W}_{ij}^{(\text{BLNY})}(b_\perp^2) = \exp\left[-g_1 b_\perp^2 - g_2 b_\perp^2 \log\left(\frac{Q}{2Q_0}\right) - g_1 g_3 b_\perp^2 \log(100 x_1 x_2)\right] \,. \tag{5.85}$$

Table 5.1 Fit values for different parameterizations of the non-perturbative function $\tilde{W}^{(\mathrm{NP})}$, taking into account a variety of TEVATRON Run I Z-boson data, from [716]. Note that more recent fits of the BLNY form tend to use $b_{\max} = 0.5\,\mathrm{GeV}^{-1}$, thereby extending the perturbative regime. These fits quite often lead to practically vanishing $g_3 \approx 0$.

	DWS [437]	LY [711]	BLNY [716]
$g_1[\mathrm{GeV}^2]$	0.016	0.02	0.21
$g_2[\mathrm{GeV}^2]$	0.54	0.55	0.68
$g_3[\mathrm{GeV}^{-1}]$		-1.50	-0.60

The parameters in some of the parameterizations above have been fitted in [716], resulting in global values of

$$Q_0 = 1.6\,\mathrm{GeV} \quad \text{and} \quad b_{\max} = 0.5\,\mathrm{GeV}^{-1}. \tag{5.86}$$

and in values for the parameters g_1, g_2, and g_3 as listed in Table 5.1. The functions F_1 and $F_{i/h}$ in the more general form of the non-perturbative function in $\tilde{W}_{ij}^{(\mathrm{CSS})}$ should, in principle, be general and would have to be extracted from data. However, by far and large, this approach has not been followed.

5.2.1.8 Practicalities of implementing the calculations

One of the main problems with the evaluation of Eq. (5.59) is that regions of large impact parameter, $b_\perp \to \infty$, contribute to the $Q_{\perp,X}$ spectrum at all values, because of the Fourier transform from $Q_\perp$ to $b_\perp$-space. This has physical and technical implications, which need to be addressed. Starting with the physics problems, due to this effect non-perturbative effects have an impact to observables in the perturbative regime. In practical implementations, typically, this problem is being solved by either adding a non-perturbative form factor in the spirit of Eqs. (5.83) – (5.85), essentially a dampening at large $b_\perp$, with parameters to be fixed through comparison with experimental data, or, by a cut-off b_*, cf. Eq. (5.84) or a combination of both. The down-side, however, of the non-perturbative dampening factor or the cut-off is that they might of course still introduce visible consequences at finite, and possibly large, $Q_{\perp,X}$, a somewhat counter-intuitive situation. Furthermore, the strong coupling is evaluated at scales proportional to $1/b_\perp$, *i.e.* at scales where the strong coupling diverges in the infrared region. To remedy this, some freezing or similar must be invoked, in order to guarantee that the α_s does not touch the Landau pole and diverges. In principle the same is also true for the PDFs, but there is no Landau pole lurking and the different scales could be simply connected through the DGLAP equation with the caveat that then, of course, problems with the scale in the strong coupling may raise their head again.

Finally, as an added technical problem, the exponential term of course oscillates: after integration over the relative angle, $\exp(i\vec{Q}_\perp \vec{b}_\perp)$ yields $J_0(b_\perp Q_\perp)$, a Bessel function. This renders the matching to the more accurate fixed-order results at large values of $Q_\perp$ a somewhat subtle and potentially numerically unstable exercise.

One way to circumvent the problem with the large $b_\perp$ integration comes from the realization [712] that the exponential leading to the Bessel function can be rewritten as

$$
\int \frac{d^2 b_\perp}{(2\pi)^2} \exp(-i\vec{b}_\perp \cdot \vec{Q}_\perp)\, f(b_\perp) \;=\; \frac{1}{2\pi} \int_0^\infty db_\perp b_\perp J_0(Q_\perp b_\perp)\, f(b_\perp)
$$

$$
\;=\; \frac{1}{4\pi} \int_0^\infty db_\perp b_\perp \left[h_1(Q_\perp b_\perp,\, v) + h_2(Q_\perp b_\perp,\, v) \right] f(b_\perp)\,, \tag{5.87}
$$

where

$$
h_1(z,\, v) \;=\; -\frac{1}{\pi} \int_{-iv\pi}^{-\pi + iv\pi} d\theta\, e^{-iz\sin\theta} \quad \text{and} \quad h_2(z,\, v) \;=\; -\frac{1}{\pi} \int_{\pi + iv\pi}^{-iv\pi} d\theta\, e^{-iz\sin\theta}\,. \tag{5.88}
$$

These functions reduce to the usual Hankel functions $H_{1,2}(z)$ for $z \to \infty$. They are finite for all finite values of z and v and fulfil

$$
h_1(z,\, v) + h_2(z,\, v) \;=\; 2J_0(z)\,, \tag{5.89}
$$

independent of v. Ultimately these functions allow the evaluation of the original $b_\perp$ integral as a function of two contours in complex space, one for each of the h_i. In so doing it is possible to deform the contours such that the Landau pole is avoided, a procedure that is equivalent to the original integral for all finite orders in perturbation theory. This is a solution that is being used in Q_T resummation, along with the more simplistic cut. Astonishingly enough, both solutions regularly yield numerically equivalent results [707].

5.2.2   Resummation in $b_\perp$-space for specific processes

In the following some example results for the Q_T-spectrum for various (singlet) final states are provided.

5.2.2.1   *Z* production

It is straightforward to arrive at expressions for all terms relevant for the production of a Z boson in $q\bar{q}$ annihilation and the respective higher-order corrections: The Born cross-section for W production given in Eq. (2.67) has to be replaced with the one for Z production. This can be achieved by adapting couplings, effectively replacing

$$
g_W^2 |V_{ij}|^2 = \frac{e^2 |V_{ij}|^2}{\sin^2\theta_W} \;\overset{W \to Z}{\longrightarrow}\; \frac{e^2 \left[\left(1 - 4|e_i|\sin^2\theta_W\right)^2 + 1 \right] \delta_{ij}}{4\sin^2\theta_W \cos^2\theta_W}\,. \tag{5.90}
$$

All other terms in the resummation part of course are not susceptible to the details of the gauge boson production, since they merely account for QCD effects in the initial state. For the hard remainder, the same holds true, after the couplings in the real-emission matrix elements have been adapted as above.

5.2.2.2 Higgs production through gluon fusion

p_T distribution of Higgs boson

gluon fusion at the 14 TeV LHC

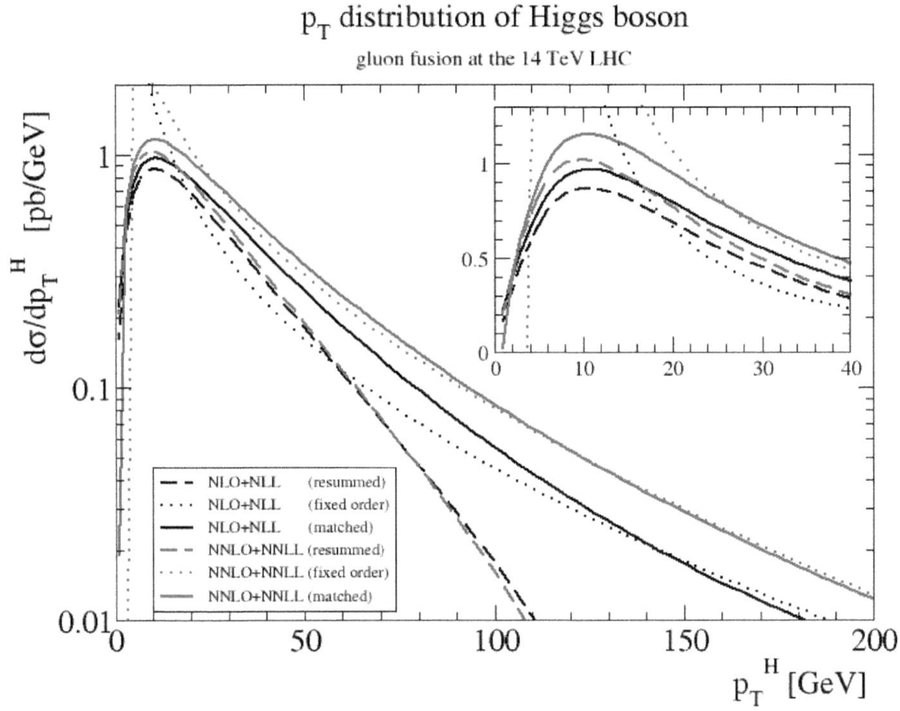

Fig. 5.5 $p_\perp$ spectrum of the Higgs boson at the 14 TeV LHC, evaluated in the $m_\perp \to infty$ approximation with the HqT code [282, 442], and ignoring contributions from b-quarks. Black lines correspond to results contributing to or accurate at NLO+NLL accuracy, while the red lines show results or contributions to the NNLO+NNLL result. Dashed and dotted lines show the resummed and fixed-order parts only, while the fully matched results are shows as straight lines. For all results the CT10 PDF NLO set with $\alpha_s = 0.118$ has been used, with $\mu_Q = \mu_F = \mu_R = m_H/2 = 62.5$ GeV. Non-perturbative effects are not included.

For Higgs production through gluon fusion mediated by the effective vertex, the structure of the result at NLO+NLL is identical to the one for the production of vector bosons in quark–anti-quark annihilation. The LO cross-section is given by

$$\sigma_{gg \to H}^{(LO)} = \frac{\sqrt{2} G_F \alpha_s(m_H^2)}{576\pi} \tag{5.91}$$

and, of course, it must be convoluted with suitable (gluon) PDFs. Similarly, in the Sudakov form factor, the terms A and B are the ones for gluon processes, taken from Eqs. (5.65) and (5.67), and the collinear terms again are those for gluons, this time taken from the corresponding item in Eq. (5.68). The hard loop correction is taken from Eq. (5.74). The same strategies already employed in the case of vector boson productions also will yield the results for the hard remainder terms R.

In Fig. 5.5 results for the $p_\perp$ spectrum of a 125 GeV Higgs boson at the 14 TeV LHC are exhibited, at NLO+NLL and NNLO+NNLL accuracy and in the large top-mass limit. The results have been obtained with HqT [282, 442]. The fixed-order correction in the matched result has two prominent effects: first, it changes the overall cross-section, and second, it fills the tail of the distribution at scales of the order of half the Higgs boson mass or above.

5.2.3 Q_T-resummation: calculating with Mellin transforms

Up to now, potentially large logarithms of the type $\log(Q_X^2/Q_{T,X}^2)$ of the transverse momentum $Q_{T,X}$ of a heavy (colour-singlet) system X with mass Q_X have been re-summed in impact-parameter or $b_\perp$-space. This variable essentially emerges from a Fourier transformation, and is conjugate to the transverse momentum. It has been introduced to guarantee that the transverse momenta $q_\perp$ of the emitted soft gluons combine to yield the overall transverses momentum of X such that

$$\vec{Q}_{\perp,X} = \sum_i \vec{q}_{\perp,i}. \qquad (5.92)$$

As a by-product of this approach, the various pieces in the master equation Eq. (5.59), and in particular the resummed and finite remainder contributions $\tilde{W}_{ij}$ and $Y_{ij\to X}$ emerge through a convolution of various contributions with the PDFs, where the natural scale for the evaluation of the latter is given by $1/b_\perp$. This cross-talk induced by the convolution sometimes renders an analysis of the individual contributions a tricky task. The somewhat unsatisfying situation can be greatly alleviated by using Mellin transforms which allow one to rewrite convolutions as simple products. The price[3] to pay for this seemingly superior procedure is the necessity to transform the results, obtained in a simpler way in Mellin space, back into x-space, which often is highly non-trivial.

5.2.3.1   Resummed cross-section in Mellin space

The resummed part of the cross-section for singlet production in Eqs. (2.166) and (5.59) can be manipulated in such a way that the Bjorken parameters x_A and x_B are not fixed anymore. To make contact with the literature, e.g., [436], this can be realized by integrating over y and considering only $d\sigma/dQ^2$. Ignoring the hard remainder,

---

[3]This is a perfect example for an important conservation law: conservation of pain.

$$\frac{Q^2 \mathrm{d}\sigma^{(\mathrm{res})}_{AB \to X}}{\mathrm{d}Q_\perp^2 \mathrm{d}Q^2} = \sum_{ij} \pi \hat{\sigma}^{(\mathrm{LO})}_{ij \to X} \int\limits_0^1 \mathrm{d}x_A \mathrm{d}x_B \int \frac{\mathrm{d}^2 b_\perp}{(2\pi)^2} J_0(Q_\perp b_\perp) \tilde{W}_{ij}(b_\perp; Q, x_A, x_B).$$

(5.93)

Introducing the rescaled invariant mass variable

$$\tau = \frac{Q^2}{S},$$

(5.94)

with S the hadronic centre-of-mass energy squared, allows the definition of the Nth moment of the cross section above with respect to τ:

$$\Sigma_{AB \to X}(N) = \int\limits_0^{1 - 2Q_\perp/Q} \mathrm{d}\tau\, \tau^N \frac{Q^2 Q_\perp^2}{\pi \sigma^{(\mathrm{LO})}_{AB \to X}} \frac{\mathrm{d}\sigma^{(\mathrm{res})}_{AB \to X}}{\mathrm{d}Q_\perp^2 \mathrm{d}Q^2}.$$

(5.95)

This constitutes a *Mellin transform* of the normalized differential cross section. Among other convenient factors, it has been multiplied by $Q_\perp^2$ to cancel the singular $1/Q_\perp^2$ behaviour. The upper limit of the integral approximates the kinematic boundary for soft particle emissions. It could be set to one under the assertion that the integrand does not have any support for larger values of τ. As discussed in some detail in Appendix A.1.4, the trick about using such Mellin moments is that the cross-section neatly factorizes into separate contributions from PDFs and partonic cross-sections without complicated convolutions of both:

$$\Sigma_{AB \to X}(N) = \sum_{ij} \left[f_{i/A}(N, \mu_F) f_{j/B}(N, \mu_F) \hat{\Sigma}_{ij}(N) \right].$$

(5.96)

Here the partonic part $\hat{\Sigma}_{ij}(N)$ collects all the terms in the Sudakov form factor and the collinear functions. After the integration over the angle between $\vec{b}_\perp$ and $\vec{Q}_\perp$ it is thus given by

$$\hat{\Sigma}_{ij}(N) = Q_\perp^2 \sum_{a,b} \int\limits_0^\infty \frac{b_\perp \mathrm{d}b_\perp}{(2\pi)} \Bigg\{ J_0(b_\perp Q_\perp) C_{ia}(N, \alpha_\mathrm{s}(b_0^2/b_\perp^2)) C_{jb}(N, \alpha_\mathrm{s}(b_0^2/b_\perp^2))$$

$$\times \exp\Bigg[- \int\limits_{b_0^2/b_\perp^2}^{Q^2} \frac{\mathrm{d}k_\perp^2}{k_\perp^2} \left(A(\alpha_\mathrm{s}(k_\perp^2)) \log \frac{Q^2}{k_\perp^2} + B(\alpha_\mathrm{s}(k_\perp^2)) \right)$$

$$- \int\limits_{b_0^2/b_\perp^2}^{\mu_F^2} \frac{\mathrm{d}k_\perp^2}{k_\perp^2} \left(\gamma_{ia}(N, \alpha_\mathrm{s}(k_\perp^2)) + \gamma_{jb}(N, \alpha_\mathrm{s}(k_\perp^2)) \right) \Bigg] \Bigg\},$$

(5.97)

where $J_0(x)$ is the Bessel function from the angular part of the integral over impact parameter space.

At this point it is useful to compare the result with the original expression in Eq. (5.59). Apart from normalization factors, there are a few differences due to the Mellin transformation. First of all, the convolution of the collinear functions C with the PDFs is replaced by a simple product of their Mellin transforms, the $C(N)$ and $f_{i,j/A,B}$, and the summation over the Mellin moments. With the PDFs at μ_F factored out, their DGLAP evolution to the scale $1/b_\perp$ is captured by the integral over their anomalous dimensions, the $\gamma(N)$. In $b_\perp$ space this was accounted for by the terms proportional to $P_{ia}^{(1)}\log(b_\perp^2\mu_F^2)$, cf. Eq. (5.68). In Mellin space these terms are not present any more. For the case of Drell–Yan-like processes they are therefore given by (remember $i = a = q$)

$$C_{ia}(N, \mu, \alpha_{\mathrm{s}}(b_0^2/b_\perp^2)) \;=\; \mathbf{M}_N\left[C\left(z, \frac{1}{b_\perp}, \mu_F\right)\right]$$

$$= \int_0^1 \mathrm{d}z\, z^N \left[\delta_{ia}\delta(1-z) + \frac{\alpha_{\mathrm{s}}(\mu)}{2\pi}\left(-P_{ia}^\epsilon(z) + \delta_{ia}\delta(1-z)C_a\frac{\pi^2}{6}\right) + \mathcal{O}\left(\alpha_{\mathrm{s}}^2\right)\right]$$

$$\xrightarrow{i=a=q} 1 + \frac{C_F\alpha_{\mathrm{s}}(\mu)}{2\pi}\left(\frac{1}{(N+1)(N+2)} + \frac{\pi^2}{6}\right) + \mathcal{O}\left(\alpha_{\mathrm{s}}^2\right). \tag{5.98}$$

Note that in the result above the constant terms are at variance with the result quoted for instance in [436]. This is because there the hard loop contribution has been absorbed into the collinear functions C. The Mellin transform of this additional part here is trivial, since the hard loop contribution is proportional to a δ-function in x-space, which yields unity.

Due to the absence of the analytically hard-to-control convolutions in x-space and their replacement by mere products due to the Mellin transform, it is now possible to try to identify logarithmic terms in such a way that the integral over the residual impact parameter $b_\perp$ can be handled with an increased level of analytic control. An important step into that direction is to replace the looming logarithms of $b_\perp$ in the Sudakov form factor, which will need to be integrated over, by logarithms of $Q_\perp$ instead.

To achieve this, the integrand is expanded in a power series of α_{s}, while collecting all logarithms of the form $\log(Q^2 b_\perp^2/b_0^2)$:

$$\hat{\Sigma}_{ij}(N) \;=\; \frac{Q_\perp^2}{2\pi}\sum_{a,b}\int_0^\infty \mathrm{d}b_\perp b_\perp\, J_0(b_\perp Q_\perp)$$

$$\times \exp\left\{\sum_{n=1}^\infty \sum_{m=0}^{n+1}\left[{}_n D_m(N, L; a, b)\left(\frac{\alpha_{\mathrm{s}}(\mu)}{2\pi}\right)^n\left(\log\frac{Q^2 b_\perp^2}{b_0^2}\right)^m\right]\right\}. \tag{5.99}$$

The term $L = \log(Q^2/\mu_F^2)$ stems from the second term in the exponential, by moving the upper limit from μ_F^2 to Q^2 to combine it with the original Sudakov form factor terms. Typically, however, this is not a large logarithm in singlet production. The other logarithms, $\log(Q^2 b_\perp^2/b_0^2)$, in contrast are potentially large, since, after integration over $b_\perp$ they will result in logarithms of the form $\log(Q^2/Q_\perp^2)$. They stem from the Sudakov form factor and the terms $\propto \gamma_{ai}(N)$ contribute some of the sub-leading logarithms.

Up to second order in α_s, the $_nD_m$, with dependence on the parton flavours a and b implicitly understood, are given by

$$_1D_2(N,L) = -\frac{1}{2}A^{(1)}$$

$$_1D_1(N,L) = -B^{(1)} - 2\gamma^{(1)}(N)$$

$$_1D_0(N,L) = 2\gamma^{(1)}(N)L + 2C^{(1)}(N)$$

$$_2D_3(N,L) = -\frac{1}{3}\beta_0$$

$$_2D_2(N,L) = -\frac{1}{2}A^{(2)} + \beta_0\left[\frac{1}{2}A^{(1)}L - \frac{1}{2}B^{(1)} - \gamma^{(1)}(N)\right]$$

$$_2D_1(N,L) = -B^{(2)} - 2\gamma^{(2)}(N) + \beta_0\left[B^{(1)}L + 2\gamma^{(1)}(N)L + 2C^{(1)}(N)\right].$$

$$(5.100)$$

In order to extract logarithms of $\log(Q^2/Q_\perp^2)$ instead of those of the modified arguments, the authors of [436] formulated a similar expansion of $\hat{\Sigma}_{ij}(N)$, namely

$$\hat{\Sigma}_{ij}(N) = \frac{1}{\pi}\left\{\sum_{n=1}^{\infty}\sum_{m=0}^{n+1}\left[_nC_m(N,L;a,b)\left(\frac{\alpha_s(\mu)}{2\pi}\right)^n\left(\log\frac{Q^2}{Q_\perp^2}\right)^m\right]\right\}. \quad (5.101)$$

The coefficients $_nC_m$ have been obtained by a direct fixed-order calculation, which explains the lacking exponential. They further differ from the $_nD_m$ due to different logarithms. The $_nC_m$ were matched in [436] to the $_nD_m$ by expansion of the exponential in Eq. (5.99).

To push this line of reasoning further, the authors of [505][4] have analysed the residual difference between the two expansions. To this end, they considered the integration over $b_\perp$ which becomes feasible by using the relation

$$\frac{d[xJ_1(x)]}{dx} = xJ_0(x) \quad (5.102)$$

between the Bessel functions and through integration by parts. The boundary terms for large impact parameters vanish due to the exponential dampening encoded in the Sudakov form factor, and therefore, after substituting $x = b_\perp Q_\perp$

---

[4]In fact, they were hoping to be able to construct a resummation procedure directly in $Q_\perp$ space rather than $b_\perp$ space. This was driven by the observation that the latter method invokes an integration over all values of $b_\perp$ inducing long-distance physics even for relatively large values of $Q_\perp$.

$$\hat{\Sigma}_{ij}(N) = -\frac{Q_\perp^2}{2\pi} \sum_{a,b} \int_0^\infty \mathrm{d}x J_1(x)$$

$$\cdot \frac{\mathrm{d}^2}{\mathrm{d}Q_\perp^2} \exp\left\{ \sum_{n=1}^\infty \sum_{m=0}^{n+1} \left[{}_nD_m(N,L) \left(\frac{\alpha_s(\mu)}{2\pi}\right)^n \left(\log\frac{Q^2 x^2}{Q_\perp^2 b_0^2}\right)^m \right] \right\}.$$

$$(5.103)$$

If $\log(x/b_0) = 0$, or, alternatively, setting $Q_\perp = b_\perp/b_0$ in Eq. (5.99), the original structure of the logarithms encoded as $\log(Q^2/Q_\perp^2)$ becomes visible. It is possible to split off the logarithms $\log(x/b_0)$ and careful analysis shows that they only contribute at the N³LL level. This can be seen by expanding the exponential in powers of this logarithm, which combined with the Bessel function yields

$$\int_0^\infty \mathrm{d}x J_1(x) \log^m(x/b_0) = \begin{cases} 1 & \text{for } m=0 \\ 0 & \text{for } m=1,2 \\ -\frac{1}{2}\zeta(3) & \text{for } m=3 \\ \dots & \text{for } m>3 \end{cases} \qquad (5.104)$$

under integration. With this in mind — and ignoring the residual sub-leading terms in the exponential — the integral over x reduces to unity and the cross-section can be written purely in $Q_\perp$ space. This basically amounts to the replacement of the $b_\perp$–integrated Sudakov form factor

$$\int_0^\infty \mathrm{d}b_\perp b_\perp J_0(b_\perp Q_\perp) \exp\left[-\int_{b_0^2/b_*^2}^{Q^2} \frac{\mathrm{d}k_\perp^2}{k_\perp^2} \left(A\log\frac{Q^2}{k_\perp^2} + B \right) \right] \tilde{W}_{ij}^{(\mathrm{non-pert.})}(b_\perp) \quad (5.105)$$

with the $Q_\perp$–space expression

$$\frac{\mathrm{d}}{\mathrm{d}Q_\perp^2} \exp\left[-\int_{Q_\perp^2}^{Q^2} \frac{\mathrm{d}k_\perp^2}{k_\perp^2} \left(\tilde{A}\log\frac{Q^2}{k_\perp^2} + \tilde{B} \right) \right]. \qquad (5.106)$$

Here, the $Q_\perp$-space parameters $\tilde{A}$ and $\tilde{B}$ coincide up to NLL accuracy with their $b_\perp$-space counterparts and the relative differences are exactly calculable. However, despite its elegance this approach ultimately could not be extended to accuracy higher than NLL and it therefore was not further pursued.

5.2.4   Threshold resummation

Soft gluon emissions off the initial state in production processes at hadron colliders do not only contribute to the transverse momentum distribution of the produced systems; they also induce large logarithmic corrections to the inclusive production cross-section. In particular, this is the case when a heavy system of mass Q is being produced. Then the emissions of gluons off the incoming partons are connected to the splitting function, which encodes a logarithmic enhancement in its soft part $\propto 1/(1-z)_+$, when $z \to 1$. The terms giving rise to these logarithms are typically of the form

$$\alpha_s^n \left[\frac{\log^k(1-z)}{1-z} \right]_+ . \tag{5.107}$$

Kinematically, this limit is approached when the combined centre-of-mass energy of the incident partons approaches the mass threshold of the system, $\sqrt{\hat{s}} \approx Q$, squeezing the phase space available for gluon emission. In this limit, therefore, the gluon emissions will not visibly change the kinematics of the produced system, their contribution however will change the production cross-section. This threshold effect on the cross-section was first found in [168, 354, 355, 863].

5.2.4.1 The factorized cross-section, once more

Consider the differential cross-section for the production of a system X

$$\frac{d\sigma_{AB \to X}}{dQ^2} = \int\limits_0^1 d\tau \int\limits_0^1 dx_i dx_j \, f_{i/A}(x_i, \mu_F) f_{j/B}(x_j, \mu_F) \, \delta\left(\tau - \frac{Q^2}{Sx_i x_j}\right) \tag{5.108}$$
$$\cdot \, W_{ij}^{(\text{thres})}(\tau, Q, \mu_R, \mu_F),$$

where Q is the mass of the system and $\sqrt{S}$ the hadronic centre-of-mass energy. As before, the renormalization and factorization scales are given by μ_R and μ_F, and $W_{ij}^{(\text{thres})}(\tau, Q, \mu_R, \mu_F)$ encodes the hard partonic cross-section, which is calculable in perturbation theory. In contrast to the function $\tilde{W}_{ij}$ from Q_T-resummation, no large logarithms explicitly show up in $W_{ij}^{(\text{thres})}$. Instead, they emerge through convolution with the PDFs. Their origin are terms inside $W_{ij}^{(\text{thres})}$ which become singular in the limit $\tau \to 1$ or $z \to 1$ and diverge like $\alpha_s^n \log^{2n-1}(1-z)/(1-z)$. Sub-dominant terms come with smaller exponents for the logarithm. Contributions from initial-state gluons splitting into quarks or initial quarks radiating a quark play no role here, since only soft gluon emissions feature the $1/(1-z)$ term in the splitting function.

After a Mellin transformation, the relevant term reads

$$\mathbf{M}_N\left[W_{ij}^{(\text{thres})}(\tau, Q, \mu_R, \mu_F)\right] \equiv \mathbf{M}_N\left[W_{ij}\right] = \int\limits_0^1 d\tau \tau^N \, W_{ij}^{(\text{thres})}(\tau, Q, \mu_R, \mu_F),$$
$$\tag{5.109}$$

with $\mu_R = Q$ as the usual choice. In the limit of soft gluon emissions,

$$\tau \longrightarrow \frac{Q^2}{Sx_i x_j} \approx 1 \tag{5.110}$$

and therefore the behaviour of $\mathbf{M}_N\left[W_{ij}\right]$ is defined by its limit for large N.

To make physical sense of it and to then evaluate this contribution, it is important to remember that soft gluon emissions can easily be treated using the eikonal approximation introduced in Eq. (2.17). This approximation is of first order in α_s, but of course the soft gluon emissions captured by the eikonal exponentiate. This is because potentially dangerous terms stemming from gluon correlations and from gluons radiating secondary gluons which may generate further large logarithms cancel due to the inclusive nature of the process [339, 863]. In any case, a simple argument can be made that typically non-exponentiating contributions are colour suppressed for sufficiently inclusive observables. This means that by evaluating the eikonal contribution to first order, as below, the leading logarithmic behaviour can be deduced.

5.2.4.2   The eikonal approximation and the emission phase space

The eikonal cross-section for the emission of a soft gluon with momentum k off incoming quark lines with momenta p_a and p_b is given by

$$\mathrm{d}w(k) = -(4\pi)\alpha_\mathrm{s}\, C_F\, \frac{\mathrm{d}^3 k}{(2\pi^3)(2\epsilon)} \left(\frac{p_a^\mu}{p_a k} - \frac{p_b^\mu}{p_b k} \right)^2, \qquad (5.111)$$

cf. Eq. (2.14), where ϵ is the energy of the emitted gluon. Soft-gluon unitarity also fixes the virtual contribution w^0 through

$$w^0 + \int \mathrm{d}w(k) = 0. \qquad (5.112)$$

The overall contribution to first order in α_s from the eikonals therefore reduces to a correction factor $\mathcal{W}_\mathrm{eik}$ to the cross-section given by

$$\mathcal{W}_\mathrm{eik} = (1 + w^0)\, \delta(1-\tau) + \int \mathrm{d}w(k)\, \delta\left(1 - \tau - \frac{\epsilon}{E}\right) \qquad (5.113)$$

where E is the energy of the incident quark. Taking the moments of this term yields the Mellin transform of $W_{ij}^{(\mathrm{thres})}$ to first order in α_s,

$$\mathbf{M}_N\left[W_{ij}^{(\mathrm{thres})}\right]^{(1)} = \frac{4C_F}{2\pi} \int_0^1 \mathrm{d}z \frac{z^N - 1}{1 - z} \int_{(1-z)Q^2}^{(1-z)^2 Q^2} \frac{\mathrm{d}k_\perp^2}{k_\perp^2}\, \alpha_\mathrm{s}(k_\perp^2). \qquad (5.114)$$

Here $1 - z = \epsilon/E$ has been identified and, as before, $k_\perp$ denotes the transverse momentum. It is related to the momentum transfer along the incoming hard propagator, which for emissions off parton a is given by

$$q^2 = |(p_a - k)^2| \approx \frac{k_\perp^2}{1 - z}. \qquad (5.115)$$

The 1 of the numerator $(z^N - 1)$ in Eq. (5.114) accounts for the virtual contribution. As the argument of the running coupling the transverse momentum of the emitted gluon is being used, *i.e.* $(1-z)q^2$. This choice is similar to the ones made before, resumming leading IR singularities.

The phase space available for the emissions is defined by the maximal interval of virtualities allowed for the intermediate propagator — the emission of the gluon with energy fraction $(1-z)$ induces a maximal virtual mass of $(1-z)Q^2$ in the propagator. The lower limit is related to the regularization of the collinear singularities. For inclusive processes, where no additional conditions are imposed on, e.g., the rapidity of the final state, the largest scale available in the process, namely Q^2, serves as a meaningful cut-off.

5.2.4.3 Leading logarithms from the eikonal

To obtain the leading logarithmic contribution only, α_s can be taken as scale-independent, reducing the otherwise tricky integral in Eq. (5.114) to

$$
\mathbf{M}_N \left[W_{ij}^{(\text{thres})} \right]^{(1,LL)} = \frac{4C_F}{2\pi} \int_0^1 dz \, \frac{z^N - 1}{1 - z} \int_{(1-z)Q^2}^{(1-z)^2 Q^2} \frac{dk_\perp^2}{k_\perp^2} \alpha_s
$$

$$
= \frac{4C_F \alpha_s}{2\pi} \int_0^1 dz \, \frac{z^N - 1}{1 - z} \log(1 - z) = \frac{4C_F \alpha_s}{2\pi} \int_0^1 dz \left[\frac{\log(1 - z)}{1 - z} \right]_+
$$

$$
= \frac{4C_F \alpha_s}{4\pi} \left\{ \psi'(N) + \zeta(2) + \left[\psi(N) + \gamma_E \right]^2 \right\}. \tag{5.116}
$$

In the limit of large N, therefore,

$$
\mathbf{M}_N \left[W_{ij}^{(\text{thres})} \right]^{(1,LL)} \xrightarrow{N \to \infty} \lim_{N \to \infty} \frac{C_F \alpha_s}{\pi} \left\{ \psi'(N) + \zeta(2) + \left[\psi(N) + \gamma_E \right]^2 \right\}
$$

$$
\frac{C_F \alpha_s}{\pi} \left[\frac{\pi^2}{6} + (\log N + \gamma_E)^2 \right]. \tag{5.117}
$$

Alternatively, directly taking the large-N limit of the integrand in Eq. (5.114) amounts to replacing

$$
z^N - 1 \approx \Theta \left(1 - \frac{1}{N} - z \right), \tag{5.118}
$$

thereby constraining the phase space for the energy integral. This results in

$$
\mathbf{M}_N \left[W_{ij}^{(\text{thres})} \right]^{(1,alt)} \approx \frac{4C_F \alpha_s}{2\pi} \int_0^1 dz \, \frac{\log(1 - z)}{1 - z} \Theta \left(1 - \frac{1}{N} - z \right)
$$

$$
= \frac{4C_F \alpha_s}{2\pi} \int_0^{1 - \frac{1}{N}} dz \, \frac{\log(1 - z)}{1 - z} = \frac{4C_F \alpha_s}{4\pi} \log^2 N, \tag{5.119}
$$

the leading logarithmic term. These terms can be exponentiated, to yield the resummed result at leading logarithmic accuracy,

$$\mathbf{M}_N \left[W_{ij}^{(\text{thres})} \right]^{(LL)} = \exp \left\{ \frac{C_F \alpha_s}{\pi} \left[\frac{\pi^2}{6} + (\log N + \gamma_E)^2 \right] \right\}. \tag{5.120}$$

5.2.4.4 Sub-leading logarithms

For sub-leading contributions, the running of α_s has to be taken into account and the global soft enhancement terms proportional to the soft pole in the splitting function must be included as well. In addition, at higher order, *i.e.*, starting at $\mathcal{O}\left(\alpha_s^2\right)$ new structures are emerging which reflect the fact that the emitted gluons skew the original colour flow by adding new directions for eikonals. Such contributions are encoded in terms $D(k_\perp^2)$. For cases, where there are also coloured particles in the *final state*, of course, also $B(k_\perp)$ terms would appear. They are absent for initial-state emissions only, since they relate to the non-soft remainders of splitting functions, once the soft, eikonal-type terms are subtracted. As such they have of course no support in the drastically reduced initial-state phase space which actually gives rise to the threshold logarithms in the first place. In other words, the phase-space constraint effectively squeezes them out in the initial state. Ignoring the case of coloured final-state particles and concentrating on colour-singlet production, therefore

$$\mathbf{M}_{N \to \infty} \left[W_{ij}^{(\text{thres})} \right] = \exp \left\{ \int_0^1 dz \, \frac{z^N - 1}{1 - z} \int_{(1-z)Q^2}^{(1-z)^2 Q^2} \frac{dk_\perp^2}{k_\perp^2} \left[A(k_\perp^2) + D(k_\perp^2) \right] \right\},$$
$$\tag{5.121}$$

similar to the case of Q_T resummation, but with the B terms replaced by the D terms.

To further stress the analogy, also for threshold resummation the functions A and D can be expanded in powers of the strong coupling as

$$A(\mu^2) = \sum_{n=1}^{\infty} A^{(n)} \left(\frac{\alpha_s(\mu^2)}{2\pi} \right)^n \quad \text{and} \quad D(\mu^2) = \sum_{n=2}^{\infty} D^{(n)} \left(\frac{\alpha_s(\mu^2)}{2\pi} \right)^n \tag{5.122}$$

with

$$A^{(1)} = 2 C_F, \quad A^{(2)} = 2 C_F K, \quad \text{and} \quad D^{(1)} = 0. \tag{5.123}$$

Not surprisingly, the global term K reads, as in the case of Q_T resummation,

$$K = C_A \left(\frac{67}{18} - \frac{\pi^2}{6} \right) - \frac{10}{9} T_R n_f, \tag{5.124}$$

cf. Eq. (5.65).

From these considerations for Drell–Yan type, *i.e.* quark-induced processes, similar expressions for gluon-initiated processes can be deduced by replacing C_F with C_A.

To obtain actual physical results in x-space from the Mellin transforms the back-transformation must be invoked, translating the expressions in terms of Mellin moments N back to centre-of-mass energies.

5.2.4.5 Comparing Q_T and threshold resummation

At first sight, the Sudakov form factor in Q_T-resummation and the threshold radiation function are remarkably similar. They have a nearly identical structure with the same coefficients A. There are some differences, though, which reflect the different physical origin of the large logarithmic corrections.

First of all, there is an explicit minus sign in front of the momentum-space integral in the Sudakov form factor of Eq. (5.60), tracing its origin to the suppression of hard radiation, while in threshold resummation the gluon contributions enhance the cross-section, as reflected by the positive exponent in the Sudakov form factor-like exponential. Second, sub-leading corrections in threshold resummation are related to either particles in the final state — a case not considered here in either of the schemes — or to the emergence of new directions after the emission of at least one parton. In Q_T resummation, in contrast, sub-leading corrections emerge already at leading order in α_s in the exponential and originate from non-soft collinear emissions. As stressed above, these have no phase space open in threshold resummation and are therefore absent there. As a consequence, even the first sub-leading corrections in threshold resummation are process-dependent while they are not in Q_T resummation.

Ultimately Q_T-resummation is being used to evaluate a kinematic distribution — the transverse momentum of a final state — without changing the fixed-order cross-section, while threshold resummation usually keeps the kinematic unchanged and alters the total partonic cross section. This latter feature actually allows to approximate higher-order corrections to the cross-section for the production of a given final state at hadron colliders that have hitherto not been evaluated, see for example [195]. These approximations do miss finite terms and rely on the assumption that the corrections in questions are dominated by the logarithmic structures in the threshold regime.

5.2.5 The BFKL equation

5.2.5.1 The high-energy limit: Large logarithms in $1/x$

Up to now, the resummation of logarithms has been discussed, where emitted gluons are collinear and/or soft and thereby alter the kinematics of a produced system or change the cross-section. The corresponding logarithms are of the form $\log(p_{T,X}^2/M_X^2)$ in the case of Q_T resummation, or $\log(M_X^2/\hat{s})$ in the case of threshold resummation, with M_X the invariant mass of the produced system X and $p_{T,X}$ its transverse momentum. In this section another class of logarithms will be considered, which appear more independent of the produced system and become important in a different kinematic situation. These logarithms originate from multiple soft emissions along t-channel propagators; they are of the form $\log(\hat{t}/\hat{s})$. They become large when the invariant mass squared $\hat{s}$ of the overall scatter system increases for fixed momentum transfers $\hat{t}$ — a limit also known as the high-energy limit.

In contrast to the DGLAP evolution, which encodes the evolution of (collinear) logarithms in transverse momentum through a strict ordering of the emissions in transverse momentum, the high-energy logarithms emerge for emissions that have all about the same transverse momentum, but are ordered in rapidities. As a consequence, applying this to PDFs for partons i in a hadron h, $f_{i/h}(x, Q^2)$, the following picture emerges. The DGLAP equation accounts for the evolution of PDFs in the (transverse) momentum scale Q while the evolution equation related to the high-energy limit accounts for their evolution in $1/x$ for $x \to 0$. This new evolution equation, the **Balitsky–Fadin–Kuraev–Lipatov (BFKL) equation** [189, 517, 708, 709], is therefore related to the rise of the gluon PDF for small x which ultimately drives the total cross-section and the structure function F_2 in this regime, see Chapter 6. In the following, the derivation of this equation will be briefly sketched, and the kinematical situation, where it is relevant, will be discussed, borrowing to a large extent from the very instructive review [451].

5.2.5.2   Connection of amplitude and cross-section: Optical theorem

In preparation for the derivation of the BFKL equation, some manipulations are necessary, which will only be sketched here. They are based on various properties of the S-matrix such as **analyticity** or more generic properties such as the **optical theorem**. It relates the total cross-section to the **elastic** or **forward scattering amplitude**. The starting point is the S **matrix**, which relates initial and final states of a reaction:

$$|f\rangle = \hat{S}|i\rangle. \tag{5.125}$$

This operator decomposes into a non-interaction part, effectively a $\delta_{fi} = \hat{\mathbf{1}}$, and an actual scattering part,

$$\hat{S} = \hat{\mathbf{1}} + i\hat{T}. \tag{5.126}$$

Demanding unitarity of the S matrix, *i.e.*, probability conservation of the theory, results in

$$1 \overset{!}{=} \hat{S}^\dagger \hat{S} = 1 + i\left(\hat{T} - \hat{T}^\dagger\right) + \hat{T}^\dagger \hat{T} \tag{5.127}$$

or, schematically,

$$i\hat{T}^\dagger \hat{T} = \left(\hat{T} - \hat{T}^\dagger\right) = \mathfrak{Im}(\hat{T}). \tag{5.128}$$

Using momentum conservation on the T matrix, and decomposing into individual entries,

$$\langle f|\hat{T}|i\rangle = T_{fi} = (2\pi)^4 \delta^4\left(\sum p_f^\mu - \sum p_i^\mu\right) \mathcal{T}_{fi} \tag{5.129}$$

implies

$$\left(\mathcal{T}_{fi} - \mathcal{T}_{if}^*\right) = i\sum_n \left[(2\pi)^4 \delta^4\left(\sum p_n^\mu - \sum p_i^\mu\right) \mathcal{T}_{fn}^* \mathcal{T}_{ni}\right] \tag{5.130}$$

summing over all possible states $|n\rangle$. Equating initial and final state — tacitly assuming that in fact the scattering processes discussed here have two initial-state particles, the two initial hadrons — and making contact with observables yields

$$\sigma_{\text{tot}} = \frac{1}{2s} \sum_n \left[(2\pi)^4 \, \delta^4 \left(\sum p_n^\mu - \sum p_i^\mu \right) T_{in}^* T_{ni} \right] \propto \frac{1}{2s} \, \mathfrak{Im}(T_{ii}) \,, \qquad (5.131)$$

the imaginary part of the forward elastic amplitude. It has to be forward and elastic since the identity $|f\rangle = |i\rangle$ also implies an identity of momenta. This relation is at the heart of the **Cutkosky rules**.

Denoting elastic $2 \to 2$ scattering amplitudes, e.g. $qq \to qq$, $qq' \to qq'$, or $gg \to gg$, with $\mathcal{A}(\hat{s}, \hat{t})$, this means that

$$\mathfrak{Im}[\mathcal{A}] = \frac{\mathcal{A}(\hat{s}, \hat{t}) - \mathcal{A}^*(\hat{s}, \hat{t})}{2i} \,. \qquad (5.132)$$

Below thresholds, e.g. from bound states, there are no imaginary contributions, implying that there is a region along the $\hat{s}$ axis where the amplitude is purely real. In this region the **Schwartz reflection principle** can be employed, which asserts that

$$\mathcal{A}^*(\hat{s}, \hat{t}) = \mathcal{A}(\hat{s}^*, \hat{t}) \qquad (5.133)$$

connecting the imaginary part of the amplitude with its s-**channel discontinuity**

$$\mathfrak{Im}[\mathcal{A}] = \lim_{\epsilon \to 0} \frac{\mathcal{A}(\hat{s} + i\epsilon, \hat{t}) - \mathcal{A}(\hat{s} - i\epsilon, \hat{t})}{2i} = \mathfrak{Disc}[\mathcal{A}(\hat{s}, \hat{t})] \,. \qquad (5.134)$$

The analyticity of the S matrix allows to rewrite the elastic amplitude through dispersion relations involving the discontinuity as

$$\mathcal{A}(\hat{s}, \hat{t}) = \int_{-\infty}^{0} \frac{\mathrm{d}s'}{2\pi i} \frac{\mathfrak{Disc}[\mathcal{A}(s', \hat{t})]}{s' - \hat{s}} + \int_{-\hat{t}}^{\infty} \frac{\mathrm{d}s'}{2\pi i} \frac{\mathfrak{Disc}[\mathcal{A}(s', \hat{t})]}{s' - \hat{s}} \,. \qquad (5.135)$$

This dispersion relation can also be rewritten as an integral in the plane of the (potentially complex) cosine of the scattering angle,

$$z_t = -\cos\theta_t = -\left(1 + \frac{2\hat{s}}{\hat{t}} \right) \,, \qquad (5.136)$$

namely,

$$\mathcal{A}(\hat{s}, \hat{t}) = \int_{-\infty}^{-1} \frac{\mathrm{d}z_t'}{2\pi i} \frac{\mathfrak{Disc}[\mathcal{A}(z_t', \hat{t})]}{z_t' - z_t} + \int_{1}^{\infty} \frac{\mathrm{d}s'}{2\pi i} \frac{\mathfrak{Disc}[\mathcal{A}(z_t', \hat{t})]}{z_t' - z_t} \,. \qquad (5.137)$$

This form documents the impact of the unphysical region of the scattering on the amplitude, which is entirely a consequence of the analytic properties of the S matrix: the integral is over the unphysical region of the scattering, *i.e.* outside the interval $z_t' \in [-1, 1]$. This consequently implies that all discontinuities or singularities in the scattering amplitude, those that impact on the cross-section, are in the unphysical regime of the scattering.

Assume that the amplitude can be decomposed according to a partial wave expansion,

$$\mathcal{A}(\hat{s}, \hat{t}) = \sum_l (2l+1)\mathcal{A}_l(\hat{s}, \hat{t}) P_l(z_t), \tag{5.138}$$

with the P_l the usual Legendre polynomials. Using Eq. (5.137) and the Legendre function

$$Q_l(z') = \frac{1}{2} \int\limits_{-1}^{1} \frac{dz'}{z'-z} P_l(z) \tag{5.139}$$

allows to rewrite the lth partial wave as

$$\mathcal{A}_l(\hat{s}, \hat{t}) = \left[1 + (-1)^{l+L}\right] \int\limits_{1}^{\infty} \frac{dz'}{2\pi i} Q_l(z') \, \mathfrak{Disc}[\mathcal{A}(z', \hat{t})]. \tag{5.140}$$

Here L is the related to the overall parity of the amplitude. In Section 7.1 the **Sommerfeld–Watson transformation** will be introduced in more detail. Here it suffices to state that it essentially replaces the sum over discrete values l of angular momentum parameter with an integral, thereby rendering l a continuous complex parameter. After the corresponding transformations the equations above, in the high-energy limit

$$z_t \longrightarrow -\frac{2\hat{s}}{\hat{t}} \longrightarrow \infty, \tag{5.141}$$

ultimately become

$$\mathcal{A}(\hat{s}, \hat{t}) = \int\limits_{\delta - i\infty}^{\delta + i\infty} \frac{dl}{2i} (2l+1) \left[1 + (-1)^{l+L}\right] \int\limits_{1}^{\infty} \frac{dz_t'}{2\pi i} \frac{P_l(-z_t) Q_l(z_t')}{\sin(\pi l)} \, \mathfrak{Disc}[\mathcal{A}(z_t', \hat{t})]$$

$$\longrightarrow -\frac{1}{4\pi} \int\limits_{\delta - i\infty}^{\delta + i\infty} dl \frac{(-1)^l + (-1)^L}{\sin(\pi l)} e^{ly} \, \mathcal{F}_l(\hat{t}). \tag{5.142}$$

Here the asymptotic values of the Legendre functions for $z \to \infty$, given by

$$P_l(z) \longrightarrow \frac{1}{\sqrt{\pi}} (2z)^l \frac{\Gamma\left(l + \frac{1}{2}\right)}{\Gamma\left(l+1\right)}$$

$$Q_l(z) \longrightarrow \frac{1}{\sqrt{\pi}} (2z)^{-(l+1)} \frac{\Gamma\left(l+1\right)}{\Gamma\left(l+\frac{3}{2}\right)} \tag{5.143}$$

have been used. The relevant quantity thus is the Laplace transform

$$\mathcal{F}_l(\hat{t}) = \int\limits_{0}^{\infty} dy\, e^{-ly} \, \mathfrak{Disc}[\mathcal{A}(z_t', \hat{t})] \tag{5.144}$$

of the discontinuity, with the Laplace parameter $y = \log(z_t/2)$. This quantity will be evaluated later on.

5.2.5.3 Kinematics of elastic parton–parton scattering

To start the discussion of how the large logarithms $\log(\hat{t}/\hat{s})$ emerge in perturbative QCD it is useful to reformulate the kinematics of elastic parton–parton scattering such that relevant limits can be inspected in a transparent way. A very handsome tool for this purpose is the **Sudakov** or **light-cone decomposition**, in which momenta are written as

$$p^\mu = \alpha P_+^\mu + \beta P_-^\mu + p_\perp^\mu \tag{5.145}$$

with two light-cone axes $P_\pm$ and the assumption that the transverse plane indicated with the subscript $\perp$ is orthogonal to both these axes. For the discussion of cross-sections at collider experiments, it is sensible to choose these two axes as the beam axes P_A and P_B. In the case of parton elastic scattering, e.g. $qq' \to qq'$, the incoming momenta are then given by

$$p_a = \sqrt{s}\left(x_A,\, 0;\, \vec{0}\right) \quad \text{and} \quad p_b = \sqrt{s}\left(0,\, x_B;\, \vec{0}\right), \tag{5.146}$$

making it explicit that $p_a \parallel P_A$ and $p_b \parallel P_B$, without any transverse momentum. The outgoing parton momenta are expressed by their two-dimensional transverse momenta $\vec{k}_{\perp,i}$ and rapidities y_i as

$$k_i^\mu = \left(k_{0,i} + k_{3,i},\, k_{0,i} - k_{3,i};\, \vec{k}_\perp\right) \stackrel{m=0}{\longrightarrow} \left(k_{\perp,i}e^{y_i},\, k_{\perp,i}e^{-y_i};\, \vec{k}_{\perp,i}\right), \tag{5.147}$$

where $k_{\perp,i}$ is the absolute value of $\vec{k}_{\perp,i}$. In these coordinates, the metric tensor is decomposed into longitudinal and transverse components according to

$$g^{\mu\nu} = 2\frac{p_a^\mu p_b^\nu + p_a^\nu p_b^\mu}{\hat{s}} - \delta_\perp^{\mu\nu}, \tag{5.148}$$

where $\delta_\perp^{\mu\nu}$ acts as a Kronecker-δ in the $x - y$ plane.

Expressed by transverse momenta and rapidities, the phase-space element reads

$$\frac{\mathrm{d}^3 k}{(2\pi)^2 (2E)} = \frac{\mathrm{d}^2 k_\perp}{(2\pi)^2} \frac{\mathrm{d}y}{4\pi}. \tag{5.149}$$

In $2 \to 2$ scattering with $p_a + p_b \to p_0 + p_1$, momentum conservation in the transverse plane requires $\vec{k}_{\perp,0} = -\vec{k}_{\perp,1}$ and therefore $|\vec{k}_{\perp,0}| = |\vec{k}_{\perp,1}| = k_\perp$. Introducing

$$\bar{y} = \frac{y_0 + y_1}{2} = \frac{1}{2}\log(x_A/x_B) \tag{5.150}$$

as the rapidity of the overall system and the rapidity distance of the outgoing partons from the centre-of-mass rapidity

$$y^* = \frac{y_0 - y_1}{2} \tag{5.151}$$

allows the rewriting

$$x_A = \frac{k_\perp}{\sqrt{s}} \left(e^{y_0} + e^{y_1} \right) = \frac{2k_\perp}{\sqrt{s}} e^{\bar{y}} \cosh y^*$$

$$x_B = \frac{k_\perp}{\sqrt{s}} \left(e^{-y_0} + e^{-y_1} \right) = \frac{2k_\perp}{\sqrt{s}} e^{-\bar{y}} \cosh y^* . \tag{5.152}$$

The Mandelstam parameters therefore read

$$\hat{s} = 4k_\perp^2 \cosh^2 y^* , \quad \hat{t} = -2k_\perp^2 \cosh y^* e^{-y^*} , \quad \text{and} \quad \hat{u} = -2k_\perp^2 \cosh y^* e^{y^*} . \tag{5.153}$$

Using these expressions on the squared leading-order matrix element yields

$$\frac{1}{(4\pi\alpha_s)^2} \left| \mathcal{M}_{qq' \to qq'} \right|^2 = \frac{C_F^2}{4} \frac{\hat{s}^2 + \hat{u}^2}{\hat{t}^2} = \frac{C_F^2}{4} \frac{4\cosh^2 y^* + e^{2y^*}}{e^{-2y^*}} = \frac{C_F^2}{4} \left(e^{y^*} \cosh y^* \right)^2 . \tag{5.154}$$

Therefore,

$$\frac{d\hat{\sigma}_{qq' \to qq'}}{d\hat{t}} = \frac{(4\pi\alpha_s)^2 \left| \mathcal{M}_{qq' \to qq'} \right|^2}{16\pi\hat{s}^2} = \frac{\pi C_F^2 \alpha_s^2}{64 k_\perp^2} e^{2y^*} . \tag{5.155}$$

The **factorization formula**, Eq. (2.52), relates the matrix element to the corresponding fully differential cross-section, in the case of elastic quark–quark scattering:

$$\sigma = \int_{\zeta_A}^{1} dx_A \int_{\zeta_B}^{1} dx_B \, f_{q/A}(x_A, \mu_F^2) f_{\bar{q}/B}(x_B, \mu_F^2) \, \hat{\sigma}_{qq' \to qq'} \left(\mu_F^2; \mu_R^2 \right) . \tag{5.156}$$

As usual, at leading order the **parton distribution function** $f_{a/A}(x_A, \mu_F^2)$ parameterizes the probability to find a parton of type a in the beam particle A, with a line-cone momentum fraction x_A, and at the **factorization scale** μ_F. In order to actually detect the two outgoing partons, they must carry some minimal energy and momentum, thus implying a minimal momentum fraction ζ w.r.t. each of the beams. Further emissions of course would make this picture more complicated and the trivial leading order partonic cross-section here would start to develop a more complicated kinematic dependence. For example, the integration over the x would allow further real emissions to be included in the partonic cross-section $\hat{\sigma}$, which would need more energy and momentum than the ζ which define the kinematics of the two outgoing quarks, and a dependence on the ratios $\zeta_{A,B}/x_{A,B}$ would emerge. However, using the relations above, Eq. (5.156) can be cast into

$$\frac{d\sigma_{qq' \to qq'}}{dk_\perp^2 dy_0 dy_1} = x_A f_{a/A}(x_A, \mu_F^2) x_B f_{b/B}(x_B, \mu_F^2)$$

$$= x_A f_{a/A}(x_A, \mu_F^2) x_B f_{b/B}(x_B, \mu_F^2) \frac{\pi\alpha_s^2}{36 k_\perp^2} e^{2y^*} . \tag{5.157}$$

This becomes large for large rapidity distances y^*. In this limit, $\hat{s} \approx -\hat{u} \gg -\hat{t}$, and in fact,

$$y^* \to \frac{1}{2} \log\left(-\frac{\hat{s}}{\hat{t}}\right), \qquad (5.158)$$

indicating that *the large rapidity distance limit is identical to the high-energy limit.* Replacing $\hat{u}$ with $-\hat{s}$ and retaining only the leading terms of order $\hat{s}^2/\hat{t}^2$, stemming from gluon exchange in the t-channel, yields the following squared matrix elements in the high-energy limit:

$$|\mathcal{M}_{qq'\to qq'}|^2 = |\mathcal{M}_{qq\to qq}|^2 = |\mathcal{M}_{q\bar{q}\to q\bar{q}}|^2 = \frac{C_F^2 g_s^4}{2} \frac{\hat{s}^2}{\hat{t}^2}$$

$$|\mathcal{M}_{qg\to qg}|^2 = \frac{C_F C_A}{2} g_s^4 \frac{\hat{s}^2}{\hat{t}^2}$$

$$|\mathcal{M}_{gg\to gg}|^2 = \frac{C_A^2 g_s^4}{2} \frac{\hat{s}^2}{\hat{t}^2}. \qquad (5.159)$$

Forming ratios shows that replacing a quark by a gluon amounts to multiplication with a factor $C_A/C_F = 9/4$, as expected.

5.2.5.4 Rewriting in the high-energy limit

To arrive at the full logarithmic structure, further emissions need to be considered. In a first step the $qq' \to qq'$ scattering amplitude will be rewritten in a fashion better suited for the following discussion. The full tree-level amplitude for $q(p_a)q'(p_b) \to q(k_0)q'(k_1)$ reads

$$\mathcal{M}_{qq'\to qq'} = \left[\bar{u}_i(k_0)(-ig_s T_{ij}^a)\gamma_\mu \bar{u}_j(p_a)\right] \frac{-ig^{\mu\nu}}{\hat{t}} \left[\bar{u}_k(k_1)(-ig_s T_{kl}^a)\gamma_\nu \bar{u}_l(p_b)\right], \quad (5.160)$$

which is not very helpful for the further discussion. The reason for this is that the form of the metric tensor in the gluon propagator suggests that the calculation is being performed in Lorenz-gauge — instead it will prove advantageous to use light-cone coordinates and a physical axial gauge.

Using the decomposition Eq. (5.148) for the metric tensor allows the identification of the leading contributions to the interaction. This is given by a structure where the helicity-conserving part of the quark–gluon coupling will be contracted with one of the light-cone-like polarizations of the t-channel gluon, while the other light-cone polarization vanishes due to the equation of motion $\not{p}_a \bar{u}(p_a) = 0$. In addition, the transverse polarizations will induce a kinematically suppressed helicity flip, usually identified with a "magnetic" interaction.[5] Ignoring these sub-leading or vanishing terms, the amplitude thus becomes

---

[5]An alternative way to see this goes as follows. In the soft limit, where the momentum transfer q along the gluon propagator is small, terms such as $\bar{u}(p_a + q)\gamma_\mu u(p_a)$ can be approximated by the helicity conserving eikonal $2p_{\mu,a}$,

$$\bar{u}(p_a + q)\gamma_\mu u(p_a) \to 2p_{\mu,a}$$

thereby eliminating the terms proportional to p_a^μ and $\delta_\perp^{\mu\nu}$.

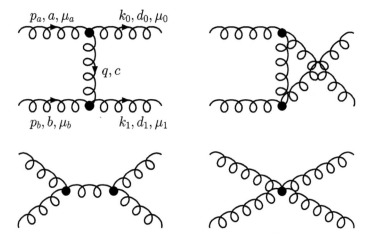

Fig. 5.6 Feynman diagrams for $gg \to gg$ scattering at leading order. In the upper left the t-channel exchange diagram, relevant in the high-energy limit, is depicted including all momentum, colour, and Lorentz-index assignments. The u-channel is shown in the upper right diagram, while the s-channel and four-gluon vertex diagrams are in the lower row.

$$\mathcal{M}_{qq' \to qq'} = \left[\bar{u}_i(k_0)(-ig_s T^a_{ij})\gamma_\mu \bar{u}_j(p_a) \right] \frac{-2ip^\mu_b p^\nu_a}{\hat{s}\hat{t}} \left[\bar{u}_k(k_1)(-ig_s T^a_{kl})\gamma_\nu \bar{u}_l(p_b) \right] .$$
(5.161)

Squaring the amplitude and summing (averaging) over final (initial) state colours and spins yields

$$\left| \overline{\mathcal{M}}_{qq' \to qq'} \right|^2 = \frac{4g_s^4 \left(\text{Tr} \left[T^a T^b \right] \right)^2}{4 \cdot 9} \frac{16 \left[2(k_0 p_b)(p_a p_b) \right] \left[2(k_1 p_a)(p_a p_b) \right]}{\hat{s}^2 \hat{t}^2}$$

$$= \frac{g_s^4 \left(T_R \delta^{ab} \right)^2}{9} \frac{4\hat{u}^2}{\hat{t}^2} = \frac{8g_s^4}{9} \frac{\hat{u}^2}{\hat{t}^2} \approx \frac{8g_s^4}{9} \frac{\hat{s}^2}{\hat{t}^2}$$
(5.162)

reproducing as anticipated the matrix element in the high-energy limit.

A similar treatment can be applied to gluon-gluon scattering, which is dominated in the high-energy limit by the t-channel diagram shown in the upper left of Fig. 5.6,

$$\mathcal{M}_{gg \to gg} = ig_s f^{ad_0 c} \left[g_{\mu_a \mu_0}(p_a + k_0)_\xi + g_{\mu_0 \xi}(-k_0 + q)_{\mu_a} + g_{\xi \mu_a}(-q - p_a)_{\mu_0} \right]$$

$$\cdot ig_s f^{bd_1 c} \left[g_{\mu_b \zeta}(p_b - q)_{\mu_1} + g_{\zeta \mu_1}(q + k_1)_{\mu_b} + g_{\mu_1 \mu_b}(-k_1 - p_b)_\zeta \right]$$

$$\cdot \frac{-ig^{\xi \zeta}}{q^2} \cdot \epsilon^{\mu_a *}_{\lambda_a}(p_a) \epsilon^{\mu_b *}_{\lambda_b}(p_b) \epsilon^{\mu_0}_{\lambda_0}(k_0) \epsilon^{\mu_1}_{\lambda_1}(k_1)$$

$$\approx -i \left(2 g_s f^{a d_0 c} g_{\mu_a \mu_0} p_{\xi,a} \right) \left(2 g_s f^{b d_1 c} g_{\mu_1 \mu_b} p_{\zeta,b} \right) \frac{2 p_a^\zeta p_b^\xi}{\hat{s} \hat{t}} \cdot \epsilon_{\lambda_a}^{\mu_a *} \epsilon_{\lambda_b}^{\mu_b *} \epsilon_{\lambda_0}^{\mu_0} \epsilon_{\lambda_1}^{\mu_1}$$

$$\approx -i g_s^2 f^{a d_0 c} f^{b d_1 c} g_{\mu_a \mu_0} g_{\mu_1 \mu_b} \frac{2 \hat{s}}{\hat{t}} \cdot \epsilon_{\lambda_a}^{\mu_a *} \epsilon_{\lambda_b}^{\mu_b *} \epsilon_{\lambda_0}^{\mu_0} \epsilon_{\lambda_1}^{\mu_1} . \tag{5.163}$$

Here the first two lines encode the triple-gluon vertices, $q = p_a - k_0$ with $\hat{t} = q^2$ denoting the momentum transfer along the t-channel propagator. In the last line before the polarization vectors the metric tensor is already rewritten in the axial gauge, omitting sub-leading or vanishing contributions following the same logic as above. To be more specific, in the limit of small momentum transfer $p_a \approx k_0$ and terms such as $k_0 \cdot \epsilon(p_a)$ or $p_a \cdot \epsilon(k_0)$ therefore are small and can be neglected. This reasoning, again, reduces the triple-gluon vertices to helicity-conserving eikonal terms to be convoluted with a single light-cone polarization of the t-channel gluon.[6] These manipulations of the elastic gluon–gluon amplitude, namely the restriction on t-channel exchange only, supplemented with rewriting the polarization vectors and the metric tensor, violate gauge invariance. This is easy to see by replacing any of the polarization vectors, for instance $\epsilon(p_a)$, with the corresponding momentum (p_a in the example) and by realising that this will not make the amplitude in Eq. (5.163) vanish. In order to restore full gauge invariance, all contributions to the amplitude and all diagrams, *i.e.*, s- and u-channel exchange and the four–gluon vertex, must be included. The complete gauge-invariant set is shown in Fig. 5.6.

However, when consistently working in the high-energy limit and in a physical gauge, the amplitude above is sufficient. With the squared colour factor given by

$$f^{a d_0 c} f^{b d_1 c} f^{d_0 a c'} f^{d_1 b c'} = C_A^2 \left(N_c^2 - 1 \right) \tag{5.165}$$

[6] In order to understand this in more detail, the sum over the **physical** helicities λ of the external gluons is cast into

$$\sum_\lambda \epsilon_\lambda^\mu (p) \epsilon_\lambda^{*\mu'} (p) = - \left(g^{\mu\nu} - \frac{n^\mu p^{\mu'} + n^{\mu'} p^\mu}{(n \cdot p)} + \frac{n^2 p^\mu p^{\mu'}}{(n \cdot p)^2} \right) ,$$

where n denotes an arbitrary four-vector, acting as the gauge vector in axial gauge. This vector can be chosen in a convenient way. For example, by setting $n = p_b$ for the incoming gluon a with momentum p_a the sum above becomes

$$\sum_{\lambda_a} \epsilon_{\lambda_a}^\mu (p_a) \epsilon_{\lambda_a}^{*\mu'} (p_a) = - \left(g^{\mu\mu'} - 2 \frac{p_b^\mu p_a^{\mu'} + p_b^{\mu'} p_a^\mu}{\hat{s}} \right) \equiv \delta_\perp^{\mu\mu'} ,$$

effectively the δ over the transverse polarizations from above. Contracting the polarization sums of the incoming gluon p_a and the outgoing gluon k_0 with a metric tensor yields the two physical polarizations plus terms that vanish in the high-energy limit $\hat{t}/\hat{s} \to 0$, thus explicitly encoding helicity conservation in this limit:

$$g_{\mu_a \mu_0} g_{\mu'_a \mu'_0} \left[\sum_{\lambda_a} \epsilon_{\lambda_a}^{\mu_a} (p_a) \epsilon_{\lambda_a}^{*\mu'_a} (p_a) \right] \left[\sum_{\lambda_0} \epsilon_{\lambda_0}^{\mu_0} (k_0) \epsilon_{\lambda_0}^{*\mu'_0} (k_0) \right] = 2 \left[1 + \mathcal{O} \left(\frac{\hat{t}}{\hat{s}} \right) \right] . \tag{5.164}$$

This means that in the high-energy limit the Lorentz structures of the triple gluon vertices can be simplified such that only the terms with the metric tensor between incoming and outgoing gluons remain.

and the relation in Eq. (5.164), the summed and averaged amplitude squared reads

$$\left|\mathcal{M}_{gg\to gg}\right|^2 = \frac{4C_A^2 g_s^4}{N_c^2 - 1} \frac{\hat{s}^2}{\hat{t}^2} = \frac{9g_s^4}{2} \frac{\hat{s}^2}{\hat{t}^2}, \tag{5.166}$$

as expected.

Turning this into a partonic cross-section necessitates folding the matrix elements squared with suitable phase-space elements; for a two-body final state

$$\begin{aligned}
d\Phi_2 &= \frac{dy_0 dk_{\perp,0}^2}{4\pi(2\pi)^2} \frac{dy_1 dk_{\perp,1}^2}{4\pi(2\pi)^2} \cdot (2\pi)^4 \delta^4 (p_a + p_b - k_0 - k_1) \\
&= \frac{1}{2\hat{s}} \frac{\cosh(y_0 - y_1) + 1}{\sinh(y_0 - y_1)} \frac{dk_{\perp,0}^2}{(2\pi)^2} \frac{dk_{\perp,1}^2}{(2\pi)^2} (2\pi)^2 \delta^2 \left(\vec{k}_{\perp,0} + \vec{k}_{\perp,1}\right) \approx \frac{1}{2\hat{s}} \frac{dk_\perp^2}{(2\pi)^2},
\end{aligned}$$
$$\tag{5.167}$$

where the large-energy limit has been used, *i.e.*, a strong ordering of rapidities $y_0 \gg y_1$, leading to the ratio involving the hyperbolic functions to approach unity. In addition, the conservation of transverse momentum leads to the $k_{\perp,0} = k_{\perp,1} = k_\perp$. Consequently, the differential partonic cross-section reads

$$\frac{d\hat{\sigma}}{dk_\perp^2} = \frac{|\mathcal{M}|^2}{16\pi\hat{s}^2}. \tag{5.168}$$

5.2.5.5   Real corrections: Multi-Regge kinematics

In a first step to an all-orders resummation, consider the kinematics of a $2 \to n+2$ gluon scattering, where the outgoing momenta are, again, labelled by k_i with $i \in [0, n+1]$. Four-momentum conservation requires

$$\vec{0} = \sum_{i=0}^{n+1} \vec{k}_{\perp,i}, \quad x_a = \sum_{i=0}^{n+1} \frac{k_{\perp,i} e^{y_i}}{\sqrt{s}}, \quad \text{and} \quad x_b = \sum_{i=0}^{n+1} \frac{k_{\perp,i} e^{-y_i}}{\sqrt{s}}, \tag{5.169}$$

as a straightforward generalization of the $2 \to 2$ case encountered above, *cf.* Eq. (5.152). In a fashion similar to Eq. (5.153), various Mandelstam invariants are given by

$$\hat{s} = x_a x_b s = \sum_{i,j=0}^{n+1} k_{\perp,i} k_{\perp,j} e^{-(y_i - y_j)} \approx k_{\perp,0} k_{\perp,n+1} e^{y_0 - y_{n+1}}$$

$$\hat{s}_{ij} = 2k_i k_j = 2k_{\perp,i} k_{\perp,j} \left[\cosh(y_i - y_j) - \cos(\phi_i - \phi_j)\right] \approx k_{\perp,i} k_{\perp,j} e^{|y_i - y_j|}$$

$$\hat{t}_{ai} = -2p_a k_i = -\sum_{j=0}^{n+1} k_{\perp,i} k_{\perp,j} e^{-(y_i - y_j)} \approx -k_{\perp,0} k_{\perp,i} e^{y_0 - y_i}$$

$$\hat{t}_{bi} = -2p_b k_i = -\sum_{j=0}^{n+1} k_{\perp,i} k_{\perp,j} e^{y_i - y_j} \approx -k_{\perp,i} k_{\perp,n+1} e^{y_i - y_{n+1}}, \tag{5.170}$$

where multi-Regge kinematics is being assumed in order to make the approximations, *i.e.* a strict ordering of rapidities while the transverse momenta are all small and of the same size,

$$y_0 \gg y_1 \gg y_2 \gg \cdots \gg y_n \gg y_{n+1} \quad \text{and} \quad k_{\perp,i} \approx k_\perp \, \forall i. \tag{5.171}$$

This leads to a strict hierarchy of scales, namely

$$\hat{s} \gg \hat{s}_{ij} \gg k_\perp^2. \tag{5.172}$$

Assuming, as before, t-channel dominance, the momentum q_1 going through the first gluon propagator is given by

$$q_1 = p_a - k_0 \approx \sqrt{s} \left(\frac{k_{\perp,0} e^{y_0} + k_{\perp,1} e^{y_1}}{\sqrt{s}}, \, 0; \, \vec{0} \right) - \left(k_{\perp,0} e^{y_0}, \, k_{\perp,0} e^{-y_0}; \, \vec{k}_{\perp,i} \right)$$

$$= \left(k_{\perp,1} e^{y_1}, \, k_{\perp,0} e^{-y_0}; \, -\vec{k}_{\perp,0} \right). \tag{5.173}$$

Therefore, the squared momentum reads

$$q_1^2 = \hat{t}_1 = -k_{\perp,1} k_{\perp,0} e^{y_1 - y_0} - k_{\perp,0}^2 \approx -k_{\perp,0}^2 = q_{\perp,1}^2, \tag{5.174}$$

and in this limit the scales in the propagators are driven by the transverse components of their four-momenta only. Of course, the same reasoning applies for the next propagator, $q_2 = p_a - k_0 - k_1$, such that, in general

$$\hat{t}_i = q_i^2 \approx -q_{\perp,i}^2. \tag{5.175}$$

5.2.5.6 Real corrections: $2 \to 3$ gluon amplitude

The $2 \to 3$ gluon scattering amplitude in the high-energy limit, *cf.* the left diagram in Fig. 5.7, is given by

$$\mathcal{M}_{gg \to ggg}$$

$$= \left(2ig_s f^{ad_0 c_1} g_{\mu_a \mu_0} p_a^{\xi_1} \right) \frac{1}{\hat{t}_1}$$

$$\cdot \left(-ig_s f^{c_1 c_2 d_1} \right) \left[g_{\xi_1 \xi_2} (q_1 + q_2)_{\mu_1} + g_{\xi_2 \mu_1} (-q_2 + k_1)_{\xi_1} + g_{\mu_1 \xi_1} (-k_1 - q_1)_{\xi_2} \right]$$

$$\cdot \left(2ig_s f^{ad_2 c_2} g_{\mu_a \mu_0} p_b^{\xi_2} \right) \frac{1}{\hat{t}_2}, \tag{5.176}$$

where the first and the last line are the by-now familiar eikonal factors for the emission of a soft gluon off the incident gluon line, and the second line stands for the additional triple-gluon vertex sandwiched between the two t-channel propagators. Multiplying this term with the "pending" four-momenta from the external vertices and suitably normalising allows to define an effective vertex, namely

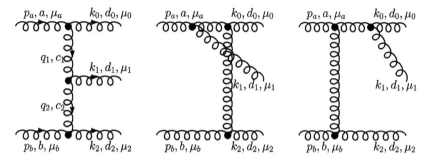

Fig. 5.7 Example Feynman diagrams for $gg \to ggg$ scattering at leading order. In the left diagram the t-channel exchange diagram, relevant in the high-energy limit, is depicted including all momentum, colour, and Lorentz-index assignments, Bremsstrahlung contributions from gluon emission off the upper gluon line are also depicted.

$$\tilde{C}_{\mu_1} = \frac{2}{\hat{s}} p_a^{\xi_1} p_b^{\xi_2} \left[g_{\xi_1 \xi_2}(q_1 + q_2)_{\mu_1} + g_{\xi_2 \mu_1}(-q_2 + k_1)_{\xi_1} + g_{\mu_1 \xi_1}(-k_1 - q_1)_{\xi_2} \right]$$

$$\approx (q_1 + q_2)_{\perp, \mu_1} - \frac{\hat{t}_{a1}}{\hat{s}} p_{\mu_1, b} + \frac{\hat{t}_{b1}}{\hat{s}} p_{\mu_1, a}, \tag{5.177}$$

where the terms proportional to the soft gluon momenta in the propagators have been neglected in the scalar products of the already sub-leading contributions, and it has been assumed that the $q_{1,2}$ are dominated by their transverse components. The Bremsstrahlung-like contributions from gluon emissions off the upper or lower gluon line, *cf.* the right two diagrams in Fig. 5.7 for emissions off the upper gluon line, can also be absorbed into the **Lipatov effective vertex**

$$C^{\mu_1}(q_1, q_2) = (q_1 + q_2)_{\perp}^{\mu_1} - \left(\frac{\hat{t}_{a1}}{\hat{s}} + \frac{2\hat{t}_2}{\hat{t}_{b1}} \right) p_b^{\mu_1} + \left(\frac{\hat{t}_{b1}}{\hat{s}} + \frac{2\hat{t}_1}{\hat{t}_{a1}} \right) p_a^{\mu_1}. \tag{5.178}$$

The amplitude in the high-energy limit thus takes a particularly simple form, namely

$$\mathcal{M}_{gg \to ggg} = 2i\hat{s} \, \epsilon^{\mu_a *}(p_a) \epsilon^{\mu_b *}(p_b) \epsilon^{\mu_0}(k_0) \epsilon^{\mu_1}(k_1) \epsilon^{\mu_2}(k_2)$$

$$\times \left[i g_s f^{a d_0 c_1} g_{\mu_a \mu_0} \right] \frac{1}{\hat{t}_1} \left[i g_s f^{c_1 d_1 c_2} C_{\mu_1}(q_1, q_2) \right] \frac{1}{\hat{t}_2} \left[i g_s f^{a d_0 c_1} g_{\mu_a \mu_0} \right]. \tag{5.179}$$

The Lipatov effective vertex is manifestly gauge-invariant, as can be seen by contracting it with k_1 instead of the polarization vector of the gluon. This allows simple contraction of this effective vertex through the ordinary metric tensor for $\epsilon(k_1)$ when squaring the amplitude, leading to

$$C^{\mu_1} C_{\mu_1} = (q_1 + q_2)_{\perp}^2 - \left(\frac{\hat{t}_{a1} \hat{t}_{b1}}{2\hat{s}} + \hat{t}_2 + \hat{t}_1 + \frac{2\hat{t}_1 \hat{t}_2 \hat{s}}{\hat{t}_{a1} \hat{t}_{b1}} \right)$$

$$\approx 2\vec{q}_{\perp,1}\vec{q}_{\perp,2} - \frac{k_{\perp,1}^2}{2} - 2\frac{q_{\perp,1}^2 q_{\perp,2}^2}{k_{\perp,1}^2} = \frac{4q_{\perp,1}^2 q_{\perp,2}^2}{k_{\perp,1}^2} \tag{5.180}$$

where the relations of Eq. (5.170) have been combined into

$$\hat{t}_{a1}\hat{t}_{b1} \approx k_{\perp,0}k_{\perp,n+1}k_{\perp,1}^2 e^{y_0-y_{n+1}} \approx k_{\perp,1}^2 \hat{s} \tag{5.181}$$

This ultimately allows the high-energy amplitude squared to be cast into the form

$$\left|\overline{\mathcal{M}}_{gg\to ggg}\right|^2 = \frac{16 C_A^3 g_s^6}{N_c^2 - 1} \frac{\hat{s}^2}{k_{\perp,0}^2 k_{\perp,1}^2 k_{\perp,2}^2} \cdot \tag{5.182}$$

In fact this result is reproduced, in the limit of strong rapidity ordering, by the **exact** squared amplitude for $gg \to ggg$,

$$\left|\overline{\mathcal{M}}_{gg\to ggg}\right|^2_{\text{exact}} = 4(\pi\alpha_s C_A)^3 \sum_{i>j} \hat{s}_{ij}^4 \sum_{\text{non-cycl}} \frac{1}{\hat{s}_{a0}\hat{s}_{01}\hat{s}_{12}\hat{s}_{2b}\hat{s}_{ba}} , \tag{5.183}$$

where $i,j \in \{a,0,1,2,b\}$ and the second sum goes over all non-cyclic permutations of this set [450].

The approximated result can also be compared with the corresponding $gg \to gg$ amplitude squared which reads

$$\left|\overline{\mathcal{M}}_{gg\to gg}\right|^2 = \frac{4 C_A^2 g_s^4}{N_c^2 - 1} \frac{\hat{s}^2}{k_{\perp,0}^2 k_{\perp,1}^2} , \tag{5.184}$$

cf. Eq. (5.166) and replacing $\hat{t}^2$ in the denominator with $k_{\perp,0}^2 k_{\perp,1}^2$, following the logic of the high-energy approximation. This gives rise to the — correct — suspicion that additional gluons emitted along the t-channel ladder will lead to factors $4C_A g_s^2/k_{\perp,i}^2$ and thereby to logarithmic corrections of the type $\log(\hat{s}/\hat{t}_i)$.

A second comment is in order here. Inspecting the three-body phase-space element, written in convenient coordinates,

$$\begin{aligned}
d\Phi_3 &= \left[\prod_{i=0}^3 \frac{dy_i d^2 k_{\perp,i}}{4\pi(2\pi)^2}\right] \cdot (2\pi)^4 \delta^4\left(p_a + p_b - \sum_{i=0}^2 k_i\right) \\
&= \frac{1}{2\hat{s}} \frac{d^2 k_{\perp,0}}{(2\pi)^2} \frac{dy_1 d^2 k_{\perp,1}}{4\pi(2\pi)^2} \frac{d^2 k_{\perp,2}}{(2\pi)^2} \cdot (2\pi)^2 \delta^2\left(\sum_{i=0}^2 \vec{k}_{\perp,i}\right) ,
\end{aligned} \tag{5.185}$$

it becomes apparent that there is not explicit rapidity dependence on the emission of the extra gluon; relieving the strong-ordering condition it will be emitted with a flat probability anywhere between the two most forward gluons, 0 and 2. This independence will remain also for further gluons, with the only constraint imposed by rapidity ordering; other than that, the probabilities for gluon emissions will be flat over rapidities. This maybe surprising pattern will change when sub-leading corrections are taken into account, a complication well beyond the scope of the discussion here. Following the findings until now, the differential partonic cross-section is given by

$$\frac{\mathrm{d}\hat{\sigma}_{gg\to ggg}}{\mathrm{d}k_{\perp,0}^2\mathrm{d}k_{\perp,2}^2\mathrm{d}\phi} = \frac{C_A^3\alpha_s^3}{4\pi} \frac{|y_0 - y_2|}{k_{\perp,0}^2 k_{\perp,2}^2 \left(k_{\perp,0}^2 + k_{\perp,0}^2 + 2k_{\perp,0}k_{\perp,2}\cos\phi_{02}\right)}, \quad (5.186)$$

where $|y_0 - y_2|$ is the rapidity interval gluon 1 is allowed to populate, and ϕ_{02} is the azimuthal angle between the gluons 0 and 2. With rapidities given by $\log(\hat{s}/k_\perp^2)$ the logarithmic enhancement of the production cross-section of three gluons with respect to the production of two gluons at the same rapidities y_0 and y_2 becomes manifest. This provides motivation for trying to resum this new class of large logarithms, which have their origin in rapidities (or as $\log(1/x)$) in contrast to the usual collinear double and single logarithms that are resummed by the DGLAP equation.

5.2.5.7   Virtual corrections

However, in order to fully appreciate the emerging logarithmic structure, virtual corrections need to be considered, too. In the large-rapidity limit, the relevant corrections are depicted in Fig. 5.8.

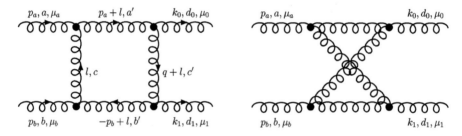

Fig. 5.8 Feynman diagrams for virtual corrections to $gg \to gg$ scattering in the high-energy limit. In the left diagram all momentum, colour, and Lorentz-index assignments have been included. The result for the crossed diagram on the right can be obtained from the one for the left diagram with suitable replacements.

Omitting from the start contributions of the form $\epsilon(p) \cdot p$, approximating $p_a \approx k_0$ and $p_b \approx k_1$, and realising that the dominant contributions stems from small loop momenta, which allows us to neglect terms $\propto l$, the amplitude can be cast into

$$\mathcal{M} = \epsilon_{\mu_a*}(p_a)\,\epsilon_{\mu_b*}(p_b)\,\epsilon_{\mu_0}(k_0)\,\epsilon_{\mu_1}(k_1)$$

$$\times \int \frac{\mathrm{d}^4 l}{(2\pi)^4} \left\{ \frac{g_s^4\, f^{aa'c}\, f^{a'd_0c'}\, f^{cb'b}\, f^{c'd_1b'}}{(p_a + l)^2 l^2 (p_b - l)^2 (q - l)^2} \cdot g_{\mu_a'\mu_0'}\, g_{\mu_a''\mu_b''}\, g_{\mu_b'\mu_1'}\, g_{\mu_0''\mu_1''} \right.$$

$$\times \left[\left(2g^{\mu_a\mu_{a'}}\, p_a^{\mu_{a''}} - 2g^{\mu_{a'}\mu_{a''}}\, l^{\mu_a} + g^{\mu_{a''}\mu_a}(l - p_a)^{\mu_{a'}} \right) \right]$$

$$\times \left. \left[\left(g^{\mu_{0'}\mu_0}(p_a + k_0 + l)^{\mu_{0''}} + g^{\mu_0\mu_{0''}}(p_a - 2k_0 + l)^{\mu_{0'}} - 2g^{\mu_{0''}\mu_{0'}}\, p_a^{\mu_0} \right) \right] \right.$$

$$\times \left[\left(g^{\mu_{b'} \mu_b} (l - 2p_b)^{\mu_{b''}} + g^{\mu_b \mu_{b''}} (p_b + l)^{\mu_{b'}} - 2g^{\mu_{b''} \mu_{b'}} l^{\mu_b} \right) \right]$$

$$\times \left[\left(g^{\mu_{1''} \mu_1} (p_a - k_0 + l + k_1)^{\mu_{1'}} + g^{\mu_1 \mu_{1'}} (l - k_1 - p_b)^{\mu_{1''}} \right) \right] \Big\}$$

$$\approx \epsilon_{\mu_a *}(p_a)\, \epsilon_{\mu_b *}(p_b)\, \epsilon_{\mu_0}(k_0)\, \epsilon_{\mu_1}(k_1)$$

$$\times \int \frac{\mathrm{d}^4 l}{(2\pi)^4} \Bigg\{ \frac{g_s^4 f^{aa'c} f^{a'doc'} f^{cb'b} f^{c'd_1 b'}}{(p_a + l)^2 l^2 (p_b - l)^2 (q - l)^2} \cdot g_{\mu'_a \mu'_0}\, g_{\mu''_a \mu''_b}\, g_{\mu'_b \mu'_1}\, g_{\mu''_0 \mu''_1}$$

$$\times \left[\left(2g^{\mu_a \mu_{a'}} p_a^{\mu_{a''}} - g^{\mu_{a''} \mu_a} p_a^{\mu_{a'}} \right) \right] \left[\left(2g^{\mu_{0'} \mu_0} p_a^{\mu_{0''}} - g^{\mu_0 \mu_{0''}} p_a^{\mu_{0'}} \right) \right]$$

$$\times \left[\left(g^{\mu_b \mu_{b''}} p_b^{\mu_{b'}} - 2g^{\mu_{b'} \mu_b} p_b^{\mu_{b''}} \right) \right] \left[\left(g^{\mu_{1''} \mu_1} p_b^{\mu_{1'}} - 2g^{\mu_1 \mu_{1'}} p_b^{\mu_{1''}} \right) \right] \Bigg\}$$

$$= g_s^4 f^{aa'c} f^{a'doc'} f^{cb'b} f^{c'd_1 b'} \cdot \mathcal{I}, \tag{5.187}$$

where in the last step the decomposition

$$g_{\mu\nu} = 2\frac{p_{\mu,a} p_{\nu,b} + p_{\mu,b} p_{\nu,a}}{\hat{s}} - \delta_{\mu\nu,\perp} \tag{5.188}$$

has been used and where the remaining integral $\mathcal{I}$ is given by

$$\mathcal{I} = 4\hat{s}^2 \int \frac{\mathrm{d}^4 l}{(2\pi^4)} \left[\frac{1}{(p_a + l)^2} \cdot \frac{1}{l^2} \cdot \frac{1}{(p_b - l)^2} \cdot \frac{1}{(q - l)^2} \right]$$

$$= 2\hat{s}^3 \int \frac{\mathrm{d}\alpha \mathrm{d}\beta \mathrm{d}^2 k_\perp}{(2\pi)^4} \left[\frac{1}{(1+\alpha)\beta\hat{s} - k_\perp^2 + i\varepsilon} \cdot \frac{1}{\alpha\beta\hat{s} - k_\perp^2 + i\varepsilon} \right.$$

$$\left. \cdot \frac{1}{\alpha(\beta - 1)\hat{s} - k_\perp^2 + i\varepsilon} \cdot \frac{1}{\alpha\beta\hat{s} - (q_\perp - k_\perp)^2 + i\varepsilon} \right]. \tag{5.189}$$

To obtain the expression in the second line, the loop momentum l has been decomposed according to

$$l^\mu = \alpha p_a^\mu + \beta p_b^\mu + l_\perp^\mu \quad \text{and} \quad \mathrm{d}^4 l = \frac{\hat{s}}{2} \mathrm{d}\alpha \mathrm{d}\beta \mathrm{d}^2 l_\perp. \tag{5.190}$$

In the final expression for $\mathcal{I}$ above, the actual pole structure of the propagators has been made explicit, since this allows the analytic structure to be analysed and ultimately the integrations to be performed. This is more or less straightforward, since closer inspection reveals that the α integration of the integral in Eq. (5.189) is not too hard: all propagators apart from the third one will have a pole in the lower complex half-plane of α and the α integration may easily be performed in the upper half-plane:

$$\mathcal{I} \approx -2i\hat{s}^2 \int \frac{\mathrm{d}\beta \mathrm{d}^2 k_\perp}{(2\pi)^4} \left[\frac{1}{\beta\hat{s} - k_\perp^2 + i\varepsilon} \cdot \frac{1}{k_\perp^2 + i\varepsilon} \cdot \frac{1}{(q_\perp - k_\perp)^2 + i\varepsilon} \right], \tag{5.191}$$

leaving a logarithmic integral over β in the range $\beta \in [k_\perp^2/\hat{s}, 1]$, resulting in

$$
\begin{aligned}
\mathcal{M} &\approx g^{\mu_a\mu_0} g^{\mu_b\mu_1} \epsilon_{\mu_a*}(p_a)\, \epsilon_{\mu_b*}(p_b)\, \epsilon_{\mu_0}(k_0)\, \epsilon_{\mu_1}(k_1) \\
&\times \frac{16\pi\alpha_s}{C_A} \cdot f^{aa'c} f^{a'd_0c'} f^{cb'b} f^{c'd_1b'} \cdot \frac{\hat{s}}{-\hat{t}} \log \frac{\hat{s}}{-\hat{t}} \alpha(\hat{t}),
\end{aligned}
\tag{5.192}
$$

with the characteristic function

$$
\alpha(\hat{t}) = \alpha_s C_A \hat{t} \int \frac{\mathrm{d}^2 k_\perp}{(2\pi)^2} \frac{1}{k_\perp^2 (q-k)_\perp^2},
\tag{5.193}
$$

which encodes the residual loop integration in the transverse plane. This function could be regularized through an infrared cut-off μ, resulting in

$$
\alpha(\hat{t}) \approx -\frac{\alpha_s C_A}{4\pi} \log \frac{q_\perp^2}{\mu^2},
\tag{5.194}
$$

showing that the amplitude is doubly logarithmic-divergent.

However, the second, crossed diagram in Fig. 5.8 needs to be considered as well. This is relatively straightforward, since the dominant term can be obtained by just replacing $\hat{s} \longrightarrow \hat{u}$, and using that $\hat{u} = -\hat{s} - \hat{t} \approx -\hat{s}$ in the usual limit $\hat{t} \to 0$. The overall amplitude thus reads

$$
\begin{aligned}
\mathcal{M} &\approx -\frac{16\pi\alpha_s}{C_A} \frac{\hat{s}}{-\hat{t}} \alpha(\hat{t})\, g^{\mu_a\mu_0} g^{\mu_b\mu_1} \epsilon_{\mu_a*} \epsilon_{\mu_b*} \epsilon_{\mu_0} \epsilon_{\mu_1} f^{aa'c} f^{a'd_0c'} \\
&\times \left[\log \frac{\hat{s}}{-\hat{t}} f^{cb'b} f^{c'd_1b'} - \left(\log \frac{\hat{s}}{-\hat{t}} + i\pi \right) f^{c'b'b} f^{cd_1b'} \right].
\end{aligned}
\tag{5.195}
$$

It is interesting to note, though, that in the calculation here, self-energy and vertex-correction diagrams have been omitted: their dominant contribution would consist of a running of α_s, which, of course, could be inserted "by hand" as well.

However, this leaves some colour algebra as a final task. In the final step for the virtual amplitude, the leading contribution corresponding to one single simple colour structure will be extracted. To this end, it is useful to recall that the underlying problem is the determination of colour structures in the scattering of two gluons, of $\mathbf{8} \otimes \mathbf{8}$ in the adjoint representation of QCD. It is important to remember that the generators of the adjoint representation, $\mathbf{8}$, indeed are anti-symmetric. Broadly speaking, therefore, two kinds of structures emerge, symmetric and asymmetric ones,

$$
\mathbf{8} \otimes \mathbf{8} = [\mathbf{8} \otimes \mathbf{8}]_{\mathbf{S}} + [\mathbf{8} \otimes \mathbf{8}]_{\mathbf{A}},
\tag{5.196}
$$

given by

$$
\begin{aligned}
[\mathbf{8} \otimes \mathbf{8}]_{\mathbf{S}} &= \mathbf{1} \oplus \mathbf{8_S} \oplus \mathbf{27} \\
[\mathbf{8} \otimes \mathbf{8}]_{\mathbf{S}} &= \mathbf{8_A} \oplus \mathbf{10} \oplus \overline{\mathbf{10}}.
\end{aligned}
\tag{5.197}
$$

The products of structure constants f^{abc} of the results are mapped onto these structures through projectors $\hat{P}^{ad_0}_{bd_1}$, given by

$$\hat{P}^{ad_0}_{bd_1}(\mathbf{1}) = \frac{1}{N_c^2 - 1}\,\delta^{ad_0}\delta_{bd_1}$$

$$\hat{P}^{ad_0}_{bd_1}(\mathbf{8_A}) = \frac{1}{N_c}\,f^{acd_0}f_{bcd_1}$$

$$\hat{P}^{ad_0}_{bd_1}(\mathbf{10} \oplus \overline{\mathbf{10}}) = \frac{1}{2}\left(\delta^a{}_b\delta^{d_0}{}_{d_1}\right) - \frac{1}{N_c}\,f^{acd_0}f_{bcd_1}\,, \tag{5.198}$$

while the other symmetric structures do not contribute any logarithmically enhanced terms, since the colour structures between s- and u-channel are antisymmetric and the two contributions differ only by finite terms. Furthermore, the projector for $\mathbf{10} \oplus \overline{\mathbf{10}}$ cancels exactly when contracted with the colour structures in the result above, Eq. (5.195). This leaves, at leading colour, only the anti-symmetric octet contribution, thereby effectively reducing the result to

$$\mathcal{M} \approx -8\pi\alpha_s\,\alpha(\hat{t})\,\frac{\hat{s}}{-\hat{t}}\,\log\frac{\hat{s}}{-\hat{t}}\,g^{\mu_a\mu_0}g^{\mu_b\mu_1}\,f^{acd_0}f^{bcd_1}\,. \tag{5.199}$$

5.2.5.8 Putting it all together: the $2 \to n$ gluon amplitude

The structure of the leading terms of the $2 \to 3$ gluon amplitude in multi-Regge kinematics, Eq. (5.179), lends itself to a generalization to the $2 \to n$ gluon case in the same kinematical situation, as advocated in Ref. [708]. This form can be further supplemented with the leading higher-order virtual corrections from above, assuming that the dominant octet structure encountered there also constitutes the leading terms for the $2 \to n$ gluon amplitude. From a Lorentz and colour structure point of view the virtual corrections look exactly like the exchange of a single t-channel gluon, which motivated the conjecture that exponentiating this correction would constitute the all-orders leading logarithmic approximation to a t-channel gluon exchange [728]. Applying this to *all* t–channel gluons amounts to replacing their simple propagators with

$$\frac{1}{\hat{t}_i} \longrightarrow \frac{1}{\hat{t}_i}\cdot\left(-\frac{\hat{s}_{i-1,i}}{\hat{t}_i}\right)^{\alpha(\hat{t}_i)} \approx \frac{1}{\hat{t}_i}\cdot\exp\left[\alpha(\hat{t}_i)(y_{i-1}-y_i)\right]\,, \tag{5.200}$$

using the approximated expressions in Eq. (5.170).

Recalling the form of the function $\alpha(t)$, the exponential on the right hand side of Eq. (5.200) has a remarkable similarity to a Sudakov form factor driven by the product of two logarithmic structures. First, there is a collinear logarithm, which manifests itself as a logarithm of the cut-off μ, $\log(q_\perp^2/\mu^2)$. Sending this cut-off to zero would of course lead to the function diverging, and, as in similar situations before, μ must be interpreted as a resolution parameter. Without any limit on resolution, the probability for not emitting a parton is zero. The other logarithm is written as rapidity differences $(y_{i-1} - y_i)$ which in fact are nothing but $\log(\hat{s}_{i-1,i}/\hat{t}_i)$. These terms are different from the ones encountered before and emerge as large contributions only in the high-energy limit.

Assembling all contributions into one expression, the $2 \to (n+2)$ gluon amplitude in the high-energy limit reads

$$\mathcal{M}_{gg \to (n+2)g} = 2i\hat{s} \left[ig_s f^{a d_0 c_1} g_{\mu_a \mu_0} \right] \epsilon^{\mu_a *}(p_a) \epsilon^{\mu_0}(k_0)$$

$$\times \frac{1}{\hat{t}_1} \exp\left[\alpha(\hat{t}_1)(y_0 - y_1)\right]$$

$$\times \left[ig_s f^{c_1 d_1 c_2} C_{\mu_1}(q_1, q_2) \right] \epsilon^{\mu_1}(k_1)$$

$$\times \frac{1}{\hat{t}_2} \exp\left[\alpha(\hat{t}_2)(y_1 - y_2)\right]$$

$$\times \left[ig_s f^{c_2 d_2 c_3} C_{\mu_2}(q_2, q_3) \right] \epsilon^{\mu_2}(k_2)$$

$$\vdots$$

$$\times \frac{1}{\hat{t}_{n+1}} \exp\left[\alpha(\hat{t}_{n+1})(y_n - y_{n+1})\right]$$

$$\times \left[ig_s f^{c_n b d_{n+1}} g_{\mu_b \mu_{n+1}} \right] \epsilon^{\mu_b *}(p_b) \epsilon^{\mu_{n+1}}(k_{n+1}). \qquad (5.201)$$

Apart from the exponential factors attached to the t-channel gluon propagators re-summing the leading virtual correction to all orders, this constitutes a straightforward generalization of the $2 \to 3$ gluon amplitude in Eq. (5.179).

5.2.5.9 Evaluating the discontinuity

In the next step the discontinuity of the elastic amplitude has to be evaluated. In this setup this translates to applying a cut on the s-channel gluons k, equivalent to putting them onto their mass-shell, *i.e.*, to replace

$$\frac{i}{k^2} \longrightarrow 2\pi\delta(k^2). \qquad (5.202)$$

Thereby the $(n+1)$-loop integral is effectively replaced with an integral over the $(n+2)$-particle phase space:

$$\Phi_{n+2} = \prod_{i=0}^{n} \int \frac{d^4 k_i}{(2\pi)^4} \prod_{j=0}^{n+1} (2\pi)\delta(k_j)^2 = \prod_{i=0}^{n+1} \int \frac{dy_i}{4\pi} \frac{d^2 k_{i,\perp}}{(2\pi)^2} (2\pi)^4 \delta^4 \left(p_+ p_b - \sum_{i=0}^{n+1} k_i \right). \qquad (5.203)$$

Being interested in the amplitude in the multi–regge regime, this can be recast into a form, where the conservation of longitudinal light-cone momentum is guaranteed by the two most forward outgoing gluons,

$$\Phi_{n+2} = \int \frac{1}{2\hat{s}} \frac{d^2 k_{0,\perp}}{(2\pi)^2} \left[\prod_{i=1}^{n} \int \frac{dy_i}{4\pi} \frac{d^2 k_{i,\perp}}{(2\pi)^2} \right] \frac{d^2 k_{n+1,\perp}}{(2\pi)^2} (2\pi)^2 \delta^2 \left(\sum_{i=0}^{n+1} k_{i,p\perp} \right). \quad (5.204)$$

To obtain the discontinuity of the elastic gluon scattering amplitude with the exchange of two t-channel gluons, one must sum over all $2 \to (n+2)$ gluon amplitudes in Eq. (5.201), which ultimately translates into

$$\mathfrak{Disc}[i\mathcal{M}^{aba'b'}_{\mu_a \mu_b \mu_{a'} \mu_{b'}}(\hat{s}, \hat{t})]$$

$$= \sum_{n=0}^{\infty} \int \frac{1}{2\hat{s}} \frac{d^2 k_{0,\perp}}{(2\pi)^2} \left[\prod_{i=1}^{n} \int \frac{dy_i}{4\pi} \frac{d^2 k_{i,\perp}}{(2\pi)^2} \right] \frac{d^2 k_{n+1,\perp}}{(2\pi)^2} (2\pi)^2 \delta^2 \left(\sum_{i=0}^{n+1} k_{i,p\perp} \right)$$

$$\times \left[-2\hat{s}\delta_{\mu_a \mu_{a'}} g_s f^{ad_0 c_1} \right] \left[-2\hat{s}\delta_{\mu_b \mu_{b'}} g_s f^{c_1' d_0 a'} \right]$$

$$\times \frac{1}{\hat{t}_1} \exp\left[\alpha(\hat{t}_1)(y_0 - y_1) \right] \frac{1}{\hat{t}_1'} \exp\left[\alpha(\hat{t}_1')(y_0 - y_1) \right]$$

$$\times \left[ig_s f^{c_1 d_1 c_2} C_{\mu_1}(q_1, q_2) \right] \left[(ig_s f^{c_1' d_1 c_2'}) C^{\mu_1}(q - q_1, q - q_2) \right]$$

$$\vdots$$

$$\times \left[ig_s f^{c_n d_n c_{n+1}} C_{\mu_n}(q_n, q_{n+1}) \right] \left[(ig_s f^{c_n' d_n c_{n+1}'}) C^{\mu_n}(q - q_n, q - q_{n+1}) \right]$$

$$\times \frac{1}{\hat{t}_{n+1}} \exp\left[\alpha(\hat{t}_{n+1})(y_n - y_{n+1}) \right] \frac{1}{\hat{t}_{n+1}'} \exp\left[\alpha(\hat{t}_{n+1}')(y_n - y_{n+1}) \right]$$

$$\times (ig_s f^{bd_{n+1} c_{n+1}})(ig_s f^{c_{n+1}' d_{n+1} b'}), \quad (5.205)$$

where $t = q^2$ and the individual $t_i = (q - q_i)^2$.

The products of Lipatov vertices $C_\mu C^\mu$ have already been obtained in Eq. (5.180). Here they are generalized to

$$C^{\mu_1}(q_i, q_{i+1}) C_{\mu_1}(q - q_i, q - q_{i+1}) = -2 \left[q_\perp^2 - \frac{(q - q_i)_\perp^2 q_{i+1,\perp}^2 + (q - q_{i+1})_\perp^2 q_{i,\perp}^2}{(q_i - q_{i+1})_\perp^2} \right]$$

$$= -2\mathcal{K}(q_i, q_{i+1}). \quad (5.206)$$

Projecting on colours — octet vs. singlet exchange along the two-gluon ladder — yields colour factors C,

$$C = \begin{cases} N_c & \text{for singlet} \\ N_c/2 & \text{for octet} \end{cases} \quad (5.207)$$

such that in total

$$\mathfrak{Disc}[i\mathcal{M}^{aba'b'}_{\mu_a \mu_b \mu_{a'} \mu_{b'}}(\hat{s}, \hat{t})] = 2i\hat{s} \sum_{n=0}^{\infty} (-g_s^2 C)^{n+2} \int \left[\prod_{i=1}^{n} \frac{dy_i}{4\pi} \right] \int \left[\prod_{j=1}^{n+1} \frac{d^2 q_{j,\perp}}{(2\pi)^2} \right]$$

$$\times \left[\prod_{k=1}^{n+1} \frac{1}{\hat{t}_k \hat{t}'_k} e^{[y_{k-1} - y_k][\alpha(\hat{t}_k) + \alpha(\hat{t}'_k)]} \right] \left[\prod_{m=1}^{n} 2\mathcal{K}(q_m, q_{m+1}) \right] .$$

$$(5.208)$$

At this point the Laplace transform prepared in Eqs. (5.142) and (5.144) allows to disentangle the multiple integrals in a very convenient way. This is achieved by

1. assuming a strong ordering of the rapidities, $y_i \gg y_{i+1}$, thereby enforcing the kinematical situation of the high-energy limit, cf. Eq. (5.171),
2. integrating over rapidity differences $y_i - y_{i+1}$ instead of rapidities, and
3. using the overall rapidity difference $y_0 - y_{n+1}$ as a Laplace parameter.

Then, the Laplace transform reads

$$\mathcal{F}_l(\hat{t}) = -2i(4\pi\alpha_s C)^2 \hat{t} \sum_{n=0}^{\infty} \left[\int \prod_{i=1}^{n+1} \frac{d^2 q_{i,\perp}}{(2\pi)^2} \right]$$

$$\times \left[\prod_{j=1}^{n} \frac{1}{\hat{t}_j \hat{t}'_j} \frac{1}{l - 1 - \alpha(\hat{t}_j) - \alpha(\hat{t}'_j)} (-2\alpha_s C) \mathcal{K}(q_j, q_{j+1}) \right]$$

$$\times \left[\frac{1}{\hat{t}_{n+1} \hat{t}'_{n+1}} \frac{1}{l - 1 - \alpha(\hat{t}_{n+1}) - \alpha(\hat{t}'_{n+1})} \right] , \qquad (5.209)$$

which can be cast into

$$\mathcal{F}_l(\hat{t}) = -2i(4\pi\alpha_s C)^2 \hat{t} \int \frac{d^2 q_{1,\perp}}{(2\pi)^2} \frac{1}{q_{1,\perp}^2 (q - q_1)_\perp^2} f_l(q_1, t) , \qquad (5.210)$$

where the function f_l is a solution of the integral equation

$$f_l(q_1, t) = \frac{1}{l - 1 - \alpha(\hat{t}_1) - \alpha(\hat{t}'_1)} \left[1 - 2\alpha_s C \int \frac{d^2 q_{2,\perp}}{(2\pi)^2} \frac{\mathcal{K}(q_1, q_2)}{q_{2,\perp}^2 (q - q_2)_\perp^2} f_l(q_2, t) \right] .$$

$$(5.211)$$

This is the celebrated BFKL equation.

5.2.5.10   Octet solution of the BFKL equation

In a first step, the octet solution of the BFKL equation is considered. Looking at Eq. (5.195) and taking into account that for **symmetric** octet exchange, the amplitude is invariant under $\hat{s} \longleftrightarrow \hat{u}$ implies that for this case $\alpha(\hat{t}) = 0$. Therefore

$$(l - 1) f_l^{\text{oct}}(q_1, \hat{t}) = 1 - \alpha_s N_c q_\perp^2 \int \frac{d^2 k_\perp}{(2\pi)^2} \frac{1}{k_\perp^2 (q - k)_\perp^2} f_l^{\text{oct}}(k, \hat{t}) \qquad (5.212)$$

which yields

$$f_l^{\text{oct}}(q_1, \hat{t}) = f_l^{\text{oct}}(k, \hat{t}) = \frac{1}{l - 1 - \alpha(\hat{t})}, \tag{5.213}$$

where the term $\alpha(\hat{t})$ in the denominator stems from the integral. Pushing this through the Laplace transform, the amplitude for octet exchange is duly obtained,

$$\mathcal{M}_{gg \to gg}^{\text{oct}}(\hat{s}, \hat{t}) = 4\pi\alpha_{\text{s}}N_c \frac{\pi\alpha(\hat{t})}{\sin[\pi\alpha(\hat{t})]} \left(1 + e^{i\pi\alpha(\hat{t})}\right) \left(\frac{\hat{s}}{-\hat{t}}\right)^{1+\alpha(\hat{t})}. \tag{5.214}$$

The exponent stems from the pole at $l = 1 + \alpha(\hat{t})$ in the complex angular momentum plane, which essentially yields the spin of the exchanged object. In Section 7.1.2 objects exchanged in the t-channel will be dubbed "**reggeons**", and the exponent indicates their "equivalent" spin — essentially the intrinsic angular momentum related to their exchange which, of course, does not need to be half-integer or integer. Using the fact that $\alpha(\hat{t}) = 0$ for $\hat{t} = 0$ shows that the **reggeized gluon** indeed has spin 1. In the high-energy limit, where $\hat{t}$ is small compared to $\hat{s}$ the amplitude thus can be approximated as

$$\mathcal{M}_{gg \to gg}^{\text{oct}}(\hat{s}, \hat{t}) = -8\pi\alpha_{\text{s}}N_c \frac{\hat{s}}{\hat{t}} e^{\alpha(\hat{t})(y_a - y_b)} \tag{5.215}$$

as anticipated.

It is worth noting that although having started the consideration with the discontinuity at $\mathcal{O}\left(\alpha_{\text{s}}^2\right)$, the overall result for the elastic amplitude is $\mathcal{O}\left(\alpha_{\text{s}}^2\right)$. This reflects the fact that of course the colour-octet exchange amplitude starts with a single gluon in the t-channel, which is at this very same perturbative order.

5.2.5.11 Singlet solution of the BFKL equation — the perturbative pomeron

After some substitutions, the BFKL equation for the exchange of a colour singlet at $\hat{t} = 0$ is given by

$$(l - 1)\tilde{f}_l^{\text{sing}}(q_1, k) = \frac{1}{2}\delta^2(\vec{q}_{1,\perp} - \vec{k}_\perp)$$
$$+ 4\alpha_{\text{s}}N_c \int \frac{\text{d}^2 q_{2\perp}}{(2\pi)^2} \frac{1}{(q_1 - q_2)_\perp^2} \left[\tilde{f}_l^{\text{sing}}(q_2, k) - \frac{q_{1,\perp}^2}{q_{2,\perp}^2 + (q_1 - q_2)_\perp^2} \tilde{f}_l^{\text{sing}}(q_1, k)\right], \tag{5.216}$$

where $\tilde{f}_l^{\text{sing}}(q_1, k)$ is the differential form of the singlet function for $\hat{t} = 0$,

$$f_l^{\text{sing}}(q_1, \hat{t} = 0) = \int \frac{\text{d}^2 k_\perp}{(2\pi)^2} \tilde{f}_l^{\text{sing}}(q_1, k, \hat{t} = 0). \tag{5.217}$$

The homogeneous part of the equation can be solved after a Fourier expansion,

$$\tilde{f}_l^{\text{sing}}(q_1,\, k) = \sum_{n=-\infty}^{\infty} \int_{-\infty}^{\infty} \mathrm{d}\nu\, a(\nu,\, n) \exp\left[i\nu \left(\log\frac{q_1^2}{\mu^2} - \log\frac{k^2}{\mu^2} \right) + in\left(\phi_1 - \phi\right) \right],$$

(5.218)

where $(\phi_1 - \phi)$ is the angle between q_1 and k in the transverse plane and μ is an arbitrary scale. Expanding the δ function in a similar way results in

$$\delta^2(\vec{q}_{1,\perp} - \vec{k}_\perp)$$

$$= \frac{1}{k_\perp q_{1,\perp}} \frac{1}{(2\pi)^2} \sum_{n=-\infty}^{\infty} \int_{-\infty}^{\infty} \mathrm{d}\nu \exp\left[i\nu \left(\log\frac{q_1^2}{\mu^2} - \log\frac{k^2}{\mu^2} \right) + in\left(\phi_1 - \phi\right) \right],$$

(5.219)

i.e. the same expansion as for the homogeneous part, but with constant coefficients $a(\nu,\, n) = 1/(4\pi^2 k_\perp q_{1,\perp})$. Inserting this expansion into the equation for $\tilde{f}$ yields an equation for the coefficients, namely

$$(l-1)a(\nu,\, n) = \frac{1}{4\pi^2 k_\perp q_{1,\perp}} + \omega(\nu,\, n)a(\nu,\, n),$$

(5.220)

and therefore

$$a(\nu,\, n) = \frac{1}{4\pi^2 k_\perp q_{1,\perp}} \frac{1}{l - 1 - \omega(\nu,\, n)}.$$

(5.221)

After a bit of algebra, the eigenvalues $\omega(\nu,\, n)$ can be written as

$$\omega(\nu,\, n) = -\frac{2\alpha_s N_c}{\pi} \mathfrak{Re} \left[\psi\left(\frac{|n|+1}{2} + i\nu \right) - \psi(1) \right].$$

(5.222)

Here $\psi(x)$ is the derivative of the logarithm of the Γ-function,

$$\psi(x) = \frac{\mathrm{d}\log\Gamma(x)}{\mathrm{d}x}.$$

(5.223)

The exchange of such a colour-singlet object in the t-channel is often identified with the exchange of a pomeron, which in turn is the driver of the total cross-section, *cf.* Sections 7.1.2 and 7.1.3. However, while in the simplistic picture advocated in this later part of the book, the pomeron is assumed to be a simple pole, this is not the case in perturbative QCD, studied here. Here, it is important to stress that the eigenvalues are continuous, which implies that for the singlet solution the idea of a simple $\hat{t}$-dependent pole must be abandoned. The perturbative pomeron is a branch cut in the complex angular momentum plane.

However, in order to analyse the behaviour of this structure in more detail, consider the leading contribution, which is located at $\nu = n = 0$. Expanding ω for small ν and $n = 0$ yields

$$\omega(\nu,\, n=0) = \frac{2\alpha_s N_c}{\pi}\left(2\log 2 - 7\zeta(3)\nu^2 + \dots \right) \approx \frac{4N_c \log 2}{\pi}\alpha_s \approx 2.65\,\alpha_s.\quad (5.224)$$

The leading contribution to $\tilde{f}_l$ therefore reads

$$\tilde{f}_l(k_a, k_b) \approx \frac{1}{4\pi^2 k_{a,\perp} k_{b,\perp}} \frac{\pi}{\sqrt{B(l-1-A)}} \exp\left[-\sqrt{\frac{l-1-A}{B}} \log \frac{k_{a,\perp}^2}{k_{b,\perp}^2}\right] \qquad (5.225)$$

with

$$A = \frac{4N_c \log 2}{\pi} \alpha_s \quad \text{and} \quad B = \frac{14\zeta(3)N_c}{\pi} \alpha_s. \qquad (5.226)$$

5.2.5.12   Total gluon-gluon cross-section

Undoing all the steps until now, Laplace transform, relation between discontinuity of the amplitude and total cross-section, *etc.*, then results in the total gluon–gluon scattering cross-section in multi-Regge kinematics

$$\hat{\sigma}_{gg\to gg}^{(\text{tot})} = \frac{8N_c^2}{N_c^2-1} \alpha_s^2 \int \frac{d^2 k_{a,\perp}}{k_{a,\perp}^2} \frac{d^2 k_{b,\perp}}{k_{b,\perp}^2} f^{\text{sing}}(k_a, k_b, |y_a - y_b|) \qquad (5.227)$$

with

$$f^{\text{sing}}(k_a, k_b, y)$$
$$= \frac{1}{4\pi^2 k_{a,\perp} k_{b,\perp}} \sum_{n=-\infty}^{\infty} \int_{-\infty}^{\infty} d\nu \exp\left[\omega(\nu, n)y + i\nu \log \frac{k_{a,\perp}^2}{k_{b,\perp}^2} + in(\phi_a - \phi_b)\right], \qquad (5.228)$$

and where k_a and k_b denote the two outermost, *i.e.* most forward, gluons. The prefactor accounts for the averaging over incident colours and spins. This is schematically depicted in Fig. 5.9.

Integrating over the azimuthal angles means that only the $n = 0$ term contributes, and thus

$$\frac{d\hat{\sigma}_{gg\to gg}^{(\text{tot})}}{dk_{a,\perp}^2 dk_{b,\perp}^2} = \frac{N_c^2 \alpha_s^2}{4k_{a,\perp}^3 k_{b,\perp}^3} \int_{-\infty}^{\infty} d\nu \exp\left[\omega(\nu, n=0)|y_a - y_b| + i\nu \log \frac{k_{a,\perp}^2}{k_{b,\perp}^2}\right]$$
$$\approx \frac{N_c^2 \alpha_s^2 \pi}{4k_{a,\perp}^3 k_{b,\perp}^3} \frac{1}{\sqrt{\pi B|y_a - y_b|}} \exp\left[A|y_a - y_b| - \frac{1}{4B|y_a - y_b|} \log^2 \frac{k_{a,\perp}^2}{k_{b,\perp}^2}\right]. \qquad (5.229)$$

As before, the leading singularity in the complex l-plane is given by $l = 1 + A$, leading to the rise of the total partonic cross-section with $\hat{s}^A$, in violation of the Froissart bound. This rise is encoded in the first term of the exponential, after realising that the rapidity difference $|y_a - y_b|$ is proportional to $\log \hat{s}$. The critical exponent A is about $A \approx 2.5\alpha_s$ at leading order, which reduces by about a factor of two at higher orders [518]. Another feature to note is that the form here includes a Gaussian in $\log(k_{a,\perp}^2/k_{b,\perp}^2)$, with the peak at balanced transverse momenta, $k_{a,\perp} = k_{b,\perp}$ and a

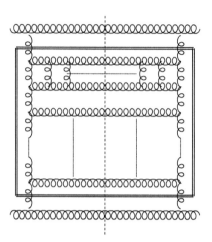

Fig. 5.9 Sketch of the scattering process $gg \to gg$ to all orders. The dou-
ble-lined box represents the exchange of a singlet gluon ladder, indicated
by a few gluons and some dotted lines hinting at further gluon, giving
rise to the function $f^{\mathrm{sing}}(k_a, k_b, y)$ in Eq. (5.227). The vertical dotted line
shows that this is actually the discontinuity of the amplitude.

width growing with the rapidity distance of the two. This is not too surprising since
the BFKL equation could also cast into the form of a diffusion equation with the
diffusion rate given by $\log(k_{a,\perp}^2/k_{b,\perp}^2)$.

Performing the integration over transverse momenta with a cutoff $k_{a,\perp}, k_{b,\perp} > p_\perp$
ultimately yields

$$
\hat{\sigma}_{gg \to gg}^{(k_\perp > p_\perp)} = \frac{N_c^2 \alpha_s^2(p_\perp^2)}{4p_\perp^2} \int_{-\infty}^{\infty} \mathrm{d}\nu \, \frac{e^{\omega(\nu,\, n=0)|y_a - y_b|}}{\nu^2 + \frac{1}{4}}
$$

$$
\approx \frac{N_c^2 \alpha_s^2(p_\perp^2)\pi}{2p_\perp^2} \frac{\exp\left(\dfrac{4N_c \alpha_s(p_\perp^2)|y_a - y_b|\log 2}{\pi}\right)}{\sqrt{\dfrac{7\zeta(3)N_c\alpha_s(p_\perp^2)|y_a - y_b|}{2}}}.
\tag{5.230}
$$

This exposes the exponential growth of the cross-section with the rapidity distance
of the two most forward partons, synonymous with a power-like growth in $\hat{s}$. Taking
$|y_a - y_b| = \log(\hat{s}/p_\perp^2)$ allows the determination of the resulting approximate K-factor
as a function of the parton centre-of-mass energy, and for different values of $p_\perp$. Simple
inspection shows that the K-factor increases steongly with the logarithm of $\hat{s}/p_\perp^2$, as
expected. The Mueller–Navelet jets [771] aim at probing exactly this regime by fixing
the x and vary the rapidities of the two most forward jets and by then measuring the
dijet cross-section. This implies that the hadron centre-of-mass energy scales with the
partonic one, *i.e.* with the exponential of $|y_a - y_b|$.

The azimuthal correlation of the two most forward jets is another interesting feature of the cross-section in the BFKL approximation. In the Born approximation for $gg \to gg$, the two gluons are back-to-back in transverse space with exactly the same transverse momentum, *i.e.* $k_{a,\perp} = k_{b,\perp}$ and $|\phi_a - \phi_b| = \pi$. This picture changes when higher-order corrections in the BFKL approximation are taken into account. Since there the leading term is given for $n = 0$, naively these two jets become entirely decorrelated. In addition, due to the diffusion property of the BFKL equation, also the correlation between their two transverse momenta fades away as their rapidity distance increases. This, the decorrelation of forward jets at large rapidity distances in transverse space, actually constitutes a possible test for the onset of BFKL dynamics.

5.3 Parton shower simulations

In this section, the simulation of QCD radiation in the production of strongly interacting particles at large scales through the parton shower picture will be introduced. Starting from an alternative, quite pictorial, introduction of the Sudakov form factor underlying this framework, details and issues of concrete implementations of this idea will be presented, and their connection to hard matrix elements at Born level will be discussed.

5.3.1 Underlying ideas

5.3.1.1 The Sudakov form factor: A simple example

In order to re-introduce the Sudakov form factor as the driving term of a probabilistic simulation of multiple parton radiation, a detour will be taken. Consider, as a toy model, the decay of a radioactive isotope with half-life τ. Ignoring trivial factors, the probability that starting with an intact nucleus at $t = 0$ the isotope is still intact at time t is given by

$$\mathcal{P}^{\text{nodec.}}(t, 0) = \exp\left[-\frac{t}{\tau}\right] = \exp\left[-\Gamma t\right]. \tag{5.231}$$

τ is related to the decay width of the isotope through $\Gamma = 1/\tau$. Of course, the probability that the isotope has decayed therefore is

$$\mathcal{P}^{\text{dec.}}(t, 0) = 1 - \mathcal{P}^{\text{nodec.}}(t, 0) = 1 - \exp\left[-\Gamma t\right]. \tag{5.232}$$

This is a very simple example of unitarity, written in the form of probability conservation: the isotope has either decayed or not.

Differentiating the decay probability thus yields the probability density that the isotope decays exactly at time t:

$$\frac{\mathrm{d}\mathcal{P}^{\text{dec.}}(t, 0)}{\mathrm{d}t} = -\frac{\mathrm{d}\mathcal{P}^{\text{nodec.}}(t, 0)}{\mathrm{d}t} = \Gamma \exp\left[-\Gamma t\right] = \Gamma \cdot \mathcal{P}^{\text{nodec.}}(t, 0). \tag{5.233}$$

To make things a bit more interesting, consider a case, where for some reason the decay width is time-dependent, $\Gamma \longrightarrow \Gamma(t)$. In this case, Eq. (5.231)) becomes

$$\mathcal{P}^{\text{nodec.}}(t,\, 0) = \exp\left[-\int\limits_{0}^{t} dt' \Gamma(t')\right], \tag{5.234}$$

and, consequently, the probability density of Eq. (5.233) is replaced with

$$\frac{d\mathcal{P}^{\text{dec.}}(t,\, 0)}{dt} = -\frac{d\mathcal{P}^{\text{nodec.}}(t,\, 0)}{dt} = \Gamma(t) \cdot \mathcal{P}^{\text{nodec.}}(t,\, 0). \tag{5.235}$$

The no-decay probability $\mathcal{P}^{\text{dec.}}(t,\, 0)$ of course is nothing but the Sudakov form factor $\Delta(t,\, t_c)$ encountered for the first time in the framework of resummation, Eq. (2.151) in Section 2.3.1, where multi-photon radiation has been discussed and the Sudakov form factor $\Sigma(Q_\perp^2)$ was also interpreted as a probability, just like here.

The question remains how these equations can be turned into a numerical simulation. The answer, on the face of it, is obvious: since the Sudakov form factor represents a probability for the isotope not to decay in some time interval, it should result in a number between 0 and 1. Due to its form as the exponential of a non-positive number, this is trivially fulfilled.

5.3.1.2   Connection with single parton splitting

The previous discussion can now be directly extended to the case of parton branching for partons of type $a \in \{q,\, g\}$, through suitably replacing the time-dependent decay width $\Gamma(t)$ of the isotope with the integrated splitting function, $\Gamma_a(T,\, t)$, previously introduced in Eq. (2.178), times the strong coupling and a propagator factor:

$$\Gamma(t) \longrightarrow \frac{\alpha_s}{\pi} \frac{\Gamma_a(T,\, t)}{t}, \tag{5.236}$$

such that the Sudakov form factor for parton splitting takes the form

$$\Delta_a(T,\, t) = \exp\left[-\int\limits_{t}^{T} \frac{dt'}{t'} \frac{\alpha_s}{\pi} \Gamma_a(T,\, t')\right]. \tag{5.237}$$

This has already been encountered in Section 2.3.2 and in Section 2.3.3. The notable difference with respect to the toy example of the radioactive decay is the fact that the time evolution from a start time $t_0 = 0$ to the decay time t_{dec} has been replaced with an evolution from a large scale T to a smaller scale t. It is no coincidence that this is identical to the form of the Sudakov form factor in Q_T resummation, as the integrated splitting functions have identical coefficients $A^{(1)}$ and $B^{(1)}$ at leading order.

5.3.1.3   Infrared cut-off, virtual contributions, and unitarity

Taking a closer look at the integrand of the Sudakov form factor, essentially given by $\Gamma(T,\, t')/t' \simeq \log(T/t')/t'$ indicates that the available scales for the parton splitting should be bounded from below in order to guarantee the convergence of the integral over the scale. In other words, introducing an **infrared cut-off** t_c, typically of the order of a few Λ_{QCD}, the additional constraint

$$t' \geq t_c > 0 \qquad (5.238)$$

must be satisfied. For any given large starting scale T, and for a scale-independent strong coupling, there is the probability

$$
\begin{aligned}
1 - \Delta_{q,g}(T, t_c) &= 1 - \exp\left[-\frac{C_{F,A}\alpha_s}{\pi}\left(\log^2\frac{T}{t_c} - \tilde{\gamma}_{q,g}^{(1)}\log\frac{T}{t_c}\right)\right] \\
&= \frac{C_{F,A}\alpha_s}{\pi}\left(\log^2\frac{T}{t_c} - \tilde{\gamma}_{q,g}^{(1)}\log\frac{T}{t_c}\right) + \mathcal{O}\left(\alpha_s^2\right)
\end{aligned}
\qquad (5.239)
$$

for a quark or gluon to emit a gluon above the resolution scale t_c, which increases with increasing T/t_c. Comparing this result with the Sudakov form factor of Q_T resummation, Eq. (2.166), the terms $A^{(1)} = C_{q,g}$ and $B^{(1)} = -\gamma_{q,g}^{(1)} = -A_{q,g}^{(1)}\tilde{\gamma}_{qg}^{(1)}$ are readily identified. The naive mismatch by a factor of two between the two formulations, parton showers and analytic resummation stems from the fact that in the former each emitter is treated individually while in the latter, and in particular in the discussion in Sections 2.3.2 and 5.2.1 the case of two incoming partons of the same type — quark or gluon — fusing into a colourless state has been considered. The corresponding combined Sudakov form factors in this case account for both of the incoming partons, explaining the relative factor of two. Note also that terms of higher order in α_s have been omitted here, although the term equivalent to $A^{(2)}$ is easily included as well.

The probability not to emit any resolvable gluon decreases with increasing ratio T/t_c,

$$
\begin{aligned}
\Delta_{q,g}(T, t_c) &= \exp\left[-\frac{C_{F,A}\alpha_s}{\pi}\left(\log^2\frac{T}{t_c} - \tilde{\gamma}_{q,g}^{(1)}\log\frac{T}{t_c}\right)\right] \\
&= 1 - \frac{C_{F,A}\alpha_s}{\pi}\left(\log^2\frac{T}{t_c} - \tilde{\gamma}_{q,g}^{(1)}\log\frac{T}{t_c}\right) + \mathcal{O}\left(\alpha_s^2\right)
\end{aligned}
\qquad (5.240)
$$

at $\mathcal{O}(\alpha_s)$. For large ratios T/t_c, the emission term $\sim \alpha_s$ can become larger than unity, which leads to this expression turning negative at $\mathcal{O}(\alpha_s)$. In such a case the higher-order terms encoded in the exponentiation will also increase and thereby guarantee that the exponent of the Sudakov form factor remains negative, rescuing the probabilistic interpretation.

Due to its probabilistic nature, the Sudakov form factor incorporates real, resolvable emissions as well as unresolvable ones, although only the former ones appear explicitly as the driving term. The unresolvable ones are taken into account only by the introduction of the cut-off and the interpretation of the Sudakov form factor. Diagrammatically speaking, such **unresolvable emissions** can be attributed to either the soft and/or collinear real gluon radiation at scales below the cut-off t_c, or to the virtual diagrams, see Fig. 5.10 for a pictorial representation. Since both exhibit infrared divergences which cancel each other, and because the expression in Eq. (5.240) is free of

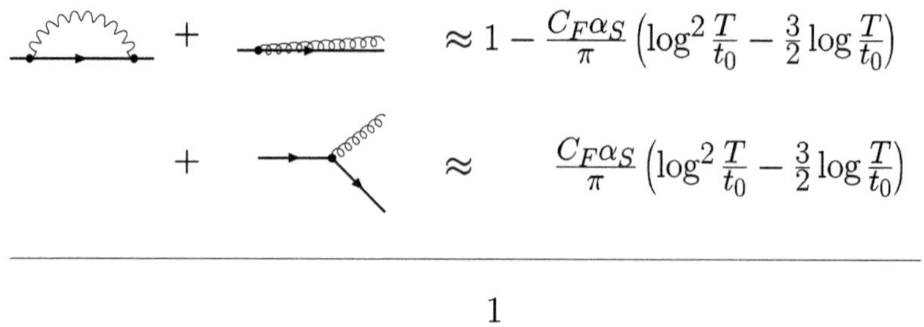

$$\approx 1 - \frac{C_F \alpha_S}{\pi}\left(\log^2\frac{T}{t_0} - \frac{3}{2}\log\frac{T}{t_0}\right)$$

$$\approx \frac{C_F \alpha_S}{\pi}\left(\log^2\frac{T}{t_0} - \frac{3}{2}\log\frac{T}{t_0}\right)$$

1

Fig. 5.10 Diagrammatic representation of the cancellation of divergences in unresolvable and virtual corrections and of large logarithms in their sum and the resolvable emissions. Here, the case of an emitting quark is shown.

pole terms $1/\epsilon$, it becomes apparent that the Sudakov form factor resums all large logarithms of the type $\log(T/t_c)$ at leading order, stemming from unresolvable and virtual diagrams, or, conversely, the same logarithms originating from the resolvable emissions. This is nothing but the celebrated Kinoshita–Lee–Nauenberg theorem [678, 724], formulated in a probabilistic way. In this fashion the unitarity-conserving character of the parton shower manifests itself: by employing a probabilistic picture virtual and unresolvable emissions are inherently accounted for.

Fixing the problems related to the divergent low-scale behaviour through the introduction of the infrared cut-off t_c, the actual scale of this cut-off needs to be discussed. There are two considerations driving this choice. For one, a well-defined description of radiation in a well-understood perturbative framework with only two parameters (α_S and t_c) is preferable over a description resting on phenomenological models with many parameters, which, at best, are understood only in qualitative terms. Such phenomenological models describing the emergence of hadrons at the end of the parton shower are discussed in more detail in the section introducing the ideas underlying hadronization models and their implementation in Section 7.3. This line of reasoning drives the infrared cut-off to minimally small scales. On the other hand, due to its confinement property, the perturbative description of strong interaction processes will eventually break down. This break-down of perturbation theory typically is related to scales of the order of Λ_{QCD}. Combining both considerations motivates choices for t_c of the order of about $1\,\mathrm{GeV}^2$.

On a similar note, also the hard scale T must be fixed. In previous sections discussing analytic resummation techniques, it became clear that this choice is quite important, as it also drives the size of the Sudakov logarithms. Bu there is no recipe based on first principles or a universal scale choice: it is process-dependent. While for simple topologies as in Drell–Yan-like processes the choice is fairly straightforward, something of the order of the invariant mass of the colour-singlet system, more intricate colour topologies will have to be described by more complicated choices. As a rule of thumb one could argue that the typical scale related to the distortion of the colour flow presents itself as a reasonable choice. It can be argued that indeed, inside the parton shower, such a scale choice is the most advantageous, as it resums certain classes of sub-leading logarithms.

5.3.1.4 Simulating a single parton branching

The determination of the full parton decay kinematics involves parameters beyond the scale related to the decay. First of all, a variable z is necessary, parameterizing the splitting of energy or, theoretically better motivated, light-cone momentum in the **branching process** $a \to bc$. For the simulation of parton splitting, the integrated splitting function $\Gamma_{a \to bc}(t, t_c)$ to first order in α_s must be replaced by the differential one $P_{a \to bc}^{(1)}(z) = P_{ba}^{(1)}(z)$, where the exact parton composition of the splitting process has been made explicit. In addition, an azimuth orientation ϕ must be included. In the single emission terms forming the Sudakov form factor of analytic resummation this parameter has been inconsequential and it appeared only as a constraint in the overall four-momentum conservation.

Taken together, the Sudakov form factor suitable for the simulation of $a \to bc$ becomes

$$\Delta_{a \to bc}(T, t) = \exp\left[-\int\limits_{t}^{T} \frac{\mathrm{d}t'}{t'} \int\limits_{z_-}^{z_+} \mathrm{d}z \int\limits_{0}^{2\pi} \frac{\mathrm{d}\phi}{2\pi} \frac{\alpha_s(p_\perp(t', z))}{\pi} P_{a \to bc}^{(1)}(z) \right] . \qquad (5.241)$$

Here, the limits of the z-integration, $z_\pm$, emerge from the need to reconstruct the kinematics of the splitting obeying exact four-momentum conservation. They therefore depend on the specifics of the scale choice, etc., which will be discussed in more detail below. It is worth stressing that the introduction of an infrared cut-off of the parton shower will guarantee that these limits cut out the divergences in the splitting function.

Another subtlety arises when actually constructing the decay kinematics parameterized by $\{t, z, \phi\}$. A massless particle on its mass-shell in general cannot decay into two other massless particles, due to four-momentum conservation. This translates into the necessity to involve other partons, which will "donate" some four-momentum, absorb the recoil of the decay and thus guarantee local four-momentum conservation. This role is typically filled by the colour partner of the decaying parton. The resulting reshuffling of momenta of course is very minor in the case of soft and collinear splittings and vanishes in the limit where the invariant mass of the produced two-body system approaches the invariant mass of the decaying parton. Therefore, the details of the momentum shuffling are beyond the intrinsic accuracy of the parton shower. For emissions away from the soft and collinear regions they start to become increasingly important. In order to characterize a parton shower, it is thus necessary not only to define the interpretation of the parameters t, z, and ϕ, but also to define precisely how four-momentum conservation is achieved. In the next section, Section 5.3.2, the actual construction of parton showers will be exemplified through some typical realizations.

5.3.1.5 Initial-state parton showering

A new problem manifests itself when simulating multiple emissions off initial state partons through the parton shower. This is due to the simple observation that, in marked difference to the radiation off final state particles, in **initial-state radiation** both the incoming hadrons at the low scale and the parton entering the hard interaction at the high scale are fixed. Of course, to circumvent this problem, a naive idea would be to just start the parton shower evolution from both hadrons and arrive at a hard scatter. A quick glance at various cross-sections relevant for phenomenology reveals that such a procedure would be prohibitively inefficient, since all interesting processes have cross-sections that are many orders of magnitude smaller than the inelastic hadronic cross-sections. Therefore, in order to arrive at any statistically significant sample of simulated events it is unavoidable that the hard interaction is fixed first and only subsequently dressed with multiple parton emissions and further stages of event simulation.

In final-state radiation, all external hadrons in the final state are equally permissible, while in the case of initial-state radiation the incoming hadron of course is fixed by the collider setup. Since the parton shower evolution preferably is described as an evolution from the large to the small scales, the **forward evolution** in the simulation of final-state radiation thus becomes a **backward evolution** for emissions off the initial state.

The theoretical motivation is as follows [851]. The parton distribution function entering the calculation of the hard process cross-section, *cf.* Eq. (2.52), already embodies an inclusive summation over all possible initial-state showers starting from a low hadronic scale and arriving at the hard factorization scale μ_F. The scaling behaviour of these PDFs is given by the DGLAP evolution equations, Eq. (2.31). Schematically,

$$
\frac{\mathrm{d} f_{b/h}(x, t)}{\mathrm{d} \log t} = \frac{\alpha_{\mathrm{s}}(t)}{2\pi} \sum_a \int \frac{\mathrm{d} x'}{x'} \, \mathcal{P} a \to bc \left(\frac{x}{x'} \right) f_{a/h}(x', t) \,.
\tag{5.242}
$$

This needs to be turned into an expression for the probability of parton b disappearing from x during a small decrease of scale, $\mathrm{d}t$. A simple way of achieving this is to divide the equation above by $f_{b/h}(x, t)$. To first order in α_{s} this leads to the following expression for the parton decay probability:

$$
\mathrm{d}\mathcal{P}_b = \frac{\mathrm{d} f_{b/h}(x, t)}{f_{b/h}(x, t)} = -\frac{\mathrm{d}t}{t} \frac{\alpha_{\mathrm{s}}(t)}{2\pi} \sum_a \int \frac{\mathrm{d} x'}{x'} \, \mathcal{P}_{ba}^{(1)} \left(\frac{x}{x'} \right) \frac{f_{a/h}(x', t)}{f_{b/h}(x, t)} \,.
\tag{5.243}
$$

Exponentiating this expression for the individual splitting, as before, yields a Sudakov form factor, this time for backward evolution. It encodes the probability for a process not to occur, where a parton b at momentum fraction x in the initial state is replaced by a parton a at $x' = x/z$ under emission of parton c.

$$
\Delta_{a \to bc}(T, t) = \exp\left\{ -\int_t^T \frac{\mathrm{d}t'}{t'} \int_{z_-}^{z_+} \frac{\mathrm{d}z}{z} \int_0^{2\pi} \frac{\mathrm{d}\phi}{2\pi} \frac{\alpha_{\mathrm{s}}(p_\perp(t', z))}{2\pi} P_{a \to bc}^{(1)}(z) \frac{f_{a/h}\left(\frac{x}{z}, t'\right)}{f_{b/h}\left(x, t'\right)} \right\} \,.
$$

$$
\tag{5.244}
$$

The only visible difference to the forward evolution is that here also ratios of PDFs enter. Their role is to assure that, starting from a hard scale, one actually arrives

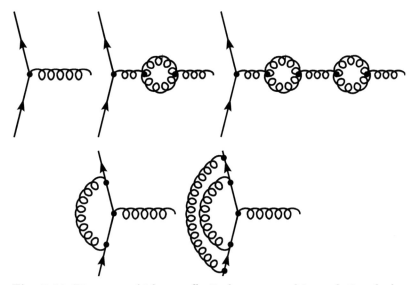

Fig. 5.11 Diagrams which are effectively resummed in analytic calculations as well as in parton showers by choosing the scale in the strong· coupling to be $\simeq k_\perp^2$. For simplicity, typically $\alpha_s(k_\perp^2)$ is assumed.

at the "right" initial hadron, and that on the way to this lower scale emissions are unfolded respecting the DGLAP evolution equation. A simple way to see that this is indeed the case is by realizing that for every splitting to take place, the PDF for parton b at its values of x_b and t_b is replaced by the PDF for the parton a at lower scale $t_a \leq t_b$ and larger $x_a \geq x_b$ as encoded in the ratio of the PDFs in Eq. (5.244). As a further welcome consequence of this backward evolution, momentum conservation is trivially guaranteed, *i.e.* no $x' > 1$ can be chosen — the PDFs just vanish there. This effectively constrains the lower limit of the z integral to be larger than x: $z_- \geq x$. In addition, the flavour symmetry of the PDFs, in principle treating all (massless) quarks on the same footing is broken. For instance, moving towards larger x' and smaller scales t, the flavour-symmetric sea components of the quark PDFs vanish and the flavour-unsymmetric valence contributions emerge.

5.3.1.6 Quantum corrections: Running of α_s

In Section 2.3.3 suitable evolution parameter definitions, related to either (scaled) angles or transverse momenta, were discussed to guarantee the resummation of leading and next-to-leading logarithms in the description of multiple emissions in the final state. In addition, the **running of α_s** demands a specification of the scale at which it is to be evaluated; as argued before in order to resum next-to-leading logarithms the correct scale is given by the transverse momentum of the splitting process. This includes a certain class of quantum corrections to the treatment of the individual emissions, see Fig. 5.11. To go even a step further, **universal higher-order corrections** may be added to **the universal soft gluon term**, which effectively amounts to adding a term $\simeq \alpha_s^2$ to the splitting kernel. Comparing with the $A^{(2)}$ terms in analytic resummation, *cf.* Eq. (5.65), motivates the replacement

$$\frac{\alpha_{\rm s}(k_\perp^2)}{2\pi} \longrightarrow \frac{\alpha_{\rm s}(k_\perp^2)}{2\pi} + \left[\frac{\alpha_{\rm s}(k_\perp^2)}{2\pi}\right]^2 \cdot \left[C_A\left(\frac{67}{18} - \frac{\pi^2}{6}\right) - \frac{5n_f}{9}\right] \tag{5.245}$$

also in the soft terms of the parton showers, see below

This choice of scale in the strong coupling poses an additional constraint on the infrared cut-off t_c. It necessitates the minimal transverse momentum in a parton split-ting, $k_\perp^{\rm min}$, to be sufficiently above the QCD scale $\Lambda_{\rm QCD}$ in order to avoid the Landau pole of $\alpha_{\rm s}$. Typically, the infrared cut-off of the parton shower is a parameter to be adjusted to data and in the region of

$$k_\perp^{\rm min} \approx 1\,{\rm GeV} > \Lambda_{\rm QCD}. \tag{5.246}$$

5.3.1.7   Quantum corrections: Angular ordering

In Section 5.1.1 the notion of **angular ordering** of subsequent emissions has been identified as an important feature of the QCD radiation pattern beyond the naive double-leading logarithmic approximation.

For parton shower simulations angular ordering implies that the exact choice of the evolution parameter t is essential for the formal accuracy in terms of leading and sub-leading logarithms. In the past, different choices for the form of the ordering parameter t have been implemented and formed the basis of parton showers successfully describing data. These choices include the following properties of the produced pair:

- invariant mass, $t = Q^2$, employed in the very first parton shower realizations [216, 533, 593, 766, 788],
- opening angle, $t = \theta^2$, taking into account quantum coherence effects [579, 750],
- and transverse momentum $t = p_\perp^2$, similarly including coherence effects [603, 729, 731, 778, 805, 856],

While all three choices exhibit logarithmic behaviour in the limit of small opening angles, the exact form of the logarithms differs for these three choices; in particular it was found that only the latter two systematically incorporate the effects of quantum coherence.

The effect of not taking into account such effects in hadronic collisions was high-lighted through an analysis performed by the CDF collaboration during Run I of the TEVATRON. The analysis studied the angular distribution of a third jet in QCD events [111], *cf.* Fig. 5.12, where its pseudo-rapidity distribution is depicted. The jets were defined by the midpoint algorithm with a radius of $R = 0.4$ and a minimal trans-verse momentum $E_\perp^{\rm (jet)} > 10\,{\rm GeV}$ in a pseudo-rapidity region given by $|\eta^{\rm (jet)}| \leq 2$. In addition, the first (hardest) jet was demanded to have at least a transverse momentum of $E_\perp^{\rm (jet_1)} > 110\,{\rm GeV}$. A number of observables are sensitive to angular ordering (QCD coherence) effects, the most intuitive ones being the η–distribution of the third jet and its spatial distance R in the η-ϕ plane from the second jet.

Both can directly be related to angular ordering, when considering the colour flow in typical jet events in hadron collisions, *cf.* Fig. 5.13. The underlying hard process can

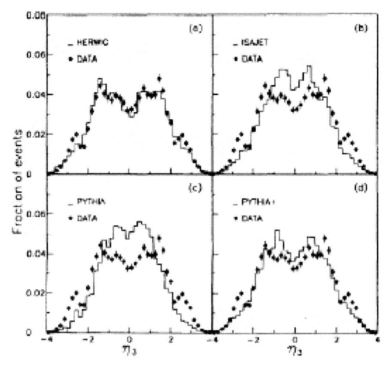

Fig. 5.12 The effect of colour coherence in QCD jet events at Run I of the TEVATRON, by the CDF collaboration [111].

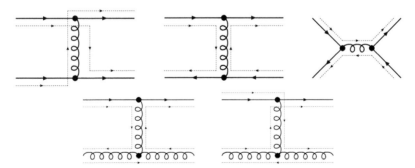

Fig. 5.13 Typical colour connections (dashed) in various parton–parton scattering processes: $qq \to qq$ and $q\bar{q} \to q\bar{q}$ scattering (top row), and $qg \to qg$ scattering (bottom row). In some cases, there are more than one possibilities in the large-N_c limit, but usually partons in the initial and final state are colour connected, thus giving rise to the angular ordering pattern discussed in the text.

be visualized as two incoming and two outgoing partons, where usually each outgoing parton is colour-connected to one of the incoming partons. This gives rise to a maximal angle $\theta_{\max}$ for the emission of the third parton, which thereby effectively is constrained inside cones of radius $\theta_{\max}$ around potential emitters — the incident or outgoing partons. Identifying partons at leading order with jets explains the features in jet production seen in the experiment. Ultimately, these findings lead to the first choice of evolution parameter, invariant mass, to be supplemented by an explicit angular veto in an improved version of the PYTHIA event generator [214, 215].

5.3.2   Realization: Parton shower algorithms

5.3.2.1   General thoughts

In this section, a number of different parton shower implementations will be briefly introduced. In general, the specific realizations of the more general ideas outlined in the previous sections will all have the same parametric accuracy, typically on the level of leading logarithms. There are a number of subtleties in each of the actual implementations, which may further enhance their accuracy or which may lead to quantitatively different behaviour.

In order to stress the general nature of the discussed parton showers, from now on the splitting kernels are denoted as $\mathcal{K}_{ij;k}(\Phi_1)$, where typically the one-particle emission phase space Φ_1 is written as a function of the ordering parameter t, the energy or light-cone momentum splitting parameter z and the azimuth angle ϕ. There are a number of requirements on the kernels and the kinematic parameters and scales which must be fulfilled in order to make contact with, e.g., the DGLAP evolution equation or analytic resummation. There are also a number of choices that ultimately define the actual implementation:

1. *The form of the splitting kernels* for the splitting of a parton (ij) into two partons i and j under the presence of a spectator k, $\mathcal{K}_{ij;k}(\Phi_1)$. These splitting kernels are subject to the requirement that they exhibit the same universal singularity structure as real-emission matrix elements at leading colour. In particular, they must reproduce the collinear and soft-collinear limits. This implies that in the collinear limit they are constrained to reduce to the usual DGLAP kernels $\mathcal{P}_{(ij)j}$. The soft limit in on the other hand is more tricky. This is because, in contrast to the collinear limit, it also receives interfering contributions at sub-leading colour which cannot be reproduced by using simple single-emitter algorithms. At leading colour, however, the soft limits must be correctly reproduced.

   Phrasing these considerations in a more formal way, the infrared limits are given by $p_i p_j \to 0$, where the soft limit in addition is characterized by $E_j \to 0$ or, equivalently, $z \to 1$. Then the kernels should satisfy

$$\mathcal{K}_{ij;k}(\Phi_1) \longrightarrow \begin{cases} \dfrac{1}{p_i p_j}\, \mathcal{P}_{(ij)i}\left(z(\Phi_1)\right) & \text{for } z \nrightarrow 1 \quad \text{(collinear)} \\[2ex] \dfrac{1}{p_i p_j} \cdot \dfrac{p_i p_k}{p_j p_k} & \text{for } z \to 1 \quad \text{(soft).} \end{cases} \tag{5.247}$$

It should be noted here that the form of the kernels in the soft limit given here may possibly not work for dipole showers, since it potentially leads to a double-counting of the soft region due to the symmetry of the eikonal. This problem can be remedied by replacing the eikonal with a similar form which will reproduce the eikonal when emissions of both parts of the dipole are added:

$$\mathcal{K}_{ij;k}(\Phi_1) \xrightarrow{z \to 1} \frac{1}{p_i p_j} \cdot \frac{p_i p_k}{(p_i + p_k) p_j} . \tag{5.248}$$

For more details, and a connection to, e.g., the Catani–Seymour subtraction method, *cf.* Section 3.3.3; for a discussion in the framework of constructing algorithms for parton showering, see for instance [807].

2. *The precise choice of evolution and splitting variable and the choice of scale in the strong coupling.* Together with the exact form of the splitting kernels this typically influences the formal accuracy of the parton shower. At this point it is worthwhile to stress, again, that in principle different evolution parameters will often lead to identical leading logarithmic (LL) accuracy, as can be seen from

$$\frac{\mathrm{d}k_{\perp ij}^2}{k_{\perp ij}^2} = \frac{\mathrm{d}m_{ij}^2}{m_{ij}^2} = \frac{\mathrm{d}\theta_{ij}^2}{\theta_{ij}^2} , \tag{5.249}$$

where $k_\perp$ is the relative transverse momentum of particle i and j after the splitting, m_{ij} is their invariant mass, and θ_{ij} their opening angle. This identity becomes apparent by realizing that in the limit of small emission angles, $k_\perp^2 = z_{ij}^2 (1 - z_{ij})^2 \theta_{ij}^2 E_{(ij)}^2$ and $m_{ij}^2 = k_\perp^2 / (z_{ij}(1 - z_{ij}))$, where $E_{(ij)}$ is the energy of parton (ij) before the splitting, and z_{ij} is the light-cone momentum fraction i retains.

Following this line of thought, differences in the formal accuracy of the parton shower manifest themselves in the sub-leading terms, *i.e.* terms of next-to-leading logarithmic (NLL) accuracy. The logic in this reasoning is related to the considerations in the framework of Q_T resummation, *cf.* Section 5.2.1. A formal treatment has been given in [356].

3. *The way the kinematics of individual splittings are constructed*, once its parameters have been fixed. This usually impacts the way in which the parton shower fills the emission phase space and therefore how it relates to matrix elements handling additional emissions. In general this can be quantified by a kinematical map, which can be characterized by four cases with splitters and spectators in the final (F) or initial state (I). For the sake of clarity, in the detailed discussion of the individual shower algorithms here and below, final-state particles will be denoted with i, j, and k, while initial state particles will be labelled with a or b.

$$\begin{aligned} \text{FF}:\ & \tilde{p}_{ij} + \tilde{p}_k \longrightarrow p_i + p_j + p_k \\ \text{IF}:\ & \tilde{p}_{aj} + \tilde{p}_k \longrightarrow p_a + p_j + p_k \\ \text{FI}:\ & \tilde{p}_{ij} + \tilde{p}_a \longrightarrow p_i + p_j + p_a \\ \text{II}:\ & \tilde{p}_{aj} + \tilde{p}_b \longrightarrow p_a + p_j + p_b \end{aligned} \tag{5.250}$$

The subscripts i, j, and k label the splitting parton, the emitted parton, and the spectator, all after the emission. The Sudakov form factor for a thus defined emission between the two scales t and T reads

$$
\Delta_{ij;k}^{(\mathcal{K})}(T,t) = \exp\left[-\int_t^T \frac{\mathrm{d}t'}{t'} \int \mathrm{d}z \int \frac{\mathrm{d}\phi}{2\pi} \, J(t', z, \phi) \, \mathcal{K}_{ij;k}(t', z, \phi)\right]
$$
$$
= \exp\left[-\int_t^T \mathrm{d}\Phi_1 \, \mathcal{K}_{ij;k}(\Phi_1)\right] .
$$

(5.251)

The quantity $J(t', z, \phi)$ in the first line of this equation takes care of any Jacobean that becomes necessary, but it is suppressed henceforth and assumed to be part of the one-particle phase-space integral in the following line.

For simplicity, parton luminosity factors as well as the specific coupling structure also including all charge factors are included in the splitting kernel. In all implementations of Eq. (5.251) the argument of the coupling is related to the transverse momentum of the splitting, given by the decay kinematics Φ_1.

5.3.2.2 Virtuality ordering

One of the first full-fledged implementations of a parton shower covering all aspects of initial- and final-state radiation was based on an ordering of the emissions by the virtual mass of the splitting parton: **virtuality ordering**, see also [216, 533, 593, 766, 788]

A parton shower with this ordering paradigm has been employed in early versions of the PYTHIA framework [852]. It is realized as a final-state shower [214, 215, 786], where the spectators are typically final-state particles as well, and as an initial-state shower [766, 851], with the other initial-state particle as spectator. The only exception is the first final-state splitting of a particle that has just been emitted from an initial-state parton, for which special arrangements are made. In general, the algorithm is built on the parton shower evolution being driven by $1 \to 2$ splittings, $(ij) \to i+j$. The spectator parton k typically is defined through the configuration of the hard process for the first emissions, or where it is related to the splitter (ij) by having a common splitter, i.e. $(ijk) \to (ij) + k$.

1. *Splitting kernels*
   In both cases, FF and II, the evolution kernels $\mathcal{K}_{ij;k}$ are given by the leading order splitting kernels of the DGLAP evolution equation for the fragmentation or parton distribution functions, $\mathcal{P}_{(ij)i}$, cf. Eq. (2.31).

2. *Evolution and splitting parameters*
   The evolution and splitting parameters t and z are given by

$$t^{(\mathrm{FF})} = t_{ij} = \tilde{p}_{ij}^2 = (p_i + p_j)^2 \quad \text{and} \quad z^{(\mathrm{FF})} = z_{ij} = \frac{E_i}{\tilde{E}_{(ij)}}$$

$$t^{(\mathrm{II})} = t_{aj} = |\tilde{p}_{aj}^2| = |(p_a - p_j)^2| \quad \text{and} \quad z^{(\mathrm{II})} = z_{aj} = \frac{x_a}{\tilde{x}_{aj}} = \frac{E_a}{\tilde{E}_{aj}}. \tag{5.252}$$

In both cases z is directly identified with an energy-splitting parameter. Different choices concerning the reference frame have been provided, with the default being the c.m.–frame of the hard scattering. It should be noted, though, that up to now the impact of kinematics has not yet been included. The limits of the energy splitting are given by

$$z_{\pm} = \frac{1}{2} \left(1 \pm \sqrt{1 - \frac{t_{ij}}{E_{(ij)}^2}} \right) \Theta(t_{ij} - m_{(ij)}^2), \tag{5.253}$$

where $m_{(ij)}$ is the on-shell mass of the splitting particle (ij), $E_{(ij)}$ its energy, fixed in the previous splitting, and t_{ij} is its virtual mass that has already been fixed by the Sudakov form factor. This choice assumes the offsprings i and j are massless — some momentum reshuffling will have to take place once they acquire a mass through their further splittings. This effectively will lead to a reinterpretation of the energy-splitting parameter of the previous parton branchings, as sketched in the following discussion (for further detail see the original literature).

The infrared cut-off scale is given by a parameter Q_0 such that it depends on the physical mass of the splitting parton, $\tilde{m}_{(ij)}$:

$$t \geq t_{\mathrm{c}}^{(ij)} = \sqrt{\tilde{m}_{(ij)}^2 + \frac{Q_0^2}{4}}. \tag{5.254}$$

Expressed by the parameters above, and assuming massless partons, the argument of the strong coupling is given as

$$\alpha_{\mathrm{s}}(k_\perp^2) \quad \text{with} \quad k_\perp^2 = \begin{cases} z_{ij}(1 - z_{ij})\, t_{ij} & (\mathrm{FF}) \\ (1 - z_{aj})\, t_{aj} & (\mathrm{II}) \end{cases} \tag{5.255}$$

while the scale argument in the PDFs in the backward evolution is given by the respective t of the splitting.

3. *Construction of kinematics*

   The kinematics of individual FF splittings are constructed in the following way. The recoil partner is selected to be either the other particle in the hard $2 \to 2$ scattering, if it is the first splitting in the process, or the other particle being generated in the splitting, or the particle emerging from them and being the colour partner. These possibilities are illustrated in Fig. 5.14. Having fixed t and z for a splitting $(ij) \to i + j$ with massless partons i and j, the decay kinematics are fixed by realizing that

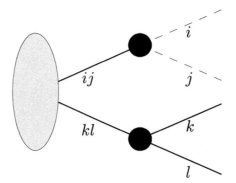

Fig. 5.14 Recoil partners in final-state parton showers. For the splitting $(kl) \rightarrow k + l$, with $t_{kl} > t_{ij}$, (ij) took the recoil, with k being colour connected with (ij). In the subsequent splitting $(ij) \rightarrow i + j$ parton k will be the recoil partner. Assuming k to be the next parton to split, it will be the turn of l to act as recoil partner in the construction of kinematics, and in turn one of the decay products of k will compensate for four-momentum imbalance in the splitting of parton l.

$$t = (p_i^{(0)} + p_j^{(0)})^2 = 2E_i^{(0)} E_j^{(0)} (1 - \cos\theta_{ij})$$
$$E = z_{ij}E + (1 - z_{ij})E = E_i^{(0)} + E_j^{(0)} \,. \tag{5.256}$$

With the offsprings becoming massive this is changed by reshuffling momenta to arrive at

$$p_{i,j} = (1 - r_{i,j})p_{i,j}^{(0)} + r_{j,i}p_{j,i}^{(0)} \,, \tag{5.257}$$

where

$$r_{i,j} = \frac{t_{ij} + t_{i,j} - t_{j,i} - \sqrt{(t_{ij} - t_i - t_j)^2 - 4t_i t_j}}{2t_{ij}} \,. \tag{5.258}$$

In emissions off initial-state particles, the energy-splitting parameter z_{ij} will lead to a rescaling of the Bjorken-x of the emitter, $x_i = x_{\text{new}} = x_{\text{old}}/z_{ij} = x_{ij}/z_{ij}$, where z is obtained from the argument of the Sudakov form factor, essentially the splitting kernel multiplied by the ratio of PDFs,

$$\int_{x_{ij}}^{1} \mathrm{d}z_{ij} \, \frac{\alpha_s(t_{ij})}{2\pi} \, \frac{x_i f_{i/h}(x_i, t_{ij})}{x_{ij} f_{ij/h}(x_{ij}, t_{ij})} \,. \tag{5.259}$$

It will always be the other initial-state particle that will account for the recoil, thereby boosting and rotating the full system. The algorithm is basically such that the Bjorken-x parameters of both initial-state particles, splitter and spectator, fix the centre-of-mass system.

To see how this works in more detail, consider a kinematical situation like the one depicted in Fig. 5.15, where two particles $\tilde{a}$ and $\tilde{b}$ collide to produce a final

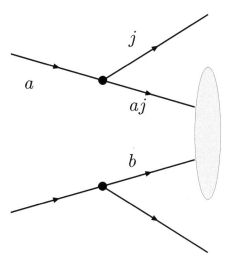

Fig. 5.15 Recoil partners in initial state parton showers.

state with total four-momentum squared $z\hat{s} = \hat{s}_{\tilde{a}\tilde{b}}$. Particle $\tilde{a}$ will eventually emit a parton j in the backward evolution, thus acquiring a negative virtual mass $p_{\tilde{a}}^2 = \tilde{p}_{aj}^2 \leq 0$, its absolute value readily identified as the evolution parameter, $|\tilde{p}_{aj}^2| = t_{aj}$. The momentum of the emitted particle, p_j, can be fixed by using local four-momentum conservation, *i.e.* $p_j = p_a - p_{aj}$. This leaves the task of constructing the four-momentum of the new initial-state parton, p_a and the emitter after emission, p_{aj}. The centre-of-mass squared of the emerging system, $\hat{s}_{ab}$, is given by rescaling the original $\hat{s}_{\tilde{a}\tilde{b}}$ by the splitting parameter,

$$\hat{s}_{a\tilde{b}} = (p_a + p_{\tilde{b}})^2 = \frac{\hat{s}_{\tilde{a}\tilde{b}}}{z_{aj}}, \tag{5.260}$$

which will provide one constraint on p_a. In the original implementation of the virtuality-ordered parton shower, it was assumed that both original incoming partons acquire a virtual mass with $t_{aj} > t_{bl} = Q_{(bl)}^2 \geq 0$. The logic of the ordering then implies that the kinematics of the backward-splitting $aj \to a + j$ is constructed first. This is achieved in the c.m.-system of the two momenta p_{aj} and p_{bl}, with energies and momenta $p^{\parallel}$ along the beam axis given by

$$
\begin{aligned}
E_{(aj),(bl)} &= \frac{\hat{s}_{\tilde{a}\tilde{b}} \mp \left(Q_{(aj)}^2 - Q_{(bl)}^2\right)}{2\sqrt{\hat{s}_{\tilde{a}\tilde{b}}}} \\[2ex]
p_{(aj),(bl)}^{\parallel} &= \pm \sqrt{\frac{\left(\hat{s}_{\tilde{a}\tilde{b}} + Q_{(aj)}^2 + Q_{(bl)}^2\right)^2 - 4Q_{(aj)}^2 Q_{(bl)}^2}{4\hat{s}_{\tilde{a}\tilde{b}}}}.
\end{aligned}
\tag{5.261}
$$

In this system p_a and p_j will have a transverse momentum with respect to the $\parallel$-axis; its azimuth angle is one of the four degrees of freedom of, say, p_a. It is chosen isotropic, and Eq. (5.260) fixes another d.o.f.. A third one is fixed by the mass of p_a, $Q^2_{(ak)} = |p^2_{ak}|$, given by the next showering step, where a parton k is being emitted. This leaves a fourth d.o.f., which is identified with the virtual mass of the outgoing parton j. The latter is fixed by the final-state parton shower, *i.e.* the further splitting of this parton, subject to kinematic constraints and, possibly, by considerations invoking quantum coherence. The kinematic constraint on the mass assumes a completely collinear branching, $\vec{p}_{(aj)} \parallel \vec{p}_a \parallel \vec{p}_j$, and reads

$$t_j = p_j^2 \leq \frac{q_{aj} q_{ak} - r_{aj} r_{ak}}{2 Q^2_{(bl)}} - Q^2_{(aj)} - Q^2_{(ak)}, \tag{5.262}$$

with

$$q_{aj} = \hat{s}_{\tilde{a}\tilde{b}} + Q^2_{(aj)} - Q^2_{(bl)}$$

$$q_{ak} = \hat{s}_{\tilde{a}\tilde{b}} + Q^2_{(ak)} - Q^2_{(bl)} = \frac{\hat{s}_{\tilde{a}\tilde{b}}}{z_{aj}} + Q^2_{(ak)} - Q^2_{(bl)}$$

$$r_{aj} = q^2_{aj} - 4 Q^2_{(aj)} Q^2_{(bl)}$$ $$\tag{5.263}$$

$$r_{ak} = q^2_{ak} - 4 Q^2_{(ak)} Q^2_{(bl)}.$$

Usually, in virtuality-ordered showers, there would be the additional constraint due to the ordering in virtual masses, demanding that $t_j \leq Q^2_{(aj)} = t_{aj}$, the value of the evolution parameter, where the parton was produced. After the splitting has been reconstructed in this way, the full system, including the full final state that has emerged so far, is boosted and rotated back on the beam axis. For more details on this and various other implementation issues, *cf.* [852].

This algorithm has been improved to approximate some of the effects of quantum coherence with an *a posteriori*-fix. This fix consists of a veto on increasing splitting angles, applied after each parton emission. For example, for final-state splittings, the opening angle of the splitting $(ij) \to i + j$ can be estimated as

$$\theta_{ij} \approx \frac{p_{\perp,i}}{E_i} + \frac{p_{\perp,j}}{E_j} \approx \frac{1}{\sqrt{z_{ij}(1 - z_{ij})}} \frac{\sqrt{t_{ij}}}{E_{(ij)}}. \tag{5.264}$$

Denoting the kinematic variants related to the splitting of parton i with subscripts $_i$ then leads to the angular ordering constraint

$$\theta_i < \theta_{ij} \quad \longrightarrow \quad \frac{z_i(1 - z_i)}{t_i} > \frac{1 - z_{ij}}{t_{ij}}, \tag{5.265}$$

cf. [214, 215].

It should be stressed, however, that implementations where the parton shower was organized through an ordering in invariant masses are not usually used any more, irrespective of whether it has been supplemented by an angular veto or not. One of the main reasons is that such showers inherently do not respect quantum coherence, with a number of implications. First of all, their emission pattern tends to generate a bit of an excess in soft radiation. Second, it is not quite clear how such showers can be improved beyond the leading logarithmic accuracy. Finally, and probably most relevant in view of the current use of parton showers, typical algorithms that match parton showers with *exact* higher-order calculations, rely on an ordering of emissions in "hardness", usually given by the transverse momentum of the splitting. In virtuality-ordered showers there is therefore a mismatch of the evolution, and quite frequently not the first but subsequent emissions off a given starting configuration are the hardest ones. This renders the matching of the hardest emission between parton shower and the fixed-order calculation a formidable task. Because of these reasons, it has become customary to use transverse-momentum or angular-ordered parton showers instead.

5.3.2.3 $p_\perp$-ordering

The original idea of ordering emissions according to the transverse momentum has been introduced in the framework of showering algorithms based on colour dipoles or colour antennae, [603, 729, 731, 805], see in the following sections. It has been adopted as a way to construct a parton shower only some time later, in [856], in the framework of the PYTHIA event generator. By now, this is the default parton showering algorithm in the latest versions of PYTHIA 6 and in its replacement, PYTHIA 8. Similar to the virtuality-ordered parton shower, again there is a distinction between final-state and initial-state parton showers.

1. *Splitting kernels*
   As in the virtuality-ordered parton showers, in both cases, FF and II, the evolution kernels $\mathcal{K}_{ij;k}$ are given by the leading-order splitting kernels of the DGLAP evolution equation, *cf.* Eq. (2.31). In addition, FI configurations are considered, again using the DGLAP splitting kernels for the evolution.

2. *Evolution and splitting parameters*
   For final-state parton showering, *i.e.* FF and FI splittings, the evolution variable is given by

$$p_{\perp,ij}^2 = z_{ij}(1 - z_{ij})\left(t_{ij} - m_{(ij)}^2\right), \tag{5.266}$$

where, as before, $m_{(ij)}$ is the physical on-shell mass of the splitting particle, and t_{ij} is its invariant mass which is now obtained after fixing the decay kinematics, *i.e.* $p_{\perp,ij}^2$ and z_{ij}. In terms of momenta after the splitting, z_{ij} is defined as

$$z_{ij} = \frac{1}{1 - k_1 - k_3}\left[\frac{x_1}{2 - x_2} - k_3\right] \tag{5.267}$$

where

$$k_1 = \frac{t_{ij} - \lambda(t_{ij}, m_i^2, m_j^2) + m_j^2 - m_i^2}{2t_{ij}} \tag{5.268}$$

$$k_3 = \frac{t_{ij} - \lambda(t_{ij}, m_i^2, m_j^2) - m_j^2 + m_i^2}{2t_{ij}} \tag{5.269}$$

$$x_{1/2} = \frac{2(p_i + p_j + p_k')p_{i/j}}{(p_i + p_j + p_k')^2} \tag{5.270}$$

$$z_{ij} = \frac{1}{1 - k_1 - k_3}\left[\frac{x_1}{2 - x_2} - k_3\right] \tag{5.271}$$

with

$$p_k' = p_k \tag{5.272}$$

for FF splittings and

$$p_k' = pk\left[1 - \frac{t_{ij} - m_{(ij)}^2}{(p_i + p_j + p_k)^2} - 2t_{ij} + 2m_{(ij)}^2\right] \tag{5.273}$$

$$\times \left[1 + \frac{t_{ij} - m_{(ij)}^2}{(p_i + p_j + p_k)^2} - 2t_{ij} + 2m_{(ij)}^2\right]^{-1} \tag{5.274}$$

for FI splittings.

For initial-state showering, the situation is a bit more complicated. Assuming a massless parton splitting into a space-like one with a virtual mass of $-Q^2$ and a time-like one with virtual mass m^2, the relative *physical* transverse momentum reads

$$p_\perp^2 = (1 - z)Q^2 - zQ^4/\hat{s}, \tag{5.275}$$

where $\hat{s}$ is the invariant mass of the system formed from the splitter and the spectator — the other parton in the initial state — before the backward splitting. The choice of this exact form of the transverse momentum would lead to an unwanted ambiguity, when mapping it onto the virtuality of the splitter *after* the emission took place. To overcome this, instead the evolution scale

$$t = (1 - z)Q^2 \equiv p_{\perp,\mathrm{evol}}^2 \tag{5.276}$$

is chosen. Close to the bottom and charm thresholds, the evolution variable is changed to

$$p_{\perp,\mathrm{evol}}^2 = (1 - z)(t_{ij} + m_{(ij)}^2). \tag{5.277}$$

In terms of momenta after the branching, z is given by the Q^2-ordered definition

$$z = \frac{2p_{(aj)}p_b}{2p_a p_b} \tag{5.278}$$

More details on the construction of this parton shower are given in [856].

3. *Construction of kinematics*

The kinematics in the implementation [856] of this parton shower algorithm in PYTHIA is constructed by mapping the quantities $p_\perp$ and z onto invariant masses and splitting parameters. The recoil of the splitting is taken by the whole final state for initial-state emissions, while the colour-connected particle acts as recoil partner for final-state splittings. This is analogous to dipole showers.

In particular, for FF splittings the branching is performed in the rest frame of splitter ij and spectator k, oriented along the positive and negative z axis. In the splitting process, the hitherto massless splitter ij receives a virtual mass t_{ij}, which induces a reduction of the energies and momenta of both splitter and spectator. For FI splittings, the construction is practically identical. For initial-state splittings, the kinematics is identical to the initial-initial kinematics described in the discussion of dipole showers below, and in [839] (up to a ϕ-rotation).

Finally, it should be noted that a veto on increasing angles (identical to the virtuality-ordered cascade) is usually applied to initial-state splittings.

5.3.2.4 Angular ordering

As discussed in Section 5.1.1, angular ordering is a quantum effect stemming from the interference of radiation from different emitters. This idea has first been successfully employed for the organization of a parton shower in [750, 886], forming the backbone of the widely used HERWIG event generator [415]. In various publications, including for example [356], it has been shown that angular-ordered parton showers are accurate up to next-to-leading logarithms, and that the first contributions that are missed are colour-suppressed by at least a factor of $1/N_c^2$.

In its original version, indeed, the opening angles of emissions, scaled by the energies of the emitting particles, have been employed as ordering parameters. In an improved version presented in [579], a modified angular variable is used instead, which allows the inclusion of mass effects in the parton shower in a more straightforward way. It thereby eliminates an artefact that is known as the "dead-cone" effect [479, 751].[7] Therefore the following discussion is based on the angular-ordered shower algorithm in its improved version, implemented in the HERWIG ++ event generator [188]. It should be noted that in this specific implementation, all outgoing partons have a minimal outgoing mass of the order of 1 GeV or below. This allows a very smooth interface with the subsequent cluster hadronization model in HERWIG ++. It also enables the parton shower to cover the transverse momentum range down to 0 GeV, while in other shower models the end of the parton shower is defined by a cut-off in transverse momentum.

1. *Splitting kernels*

    The splitting kernels for emissions off both initial- and final-state particles are given by DGLAP kernels involving masses, in particular

---

[7]This artefact is based on the observation that the masses of the emitting or emitted particles shield the collinear divergence in the emission pattern, due to simple four-momentum conservation. In the original angular-ordered showers the corresponding suppression of radiation was approximated by a hard cut on the emission angle in $q \to qg$ splitting, given by $\theta > m/E$, with m the quark mass and E its energy. This cut is too hard, and to obtain a better description, massive splitting kernels like the ones below have to be employed. They still exhibit a substantial depletion of radiation in the low-angle region, but the transition is smooth.

$$\hat{P}_{qq}(z, p_\perp^2) = C_F \left[\frac{1+z^2}{1-z} - \frac{2z(1-z)m^2}{p_\perp^2 + (1-z)^2 m^2} \right].$$

$$\hat{P}_{qg}(z, p_\perp^2) = T_R \left[1 - 2z(1-z)\frac{p_\perp^2}{p_\perp^2 + m^2} \right],$$

(5.279)

with the $g \to gg$ splitting of course still given by the form in Eq. (2.31).

2. *Evolution and splitting parameters*
The evolution parameter is given by a generalized flavour-dependent angle, optimized for mapping out the singular structures in the respective splitting process. For final-state emissions,

$$t = \frac{\vec{k}_\perp^2}{z^2(1-z)^2} - \frac{\mu_{ij}^2}{z(1-z)} + \frac{\mu_i^2}{z^2(1-z)} + \frac{\mu_j^2}{z(1-z)^2}$$

$$\to \begin{cases} \dfrac{\vec{k}_\perp^2}{z^2(1-z)^2} + \dfrac{\mu^2}{z^2} + \dfrac{Q_{g,\mathrm{min}}^2}{z(1-z)^2} & \text{for } q \to qg \\[2ex] \dfrac{\vec{k}_\perp^2 + \mu^2}{z^2(1-z)^2} & \text{for } g \to q\bar{q} \\[2ex] \dfrac{\vec{k}_\perp^2 + Q_{g,\mathrm{min}}^2}{z^2(1-z)^2} & \text{for } g \to gg, \end{cases}$$

(5.280)

while the argument in the running coupling is given by

$$\mu_R^2 = z^2(1-z)^2 t.$$

(5.281)

The actual angular ordering condition in terms of scales t (and $\bar{t}$) for subsequent emissions i and $i+1$ is given by

$$t_{i+1} < z_i^2 t_i \quad \text{and} \quad \bar{t}_{i+1} < (1-z_i)^2 t_i,$$

(5.282)

where the splitting factors enter because of the rescaling of the momenta by them. In all cases above, $Q_{g,\mathrm{min}}$ is the minimal virtual mass for gluons and light quarks at the end of the parton shower, and

$$\mu = \min\{m, Q_{g,\mathrm{min}}\}$$

(5.283)

with m the mass of the light or heavy quark. The splitting parameter z is the ratio of light-cone momentum fractions along the direction of the splitting parton before and after the splitting took place, and $\vec{k}_\perp$ is the transverse momentum in the splitting with respect to this axis. In its original version [579], the two light-cones were fixed by the two hardest partons, for instance the quark and the anti-quark directions in $e^- e^+ \to q\bar{q}$. In an improvement this was replaced by using the splitter and spectator axes of motion to fix the light-cone momenta.

In initial-state branchings, all partons are assumed to be massless (or light) and the evolution parameter is given by

$$t = \frac{\vec{k}_\perp^2 + z Q_{g,\min}^2}{(1-z)^2}$$

(5.284)

with the scale in the strong coupling given by

$$\mu_R^2 = (1-z)^2 t.$$

(5.285)

The splitting parameter z again is related to the splitting in light-cone momentum.

3. *Construction of kinematics*

The construction of the kinematics in angular-ordered showers is somewhat tricky to describe, compared to the other showering mechanisms. This is because the actual four-momenta of the particles are fixed only at the end of the parton shower evolution, and the kinematics of the intermediate partons is recovered recursively. In the final-state case, this recursion is based on writing the momentum of particle i as

$$q_i^\mu = \alpha_i p^\mu + \beta_i n^\nu + q_{\perp,i}^\mu,$$

(5.286)

where p and n are the two axes of the splitter and spectator directions, with $p^2 = m^2$, the mass of the particle, $n^2 = 0$, and $n \cdot q_{\perp,i} = p \cdot q_{\perp,i} = 0$. The splitting parameters z are given by the ratios of subsequent α's, $z_i = \alpha_i/\alpha_{i-1}$ and

$$\vec{p}_{\perp,i} = \vec{q}_{\perp,i} - z_i \vec{q}_{\perp,i-1}.$$

(5.287)

With this in mind, the virtuality of parton $i-1$ is given by its successors i through

$$q_{i-1}^2 = \frac{q_i^2}{z_i} + \frac{k_i^2}{1-z_i} + \frac{\vec{p_{\perp,i}}^2}{z_i(1-z_i)},$$

(5.288)

which is successively applied until the full final-state kinematics is recovered. Similar reasoning also allows the reconstruction of the kinematics of the initial state shower. The reader is referred to the original literature for a more detailed description.

One feature of angular-ordered parton showers is that they do not usually fill the full phase space available for emission. These gaps in the radiation pattern emerge in the hard, wide-angle regime and must be filled. In all practical implementations this is achieved by supplementing the first emission with **hard matrix-element corrections** filling these gaps. This "dead region" is shown in Fig. 5.18, Section 5.4.2, when the generic method of matrix-element corrections is discussed. In addition, it is worthwhile to stress that the parton shower fills the available phase space in such a way that typically the first emissions are the large-angle soft ones at relatively low transverse momentum, while harder emissions are usually appearing later in the process. This is in contrast to the other parton shower algorithms based on an ordering of the emissions by their transverse momentum.

5.3.2.5   Dipole showers

In a more recent development, parton showers have emerged that are based on the dipoles forming the Catani–Seymour subtraction kernels. This construction has for the first time been suggested in [778], and has been implemented in [466, 807, 839]. The underlying idea is to retain the notion of a well-defined splitting parton but to also explicitly identify the spectator parton in the splitting, whose kinematics in turn impacts on the splitting kernels.

1. *Splitting kernels*

   This **dipole shower** is further specified by employing (Catani–Seymour) dipole splitting kernels. For the case of massless partons, the splitting kernels for the FF case (final-state splitter and final-state spectator) read

   $$
   \begin{aligned}
   \mathcal{K}_{qg,k}^{(FF)} &= C_F \left[\frac{2}{1 - z_i(1 - y_{ij;k})} - (1 + z_i) \right] \\
   \mathcal{K}_{gg,k}^{(FF)} &= 2C_A \left[\frac{1}{1 - z_i(1 - y_{ij;k})} + \frac{1}{1 - (1 - z_i)(1 + y_{ij;k})} - 2 + z_i(1 - z_i) \right] \\
   \mathcal{K}_{q\bar{q},k}^{(FF)} &= T_R \left[1 - 2z_i(1 - z_i) \right] ,
   \end{aligned}
   $$

   $$(5.289)$$

   cf. also Eq. (3.168). They depend on a recoil parameter $y_{ij;k}$ and a splitting parameter z_i,

   $$
   \begin{aligned}
   y_{ij;k} &= \frac{p_i p_j}{p_i p_j + p_i p_k + p_j p_k} \\
   z_i &= \frac{p_i p_k}{p_i p_k + p_j p_k} = 1 - z_j .
   \end{aligned}
   $$

   $$(5.290)$$

   The FI, IF, and II expressions emerge from the FF case by replacing the final–state splitter momentum p_i or spectator momentum p_k with corresponding initial-state ones: $p_{i,k} \leftrightarrow -p_{a,b}$. In addition, for initial-state splitters, recoil and splitting parameter change their roles, *cf.* also Table 5.2. Splitting kernels for them can be found in Appendix C.2; a modification for the FI and IF kernels reproducing exact fixed-order matrix elements has been proposed in [330]. This modification essentially consists of adding some non-singular terms which vanish in the soft and collinear limits and therefore do not change the logarithmic accuracy of the shower.

   It should be mentioned here that the splitting kernels above come with a factor of $1/(p_i p_j)$, a suitably normalized strong coupling factor and a normalization taking into account the number of spectators, $1/N_{\mathrm{spec}}$. The choices of relevant kinematic quantities of course also depend on the specific case. However, in the original publications they are identical to the corresponding expressions used in Catani–Seymour subtraction. For a massless shower, they are listed in Appendix C.2.

2. *Evolution and splitting parameters*

   The formalism in principle allows different evolution parameters to be employed,

Table 5.2 Recoil and splitting parameters for various dipoles in a Catani–Seymour shower. The FI, IF, and II expressions emerge from the FF case by replacing the final-state splitter momentum p_i or spectator momentum p_k with corresponding initial-state ones: $p_{i,k} \leftrightarrow -p_{a,b}$. In addition, for initial-state splitters, recoil and splitting parameters change roles.

case	recoil parameter	splitting parameter
FF	$y_{ij;k} = \dfrac{p_i p_j}{p_i p_j + p_i p_k + p_j p_k}$	$z_i = \dfrac{p_i p_k}{(p_i + p_j)p_k} = 1 - z_j$
FI	$x_{ij,a} = \dfrac{p_i p_a + p_j p_a - p_i p_j}{(p_i + p_j)p_a}$	$z_i = \dfrac{p_i p_a}{(p_i + p_j)p_a} = 1 - z_j$
IF	$u_i = \dfrac{p_i p_a}{(p_i + p_k)p_a} = 1 - u_k$	$x_{ik,a} = \dfrac{p_i p_a + p_k p_a - p_i p_k}{(p_i + p_k)p_a}$
II	—	$x_{i,ab} = \dfrac{p_a p_b - p_i p_a - p_i p_b}{p_a p_b}$

including the invariant mass s_{ij} of the pair after the splitting or its transverse momentum. This latter choice was the default in the original publications, *i.e.* $t = k_\perp^2$, given by

$$t = k_\perp^2 = \begin{cases} 2p_i p_j z_i(1 - z_i) & \text{for final-state emissions} \\ 2p_a p_j(1 - x_a) & \text{for initial-state emissions} \end{cases} \tag{5.291}$$

for massless partons and more complicated expressions for massive ones. Irrespective of the evolution parameter t, the transverse momentum squared $k_\perp^2$ is used as the renormalization scale for the strong coupling and as the factorization scale, at which PDFs are evaluated in initial-state splitting processes. Various refinements to the definition of the transverse momentum variable with respect to the original proposal in [778] and its first implementations [466, 839] have been suggested, the most recent one in [626, 628]. There, a variation of the original evolution parameter $t = k_\perp^2$ in Eq. (5.291), making it flavour-specific, has been introduced, in order to better capture the singular behaviour of the splitting kernel. In particular, for massless final-state splittings $(\tilde{ij}) + \tilde{k} \to i + j + k$ the refined evolution parameter t reads

$$t = 2p_i p_j \cdot \begin{cases} z_i(1 - z_i) & \text{if } i, j = g \\ (1 - z_i) & \text{if } i \neq g \text{ and } j = g \\ z_i & \text{if } i = g \text{ and } j \neq g \\ z_i(1 - z_i) & \text{if } i, j \neq g, \end{cases} \tag{5.292}$$

while for initial-state splittings it is given by

$$t = 2p_a p_j \cdot \begin{cases} 1 - x_{aj,k} & \text{if } j = g \\ 1 & \text{if } j \neq g. \end{cases} \tag{5.293}$$

3. *Construction of kinematics*

In the original implementations, for the construction of the kinematics a recoil scheme has been used, which essentially is the inverse of the Catani–Seymour mappings. For the example of massless FF splittings, the kinematics is therefore constructed according to the inverse of the mapping in Eq. (3.160):

$$
\begin{aligned}
p_i &= & z_i \tilde{p}_{ij} + (1 - z_i)\, y_{ij;k}\, \tilde{p}_k + \vec{k}_\perp \\[4pt]
p_j &= (1 - z_i)\, \tilde{p}_{ij} + & z_i\, y_{ij;k}\, \tilde{p}_k - \vec{k}_\perp \qquad\qquad (5.294) \\[4pt]
p_k &= & (1 - y_{ij;k})\, \tilde{p}_k.
\end{aligned}
$$

The other cases are detailed in Appendix C.2. Of course, $k_\perp^2 = -\vec{k}_\perp^2$ with the transverse momentum $k_\perp$ sitting in the plane orthogonal to both $\tilde{p}_{ij}$ and $\tilde{p}_k$. There is, however, one caveat concerning the way kinematics is constructed. In the FF case, the Catani–Seymour simply had been inverted. For initial-state showering the situation is not quite as obvious. Merely employing the inverse of the Catani–Seymour kinematics, as in the FF case detailed above, would lead to a violation of the NLL accuracy. Applying the logic inherent to the Catani–Seymour language, consider the case where, say, an incident quark–anti-quark pair producing a gauge boson emits a gluon. In the first gluon emission, there is no other final-state particle, and therefore the gauge boson must compensate for the transverse momentum by recoiling against this gluon and thereby experiencing a $k_\perp$-"kick". This first gluon would decouple the quark–anti-quark pair in colour space. Thus, for the next emission off one of the incoming quarks, the gluon emitted first would act as the spectator in this IF splitting and naively take all the recoil and, in particular, compensate for the transverse momentum. This is to be contrasted with the master formula for Q_T resummation in the hadro-production of colour-singlets, *cf.* Eq. (2.166) and Eq. (5.59). There, the overall transverse momentum of the singlet results through the coherent sum of all gluon emissions in the Sudakov form factor, plus, eventually, some hard corrections. It is therefore mandatory to subject the complete final state — and not only the spectator — to recoils in transverse momentum in initial-state splittings with final-state spectators. This has very explicitly been worked out in [807], where in addition the issue of the potential double-counting of the soft region when using naive splitting kernels has been addressed. As a result, the authors explicitly showed that parton showers based on Catani–Seymour splitting kernels with transverse momentum as the evolution parameter and a good recoil strategy exhibit the correct logarithmic behaviour. As a consequence, by now, the kinematics of emissions off initial state particles with a final-state spectator is constructed in such a way that the full final state receives a transverse momentum "kick". Further details are given in [807].

5.3.2.6 Antenna showers

Finally, there are **antenna showers** (sometimes also called "dipole showers"), which in contrast to the dipole showers discussed above do not make any distinction between splitter and spectator parton but rather consider all secondary emissions as coherent radiation of an antenna consisting of two, typically colour-connected partons [602, 603]. This physical picture was first realized for final-state radiation in the ARIADNE program [729]. The absence of defined splitter and spectator partons translates into a notion of splitting kinematics as

$$\tilde{p}_i + \tilde{p}_k \rightarrow p_i + p_j + p_k \tag{5.295}$$

for final state splittings. This leads to light-cone fractions x_l (with $l \in \{i, j, k\}$) of the outgoing partons given by

$$x_l = \frac{2p_l Q}{Q^2}. \tag{5.296}$$

As before, $Q^\mu = \tilde{p}_i^\mu + \tilde{p}_k^\mu = p_i^\mu + p_j^\mu + p_k^\mu$ is the total momentum of the antenna — the splitter and spectator — and $Q^2 = s_{ijk}$. For massless partons the relevant phase space can be parameterized by a generic Lorentz-invariant quantity which reduces to transverse momentum in a suitable frame and a generalized rapidity of the emitted parton:

$$k_\perp^2 = \frac{s_{ij} s_{jk}}{s_{ijk}} \quad \text{and} \quad y = \frac{1}{2} \log \frac{s_{ij}}{s_{jk}}. \tag{5.297}$$

Note that with the evolution parameters $k_\perp$ and y as defined in Eq. (5.297), phase-space boundaries in $k_\perp$, $k_\perp > k_{\perp,c}$, translate into limits on y, ensuring finite integration volumes and, hence, a Sudakov form factor that can be evaluated in a fairly straightforward way.

1. *Splitting kernels*
   Writing unnormalized differential splitting probabilities in striking similarity with Eq. (2.2) as

$$d\mathcal{P}_{\tilde{i}\tilde{k}\rightarrow ijk} = \frac{dk_\perp^2}{k_\perp^2} \, dy \, \frac{d\phi}{2\pi} \, \mathcal{K}_{\tilde{i}\tilde{k}\rightarrow ijk} \tag{5.298}$$

   allows corresponding splitting kernels to be defined. In antenna showers this is typically achieved by deducing them from suitably chosen matrix elements through

$$d\mathcal{P}_{\tilde{i}\tilde{k}\rightarrow ijk} = dk_\perp^2 \, dy \, \frac{d\phi}{2\pi} \, \frac{|\mathcal{M}_{X\rightarrow ijk}|^2}{|\mathcal{M}_{X\rightarrow \tilde{i}\tilde{k}}|^2}, \tag{5.299}$$

   thereby extracting the $1/k_\perp^2$ singularity from the real-emission matrix element.[8]

---

[8]Ultimately, such an approach means that instead of exponentiating the singular terms only in the Sudakov form factor, full matrix elements are used and exponentiated in a way similar to what will be introduced as a matrix-element correction in Section 5.4.2. It is not hard to imagine that this is the source of the success ARIADNE enjoyed in describing QCD data from LEP.

In the case of FF splittings, and for gluon emission off a quark–anti-quark antenna, typically the matrix elements for $\gamma^* \to q\bar{q}(g)$ are used. Using that for any combination $\{l, m, n\}$ of the three massless final-state vectors

$$s_{lm} = 2p_l p_m = Q^2(1 - x_n) \tag{5.300}$$

this leads to

$$
\begin{aligned}
d\mathcal{P}_{\tilde{i}\tilde{k}\to ijk} &= dx_q d_{\bar{q}} \frac{d\phi}{2\pi} \frac{C_F \alpha_s}{2\pi} \frac{x_q^2 + x_{\bar{q}}^2}{(1 - x_q)(1 - x_{\bar{q}})} \\
&= \frac{ds_{\bar{q}g}}{Q^2} \frac{ds_{qg}}{Q^2} \frac{d\phi}{2\pi} \frac{C_F \alpha_s}{2\pi} \frac{(Q^2 - s_{\bar{q}g})^2 + (Q^2 - s_{qg})^2}{s_{\bar{q}g} s_{qg}} \\
&= \frac{dk_\perp^2}{k_\perp^2} dy \frac{d\phi}{2\pi} \frac{2C_F \alpha_s}{2\pi} \frac{(1 - x_\perp e^y)^2 + (1 - x_\perp e^{-y})^2}{2}.
\end{aligned} \tag{5.301}
$$

Here, in the last line,

$$s_{\bar{q}g,qg} = \sqrt{k_\perp^2 Q^2} e^{\pm y} = Q^2 x_\perp e^{\pm y} \tag{5.302}$$

has been employed. With similar considerations, and using slightly different processes, splitting kernels for other splittings such as $qg \to qgg$, $qg \to q\bar{q}'q'$ etc. have been obtained for the FF case, for instance in [805].

For splittings involving initial-state particles, two approaches have been pursued. First of all, in the original implementation of an antenna shower in ARIADNE, initial-state radiation has been re-interpreted as final-state radiation. This is achieved by replacing the initial-state partons of a given antenna by the corresponding hadron remnants in the final state, with the typically enhanced phase-space due to their larger momenta being compensated by tunable phase-space cuts. The origin of this idea can be traced back to [164], where emissions off an extended colour source, such as a hadron remnant in a collision induced by incoming hadrons, have been discussed.

An alternative approach, in parallel to the treatment in the other parton shower algorithms, has been worked out in [895] and, in the framework of the VINCIA code [577], in [826]. In this approach, emissions from the IF and II antenna are treated in a standard perturbative language with suitably defined splitting kernels obtained in a way very similar to the FF case.

2. *Evolution and splitting parameters*
   The customary evolution parameter is the transverse momentum given in the form of Eq. (5.297) or analogous for antennae with initial-state particles. Instead of an explicit energy splitting parameter, usually, the rapidity of the emitted parton in the c.m.- or Breit-frame of the emitting antennae is used, which of course can be translated into light-cone splitting parameters.

3. *Construction of kinematics*
   In the FF case, the construction of the splitting kinematics is best understood in the rest frame of the antenna. In this frame the parameters x_i, x_j, and x_k yield the

energy fractions of the outgoing partons with respect to the full antenna. Orienting the original partons along the z axis and assuming them to be massless,

$$\tilde{p}_{i,k} = \frac{Q}{2}. \tag{5.303}$$

The first way, usually implemented, for instance in ARIADNE, is to have one of the partons i and k keep its direction (given by $\tilde{i}$ or $\tilde{k}$) in the rest-frame of the antenna and just rescale its momentum by the corresponding x. With the absolute value of the transverse momentum fixed by $k_\perp$, only the azimuthal angle ϕ with respect to the original axis must be selected. The parton which keeps its direction typically is chosen in the following way. In the case of gluon emissions by a $q\bar{q}$ or a gg antenna, the less energetic of the two partons, *i.e.* the one with the smaller x takes the transverse momentum, while the more energetic one, the one with the larger x, just has its momentum compressed by its x. If the gluon is radiated off a qg or a $\bar{q}g$ antenna, it is always the quark that retains its direction. Finally, in the case of a gluon splitting into quarks it is always the other parton that keeps its direction. Alternatively, phase-space mappings used in antenna subtraction for NLO calculations [430, 561–563] could be employed. Emissions from IF and II antennae are treated in very different ways in different realizations of the antenna shower paradigm, although the same reasoning as for the dipole showers applies. Care must be taken to ensure that these splittings continue to transfer transverse momentum to the full final state in order to ensure the logarithmic accuracy in the description of, say, the transverse momentum of Drell–Yan pairs or similar.

5.3.2.7   Accuracy of shower algorithms

In general, it is a non-trivial task to work out the formal accuracy of parton showers, and for a variety of reasons.

First of all, it is important to realize that the formal accuracy of a parton shower is related to the logarithmic order that it controls. Formally speaking, it is synonymous with the question of which towers of $\alpha_s^n \log^{2n-m}$ are correctly described by the parton shower. But this, of course, is only a very simplistic way of looking at it: it is of course also relevant *which* logarithms, *i.e.*, *which* arguments, are thus treated. With the parton showers providing completely exclusive partonic final states it is obvious that not all logarithmic observables can be described at the same formal accuracy level. One could therefore easily imagine situations where one class of fairly inclusive logarithms, like, for example, logarithms in the transverse momentum of the lepton-pair in Drell–Yan production or logarithms in the thrust observable in electron–positron annihilations to hadrons are described at a fairly large accuracy (*i.e.* relatively large m). At the same time other, more exclusive observables, such as the correlation of the two leading jets in Drell–Yan processes or the thrust minor distribution in $e^-e^+ \rightarrow$ hadrons are not well described at all. When discussing the formal logarithmic accuracy of a given parton shower algorithm it is therefore mandatory to specify not only the level $N^m LL$ but also the arguments of the logarithms.

At this point, another comment is in order. It is true that using $k_\perp^2$ as an evolution parameter in splitting processes where a potentially soft gluon is emitted correctly captures NLL effects in the parton shower evolution. The issue of formal accuracy is not as clean-cut for the case of a gluon splitting into a quark–anti-quark pair. This is because, formally speaking, this kind of process first appears at NLL accuracy such that the freedom in the evolution parameter may yield a sub-leading effect related to that choice. The same reasoning also applies to the choice of renormalization scale. The effect of different choices on the invariant mass distribution of secondary quark pairs is quite large, and especially so for the gluon splitting into heavy quarks, where the differences of relative transverse momentum of the two quarks and their invariant mass can be quite large.

Furthermore, there are a number of other obstacles when trying to analyse the formal accuracy in the radiation pattern produced by a given parton shower. There is a subtle way recoil schemes — the way the splitting kinematics is constructed for each branching — impacts on observables and, possibly, on the logarithmic accuracy. The most obvious example has already been discussed. For dipole or antenna showers, systems made from an initial-state splitter with a final-state spectator naively may decouple their kinematics from the rest of the parton ensemble and, in particular, from the other final-state particles. As a consequence, in this case the transverse momentum of the individual splitting is no longer transferred to the other final-state particles, and, consequently, the resummation of logarithms in transverse momentum breaks down.

Correspondingly, it must be stressed that parton showers automatically implement detailed four-momentum conservation in each emission of an additional parton. In contrast, by far and large, this is not the case in analytic resummation. In most cases the effects of momentum conservation induce non-logarithmic corrections ("power–corrections") of the form $k_\perp/Q$ with Q a typical scale related to the splitter–spectator pair. It is not entirely clear if (and how) such effects produce additional logarithms when convoluted over many emissions.

5.3.3 Emissions off a Born-level configuration

5.3.3.1 Hardest/first emission

To understand the dynamics provided by the parton showers in more detail, and to develop the formalism further, consider the differential cross-section for the emission of the first — typically the hardest — parton off a core process, modelled at the Born level. Note that, due to its probabilistic nature, the parton shower does not change the Born-level cross-section for a state with N external particles given by Eq. (3.1). However, the process at hand and the specifics of its parton configuration, given by their flavours and momenta, will influence the parton shower by providing it with the scale μ_Q, the upper limit for further parton emissions. Thereby, this scale μ_Q also defines the hard scale for the logarithms that are resummed by the parton shower evolution.

The radiation pattern up to the first emission is given by the parton shower as

$$
\mathrm{d}\sigma_N^{(\mathrm{Born})} \;=\; \mathrm{d}\Phi_{\mathcal{B}}\,\mathcal{B}_N(\Phi_{\mathcal{B}}) \left\{ \Delta_N^{(\mathcal{K})}(\mu_Q^2, t_{\mathrm{c}}) + \int\limits_{t_{\mathrm{c}}}^{\mu_Q^2} \mathrm{d}\Phi_1 \left[\mathcal{K}_N(\Phi_1)\,\Delta_N^{(\mathcal{K})}(\mu_Q^2,\, t(\Phi_1)) \right] \right\}.
$$
$$(5.304)$$

Here, combined splitting kernels for emissions off an N-body state, $\mathcal{K}_N(\Phi_1)$, are introduced and read

$$
\mathcal{K}_N(\Phi_1) \;=\; \sum_{\{ij;k\}\in N} \mathcal{K}_{ij;k}(\Phi_1).
$$
$$(5.305)$$

In this equation, of course, the sum over $\{ij;k\}$ covers all viable combinations of emitting, emitted, and spectator partons. Their very nature as an exponential allows the introduction of compound Sudakov form factors,

$$
\Delta_N^{(\mathcal{K})}(T,\, t) \;=\; \prod_{\{ij;k\}\in N} \Delta_{ij;k}^{(\mathcal{K})}(T,\, t).
$$
$$(5.306)$$

Furthermore, $t(\Phi_1)$ is the parton shower scale associated with the emission phase space given by Φ_1.

The first term in the curly bracket above gives the no-emission probability down to the parton shower cut-off scale t_{c}, while the second term takes into account the first (hardest) emission at t with the no-emission probability at higher scales again encoded in the Sudakov form factor. The sum of the two terms integrates to unity reflecting the Born configuration to either emit a parton or not. This simple probabilistic reasoning typically is dubbed the "**unitarity**" of the parton shower. As a consequence, the cross-section thus simulated is identical with the Born level one, while the pattern of the first emission is determined by the parton shower.

Looking at the second term, the emission term, however, it is clear that further emissions at lower scales should also be included. In fact, they emerge by iterating the curly bracket in an appropriate fashion, leading to an expression of the form

$$
\mathrm{d}\sigma_N^{(\mathrm{Born})} = \mathrm{d}\Phi_{\mathcal{B}}\,\mathcal{B}_N(\Phi_{\mathcal{B}}) \left\{ \Delta_N^{(\mathcal{K})}(\mu_Q^2, t_{\mathrm{c}}) + \int\limits_{t_{\mathrm{c}}}^{\mu_Q^2} \mathrm{d}\Phi_1 \left[\mathcal{K}_N(\Phi_1)\,\Delta_N^{(\mathcal{K})}(\mu_Q^2, t(\Phi_1)) \right.\right.
$$
$$
\times \left\{ \Delta_{N+1}^{(\mathcal{K})}(t, t_{\mathrm{c}}) + \int\limits_{t_{\mathrm{c}}}^{t} \mathrm{d}\Phi_1' \left[\mathcal{K}_{N+1}(\Phi_1')\,\Delta_{N+1}^{(\mathcal{K})}(t(\Phi_1), t'(\Phi_1')) \right.\right.
$$
$$
\left.\left.\left.\left. \times \left\{ \Delta_{N+2}^{(\mathcal{K})}(t', t_{\mathrm{c}}) + \ldots \right.\right.\right.\right.\right\} \Big]\Big\}\Big]\Big\}\right\}.
$$
$$(5.307)$$

This is the explicit manifestation of the idea that in the soft and collinear limits multiple emissions off a given parton configuration can be constructed recursively, taking advantage of the factorization of matrix elements and the corresponding phase space,

$$\mathrm{d}\Phi_{N+1}\mathcal{B}_{N+1}(\Phi_{N+1}) = \mathrm{d}\Phi_N \mathcal{B}_N(\Phi_N)\, \mathcal{K}_N \,\mathrm{d}\Phi_1\, \Theta\left(\mu_Q^2(\Phi_N) - t(\Phi_1)\right), \qquad (5.308)$$

which of course leads to

$$\mathrm{d}\Phi_{N+m}\mathcal{B}_{N+m}(\Phi_{N+m}) = \mathrm{d}\Phi_N \mathcal{B}_N(\Phi_N) \prod_{i=1}^{m} \left[\mathcal{K}_{N+i-1}\,\mathrm{d}\Phi_1^{(i)}\, \Theta(t^{(i-1)} - t^{(i)})\right]. \quad (5.309)$$

Here, $t_0 = \mu_Q^2$ is defined by the N-particle Born configuration. This equation in fact represents the expression for the leading order cross-section for $(n+m)$ particles extracted from Eq. (5.307), *i.e.*, where the higher-order contributions encoded in the Sudakov form factors of Eq. (5.307) have been ignored.

To allow a more compact notation, a parton shower all-emission operator $\mathcal{E}_n^{(\mathcal{K})}(\mu_Q^2, t_c)$ is introduced defined recursively through

$$\mathcal{E}_n^{(\mathcal{K})}(\mu_Q^2, t_c) = \Delta_n^{(\mathcal{K})}(\mu_Q^2, t_c) + \int\limits_{t_c}^{\mu_Q^2} \mathrm{d}\Phi_1 \left[\mathcal{K}_n(\Phi_1)\, \Delta_n^{(\mathcal{K})}(\mu_Q^2, t(\Phi_1)) \otimes \mathcal{E}_{n+1}^{(\mathcal{K})}(t(\Phi_1), t_c)\right],$$

$$(5.310)$$

which, due to its recursive nature, also takes care of all further emissions. Applied to Eq. (5.307), this yields the compact expression

$$\mathrm{d}\sigma_N^{(\mathrm{Born})} = \mathrm{d}\Phi_{\mathcal{B}}\, \mathcal{B}_N(\Phi_{\mathcal{B}})\, \mathcal{E}_N^{(\mathcal{K})}(\mu_Q^2, t_c) \qquad (5.311)$$

for the Born-level cross-section and parton configuration, dressed with all possible emissions.

5.4   Matching parton showers and fixed-order calculations

In this section, techniques will be reviewed, which allow the combination of fixed-order results for exactly calculated matrix elements with parton showering algorithms.

5.4.1   Motivation

Results from fixed-order matrix elements and resummation incorporated in the parton shower provide good descriptions of essential event characteristics in complementary regions of phase space. The same also holds true for analytic resummation techniques, like the Q_T resummation scheme discussed in Section 5.2. There this complementarity results in supplementing the resummation part in $\tilde{W}_{ij}$, Eq. (5.60), with a hard remainder part Y_{ij}. It encodes the difference between the logarithmically enhanced contributions from the Sudakov form factor and the full fixed-order extra emission part of the real correction. In addition, virtual corrections can be encoded by adding the loop correction; in Q_T resummation this is the term H_{ab}, which quite often is absorbed as a part of $\tilde{W}_{ij}$. It thus allows the systematic correction of the approximate resummation result, order by order, to the full result. In particular, the total cross-section of the process can be recovered, order-by-order, and the radiation pattern of hard emissions approaches the fixed-order result.

This pattern can also be found in parton shower simulations, where the effect of corrections to the fixed-order results becomes most prominent in observables sensitive to additional emissions, real or virtual. For example, due to its probabilistic nature, the parton shower fails to account for any higher-order effects impacting on the total cross-section of the simulated process. This could be cured trivially by multiplying the cross-section fed into the parton shower with a suitable global K-factor. Such a treatment would provide a fairly satisfying solution to the problem, provided the patterns of additional particle radiation of the underlying fixed-order calculation and the parton shower were not visibly different. This, however, is not necessarily the case, and typically parton showers and fixed-order calculations differ substantially in large parts of the emission phase space. In fact, fixed-order calculations are primed to correctly describe the emission of additional, highly energetic particles at large angles and typically fail to describe the softer and more collinear emissions due to the occurrence of associated large logarithms which eventually may overcome the smallness of the perturbative parameter, the coupling constant. In contrast, the parton shower, constituting an expansion around the soft and collinear limits of particle radiation excels at describing such emissions, while it typically is incapable of taking into account the more complicated pattern of hard emissions. Stated in a slightly more extreme way: in order to capture the full effect of the quantum nature of particle emission, quantum field theoretical methods have to be applied — in practical terms this typically means that full matrix elements must be calculated. They can then be used to correct the classical parton shower picture in a systematic way, similar to the way this is achieved in analytic resummation.

Broadly speaking, methods to combine parton showers and fixed order matrix elements aim at combining the best features of both: of exact order-by-order calculations, which capture all quantum interferences and, possibly, higher-order effects due to virtual corrections, and of the parton shower, which, depending on its formulation, provides a simulation of further soft and collinear emissions to leading or next-to-leading order accuracy. The aim of such a combination exercise always is to maintain the fixed-order accuracy for the overall cross-section. At the same time, the fixed-order accuracy of the hardest emissions should be guaranteed, but supplemented with those leading logarithms that are captured by the parton shower's Sudakov form factor. In addition, all further, softer or more collinear emissions must still be described at the intrinsic accuracy of the parton shower.

The persistent problem in any combination procedure, however, is that both the matrix elements and the parton shower may allow for the emission of additional partons off a core process, which would lead to an unwanted double-counting if not properly taken into account.

In order to appreciate fully the difference of fixed-order calculations and parton showers and the problem underlying any combination of the two, consider the diagram in Fig. 5.16, where the orders of α_s and the accompanying logarithms L are depicted for the case of resolvable parton emissions in $e^- e^+ \to q\bar{q} + X$. One could think of

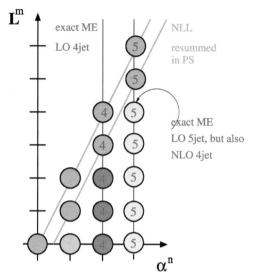

Fig. 5.16 Orders in α_s and large logarithms for emissions in $e^- e^+$ annihilations to hadrons.

such resolvable partons as jets, defined by a suitable algorithm. Obviously, for every additional jet emission, the order of α_s is incremented by one, and up to two new logarithms could emerge. This is the pattern of orders in α_s and L at leading order; inclusive higher-order corrections typically only add an order in the coupling without necessarily introducing new logarithms. Therefore, in this diagram the fixed-order matrix elements, being of a fixed order in α_s, live on one (leading order only) or more (higher-order corrections) vertical lines; in contrast, the parton shower, resumming terms of the form $\alpha_s L^{2n}$ and possibly also terms $\alpha_s L^{2n-1}$, occupies the diagonals. These lines, vertical and diagonal, cross. This indicates a double-counting, which could either be positive, as an over-counting, when contributions from both are wrongly added, or negative, when contributions are missed in both.

In this and the following sections, a variety of existing methods to include higher fixed-order terms into the parton shower will be presented, starting with a discussion of **matrix-element corrections (MEC)** [215, 416, 417, 786, 843] in this section. This method effectively allows the inclusion of the full $\mathcal{O}(\alpha_s)$ kinematics into the parton shower, but without including the effect of the $\mathcal{O}(\alpha_s)$ on the total rate. This will be achieved in the next section by introducing existing NLO matrix-element parton-shower **matching algorithms (NLoPs)** [543, 546, 630, 782]. An alternative approach, aiming at combining multiple fixed-order calculations into one inclusive sample is by now known as **multijet merging methods at leading order (MePs)** [351, 695, 730, 744] and **at next-to-leading order (MePs@Nlo)** [535, 558, 631, 737]. Finally, an outlook is given to recently devised methods to even include NNLO matrix elements for simple processes into the parton shower simulation (**NNLoPs**) [610, 633, 634, 652]. In most cases technical details will be ignored in favour of clarity of the presentation. Readers interested in more technical and implementation details and proofs of the respective accuracies are referred to the vast literature on this subject.

5.4.2 Matrix-element corrections

5.4.2.1 General idea

The first approach to improve the radiation pattern of the parton shower in the region of hard emissions concentrated on the first (hardest) emission [215, 416, 417, 786, 843], correcting it by the exact matrix element. This technique uses the fact that in many processes the differential emission cross-section for one additional parton as given by the parton shower exceeds its exact counterpart, given by the matrix element, in the full emission phase space. Examples for such a behaviour, where the technique sketched below has traditionally been implemented, include the processes $e^-e^+ \to q\bar{q}g$, $t \to bWg$ or the emission of an additional parton in the production of vector bosons at hadron colliders. In other words,

$$\mathcal{R}_N(\Phi_\mathcal{B} \times \Phi_1) \leq \mathcal{B}_N(\Phi_\mathcal{B}) \otimes \mathcal{K}_N(\Phi_1), \tag{5.312}$$

where $\mathcal{R}_N$ denotes the real correction to the process with N external particles, *i.e.* a matrix element for $(N+1)$ external particles at Born level. The trick of the method is to assume a modified splitting kernel $\tilde{\mathcal{K}}_N$, given by

$$\tilde{\mathcal{K}}_N(\Phi_1) = \mathcal{R}_N(\Phi_\mathcal{B} \times \Phi_1)/\mathcal{B}_N(\Phi_\mathcal{B}) \tag{5.313}$$

and to use this kernel for the description of the first emission in the parton shower. It should be stressed, though, that rather than evaluating the real-emission term $\mathcal{R}_N$ at a fixed scale, it is customary to use the same kinematics-dependent scale as in the parton shower.

This implies that the equation for the Born cross-section including the first emission through the parton shower of Eq. (5.304) in the section above becomes

$$d\sigma_N^{(\text{Born})} = d\Phi_\mathcal{B}\,\mathcal{B}_N(\Phi_\mathcal{B}) \left\{ \Delta_N^{(\tilde{\mathcal{K}})}(\mu_Q^2, t_c) + \int_{t_c}^{\mu_Q^2} d\Phi_1 \left[\tilde{\mathcal{K}}(\Phi_1)\, \Delta_N^{(\tilde{\mathcal{K}})}(\mu_Q^2, t(\Phi_1)) \right] \right\}$$

$$= d\Phi_\mathcal{B}\,\mathcal{B}_N(\Phi_\mathcal{B}) \left\{ \Delta_N^{(\mathcal{R}/\mathcal{B})}(\mu_Q^2, t_c) + \int_{t_c}^{\mu_Q^2} d\Phi_1 \left[\frac{\mathcal{R}_N(\Phi_\mathcal{B} \times \Phi_1)}{\mathcal{B}_N(\Phi_\mathcal{B})}\, \Delta_N^{(\mathcal{R}/\mathcal{B})}(\mu_Q^2, t(\Phi_1)) \right] \right\}. \tag{5.314}$$

Again, the terms in the curly brackets integrate to unity, indicating that the simulated cross-section is identical to the Born level one. This time, however, the radiation pattern is determined by the full real radiation matrix element as encoded in $\mathcal{R}_N$ rather than by the parton shower. This can trivially be seen by expanding the emission term in Eq. (5.314) up to first order in the coupling, keeping in mind that $\mathcal{R}_N$ has one more power in α_s than the corresponding Born term $\mathcal{B}_N$, and that effects of running

α_s are of higher order as well. All further emissions are still driven by the original parton shower kernels.

In practical implementations, usually the original splitting kernels and parton shower evolution parameters are used, and the parton shower is corrected to the modified splitting kernel through simple reweighting. Algorithmically, this means that first test emissions are generated by the original parton shower and accepted with a probability given by

$$\mathcal{P}_{\mathrm{MEC}} = \frac{\tilde{\mathcal{K}}_N(\Phi_1)}{\mathcal{K}_N(\Phi_1)} = \frac{\mathcal{R}_N(\Phi_{\mathcal{B}} \times \Phi_1)}{\mathcal{B}_N(\Phi_{\mathcal{B}}) \times \mathcal{K}_N(\Phi_1)}. \tag{5.315}$$

5.4.2.2 Example: First emission in $e^- e^+ \to q\bar{q}$

Using $e^- e^+ \to q\bar{q}g$ to further illustrate the idea, the first step is to compare the differential cross-sections for this process in the matrix element and parton shower approach. The matrix element given by the Feynman diagrams of Fig. 5.17 yields [498]

$$\mathrm{d}\Phi_g \, \mathcal{R}_{q\bar{q}}(\Phi_{q\bar{q}} \times \Phi_g) = \mathcal{B}_{q\bar{q}}(\Phi_{q\bar{q}}) \times \frac{C_F \alpha_s}{2\pi} \frac{x_1^2 + x_3^2}{(1 - x_1)(1 - x_3)} \mathrm{d}x_1 \mathrm{d}x_3 \tag{5.316}$$

where

$$x_{1,3} = 2E_{q,\bar{q}}/E_{\mathrm{c.m.}} \in [0, 1] \tag{5.317}$$

are the energy fractions the massless quark and anti-quark carry after gluon emission. The gluon emission phase space, after performing the azimuthal integration, is given by $\mathrm{d}\Phi_g \propto \mathrm{d}x_1 \, \mathrm{d}x_3$.

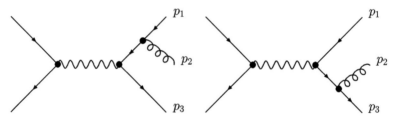

Fig. 5.17 Feynman diagrams for the emission of an additional gluon in quark–anti-quark pair production in lepton annihilations.

The parton shower expression has to be obtained from the details of the map relating its variables to the splitting kinematics. For the case of a virtuality-ordered parton shower with z defining the energy components in the splitting, the virtual mass and energy-splitting variables are given by

$$\begin{aligned} t_q &= m_{qg}^2 = (p_1 + p_2)^2 = E_{\mathrm{c.m.}}^2 (1 - x_3) \\ z_q &= \frac{E_1}{E_1 + E_2} = \frac{x_1}{2 - x_3}. \end{aligned} \tag{5.318}$$

Then the $\mathcal{O}(\alpha_s)$ expression for the gluon emission by the parton shower becomes [215]

$$d\Phi_g \left[\mathcal{B}_{q\bar{q}}(\Phi_{q\bar{q}}) \times \mathcal{K}(\Phi_g) \right] = \mathcal{B}_{q\bar{q}}(\Phi_{q\bar{q}}) \times \sum_{i \in \{q,\bar{q}\}} \frac{dt_i}{t_i} dz_i \frac{C_F \alpha_s}{2\pi} \frac{1 + z_i^2}{1 - z_i}$$

$$= \mathcal{B}_{q\bar{q}}(\Phi_{q\bar{q}}) \times \frac{C_F \alpha_s}{2\pi} \frac{dx_1 dx_3}{(1 - x_1)(1 - x_3)}$$

$$\times \left\{ \frac{1 - x_1}{x_2} \left[1 + \left(\frac{x_1}{2 - x_3} \right)^2 \right] + \frac{1 - x_3}{x_2} \left[1 + \left(\frac{x_3}{2 - x_1} \right)^2 \right] \right\}. \quad (5.319)$$

For the angular-ordered parton shower in HERWIG, the situation is a bit more complicated due to the somewhat non-trivial phase-space limits, leading to

$$d\Phi_g \left[\mathcal{B}_{q\bar{q}}(\Phi_{q\bar{q}}) \times \mathcal{K}(\Phi_g) \right] = \mathcal{B}_{q\bar{q}}(\Phi_{q\bar{q}}) \times \frac{C_F \alpha_s}{2\pi} \frac{dx_1 dx_3}{(1 - x_1)(1 - x_3)}$$

$$\times \left\{ \left[1 + \left(\frac{x_1 + x_3 - \frac{1}{2}}{x_1} \right)^2 \right]_{\left[\begin{matrix} x_1 > 1 - z(1 - z) \\ x_3 > 1 - x_1 + zx_1 \end{matrix} \right]} + x_1 \leftrightarrow x_3 \right\}, \quad (5.320)$$

where the splitting parameter z is given by

$$z = \frac{1}{2} + \frac{x_1 + 2x_3 - 1}{2x_1} \quad (5.321)$$

for massless quarks.

If instead a parton shower based on Catani–Seymour splitting kernels was used, the evolution parameters are $\vec{k}_\perp^2$ and z_i with

$$\begin{aligned}
\vec{k}_\perp^2 &= E^2 y_{ij;k} z_i (1 - z_i) \\
y_{ij;k} &= \frac{p_i p_j}{p_i p_j + p_j p_k + p_k p_i} = 1 - x_k \\
z_i &= \frac{p_i p_k}{(p_i + p_j) p_k} = \frac{1 - x_j}{2 - x_j - x_i} = \frac{x_i + x_k - 1}{x_k}.
\end{aligned} \quad (5.322)$$

Using the dimensionless parameter $y_{ij;k}$, with

$$\frac{d\vec{k}_\perp^2}{\vec{k}_\perp^2} = \frac{dy_{ij;k}}{y_{ij;k}}, \quad (5.323)$$

yields as the Catani–Seymour parton shower result for gluon emission in $e^- e^+ \to q\bar{q}$

$$d\Phi_g \left[\mathcal{B}_{q\bar{q}}(\Phi_{q\bar{q}}) \times \mathcal{K}(\Phi_g) \right] = \mathcal{B}_{q\bar{q}}(\Phi_{q\bar{q}}) \times \frac{C_F \alpha_s}{2\pi} \frac{dx_1 dx_3}{(1 - x_1)(1 - x_3)}$$

$$\times \left\{ \left[x_1^2 + x_3^2 \right] + \left[\frac{(1 - x_1)^2 (1 - x_3)}{x_3} + \frac{(1 - x_1)(1 - x_3)^2}{x_1} \right] \right\}. \quad (5.324)$$

In all cases, the acceptance weight is given by the ratio of $x_1^2 + x_3^2$ and the terms in the

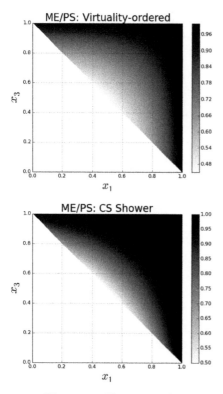

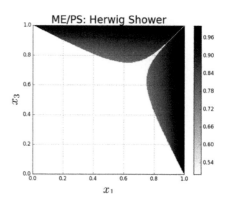

Fig. 5.18 The ratio of the matrix element and the parton shower expression for the differential emission cross section of an additional gluon in quark–anti-quark pair production in lepton annihilations, Eq. (5.319) (virtuality-ordered parton shower, upper left), Eq. (5.320) (angular ordered parton shower, upper right) and Eq. (5.324) (Catani-Seymour parton shower, lower left).

curly brackets. Of course, further parton shower formulations may lead to yet different weights, depending on the details of the map between the parton shower parameters and the kinematical quantities parameterizing the matrix element.

The algorithm then is to generate an emission through the parton shower, and to accept it with a probability given by the ratio of the two expressions above. Profiles for this ratio for different parton shower implementations are depicted in Fig. 5.18.

5.4.2.3 Limitations

While this technology is fairly transparent and straightforward in its implementation, it is limited in its applicability. First of all, it depends on the parton shower expression (or a suitable multiple of it) to be larger than the corresponding exact matrix element in all emission phase space, which is not always the case. This is especially true for production processes at hadron colliders, where the huge phase space for emissions off the initial state is not necessarily filled by the parton shower, see below. In addition, going to higher multiplicities the analytic parton shower expression becomes increasingly untraceable, which translates into the problem that the rejection weight cannot be constructed any more.

As already mentioned, there are situations, where the parton shower does not entirely fill the phase space. There are two possible reasons:

1. In cases like the radiation of extra partons in production processes at hadron colliders, the emission phase space is typically constrained by an upper scale given by the kinematics of the hard process. For example, in the production of vector bosons, such a scale would be given by the (virtual) mass of the vector boson. Since the transverse momenta in any parton emission generated by the parton shower are bounded from above by the scale of the previous process, as a consequence, only transverse momenta below the mass of the vector boson would be generated — the parton shower would completely miss the large-$p_\perp$ tail of the boson kinematics. This problem naively could be cured by opening up the emission phase space through a harder starting scale of the parton shower and a suitable matrix element reweighting. This is sometimes referred to as "**power shower**" in the literature, while sticking to the boson mass is dubbed "**wimpy shower**" [858]. However, the natural question then arises which scale-logarithms would actually be resummed in the Sudakov form factor in each case, and it quickly becomes apparent that in the power shower case the link to the standard analytic resummation techniques is broken. There, the upper limit of the $k_\perp$–integral in the Sudakov form factor is given by the boson mass, at odds with the power shower, where the upper limit is given by the energy scale of the hadronic collision, a violation of factorization theorems underlying all perturbative calculations.

2. Secondly, it is possible that the parton shower, while perfectly in line with resummation techniques, misses some regions in phase space by construction. This is particularly true for angular-ordered parton showers like the ones implemented in HERWIG [415, 750] and HERWIG++ [188, 579], see the upper-right panel of Fig. 5.18 for the case of gluon radiation in $e^-e^+ \rightarrow q\bar{q}$. In this case the **soft matrix-element correction** discussed so far needs to be supplemented with a **hard matrix-element correction**. This essentially boils down to using the matrix element to fill the phase-space region omitted by the parton shower. To guarantee a smooth transition, in this hard correction renormalization and factorization scale definitions as in the parton shower are used as well as some Sudakov weight for the intermediate quark line.

5.4.3   Next-to-leading order matching – the POWHEG method

5.4.3.1   Underlying idea

To combine an NLO calculation with a parton shower, it is essential to ensure that the result inherits the total cross-section from the fixed-order (NLO) calculation and that the radiation pattern to first order follows the real emission part of the calculation. In addition, from a parton shower point of view, it is also important to maintain its intrinsic logarithmic accuracy, which is substantially harder to achieve and to prove.

The fixed-order part of these requirements could trivially be achieved by multiplying a parton shower simulation with a suitable **global K-factor** and by applying a matrix-element correction in the style discussed in Section 5.4.2, to the first emission. The trick to improve this simple recipe, and the idea underpinning the POWHEG method, however, is to define a **local K-factor** in such a way that the integral over the phase space of the Born-level configurations yields the full result at next-to-leading order. To see how this works in more detail, remember the form of the total NLO cross-section

$$
\begin{aligned}
\sigma_N^{(\mathrm{NLO})} &= \int \mathrm{d}\Phi_\mathcal{B} \left[\mathcal{B}_N(\Phi_\mathcal{B}) + \mathcal{V}_N(\Phi_\mathcal{B}) + \mathcal{I}_N^{(\mathcal{S})}(\Phi_\mathcal{B}) \right] \\
&+ \int \mathrm{d}\Phi_\mathcal{R} \left[\mathcal{R}_N(\Phi_\mathcal{R}) - \mathcal{S}_N(\Phi_\mathcal{R}) \right]
\end{aligned}
\tag{5.325}
$$

from Eqs. (3.120) and (3.130).

5.4.3.2 Local K-factors

Using the logic of the subtraction method and the implicit factorization of the real emission phase space into a Born-like phase space and a one-particle emission phase space, it is simple to construct expressions for Born-like configurations $\tilde{B}$, which yield the cross-section fully accurate at next-to-leading order level. Symbolically they are given by

$$
\bar{\mathcal{B}}_N(\Phi_\mathcal{B}) = \mathcal{B}(\Phi_\mathcal{B}) + \tilde{\mathcal{V}}_N(\Phi_\mathcal{B}) + \int \mathrm{d}\Phi_1 \left[\mathcal{R}_N(\Phi_\mathcal{B} \otimes \Phi_1) - \mathcal{S}_N(\Phi_\mathcal{B} \otimes \Phi_1) \right] , \tag{5.326}
$$

where the sum over different subtraction terms is understood, and where the renormalized and infrared-subtracted virtual contribution has been combined into

$$
\tilde{\mathcal{V}}_N(\Phi_\mathcal{B}) = \mathcal{V}_N(\Phi_\mathcal{B}) + \mathcal{I}_N^{(\mathcal{S})}(\Phi_\mathcal{B}) \tag{5.327}
$$

This construction only works out, if the full real-emission phase space $\Phi_\mathcal{R}$ can be written in the factorized form as

$$
\Phi_\mathcal{R} = \Phi_\mathcal{B} \otimes \Phi_1 . \tag{5.328}
$$

If this is not the case, the additional phase space is guaranteed to be infrared finite and the correct result of Eq. (5.325) can be recovered by merely adding the difference, a difficulty which will be ignored in the following. In their absence Eq. (5.326) indeed yields the full NLO cross-section upon integration over the Born-level phase space $\Phi_\mathcal{B}$.

The terms $\bar{B}$ therefore can be interpreted as fully differential cross-sections of Born-level configurations with a next-to-leading order weight, or, stated slightly differently, as Born-level configurations modified by a **local K factor**. The next-to-leading order accuracy of course would not be spoilt by any unitary parton shower added to it, so one just has to ensure that the pattern of the first emission is correct up to first order in α_s in order to arrive at a fully NLO accurate simulation. Going back to the technology of matrix-element corrections introduced in Section 5.4.2 indicates how this can be achieved: the full real-emission matrix element for the first emission must be used, this is easily achieved by essentially replacing the parton shower kernels $\mathcal{K}$ with $\mathcal{R}/\mathcal{B}$. Therefore, combining the $\bar{B}$ terms with the first-order correct radiation pattern of Eq. (5.314) results in a simulation which is correct to first order of the coupling for both the inclusive cross-section and for the emission of the hardest parton. This essentially is the core of the POWHEG method introduced for the first time in [543, 782] and in heavy use ever since.

In this matching method, the differential rate up to the hardest emission is given by

$$
d\sigma_N^{(\mathrm{NLO})} = d\Phi_\mathcal{B}\, \bar{\mathcal{B}}_N(\Phi_\mathcal{B})
$$
$$
\times \left\{ \Delta_N^{(\mathcal{R}/\mathcal{B})}(\mu_Q^2, t_\mathrm{c}) + \int\limits_{t_\mathrm{c}}^{\mu_Q^2} d\Phi_1 \left[\frac{\mathcal{R}_N(\Phi_\mathcal{B} \times \Phi_1)}{\mathcal{B}_N(\Phi_\mathcal{B})} \Delta_N^{(\mathcal{R}/\mathcal{B})}(\mu_Q^2, t(\Phi_1)) \right] \right\}.
$$
$$(5.329)$$

It is straightforward to prove that this yields the correct NLO cross-section, since the term in the second line integrates, again, to unity. In order to see that the radiation of the first/hardest emission follows the exact real correction matrix element at first order in the coupling, it suffices to note that the term $\mathcal{R}_N/\mathcal{B}_N$ already is at first order in the coupling. This allows to ignore, at this accuracy level, all next-to-leading terms in $\bar{\mathcal{B}}_N$, *i.e.* all terms stemming from real or virtual corrections or the corresponding subtractions, since they would yield terms of order $\mathcal{O}(\alpha_\mathrm{s}^2)$.

One subtlety ignored so far is the choice of the "right" renormalization and factorization scales in both $\bar{\mathcal{B}}_N$ and in $\mathcal{R}_N/\mathcal{B}_N$. By far and large it is advantageous to keep to the choices made in the parton shower in the emission term, *i.e.* $\mathcal{R}_N/\mathcal{B}_N$. For the integrated emissions in $\mathcal{B}_N$, on the other hand, it is probably better to keep choices in line with the choices also made in $\bar{\mathcal{B}}_N$ and especially in the virtual part $\mathcal{V}_N$ in order not to spoil the exact cancellation of infrared divergences. In all cases, such choices are essentially of higher order in α_s and therefore do not hamper the fixed-order (NLO) accuracy of the approach.

Similar to the case of matrix-element corrections, however, there are a number of pitfalls beyond fixed-order accuracy. This is especially true for processes at hadron colliders; as an example consider the case of Higgs boson production in gluon fusion.

First of all, as in the case of matrix-element corrections, it is not clear what scale to pick as upper scale for the parton shower evolution. Standard resummation technology suggests choosing a scale of the order of the Higgs boson mass m_H as an argument in the logarithms, *i.e.* $\mu_Q \approx m_H$. This choice however does not allow a description of transverse momenta of the Higgs boson in the high-$p_\perp$ tail, at scales above m_H. Conversely, choosing $\mu_Q = m_H$ in the equation above, Eq. (5.329), will automatically constrain the phase space available for the hardest emissions to scales below m_H. This is at odds with maintaining $\mathcal{O}(\alpha_\mathrm{s})$ accuracy over the *full* emission phase space. Without modifying the algorithm outlined up to now, therefore a choice must be made between logarithmic and fixed order accuracy of the approach in the high-$p_\perp$ tails of additional emissions.

Assuming that the full phase space is opened up for the hardest emission, *i.e.* $\mu_Q \to E_{\text{cms}}$ of the hadronic collision, a second question naturally arises. The local K-factor encoded in $\bar{\mathcal{B}}_N$ corresponds to the *inclusive* production of the n-body final state, and in particular after integrating out additional partons in $\mathcal{R}_N$. By construction it is applied to *all* events, and in particular to those where the hardest emission is harder than the typical scale related to the n-particle state. This of course is questionable, since *a priori* the K-factors for n-particle and for $(n+1)$-particle final states do not coincide. It is of course possible that such discrepancies are not so large, and that therefore the tails of large-$p_\perp$ of the produced system (in the example here the Higgs boson) are well described. An illustrative study of this problem is depicted in Fig. 5.19. However, as can be seen in the left panel of this figure, the tail of the distribution differs significantly from the fixed-order, *i.e.* NLO, result. Even more, comparing with another NLO matching method, Mc@Nlo discussed in the next section, Section 5.4.4, it appears as if the latter interpolates between a low-$p_\perp$ regime, where a K factor is applied for both, which therefore agree with each other, and the region of large $p_\perp$, where only the Powheg simulation is modified by the K factor. In the right panel, this difference is unambiguously traced back to the influence of the K-factor. Replacing the Born-term in the denominator of the emission kernel $\mathcal{R}/\mathcal{B}$ in the Sudakov form factor with $\bar{\mathcal{B}}$, Eq. (5.329) schematically becomes

$$
d\sigma_N^{(\text{NLO})} \longrightarrow d\Phi_{\mathcal{B}}\,\bar{\mathcal{B}}_N \left\{ \Delta_N^{(\mathcal{R}/\bar{\mathcal{B}})}(\mu_Q^2, t_c) + \int\limits_{t_c}^{\mu_Q^2} d\Phi_1 \left[\frac{\mathcal{R}_N}{\bar{\mathcal{B}}_N} \, \Delta_N^{(\mathcal{R}/\bar{\mathcal{B}})} \right] \right\}, \qquad (5.330)
$$

where the by-now familiar phase-space arguments in the individual pieces have been omitted. In this form, the higher-order enhancement is cancelled, and the $p_\perp$ spectrum of the NLO result is recovered. Keeping, on the other hand, the original form of the emission kernel in Eq. (5.329), the result resembles more the NNLO result. It appears however that this is purely a coincidence and not related to any systematic improvement.

5.4.3.3 Improving the Powheg method

One way of solving both potential pitfalls outlined above is to decompose the real emission phase space into a "soft" and a "hard" part. This is essentially achieved by defining soft and hard real emission matrix elements $\mathcal{R}^{(\text{S})}$ and $\mathcal{R}^{(\text{H})}$ with support only in the corresponding phase-space regions. The former would contain the infrared-divergent parts and thus have to be suitably subtracted, while the latter would be free of infrared divergences. As a further result, the soft part only would be employed for the definition of the local K factors modifying $\mathcal{B}_N$ to obtain $\bar{\mathcal{B}}_N$ and for the parton shower kernel for the hardest emission off such configurations, while the hard part would be added separately, with the parton shower defined for all emissions through its usual kernel.

As an improvement of the Powheg implementation this was first discussed in [139], and the same idea has been used for the Powheg simulation of various other processes

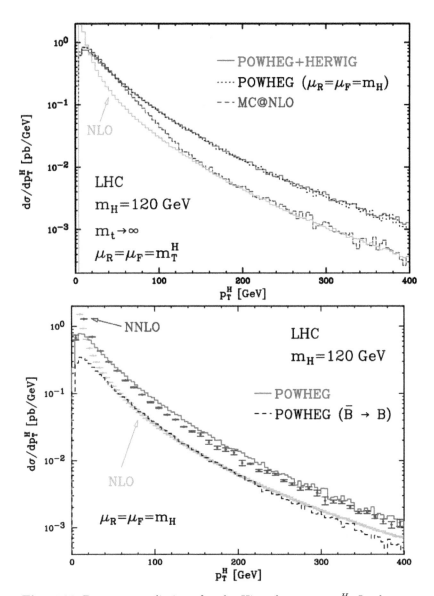

Fig. 5.19 POWHEG predictions for the Higgs boson $p_\perp$, $p_\perp^H$. In the top panel the inclusion of the parton shower on $p_\perp^H$ (POWHEG +HERWIG) is compared with the NLO result and with the distribution where only the first emission is simulated (POWHEG). In the lower panel a comparison is made where $\mathcal{R}/\mathcal{B}$ is replaced by $\mathcal{R}/\bar{\mathcal{B}}$ and with the NNLO result. (Figures taken from [139].)

since then. Specifically, the decomposition of phase space is achieved with a smooth function. For the example case of Higgs boson production in gluon fusion it reads

$$\mathcal{R}_N = \mathcal{R}_N \left(\frac{h^2}{p_\perp^2 + h^2} + \frac{p_\perp^2}{p_\perp^2 + h^2} \right) = \mathcal{R}_N^{(\mathrm{S})} + \mathcal{R}_N^{(\mathrm{H})} \tag{5.331}$$

with $p_\perp$ the transverse momentum of the Higgs boson. The new parameter h will typically be of the order of the relevant resummation scale of the underlying Born configuration, in the example here therefore $h \approx m_H$. Alternatively it could be "tuned" to exact higher-order calculations or calculations involving the resummation at higher logarithmic accuracy.[9]

Omitting again the phase-space arguments in the various parts, the differential rate up to the first emission in this improved formalism therefore reads

$$d\sigma_N^{(\mathrm{NLO})} = d\Phi_\mathcal{B} \, \tilde{\mathcal{B}}_N \left\{ \Delta_N^{(\mathcal{R}^{(\mathrm{S})}/\mathcal{B})}(\mu_Q^2, t_c) + \int_{t_c}^{\mu_Q^2} d\Phi_1 \left[\frac{\mathcal{R}_N^{(\mathrm{S})}}{\mathcal{B}_N} \Delta_N^{(\mathcal{R}^{(\mathrm{S})}/\mathcal{B})}(\mu_Q^2, t) \right] \right\}$$

$$+ \, d\Phi_\mathcal{R} \, \mathcal{R}^{(\mathrm{H})} \tag{5.332}$$

with a modified local K-factor defined through

$$\tilde{\mathcal{B}} = \mathcal{B} + \tilde{\mathcal{V}}_N + \int d\Phi_1 \left[\mathcal{R}^{(\mathrm{S})} - \mathcal{S} \right]. \tag{5.333}$$

The first line describes emissions in the soft regime, multiplied by the modified K-factor. For the example of Higgs boson production, it thereby experiences an enhancement. The second line describes the radiation in the hard regime, and it is not modified by any K-factor.

5.4.3.4 The interface to the parton shower

In its inception the POWHEG method was presented as an algorithm to promote arbitrary showers to higher fixed order, and in particular NLO, accuracy. There are, however, some caveats to this, which have to do with the fact that POWHEG works for ordering the emissions through their hardness, typically identified through a quantity like transverse momentum. However, as has already been seen in Section 5.3.2, there are about as many definitions of such a quantity as there are parton shower algorithms. For instance, in many cases the transverse momentum of a particular splitting is defined with respect to a well-specified spectator parton; but in the commonly used interface structures the information of a spectator parton is not provided, and the parton shower algorithm may make other choices than the matching code feeding parton-level configurations into it. In addition, for many showers, only one such hard

---

[9]This indeed is the case for the illustrative example, Higgs boson production in gluon fusion, where h has been fixed to a value of $h = 1.2 m_H$ by comparison with the result of NNLL resummation [442, 596].

scale is provided globally for the full parton ensemble, which may therefore have a meaning which in extreme cases may lead to sizable mismatches. Consequently, this leads to an additional source of uncertainty purely related to the matching, which could and possibly should be systematically assessed.

The obvious way out to is to ensure that the definition of the hardness scale or transverse momentum in the fixed-order part of the simulation is identical to the evolution parameter in the parton shower. Alternatively, if this cannot be achieved, one could invoke **truncated showering**, already introduced in [782]. There, it was noticed that the parton shower implementation in HERWIG uses angles instead of transverse momenta as the evolution parameter. This leads to a situation where the first emissions in the parton shower typically are large-angle emissions of rather soft partons which will often result in a relatively small transverse momentum. At the same time, the more collinear splittings into partons with larger energies quite often appear towards the end of the shower evolution — implying that the hardest splitting, *i.e.* the one with the largest transverse momentum, can essentially happen at all stages of the evolution. This has to be accounted for in the matching, in order not to upset the resummation of logarithms in the parton shower. The only way to achieve this is to allow the parton shower to emit partons at larger angles, but lower transverse momenta relative to the hardest emission fixed by the POWHEG formalism, which in turn must be inserted at some point into the parton shower evolution. Such a strategy in principle could be employed whenever there is a mismatch of evolution and hardness scales in the parton shower and fixed-order parts of the simulation.

5.4.4   Next-to-leading order matching — the MC@NLO method

5.4.4.1   Underlying idea

Historically, the first solution of how to match NLO matrix elements with the parton shower has been provided by the MC@NLO method, pioneered in [546]. While, somewhat loosely speaking, the POWHEG method is nothing but a matrix-element correction method supplemented with local K-factors, the MC@NLO method is closer in spirit to analytic resummation. Similar to the way the calculation is organized there, the real emission correction is decomposed into a part driven by Sudakov form factors and realized by the parton shower, and a hard remainder. And, while the former will experience higher-order corrections, like in Q_T resummation at NLO+NLL accuracy, the latter will not. In particular, the decomposition in MC@NLO is given by

$$\mathcal{R}_N(\Phi_\mathcal{R}) \;=\; \mathcal{R}_N^{(\mathrm{S})}(\Phi_\mathcal{R}) + \mathcal{R}_N^{(\mathrm{H})}(\Phi_\mathcal{R}) \;=\; \mathcal{S}_N(\Phi_\mathcal{B} \otimes \Phi_1) + \mathcal{H}_N(\Phi_\mathcal{R})\,. \tag{5.334}$$

The catch in MC@NLO is to identify the subtraction terms with the shower kernels such that, symbolically,

$$\mathcal{S}_N(\Phi_\mathcal{B} \otimes \Phi_1) \;\equiv\; \sum_{ijk} \mathcal{B}_N(\Phi_\mathcal{B}) \otimes \mathcal{K}_{ij;k}(\Phi_1) = \mathcal{B}_N(\Phi_\mathcal{B}) \otimes \mathcal{K}(\Phi_1)\,. \tag{5.335}$$

The MC@NLO version for the differential rate up to the first emission thus is given by

$$\mathrm{d}\sigma_N^{(\mathrm{NLO})} = \mathrm{d}\Phi_\mathcal{B}\,\tilde{\mathcal{B}}_N(\Phi_\mathcal{B})\left\{\Delta_N^{(\mathcal{K})}(\mu_Q^2, t_\mathrm{c}) + \int_{t_\mathrm{c}}^{\mu_Q^2} \mathrm{d}\Phi_1\,\mathcal{K}(\Phi_1)\,\Delta_N^{(\mathcal{K})}(\mu_Q^2, t(\Phi_1))\right\}$$

$$+ \,\mathrm{d}\Phi_\mathcal{R}\,\mathcal{H}_N(\Phi_\mathcal{R})\,,$$

(5.336)

where this time the NLO-modified Born term is

$$\tilde{\mathcal{B}}_N(\Phi_\mathcal{B}) = \mathcal{B}_N(\Phi_\mathcal{B}) + \tilde{\mathcal{V}}_N(\Phi_\mathcal{B})\,.$$

(5.337)

The virtual part of the NLO correction is applied only to those emissions that follow the kinematics given by the parton shower. In turn the phase space of such emissions is guaranteed to be in line with its leading logarithmic pattern. Since, by construction, the hard remainder $\mathcal{H}_N$ does not contribute any terms at the same logarithmic accuracy as the parton shower, the logarithmic accuracy of the latter is automatically maintained. Similar to its counterpart in Q_T resummation, the hard emission term in the second line of Eq. (5.336) therefore has a dual role. It firstly corrects the hardest emissions from the parton shower such that they follow at fixed order the exact matrix element. At the same time it fills emissions in those regions which are inaccessible by the parton shower with the exact leading-order pattern. Therefore, also the MC@NLO method fulfills the fixed-order requirements for a successful NLO matching of matrix elements and the parton shower. It can trivially be seen that it also maintains the logarithmic accuracy of the parton shower, and potential problems with the arguments in the resummation related to the choice of parton shower starting scale are avoided by construction: the parton shower starts exactly at the same scale as it would without the matching: at all orders and with help of the all-emissions operator $\mathcal{E}_n^{(\mathcal{K})}$ from Eq. (5.310) the expression in Eq. (5.336) can be rewritten as

$$\mathrm{d}\sigma_N^{(\mathrm{NLO})} = \mathrm{d}\Phi_\mathcal{B}\,\tilde{\mathcal{B}}_N(\Phi_\mathcal{B})\,\mathcal{E}_N^{(\mathcal{K})}(\mu_Q^2, t_\mathrm{c}) + \mathrm{d}\Phi_\mathcal{R}\,\mathcal{H}(\Phi_\mathcal{R})\,\mathcal{E}_{N+1}^{(\mathcal{K})}(\mu_H^2, t_\mathrm{c})\,,$$

(5.338)

where μ_H is the starting scale related to the hard remainder.

5.4.4.2   Treatment of colour

A potential pitfall in the MC@NLO method is related to the fact that the parton shower is a leading colour approximation. This translates into the fact that for general processes it is impossible to subtract all soft divergences, since sub-leading colour configuration usually do exhibit soft singularities at next-to-leading order (but of course no collinear ones).

   This is merely a technical problem — the NLO accuracy remains unharmed by such potentially uncancelled singularities at sub-leading colour. This is because in principle they could be cancelled after averaging over the various directions in the eikonals and identifying this suitably in the splitting kernels of the parton shower. This is how this problem indeed was cured in the original MC@NLO algorithm, when applied to $t\bar{t}$-production at hadron colliders in [544, 546]. In practical terms, a dampening function was introduced there, which essentially redistributes the soft sub-leading colour divergences by modifying the leading colour subtractions. However, at fixed order, the hard remainder term $\mathcal{H}_N$ will modify this averaging and it will thereby guarantee that the correct resolvable radiation pattern at NLO accuracy is always recovered.

Alternatively, it is also possible to modify the emission kernels for the first emission in such a way that the full colour structure is directly recovered [630]. To do this, the summing over colour structures inherent in the construction of the splitting kernels usually employed in parton showers has to be undone allowing to recover all colour terms. In the framework of a parton shower based on Catani–Seymour splitting kernels [466, 778, 839], this translates into replacing the colour-averaged kernels discussed in Section 5.3.2 with a sum over *all* dipole terms $\tilde{\mathcal{D}}_{ij;k}$. Phrased in other words, the sum must go over *all* splitters ij and *all* spectators k irrespective of whether they are colour-connected or not. Thereby structures like the colour-insertion operators $\mathbf{T}_{ij} \cdot \mathbf{T}_k$ will appear, made explicit for instance in Eq. (3.167) for the case of two final-state particles.

Upon evaluation in the explicit case, these operators however may become negative, leading to a negative weight for a splitting in this particular configuration. By construction, the sum over all $\mathbf{T}_k$ will lead to $\mathbf{T}_{ij}^*$ due to colour conservation in any physical parton ensemble, and thus the splitting weights will be positive overall. However, in certain directions given by certain spectators k, the admixture of positive and negative contributions will result in a negative contribution. At fixed order, such terms do not pose a problem. But in addition to the correct fixed-order result, in the implementation in [630], these sub-leading colour structures are also part of the Sudakov form factor, thereby accounting to some degree also for effects beyond fixed order. The technical problem there is related to the fact that negative splitting weights lead to negative arguments in the Sudakov form factors. This apparent violation of a probabilistic interpretation manifests itself naturally in Sudakov form factors larger than unity. Such "anti-probabilistic" features necessitate a modification of the showering algorithm, *cf.* [630, 809] for technical details.

While in most cases the effects of including such sub-leading terms are small, there are some observables for which they are surprisingly large. An example is provided by the case of the forward–backward asymmetry A_{FB} in $t\bar{t}$ production at the TEVATRON, where such effects have been studied for instance in [627].

5.4.4.3 The interface to the parton shower

Similar to what has been discussed already in the previous section, the K factor — essentially the correction related to the virtual part $\tilde{V}$ — is only applied to those radiation configurations that are produced by the parton shower, while the other configurations, produced by the hard correction term, essentially come with their tree-level weight. This leads to an NLO modification of the radiation pattern below the parton shower starting scale, and an LO distribution above. In fact, this is exactly the behaviour one would expect, and it is more or less a matter of taste whether a smooth transition like in the POWHEG case or a steep one like in the MC@NLO case takes this into account in a cleaner, better, or more transparent way.

However, this behaviour leads to yet another subtlety related to the practical implementations of this method. As already seen in Section 5.3.2, there are parton shower algorithms that may not fill the full emission phase space, like, for instance, the angular-ordered parton showers implemented in HERWIG [415] and HERWIG ++ [188]. Typically, these "holes" in the emission phase space are filled through the hard matrix element corrections, which, by a combination of clever scale choices in α_s analogous to the parton shower and, eventually, some Sudakov weights will smoothly connect to the parton shower description in the transition between the two regimes.

In MC@NLO, life may not be quite as simple. As the cross-section in the regime of soft emissions, treated by the parton shower is modified by the local K factor — essentially by the term $\tilde{V}$ — this smooth transition may be lost and a mismatch in the radiation pattern may emerge. The emergence of a similar pattern can also be observed in the MC@NLO method as implemented in SHERPA [630] by constraining the real emission phase space by measures that are incompatible with the parton shower evolution.

5.4.4.4   Comparison with analytic resummation methods

At this point it is worth comparing the MC@NLO method with analytic resummation methods, and in particular Q_T resummation.

In the master formula for Q_T resummation of heavy singlets in hadron–hadron collisions, Eq. (5.59), the hard remainder parts $Y_{ij \to X}$ are not supplemented with any further emissions encoded in Sudakov form factors. However, the construction of the R-terms, exemplified for W production in Eq. (5.78), follows the same principle as the construction of $\mathcal{H}_N$ here: It results from the fixed-order real-emission cross-section based on the matrix element $\mathcal{R}_N$ through subtraction of the $\mathcal{O}(\alpha_s)$ terms in the resummed part.

In a similar way, analysing the structure of the resummed part in the MC@NLO equation above, it is fairly straightforward to realize that the subtracted virtual term, $\tilde{V}_N$ includes not only the hard loop correction H_{ab} of Eq. (5.60) to first order in α_s, but also the two collinear terms C_{ia} to the same order; in particular terms such as $P_{ia}^\epsilon(z)$ are part of the underlying standard NLO subtraction procedure. This leaves the Sudakov form factors. In standard Q_T resummation overall momentum conservation is guaranteed through the Fourier transform to impact-parameter space ($b_\perp$-space). This of course is not necessary in parton showers, which have momentum conservation included for each individual emission. The only subtlety here is to realize that along the initial-state parton shower, *all* emissions must contribute to the overall $Q_\perp$ of the singlet system. In the original proposal for the construction of a parton shower based on the Catani–Seymour splitting kernels and kinematics [778] this is not quite the case: there the recoil, *i.e.* the transverse momentum "kick", of a splitting is only absorbed by the colour partner of the system and, in particular in the case of initial-state splittings, not by the full final state. This is the reason why this proposal has later been augmented with kinematical mappings where initial-state legs never decouple from the rest of the system during parton showering [626, 780, 807]. However, when comparing the terms in the Sudakov form factor of analytic resummation, Eq. (5.60) and Eqs. (5.65) – (5.67), it becomes apparent that the same terms $A^{(1,2)}$ and $B^{(1)}$ — possibly up to power corrections of the type $Q_\perp^2/Q^2$ involving no logarithms — are also present in the parton shower Sudakov form factors. This suggest that parton showers that are ordered in $k_\perp$ or opening angles correctly resum those next-to-leading logarithms $\log(Q_\perp^2/Q^2)$ in singlet production, which are also present in the Sudakov form factors found in analytic resummations of the same quantity.

In addition, a successful matching with fixed-order calculations at next-to-leading order also includes all terms appearing in the collinear and hard contributions — the formal accuracy of such samples therefore is NLO+NLL in the language of Q_T resummation based on the Collins–Soper–Sterman approach.

5.5 Multijet merging of parton showers and matrix elements

5.5.1 Multijet merging at leading order

5.5.1.1 Underlying idea

The **merging of multijet matrix elements** at leading order recognizes the fact that for all emissions leading to a sufficiently hard new jet matrix elements provide the best description, while the strength of the parton shower lies in accounting for the softer or more collinear radiation inside such jets. The idea therefore is to decompose the emission phase space into two complementary regimes: one hard regime of jet production, and a soft regime of jet evolution. In general such a division can be achieved by introducing a jet resolution criterion Q_{cut}, typically related to the transverse momentum of emissions. In the first multijet merging algorithm presented in [351], Q_{cut} indeed was identified with a $k_\perp$-jet measure from the Durham jet algorithm [345]. Consequently, emissions harder than Q_{cut} are described by the appropriate matrix elements while softer emissions are left to the parton shower.

Naively, thus, it would be preferable to push Q_{cut} to the lowest possible values, of the order of the parton shower cut-off. In practice, however, this turns out to be potentially tricky. First of all, there are technical problems related to the increasingly poorer convergence of the multijet matrix elements with increasingly softer cuts, rendering event generation, especially of unweighted events, increasingly or even prohibitively CPU consuming. In addition, and as seen in previous chapters, the introduction of such a jet measure in fixed-order calculations for processes characterized by a typical hard scale Q_{hard} introduces logarithms (or their squares) of the form $\log(Q_{\mathrm{hard}}/Q_{\mathrm{cut}})$. These logarithms will become larger for increasingly disparate scales Q_{cut} and Q_{hard}, and therefore they may overcome the smallness of the perturbation parameter, typically α_{s}/π or similar. This signals a collapse of the validity of fixed-order perturbation theory and the need for resummation. While the latter is partially achieved by the parton shower, at least at leading or next-to-leading logarithmic accuracy, the residual sub-leading logarithmic terms may still introduce unwanted and dangerous corrections that will not be accounted for. It is therefore important that the behaviour of the calculation for small Q_{cut} be carefully monitored, to thereby protect the calculation against such unavoidable sub-leading contributions becoming numerically large. In practice this can be achieved by varying Q_{cut} and evaluating the effect on critical distributions.

However, directly including the parton shower Sudakov form factors, maybe in their analytic form, into the fixed-order calculation will lead to a stabilization of their behaviour and a vastly increased level of convergence for low emission scales, *i.e.*

in the **Sudakov region**. This in fact has been worked out in more detail by the authors of the MINLO method in [609, 611], see Section 5.6.1, who combined the scale-setting prescription of multijet merging methods discussed below with a corresponding Sudakov reweighting, allowing them to push Q_{cut} to values around or below $\mathcal{O}\,(1\,\text{GeV})$, the infrared cut-off of the parton shower.

Going back to Fig. 5.16, the picture there as applied to multijet merging at leading order translates into using the vertical lines, corresponding to fixed-order expressions for jet production, with the number of jets increasing with the order of α_{s} and to combine them with the terms populating the diagonals. Immediately, the problem of double-counting of some terms becomes apparent; it should be stressed that this double-counting in general can be constructive, *i.e.* the same terms are attributed twice, or it can be destructive, by not covering them at all. In order to remedy this problem, a dual strategy comes into play. First of all, the matrix-element expressions are evaluated with suitable scale choices for the strong coupling, in such a way that their counterparts in the parton shower are emulated. In addition, they are weighted with suitable Sudakov form factors. With the interpretation of the Sudakov form factors as no-emission probabilities in mind, this second step transforms the matrix elements, which describe the *inclusive* production of an N-jet system *plus anything else* into matrix elements that describe the *exclusive* production of an N-jet system *only*, with no further resolvable emission above Q_{cut}. At the same time, the parton shower is modified such that any further jet emissions are vetoed. There are various ways of achieving this, which impact differently on the parametric accuracy provided by the matrix elements and, in particular, by the parton shower. While maintaining the fixed order accuracy of the former is fairly straightforward, the logarithmic accuracy of the parton shower is harder to conserve.

5.5.1.2 Reweighting matrix elements

The idea underlying actual algorithms can easily be understood going back to Section 2.3.3, where $k_\perp$-jet rates in electron–positron annihilations to hadrons were analytically resummed, *cf.* Eqs. (2.184) and (2.186). With the interpretation of Sudakov form factors as no–emission probabilities, the two- and three-jet rates, *i.e.* the probability for emitting no or only one jet, can be approximated, at logarithmic accuracy, as products of such no-emission probabilities and the terms relevant for the emission of a single parton,

$$\Re_2(Q_{\text{cut}}) = \left[\Delta_q(\mu_Q^2, Q_{\text{cut}}^2)\right]^2$$

$$\Re_3(Q_{\text{cut}}) = 2\Delta_q(\mu_Q^2, Q_{\text{cut}}^2) \int\limits_{Q_{\text{cut}}^2}^{\mu_Q^2} \frac{dq_\perp^2}{q_\perp^2} \left[\left(\frac{\alpha_{\text{s}}(q_\perp^2)}{\pi} C_F\, \Gamma_q(\mu_Q^2, q_\perp^2)\, \frac{\Delta_q(\mu_Q^2, Q_{\text{cut}}^2)}{\Delta_q(\mu_Q^2, q_\perp^2)}\right)\right.$$

$$\left. \times \Delta_q(q_\perp^2, Q_{\text{cut}}^2)\Delta_g(q_\perp^2, Q_{\text{cut}}^2)\right].$$

$$(5.339)$$

The integrated splitting kernels $\Gamma_{q,g}$ have already been introduced in Eq. (2.181), leading to Sudakov form factors $\Delta_{q,g}$, which, depending on whether α_s is taken as fixed or running assume the form of Eq. (2.182) or Eq. (2.183). The upper limit μ_Q of the Sudakov form factors, and correspondingly of the integration over the transverse momentum of the emitted gluon, is usually identified with the only hard scale of the process, the centre-of-mass energy E_{cms} of the e^-e^+ pair. Since $\Re_2$ and $\Re_3$ are jet *rates*, they are normalized to the total hadron production cross-section. Therefore the production of the original quark–anti-quark pair in the electron–positron annihilation, proceeding through electroweak interactions, has been factored out. The relationship of the corresponding approximate cross-sections at $\mathcal{O}(\alpha_s)$ is given by

$$\Re_3(Q_{cut}) = 1 - \Re_2(Q_{cut}) = \int\limits_{Q_{cut}^2}^{\mu_Q^2} \frac{dq_\perp^2}{q_\perp^2} \left[\frac{2\,C_F\,\alpha_s(q_\perp^2)}{\pi}\,\Gamma_q(\mu_Q^2, q_\perp^2) \right] + \mathcal{O}(\alpha_s^2). \quad (5.340)$$

Remembering that Γ_q is nothing but the integrated splitting kernel it becomes quickly apparent that this expression indeed is the $\mathcal{O}(\alpha_s)$–approximation of a $k_\perp$-ordered parton shower to the respective cross-section.[10]

In the parton shower language, Eq. (5.340) can be cast into

$$\Re_3(Q_{cut}) = \int\limits_{Q_{cut}^2}^{\mu_Q^2} d\Phi_1 \left[\mathcal{K}_{qg;\bar{q}}(\Phi_1) + \mathcal{K}_{\bar{q}g;q}(\Phi_1) \right] + \mathcal{O}(\alpha_s^2) = \frac{d\sigma_3^{(res)}(Q_{cut})}{\sigma^{(tot)}}. \quad (5.341)$$

In order to improve this description of the three-jet rate as provided by the parton shower with exact tree-level matrix elements, the expressions in the integral of Eqs. (5.340) and (5.341) merely have to be replaced with the exact matrix elements. Choosing the scale of α_s in the matrix element as in the parton shower also includes the corresponding logarithmic terms. This idea has already been encountered in Section 5.4.2, where matrix-element corrections to the hardest emission were discussed. Using them, the exact radiation pattern up to $\mathcal{O}(\alpha_s)$ was recovered by reweighting the parton shower with the fixed-order matrix element at the same perturbative order.

In contrast, multijet merging algorithms start from a matrix element, with an inverted logic: instead of the parton shower being reweighted with the matrix elements, the matrix elements are modified by terms stemming from the resummed expression. This is achieved through a combination of adjusted scales of α_s and reweighting with the Sudakov suppression factors which have been omitted in the fixed-order expressions in Eqs. (5.340) and (5.341). As a result, the procedure resums the same logarithms as the parton shower does, but, in addition, reproduces the *exact* fixed-order result provided by the tree-level matrix elements.

The remaining problem is to determine the adjusted scales for both the α_s and the

---

[10]The Sudakov suppression factors, being exponentials, could in principle be Taylor-expanded to approximate higher order terms. While this is not relevant for merging technology at leading order, it becomes relevant for fixed order matrix elements beyond tree-level, see Sections 5.5.3 and 5.6.1.

Sudakov reweighting. To this end, a **parton-shower emission history** is constructed from the matrix-element configuration by a recursive algorithm. In each recursion step r, all allowed parton pairs i and j, which may be combined into a joint emitter $\widetilde{ij}$, together with all possible spectators k are considered. The respective flavours and momenta define a splitting kernel $\mathcal{K}_{ij,k}$ and the corresponding kinematics. Following the logic of parton showers, the latter is characterized through the ordering parameter $t^{(r)}$, the splitting variable $z^{(r)}$ and the azimuth angle $\phi^{(r)}$. From them, the nodal scale value $\mu_R^{(r)}$ entering the strong coupling is readily read off. The momenta after undoing the splitting $\widetilde{ij} \to i+j$ are given by the inverse of the parton shower phase-space map. Repeating this procedure until a core process — typically a $2 \to 2$ scattering process — is reached, thus yields a sequence of splitting kernels and nodal values for the coupling constants.

There are two ways of selecting the actual parton shower history from such a procedure: Either in a "winner-takes-all" strategy, where step-by-step only the most probable clustering, typically the one with the smallest transverse momentum, is retained, or in a probabilistic strategy where the "winning" parton history is selected according to overall weights given by the product of the respective splitting kernels.

In general, irrespective of such details of the final selection strategy, the construction of the parton shower emission history leads to a sequence of m nodal values of hardness scales — typically transverse momenta — for the backwards clustering,

$$Q^{(m)} \le Q^{(m-1)} \le \ldots \le Q^{(2)} \le Q^{(1)} \le Q^{(\text{core})} = Q^2 \tag{5.342}$$

the respective values of the parton shower evolution parameter

$$t^{(m)} \le t^{(m-1)} \le \ldots \le t^{(2)} \le t^{(1)} \le t^{(\text{core})} = \mu_Q^2 = \approx Q^2 \tag{5.343}$$

and, correspondingly, for the scales entering the strong couplings

$$\mu_R^{(m)} \le \mu_R^{(m-1)} \ldots \le \mu_R^{(2)} \le \mu_R^{(1)} \le \mu_R^{(\text{core})}. \tag{5.344}$$

In the discussion up to now, the values of t have been identified with the corresponding Q^2, which would usually be the case; it is worthwhile to mention, however, that for instance for angular ordered showers this is not quite the case. There an ordering in hardness/transverse momentum Q does not usually also manifest itself in an ordering in the emission angles. As in the case of next-to-leading order matching, **truncated showering** as defined in [782], must be applied in such circumstances, see below for a more detailed discussion of its effect. But, indeed, there are some further subtleties, which, as already hinted at, most notably concern strategies for dealing with unordered emissions. Such rather pathological cases will not be discussed here.

Nevertheless, with the information encoded in the parton history at hand, the overall scale μ_R of a matrix-element configuration is determined as

$$\alpha_s^{M+m}(\mu_R^2) = \alpha_s^M(\mu_{R,(\text{core})}^2) \cdot \prod_{i \in m} \alpha_s(\mu_{R,(i)}^2), \tag{5.345}$$

where M is the power of α_s in the hard core process, $\mu_R^{(\mathrm{core})}$ is the scale choice associated to this process, and where the $\mu_R^{(i)}$ are the scales of the m QCD splitting nodes. In a similar way, the evolution scales $t^{(i)}$ and the core scale $t^{(\mathrm{core})}$ define the Sudakov suppression factors.

5.5.1.3   Interlude: A scale-setting prescription for LO and NLO calculations

One of the reasons for the phenomenological success of multijet merging can be traced back to the fact that multi-parton cross-sections that have genuinely been calculated at leading order look surprisingly similar in shape to the corresponding next-to-leading order ones, if the former have been subjected to the scale-setting prescriptions described previously. Of course, this prescription simply generalizes to the case of NLO calculations, where, after ignoring the softest parton in the real emission contribution the overall scale can be deduced from Eq. (5.345) as well. These ideas have also been adopted in the MINLO framework [611], *cf.* Section 5.6.1.

5.5.1.4   Vetoed parton showers

Moving to the jet evolution part of the simulation, the interplay of the two expressions for the two- and three-jet rates, Eq. (5.339), will also fix the way the parton shower must supplement the radiation pattern below Q_{cut} in such merged simulations. Simply put, the correct jet rates must be recovered at the same logarithmic accuracy for all values of a jet resolution Q_J, above and below Q_{cut}.

The cross-sections for the various jet multiplicities have been made exclusive with respect to the emission of additional jets, by multiplying them with the Sudakov form factors. This is, yet again, a manifestation of the interpretation of the Sudakov form factors as encoding no-emission probabilities at their logarithmic accuracy. The modified cross-section expressions can thus be combined into an inclusive sample, where each jet multiplicity covered by the fixed-order matrix elements is described with differential and total cross-sections at tree-level accuracy with an improved scale choice. From the matrix element point of view, there is no double-counting present in such a sample.[11] However, up to now each of the jets in this sample consists of exactly one parton only, and therefore must be further populated by invoking the parton shower.

There are two caveats related to naively applying the parton shower. First, the parton shower is not allowed to produce any unwanted jets, in order not to spoil the description of jet rates provided by the reweighted matrix elements, which is accurate at fixed order (supplemented with a resummation of leading and, eventually, subleading logarithms). Naively, this could be achieved by choosing t_{cut} as the starting scale of the parton shower evolution of all external partons, thus explicitly disallowing all harder emissions, which typically would become additional jets. As will be seen,

---

[11]A fixed-order version of this idea has been provided in the BLACKHAT +SHERPA framework, dubbed "exclusive sums". There, NLO matrix elements for $V+$ jets production with increasing jet multiplicity are added, with the phase space for the real emission correction constrained to inner-jet radiation only, *i.e.* such that it does not produce additional jets; see also a short description in [134].

this conflicts with the second potential problem of evolving the jets through the parton shower, namely the condition to maintain the intrinsic accuracy of the latter.

To see how this works out, consider, as an example the emission of a parton off a two-parton configuration in $e^-e^+ \to$ jets. Such a configuration has been identified with a two-jet event, and thus contributes with a rate given by $\mathfrak{R}_2(Q_{\text{cut}})$ as given in Eq. (5.339). The event therefore already has been weighted with a Sudakov suppression factor

$$
\mathfrak{R}_2(Q_{\text{cut}}) = \left[\Delta_q(\mu_Q^2, Q_{\text{cut}}^2)\right]^2 = \exp\left[-2\int_{Q_{\text{cut}}}^{\mu_Q} \frac{dq_\perp}{q_\perp} \frac{\alpha_s C_F}{\pi}\left(\log\frac{\mu_Q}{q_\perp} - \frac{3}{4}\right)\right]
$$

$$
\xrightarrow{\alpha_s \to \text{const.}} \exp\left[-\frac{\alpha_s C_F}{\pi}\left(\log^2\frac{\mu_Q}{Q_{\text{cut}}} - \frac{3}{2}\log\frac{\mu_Q}{Q_{\text{cut}}}\right)\right]
$$

(5.346)

where, for simplicity, effects of the running of the strong coupling are ignored. What would the two-jet rate be for values $Q_J \leq Q_{\text{cut}}$? Applying the same logic, the result should be given by $\mathfrak{R}_2(Q_J)$. However, if the suppression factor above was combined with the corresponding Sudakov suppression in a parton shower starting at Q_{cut}

$$
\mathfrak{R}_2'(Q_J) = \left[\Delta_q(\mu_Q^2, Q_{\text{cut}}^2) \cdot \Delta_q(Q_{\text{cut}}^2, Q_J^2)\right]^2
$$

$$
\xrightarrow{\alpha_s \to \text{const.}} \exp\left[-\frac{\alpha_s C_F}{\pi}\left(\log^2\frac{\mu_Q}{Q_{\text{cut}}} - \frac{3}{2}\log\frac{\mu_Q}{Q_{\text{cut}}} + \log^2\frac{Q_{\text{cut}}}{Q_J} - \frac{3}{2}\log\frac{Q_{\text{cut}}}{Q_J}\right)\right]
$$

$$
= \exp\left[-\frac{\alpha_s C_F}{\pi}\left(\log^2\frac{\mu_Q}{Q_J} - \frac{3}{2}\log\frac{\mu_Q}{Q_J} + 2\log\frac{\mu_Q}{Q_{\text{cut}}}\log\frac{Q_J}{Q_{\text{cut}}}\right)\right]
$$

$$
\neq \mathfrak{R}_2(Q_J)|_{\alpha_s \to \text{const.}}
$$

(5.347)

results. This simple consideration shows that a naive treatment leads to an unwanted and unphysical dependence on Q_{cut} in the two-jet rate at Q_J. In addition, it is apparent that a parton shower starting at Q_{cut} only yields very limited radiation just below this scale — in fact, for a scale $q \to Q_{\text{cut}}$ the radiation vanishes completely. This of course would yield a completely unphysical radiation dip just below Q_{cut}. Therefore, simply starting the parton shower at Q_{cut} is not an option.

The solution to this problem presents itself, when analysing the structure of emissions mediated by the parton shower. Starting the parton shower at μ_Q and vetoing every emission above Q_{cut} yields an expression that reads, for a single quark leg,

$$
1 + \int_{Q_{\text{cut}}}^{\mu_Q} \frac{dq_\perp}{q_\perp} \frac{\alpha_s C_F}{\pi} \Gamma_q(\mu_Q, q_\perp)
$$

$$
+ \int_{Q_{\text{cut}}}^{\mu_Q} \frac{dq_\perp}{q_\perp} \frac{\alpha_s C_F}{\pi} \Gamma_q(\mu_Q, q_\perp) \int_{Q_{\text{cut}}}^{q_\perp} \frac{dr_\perp}{r_\perp} \frac{\alpha_s C_F}{\pi} \Gamma_q(q_\perp, r_\perp) + \dots
$$

$$
= 1 + \int\limits_{Q_{\mathrm{cut}}}^{\mu_Q} \frac{\mathrm{d}q_\perp}{q_\perp} \frac{\alpha_{\mathrm{s}} C_F}{\pi} \Gamma_q(\mu_Q, q_\perp) + \frac{1}{2} \left[\int\limits_{Q_{\mathrm{cut}}}^{\mu_Q} \frac{\mathrm{d}q_\perp}{q_\perp} \frac{\alpha_{\mathrm{s}} C_F}{\pi} \Gamma_q(\mu_Q, q_\perp) \right]^2 + \dots
$$

$$
= \exp\left[\int\limits_{Q_{\mathrm{cut}}}^{\mu_Q} \frac{\mathrm{d}q_\perp}{q_\perp} \frac{\alpha_{\mathrm{s}} C_F}{\pi} \Gamma_q(\mu_Q, q_\perp) \right] = \Delta_q^{-1}(\mu_Q, Q_{\mathrm{cut}}), \tag{5.348}
$$

thus compensating the Sudakov suppression on the matrix element and eliminating the dependence on Q_{cut} for the two-jet rate at scales $Q_J \le Q_{\mathrm{cut}}$. This interplay of Sudakov rejection and vetoed parton showering was at the core of the proof of the logarithmic accuracy of multi-jet merging in [351]. The original proof has later been extended to also include initial state radiation, showing that a merging prescription can be formulated which exactly maintains the logarithmic accuracy of the parton shower, irrespective of implementation details [632].

In general this reasoning translates into an algorithm, where the parton shower evolution of each parton starts at the scale where it was first produced, as constructed from the parton history. This also gives rise to the Sudakov suppression weight on the matrix element. When running the parton shower, however, all emissions that would lead to the production of a jet above Q_{cut} are vetoed. The corresponding algorithm is also known as **vetoed parton shower**. This compensates, by construction, the Q_{cut}-dependence in the event inherited from the suppression applied on the matrix element.

Two comments are in order here. First of all, while this compensation is, in principle, accurate at the logarithmic accuracy of the Sudakov form factors employed, there may be mismatches. This occurs if the analytic Sudakov form factors are not exactly reflected in the parton shower, which typically is the case. Reasons for this, of course, range from an ordering in an evolution parameter different from the transverse momentum used in the $k_\perp$ algorithm employed to construct the resummed jet rates above, over non-logarithmic contributions emerging from finite terms in the z–integral of the splitting functions, to the exact inclusion of recoil effects inside the parton shower. These mismatches, despite typically being of sub-leading logarithmic accuracy, may become numerically important and would then manifest themselves in observables such as differential jet rates or similar.

In addition, when using parton showers with a sufficiently different ordering such as, for instance, angular ordering, it quite often happens that the first emissions are not the hardest ones according to the jet criterion. In such a case, vetoed showering alone will not be sufficient, and measures must be taken not to upset the logarithmic and colour structure produced by the parton shower. The solution to this problem of a mismatch of the parton shower evolution parameter and the hardness ordering consists of employing what is known as **truncated showering** [782]. In this formalism, the parton shower is allowed to emit partons at a transverse momentum *below* the relevant cut in hardness but with an evolution parameter *above* the one related to the hard emission providing this cut. In this way a radiation pattern is generated that is ordered in the evolution parameter of the parton shower but unordered in the hardness

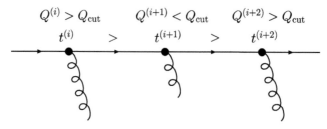

Fig. 5.20 Sketch to illustrate the idea underlying truncated showering. While the hardness of an emission according to the jet definition is given by Q, the parton shower evolution parameter squared is denoted by t.

parameter of the jet criterion. This is illustrated in Fig. 5.20.

5.5.1.5   Compact description

There remains a potential problem, namely a mismatch of analytic expressions employed for the Sudakov suppression weight on the matrix elements and the Sudakov form factors employed in the parton shower. The former encompass logarithmically enhanced terms only, which emerge when integrating over the energy splitting parameter z in its limits from 0 to 1. The latter, on the other hand also have finite terms of the type q^2/Q^2 related to the limits in z, essentially structures that look like power corrections in the transverse momentum of the emission.

This slight mismatch, however, can easily be overcome by directly using the parton shower to generate the Sudakov suppression terms of the matrix element. This was first noted [730] in the framework of multijet merging with the dipole shower implemented in ARIADNE [729]. The logic behind this is very simple: using that the Sudakov form factor in any case represents a no-emission probability, vetoing every event where the parton shower produces an unwanted emission with a transverse momentum $k_\perp > Q_{\mathrm{cut}}$ automatically generates the correct Sudakov suppression, thus playing the same role as the analytic weights. Of course, the initial parton configurations produced by the matrix elements still need to be clustered back to a core process in order to construct a parton shower history, providing the nodal scale values entering α_{s} and yielding the starting conditions of the parton shower.

This also enables a very compact way of analysing the structure of emissions in multijet merging. To see how this works in principle, consider first the case where matrix elements corresponding to the production of N and $(N+1)$ jets are merged into one inclusive sample. Up to the first emission of the N-parton configuration, the differential cross-section for the production of $(N+1)$ particles reads

$$
\mathrm{d}\sigma = \mathrm{d}\Phi_N\,\mathcal{B}_N\left[\Delta_N^{(\mathcal{K})}(\mu_N^2,\,t_{\mathrm{c}}) + \int\limits_{t_{\mathrm{c}}}^{\mu_N^2}\mathrm{d}\Phi_1\,\mathcal{K}_N\Delta_N^{(\mathcal{K})}(\mu_N^2,\,t_{N+1})\Theta(Q_{\mathrm{cut}}-Q_{N+1})\right]
$$

$$
+\,\mathrm{d}\Phi_{N+1}\,\mathcal{B}_{N+1}\,\Delta_N^{(\mathcal{K})}(\mu_{N+1}^2,\,t_{N+1})\Theta(Q_{N+1}-Q_{\mathrm{cut}})
$$

$$(5.349)$$

Here, Q_{N+1} is the hardness scale related to the emission of the $(N+1)$th particle. As advertised before, this scale defines two different regimes, namely the parton shower region $Q_{N+1} < Q_{\text{cut}}$ and the matrix-element region $Q_{N+1} > Q_{\text{cut}}$. While in the parton shower only expression, Eq. (5.304), such a differentiation is not being made, in contrast here the description of the emission of the $(N+1)$th particle is susceptible to this difference.

This of course has some consequences. Taking a closer look at the square bracket in the first line it becomes apparent that it does not integrate to unity any more, due to the phase-space constraint — encoded through $\Theta(Q_{\text{cut}} - Q_{N+1})$ — present in the second term, the emission term. The missing hard emissions are of course supplied through the second line, explicit through the complementary constraint $\Theta(Q_{N+1} - Q_{\text{cut}})$. There are two slight mismatches which prevent this term to also account for the pieces lacking in the first line to integrate to unity. First of all, there is a mismatch in the form of the emission term: the exact matrix element for the $(N + 1)$-particle state is different from the exact matrix element for the N-particle state convoluted with the parton shower kernel. This does not come as a surprise, as this was the very reason multijet merging has been introduced in the first place.

On top of this, the phase space available for this additional emission, the $(N+1)$th particle, differs in both lines. In the first line the upper limit on the available phase space is given by the relevant parton shower starting or resummation scale defined through the N particle kinematics, μ_N, while in the second line it is the potentially different $(N + 1)$-particle kinematics, which defines the corresponding scale μ_{N+1}. While in processes with a fixed hardest scale, such as electron–positron annihilations to jets it is safe to assume that typically μ_N and μ_{N+1} are very similar or even identical, this is not true for processes at hadron colliders such as the production of lepton pairs (the Drell-Yan process) in association with jets.

There, the emission of additional jets typically may offer phase-space regions associated with larger scales than processes without these additional jets — in the example of Drell–Yan-type processes this would correspond to jet emissions taking place at transverse momentum scales above the invariant mass of the lepton pair. Taken together, this leads to the fact that the combined contributions from the hard and soft first emissions will not exactly combine with the no-emission term to yield unity. This has been, in a somewhat sloppy use of language been coined "unitarity violation". It consequently leads to a variation of the total cross-section related to the inclusive sample produced with respect to the Born result. Postponing details to later parts of this section, it should be noted that this effect, while present in some of the more widely used multijet merging implementations, has been taken care of by the UMEPS algorithm [733, 806], which conserves inclusive cross-sections by suitably reshuffling different contributions.

Taking into account first emissions only, but combining many Born matrix elements up to a maximal number N_{max} of external legs promotes the simple expression of Eq. (5.349) to

$$
\overbrace{\phantom{\prod_{j=N}^{n} \Theta(Q_j - Q_{\rm cut})}}^{(n-N)\ \text{extra jets}} \quad \overbrace{\phantom{\prod_{j=N}^{n-1} \Delta_j^{(\mathcal{K})}(t_j, t_{j+1})}}^{\text{no emissions off internal lines}}
$$

$$
d\sigma \;=\; \sum_{n=N}^{N_{\rm max}-1} \left\{ d\Phi_n\, \mathcal{B}_n \left[\prod_{j=N}^{n} \Theta(Q_j - Q_{\rm cut}) \right] \left[\prod_{j=N}^{n-1} \Delta_j^{(\mathcal{K})}(t_j,\, t_{j+1}) \right] \right.
$$

$$
\times \left[\underbrace{\Delta_N^{(\mathcal{K})}(t_N, t_{\rm c})}_{\text{no emission}} + \underbrace{\int_{t_{\rm c}}^{t_N} d\Phi_1\, \mathcal{K}_N \Delta_N^{(\mathcal{K})}(t_N, t_{N+1}) \Theta(Q_{\rm cut} - Q_{N+1})}_{\text{next emission no jet \& below last ME emission}} \right] \Bigg\}
$$

$$
+\; d\Phi_{N_{\rm max}}\, \mathcal{B}_{N_{\rm max}} \left[\prod_{j=N}^{N_{\rm max}} \Theta(Q_{j+1} - Q_{\rm cut}) \right] \left[\prod_{j=N}^{N_{\rm max}-1} \Delta_j^{(\mathcal{K})}(t_j,\, t_{j+1}) \right]
$$

$$
\times \left[\Delta_{N_{\rm max}}^{(\mathcal{K})}(t_{N_{\rm max}}, t_{\rm c}) + \int_{t_{\rm c}}^{t_{N_{\rm max}}} d\Phi_1\, \mathcal{K}_{N_{\rm max}} \Delta_{N_{\rm max}}^{(\mathcal{K})}(t_{N_{\rm max}}, t_{N_{\rm max}+1}) \right.
$$

$$
\left. \cdot\, \Theta(Q_{N_{\rm max}} - Q_{N_{\rm max}+1}) \right].
$$

$$
(5.350)
$$

Note that in the contribution from the $N_{\rm max}$ configuration, the phase space filled by the parton shower is not constrained by the jet cut $Q_{\rm cut}$ but by the jet measure of the last emission filled by the matrix element. This allows the parton shower to account — within its limitations — for even higher jet multiplicities.

At this point it should of course be stressed that further emissions are based on splitting kernels supplemented with the phase-space veto:

$$
\mathcal{K}_N(\Phi_1) \xrightarrow{\text{MEPS}} \mathcal{K}_N^{<Q}(\Phi_1) = \mathcal{K}_N(\Phi_1)\Theta(Q - Q_{N+1}). \tag{5.351}
$$

Taking this into account and applying it to the subsequent emissions encoded in the parton shower evolution operator, *cf.* Eq. (5.310), $\mathcal{E}_N^{(\mathcal{K})}$ becomes

$$
\mathcal{E}_N^{(\mathcal{K}, <Q)}(\mu_Q^2, t_{\rm c})
$$

$$
= \Delta_N^{(\mathcal{K})}(\mu_Q^2, t_{\rm c}) + \int_{t_{\rm c}}^{\mu_Q^2} d\Phi_1 \left[\mathcal{K}_N^{<Q}(\Phi_1)\, \Delta_N^{(\mathcal{K})}(\mu_Q^2, t(\Phi_1)) \otimes \mathcal{E}_{N+1}^{(\mathcal{K}, <Q)}(t(\Phi_1), t_{\rm c}) \right].
$$

$$
(5.352)
$$

Inserting this into the merging equation Eq. (5.350) above results in

$$d\sigma = \sum_{n=N}^{N_{\max}-1} d\Phi_n \, \mathcal{B}_n \, \Theta(Q_n - Q_{\text{cut}}) \, \mathcal{E}_n^{(\mathcal{K}, <Q_{\text{cut}})}(\mu_n^2, t_{\text{c}})$$

$$+ \, d\Phi_{N_{\max}} \, \mathcal{B}_{N_{\max}} \, \Theta(Q_{N_{\max}} - Q_{\text{cut}}) \, \mathcal{E}_{N_{\max}}^{(\mathcal{K}, <Q_{N_{\max}})}(\mu_{N_{\max}}^2, t_{\text{c}})$$

(5.353)

This improved description has been implemented for a variety of parton showers, most of which use an ordering parameter related to the transverse momentum of the emissions and thus do not exhibit any real mismatch of evolution and hardness parameters. For a more in-depth discussion of the various realizations the reader is referred to the literature [632, 732]. Multijet merging with an angular-ordered parton shower in the framework of HERWIG++ has been explored in [612], where the impact of not employing truncated showering in the merging has been analysed, too.

5.5.1.6 Alternative: The MLM method

The MLM prescription for multijet merging [147, 744] was introduced and further refined in parallel to the method discussed so far. It is based on slightly different ideas: most notably it aims at using traditional parton shower routines like the ones implemented in PYTHIA [853] (virtuality or $p_\perp$-ordered shower) or in HERWIG [414] (angular-ordered shower) which are accessed through a standardized interface between general matrix elements and parton showers [145] without any direct alteration in the code. The structure of the resulting algorithm implies that vetoes on unwanted emissions can only be applied *a posteriori*, i.e. after the parton shower evolution has finished. In practice this is realized by reclustering the partons after the parton shower, i.e. at the hadronization scale t_{c}, into jets and by comparing these jets with the original ones, from the parton level at scale Q_{cut}. If an additional jet with respect to the original ones has been produced, or, conversely, if a jet has been lost, the event then is rejected. This algorithm of course incorporates a slightly different way of treating emissions off intermediate legs, which here are re-interpreted as radiation off external legs, by a suitable definition of parton shower starting conditions.[12]

In its original version, the MLM prescription uses a simple cone definition as the jet criterion to generate the parton configurations at the matrix-element level. The scales of strong couplings are reconstructed by using a backward clustering with a $k_\perp$ measure. Then the accepted configurations are passed on to the parton-shower routines. They in turn typically reconstruct the parton-shower starting scales of a multi-parton configuration by directly inspecting colour connections, without any backward clustering and therefore partially neglecting intermediate legs and the radiation originating from them. Having the starting conditions at hand, the parton shower is invoked without any constraint. After it has terminated at the hadronization scale t_{c}, the original partons stemming from the matrix element are exactly matched to the jets present at parton-shower level, again defined by the cone algorithm with parameters, which in principle may slightly differ from the ones applied at matrix-element level. If such a

---

[12]In such a treatment, the full capture of truncated showering effects is not guaranteed, and as a consequence some residual sub-leading logarithmic terms may be left out [632], which would be present in the original parton shower algorithm.

one-to-one match is not possible, either due to extra, unwanted jets being produced in the parton shower or due to "losing" jets in the parton shower, the event is rejected. This means that the Sudakov rejection factors that are applied locally, either through analytic Sudakov form factors or by using the shower, as explained above, are applied inclusively over the full parton configuration. As a by-product of this treatment, the MLM approach penalizes "losing" jets which is not the case in the original merging prescriptions. This algorithm has originally been implemented in ALPGEN [743, 744]; a variant of it using the Durham $k_\perp$-algorithm has later been provided in the MADGRAPH framework [146].

Despite the subtle differences in the different approaches, by far and large a good agreement of predictions obtained between both methods can be observed. Respective results have been reported for the case of W+jets production at the TEVATRON and the LHC for example in [147].

5.5.1.7   Extensions: Dealing with photons

The extension of the merging algorithm to also include photons is fairly straightforward: treating the emission of photons "democratically", *i.e.* on the same footing as QCD emissions allows to embed them into a multijet merging such that both the number of QCD particles and photons will vary. Evaluating matrix elements at the tree-level with $N_{\rm QCD}$ QCD particles and N_γ photons is not a problem and can be dealt with standard technology. There is also no difficulty in supplementing the jet definition encoded in Q_J and a corresponding cut $Q_{\rm cut}$ with an isolation criterion Q_γ and a corresponding cut $Q_{\rm iso}$ for the emission of photons. From the parton shower side, things are similarly trivial and basically amount to supplementing $q \to q\gamma$ splitting kernels, which typically can be obtained from the $q \to qg$ ones by suitably replacing coupling factors. In principle it is also possible — and probably even desirable — to also include $\gamma \to f\bar{f}$ splittings and to have different infrared cut-offs for the QCD and the QED part of the parton shower, in other words to supplement t_c with a $t_{c,({\rm QED})}$. The latter typically would be chosen on scales of the order of the π^0 mass or similar, which is of course possible due to the absence of a Landau pole in QED in the soft regime.

Of course, the same logic could also be applied to extend this treatment to, e.g., the emission of the weak vector bosons, the $W^\pm$ and Z bosons. There is one caveat, though, related to the fact that the coupling of, say, W bosons to fermions is chiral and therefore highly sensitive to the spin of the fermions and, consequently, introduces non-trivial spin correlations in the parton shower. Including them in a systematic way would imply that the easily implemented structure of more or less independent emissions in a simple probabilistic fashion would have to be augmented with spin-correlation matrices of the kind discussed in [825] or a treatment similar to the one discussed in [779]. However, for the emission of one boson only some matrix-element corrections could be applied which would by far and large capture such effects. This has been studied in more detail in the framework of the PYTHIA event generator [396].

5.5.1.8   Aside: unitarization of multijet merging: UMEPS

The idea underlying the "unitarization" of multijet merging is to conserve the fixed-order cross-section of the inclusive process with lowest multiplicity — in the MEPS master equation, Eq. (5.353), this would be the cross-section of the N-particle process,

$$\int d\sigma^{(LO)} \;=\; \int d\Phi_N \mathcal{B}_N \;\overset{!}{=}\; \int d\sigma^{(\text{UMEPS})} . \tag{5.354}$$

This is in principle achieved by shuffling contributions from higher multiplicities to lower ones, thus compensating for the mismatch of fixed-order real emission cross-sections and their parton-shower approximation above Q_{cut}. Such a mismatch manifested itself in Eq. (5.349) as a difference between $\mathcal{B}_{N+1}$ and $\mathcal{B}_N \otimes \mathcal{K}_N$.

To see in more detail how this works, consider first the expression for the Sudakov form factor, Eq. (5.251), which can be decomposed as

$$\Delta_N^{(\mathcal{K})}(T, t) \;=\; \Delta_N^{(\mathcal{K})}(T, t; Q_{N+1} \geq Q_{\text{cut}}) \,\times\, \Delta_N^{(\mathcal{K})}(T, t; Q_{N+1} < Q_{\text{cut}})$$

$$= \Delta_N^{\geq Q_{\text{cut}},(\mathcal{K})}(T, t) \,\times\, \Delta_N^{< Q_{\text{cut}},(\mathcal{K})}(T, t)$$

$$= \exp\left[-\int_t^T d\Phi_1 \, \mathcal{K}_N^{\geq Q_{\text{cut}}}(\Phi_1) \right] \,\times\, \exp\left[-\int_t^T d\Phi_1 \, \mathcal{K}_N^{< Q_{\text{cut}}}(\Phi_1) \right] . \tag{5.355}$$

Using this in Eq. (5.349) allows to rewrite the first line as

$$d\Phi_N \, \mathcal{B}_N \left[\Delta_N^{(\mathcal{K})}(\mu_N^2, t_{\text{c}}) + \int_{t_{\text{c}}}^{\mu_N^2} d\Phi_1 \, \mathcal{K}_N \Delta_N^{(\mathcal{K})}(\mu_N^2, t_{N+1}) \Theta(Q_{\text{cut}} - Q_{N+1}) \right]$$

$$= d\Phi_N \, \mathcal{B}_N \, \Delta_N^{\geq Q_{\text{cut}},(\mathcal{K})}(\mu_N^2, t_{\text{c}})$$

$$\times \left[\Delta_N^{< Q_{\text{cut}},(\mathcal{K})}(\mu_N^2, t_{\text{c}}) + \int_{t_{\text{c}}}^{\mu_N^2} d\Phi_1 \mathcal{K}_N^{< Q_{\text{cut}}} \Delta_N^{< Q_{\text{cut}},(\mathcal{K})}(\mu_N^2, t) \right] \tag{5.356}$$

Again, because of the identical kernels in the individual terms and the identical phase-space constraints, the square bracket integrates to unity. Using the probabilistic nature of the parton shower, the Sudakov form factor, being interpreted as the no-emission probability, can be recast as unity minus the emission probability. In other words, in

$$\Delta_N^{\geq Q_{\text{cut}},(\mathcal{K})}(\mu_N^2, t_{\text{c}}) = 1 - \int_{t_{\text{c}}}^{\mu_N^2} d\Phi_1 \, \mathcal{K}^{\geq Q_{\text{cut}}} \, \Delta_N^{\geq Q_{\text{cut}},(\mathcal{K})}(\mu_N^2, t) \tag{5.357}$$

the integrated parton shower emission rate above Q_{cut} is given by the second term.

The unitarization of the cross-section is now achieved by suitably replacing the kernel in the integrand of Eq. (5.357),

$$\tilde{\Delta}_N^{\geq Q_{\mathrm{cut}},(\mathcal{K})}(\mu_N^2, t_c) = 1 - \int_{t_c}^{\mu_N^2} d\Phi_1 \frac{\mathcal{B}_{N+1}}{\mathcal{B}_N} \Theta(Q_{N+1} - Q_{\mathrm{cut}}) \Delta_N^{\geq Q_{\mathrm{cut}},(\mathcal{K})}(\mu_N^2, t) \quad (5.358)$$

Merging now the first emission through the corresponding Born matrix element yields

$$
\begin{aligned}
d\sigma = {} & d\Phi_N \mathcal{B}_N \tilde{\Delta}_N^{\geq Q_{\mathrm{cut}},(\mathcal{K})}(\mu_N^2, t_c) \\
& \times \left[\Delta_N^{< Q_{\mathrm{cut}},(\mathcal{K})}(\mu_N^2, t_c) + \int_{t_c}^{\mu_N^2} d\Phi_1 \mathcal{K}_N^{< Q_{\mathrm{cut}}} \Delta_N^{< Q_{\mathrm{cut}},(\mathcal{K})}(\mu_N^2, t) \right] \\
& + d\Phi_{N+1} \mathcal{B}_{N+1} \Delta_N^{\geq Q_{\mathrm{cut}},(\mathcal{K})}(\mu_N^2, t) \, \Theta(Q_{N+1} - Q_{\mathrm{cut}}) \, .
\end{aligned}
\quad (5.359)
$$

Similar replacements are applied in Eq. (5.350) to all brackets encoding the parton shower evolution, with the exception of the N_{max}-term. Consequently, all N-parton inclusive cross-sections are governed by the input cross-sections associated with the $\mathcal{B}_N$. Some comments are in order here. First of all, it is possible that a parton shower cannot produce all particles or phase-space configurations entering these cross-sections. In such a case the corresponding fixed-order cross-sections must be regarded as genuine corrections to the parton shower and therefore they will be just added. This is also true for the impact of the phase-space constraints given by the Θ functions. It is not always guaranteed that integrating over the one-particle phase space will yield a lower-multiplicity state that passes the Q_{cut} criterion — sometimes one loses more than one jet. Handling such contributions is ambiguous, since they could be regarded as genuine fixed-order corrections as discussed in [806] or as contributions to even lower multiplicity states as in [733].

5.5.2 Hybrids of merging and matching: MENLOPS

In a first step towards a full multijet merging on the basis of NLO multijet matrix elements, and with NLO matching and LO merging methods well established it is worth discussing a hybrid between these two methods. By now this has been established as the MENLOPS method, and it has been introduced and implemented for various processes in [608, 629]. It combines matching, either with the POWHEG or the MC@NLO method, of QCD NLO matrix elements for the lowest-multiplicity final states with an LO multijet merging of higher-multiplicity matrix elements. This implies that the NLO matching has to be performed in such a way that the real-emission part of the NLO correction does not produce any additional jets; fully in line with the LO merging paradigm therefore the emission phase space has to be constrained through a jet measure Q_J and a corresponding cut Q_{cut}. This requirement is realized by suitably multiplying with a Sudakov form factor. In addition, the no-jet emission constraint must act on the phase space available for both the real-emission correction encoded through an exact matrix element and on the parton shower parts.

While this is, by far and large, not too hard to implement, there is another some-what more nagging problem. The jet-exclusive cross-section for the lowest-multiplicity final state will be modified by the local K-factor discussed in Sections 5.4.3 and 5.4.4, while the higher multiplicities will not be corrected and be at Born level only. This can lead to discontinuities in the radiation pattern, in particular of the first, hardest emission, and especially in those cases, where the K factors are fairly different from unity, like, for instance, in the case of Higgs production through gluon fusion. In order to remedy this situation, the higher-multiplicity part could be multiplied by an inter-polating K-factor, k_N, capturing the NLO correction to the lowest multiplicity N. For example, for the MENLOPS method, where the NLO matching part is realized through the MC@NLO algorithm, this interpolating K-factor may be given by something like

$$k_N(\Phi_{N+1}) = \frac{\tilde{\mathcal{B}}_N}{\mathcal{B}_N}\left(1 - \frac{\mathcal{H}_N}{\mathcal{B}_{N+1}}\right) + \frac{\mathcal{H}_N}{\mathcal{B}_{N+1}} \longrightarrow \begin{cases} \tilde{\mathcal{B}}_N/\mathcal{B}_N & \text{for soft emissions} \\ 1 & \text{for hard emissions.} \end{cases} \tag{5.360}$$

Such an interpolation works fairly well, since in the soft limit the real correction term and its subtraction become very similar and therefore the hard remainder is given by

$$\mathcal{H}_N = \mathcal{R}_N - \mathcal{S}_N \xrightarrow{\text{soft}} 0 \, , \tag{5.361}$$

while in the hard limit the subtraction term is not very prominent and therefore

$$\mathcal{H}_N = \mathcal{R}_N - \mathcal{S}_N \xrightarrow{\text{hard}} \mathcal{R}_N = \mathcal{B}_{N+1} \, . \tag{5.362}$$

Concentrating again on contributions up to the first emissions only, the differential cross-section for such a sample is given by

$$\begin{aligned} d\sigma = \; & d\Phi_N \, \Theta(Q_N - Q_{\text{cut}}) \, \tilde{\mathcal{B}}_N \\ & \times \left[\Delta_N^{(\mathcal{K})}(\mu_N^2, t_{\text{cut}}) + \int_{t_{\text{cut}}}^{\mu_N^2} d\Phi_1 \, \mathcal{K}_N \Delta_N^{(\mathcal{K})}(\mu_N^2, t_{N+1}) \Theta(Q_{\text{cut}} - Q_{N+1}) \right] \\ & + d\Phi_{N+1} \, \Theta(Q_N - Q_{\text{cut}}) \Theta(Q_{\text{cut}} - Q_{N+1}) \, \mathcal{H}_N \Delta_N^{(\mathcal{K})}(\mu_{N+1}^2, t_{N+1}) \\ & + d\Phi_{N+1} \, k_N(\Phi_{N+1}) \, \Theta(Q_{N+1} - Q_{\text{cut}}) \, \mathcal{B}_{N+1} \Delta_N^{(\mathcal{K})}(\mu_{N+1}^2, t_{N+1}) \\ & \times \left[\Delta_{N+1}^{(\mathcal{K})}(t_{N+1}, t_{\text{cut}}) + \int_{t_{\text{cut}}}^{t_{N+1}} d\Phi_1 \, \mathcal{K}_{N+1} \Delta_{N+1}^{(\mathcal{K})}(t_{N+1}, t_{N+2}) \Theta(Q_{\text{cut}} - Q_{N+2}) \right] \\ & + d\Phi_{N+2} \, k_N(\Phi_{N+2}) \, \Theta(Q_{N+2} - Q_{\text{cut}}) \, \mathcal{B}_{N+2} \Delta_{N+1}^{(\mathcal{K})}(\mu_{N+2}^2, t_{N+2}) \Delta_N^{(\mathcal{K})}(t_{N+2}, t_{N+1}) \\ & \times \left[\Delta_{N+1}^{(\mathcal{K})}(t_{N+2}, t_{\text{cut}}) + \dots \right. \end{aligned} \tag{5.363}$$

Here, the first three lines correspond to an MC@NLO simulation, where the phase space for the $(N + 1)$th particle is constrained such that it does not yield another jet — this is what the Θ–functions $\Theta(Q_{\mathrm{cut}} - Q_{N+1})$ are there for. In order to make the jet counting more explicit, another Θ-function, $\Theta(Q_N - Q_{\mathrm{cut}})$ has been added to highlight that all other QCD particles in the N-particle Born-level final state should be jets. Similarly, the fourth and fifth lines encapsulate the first additional tree-level matrix element, merged into the sample, and supplemented with the interpolating K factor k_N. Again, no further jet emission is allowed. This could now continue like indicated by the sixth line, to include higher and higher jet multiplicities. In any case, any further emissions, as discussed already for the multijet merging at leading order, cannot result in additional unwanted jets.

5.5.3   Multijet merging at next-to-leading order: MEPS@NLO

5.5.3.1   Basic idea

The idea underlying multijet merging with next-to-leading order matrix elements, MEPS@NLO, is identical to multijet merging at leading order: towers of matrix elements with increasing jet multiplicities are combined into one inclusive sample in such a way that no double-counting occurs. Of course, as before, it is important to maintain the accuracy of both the matrix elements, *i.e.* the cross-sections related to the processes as well as the fixed-order accuracy of the first emission, and the parton shower, *i.e.* the resummation of leading and the next-to-leading logarithms encoded in the showering.

This has been first achieved in [558, 631], where the first missing terms have explicitly been shown to be of order $\alpha_{\mathrm{s}}^2 L^3 / N_c^2$ — the colour-suppressed sub-leading logarithms beyond shower accuracy. Alternative methods and implementations have been presented in [535], and in [737, 806]. The latter two are closely related to one another, and both guarantee the proper "unitarization" of the emissions in line with the leading-order treatment by the same authors [733, 806].

5.5.3.2   First emission(s), once more

Following the original presentation of a multijet-merging method of next-to leading-order matrix elements in [558, 631], this MEPS@NLO method can be understood as a merging of individual MC@NLO simulations. Some additional terms need to be included, though, to guarantee the fixed-order and logarithmic correctness. Concentrating first on the merging of matrix elements for the production of an N and an $(N + 1)$-particle state only, and taking into account the first emissions only, a naive addition of two MC@NLO simulations for both would yield the (wrong) cross-section

$$\mathrm{d}\sigma^{(\mathrm{wrong})} = \mathrm{d}\Phi_N \, \Theta(Q_N - Q_{\mathrm{cut}}) \, \tilde{\mathcal{B}}_N$$

$$\times \left[\Delta_N^{(\mathcal{K})}(\mu_N^2, t_{\mathrm{c}}) + \int\limits_{t_{\mathrm{c}}}^{\mu_N^2} \mathrm{d}\Phi_1 \, \mathcal{K}_N \Delta_N^{(\mathcal{K})}(\mu_N^2, t_{N+1})\Theta(Q_{\mathrm{cut}} - Q_{N+1}) \right]$$

$$+ \, \mathrm{d}\Phi_{N+1} \, \Theta(Q_N - Q_{\mathrm{cut}})\Theta(Q_{\mathrm{cut}} - Q_{N+1}) \, \mathcal{H}_N \Delta_N^{(\mathcal{K})}(\mu_N^2, t_{N+1})$$

$$+ \, \mathrm{d}\Phi_{N+1} \, \Theta(Q_{N+1} - Q_{\mathrm{cut}}) \, \tilde{\mathcal{B}}_{N+1}$$

$$\times \left[\Delta_{N+1}^{(\mathcal{K})}(t_{N+1}, t_{\mathrm{c}}) + \int\limits_{t_{\mathrm{c}}}^{t_{N+1}} \mathrm{d}\Phi_1 \, \mathcal{K}_{N+1} \Delta_{N+1}^{(\mathcal{K})}(t_{N+1}, t_{N+2}) \right]$$

$$+ \, \mathrm{d}\Phi_{N+2} \, \Theta(Q_{N+1} - Q_{\mathrm{cut}}) \, \mathcal{H}_{N+1} \Delta_N^{(\mathcal{K})}(\mu_N^2, t_{N+1})\Delta_{N+1}^{(\mathcal{K})}(t_{N+1}, t_{N+2}) \,.$$

$$(5.364)$$

The first three lines would correspond to an MC@NLO simulation, where, in complete agreement with Eq. (5.363), the phase space for the $(N+1)$th particle is constrained such that it does not yield another jet, realized by $\Theta(Q_{\mathrm{cut}} - Q_{N+1})$ in the two emission terms, soft and hard. The next three lines would then stand for the next MC@NLO simulation, for a process with one more particle in the final state.

Naively, it seems as if there is no double counting present here. However, this is not entirely true. To see this, consider the emission term in the first, soft part of the lowest order MC@NLO simulation, the term $\mathcal{K}_N \Delta_N^{(\mathcal{K})}(\mu_N^2, t_{N+1})$ in the first square bracket and the corresponding hard emission part in the line below. Closer inspection reveals that at first order in α_{s}, there is some unwanted contribution to the phase space of the next MC@NLO simulation. It stems from the expansion of the Sudakov form factor accounting for no emissions harder than t_{N+1} in the combined soft and hard radiation pattern, ranging over the full emission phase space from t_{N+1} up to μ_N^2:

$$\left[\mathrm{d}\Phi_N \int\limits_{t_{\mathrm{c}}}^{\mu_N^2} \mathrm{d}\Phi_1 \, \tilde{\mathcal{B}}_N \mathcal{K}_N + \mathrm{d}\Phi_{N+1} \mathcal{H}_N \right] \Theta(Q_{\mathrm{cut}} - Q_{N+1}) \Delta_N^{(\mathcal{K})}(\mu_N^2, t_{N+1})$$

$$= \left[\mathrm{d}\Phi_N \int\limits_{t_{\mathrm{c}}}^{\mu_N^2} \mathrm{d}\Phi_1 \, (\mathcal{B}_N \otimes \mathcal{K}_N + \mathcal{H}_N) + \mathcal{O}\left(\alpha_{\mathrm{s}}^2\right) \right] \Theta(Q_{\mathrm{cut}} - Q_{N+1}) \Delta_N^{(\mathcal{K})}(\mu_N^2, t_{N+1})$$

$$= \mathrm{d}\Phi_{N+1} \mathcal{B}_{N+1} \Theta(Q_{\mathrm{cut}} - Q_{N+1}) \left[1 - \int\limits_{t_{N+1}}^{\mu_N^2} \mathrm{d}\Phi_1 \mathcal{K}_N + \mathcal{O}\left(\alpha_{\mathrm{s}}^2\right) \right] \,. \qquad (5.365)$$

The second part of the bracket in the last line therefore interferes with the emissions of the MC@NLO simulation of the incremented multiplicity. This double-counting of emissions must obviously be avoided. There are various way of achieving this, the simplest one is by adding the second term in the square bracket above to the $(N+1)$-

MC@NLO simulation:

$$
\begin{aligned}
d\sigma = \quad & d\Phi_N \, \Theta(Q_N - Q_{\text{cut}}) \, \tilde{\mathcal{B}}_N \\
& \times \left[\Delta_N^{(\mathcal{K})}(\mu_N^2, t_c) + \int_{t_c}^{\mu_N^2} d\Phi_1 \, \mathcal{K}_N \Delta_N^{(\mathcal{K})}(\mu_N^2, t_{N+1}) \Theta(Q_{\text{cut}} - Q_{N+1}) \right] \\[1ex]
& + d\Phi_{N+1} \, \Theta(Q_N - Q_{\text{cut}}) \Theta(Q_{\text{cut}} - Q_{N+1}) \, \mathcal{H}_N \, \Delta_N^{(\mathcal{K})}(\mu_N^2, t_{N+1}) \\[1ex]
& + d\Phi_{N+1} \Theta(Q_{N+1} - Q_{\text{cut}}) \, \tilde{\mathcal{B}}_{N+1} \, \Delta_N^{(\mathcal{K})}(\mu_{N+1}^2, t_{N+1}) \left(1 + \frac{\mathcal{B}_{N+1}}{\tilde{\mathcal{B}}_{N+1}} \int_{t_{N+1}}^{\mu_N^2} d\Phi_1 \, \mathcal{K}_N \right) \\[1ex]
& \times \cdot \left[\Delta_{N+1}^{(\mathcal{K})}(t_{N+1}, t_c) + \int_{t_c}^{t_{N+1}} d\Phi_1 \, \mathcal{K}_{N+1} \Delta_{N+1}^{(\mathcal{K})}(t_{N+1}, t_{N+2}) \right] \\[1ex]
& + d\Phi_{N+2} \, \Theta(Q_{N+1} - Q_{\text{cut}}) \Theta(Q_{\text{cut}} - Q_{N+2}) \, \mathcal{H}_{N+1} \\
& \quad \times \Delta_N^{(\mathcal{K})}(\mu_{N+1}^2, t_{N+1}) \Delta_{N+1}^{(\mathcal{K})}(t_{N+1}, t_{N+2}) \,.
\end{aligned}
$$

$$\tag{5.366}$$

As in the MENLOPS method described in Section 5.5.2, the first three lines merely describe an MC@NLO simulation of the lowest multiplicity sample with an additional jet veto applied to all emissions. In a similar way, the next lines describe an MC@NLO simulation for the next higher multiplicity, modified by the term in round brackets in the fourth line compensating for a double-counting of terms stemming from the lowest multiplicity MC@NLO.

As already worked out, this term compensates contributions stemming from the Sudakov form factor encoding the veto of hard emissions from the lower multiplicity, the two first lines. Such a compensation must be introduced into the formalism in order to maintain the NLO accuracy of the overall procedure. The fact that this double-counting of NLO terms is introduced by the parton shower (or analytic Sudakov form factors encoding the jet veto) renders these contributions potentially hard to understand at first. Realizing, however, that Sudakov form factors encode some of the NLO corrections in a leading-logarithmic approximation, this should not come as a surprise. In fact, the treatment here actually is fully analogous to the one of the Sudakov form factor in its interaction with higher-order matrix elements, presented later in Section 5.6.1, and made explicit in Eq. (5.372) there.

Naively, such terms also seem to be hard to implement when generating the Sudakov rejection directly from the parton shower rather than analytically. This turns out not to be entirely true. In close analogy to the discussion of vetoed emissions, *cf.* Eq. (5.348), it can be seen that these terms correspond to a vetoed emission off the N-particle state only. Instead of vetoing an event when an unwanted emission takes place, it is merely the hard emission itself that should be vetoed.

Of course, the very same logic could be iterated to include even higher multiplicities. In addition, further leading-order matrix elements could be added following the same reasoning with an interpolating K-factor as in the MENLOPS algorithm.

5.5.3.3 Scale setting in MEPS@NLO

One remaining problem is the definition of scales, and in particular the nodal values for the individual emissions and, correspondingly, the renormalization scales of the strong coupling in the various terms. Simply put, the same logic as in the LO case, Section 5.5.1, will apply. Starting from Born-type configurations, the partons are clustered backwards, yielding, as before, a parton shower history for the configuration. Its nodal values, t_i will be used in the Sudakov rejection factors, and, together with the rest of the kinematics of the ith splitting the corresponding renormalization scale $\mu_R^{(i)}$ can be deduced. As in Eq. (5.345), then the overall scale for a process with M powers of the strong coupling in the $2 \to 2$ core process and with m additional QCD emissions will be given as

$$\alpha_{\mathrm{s}}^{M+m}(\mu_R^2) = \alpha_{\mathrm{s}}^n(\mu_R^2) = \alpha_{\mathrm{s}}^M(\mu_{R(\text{core})}^2) \cdot \prod_{i \in m} \alpha_{\mathrm{s}}(\mu_{R,(i)}^2). \tag{5.367}$$

This scale is also used for the strong coupling related to loop corrections or to the soft emissions below the jet-threshold encoded for instance in the terms $\mathcal{H}_N$. They are effectively ignored in the construction of the parton shower history, which relies on tree-level configurations only.

When performing a scale variation of the renormalization scale in order to assess the corresponding uncertainty, one must therefore modify the Born matrix element by a factor such that

$$\alpha_{\mathrm{s}}^n(\mu_R^2) \longrightarrow \alpha_{\mathrm{s}}^n(\tilde{\mu}_R^2)\left(1 - \frac{\alpha_{\mathrm{s}}(\tilde{\mu}_R^2)}{2\pi}\beta_0 \sum_{i=1}^{n} \log\frac{\mu_i^2}{\tilde{\mu}_R^2}\right), \tag{5.368}$$

while all higher order terms are just evaluated with the coupling taken at the new scale $\tilde{\mu}_R^2$. This is slightly different from the scale setting prescription in MINLO, where the value of α_{s} chosen for the additional real or virtual correction is fixed to be the average of all other values of the strong coupling, *cf.* Eq. (5.375).

In a similar way, a new factorization scale can be chosen for the matrix elements, but must be compensated for by a term of the form

$$\mathcal{B}_N \log\frac{\tilde{\mu}_F^2}{\mu_F^2}\left[\sum_{c=q,g}\int_{x_a}^1 \frac{\mathrm{d}z}{z}\mathcal{P}_{ac}(z)\,f_{c/h_a}\left(\frac{x_a}{z},\tilde{\mu}_F^2\right) + \sum_{d=q,g}\int_{x_b}^1 \frac{\mathrm{d}z}{z}\mathcal{P}_{bd}(z)\,f_{d/h_b}\left(\frac{x_b}{z},\tilde{\mu}_F^2\right)\right].$$
$$\tag{5.369}$$

5.6 NNLO and parton showers

Building on the algorithms and technologies discussed so far, the first implementations of combinations of calculations at next-to-next-to leading order (NNLO) in the strong coupling for simple processes — essentially the production of simple colour–singlet systems — and parton showers have emerged. The first of them, dubbed MINLO, builds on a scale–setting prescription similar to the one employed in MEPS@NLO; its alternative, UNNLOPS in turn employs technology imported from the UNLOPS approach to multijet merging of NLO-QCD matrix elements. Both will briefly be sketched in the remainder of this chapter.

5.6.1 Multijet merging without scales: MINLO

5.6.1.1 Underlying ideas

The MINLO ideas have been worked out in [611] (MINLO-1) and [609] (MINLO-2).

The first instance focuses on improving the behaviour of fixed-order calculations in the soft region of parton emission, where they usually show unphysical, *i.e.* diverging, behaviour. This is achieved by a combination of clever choices for the scale of the strong coupling and the modification of the matrix elements by Sudakov form factors, similar to the programme already introduced in multijet merging at leading order, *cf.* Section 5.5.1. The main complication, though, that needed to be addressed stems from the fact that in contrast to the leading-order matrix elements, on which the CKKW method [351], operated, MINLO-1 aims at modifying next-to-leading order matrix elements.

MINLO-2 pushes the ideas in MINLO-1 one step further, by reweighting the matrix elements with Sudakov form factors not at next-to-leading logarithmic accuracy (NLL) but actually at the next-to-next-to-leading logarithmic accuracy (NNLL). This ingenious idea allows the transverse momentum of the softest parton of the Born-level configuration to include scales of the order the parton shower cut-off t_c, around 1 GeV. In turn MINLO-2 feeds such configurations directly into the parton shower through the POWHEG method, thereby essentially taking into account NLO corrections for processes such as Drell–Yan or Higgs production in association with one potentially very soft parton in a full hadron-level simulation.

5.6.1.2 MINLO-1

As already stated, the primary aim of MINLO-1 is to improve the behaviour of single fixed–order calculations at NLO accuracy in the strong coupling in the Sudakov region, without upsetting their formal NLO accuracy. This is an important step towards improved stability of NLO calculations where a heavy system such as, e.g., a gauge or a Higgs boson is produced in association with light jets. In such cases, the stability of the calculation suffers with decreasing transverse momenta of the heavy system, or, correspondingly, of the light jets, due to the emergence of increasingly large logarithms, which must be resummed to all orders. As already encountered in previous sections, a convenient way to achieve this stabilization is through inclusion of now familiar Sudakov form factors. It should thus not be a big surprise that successful algorithms achieving this aim, such as the MINLO-1 method, are further developing ideas imported from multijet merging, presented in Section 5.5.3, but usually without the additional intricacies related to the combination of various NLO calculations for increasing multiplicities.

There are two ingredients in MINLO-1, already familiar from multijet merging. First of all, the renormalization and factorization scales are chosen to capture better the interplay of various scales in the process, and, secondly, the introduction of Sudakov weights which tame the instabilities emerging in the regime of small transverse momenta. This proceeds by constructing a parton shower history for the underlying Born-level configuration, omitting the softest emission in the real-emission contribution. As before, a strong ordering of the N nodal values in the hardness scale for the emissions is assumed,

$$Q_0 \geq Q_1 \geq Q_2 \ldots \geq Q_{N-1} \geq Q_N = Q_{\text{cut}}, \tag{5.370}$$

where Q_0 denotes the hardest scale of the core process. For instance, in the case of the production of a heavy singlet in association with a number of jets, Q_0 typically is the mass of the singlet. The weight corresponding to the individual phase-space point is multiplied with an overall suppression factor, given by a product of Sudakov form factors of all *internal lines i of the Born configuration* and a product of all Sudakov form factors of all *outgoing partons k of the Born configuration*, which have been produced through a splitting at the scale Q_k,

$$\mathcal{S} = \prod_{i=1}^{N} \Delta_i(Q_{i-1}^2, Q_i^2) \prod_k \Delta_k(Q_k^2, Q_{\text{cut}}^2). \tag{5.371}$$

Here, the subscripts i and k in Δ_i and Δ_k denote the flavour of the internal and external lines and the analytic Sudakov form factors from Eq. (2.183) are employed. The two external partons emerging at the last branching, *i.e.* at scale Q_N, will be associated with a Sudakov factor of $\Delta_k(Q_{\text{cut}}^2, Q_{\text{cut}}^2) = 1$.

In order to compensate for higher-order effects induced by the Sudakov form factors, the Born-term is modified by a factor which captures the first order in the α_s-expansion of the analytic Sudakov form factors above. For each of the Sudakov form factors, the corresponding first-order term in its expansion is denoted by $\Delta^{(1)}$. With this notation, the correction factor is given by

$$1 - \sum_i \Delta^{(1)}(Q_{i-1}^2, Q_i^2) - \sum_k \Delta_k^{(1)}(Q_k^2, Q_{\text{cut}}^2)$$

$$= 1 + \sum_i \int\limits_{Q_i^2}^{Q_{i-1}^2} \frac{\mathrm{d}q_\perp^2}{q_\perp^2} \frac{\alpha_s(q_\perp^2)}{2\pi} \Gamma_i(Q_{i-1}^2, q_\perp^2) + \sum_k \int\limits_{Q_{\text{cut}}^2}^{Q_k^2} \frac{\mathrm{d}q_\perp^2}{q_\perp^2} \frac{\alpha_s(q_\perp^2)}{2\pi} \Gamma_i(Q_k^2, q_\perp^2),$$

$$\tag{5.372}$$

where the Γ_i are the integrated splitting functions introduced in Eq. (2.178), and

$$\Delta_k^{(1)}(Q^2, Q_0^2) = -\int\limits_{Q_{\text{cut}}^2}^{Q_k^2} \frac{\mathrm{d}q_\perp^2}{q_\perp^2} \frac{\alpha_s(t)}{2\pi} \Gamma_i(Q_k^2, q_\perp^2). \tag{5.373}$$

Here, the first sum ranges over all internal lines i, while the second sum includes all outgoing parton of the Born configuration. This factor is in complete analogy to the one in Eq. (5.365), see also Section 5.5.3.

To see how the scale setting works out for the MINLO method, consider again a Born-level configuration with M powers of the strong coupling related to the core process and with m QCD emissions with corresponding scales $\mu_{R,(i)}$. In tree-level multijet merging methods, the overall scale μ_R is implicitly given by

$$\alpha_s^{M+N}(\mu_R^2) = \alpha_s^M(\mu_{R,(\text{core})}) \cdot \prod_{i \in N} \alpha_s(\mu_{R,(i)}), \tag{5.374}$$

cf. Eq. (5.345).

It may seem natural to use this scale also as the argument of the strong coupling in the real or virtual correction, and in fact, this is the way multijet merging methods at NLO choose the scale, *cf.* Section 5.5.3. In slight contrast, in MINLO-1 the average value of α_s,

$$\alpha_s^{(M+N+1)} = \frac{1}{M+n} \left[M\alpha_s(\mu_{R,(\text{core})}^2) + \sum_{i=1}^N \alpha_s(\mu_{R,(i)}) \right], \tag{5.375}$$

is employed for them; the rationale for this choice is detailed in the publication [611].

5.6.1.3   MINLO-2: first implementation of NNLOPS

Up to now, the MINLO-2 method has been formulated for the production of singlet systems only, for reason that will become obvious. The idea in it is to basically feed the next-to-leading order expressions in MINLO-1 for the production of such a singlet system S in association with a parton j into the POWHEG formalism, thereby providing the link to the parton shower. By allowing the parton j to become as soft or collinear as the parton shower cut-off t_c allows, the full one-parton emission phase space is filled at NLO accuracy, and the second parton is distributed according to LO accuracy. This is possible by reweighting the Born-level singlet plus jet configuration with a Sudakov form factor at $\mathcal{O}\left(\alpha_s^2\right)$ accuracy, *i.e.* including terms A_2 and B_2 from Q_T-resummation, thereby compensating all dangerous logarithms of the low scale. The catch then is to reweight the emergent event sample to a suitable distribution of the singlet at NNLO accuracy, to achieve the overall NNLO+PS accuracy.

Taking as an illustrative example for this procedure the production of a Higgs boson in conjunction with a parton, the $\bar{\mathcal{B}}$ term for $H + j$ production can be written in a similar way as before, including the Born term plus the real and virtual corrections. Following the reasoning of MINLO-1 above, however, Sudakov form factors are invoked to account for higher-order corrections which become large for small transverse momenta, and their interplay with the genuine higher-order terms must be accounted for by subtracting out their respective first order expansion. Taken together, and making all factors of α_s explicit,

$$\bar{\mathcal{B}}(\Phi_{\mathcal{B}}) = \alpha_s(m_H^2)\,\alpha_s(q_\perp)\,\Delta_g^2(m_H^2, Q_\perp^2) \cdot \left\{ \mathcal{B}(\Phi_{\mathcal{B}}) \left[1 - 2\Delta^{(1)}(m_H^2, Q_\perp^2) \right] \right.$$
$$\left. + \alpha_s(Q_\perp^2) \left[\tilde{\mathcal{V}}(\Phi_{\mathcal{B}}) + \int d\Phi_1\, \mathcal{R}(\Phi_{\mathcal{B}} \times \Phi_1) \right] \right\},$$

(5.376)

where $Q_\perp$ is the transverse momentum of the Higgs boson (and therefore, at Born-level, the extra parton) with $Q_\perp^2 \geq t_c$.

Integrating Eq. (5.376) over the full phase space of the Born-level $H + j$ configuration will therefore, by construction, yield the cross-section for $H + j$ at NLO accuracy, including all terms that are singular in the limit $Q_\perp \to 0$, giving rise to logarithms of $Q_\perp$. All singular terms, formally up to $\mathcal{O}\left(\alpha_s^2\right)$ with respect to the inclusive Higgs boson production, can of course also be obtained through the Q_T-resummation formalism at NNLL accuracy, which recovers *all* these logarithms. These are all terms of the form $\alpha_s L^2$ and $\alpha_s L$ as well as $\alpha_s^2 L^4$, $\alpha_s^2 L^3$, $\alpha_s^2 L^2$, and $\alpha_s^2 L$, where the short-hand notation $L = \log(Q^2/Q_\perp^2)$ has been used. In Q_T resummation the resummation scale Q typically is identified to be of order of the singlet mass, in this case therefore $Q = \mathcal{O}\left(m_H\right)$.

Therefore, including also the A_2 and B_2 terms into the reweighting Sudakov form factors above, and fixing the scale of α_s in the real and virtual contributions to $Q_\perp^2$ guarantees that the result not only is NLO-accurate for $H + j$, but also for inclusive H production.

There is one minor problem remaining, though, namely that the Sudakov form factors and coefficients given in Section 5.2.1, Eq. (5.65), Eq. (5.66), and Eq. (5.72), are for Q_T resummation in the conjugate $b_\perp$-space. Here, however, the Sudakov form factors are meant to be directly applied in transverse momentum space, and therefore a translation between both must be applied. As noted in [609], this actually has been worked out in [505] and essentially leads to adding a term

$$\Delta B_{2,b_\perp \to q_\perp}^{(q,g)} = 4\zeta(3) \left(A_1^{(q,g)} \right)^2,$$

(5.377)

with $A_1^{(q,g)}$ the corresponding soft coefficient, while the A_2 remain unaltered. The B_2 coefficients in the case of resummation directly in Q_T-space therefore are given by

$$B_{2,Q_T}^{(q,g)} = B_{2,b_\perp}^{(q,g)} + \Delta B_2^{(q,g)}.$$

(5.378)

These terms are used in the analytic NNLL Sudakov form factors, multiplied to the matrix element, of the MINLO method such that

$$\Delta_k^{(\mathrm{NNLL})}(Q^2, Q_0^2) = \exp\left\{ -\int_{Q_0^2}^{Q^2} \frac{dq_\perp^2}{q_\perp^2} \left[A(q_\perp^2) \log \frac{Q^2}{q_\perp^2} + B(q_\perp^2) \right] \right\},$$

(5.379)

where

$$A(q_\perp^2) = \frac{\alpha_s(q_\perp^2)}{2\pi} A_1 + \left(\frac{\alpha_s(q_\perp^2)}{2\pi}\right)^2 A_2$$

$$B(q_\perp^2) = \frac{\alpha_s(q_\perp^2)}{2\pi} B_1 + \left(\frac{\alpha_s(q_\perp^2)}{2\pi}\right)^2 B_{2,Q_T},$$

(5.380)

with the coefficients A from Eq. (5.65), B_1 from Eq. (5.67), and the term B_2 from Eq. (5.378) with the B_2 term in $b_\perp$-space given in Eq. (5.72). Note that the latter also include finite loop-correction terms, making them process-dependent.

Finally, with both the inclusive process — in the example, inclusive Higgs production — and the process with an additional jet — $H+$ jet production — NLO correct, a seamless merging of the two multiplicities has been successfully achieved. This happened, and it is important to stress this, without any jet cut like in the usual multijet merging. Instead, by carefully adjusting Sudakov weights and scales in α_s it was possible to modify the $H+$ jet part in such a way that it automatically also accounts for the inclusive part.

In [610] it was then further realized that this could be turned into a full parton-level simulation, but accurate not only at NLO for both H and $H+$ jet production, but in fact accurate at NNLO in the inclusive cross-section. In order to achieve this, the rapidity distribution of the Higgs boson in the seamless merged sample merely has to be reweighted to the NNLO distribution. The same algorithm was also applied in a follow-up study, concerning Drell–Yan production [652]. There, the reweighting proceeds in three dimensions, thereby capturing the dynamics of the lepton system.

5.6.2   An alternative NNLO+PS implementation: UNNLOPS

To see how the UNNLOPS method [633, 634] works, consider first the simpler NLO case, known as UNLOPS. The idea there is to decompose the radiation pattern of *all* emissions into one part, where **no** emission happens and a complementary part capturing all emissions, and in particular at least one. Omitting the by-now obvious arguments of phase space for the different fixed-order contributions, the expectation value of an observable $\mathcal{O}$ is given by

$$\langle \mathcal{O} \rangle = \left\{ \int d\Phi_{\mathcal{B}} \left[\bar{\mathcal{B}}_N - \int_{t_c} d\Phi_1 \mathcal{R}_N \right] + \int_{t_c} d\Phi_{\mathcal{R}} \left[1 - \Delta_N(t_1, \mu_Q^2) \right] \mathcal{R}_N(\Phi_{\mathcal{R}}) \right\} \mathcal{O}(\Phi_{\mathcal{B}})$$

$$+ \int_{t_c} d\Phi_{\mathcal{R}} \Delta_N(t_1, \mu_Q^2) \mathcal{R}_N \mathcal{E}_N^{(\mathcal{K})}(t_1, t_c; \mathcal{O}).$$

Here, $\bar{\mathcal{B}}$ is the differential NLO cross-section for Born-level kinematics from Eq. (5.326),

$$\bar{\mathcal{B}}_N(\Phi_{\mathcal{B}}) = \mathcal{B}(\Phi_{\mathcal{B}}) + \tilde{\mathcal{V}}_N(\Phi_{\mathcal{B}}) + \int d\Phi_1 \left[\mathcal{R}_N(\Phi_{\mathcal{B}} \otimes \Phi_1) - \mathcal{S}_N(\Phi_{\mathcal{B}} \otimes \Phi_1) \right].$$ (5.381)

The argument of the observable indicates the particle composition and phase space from which it is evaluated. In view of this, the first line contains the observable after no emission has taken place — the first term is the fixed-order result for a Born-level configuration where all fixed-order parton emissions above t_c in the real correction have been subtracted, such that the one potentially remaining emission is below the parton shower cut-off. The second term in the first line refers to those events, which, starting from a real correction with a parton above the parton shower cut-off where vetoed due to an unwanted emission harder than t_1. In this case the parton configuration is projected onto the underlying Born configuration. The second line refers to those events which starting from a real-emission configuration experienced a parton shower evolution without unwanted emissions. This is signalled by the parton shower evolution operator $\mathcal{E}_n^{(\mathcal{K})}(t, t_c; \mathcal{O})$, defined in full analogy to the Eq. (5.310),

$$
\begin{aligned}
\mathcal{E}_n^{(\mathcal{K})}(t, t_c; \mathcal{O}) \;=\; & \Delta_n^{(\mathcal{K})}(t, t_c)\, \mathcal{O}(\Phi_n) \\
& + \int_{t_c}^{t} \mathrm{d}\Phi_1 \left[\mathcal{K}_n(\Phi_1)\, \Delta_n^{(\mathcal{K})}(t, t(\Phi_1)) \otimes \mathcal{E}_{n+1}^{(\mathcal{K})}(t(\Phi_1), t_c; \mathcal{O}) \right] .
\end{aligned}
$$

While the expression in Eq. (5.381) captures all relevant terms to maintain fixed-order accuracy, it does not account for the dependence of the real-emission contribution $\mathcal{R}_N$ on the renormalization and factorization scales. This can trivially remedied, however, by suitable weights, see also the original literature.

The same logic has been extended to match the parton shower to NNLO calculations for the production of colour singlets, performed with q_T subtraction. There, contributions with exactly zero, exactly one, and more than one emissions above t_c have been identified, leading to a somewhat lengthy expression, for which we refer to the literature.

6

Parton Distribution Functions

Parton Distribution Functions (PDFs) are a necessary ingredient in the calculation of particle cross-sections at collider experiments with hadron beams.

In Chapter 2, the master formula Eq. (2.52)

$$\sigma \;=\; \sum_{a,b} \int_0^1 \mathrm{d}x_a \mathrm{d}x_b \int f_{a/h_1}(x_a, \mu_F) f_{b/h_2}(x_b, \mu_F)\, \mathrm{d}\hat{\sigma}_{ab \to n}(\mu_F, \mu_R) \qquad (6.1)$$

was introduced to describe the calculation of the production of n-parton final states in processes of the type $h_1 h_2 \to n + X$ with incoming hadrons h_1 and h_2 – usually protons. The calculation of such a hard (parton-level) cross-section within perturbative QCD relies on also using partons in the inital state and thus requires the knowledge of the distribution of partons i in the hadrons h. This is most often achieved by using a scheme called collinear factorization in which this distribution depends on the longitudinal momentum fraction x of the partons with respect to the hadron, taken at the **factorization scale** μ_F. The resulting distributions $f_{i/h}(x, \mu_F)$ are known as parton distribution functions (**PDFs**), and parameterize the transition of incident hadrons to incident partons, thereby absorbing all emissions at scales below μ_F, cf. Chapter 2. In turn, the PDFs obviously depend upon this measure of the hardness of the parton–level interaction, which will be of the same order as the renormalization scale μ_R. As a consequence of the factorization of possible parton emissions into a soft and collinear part, below μ_F, and a hard part, above μ_F, the latter are explicitly described by the matrix element. Factorization theorems, proven for deep–inelastic lepton–proton scattering and for Drell-Yan production of gauge bosons in hadron collisions, assert that the PDFs in collinear factorization are process–independent. There are however terms of higher dimension – so called higher–twist contributions – that are process dependent. These terms are ignored throughout this book, since they are suppressed by factors of the type m_p^2/μ_F^2. Note that in this chapter both μ_R and μ_F will often be represented just as Q.

Pictorially, as was shown in Fig. 2.5 (left), more of the quantum fluctuations can be resolved as the hardness scale, corresponding to the inverse of the time scale, increases. Low–scale processes can only resolve longer time intervals, and with increasing scale,

The Black Book of Quantum Chromodynamics. John Campbell, Joey Huston and Frank Krauss, Oxford University Press. © John Campbell, Joey Huston, and Frank Krauss 2018, First published in paperback 2022. DOI: 10.1093/oso/9780192871961.001.0006

smaller time intervals are being probed, and more of the quantum fluctuations inside the hadron are resolved, which transfer momentum from high–x partons to low–x partons. Thus, at higher μ_F one expects to find more of the momentum of the proton given to gluons and sea quarks with relatively low values of the momentum fraction x, as the quantum fluctuations creating these partons can be resolved. Conversely, the population of partons containing a large fraction of the parent hadron's momentum will decrease.

The PDFs describing this dynamics are treated as being universal, *i.e.* they can be determined by one set of processes and used to predict the cross-section for any process, using the factorized form shown above. These parton distribution functions currently can not be calculated perturbatively; ultimately, however, it may become possible in the future to calculate these PDFs non-perturbatively, using lattice gauge theory [310]. On the other hand, the evolution of PDFs with Q^2 can be calculated with a perturbative treatment using the **DGLAP evolution equations**, as sketched in Section 2.1.3 and discussed in more detail in Section 6.1.

PDFs have been determined by global fits to data from a plethora of different data. Modern global PDF fits use of the order of 3000 data points from a number of processes from fixed target experiments, the TEVATRON, HERA, and the LHC. In particular, data from deep-inelastic scattering (DIS), Drell-Yan (DY) and jet production processes, have played a dominant role in the past. With the advent of the LHC and its huge amount of data, also processes such as γ+jet, W+jets, and heavy flavour production (including $t\bar{t}$) have become increasingly important as the statistical and systematic errors of the data sets have improved [827]. Many of these processes have recently been calculated to NNLO, allowing the data to be used in PDF fits at that order. At the moment this endeavour is actively pursued by a number of different groups: ABM [136], CTEQ/CT [489, 551], HERAPDF [115, 413, 819], JR [649], MSTW/MMHT [614, 755] and NNPDF [192, 194], which provide semi-regular updates to their fits of parton distributions, when new data and/or theoretical developments become available. Similarities and differences in the fitting procedure performed by the different groups will be discussed in some detail in Section 6.2.

The most commonly used are those PDFs which perform a global analysis on a broad variety of data using a variable flavour number scheme; the most recent updates (CT14, MMHT2014, NNPDF3.0) are given in [194, 489, 614]. The resulting PDFs are available at leading order (LO), next-to-leading order (NLO), and next-to-next-to-leading order (NNLO) in the strong coupling constant α_S, depending on the order(s) at which the global PDF fits have been carried out. Some PDFs have also been produced at what has been termed modified leading order [715, 846] (LO* or similar), in an attempt to reduce some of the problems that result from the use of LO PDFs in parton shower Monte Carlo programs. These PDFs are no longer in wide use, however. The choices of parameterization for these PDFs are then discussed, along with the impact of the use of particular renormalization and factorization schemes. Given the wide kinematic coverage of the data, the parameterization of the PDFs must be flexible enough to describe the parton distributions over a wide range of x and Q^2, and to not introduce artificial correlations between different x regions. Modern PDFs take into account the finite charm and bottom quark masses (using a variety of heavy quark mass schemes) in their fits to data, particularly to data from deep-inelastic scattering. This has consequences for low-mass cross-sections at the LHC that will be examined. Technical aspects related to the different orders, schemes and parameterizations are highlighted in Section 6.2.2.

There are two different classes of technique employed for the PDF determination: those based on the Hessian approach, and those using a Monte Carlo approach. The background behind both classes will be discussed in this chapter, and some of the actual procedures employed in the fitting of the PDFs and the determination of the respective errors will be highlighted. An important consequence of the method chosen is the way uncertainties in the fits are handled. These uncertainties also have implications on the accuracy of overall cross-section calculations. This is followed by a similar discussion of the choice of the value of $\alpha_s(m_Z)$ in the global fits. The data in the global fits are mostly from strong interaction physics, and they are thus sensitive to the exact value of $\alpha_s(m_Z)$. As a consequence, the global fit itself can be used to determine it. Since $\alpha_s(m_Z)$ is a universal parameter, however, another approach is to assume the world average value, and to not let α_s be a free parameter in the fit. Both approaches can be and have been used. Recently, most groups have provided PDFs at the world average value of $\alpha_s(m_Z)$, along with PDFs at alternate values of $\alpha_s(m_Z)$. Issues related to the fitting technology, impacting on estimates of the intrinsic uncertainties of PDFs and their impact on cross-sections are discussed in Section 6.3.

The next section is devoted to a discussion of PDF correlations. The examination and understanding of correlations between two fitted PDFs, or between a PDF and a cross-section, or between two cross-sections is crucial to a detailed appreciation of PDFs at the LHC. In adition, such correlations can be used to decrease the PDF uncertainties for the ratios of such quantities at the LHC. In Section 6.4, the resultant PDFs at LO (modified LO), NLO, and NNLO are presented. Parton luminosities are defined and are shown for several important initial states at the LHC in Section 6.5. PDF uncertainties using inputs from several PDF groups are then defined, using the PDF4LHC accords. Correlations among LHC processes are examined using these PDF combinations, and finally, in Section 6.6 several useful PDF tools that are currently available are described.

6.1 PDF evolution: the DGLAP equation revisited

6.1.1 PDF evolution at leading order

An essential feature of the PDFs is the fact that their evolution with energy can be calculated within the framework of perturbative QCD, as has already been outlined in Chapter 2. In that chapter the equation governing this evolution – the QCD **DGLAP equation** – was first introduced. At the leading order it reads,

$$
\frac{\partial}{\partial \log Q^2} \begin{pmatrix} f_{q/h}(x, Q^2) \\ f_{g/h}(x, Q^2) \end{pmatrix}
$$

$$
= \frac{\alpha_s(Q^2)}{2\pi} \int_x^1 \frac{\mathrm{d}z}{z} \begin{pmatrix} \mathcal{P}_{qq}^{(1)}\left(\frac{x}{z}\right) & \mathcal{P}_{qg}^{(1)}\left(\frac{x}{z}\right) \\ \mathcal{P}_{gq}^{(1)}\left(\frac{x}{z}\right) & \mathcal{P}_{gg}^{(1)}\left(\frac{x}{z}\right) \end{pmatrix} \begin{pmatrix} f_{q/h}(z, Q^2) \\ f_{g/h}(z, Q^2) \end{pmatrix}, \tag{6.2}
$$

The leading order splitting functions $\mathcal{P}_{ij}^{(1)}$ have previously been given in Eq. (2.33). They represent the kernel of this evolution equation and clearly couple different PDFs together. Note that the symmetries of QCD mean that, at this order, splitting functions involving anti-quarks are simply equal to their quark counterparts. Before turning to the form of the DGLAP equation and the splitting kernels at higher orders in the next section, the calculation of the kernels $\mathcal{P}_{ij}$ at leading order will be sketched below.

6.1.1.1 Factorization of initial state emissions in the collinear limit

The starting point for the calculation of the splitting kernels is the collinear limit of particle emission off initial-state partons. In order to see how this works, assume the production of a system X with mass M_X in the scattering of two incident partons j and j' with momenta p and p',

$$
p^\mu = \frac{\sqrt{\hat{s}}}{2}\,(1, 0, 0, 1) \quad \text{and} \quad p'^\mu = \frac{\sqrt{\hat{s}}}{2}\,(1, 0, 0, -1)\,, \tag{6.3}
$$

where $\hat{s} = (p + p')^2$. The differential parton–level cross-section for this process, at leading order, is

$$
\mathrm{d}\hat{\sigma}_{jj'\to X}^{(0)}(p, p') = \frac{1}{2\hat{s}}\,\mathrm{d}\Phi_X\,\left|\mathcal{M}_{jj'\to X}^{(0)}(p, p')\right|^2\,. \tag{6.4}
$$

Assume now that parton j emits a parton k with momentum k into the final state, turning into a parton i with reduced four momentum $p - k$, with $(p - k)^2 < 0$, which together with j' produces X. The momentum k is then given by

$$
k^\mu = (1 - x)\,p^\mu + \beta\,p'^\mu + k_\perp^\mu\,, \tag{6.5}
$$

where $k_\perp$ is perpendicular to both p and p'. β is fixed by the on-shell condition for k, $k^2 = 0$, and hence

$$
k^\mu = (1 - x)\,p^\mu + \frac{\mathbf{k}_\perp^2}{(1 - x)\hat{s}}\,p'^\mu + k_\perp^\mu\,. \tag{6.6}
$$

Here $\mathbf{k}_\perp^2$ denotes the positive square of the transverse momentum. Similarly,

$$
q^2 = (p - k)^2 = \frac{\mathbf{k}_\perp^2}{1 - x}\,, \tag{6.7}
$$

showing that the propagator of i is proportional to $\mathbf{k}_\perp^2$. As a consequence, the corresponding matrix element for the process under emission of k diverges with $\mathbf{k}_\perp \to 0$. This is the collinear limit, where the transverse momentum of the emitted particle vanishes. This singular behaviour also triggers divergences in the phase space integration over k. All these divergences however cancel, according to the **Kinoshita–Lee–Nauenberg theorem**, and as seen previously, when virtual corrections are added.

To proceed, the phase space integral over particle k has to be cast into a form useful for further consideration. With a bit of algebra it is easy to show that

$$d\Phi_k = \frac{d^4 k}{(2\pi)^4}(2\pi)\delta(k^2) = \frac{1}{(2\pi)^3}\,dx d\beta d^2 \mathbf{k}_\perp\, \delta\left((1-x)\beta - \frac{\mathbf{k}_\perp^2}{\hat{s}}\right)$$
$$= \frac{dx d\mathbf{k}_\perp^2}{16\pi^2(1-x)}. \tag{6.8}$$

Adding in the emission of the final state parton k, and ignoring all other possible emissions, yields a higher–order contribution to the cross-section of X production, and Eq. (6.4) becomes

$$d\hat{\sigma}_{jj'\to X}(p, p') = \frac{1}{2\hat{s}}\, d\Phi_X\, \left|\mathcal{M}_{jj'\to X}^{(0)}(p, p')\right|^2$$
$$+ \frac{1}{2\hat{s}}\frac{1}{16\pi^2}\, d\Phi_X \int_0^1 \frac{dx}{1-x}\int d^2\mathbf{k}_\perp^2 \left|\mathcal{M}_{jj'\to ikj'\to kX}^{(0)}(p, p')\right|^2. \tag{6.9}$$

The leading-order splitting functions $P_{ji}^{(1)}(x)$ in **collinear factorization** are then defined as the part of the second line of the equation above that diverges in the collinear limit:

$$\frac{1-x}{x}\frac{\alpha_s}{2\pi}P_{ji}^{(1)}(x)\left|\mathcal{M}_{ij'\to X}^{(0)}(xp, p')\right|^2 = \frac{1}{16\pi^2}\lim_{\mathbf{k}_\perp\to 0}\left[\mathbf{k}_\perp^2\left|\mathcal{M}_{jj'\to ikj'\to kX}^{(0)}(p, p')\right|^2\right]. \tag{6.10}$$

The x in the denominator of the left-hand side of the equation arises from the fact that the matrix element for $ij' \to X$ has an incoming flux given by $2x\hat{s}$ instead of the original $2\hat{s}$.

6.1.1.2 Calculating the kernels: $P_{qq}^{(1)}$

To calculate the splitting function $P_{qq}^{(1)}$, i and j are quark lines and k is a gluon. Clearly, gauge invariance of the overall matrix element for the associated production of X and the gluon k is only guaranteed if amplitudes for the emission of the gluon off all coloured lines are included. The question is whether these contributions exhibit collinear divergent behaviour.

It can be shown that this is not the case when choosing a smart physical gauge where the gluon's polarization vector ϵ^μ is transverse to both p' and k. To see this, an additional axis $n_\perp \perp k_\perp$ with $n_\perp^2 = 1$ is introduced, which allows the explicit construction of the polarization vectors as

$$\epsilon_\pm^\mu = \frac{\sqrt{2}\,\mathbf{k}_\perp}{(1-x)\hat{s}}p'^\mu + \frac{1}{\sqrt{2}\,\mathbf{k}_\perp}k_\perp^\mu \pm in_\perp^\mu. \tag{6.11}$$

In addition to the orthogonality requirements underlying their construction,

$$\epsilon \cdot p' = \epsilon \cdot k = 0 . \tag{6.12}$$

they satisfy

$$\epsilon_\pm \cdot k_\perp = \frac{\mathbf{k}_\perp}{\sqrt{2}} \quad \text{and} \quad \epsilon_\pm \cdot p = \frac{\mathbf{k}_\perp}{\sqrt{2}(1-x)} . \tag{6.13}$$

While the square of the graph with the propagator of i boasts two propagators and is thus proportional to $1/k_\perp^4$, all other graphs have at best one of these propagators. And since in the squared amplitude there are two scalar products involving the polarization vector, the numerator of the squared amplitude is proportional to $k_\perp^2$. This leaves no overall divergence in $1/k_\perp^2$ the squared amplitude apart from the one stemming from the graph with the i-propagator.

Analysing the structure of the reduced amplitude $\mathcal{M}_{ij' \to X}^{(0)}$, it becomes clear that its square can be written as

$$\left| \mathcal{M}_{ij' \to X}^{(0)} \right|^2 = \bar{u}(xp, \lambda) \left[\gamma_\mu M^\mu \right] u(xp, \lambda) , \tag{6.14}$$

where the spinors refer to the intermediate quark i. Because it is a massless quark, and since incoming and outgoing spinors come with the same helicity, this is the only allowed structure. Decomposing M according to

$$M^\mu = a p^\mu + b p'^\mu + m_\perp^\mu \tag{6.15}$$

with $m_\perp$ a vector in the same transverse plane as $k_\perp$ and using the Dirac equatoin elimiates the term proportional to p. Also, since the helicities are the same, the spin-flip operator $\gamma \cdot m$ vanishes when sandwiched between the two spinors. In the end, therefore

$$\left| \mathcal{M}_{ij' \to X}^{(0)} \right|^2 = \bar{u}(xp, \lambda) \left[\gamma_\mu b p'^\mu \right] u(xp, \lambda) . \tag{6.16}$$

realizing that

$$\left| \bar{u}(xp, \lambda) \left(\gamma \cdot p' \right) u(xp, \lambda) \right|^2 = (2xpp')^2 = x^2 \hat{s}^2 \tag{6.17}$$

identifies b and therefore

$$M^\mu = \frac{p'^\mu}{2x\hat{s}} \left| \mathcal{M}_{ij' \to X}^{(0)} \right|^2 . \tag{6.18}$$

These identities can now be rolled out to the calculation of the emission matrix element in the collinear limit. Including factors of $1/2$ and $1/N_c$ for the average over incoming quark helicities and colours,

$$\left| \mathcal{M}_{jj' \to ikj' \to kX}^{(0)}(p, p') \right|^2$$

$$= \frac{g^2 \operatorname{Tr} [T^a T^a]}{2 N_c} \sum_{\lambda, \pm} \bar{u}(p, \lambda) \not{\epsilon}_\pm \frac{\not{p} - \not{k}}{(p-k)^2} \left[\frac{\not{p}'}{x\hat{s}} \left| \mathcal{M}_{ij' \to X}^{(0)} \right|^2 \right] \frac{\not{p} - \not{k}}{(p-k)^2} \not{\epsilon}_\pm^* u(p, \lambda)$$

$$= \frac{g^2 \operatorname{Tr} [T^a T^a]}{2 N_c x \hat{s} (-2p \cdot k)^2} \sum_\pm \operatorname{Tr} \left[\not{p} \not{\epsilon}_\pm (\not{p} - \not{k}) \not{p}' (\not{p} - \not{k}) \not{\epsilon}_\pm^* \right] \left| \mathcal{M}_{ij' \to X}^{(0)} \right|^2$$

$$= \frac{2g^2 C_F}{x \mathbf{k}_\perp^2} (1 + x^2) \left| \mathcal{M}_{ij' \to X}^{(0)} \right|^2 . \tag{6.19}$$

Here, the following identities have been used:

$$2kp = \frac{\mathbf{k}_\perp^2}{1 - x}$$
$$2kp' = \hat{s}(1 - x)$$
$$\sum_\pm \epsilon_\mu^\pm \epsilon_\nu^{*\pm} = -g_{\mu\nu} + \frac{k_\mu p'_\nu + k_\nu p'_\mu}{kp'} . \tag{6.20}$$

Inserting this into Eq. (6.10) yields

$$\frac{1 - x}{x} \frac{\alpha_s}{2\pi} P_{qq}^{(1)}(x) = \frac{1}{16\pi^2} \times \frac{2g^2 C_F}{x} (1 + x^2) = \frac{8\pi \alpha_s C_F (1 + x^2)}{16\pi^2 x} \tag{6.21}$$

and therefore

$$P_{qq}^{(1)}(x) = C_F \frac{1 + x^2}{1 - x} , \tag{6.22}$$

as expected from Eq. (2.33).

6.1.1.3 Virtual contributions: The "+"-function and all that

Having arrived at the expression in Eq. (6.22) is not the end of the story, though. Taking a closer look at the expression for $P_{qq}^{(1)}$ shows the by now well–anticipated divergent behaviour for $x \to 1$, or, for vanishing gluon energies. In this limit of course also the transverse momentum vanishes, leading, ultimately, to the usual double soft and collinear divergence. Of course, these divergences will cancel, when taking into account the virtual contributions, a feature that consistently appeared throughout the book. Formalizing this idea results in

$$\text{finite} = \int_0^1 dx \left[\frac{\alpha_s}{2\pi} P_{qq}^{(1)}(x) |\mathcal{M}_{qj' \to X}(xp)|^2 + (1 + V) |\mathcal{M}_{qj' \to X}(xp)|^2 \right]$$
$$= \int_0^1 dx \left[\frac{\alpha_s}{2\pi} P_{qq}^{(1)}(x) + \delta(1 - x)(1 + V) \right] |\mathcal{M}_{qj' \to X}(xp)|^2 , \tag{6.23}$$

where V is the virtual contribution at order α_s. This contribution can be obtained by invoking a simple probabilistic idea, similar to what has been seen in the construction of parton showers. To order α_s, the probability of a gluon emission by the quark plus the probability of no emission must add to unity. Hence,

$$\int_0^1 dx \left[\frac{\alpha_s}{2\pi} P_{qq}^{(1)}(x) + \delta(1 - x)(1 + V) \right] = 1 \tag{6.24}$$

and therefore the virtual part must compensate the real emission encoded in the splitting function:

$$V = -\frac{\alpha_s}{2\pi} \lim_{\delta \to 0} \int_0^{1-\delta} \mathrm{d}x \, P_{qq}^{(1)}(x)$$

$$= \frac{\alpha_s C_F}{2\pi} \lim_{\delta \to 0} \int_0^{1-\delta} \mathrm{d}x \left(\frac{2}{1-x} - (1+x) \right) = \frac{\alpha_s C_F}{2\pi} \left(\frac{3}{2} + 2 \log \delta \right). \quad (6.25)$$

This implies that the complete splitting function is given by

$$\mathcal{P}_{qq}(x) = C_F \lim_{\delta \to 0} \left[\frac{1+x^2}{1-x} \Theta(1-x-\delta) + \delta(1-x) \left(\frac{3}{2} + 2 \log \delta \right) \right] \quad (6.26)$$

This looks quite cumbersome. A solution, however, presents itself by realizing that the "splitting functions" $\mathcal{P}_{ij}$ actually are not really functions but in fact distributions that will always act on some other functions $f(x)$ such as PDFs or similar. These functions are typically regular at $x = 1$ and it is therefore meaningful to define a way to regulate the divergent behaviour of the $\mathcal{P}_{ij}$ at $x = 1$ independent of the limit-procedure above. Introducing the "+" prescription, applicable for distributions that diverge at $x = 1$, such as $1/(1-x)$,

$$\left(\frac{1}{1-x} \right)_+ f(x) \stackrel{!}{=} \frac{f(x) - f(1)}{1-x}, \quad (6.27)$$

cf. Eq. (2.10). This prescription will take care of the $\log \delta$-term, located at $x = 1$ by virtue of the δ-function and therefore, finally,

$$\mathcal{P}_{qq}(x) = C_F \left[\left(\frac{1+x^2}{1-x} \right)_+ + \frac{3}{2} \delta(1-x) \right]. \quad (6.28)$$

This actually implies that the x–integral over $\mathcal{P}_{ij}$ vanishes exactly,

$$\int_0^1 \mathrm{d}x \, \mathcal{P}_{qq}(x) = 0, \quad (6.29)$$

thereby incorporating the exact cancellation of real and virtual contributions.

6.1.2   PDF evolution at higher orders

PDF evolution at the next order in perturbation theory is achieved by expanding the splitting functions as a series in the strong coupling,

$$\mathcal{P}_{ij}(x) = \mathcal{P}_{ij}^{(1)} + \frac{\alpha_s}{2\pi} \mathcal{P}_{ij}^{(2)} + \mathcal{O}(\alpha_s^2) \quad (6.30)$$

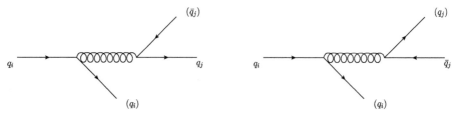

Fig. 6.1 Processes corresponding to the NLO splitting functions $\mathcal{P}^{(2)}_{q_i q_j}$ (left) and $\mathcal{P}^{(2)}_{q_i \bar{q}_j}$ (right). Unobserved partons are indicated in the figure by parentheses.

The inclusion of higher–order terms, $\mathcal{P}^{(2)}_{ij}$, introduces several subtleties that did not arise at leading order. The first is that, at leading order, the flavour structure of the evolution is trivial: for two flavours of quark, q_i and q_j, the splitting function trivially vanishes unless $i = j$, *i.e.* $\mathcal{P}^{(1)}_{q_i q_j} \propto \delta_{ij}$. This is no longer true at the next order, as indicated in Fig. 6.1 (a). Contributions originating from the splitting of a virtual gluon into a quark-antiquark pair of different flavour, $q_i \to q_j(q_i\bar{q}_j)$, give rise to kernels that are no longer diagonal in flavour space. Moreover, the same process also gives rise to kernels that directly couple quarks and antiquarks, $q_i \to \bar{q}_j(q_i q_j)$, *cf.* Fig. 6.1 (b).

Therefore Eq. (6.2) must be generalized in order to account for effects at higher orders. The evolution can be written as,

$$
\frac{\partial}{\partial \log Q^2}
\begin{pmatrix}
f_{q_i/h}(x, Q^2) \\
f_{\bar{q}_i/h}(x, Q^2) \\
f_{g/h}(x, Q^2)
\end{pmatrix}
$$
$$
= \frac{\alpha_s(Q^2)}{2\pi} \int_x^1 \frac{dz}{z}
\begin{pmatrix}
\mathcal{P}_{q_i q_k}\left(\frac{x}{z}\right) & \mathcal{P}_{q_i \bar{q}_k}\left(\frac{x}{z}\right) & \mathcal{P}_{qg}\left(\frac{x}{z}\right) \\
\mathcal{P}_{\bar{q}_i q_k}\left(\frac{x}{z}\right) & \mathcal{P}_{\bar{q}_i \bar{q}_k}\left(\frac{x}{z}\right) & \mathcal{P}_{\bar{q}g}\left(\frac{x}{z}\right) \\
\mathcal{P}_{gq_k}\left(\frac{x}{z}\right) & \mathcal{P}_{g\bar{q}_k}\left(\frac{x}{z}\right) & \mathcal{P}_{gg}\left(\frac{x}{z}\right)
\end{pmatrix}
\begin{pmatrix}
f_{q_k/h}(z, Q^2) \\
f_{\bar{q}_k/h}(z, Q^2) \\
f_{g/h}(z, Q^2)
\end{pmatrix} , \tag{6.31}
$$

with all of the distributions coupled together. The presence of some amount of anti-quarks in the parent quark means that there is no physically-meaningful separation into "valence" and "sea" contributions beyond leading order. However it is possible to identify combinations of quark PDFs that can be identified with the valence contribution as follows. The nature of the leading order splitting functions suggests that they be decomposed according to,

$$
\mathcal{P}_{q_i q_j} = \delta_{ij}\mathcal{P}^V_{qq} + \mathcal{P}^S_{qq} ,
$$
$$
\mathcal{P}_{q_i \bar{q}_j} = \delta_{ij}\mathcal{P}^V_{q\bar{q}} + \mathcal{P}^S_{q\bar{q}} . \tag{6.32}
$$

In order to solve the coupled evolution equations, Eq. (6.31), it is useful to introduce the following combinations of the splitting functions [550],

$$
\mathcal{P}_{(\pm)} = \mathcal{P}^V_{qq} \pm \mathcal{P}^V_{q\bar{q}} ,
$$
$$
\mathcal{P}_{QQ} = \mathcal{P}_{(+)} + 2n_f \mathcal{P}^S_{qq}
$$

$$\mathcal{P}_{Qg} = 2n_f \mathcal{P}_{qg}$$
$$\mathcal{P}_{gQ} = \mathcal{P}_{gq} \,, \tag{6.33}$$

for n_f flavours of light quarks. Similarly it is convenient to express the parton distributions in terms of the following quantities,

$$f_{q_i^{(\pm)}/h}(x, Q^2) = f_{q_i/h}(x, Q^2) \pm f_{\bar{q}_i/h}(x, Q^2) \,, \tag{6.34}$$

$$f_{Q^{(\pm)}/h}(x, Q^2) = \sum_{i=1}^{n_f} f_{q_i^{(\pm)}/h}(x, Q^2) \tag{6.35}$$

that is, into either a sum or difference of quark and antiquark distributions. The evolution expressed in Eq. (6.31) can now be written in a much simpler form in which the components decouple significantly. Two combinations satisfy a particularly simple form of the evolution,

$$\frac{\partial}{\partial \log Q^2} f_{q_i^{(-)}/h}(x, Q^2) = \frac{\alpha_s(Q^2)}{2\pi} \int_x^1 \frac{dz}{z} \mathcal{P}_{(-)} f_{q_i^{(-)}/h}(z, Q^2) \tag{6.36}$$

$$\frac{\partial}{\partial \log Q^2} S_i(z, Q^2) = \frac{\alpha_s(Q^2)}{2\pi} \int_x^1 \frac{dz}{z} \mathcal{P}_{(+)} S_i(z, Q^2) \,, \tag{6.37}$$

where the quantity $S_i(x, Q^2)$ is given by,

$$S_i(x, Q^2) = n_f f_{q_i^{(+)}/h}(x, Q^2) - f_{Q^{(+)}/h}(x, Q^2) \,. \tag{6.38}$$

These evolution equations can be solved separately; the solutions for $f_{q_i^{(-)}/h}(x, Q^2)$ and $S_i(x, Q^2)$ are referred to as **non-singlet** contributions. The remaining quark **singlet** combination, $f_{Q^{(+)}/h}(x, Q^2)$, remains coupled to the gluon contributions, but in a fashion akin to the leading order evolution (*cf.* Eq. (6.2)),

$$\frac{\partial}{\partial \log Q^2} \begin{pmatrix} f_{Q^{(+)}/h}(x, Q^2) \\ f_{g/h}(x, Q^2) \end{pmatrix}$$
$$= \frac{\alpha_s(Q^2)}{2\pi} \int_x^1 \frac{dz}{z} \begin{pmatrix} \mathcal{P}_{QQ}\left(\frac{x}{z}\right) & \mathcal{P}_{Qg}\left(\frac{x}{z}\right) \\ \mathcal{P}_{gQ}\left(\frac{x}{z}\right) & \mathcal{P}_{gg}\left(\frac{x}{z}\right) \end{pmatrix} \begin{pmatrix} f_{Q^{(+)}/h}(z, Q^2) \\ f_{g/h}(z, Q^2) \end{pmatrix} \,. \tag{6.39}$$

Eqs. (6.36), (6.37), and (6.39) are sufficient to determine the evolution of all of the individual PDFs at a given order, if the corresponding splitting functions are known to the same accuracy.

The first corrections to the leading-order splitting functions were originally computed in Refs. [421, 549]. Their forms are reproduced here for completeness. The two-loop gluon splitting function receives contributions from two different colour structures,

$$\mathcal{P}_{gg}^{(2)}(z) = C_A \mathcal{P}_{ggA}^{(2)}(z) + T_R n_f \mathcal{P}_{ggF}^{(2)}(z), \tag{6.40}$$

that are given by,

$$\mathcal{P}_{ggA}^{(2)}(z) = \frac{\Gamma_1}{8} \frac{1}{C_A} \left[P_{gg}^{(1)}(z) \right]_+ + \delta(1-z) \left[C_A(-1 + 3\zeta_3) + \beta_0 \right]$$
$$+ \left[P_{gg}^{(1)}(z) \right]_+ \left(-2\ln(1-z) + \frac{1}{2}\ln z \right) \ln z + \left[P_{gg}^{(1)}(-z) \right]_+ \left(S_2(z) + \frac{1}{2}\ln^2 z \right)$$
$$+ C_A \left(4(1+z)\ln^2 z - \frac{4(9+11z^2)}{3}\ln z - \frac{277}{18z} + 19(1-z) + \frac{277}{18}z^2 \right)$$
$$+ \beta_0 \left(\frac{13}{6z} - \frac{3}{2}(1-z) - \frac{13}{6}z^2 + (1+z)\ln z \right), \tag{6.41}$$

$$\mathcal{P}_{ggF}^{(2)}(z) = C_F \left[-\delta(1-z) + \frac{4}{3z} - 16 + 8z + \frac{20}{3}z^2 - 2(1+z)\ln^2 z - 2(3+5z)\ln z \right],$$

where the plus-part of the first-order splitting functions has been defined previously in Eq. (2.33). This equation also introduces both the two-loop cusp anomalous dimension,

$$\Gamma_1 = \frac{4}{3} \left(C_A(4 - \pi^2) + 5\beta_0 \right) \tag{6.42}$$

as well as the function,

$$S_2(z) = -2\operatorname{Li}_2(-z) - 2\ln(1+z)\ln z - \frac{\pi^2}{6}. \tag{6.43}$$

The corresponding result for $\mathcal{P}_{gq}^{(2)}(z)$ is,

$$\mathcal{P}_{gq}^{(2)}(z) = C_A \left\{ P_{gq}^{(1)}(z) \left[\ln^2(1-z) - 2\ln(1-z)\ln z - \frac{101}{18} - \frac{\pi^2}{6} \right] + P_{gq}^{(1)}(-z)S_2(z) \right.$$
$$\left. + C_F \left(2z\ln(1-z) + (2+z)\ln^2 z - \frac{36 + 15z + 8z^2}{3}\ln z + \frac{56 - z + 88z^2}{18} \right) \right\}$$
$$- C_F \left\{ P_{gq}^{(1)}(z)\ln^2(1-z) + \left[3P_{gq}^{(1)}(z) + 2zC_F \right]\ln(1-z) \right.$$
$$\left. + C_F \left(\frac{2-z}{2}\ln^2 z - \frac{4+7z}{2}\ln z + \frac{5+7z}{2} \right) \right\}$$
$$+ \beta_0 \left\{ P_{gq}^{(1)}(z) \left[\ln(1-z) + \frac{5}{3} \right] + z \right\}. \tag{6.44}$$

The quark splitting functions employ the decomposition of Eq. (6.32). The components are,

$$\mathcal{P}_{qqV}^{(2)}(z) = \frac{\Gamma_1}{8} C_F \frac{(1+z^2)}{(1-z)_+}$$

$$+ \delta(1-z)C_F\left[C_F\left(\frac{3}{8} - \frac{\pi^2}{2} + 6\zeta_3\right) + C_A\left(\frac{1}{4} - 3\zeta_3\right) + \beta_0\left(\frac{1}{8} + \frac{\pi^2}{6}\right)\right]$$

$$- C_F^2\left\{\frac{1+z^2}{1-z}\left[2\ln(1-z) + \frac{3}{2}\right]\ln z + \frac{1+z}{2}\ln^2 z + \frac{3+7z}{2}\ln z + 5(1-z)\right\}$$

$$+ C_A C_F\left[\frac{1}{2}\frac{1+z^2}{1-z}\ln^2 z + (1+z)\ln z + 3(1-z)\right]$$

$$+ \beta_0\left[\frac{1}{2}\frac{1+z^2}{1-z}\ln z + 1 - z\right],$$

$$\mathcal{P}_{q\bar{q}V}^{(2)}(z) = (2C_F - C_A)C_F\left\{\frac{1+z^2}{1+z}\left[S_2(z) + \frac{1}{2}\ln^2 z\right] + (1+z)\ln z + 2(1-z)\right\},$$

$$\mathcal{P}_{qqS}^{(2)}(z) = T_R C_F\left[-(1+z)\ln^2 z + \left(1 + 5z + \frac{8}{3}z^2\right)\ln z + \frac{20}{9z} - 2 + 6z - \frac{56}{9}z^2\right],$$

$$\tag{6.45}$$

and

$$\mathcal{P}_{qg}^{(2)}(z) = C_F T_R\left[\left(z^2 + (1-z)^2\right)\left(\ln^2\frac{1-z}{z} - 2\ln\frac{1-z}{z} - \frac{\pi^2}{3} + 5\right) + 2\ln(1-z)\right.$$

$$\left. - \frac{1-2z}{2}\ln^2 z - \frac{1-4z}{2}\ln z + 2 - \frac{9}{2}z\right]$$

$$+ C_A T_R\left\{\left(z^2 + (1-z)^2\right)\left[-\ln^2(1-z) + 2\ln(1-z) + \frac{22}{3}\ln z - \frac{109}{9} + \frac{\pi^2}{6}\right]\right.$$

$$+ \left(z^2 + (1+z)^2\right)S_2(z) - 2\ln(1-z) - (1+2z)\ln^2 z$$

$$\left. + \frac{68z - 19}{3}\ln z + \frac{20}{9z} + \frac{91}{9} + \frac{7}{9}z\right\}.$$

$$\tag{6.46}$$

The solution of the PDF evolution equations given earlier, in the presence of these splitting functions, is obtained numerically. Some examples of the effects of DGLAP evolution on the PDFs will be given later in this chapter.

6.2 Fitting parton distribution functions

6.2.1 Processes involved in global analysis fits

6.2.1.1 Interlude: Deep-inelastic scattering

The parton distribution functions were introduced in Chapter 2 with reference to **deep-inelastic scattering (DIS)** of leptons and protons. These processes play an important role in the extraction of PDFs since, within the parton model, the PDFs are related in a fairly straightforward manner to cross-sections that can be measured experimentally. The kinematics of a DIS process can be described by the following variables: Q^2, the square of the four-momentum transferred to the proton in the exchange; x, the fraction of the proton's momentum carried by the struck quark, y, the fraction of the incident lepton's energy, in the proton's rest frame, that is lost in the collision. In terms of these variables the doubly-differential DIS cross-section can be written as,

$$\frac{d^2\sigma}{dxdy} = \frac{2\pi\alpha^2}{xyQ^4} \left[\left(1 + (1-y)^2\right) F_2 - \left(1 - (1-y)^2\right) xF_3 - y^2 F_L \right] . \tag{6.47}$$

This introduces a parameterization of the cross-section in terms of the **structure functions** F_2, F_3, and F_L. This formula is correct for **neutral current** (**NC**) processes where the exchanged particle is a photon or a Z-boson. For **charged current** (**CC**) interactions the particle exchanged is a W-boson and, for a beam of the appropriate helicity lepton to facilate the weak interaction, Eq. (6.47) must be multiplied by an additional coupling and propagator factor,

$$2 \left(\frac{G_F m_W^2}{4\pi\alpha} \frac{Q^2}{Q^2 + m_W^2} \right)^2 . \tag{6.48}$$

In the quark-parton model the relationship between the structure functions and the PDFs is:

$$F_2^{NC} = x \sum_q C_q \left[f_q(x, Q^2) + f_{\bar{q}}(x, Q^2) \right] , \tag{6.49}$$

$$F_3^{NC} = \sum_q C_q' \left[f_q(x, Q^2) - f_{\bar{q}}(x, Q^2) \right] , \tag{6.50}$$

$$F_2^{CC(-)} = 2x \left[f_u(x, Q^2) + f_{\bar{d}}(x, Q^2) + f_{\bar{s}}(x, Q^2) + f_c(x, Q^2) + \ldots \right] , \tag{6.51}$$

$$F_3^{CC(-)} = 2 \left[f_u(x, Q^2) - f_{\bar{d}}(x, Q^2) - f_{\bar{s}}(x, Q^2) + f_c(x, Q^2) + \ldots \right] . \tag{6.52}$$

Here, C_q and C_q' represent the combinations of couplings and propagators necessary to account for the separate photon and Z-boson contributions to the neutral current process. In particular, F_3 would be zero for the case of pure photon exchange. The charged-current process corresponds to the case of an incident electron, so that a W^- is exchanged, with the corresponding result for W^+ obtained by interchanging $d \leftrightarrow u$ and $s \leftrightarrow c$. Since the contribution of sea quarks and anti-quarks is equal, measurements of F_3^{NC} are particularly useful probes of the valence quark distributions.

6.2.1.2 Interlude: End

Measurements of deep-inelastic scattering (DIS) structure functions (F_2, F_3), or of the related cross-sections, in lepton-hadron scattering and of lepton pair production cross-sections in hadron-hadron collisions provide the main source of information on quark distributions $f_{q/p}(x, \mu_F^2)$ inside hadrons. For simplicity, for the rest of this chapter, the scale μ_F will be replaced by the scale Q, representing both the renormalization and factorization scales.

At leading-order, the gluon distribution function $f_{g/p}(x, Q^2)$ enters directly in hadron-hadron scattering processes with jet final states. Modern global parton distribution fits are carried out to NLO and NNLO, which allows $\alpha_S(Q^2)$, $f_{q/p}(x, Q^2)$ and $f_{g/p}(x, Q^2)$ to all mix and contribute in the theoretical formulae for all processes.

Nevertheless, the broad picture described above still holds to some degree in global PDF analyses. Again, Q here refers to a scale representing the hardness of the interaction, as for example the jet transverse momentum in a calculation of the inclusive jet cross-section.

A NLO (NNLO) global PDF fit requires thousands of iterations and thus thousands of estimates of NLO (NNLO) matrix elements. The NLO (NNLO) matrix elements require too much time for evaluation to be used directly in global fits. Previously, a K-factor (NLO/LO or NNLO/LO) was calculated for each data point used in the global fit, and the LO matrix element (which can be calculated very quickly) was changed in the global fit (multiplied by the K-factor). Currently, a routine such as fastNLO [686] or Applgrid [329] is often used for fast evaluation of the NLO matrix element with the new iterated PDF. Practically speaking, both provide the same order of accuracy. Even when fastNLO or Applgrid is used at NLO, a K-factor approach (NNLO/NLO) is still needed at NNLO, at least until the fastNLO/Applgrid technique can be adapted for NNLO calculations (in progress at the completion of this book).

The data from DIS, DY, and jet processes utilized in PDF fits cover a wide range in x and Q^2. HERA data [77, 115] are predominantly at low x, while the fixed target DIS [169, 170, 217, 900] and DY [769, 874] data are at higher x. Collider jet data at both the TEVATRON and LHC [9, 61, 86, 103, 116, 117, 119, 126, 173, 360] cover a broad range in x and Q^2 by themselves and are particularly important in the determination of the high x gluon distribution. Jet data from the LHC have now been used in global PDF fits, and their importance will increase as high statistics data, and their detailed systematic error information, are published. In addition, jet production data from HERA have been used in the HERAPDF global PDF fits [78, 79, 389, 390].

As an example, the kinematic coverage of the data used in the NNPDF2.3 fit [192] is shown in Fig. 6.2.

There is a tradeoff between the size and the consistency of a data set used in a global PDF fit, in that a wider data set contains more information, but information coming from different experiments may be partially inconsistent. Most of the fixed target data have been taken on nuclear targets and suffer from uncertainties in the nuclear corrections that must be made [693]. This is unfortunate as it is the neutrino fixed target data that provide most of the quark flavour differentiation, for example between up, down, and strange quarks. As LHC collider data become more copious, it may be possible to reduce the reliance on fixed target nuclear data. For example, the rapidity distributions for W^+, W^-, and Z production at the LHC (as well as the TEVATRON) are proving to be very useful in constraining u and d valence and sea quarks, as described in Chapter 9.

There is considerable overlap, however, for the kinematic coverage among the datasets with the degree of overlap increasing with time as the full statistics of the HERA experiments have been published. Parton distributions determined at a given x and Q^2 'feed-down' or evolve to lower x values at higher Q^2 values, as discussed in Chapter 2. DGLAP-based NLO and NNLO pQCD should provide an accurate description of the data (and of the evolution of the parton distributions) over the entire kinematic range present in current global fits. At very low x and Q^2, DGLAP evolution is believed to be no longer applicable and a BFKL [189, 517, 708, 709] description

NNPDF2.3 dataset

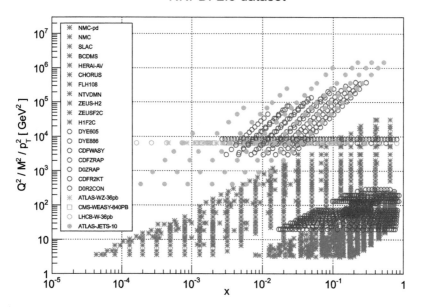

Fig. 6.2 The kinematical coverage in x and Q^2 for the data sets included in the NNPDF2.3 global PDF fit. Reprinted with permission from Ref. [192].

should be used. No clear evidence of BFKL physics is seen in the current range of data; thus all global analyses use conventional DGLAP evolution of PDFs.

There is a remarkable consistency between the data in the PDF fits and the perturbative QCD theory fit to them. The CT,MMHT, and NNPDF groups use over 3000 data points in their global PDF analyses and the χ^2/DOF for the fit of theory to data is on the order of unity, for both the NLO and NNLO analyses. For most of the data points, the statistical errors are smaller than the systematic errors, so a proper treatment of the systematic errors and their bin-to-bin correlations is important. All modern day experiments provide the needed correlated systematic error information. The H1 and ZEUS experiments have combined the data from the two experiments from Run 1 at HERA (and now Run 2) in such a way as to reduce both the systematic and statistical errors, providing errors of both types of the order of a percent or less over much of the HERA kinematics [77]. In the Run 1 combination, for example, 1402 data points are combined to form 742 cross-section measurements (including both neutral current and charged current cross-sections). The combined data sets, with their small statistical and systematic errors, form a very strong constraint for all modern global PDF fits. Thus, it can be hard for other data sets, for example from the LHC, to match the statistical and systematic errors of the HERA data. The manner of using the systematic errors in a global fit will be discussed later in Section 6.3.

The accuracy of the extrapolation to higher Q^2 depends on the accuracy of the original measurement, any uncertainty on $\alpha_S(Q^2)$ and the accuracy of the evolution code. Most global PDF analyses are carried out at NLO and NNLO. Both the NLO and the NNLO evolution codes have now been benchmarked against each other and found to be consistent [190, 191, 251, 262, 470, 572]. Most processes of interest have been calculated to NLO and there is the possibility, as discussed previously, of including data from these processes in global fits. Fewer processes have been calculated at NNLO [134, 296]. The processes that have been calculated include DIS, DY, diphoton [340], $t\bar{t}$ production [427], and inclusive W,Z, and Higgs boson + jet production [269, 271, 272, 274]. Late in the writing of this book, the complete NNLO inclusive jet production cross-section has been completed (a monumental feat), but the results are not yet in the form to be easily used in global PDF fits [422]. Typically, jet production has been included in global PDF fits using NLO matrix elements. Threshold corrections [445, 676] can be used to make an approximate NNLO prediction, but the corrections are valid only over a limited phase space at the LHC, thus greatly reducing the size and power of the jet data in the global fits. Thus, any of the NNLO global PDF analyses discussed here are still approximate for this reason, but in practice the approximation should work reasonably well. The NNLO corrections for the inclusive jet cross-section have been found to be small (and relatively constant for the LHC phase space), if a scale equal to the transverse momentum of the jet is used [335].[1] The CT14, MMHT2014, and NNPDF3.0 PDFs follow different philosophies regarding the use of LHC jet data in NNLO fits. CT14 makes no cuts on the LHC jet data, MMHT2014 doesn't include the LHC jet data, and NNPDF3.0 uses only the jet data for which threshold resummation provides a reasonable prediction. Current evolution programmes should be able to carry out the evolution using NLO and NNLO DGLAP to an accuracy of a few percent over the hadron collider kinematic range, except perhaps at very large and very small x.

The kinematics appropriate for the production of a state of mass M and rapidity y at the LHC is shown in Fig. 6.3 [320]. For example, to produce a state of mass 100 GeV and rapidity 2 requires partons of x values 0.05 and 0.001 at a Q^2 value of 1×10^4 GeV^2. Compare this figure to the scatterplot of the x and Q^2 range included in the recent NNPDF2.3 fit and it is clear that an extrapolation to higher Q^2 (M^2) is required for predictions for many of the LHC processes of interest. As more Standard Model processes are included in global PDF fits, the need for extrapolation will be reduced.

6.2.2 Parameterizations and schemes

A global PDF analysis carried out at NLO or NNLO needs to be performed in a specific renormalization and factorization scheme. The evolution kernels are calculated in a specific scheme and to maintain consistency, any hard scattering cross section calculations used for the input processes or utilizing the resulting PDFs need to have been implemented in that same renormalization scheme. As we saw earlier in Chapter 2, one needs to specify a scheme or convention in subtracting the divergent terms from the PDFs. These divergent terms result from collinear gluon emission from the initial

---

[1]The NNLO cross-section is higher than the NLO one at smaller jet transverse momenta if a scale equal to that of the largest transverse momentum jet in the event is used.

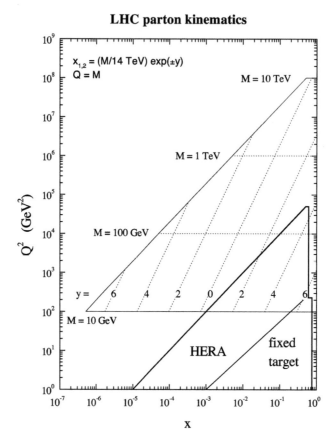

Fig. 6.3 A plot showing the x and Q^2 values needed for the colliding partons to produce a final state with mass M and rapidity y at the LHC (14 TeV).

state partons. The collinear emissions result in pole terms of the form $1/\epsilon$, where ϵ is the dimensional regularization parameter. Basically the scheme definition specifies how much of the finite corrections to subtract along with the divergent pieces. Almost universally, the $\overline{MS}$ scheme is used; using dimensional regularization, in this scheme the pole terms and accompanying $log\ 4\pi$ and Euler constant terms are subtracted.[2] PDFs are also available in the DIS scheme (where the full order α_s corrections for F_2

---

[2] Within the $\overline{MS}$ scheme, PDFs can also be defined for a fixed number of flavours, which then have a validity over the kinematic range for which that number (and only that number) of flavours can be present in the proton.

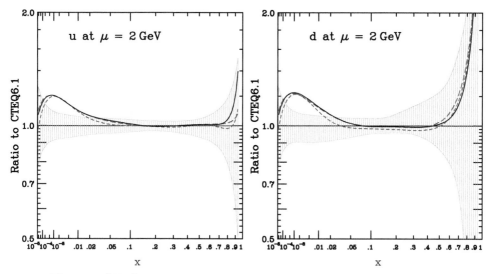

Fig. 6.4 CTEQ6.5 up and down quark distributions normalized to those of CTEQ6.1, showing the impact of the heavy quark mass corrections. Reprinted with permission from Ref. [876].

are absorbed into the quark PDFs).

Basically all modern PDFs now incorporate a treatment of heavy quark effects in their fits, either via the ACOT general-mass (GM) variable flavour number scheme [131] (supplemented by by a unified treatment of both kinematical and dynamical effects using the S-ACOT [694] and ACOT-χ [698, 875] concepts), used by CTEQ/CT, or by the Thorne-Roberts scheme [872, 873], used by both MSTW and HERAPDF, and the FONLL scheme, used by NNPDF [302, 532].

Incorporation of the full heavy-quark mass effects in the general-mass formalism suppresses the heavy flavour contributions to the DIS structure functions, especially at low x and Q^2. In order for the theoretical calculations in the global fits to agree with the data in these kinematic regions, the contributions of the light quark and anti-quark PDFs must increase accordingly. This has a noticeable impact, especially on predictions for W and Z cross-sections at the LHC.

Fig. 6.4 shows the impact of the heavy quark mass corrections on the up and down quark distributions for CTEQ6.5, at a Q value of 2 GeV [876]. The CTEQ6.5 up and down quark distributions are normalized to the corresponding ones from CTEQ6.1 (which does not have the heavy quark mass corrections). The shaded areas indicate the CTEQ6.1 PDF uncertainty. The dashed curves represent slightly different parameterizations for the CTEQ6.5 PDFs. The heavy quark mass corrections have a strong effect (larger than the PDF uncertainty for CTEQ6.1) at low x, in a region sensitive to W and Z production at the LHC.

The impact of general-mass variable flavour number schemes (GM-VFNS) lies mostly in the low x and Q^2 regions. Aside from modifications to the fits to the HERA data, and the commensurate change in the fitted PDFs, there is basically no modification for predictions at high Q^2 at the LHC.

It is also possible to use only leading-order matrix element calculations in the global fits which results in leading-order parton distribution functions, which have been made available, for example, by the CTEQ [489, 818], MSTW/MMHT [614, 755] and NNPDF [191, 194] groups. For many hard matrix elements for processes used in the global analysis, there exist K factors significantly different from unity. Thus, one expects there to be noticeable differences between the LO and NLO parton distributions (and indeed this is often the case, especially at low x and high x).

Global analyses have traditionally used a generic form for the parameterization of both the quark and gluon distributions at some reference value Q_0:[3]

$$F(x, Q_0) = x^{A_1}(1-x)^{A_2} P(x; A_3, A_4...). \tag{6.53}$$

The reference value Q_0 is usually chosen in the range of 1–2 GeV. The parameter A_1 is associated with small-x Regge behaviour while A_2 is associated with large-x valence counting rules. We expect A_1 to be approximately -1 for gluons and anti-quarks, and of the order of $1/2$ for valence quarks, from the Regge arguments mentioned in Chapter 2. Counting rule arguments tell us that the A_2 parameter should be related to $2n_s - 1$, where n_s is the minimum number of spectator quarks. So, for valence quarks in a proton, there are two spectator quarks, and we expect $A_2 = 3$. For a gluon, there are three spectator quarks, and $A_2 = 5$; for anti-quarks in a proton, there are four spectator quarks, and thus $A_2 = 7$. Such arguments are useful, for example in telling us that the gluon distribution should fall more rapidly with x than quark distributions, but it is not clear exactly at what value of Q that the arguments made above are valid.

The first two factors, in general, are not sufficient to completely describe either quark or gluon distributions. The term $P(x; A_3, ...)$ is a suitably chosen smooth function, depending on one or more parameters, that adds more flexibility to the PDF parameterization. $P(x; A_3, ...)$ is chosen so as to tend towards a constant for x approaching either 0 or 1, so that the limiting behaviour is given by the first two terms.

In general, both the number of free parameters and the functional form can have an influence on the global fit. A too-limited parameterization not only can lead to a worse description of the data, but also to PDFs in different kinematic regions being tied together not by the physics, but by the limitations of the parameterization. Note that the parameterization forms shown here imply that PDFs are positive-definite. As they are not physical objects by themselves, it is possible for them to be negative, especially at low Q^2. Some PDF groups (such as CT) use a positive-definite form for the parameterization; others do not. For example, the MSTW2008 gluon distribution is negative for $x < 0.0001, Q^2 = 2 GeV^2$. Evolution quickly brings the gluon into positive territory.

The CT14 fit uses 28 free parameters (many of the PDF parameters are either fixed at reasonable values, or are constrained by sum rules). There are a total of 8 free

---

[3]Recently, there has been a trend towards the use of more sophisticated forms of parameterization in global fits, but the physics arguments listed here are still valid. For example, in the CT14 global fit [489], P(x) is defined by a fourth-order polynomial in $\sqrt{x}$; the polynomial is then re-expressed in terms of Bernstein polynomials in order to reduce correlations among the coefficients.

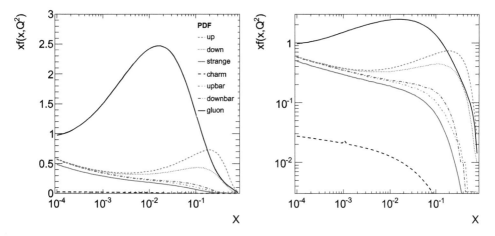

Fig. 6.5 The CT14 NNLO parton distribution functions evaluated at a Q^2 value of 2 GeV^2, with a linear (left) and logarithmic (right) scale.

parameters for the valence quarks, 5 for the gluon and 15 for the sea quarks.

The MMHT2014 fit uses 20 free parameters, while the NNPDF fits effectively has 259 free parameters. The NNPDF approach attempts to minimize the parameterization bias by exploring global fits using a large number of free parameters in a Monte Carlo approach. The general form for NNPDF can be written as $f_i(x, Q_o) = c_i(x) NN_i(x)$, where $NN_i(x)$ is a neural network, and $c_i(x)$ is a "pre-processing function".

In the past, PDFs were often made available to the world in a form where the x and Q^2 dependence was parameterized. Now, almost universally, the PDFs for a given x and Q^2 range can be interpolated from a grid that is provided by the PDF groups, or the grid can be generated given the starting parameters for the PDFs (see the discussion on LHAPDF in Section 6.6). All techniques should provide an accuracy on the output PDF distributions on the order of a few percent or better.

The parton distributions from the CT14 NNLO PDFs are plotted in Fig. 6.5 at a Q^2 value of 2 GeV^2 (near the starting point of evolution) and in Fig. 6.6 at a Q^2 value of 10000 GeV^2 (more typical of LHC processes). At the lower Q^2 value, the up quark and down quark distribution peak near x values of 1/3, the remnant of the spike in the primitive model described in Chapter 2. The charm PDF is very suppressed as it is produced entirely by evolution, and the Q^2 value is near the starting scale for the evolution. There is no bottom quark distribution since it is below threshold. At higher Q^2 values, the up and down quark peaks become shoulders due to the effects of evolution. At high Q^2, the gluon distribution is dominant at x values of less than 0.1 with the valence quark distributions dominant at higher x. One of the major influences of the HERA data has been to steepen the gluon distribution at low x.

The CT14 up quark, up-bar quark, b-quark, and gluon distributions are shown as a function of Q^2 for x values of 0.001, 0.01, 0.1 and 0.3 in Figs. 6.7 and 6.8. At low x, the PDFs increase with Q^2, while at higher x, the PDFs decrease with Q^2. Both effects are due to DGLAP evolution, as discussed previously. An x value of approximately

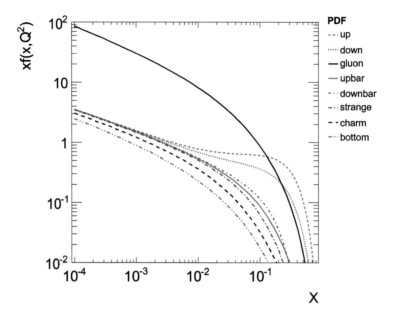

Fig. 6.6 The CT14 NNLO parton distribution functions evaluated at a Q^2 of 10000 GeV^2.

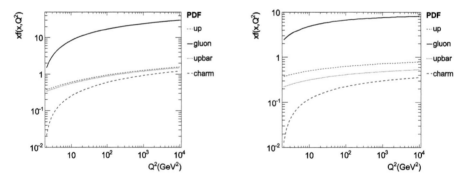

Fig. 6.7 The CT14 NNLO up quark, up-bar quark, c quark, and gluon parton distribution functions evaluated as a function of Q^2 at x values of 0.001 (left) and 0.01 (right).

0.1 is the *pivot − point* for gluon evolution; at this x value, the gluon distribution changes little as the value of Q^2 increases. It can also be seen that both the charm (and bottom) quark distributions are generated perturbatively from gluon splitting, and thus the distributions are zero below threshold for heavy quark pair production and rise rapidly thereafter with increasing Q^2. It is also possible for there to be intrinsic charm, where charm quarks are present below threshold. However, no strong evidence has been observed to date for the existence of intrinsic charm.

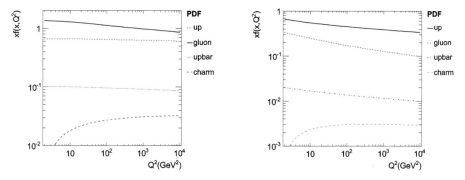

Fig. 6.8 The CT14 NNLO up quark, up-bar quark, c quark and gluon parton distribution functions evaluated as a function of Q^2 at an x value of 0.1 (left) and 0.3 (right).

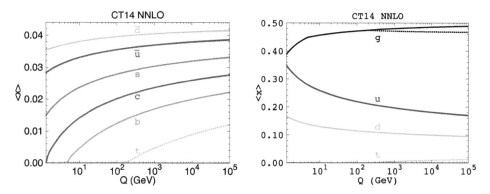

Fig. 6.9 The momentum fractions carried by the CT14 NNLO quark and gluon distributions, as a function of Q. The gluon distribution in the right figure is shown without (solid) and with (dotted) the presence of a top quark PDF.

The average proton momentum carried by each parton species is shown in Fig. 6.9. As Q^2 increases, the momentum carried by up and down valence quarks decreases, while the momentum carried by the gluon and by sea quarks increases. For typical LHC hard-scattering scales, the gluon carries slightly less than 50% of the parent proton's momentum. Note that a 5-flavour scheme is most commonly used, i.e. the charm and bottom (but not top) can appear as sea quarks once the Q value is sufficient to pair produce them from gluon splitting. It is also possible to allow top quarks in the sea in a 6-flavour scheme; even at the highest Q values, only about 1% of the proton's momentum is carried by top quarks. The momentum added to the top quark distribution comes primarily from the gluon distribution.

The photon is also a parton constituent of the proton, just as a quark or gluon is, and can be produced from QED radiation from quark lines [754]. This source of photons is known as the inelastic component. There is also a (well-known) elastic component, which cannot be ignored, resulting from coherent electromagnetic radiation from the proton as a whole, leaving the proton intact [292, 587]. Both components contribute to photon-induced processes at the LHC. The inelastic component evolves with Q^2, while the elastic component is relatively constant (varying with the running of the QED coupling). To include the photon PDF, the QCD evolution of the partons in the proton now has to be expanded to a QCD+QED evolution, where the QED aspect involves the electromagnetic coupling $\alpha(Q^2)$ instead of $\alpha_s(Q^2)$ and the corresponding splitting function is Abelian rather than non-Abelian. At the LHC, especially for the 13-14 TeV running, processes involving photons in the initial state will become increasingly important, *c.f.* $\gamma\gamma \to WW$, or photon-initiated production of WH. There is little hadronic data to directly constrain the photon PDF, but it has been known that less than 1% of the proton's momentum is carried by photons. The first attempt (MRST2004QED) to model the inelastic component just considered photon emission from quark lines, using the known quark distributions, and using either the current quark mass (few MeV) or the constituent quark mass (few hundred MeV) as a cutoff [754] (see also Refs. [587, 772]). The disparity in the cutoffs leads to a wide range in the possible size of the photon PDF. Other attempts to determine the photon distribution used Drell-Yan data from the LHC [193, 194, 275](NNPDF2.3qed followed by NNPDF3.0qed) to fit for the photon PDF.[4]

Another approach [80, 837] (CT14qed,CT14qed_inc) used data from the scattering process $ep \to e\gamma X$ measured by the ZEUS collaboration [391], leading to an upper constraint on the total inelastic photon PDF momentum, at a scale of 1.3 GeV, of approximately 0.14% (and a lower constraint of 0). This is to be compared to the total elastic photon PDF momentum fraction of 0.15%. The photon PDFs for the inelastic component only, assuming the maximum intrinsic momentum fraction of 0.14%, and the total photon PDF, including the elastic component as well, are shown in Fig. 6.10 for $Q=1.3$ GeV (left) and $Q=85$ GeV (right). The dominance of the inelastic component at high Q can be observed.

Recently, the photon PDF has been determined to a high precision (1-2%) using electron-proton scattering data, considering an equivalence between a template cross-section calculated either using proton structure functions, or a photon PDF [746]. The resulting PDF (LUXqed) is in good agreement with CT14qed_inc and is at the lower edge of the NNPDF2.3qed photon PDF uncertainty band at high x, as shown in Fig. 6.11.

The charm quark distribution also has a dynamic component, generated through gluon splitting into a $c\bar{c}$ pair.[5] The photon/charm ratio increases with increasing Q and increasing x value. One reason for the variation in Q is that while α_s decreases with Q, α remains approximately constant (actually, it rises slightly). At low x, the photon/charm ratio is of the order of 5-10%, due to the difference in coupling constants

---

[4]NNPDF3.0qed improves on the NNPDF2.3qed photon PDF with a correct treatment of $\alpha(\alpha_s L)^n$ terms in the evolution. These resulted in a large uncertainty for the photon PDF, as the errors in the reasonably precise high-mass Drell-Yan data are still large compared to the expected contributions from photon-initiated processes.

[5]There may also be an intrinsic charm component at low Q, but the evidence is not convincing.

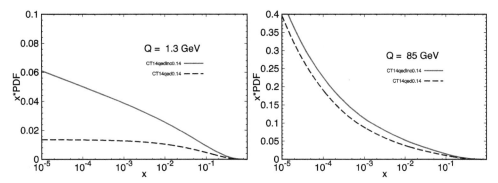

Fig. 6.10 The photon PDF from the inelastic component only, with a photon momentum fraction of 0.14%, and the total photon PDF, including the elastic component, at a Q value of 1.3 GeV (left) and 85 GeV (right).

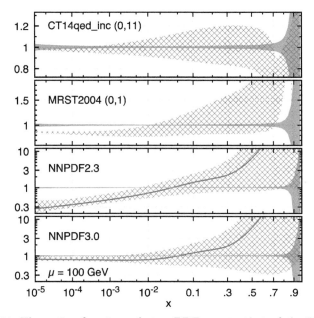

Fig. 6.11 The ratio of various photon PDF sets to that of the LUXqed photon PDF set, all evaluated at a scale of 100 GeV. Note that the vertical axis scales are different for the sub-plots. Reprinted with permission from Ref. [746].

(α_s vs. α) and the larger gluon than quark (primarily up quark) distribution at low x. (Remember that the photon does have a substantial elastic component at small x which is included in this ratio.) At high x, the dominance of the valence up quark over the gluon results in the photon distribution becoming larger than the charm distribution.

6.3 PDF uncertainties

In addition to having the best estimates for the values of the PDFs in a given kinematic range, it is also important to understand the allowed range of variation of the PDFs, i.e. their uncertainties. A prior conventional method of estimating parton distribution uncertainties had been to compare different published parton distributions. Although in some cases, this procedure does provide useful information, in general this is unreliable since most published sets of parton distributions adopt similar assumptions and the differences between the sets do not fully explore the uncertainties that actually exist. In addition, some PDF fits may use only a limited set of data in their fits, and they may result in PDFs that (a) have larger intrinsic uncertainties and (b) differ in their central values from PDF sets which use a more global set of input data. Some comparisons of PDFs and PDF predictions at the LHC will be given later in this chapter.

The sum of the quark distributions $(\Sigma f_{q/p}(x, Q^2) + f_{g/p}(x, Q^2))$ is, in general, well determined over a wide range of x and Q^2. As stated earlier, the quark distributions are predominantly determined by the DIS and DY data sets which have large statistics, and systematic errors in the few percent range ($\pm 3\%$ for $10^{-4} < x < 0.75$). Thus the sum of the quark distributions is basically known to a similar accuracy. The individual quark flavours, though, may have a greater uncertainty than the sum. This can be important, for example, in predicting distributions that depend on specific quark flavours, like the W lepton asymmetry distribution and the W and Z rapidity distributions.

The largest uncertainty of any parton distribution, however, is that on the gluon distribution. The gluon distribution can be determined indirectly at low x by measuring the scaling violations in the quark distributions, but a direct measurement is necessary at moderate to high x. About 40-50% of the momentum of the proton is carried by gluons, and most of that momentum is at relatively small x (16% of the momentum of the proton, for example, is carried by gluons in the x range from 0.01 to 0.1.) The best direct information on the gluon distribution at moderate to high x comes from jet production at the TEVATRON and the LHC, although new processes (photon production, top pair production, etc.) will increasingly contribute.

There has been a great deal of activity on the subject of PDF uncertainties. Two techniques in particular, the Lagrange Multiplier and Hessian techniques, have been used by CTEQ/CT, MSTW/MMHT, and HERAPDF to estimate PDF uncertainties [413, 753, 817, 819, 868]. The Lagrange Multiplier technique is useful for probing the PDF uncertainty of a given process, such as the Higgs boson cross-section, while the Hessian technique provides a more general framework for estimating the PDF uncertainty for any cross-section. In addition, the Hessian technique results in tools more accessible to the general user, such as error PDFs. Both techniques are described in the following. The Monte Carlo technique used by the NNPDF group for PDF uncertainty estimation will be described later in this section.

6.3.1 The Hessian method

The Hessian method for the determination of a central PDF and/or determination of PDF uncertainties involves minimizing a suitable log-likelihood function. The χ^2

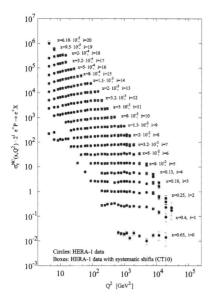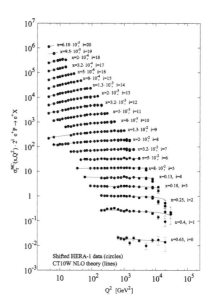

Fig. 6.12 A comparison of the unshifted and shifted HERA1 combined neutral current data (left) and the comparison of the NLO CT10 predictions to the shifted data (right). Reprinted with permission from Ref. [713].

function may contain the full set of correlated errors, or only a partial set. The correlated systematic errors may be accounted for using a covariance matrix, or as a shift to the data, adopting a χ^2 penalty proportional to the size of the shift divided by the systematic error. The two methods should be equivalent. Below we discuss the shift method.

In the following description, CT10 NLO PDFs [713] are used (similar considerations apply at NNLO), along with the HERA Run I combined (H1+ZEUS) neutral current (e^+p) cross sections [77], to discuss the use of the Hessian formalism. A comparison of the HERA data and the NLO predictions using the CT10 PDFs is shown in Fig. 6.12. On the left, the data are presented in unshifted form, and on the right, optimal systematic error shifts have been applied in the manner detailed below. There is good agreement between the combined HERA Run I data and the CT10 NLO predictions with a global χ^2 of about 680 for the 579 data points, which is typical for the global fit PDFs.

The HERA Run I combined data have $N_\lambda = 114$ independent sources of experimental systematic uncertainty, with parameters λ_α that should obey a standard Gaussian or normal distribution. The contribution of the HERA dataset to the χ^2 can be written as

$$\chi^2(\{a\}, \{\lambda\}) = \sum_{k=1}^{N} \frac{1}{s_k^2} \left(D_k - T_k(\{a\}) - \sum_{\alpha=1}^{N_\lambda} \lambda_\alpha \beta_{k\alpha} \right)^2 + \sum_{\alpha=1}^{N_\lambda} \lambda_\alpha^2, \qquad (6.54)$$

where N is the total number of points and $T_k(a)$ is the theory value for the kth data

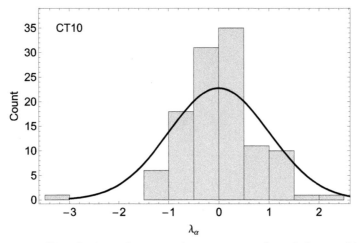

Fig. 6.13 Distribution of systematic parameters λ_α of the combined HERA Run I data set [871] in the CT10 best fit (CT10.00). Reprinted with permission from Ref. [713].

point, dependent on the PDF parameters a. Furthermore, s_k is the total uncorrelated error on the measurement D_k that is obtained by summing the statistical and uncorrelated systematic errors on D_k in quadrature,

$$s_k = \sqrt{s_{k,\text{stat}}^2 + s_{k,\text{uncorr sys}}^2} \tag{6.55}$$

The χ^2 function is minimized with respect to the size of the systematic error shifts λ_α using the algebraic procedure described. It is also possible to add a penalty term to the χ^2 function that prevents relatively unconstrained PDF parameters from reaching values that might lead to unphysical predictions in regions where experimental data are sparse (e.g. very low x).

As expected, a better agreement of data with theory is observed when the systematic error shifts are allowed. It is important to check that the systematic error parameters $\lambda_\alpha(a)$ contribute to the χ^2 an amount on the order of the total number of systematic errors (114) and (b) that the sizes of the parameters follow a Gaussian distribution. For the case of CT10 and the HERA Run I data, the systematic error contribution to the total χ^2 is 65, or somewhat better than expected, and their distribution is approximately Gaussian-distributed, as shown in Fig. 6.13.

All systematic errors are not equally important, though, and it is also crucial to verify that no 'major' systematic error needs to be shifted by several sigma. Given the precision of the HERA Run I data, the size of the systematic error shifts required are relatively small. This need not be the case, for example, for the case of inclusive jet production, either at the TEVATRON or the LHC.

The Hessian method results in the production of a central (best fit) PDF, and a set of error PDFs. In this method, a large matrix (26×26 for CT10 and 28×28 for CT14), with dimension equal to the number of free parameters in the fit, has to

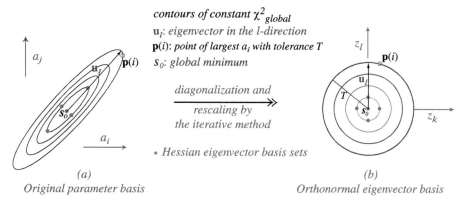

Fig. 6.14 A schematic representation of the transformation from the PDF parameter basis to the orthonormal eigenvector basis. Reprinted with permission from Ref. [818].

be diagonalized.[6] The result is 26 (28) orthonormal eigenvector directions for CT10 (CT14) which provide the basis for the determination of the PDF error for any cross-section.

This process is shown schematically in Fig. 6.14. The eigenvectors are now admixtures of the PDF parameters left free in the global fit. There is a broad range for the eigenvalues, over a factor of one million. The eigenvalues are distributed roughly linearly as $\log \epsilon_i$, where ϵ_i is the eigenvalue for the i-th direction. The larger eigenvalues correspond to directions which are well-determined; for example, eigenvectors 1 and 2 are sensitive primarily to the valence quark distributions at moderate x, a region where they are well-constrained. The theoretical uncertainty on the determination of the W mass at the TEVATRON depends primarily on these 2 eigenvector directions, as W production at the TEVATRON proceeds primarily through collisions of valence quarks. The most significant eigenvector directions for determination of the W mass at the LHC correspond to larger eigenvector numbers, which are primarily determined by sea quark distributions. In most cases, the eigenvector can not be directly tied to the behaviour of a particular PDF in a specific kinematic region. There are exceptions, such as eigenvector 15 in the CTEQ6.1 fit, discussed in the following.

In the past, one of the most controversial aspect of PDF uncertainties has been the determination of the $\Delta\chi^2$ excursion from the central fit that is representative of a reasonable error. Nominally, a $\Delta\chi^2 = T^2$(**tolerance**) would correspond to a $1 - \sigma(68\%CL)$ error. PDF fits performed with a limited number of experiments may be able to maintain that criterion. For example, HERAPDF uses a χ^2 excursion of 1 for a 1σ error.[7] For general global fits, such as from CT and MMHT, however, a χ^2

---

[6]As more data is included, more PDF parameters in the global fit can be set free, resulting in a larger number of eigenvectors.

[7]But the total error also includes other sources of uncertainty, for example from possible parameterization bias.

excursion of 1 (for a 1σ error) is too low of a value in a global PDF fit. These global fits use data sets arising from a number of different processes and different experiments; there is a non-negligible tension between some of the different data sets. In addition, the finite number of PDF parameters used in the parameterizations (parameterization bias) also leads to the need for a larger tolerance. Thus, a larger variation in $\Delta\chi^2$ is required for a 68% CL. For example, CT10 uses a tolerance T=10 for a 90% CL error, corresponding to T=6.1 for a 68% CL error,[8] while MSTW uses a dynamical tolerance (varying from 1 to 6.5) for each eigenvector.

The uncertainties for all predictions should be linearly dependent on the tolerance parameter used; thus, it should be reasonable to scale the uncertainty for an observable from the 90% CL limit provided by the CT error PDFs to a one-sigma error by dividing by a factor of 1.645. Such a scaling will be a better approximation for observables more dependent on the lower number eigenvectors, where the χ^2 function is closer to a quadratic form.

Even though, the data sets and definitions of tolerance are different among the different PDF groups, we will see later in this chapter that the PDF uncertainties at the LHC are fairly similar. Note that relying on the errors determined from a single PDF group may be an underestimate of the *true* PDF uncertainty, as the central results among the PDF groups can in some cases differ by an amount similar to this one-sigma error. (See the discussion later in this chapter regarding benchmarking comparisons of predictions and uncertainties for the LHC.)

Each error PDF results from an excursion along the "+" and "−" directions for each eigenvector. Consider a variable X; its value using the central PDF for an error set (say CT14) is given by X_0. X_i^+ is the value of that variable using the PDF corresponding to the "+" direction for eigenvector i and X_i^- the value for the variable using the PDF corresponding to the "−" direction. The excursions are symmetric for the larger eigenvalues, but may be asymmetric for the more poorly determined directions. In order to calculate the PDF error for an observable, a **Master Equation** should be used:

$$\Delta X_{max}^+ = \sqrt{\sum_{i=1}^{N}[max(X_i^+ - X_0, X_i^- - X_0, 0)]^2}$$

$$\Delta X_{max}^- = \sqrt{\sum_{i=1}^{N}[max(X_0 - X_i^+, X_0 - X_i^-, 0)]^2}. \qquad (6.56)$$

ΔX^+ adds in quadrature the PDF error contributions that lead to an increase in the observable X and ΔX^- the PDF error contributions that lead to a decrease. The addition in quadrature is justified by the eigenvectors forming an orthonormal basis. The sum is over all N eigenvector directions, or 20 in the case of CTEQ6.1 and 26 (28) in the case of CT10 (CT14). Ordinarily, $X_i^+ - X_0$ will be positive and $X_i^- - X_0$

---

[8]A penalty is applied in the CT10 approach, if a disproportionate fraction of the increase in χ^2 along a particular eigenvector direction is concentrated in one, or a few, experiments; this results in the tolerance being effectively smaller for those directions.

will be negative (or vice versa), and thus it is trivial as to which term is to be included in each quadratic sum. For the higher number eigenvectors, however, the "+" and "−" contributions may be in the same direction (see for example eigenvector 17 in Fig. 6.15). In this case, only the most positive term will be included in the calculation of ΔX^+ and the most negative in the calculation of ΔX^-. Thus, there may be less than N terms for either the "+" or "−" directions. There are other versions of the Master Equation in current use but the version listed above is the "official" recommendation of the authors.

There are two things that can happen when new PDFs (eigenvector directions) are added: a new direction in parameter space can be opened to which some cross-sections will be sensitive to (an example of this is eigenvector 15 in the CTEQ6.1 error PDF set, which is sensitive to the high x gluon behaviour and thus influences the high p_T jet cross-section at the TEVATRON and LHC). This particular eigenvector direction happens to be dominated by a parameter which affects mostly the large x behaviour of the gluon distribution.

In this case, a smaller parameter space is an underestimate of the true PDF error since it did not sample a direction important for some physics. In the second case, adding new eigenvectors does not appreciably open new parameter space and the new parameters should not contribute much PDF error to most physics processes (although the error may be redistributed somewhat among the new and old eigenvectors).

In Fig. 6.15, the PDF errors are shown in the "+" and "−" directions for the 20 CTEQ eigenvector directions for predictions for inclusive jet production at the TEVATRON from the CTEQ6.1 PDFs. The excursions are symmetric for the first 10 eigenvectors but can be asymmetric for the last 10, as they correspond to less well-determined directions.

Either X_0 and $X_i^\pm$ can be calculated separately in a matrix element/Monte Carlo program (requiring the program to be run $2N+1$ times) or X_0 can be calculated with the program and at the same time the ratio of the PDF luminosities (the product of the two PDFs at the x values used in the generation of the event) for eigenvector i ($\pm$) to that of the central fit can be calculated and stored. This results in an effective sample with $2N+1$ weights, but identical kinematics, requiring a substantially reduced amount of time to generate. PDF re-weighting will be discussed later in this chapter.

As an example of PDF uncertainties using the Hessian method, the CT10 and MSTW2008 NLO uncertainties for the up quark and gluon distributions are shown in Figs. 6.16 and 6.17. While the CT10 and MSTW2008 PDF distributions and uncertainties are reasonably close to each other, some differences are evident, especially at low and high x.

After the initial diagonalization of the Hessian matrix, it is also possible to diagonalize any one chosen function of the fitting parameters while maintaining a diagonal form for the χ^2 function [816]. This additional function could be a particular cross-section, say for Higgs boson production through gg fusion. It may be that such an observable may be dominated by a few eigenvector directions, something which will be illuminated by the additional diagonalization. It is also possible to determine the particular direction in eigenvector space that has the greatest sensitivity to a particular observable, i.e. the steepest gradient. This will become important when looking

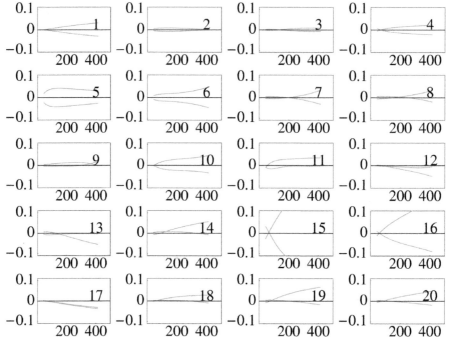

Fig. 6.15 The PDF errors for the CDF inclusive jet cross-section in Run I for the 20 different eigenvector directions contained in the CTEQ6.1 PDF error set. The vertical axes show the fractional deviation from the central prediction and the horizontal axes the jet transverse momentum in GeV. Reprinted with permission from Ref. [867].

for PDF correlations among cross-sections, and in the discussion of meta-PDFs.

6.3.2 The Lagrange Multiplier method

Another technique for determining the uncertainty on a physical observable is the Lagrange Multiplier (LM) method [868]. The LM method can be considered as an extension of the χ^2 minimization procedure that relates the uncertainty of the physical observable (which depends on the PDFs) to the variation of the χ^2 function used in the global fitting. A Lagrange Multiplier variable λ is introduced and the function $\psi(\lambda, a) = \chi^2_{global}(a) + \lambda X(a)$ is minimized for different values of the parameter λ. Here, a refers to the original set of (d) parameters determined in the PDF fit, and $X(a)$ is the physical observable dependent upon those PDF parameters. The Lagrange Multiplier method provides optimal PDFs tailored for a specific study. A representation of the LM method is shown in Fig. 6.18 where a hypothetical mapping of the set of d PDF parameters corresponding to various values of the LM parameter λ are mapped onto the global χ^2 plotted as a function of the possible values of the physical observable X.

An example Lagrange Multiplier analysis for the production of a Higgs boson

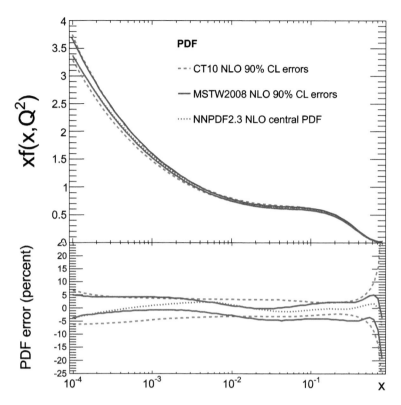

Fig. 6.16 A comparison of the CT10 and MSTW2008 up quark PDF uncertainty bands at $Q^2 = 10^4$ GeV^2. The NNPDF2.3 central up PDF is also shown for comparison.

through gg fusion at 8 and 13 TeV using the CT14 PDFs is shown in Fig. 6.19. The parabolic curve has been determined using the Lagrange Multiplier method, while the points indicate the results of the Hessian analysis for 90% CL ($\Delta\chi^2 = 100$). (The dashed curve shows the results of applying the Tier-2 penalty (a penalty that prevents the agreement with any particular experiment from degrading too greatly) when the χ^2 of any experiment starts seriously degrading.

Table 6.1 is reproduced below from Ref. [489], showing both the PDF and the PDF+$\alpha_s(m_Z)$ uncertainties determined from the Hessian and the Lagrange Multiplier methods for Higgs boson production through gg fusion at NNLO. The results indicate both the agreement between the Hessian and Lagrange Multiplier techniques and the efficacy of scaling the 90%CL Hessian uncertainty by a factor of 1.645 to get the 68% CL uncertainty.

6.3.3 The NNPDF approach

For predictions using NNPDF PDFs, a Monte Carlo sample of PDFs is given, such that the expectation value of any observable $F[q]$ depending on the PDFs is calculated

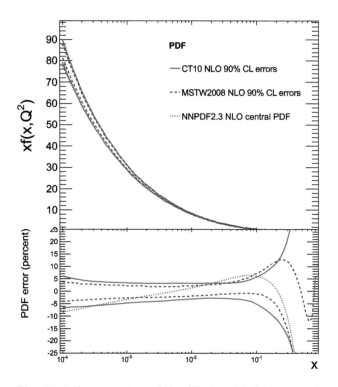

Fig. 6.17 A comparison of the CT10 and MSTW2008 gluon PDF uncertainty bands at $Q^2 = 10^4$ GeV^2. The NNPDF2.3 central gluon PDF is also shown for comparison.

Table 6.1 Uncertainties of $\sigma_H(gg \to H)$ at NNLO computed by the LM method and by the Hessian method, with the Tier-2 penalty included. The 68% C.L. errors are given as percentage of the central value, and the PDF-only uncertainties are for $\alpha_s = 0.118$.

$gg \to H$ (pb), PDF unc., $\alpha_s = 0.118$	8 TeV	13 TeV
68% C.L. (Hessian)	$18.7 + 2.1\% - 2.3\%$	$42.7 + 2.0\% - 2.4\%$
68% C.L. (LM)	$+2.3\% - 2.3\%$	$+2.4\% - 2.5\%$
$gg \to H$ (pb), PDF+α_s	8 TeV	13 TeV
68% C.L. (Hessian)	$18.7 + 2.9\% - 3.0\%$	$42.7 + 3.0\% - 3.2\%$
68.0% C.L. (LM)	$+3.0\% - 2.9\%$	$+3.2\% - 3.1\%$

over an ensemble of PDF replicas using the formula

$$\langle \mathcal{F}[\{q\}] \rangle = \frac{1}{N_{\mathrm{rep}}} \sum_{k=1}^{N_{\mathrm{rep}}} \mathcal{F}[\{q^{(k)}\}], \qquad (6.57)$$

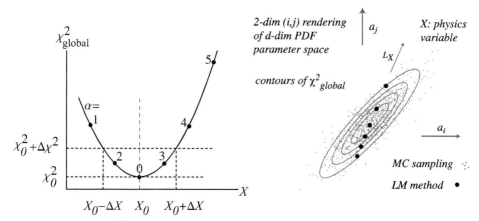

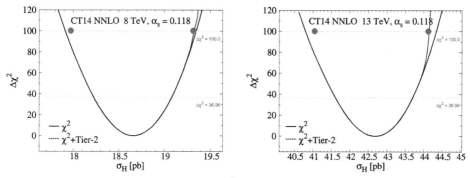

Fig. 6.18 On the left is an illustration of how the Lagrange Multiplier Method provides sample points along a curve L_X in the multi-dimensional parameter space, where X is the observable of interest. On the right is an illustration of how these sample points are mapped onto a global χ^2 distribution, plotted as a function of the value of the cross-section of the observable x. Reprinted with permission from Ref. [868].

Fig. 6.19 A calculation of the χ^2 distribution vs the NNLO Higgs boson cross-section in the CT14 global analysis at 8 TeV (left) and 13 TeV (right). The Lagrange Multiplier (curve) and Hessian approaches (dots) are compared. Reprinted with permission from Ref. [490].

where $N_{\rm rep}$ is the number of replicas of PDFs in the Monte Carlo ensemble. The uncertainty for any observable is calculated as the standard deviation of the sample.

$$\sigma_{\mathcal{F}} = \left(\frac{N_{\rm rep}}{N_{\rm rep} - 1} \left(\langle \mathcal{F}[\{q\}]^2 \rangle - \langle \mathcal{F}[\{q\}] \rangle^2 \right) \right)^{1/2}$$

$$= \left(\frac{1}{N_{\text{rep}} - 1} \sum_{k=1}^{N_{\text{rep}}} \left(\mathcal{F}[\{q^{(k)}\}] - \langle \mathcal{F}[\{q\}] \rangle \right)^2 \right)^{1/2} . \tag{6.58}$$

This equation provides the 1-sigma error on any observable; one advantage of the Monte Carlo approach is that any confidence-level can be calculated by removing the appropriate upper and lower PDF outliers. The NNPDF collaboration provides sets of N_{rep}=100 and 1000 replicas. For most applications, the smaller replica set is sufficient. A central set corresponding to the average of the replicas

$$q^{(0)} \equiv \langle q \rangle = \frac{1}{N_{\text{rep}}} \sum_{k=1}^{N_{\text{rep}}} q^{(k)} . \tag{6.59}$$

is also provided[190–192, 194].

6.3.4   Meta-PDFs

It is possible to re-fit and to re-parameterize, in a common functional form, the error PDFs from a number of PDF fitting groups. The result is an ensemble of PDFs that encompasses the uncertainty of all of the PDF error sets included. The ensemble can also be expanded to cover the combined PDF+α_{s} uncertainties, instead of just the PDF uncertainties alone. The ensemble can be transformed into a Hessian basis, and then only a limited number of the (important) eigenvectors can be retained, leading to a smaller ensemble that provides PDFs corresponding to both the central behaviour of the group of PDFs (for example, CT14, MMHT14, and NNPDF3.0) and to the full uncertainty range. Such PDFs are known as meta-PDFs [552] and they can make it easier to calculate PDF($+\alpha_{\text{s}}$) uncertainties for any observable at the LHC. In addition, by using the technique of data set diagonalization, the number of error PDFs needed to describe the $PDF + \alpha_{\text{s}}$ uncertainties for all Higgs production processes for all LHC energies can be reduced to 8.

6.3.5   PDF uncertainties and evolution

Evolution is the great equalizer. Parton uncertainties tend to decrease as the factorization scale increases. This can be seen for example for the case of the gluon distribution in Figs. 6.20, 6.21 6.22 and 6.23 where the gluon uncertainties for the CT10 and MSTW2008 NLO PDFs are shown for Q^2 values of 2, 10, 100, and 10000 GeV^2. Aside from the high x region, a significant decrease in the uncertainty is observed.

6.3.6   PDF uncertainties and Sudakov form factors

As discussed in the above section, it is often useful to use the error PDF sets with parton shower Monte Carlos. The caveat still remains that a true test of the acceptances would use a NLO MC. Similar to their use with matrix element calculations, events can be generated once using the central PDF and the PDF weights stored for the error PDFs. These PDF weights then can be used to construct the PDF uncertainty for any

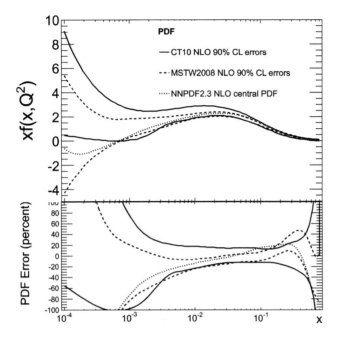

Fig. 6.20 A comparison of the uncertainties of the CT10 and MSTW2008 NLO gluon distributions, along with the central PDF of NNPDF2.3, for a Q^2 value of 2 GeV^2.

observable. One additional complication with respect to their use in matrix element programmes is that the parton distributions are used to construct the initial state parton showers through the backward evolution process. The space-like evolution of the initial state partons is guided by the ratio of parton distribution functions at different x and Q^2 values. Thus the Sudakov form factors in parton shower Monte Carlos will be constructed using only the central PDF and not with any of the individual error PDFs and this may lead to some errors for the calculation of the PDF uncertainties of some observables. However, it was demonstrated in Ref. [578] that the PDF uncertainty for Sudakov form factors in the kinematic region relevant for the LHC is minimal, and the weighting technique can be used just as well with parton shower Monte Carlos as with matrix element programmes.

6.3.7 Choice of $\alpha_s(m_Z)$ and related uncertainties

Global PDF fits are sensitive to the value of the strong coupling constant α_s, explicitly through the QCD cross-sections used in the fits, and implicitly through the scaling violations observed in DIS. In fact, a global fit can be used to determine the value of $\alpha_s(m_Z)$, albeit less accurately than provided by the world average. Historically, some PDF groups have used the world average value of $\alpha_s(m_Z)$ [239, 240, 791] as a fixed constant in the global fits, while other groups have allowed $\alpha_s(m_Z)$ to be a free parameter in the fit. It is also possible to explore the effects of the variation of

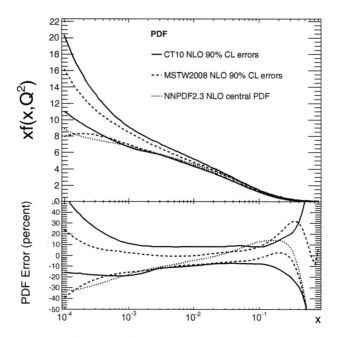

Fig. 6.21 A comparison of the uncertainties of the CT10 and MSTW2008 NLO gluon distributions, along with the central PDF of NNPDF2.3, for a Q^2 value of 10 GeV^2.

$\alpha_{\rm s}(m_Z)$ by producing PDFs at different fixed $\alpha_{\rm s}(m_Z)$ values. There is now a consensus to use $\alpha_{\rm s}(m_Z) = 0.118$ as a central value (basically an approximation/truncation of the current world value) in global PDF fits,at both NLO and NNLO, and to publish alternative fits with $\alpha_{\rm s}(m_Z)$ values in intervals of ± 0.001 around that central value. It is expected that the LO value of $\alpha_{\rm s}(m_Z)$ is considerably larger than the NLO value (0.130 compared to 0.118 for the CTEQ/CT PDFs, for example).

There is a correlation/anti-correlation between the value of $\alpha_{\rm s}(m_Z)$ used in the global PDF fit and the gluon distribution; whether there is a correlation or anti-correlation depends on the gluon x range being considered. At low x (less than 0.1), a decrease in the value of $\alpha_{\rm s}(m_Z)$ results in an increase in the gluon distribution and vice versa, i.e. there is an anti-correlation. The net impact is to reduce the sensitivity of cross-sections that depend on both the value of $\alpha_{\rm s}(m_Z)$ and the gluon distribution in this x range to variations in the value of $\alpha_{\rm s}(m_Z)$. The sensitivity becomes smaller as the x value approaches 0.1. In the x range from 0.1 to 0.8, there is a correlation between the value of $\alpha_{\rm s}(m_Z)$ and the gluon distribution, with the correlation becoming larger as the x value increases.

The diagonalization technique can also be used with respect to the value of $\alpha_{\rm s}(m_Z)$; in fact, it can be shown, using this technique, that, within the quadratic approximation, the uncertainty in $\alpha_{\rm s}(m_Z)$ is uncorrelated with the PDF uncertainty [714]. Thus the combined PDF+$\alpha_{\rm s}$ can be calculated by computing the $1 - \sigma$ PDF uncertainty with

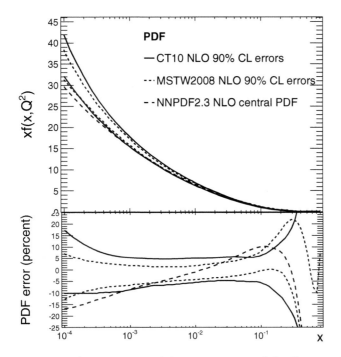

Fig. 6.22 A comparison of the uncertainties of the CT10 and MSTW2008 NLO gluon distributions, along with the central PDF of NNPDF2.3, for a Q^2 value of 100 GeV^2.

$\alpha_{\rm s}(m_Z)$ fixed at its central value, and adding in quadrature the $1 - \sigma$ uncertainty in $\alpha_{\rm s}(m_Z)$ (and of course, this can also be done for any other desired confidence level).

6.3.8 PDF correlations

The uncertainty analysis may be extended to define a *correlation* between the uncertainties of two variables, say $X(\vec{a})$ and $Y(\vec{a})$. As for the case of PDFs, the physical concept of PDF correlations can be determined both from PDF determinations based on the Hessian approach and on the Monte Carlo approach.

6.3.8.1 PDF correlations in the Hessian approach

Consider the projection of the tolerance hypersphere onto a circle of radius 1 in the plane of the gradients $\vec{\nabla}X$ and $\vec{\nabla}Y$ in the parton parameter space [776, 817]. The circle maps onto an ellipse in the XY plane. This "tolerance ellipse" is described by Lissajous-style parametric equations,

$$X = X_0 + \Delta X \cos\theta, \tag{6.60}$$
$$Y = Y_0 + \Delta Y \cos(\theta + \varphi), \tag{6.61}$$

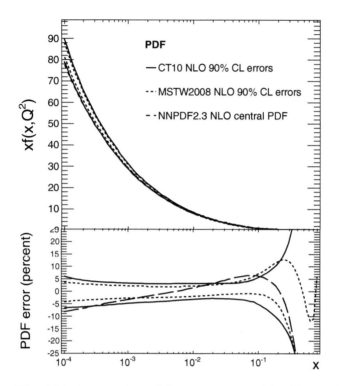

Fig. 6.23 A comparison of the uncertainties of the CT10 and MSTW2008 NLO gluon distributions, along with the central PDF of NNPDF2.3, for a Q^2 value of 10000 GeV^2.

where the parameter θ varies between 0 and 2π, $X_0 \equiv X(\vec{a}_0)$, and $Y_0 \equiv Y(\vec{a}_0)$. ΔX and ΔY are the maximal variations $\delta X \equiv X - X_0$ and $\delta Y \equiv Y - Y_0$ evaluated according to the *Master* Equation, and φ is the angle between $\vec{\nabla} X$ and $\vec{\nabla} Y$ in the $\{a_i\}$ space, with

$$\cos\varphi = \frac{\vec{\nabla} X \cdot \vec{\nabla} Y}{\Delta X \Delta Y} = \frac{1}{4\Delta X \, \Delta Y} \sum_{i=1}^{N} \left(X_i^{(+)} - X_i^{(-)} \right) \left(Y_i^{(+)} - Y_i^{(-)} \right). \qquad (6.62)$$

The quantity $\cos\varphi$ characterizes whether the PDF degrees of freedom of X and Y are correlated ($\cos\varphi \approx 1$), anti-correlated ($\cos\varphi \approx -1$), or uncorrelated ($\cos\varphi \approx 0$). If units for X and Y are rescaled so that $\Delta X = \Delta Y$ (e.g., $\Delta X = \Delta Y = 1$), the semimajor axis of the tolerance ellipse is directed at an angle $\pi/4$ (or $3\pi/4$) with respect to the ΔX axis for $\cos\varphi > 0$ (or $\cos\varphi < 0$). In these units, the ellipse reduces to a line for $\cos\varphi = \pm 1$ and becomes a circle for $\cos\varphi = 0$, as illustrated by Fig. 6.24. These properties can be found by diagonalizing the equation for the correlation ellipse. Its semi-minor and semi-major axes (normalized to $\Delta X = \Delta Y$) are

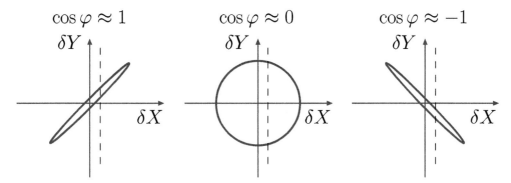

$$\cos\varphi \approx 1 \qquad\qquad \cos\varphi \approx 0 \qquad\qquad \cos\varphi \approx -1$$

Fig. 6.24 Correlations ellipses for a strong correlation (left), no correlation (centre) and a strong anti-correlation(right). Reprinted with permission from Ref. [774].

$$\{a_{minor}, a_{major}\} = \frac{\sin\varphi}{\sqrt{1 \pm \cos\varphi}}. \tag{6.63}$$

The eccentricity $\epsilon \equiv \sqrt{1 - (a_{minor}/a_{major})^2}$ is therefore approximately equal to $\sqrt{|\cos\varphi|}$ as $|\cos\varphi| \to 1$.

The ellipse itself is described by

$$\left(\frac{\delta X}{\Delta X}\right)^2 + \left(\frac{\delta Y}{\Delta Y}\right)^2 - 2\left(\frac{\delta X}{\Delta X}\right)\left(\frac{\delta Y}{\Delta Y}\right)\cos\varphi = \sin^2\varphi. \tag{6.64}$$

A magnitude of $|\cos\varphi|$ close to unity suggests that a precise measurement of X (constraining δX to be along the dashed line in Fig. 6.24) is likely to constrain tangibly the uncertainty δY in Y, as the value of Y shall lie within the needle-shaped error ellipse. Conversely, $\cos\varphi \approx 0$ implies that the measurement of X is not likely to constrain δY strongly.[9]

The values of ΔX, ΔY, and $\cos\varphi$ are also sufficient to estimate the PDF uncertainty of any function $f(X,Y)$ of X and Y by relating the gradient of $f(X,Y)$ to $\partial_X f \equiv \partial f/\partial X$ and $\partial_Y f \equiv \partial f/\partial Y$ via the chain rule:

$$\Delta f = \left|\vec{\nabla} f\right| = \sqrt{(\Delta X\, \partial_X f\,)^2 + 2\Delta X\, \Delta Y\, \cos\varphi\, \partial_X f\, \partial_Y f + (\Delta Y\, \partial_Y f)^2}. \tag{6.65}$$

Of particular interest is the case of a rational function $f(X,Y) = X^m/Y^n$, pertinent to computations of various cross-section ratios, cross-section asymmetries, and statistical significance for finding signal events over background processes [776]. For rational functions Eq. (6.65) takes the form

---

[9]The allowed range of $\delta Y/\Delta Y$ for a given $\delta \equiv \delta X/\Delta X$ is $r_Y^{(-)} \le \delta Y/\Delta Y \le r_Y^{(+)}$, where $r_Y^{(\pm)} \equiv \delta\cos\varphi \pm \sqrt{1-\delta^2}\sin\varphi$.

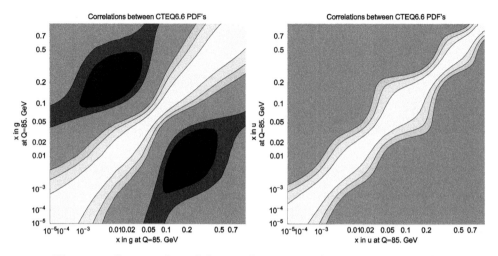

Fig. 6.25 Contour plots of the correlation cosine between two PDFs, for the up quark (left) and the gluon (right).

$$\frac{\Delta f}{f_0} = \sqrt{\left(m\frac{\Delta X}{X_0}\right)^2 - 2mn\frac{\Delta X}{X_0}\frac{\Delta Y}{Y_0}\,\cos\varphi\, + \left(n\frac{\Delta Y}{Y_0}\right)^2}. \qquad (6.66)$$

For example, consider a simple ratio, $f = X/Y$. Then $\Delta f/f_0$ is suppressed ($\Delta f/f_0 \approx |\Delta X/X_0 - \Delta Y/Y_0|$) if X and Y are strongly correlated, and it is enhanced ($\Delta f/f_0 \approx \Delta X/X_0 + \Delta Y/Y_0$) if X and Y are strongly anticorrelated.

As would be true for any estimate provided by the Hessian method, the correlation angle is inherently approximate. Eq. (6.62) is derived under a number of simplifying assumptions, notably in the quadratic approximation for the χ^2 function within the tolerance hypersphere, and by using a symmetric finite-difference formula for $\{\partial_i X\}$ that may fail if X is not monotonic. Even with these limitations in mind, the correlation angle is a convenient measure of the interdependence between quantities of diverse nature, such as physical cross-sections and parton distributions themselves.

Correlations can be calculated between two PDFs, $f_{a1}(x_1, \mu_1)$ and $f_{a2}(x_2, \mu_2)$ at a scale $\mu_1 = \mu_2 = 85$ GeV. In the figure below, the self-correlations for the up quark (left) and the gluon (right) are shown. Light (dark) shades of grey correspond to $cos\phi$ close to 1 (-1). Each self-correlation includes a trivial correlation ($cos\phi = 1$) when x_1 and x_2 are approximately the same (along the $x_1 = x_2$ diagonals). For the up quark, this trivial correlation is the only pattern present. The gluon distribution, however, also shows a strong anti-correlation when one of the x values is large and the other small. This arises as a consequence of the momentum sum rule.

PDF correlations for physics processes at the LHC will be discussed later in this chapter.

6.3.8.2 Correlations within the Monte Carlo approach

Correlations can also be calculated using the Monte Carlo approach, as practiced for example by NNPDF [135]. The correlation cosine between two observables A and B can be calculated in this approach as

$$cos\phi[A, B] = \frac{N_{rep}}{(N_{rep} - 1)} \frac{\langle AB \rangle_{\rm rep} - \langle A \rangle_{\rm rep} \langle B \rangle_{\rm rep}}{\sigma_A \sigma_B} \tag{6.67}$$

where the averages are taken over the ensemble of the n_{rep} values of the observables computed with the different replicas of the NNPDF set, and $\sigma_{A,B}$ are the standard deviations of the ensembles.

6.4 Resulting parton distribution functions

6.4.1 LO, NLO and NNLO PDFs

Global PDF fitting groups have also traditionally produced sets of PDFs in which leading order rather than next-to-leading order (or next-to-next-to-leading order) matrix elements, along with the 1-loop α_S rather than the 2-loop α_S, have been used to fit the input datasets. The resultant leading order PDFs have most often been used in conjunction with leading order matrix element programmes or parton shower Monte Carlos. However, the leading order PDFs of a given set will tend to differ from the central PDFs in the NLO fit, and in fact will most often lie outside the PDF error band. Such is the case for the up quark distribution and the gluon distribution from the CTEQ6.1 set of PDFs shown in Fig. 6.26 where the LO PDFs are plotted along with the NLO PDF error bands.[10] The LO up quark distribution is considerably larger than its NLO counterpart at both small x and large x. This is due to (1) the larger gluon distribution at small x for the LO PDF and (2) the influence of missing $log(1-x)$ terms in the LO DIS matrix element. The gluon distribution is outside of the NLO error band basically for all x. It is higher than the NLO gluon distribution at small x due to missing $log(1/x)$ terms in the LO DIS matrix element. It is smaller than the NLO gluon distribution at large x basically due to the momentum sum rule and the lack of constraints at high x.

The global PDF fits are dominated by the high statistics, low systematic error deep inelastic scattering data, and the differences between the LO and NLO PDFs are determined most often by the differences between the LO and NLO matrix elements for deep inelastic scattering. This is especially true at low x and at high x, due to missing terms that first arise in the hard matrix elements for DIS at NLO. As the NLO corrections for most processes of interest at the LHC are reasonably small, the use of NLO PDFs in conjunction with LO matrix elements will most often give a closer approximation of the full NLO result (although the result remains formally LO). In many cases in which a relatively large K-factor results from a calculation of collider processes, the primary cause is the difference between LO and NLO PDFs, rather than the differences between LO and NLO matrix elements.

---

[10]These observations are true in general for comparison of any sets of LO and NLO PDFs.

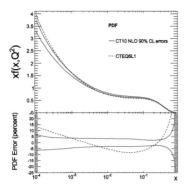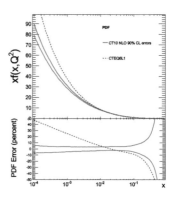

Fig. 6.26 The CTEQ6L1 up quark and gluon PDFs, evaluated at $Q^2 = 10^4$ GeV^2 compared to the CT10 NLO PDF error bands for the same.

In most cases, LO PDFs will be used not in fixed order calculations, but in programmes where the LO matrix elements have been embedded in a parton shower framework. In the initial state radiation algorithms in these frameworks, shower partons are emitted at non-zero angles with finite transverse momentum, and not with a zero k_T implicit in the collinear approximation. It might be argued that the resulting kinematic suppression due to parton showering should be taken into account when deriving PDFs for explicit use in Monte Carlo programmes. Indeed, there is substantial kinematic suppression for production of a low-mass (10 GeV) object at forward rapidities due to this effect, but the suppression becomes minimal once the mass rises to the order of 100 GeV [715].

6.4.2 Modified LO PDFs

Due to the inherent differences between LO and NLO PDFs, and the relatively small differences between LO and NLO matrix elements for processes of interest at the LHC, LO calculations at the LHC using LO PDFs often lead to erroneous predictions. This is true not only of the normalization of the cross-sections, but also for the kinematic shapes. This can be seen for example in the predictions for the $W^{+/-}/Z$ and Higgs rapidity distributions seen in Fig. 6.27, where the wrong shapes for the vector boson rapidity distributions result from the deficiencies of the LO DIS matrix elements used in the fit. This can have an impact, for example, if the LO predictions are used to calculate final-state acceptances.

In an attempt to reduce the size of the errors obtained using LO PDFs with LO predictions, modified LO PDFs have been produced. The techniques used to produce these modified PDFs include (1) relaxing the momentum sum rule in the global fit and (2) using NLO pseudo-data in order to try to *steer* the fit towards the desired NLO behaviour. Both the CTEQ [715] and MRST [846] modified LO PDFs use the first technique, while the CTEQ PDFs use the second technique as well.

Of course, the desired behaviour can also be obtained (in most cases) by the use of NLO PDFs in the LO calculation. Here, care must be taken that only positive-definite

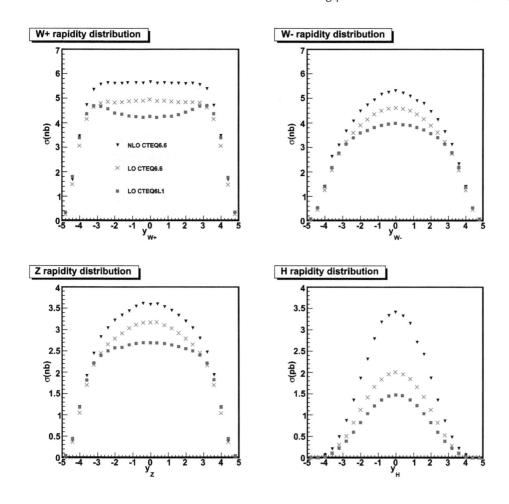

Fig. 6.27 A comparison of NLO predictions for SM boson rapidity distributions to LO predictions for the same, using CTEQ6.6 and CTEQ6L1 PDFs respectively. Reprinted with permission from Ref. [715].

NLO PDFs be used. Increasingly, most processes of interest have been included in NLO parton shower Monte Carlos. Here, the issue of LO PDFs becomes moot, as NLO PDFs must be used in such programs for consistency with the matrix elements. As a result, the use of modified LO PDFs has been decreasing.

6.4.3 NNLO PDFs, and beyond

All of the PDF groups now have PDFs determined at NNLO as well as at NLO. The transition from NLO to NNLO results in much smaller changes to the PDFs than for the transition from LO to NLO. Although the changes from NLO to NNLO are much smaller than those from LO to NLO, they can still be observed, especially at

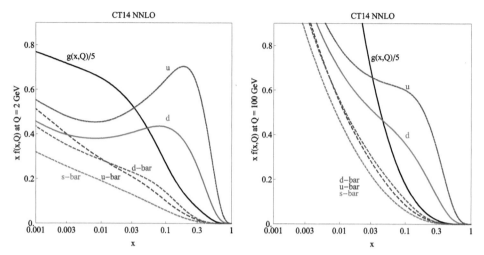

Fig. 6.28 CT14 NNLO PDFs as a function of x for $Q = 2$ GeV (left) and $Q = 100$ GeV (right). Reprinted with permission from Ref. [489].

low Q values.[11] At higher Q values, though, the differences are reduced. As the PDF uncertainties are dominated by the experimental errors of the data included in the PDF fits, the uncertainties at NNLO will be similar to those determined at NLO.

As mentioned previously, the most recent PDF set from the CTEQ-TEA group is CT14 [489]. The NNLO PDFs from CT14 are shown in Fig. 6.28 for a Q values of 2 GeV and 100 GeV.

Differences between the CT14 and the CT10 PDFs for the up quark and the gluon distributions are shown in Fig. 6.29. The differences are relatively small, within the error bands of either PDF set, and tend to be the most significant at low x and high x where the PDFs are the most unconstrained. One of the most important changes is not easily visible in these plots; that is of the gluon distribution in the x region around 0.01. The changes from CT10 to CT14 are small, but have an impact on the PDF uncertainty for Higgs boson production at the LHC, as discussed in Section 6.5.1.

As described in Chapter 2, the gg fusion cross-section for Higgs production to NNNLO has been completed. Since it will be quite some time before any PDFs are produced at this order, there arises the question as to the level of error produced when NNLO PDFs are used with such NNNLO calculations. It has been shown [531] that the error should be much smaller than the level of the difference between the NLO and the NNLO matrix elements.

6.5 CT14 and parton luminosities

It is useful to introduce the idea of differential parton-parton luminosities. Such luminosities, when multiplied by the dimensionless cross-section $\hat{s}\hat{\sigma}$ for a given process,

---

[11]For example, at low Q, the order α_s^2 evolution in the NNLO PDF suppresses $g(x, Q)$ and increases $q(x, Q)$ relative to NLO.

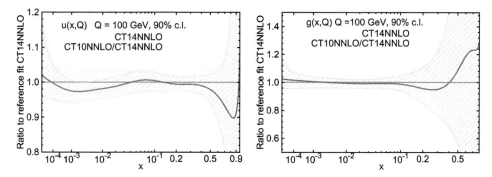

Fig. 6.29 A comparison of the CT10 and CT14 up quark (left) and gluon (right) distributions. Reprinted with permission from Ref. [489].

provide a useful estimate of the size of an event cross-section at the LHC. Below we define the differential parton-parton luminosity $dL_{ij}/d\hat{s}\,dy$ and its integral $dL_{ij}/d\hat{s}$:

$$\frac{dL_{ij}}{d\hat{s}\,dy} = \frac{1}{s}\frac{1}{1+\delta_{ij}}\left[f_i(x_1,\mu)f_j(x_2,\mu) + (1 \leftrightarrow 2)\right]. \tag{6.68}$$

The prefactor with the Kronecker delta avoids double-counting in case the partons are identical. The generic parton-model formula

$$\sigma = \sum_{i,j}\int_0^1 dx_1\,dx_2\,f_i(x_1,\mu)\,f_j(x_2,\mu)\,\hat{\sigma}_{ij} \tag{6.69}$$

can then be written as[12]

$$\sigma = \sum_{i,j}\int\left(\frac{d\hat{s}}{\hat{s}}\,dy\right)\left(\frac{dL_{ij}}{d\hat{s}\,dy}\right)(\hat{s}\,\hat{\sigma}_{ij}). \tag{6.70}$$

Fig. 6.30 shows a plot of the luminosity function integrated over rapidity, $dL_{ij}/d\hat{s} = \int (dL_{ij}/d\hat{s}\,dy)\,dy$, at the LHC $\sqrt{s} = 13\,\text{TeV}$ for various parton flavour combinations, for the CT14 PDFs. The gluon-gluon PDF luminosity dominates at low mass, the gluon-quark PDF luminosity for masses from 300 GeV to approximately 2 TeV, with the quark-quark luminosity being largest for all masses above 2 TeV.

6.5.1 Comparison of PDFs at the LHC

As mentioned earlier in this chapter, the three major PDF fitting groups (with the latest PDF sets being CT14, MMHT2014, and NNPDF3.0) fit to basically the same data sets (albeit with different kinematic cuts in some cases). However, there can still be differences in the resultant PDF fits and uncertainties due to differences in the fitting procedures and to details such as the parameterizations and heavy quark

---

[12]Note that this result is easily derived by defining $\tau = x_1\,x_2 = \hat{s}/s$ and observing that the Jacobian $\partial(\tau,y)/\partial(x_1,x_2) = 1$.

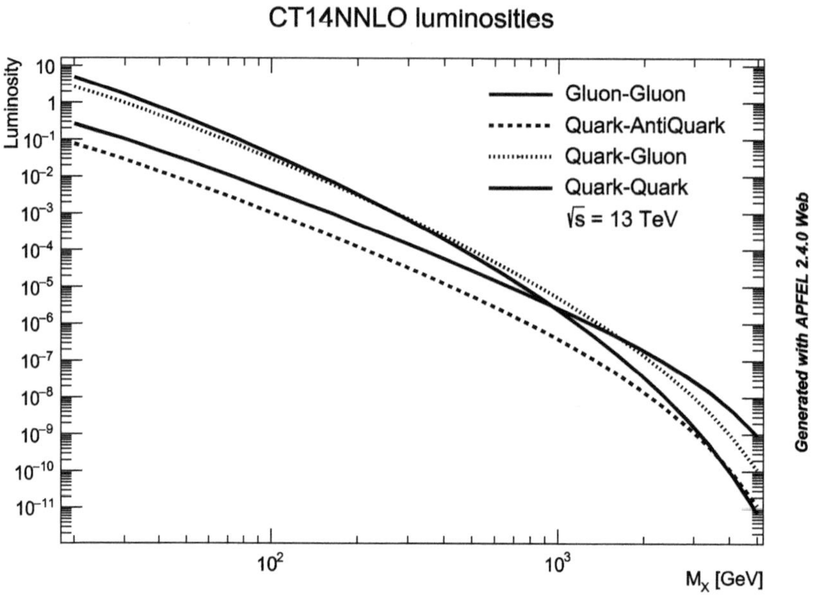

CT14NNLO luminosities

Fig. 6.30 The parton-parton luminosities for CT14 for *pp* collisions at 13 TeV plotted as a function of mass ($M_X \equiv \sqrt{\hat{s}}$).

flavour schemes used, etc. However, as time has progressed, the tendency has been for the three groups to have results that are in good agreement with each other.

Consider for example the quark-antiquark and gluon-gluon PDF luminosity uncertainties as a function of mass, at NNLO, for CT14, MMHT2014, and NNPDF3.0 (and the prior, to be defined in the following) for an LHC centre-of-mass energy of 13 TeV, shown in Fig. 6.31 from Ref. [196]. The central values and the size of the uncertainties can vary among the 3 PDF groups at low-mass and at high-mass, but in the precision mass region (say from 50-500 GeV), both the central values and the uncertainties are in remarkably good agreement with each other (especially so for the gluon-gluon case). This was not the situation for the gluon-gluon luminosity for the previous round of PDF fitting (CT10, MSTW08 and NNPDF2.3), where the envelope of the uncertainty bands for the 3 PDF groups yielded a PDF uncertainty for Higgs boson production through *gg* fusion that was about a factor of 2.5 larger than the PDF uncertainty band for any of the individual PDFs [196].[13] The resultant PDF uncertainty was similar in size to the NNLO scale uncertainty for the cross-section [618]. As the scale uncertainty at NNNLO has shrunk to the order of 2-3% [156], the use of the older generation of PDFs would have left the PDF($+\alpha_s(m_Z)$) uncertainty as the largest source of uncertainty for that cross-section, and implicitly for the determination of the Higgs couplings and other parameters that depend on the absolute knowledge of the

---

[13]The CT14 *gg* PDF luminosity increased by about 1% in the Higgs region at 13 GeV (compared to the older generation CT10), the MMHT2014 decreased by about 0.5%, and the NNPDF3.0 PDF luminosity decreased by about 2-2.5%.

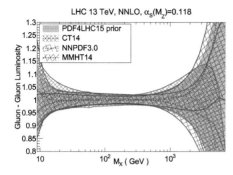
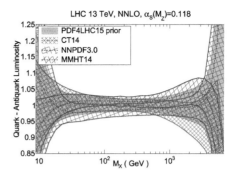

Fig. 6.31 A comparison of the PDF luminosities for the prior, CT14, MMHT2014 and NNPDF3.0 are shown for the gg initial state (left) and the $q\bar{q}$ initial state (right) for a centre-of-mass energy of 13 TeV. Reprinted with permission from Ref. [295].

Standard Model cross-sections for Higgs boson production.

The PDF4LHC working group has been responsible for the determination of appropriate recommendations for both the PDFs and the PDF uncertainties for the LHC. Implicitly, this also includes recommendations for the value of $\alpha_s(m_Z)$ and its uncertainties. Through its history, the group has performed a number of benchmarking exercises and recommendations [135, 196, 268],[14] with the most recent being [295], using the 3 global PDF fits discussed above. Prior to this most recent document, the recommendation for $\alpha_s(m_Z)$ was to use a central value of 0.118, with an uncertainty (at 90% CL) of ± 0.002, corresponding to a 68% CL uncertainty of ± 0.0012. The central value was chosen as a truncation of the world average value of $\alpha_s(m_Z)$ (0.1185), and the uncertainty was an enlargement of the value obtained in the world average (± 0.0006) [799]. All of the global PDF groups (as well as HERAPDF) use this central value for their central fits, as well as producing alternate PDF sets with the value of $\alpha_s(m_Z)$ typically varying in increments of ± 0.001. In the recent PDF4LHC recommendation, the central value of $\alpha_s(m_Z)$ has remained the same (0.118), but the 68% CL uncertainty was changed to ± 0.0015, partially to reflect the increased level of uncertainty adopted by the Particle Data Group [791], and partially to reflect that the same value of $\alpha_s(m_Z)$ is recommended for both NLO and NNLO. (There is a question whether the value should be the same for the two orders, or whether the NNLO value should be slightly smaller.)

Previous PDF4LHC recommendations have determined the uncertainty for a given cross-section by using the envelope of the error bands of PDF luminosities from the 3 global PDF groups. This has the drawback that it lacks a rigorous statistical interpretation (the envelope can be determined by a few extreme sets). In the new recommendation, the uncertainty is obtained from the combination of error PDFs from the 3 groups using Monte Carlo replicas [885]. For CT14 and MMHT2014, an arbitrary number of Monte Carlo replicas can be generated from the Hessian eigenvectors such that the replicas represent the underlying probability density. It is possible

---

[14]See also http://www.hep.ucl.ac.uk/pdf4lhc/

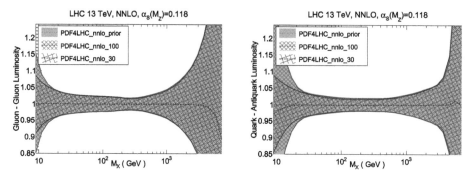

Fig. 6.32 A comparison of the PDF luminosities for the prior, the 30 PDF set and the 100 PDF set, is shown for the gg initial state (left) and the $q\bar{q}$ initial state (right) for a centre-of-mass energy of 13 TeV. Reprinted with permission from Ref. [295].

to convert from a Hessian representation to a Monte Carlo representation. The use of a Monte Carlo representation allows a straightforward combination of PDF sets and a more rigorous determination of the 68% CL.[15] The new PDF4LHC recommendation is based on a 900 Monte Carlo replica set, 300 derived each from the eigenvectors of CT14, MMHT2014, and NNPDF3.0. The 900 PDF set can be reduced to either 30 error PDFs or 100 error PDFs (the latter with either symmetric or asymmetric errors) using 3 separate techniques [332, 334, 552]. Each adequately represents (1) the central PDF and (2) the PDF uncertainties of the 900 Monte Carlo replica prior. Details on each of these techniques and their specific applications are given in the PDF4LHC recommendation document and in the forthcoming Higgs Yellow Report 4. Each technique also includes 2 member sets with $\alpha_s(m_Z)$ 0.0015 higher or lower than the nominal value of 0.118. The PDF+$\alpha_s(m_Z)$ error can be calculated by adding these sets in quadrature with the PDF error sets.

The PDF luminosity uncertainties for the 900 PDF Monte Carlo replica set, along with those from the PDF4LHC15 error set containing 30 error PDFs and from the error set containing 100 error PDFs, are shown in Fig. 6.32, for the gg PDF luminosity (left) and the $q\bar{q}$ PDF luminosity (right). The 900 PDF Monte Carlo replica set represents the best estimate by the PDF4LHC group for the PDF uncertainty, but its absolute accuracy is unknown. Both the 30 and 100 PDF sets are capable of reproducing the uncertainty of the prior in the precision mass range, and in the high mass range; the 100 PDF set better reproduces the uncertainty for very low mass.[16] The resultant PDF uncertainty for gg fusion production of a Higgs boson at 13 TeV is 2%, and the $\alpha_s(m_Z)$ uncertainty is 2%, both comparable to the NNNLO scale uncertainty.

A comparison of the predictions for Higgs boson production at 13 TeV relative to the prediction derived from the prior, for gg fusion (left) and vector boson fusion

---

[15]The NNPDF3.0 set is already in this formulation and the CT14 and MMHT2014 Hessian sets can be converted to such.

[16]The low-mass difference is basically due to low-mass final states produced at rapidities beyond the acceptance of the LHC detectors, which were not included in the construction of the 30 PDF set.

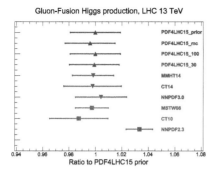 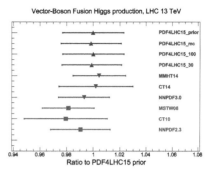

Fig. 6.33 A comparison of the predictions for Higgs boson production through gg fusion (left) and vector boson fusion (right) for a centre-of-mass energy of 13 TeV. Reprinted with permission from Ref. [441].

Table 6.2 The correlation coefficients between various Higgs production cross-sections at 13 TeV. In each case, the PDF4LHC15 NNLO prior set is compared to the Monte Carlo and with the two Hessian reduced sets, and the results from the three individual sets, CT14, MMHT14 and NNPDF3.0.

PDF Set	correlation coefficient					
	$t\bar{t}, Ht\bar{t}$	$t\bar{t}, hW$	$t\bar{t}, hZ$	$ggh, ht\bar{t}$	ggh, hW	ggh, hZ
PDF4LHC15_nnlo_prior	0.87	-0.23	-0.34	-0.13	-0.01	-0.17
PDF4LHC15_nnlo_mc	0.87	-0.27	-0.35	-0.10	0.07	-0.01
PDF4LHC15_nnlo_100	0.87	-0.24	-0.34	-0.13	-0.02	-0.17
PDF4LHC15_nnlo_30	0.87	-0.27	-0.43	-0.13	-0.04	-0.23
CT14	0.09	-0.32	-0.44	-0.26	-0.03	-0.18
MMHT14	0.90	-0.22	-0.52	0.08	-0.18	-0.33
NNPDF3.0	0.90	-0.17	-0.21	0.18	0.52	0.49

(right), is shown in Fig. 6.33 [441]. From these plots, the level of agreement of the predictions from the PDF sets provided by the PDF4LHC group with each other, and with the current and previous generation of global PDF sets, can be determined.

The correlation coefficients between various Higgs boson production processes at 13 TeV are shown in Table 6.2 [295]. Note the spread in correlation coefficients among the three global PDFs. No more than a single digit accuracy should be ascribed to the correlation numbers.

A small number of error PDFs can be very useful in instances where the PDF uncertainties are used as nuisance parameters. It is possible, for example, to reduce the number of error PDFs needed to describe Higgs physics and backgrounds down to a smaller number, on the order of 7, using the METAPDF technique, without significant loss of precision [552]. Other techniques are available which also reduce the number of error PDFs needed [333].

6.6   LHAPDF and other tools

6.6.0.1   LHAPDF, Durham PDF plotter, and APFEL

Libraries such as PDFLIB [812] were established to maintain a large collection of available PDFs. However, PDFLIB is no longer supported, making it more difficult for easy access to the most up-to-date PDFs. In addition, the determination of the PDF uncertainty of any cross-section typically involves the use of a large number of PDFs (up to several hundred) and PDFLIB was not set up for easy accessibility for a large number of PDFs.

At Les Houches in 2001, representatives from a number of PDF groups were present and an interface (Les Houches Accord 2, or LHAPDF) [572] that allows the compact storage of the information needed to define a PDF was defined. Each PDF can be determined either from a grid in x and Q^2, or by a few lines of information – essentially the starting values of the parameters at $Q = Q_0$. The interface then carries out the evolution to any x and Q value, at either LO, NLO, or NNLO, as appropriate for each PDF.

The interface is as easy to use as PDFLIB and consists essentially of 3 subroutine calls:

- `call InitPDFset(`*name*`)`: called once at the beginning of the code; *name* is the file name of the external PDF file that defines the PDF set (for example, CTEQ, MSTW, or NNPDF).
- `call Initpdf(`*mem*`)`: *mem* specifies the individual member of the PDF set.
- `call evolvepdf(`*x,Q,f*`)`: returns the PDF momentum densities for flavour *f* at a momentum fraction *x* and scale *Q*.

Responsibility for LHAPDF has been taken over by the Durham HEPDATA project [891] and regular updates and improvements have been produced. Interfaces with LHAPDF are now included in most matrix element programs. Recent modifications make it possible to include all error PDFs in memory at the same time. Such a possibility reduces the amount of time needed for PDF error calculations on any observable, as discussed below.

One very useful tool is the Durham PDF plotter[17] which allows the fast plotting and comparisons of PDFs, including their error bands. Recently **APFEL Web**[18] [238, 331], a web-based application for the graphical visualization of PDFs was developed that allows a very fast online calculation of PDFs and their associated errors, as well as PDF luminosities. Many of the PDF plots in this book were made using these two routines.

6.6.1   PDF event re-weighting, Applgrid, and fastNLO

NLO and NNLO programmes are notoriously slow. Thus, it can be very time-consuming to generate a higher order cross-section with one PDF, and then have to re-run the program as well for the $2N$ (where N is the number of PDF eigenvectors) error PDFs.

---

[17]http://hepdata.cedar.ac.uk/pdf/pdf3.html
[18]http://apfel.mi.infn.it

Such a step is in fact unnecessary, and most programs have the ability to use PDF event re-weighting to substitute a new PDF for the PDF used in the original generation.

For each event generated with the central PDF from the set, PDF weights, for the error PDFs, can also be determined. Only one Monte Carlo event sample is generated but $2N + 1$ (e.g. 57 for CT14) PDF weights are obtained using

$$W_n^0 = 1, W_n^i = \frac{f(x_1, Q; S_i) f(x_2, Q; S_i)}{f(x_1, Q; S_0) f(x_2, Q; S_0)} \tag{6.71}$$

where $n = 1, \ldots, N_{events}$, $i = 1, \ldots, 2N+1$ [280]. Any statistical error due to the finite event size of the original event distribution is cancelled in the ratios that determine the error PDF relative weights. Such a PDF re-weighting has been shown to work both for exact matrix element calculations as well as for matrix element+parton shower calculations (modulo subtle effects in regions where Sudakov suppression is large). The PDF error weights can either be stored at the time of generation, or can be generated on the fly by the program.

Since the PDF-dependent information in a QCD calculation can be factorized from the rest of the hard scattering terms, it is possible to calculate the non-PDF terms one time and then to store the PDF information on a grid (in terms of the PDF x values and their μ_r and μ_f dependence). This allows for the fast calculation of any hard scattering cross-section and the a posteriori inclusion of PDFs and the strong coupling constant α_s in higher order QCD calculations. The technique also allows the a posteriori variation of the renormalization and factorization scales. This is the working principle of the two programmes fastNLO [686] and Applgrid [329]. These programmes are greatly used for calculation of the NLO matrix elements used in PDF fits, and as discussed earlier in Section 6.2.1 are now being adapted for use with NNLO matrix elements.

6.7 Summary

An accurate knowledge of parton distribution functions is crucial for precision LHC phenomenology. In this chapter, the techniques for the determination of PDFs were described, as well as techniques for the determination of the uncertainties of the PDFs. Global PDF fits involve data from deep-inelastic scattering, Drell-Yan and inclusive jet production, with increasing contributions from single photon and top production. Previous generations of fits used only data from the TEVATRON, HERA, and fixed target experiments, but the copious data from Run 1 (and now Run 2) at the LHC are starting to have a significant impact.

The evolution PDFs using the DGLAP equation was revisited in more detail. Evolution is the great equalizer, as differences among PDFs from different groups, and the sizes of PDF uncertainties, decrease with increasing Q^2. In general, there is a larger difference between PDFs determined with a leading order framework and a next-to-leading order framework, with a much smaller difference when going from NLO to NNLO. LHC predictions carried out with LO PDFs can differ greatly both in normalization and shape from those carried out in a purely NLO framework. NLO shapes can often be recovered with LO matrix elements by using NLO PDFs.

Parton PDF luminosities were defined and a new framework (PDF4LHC15) for combining PDFs from the three global PDF fitting groups was described. The PDF uncertainties for the three PDF groups can be summarized with a limited number of error PDFs, as low as 30. Lastly, a number of useful PDF tools were described.

PDFs will continue to improve as more data from the LHC is included, and as more crucial processes are calculated at NNLO. As for the theory, and for the LHC data, what is presented here is just a snapshot of a rapidly developing field.

7
Soft QCD

When looking at event displays at hadron colliders, it becomes apparent that the perturbative picture developed so far does not yet cover all aspects of what can be seen. First of all, there are many events with only a few — if any — particles hitting the central regions of the detector, either as charged tracks or as energy deposits in the calorimeters. Quite often these particles are relatively soft, with transverse momenta at around or below $1\,\mathrm{GeV}$. At the beginning of this chapter, in Section 7.1, the soft inclusive physics underlying such events will briefly be discussed. Building on a short presentation of the ideas behind typical models for soft strong interactions, which are quite often based on Pomeron and Regge pole physics, total hadronic cross-sections and their parametrizations will be introduced. This will extend to elastic and diffractive processes, which populate the forward regions of the detector, usually with low-$p_\perp$ particles.

Increasing the energy scales, the next section, Section 7.2 will focus on multiple parton scattering. This phenomenon is closely related to the fact that hadrons are extended objects, containing many partons. In contrast to the usual factorization theorems underpinning the perturbative machinery discussed at great length in the first chapters of this book, this may lead to more than one parton pair interacting with each other. With increasing available energies, secondary parton–parton scatters may start populating regions of phase space usually considered to be mainly driven by perturbative physics; this is true in particular for the multiple production of hard objects, such as gauge bosons or jets. But even without such relatively spectacular manifestations of this phenomenon, multiple parton scatterings contribute to the overall particle yield in collisions, to the overall energies of jets, etc.. This manifestation of multiple scattering is often called the "Underlying Event". Models describing this part of the overall event structure will also be introduced in Section 7.2.

In the penultimate section of this chapter, Section 7.3, some light will be shed on the transition of the partons produced in the hard interaction, the parton showering, and the underlying event into the observable hadrons. While to date this transition can quantitatively be described by phenomenological models only, some qualitative ideas underlying their construction could be tested, and this interplay will be discussed.

Finally, Section 7.4 rounds off this chapter with a very brief description of the

The Black Book of Quantum Chromodynamics. John Campbell, Joey Huston and Frank Krauss, Oxford University Press. © John Campbell, Joey Huston, and Frank Krauss 2018, First published in paperback 2022. DOI: 10.1093/oso/9780192871961.001.0007

technology used in the understanding or parameterization of the decays of unstable particles.

7.1 Total cross-sections and all that

In this section, inclusive quantities such as the total, elastic, and inelastic cross-sections are discussed. After some definitions and general thoughts on how these quantities manifest themselves in the experiment, the optical theorem is introduced, which connects the total cross-section with the elastic scattering amplitude and, thereby, with elastic scattering. This motivates a closer look at the elastic scattering amplitude and its analytic behaviour. This is known as S-matrix theory and, by today's standards, seems old-fashioned. However, the resulting Regge calculus and the pomeron [815, 823, 824] embedded in it still serve as the basis for nearly all parameterizations of the total, elastic, and inelastic cross-sections used today. After discussing the properties of these parameterizations, the concept of opacity is presented, which helps to maintain the unitarity of the total cross-section. It will also be used in the very short introduction to diffraction in hadronic collisions. The section closes by connecting the Regge picture of strong interactions with perturbative QCD and in particular the BFKL pomeron and its relationship to the pomeron introduced previously.

7.1.1 Setting the scene

7.1.1.1 Definitions

By definition, the total cross-section of particle interactions in collisions is given by the number of events where the initial-state experiences some interaction leading to an observable change. From a theoretical perspective there are three types of processes contributing to the total cross-section:

1. Elastic interactions, where the particles in the final state are identical to the ones in the initial state, but just experience a "kick" such that their direction changes.
2. Diffractive interactions, where the final states can be seen as excitations of the initial-state particles, which populate the forward regions of the detector. This new final state might be roughly equal in mass to the incident particle and thereby would be understood as an excited particle; for example, a proton could become one of the N^* resonances such as the $N(1440)$. But the final state could also have a mass decidedly different from the typical hadronic mass scale. In such a case a "**rapidity gap**", the absence of any QCD radiation over sufficiently many units of rapidity, usually more than three, between this state and any other particle emerging from the radiation is generally thought to signal a diffractive interaction. From a theoretical point of view, this is connected with the exchange of a (hadronic) colour-singlet state interacting with the diffracted initial-state particle. It should be noted that quite often also single and double diffractive (SD and DD) events are distinguished, depending on how many initial-state particles experience this kind of interaction. In addition, there are also events with more than one rapidity gap. Probably the most prominent representatives are central exclusive production (CXP) events, where a heavy system is produced at mid-rapidity with rapidity gaps on both sides.

3. Inelastic interactions, where the initial state gets seriously distorted, typically due to sufficiently hard QCD interactions leading to a number of final-state particles not only in the forward direction but also in the central detector region.

By far and large, the total, inelastic, elastic, and diffractive cross-sections increase with energy, where typically

$$\sigma_{\text{tot}} > \sigma_{\text{inel}} > \sigma_{\text{el}} > \sigma_{\text{SD}} > \sigma_{\text{DD}} > \sigma_{\text{CXP}} \tag{7.1}$$

and all being of the order of 1-100 mb at hadronic centre-of-mass energies that have been accessed until now.

7.1.1.2   Measurement, selection bias, and minimum bias

In this context it is worthwhile to discuss briefly the concept of **minimum bias**, typically applied to a class of events analysed at collider experiments. The idea is that, by construction in every such experiment, events must pass certain thresholds in order to be further analysed and thereby are subjected to a **selection bias**. In minimum bias events, this selection bias is reduced as much as possible. This is limited by the realities of detector construction and data acquisition. In most cases, modern detectors have an inner magnetic field to improve the tracking performance of charged particles, and this, together with a limited efficiency for individual components, induces a minimal transverse momentum for particles to be detected with appreciable probability. This induces a non-trivial and certainly non-negligible bias in the selection of events, which must be properly treated and documented. It has therefore become customary at experiments such as the ones at the LHC to define different minimum bias characteristics based on a number of charged tracks with a minimal transverse momentum in the central detector. Similar reasoning of course also holds true for other process classes, where in modern analyses the definitions are increasingly based on visible objects in the detector rather than supposed production mechanisms.

7.1.2   Reggeons and pomerons

7.1.2.1   Complex angular momenta

Consider the amplitude for a $2 \to 2$ scattering process ($\mathcal{A}$) which, supposing symmetry of the interaction around the azimuth, can be expanded in Legendre polynomials $P_l(\cos\theta)$ as,

$$\mathcal{A}_{ab\to cd}(s,\, t) = \sum_{l=0}^{\infty} (2l+1)\, a_l(s)\, P_l(\cos\theta)\,. \tag{7.2}$$

Here l labels the angular momentum and the polar angle θ can be expressed through the Mandelstam variables s and t as

$$\cos\theta = 1 + \frac{2t}{s}\,. \tag{7.3}$$

The a_l in the equation above are called **partial wave amplitudes**, and, correspondingly, this expansion is also called **partial wave expansion**. It has already been

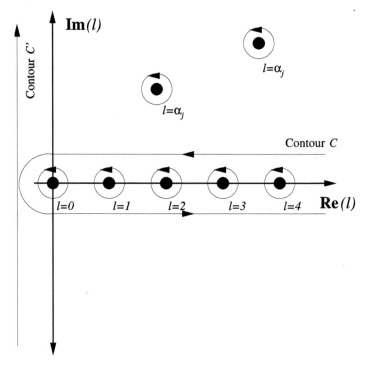

Fig. 7.1 The Sommerfeld–Watson contours C and C'.

encountered before, in Section 5.2.5, where also the connection of amplitudes and cross-sections through the **optical theorem** has been discussed in detail.

By continuing the angular momentum l to the complex plane, it is possible to rewrite the summation in the expansion above as an integral, namely

$$\mathcal{A}_{ab \to cd}(s,\,t) \;=\; \frac{1}{2i} \oint_C \mathrm{d}l \, (2l+1) \, a(l,\,t) \, \frac{P\left(l,\, 1+\frac{2t}{s}\right)}{\sin(\pi l)} \,. \tag{7.4}$$

This is known as the **Sommerfeld–Watson transformation**. The contour C surrounds the positive real l axis, excluding the poles at integer values of l, as shown in Fig. 7.1. In order to close the contour for large l ($|l| \to \infty$), the partial wave amplitudes must fulfil

$$a(l,\,t) < \exp(\pi l) \quad \text{for} \quad |l| \to \infty. \tag{7.5}$$

These poles come with factors $(-1)^l$ from the sine, thus violating the inequality along the imaginary axis. In order to guarantee convergence for infinitely large l, two analytic functions $a^{(\eta=\pm)}(l,\,t)$ are introduced, such that the integral is separately finite for either of them. The $\eta = \pm 1$ are called "signatures" of the corresponding partial waves, and

$$\mathcal{A}_{ab \to cd}(s,\,t) \;=\; \frac{1}{2i} \oint_C dl\,(2l+1) \sum_{\eta=\pm} \left[\frac{\eta + e^{-i\pi l}\, P\left(l,\, 1 + \frac{2t}{s}\right)}{2\sin(\pi l)}\, a^{(\eta)}(l,\,t) \right]. \tag{7.6}$$

In a next step, the contour C surrounding the positive real l axis is closed in the complex-l plane by adding a half-circle and the line $-C'$, running in parallel to the imaginary l axis at $\mathcal{R}l = -1/2$, thus ranging from $l = -1/2 + i\infty$ to $l = -1/2 - i\infty$. The overall result equals the residues of all poles inside this closed integration path, labelled by n_η, thus reflecting their signature. Assuming the integral to vanish for large absolute values of l, $|l| \to \infty$, this leaves only the poles and the integration along C'. The various pieces of this manipulation are sketched in Fig. 7.1. After it, the amplitude reads

$$\begin{aligned}
\mathcal{A}_{ab \to cd}(s,\,t) \;=\; &\frac{1}{2i} \int_{-\frac{1}{2}-i\infty}^{-\frac{1}{2}+i\infty} dl\,(2l+1) \sum_{\eta=\pm} \left[\frac{\left(\eta + e^{-i\pi l}\right) P\left(l,\, 1 + \frac{2t}{s}\right)}{2\sin(\pi l)}\, a^{(\eta)}(l,\,t) \right] \\
&+ \sum_{\eta=\pm} \sum_{j \in n_\eta} \left[\frac{\left(\eta + e^{-i\pi\alpha_j(t)}\right) P\left(\alpha_j(t),\, 1 + \frac{2t}{s}\right)}{2\sin(\pi\alpha_j(t))}\, \beta_j(t) \right].
\end{aligned} \tag{7.7}$$

The new poles reside at positions $\alpha_j(t)$ in the complex-l plane, replacing the values l in the "regular" real poles and contribute with a strength according to their partial wave amplitudes, denoted as $\beta_j(t)$. They are called **Regge poles** with even and odd signatures, depending on the value $\eta = \pm$. There may be additional, more complicated analytic structures in the complex-l plane, like branch cuts, etc., but they are beyond the scope of this very brief introduction.

7.1.2.2 Reggeons and pomerons

The remarkable thing about this simple picture is that in the region of $s > |t|$ the Legendre polynomial is dominated by the first term. Furthermore, at large energies, $s \to \infty$, the integral along the new contour C' vanishes, and only the poles drive the behaviour of the scattering amplitude. In this limit the pole with the largest real value of $\alpha(t) = \alpha_j(t)$ will yield the dominant contribution,

$$\mathcal{A}_{ab \to cd}(s,\,t) \;\overset{s \to \infty}{\longrightarrow}\; \frac{\eta + e^{-i\pi\alpha_j(t)}}{2}\, \beta(t)\, s^{\alpha_j(t)}. \tag{7.8}$$

This can be viewed as the t-channel exchange of something with angular momentum $\alpha(t)$. This "something" defies usual particle definitions, as its spin is t-dependent and thereby it cannot be identified with an integer or half-integer number. Such an object is called a "**Reggeon**" [815, 823, 824]. One way to look at it is to identify it with the superposition of amplitudes for a variety of different particles or mechanisms exchanged in the t channel.

However, assuming the exchange of one Reggeon only, in a factorized picture, the amplitude can thus be written as

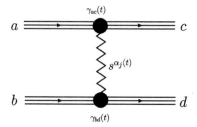

Fig. 7.2 Factorized amplitude with a Reggeon exchange in the t channel; the couplings $\gamma(t)$ of the Reggeon to the external particles a, b, c, and d are made explicit, and the "propagator" term behaves like s^{α}.

$$\mathcal{A}_{ab \to cd}(s, t) \xrightarrow{s \to \infty} \frac{\eta + e^{-i\pi\alpha(t)}}{2} \frac{\gamma_{ac}(t)\gamma_{bd}(t)}{\sin[\pi\alpha(t)]\,\Gamma(\alpha(t))} s^{\alpha(t)}, \tag{7.9}$$

where the γ are the "couplings" of the Reggeon to the incoming and outgoing particles, *cf.* Fig. 7.2. If $\alpha(t)$ assumes an integer value, $\sin[\alpha(t)\pi]$ will vanish, and the amplitude will develop a pole. For positive integers this can be understood as the resonant exchange of a physical particle, for negative integers the Γ function will lead to a cancellation of the contributions. This is to be compared with the exchange of particles with mass m and spin $J = \alpha(m^2)$ for positive t (or in the s channel), which also exhibits a resonant structure at $0 < t, s = m^2$. This observation led Chew and Frautschi to plot the spins of hadrons against their masses squared, discovering straight lines, as shown in Fig. 7.3. Such graphs are also known as **Chew–Frautschi plots**, and they provide a motivation to parameterize the **Regge trajectories** $\alpha(t)$ as

$$\alpha(t) = \alpha(0) + \alpha' \cdot t \tag{7.10}$$

for *all* values of t, and in particular for the physical region of negative t. For a more detailed discussion, the reader is referred to the literature, for example [548].

The Reggeon amplitude will be equal to the forward elastic amplitude for $t \to 0$, if $a = c$ and $b = d$. In this case, the optical theorem expressed in Eq. (5.131) links the total cross-section to such an amplitude, and therefore

$$\sigma_{\text{tot}} \propto s^{\alpha(0)-1}. \tag{7.11}$$

In the specific case of the ρ-ω trajectory, which is related to processes where isospin is exchanged ($\Delta I = 1$ processes), the exponent of s is smaller than 0, and therefore the cross-section decreases with increasing s. This in fact has been observed experimentally, and it is in line with the **Pomeranchuk theorem** asserting that the cross-sections for all scattering processes involving any charge exchange vanish asymptotically [790, 814]. Conversely, at asymptotically high energies, processes with the exchange of the quantum numbers of the vacuum dominate the cross-section [530]. In fact, all experiments to-date exhibit a total hadronic scattering cross-section that increases slowly with the centre-of-mass energy of the hadronic collision. Attributing this behaviour to a single Regge pole, then it must carry the quantum numbers of the

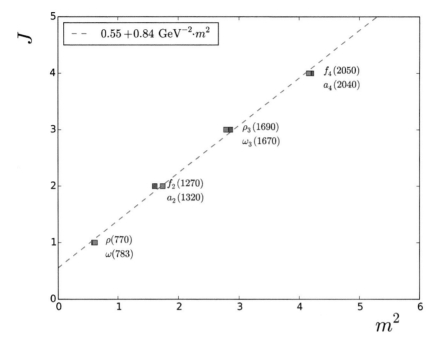

Fig. 7.3 The Chew–Frautschi plot of the ρ-ω trajectory, including a fit to the linear form of $\alpha(t)$.

vacuum: this specific Regge trajectory is called the **pomeron** or **Pomeranchuk pole**. The increase of the cross-section with energy means that its intercept $\alpha_{\mathbf{P}}(0) > 1$.

At this point, it is important to stress that reggeons in general and the pomeron in particular are, *per se*, poles or cuts in the complex plane of the scattering amplitude, and *not* particles. The identification of reggeon exchange with the exchange of a tower of physical particles is not neccessarily a coincidence, since physical particles can and will be exchanged in scattering processes, but this relation is not bi-directional: there may be poles/cuts that cannot be identified with known physical particles. Incidentally, the pomeron is a prime example for it: it *cannot* be related to any known physical particle.

To rephrase this in a different way: the pomeron is *not* a particle and it thus actually has a nature different from Regge trajectories like the one in the example above, the ρ-ω trajectory, for two reasons. First of all, to-date, there are no strongly interacting particles with integer spins that could serve as manifestations of the trajectory for $t > 0$ or, stated differently, as resonances in s-channel scattering. This relates to the fact that in a picture inspired by perturbative QCD, the pomeron is thought of as the exchange of gluonic degrees of freedom, arranged in a color-singlet state. The very existence of purely gluonic bound states, essentially hadrons made of gluons only, also known as **glueballs**, remains a subject of speculation. Secondly, in the QCD picture, however, the exchange of gluons does not lead to a simple pole, but rather to a *branch cut*. This indicates that a simple particle interpretation like for other Reggeons is not obvious: there is just no unique relation between mass and spin, a hallmark of "proper" particles.

7.1.3   Simple parameterizations

7.1.3.1   The total cross-section in simple fits

The total cross-section in hadronic collisions increases with the centre-of-mass energy of the colliding particles. Comparison with scattering data up to 13 TeV c.m.-energy that have been reached at Run I of the LHC and up to about 50 TeV in cosmic rays suggest a total proton–proton cross-section that for large energies scales like

$$\sigma_{\text{tot}}^{(pp)}(s_{pp}) \;=\; \sigma_{\mathbf{P}} \left(\frac{s_{pp}}{\text{GeV}^2} \right)^{\epsilon}, \tag{7.12}$$

where as usual $s_{pp} = E_{\text{cms}}^2$, the square of the centre-of-mass energy of the incoming protons. In typical fits [484, 485]

$$\sigma_{\mathbf{P}} \;=\; 21.7\,\text{mb} \quad \text{and} \quad \epsilon \;=\; 0.0808. \tag{7.13}$$

In this "**Donnachie–Landshoff fit**", the exponent ϵ typically is related to the **soft Pomeron intercept**, $\alpha_{\mathbf{P}}$,

$$\epsilon \;\overset{<}{\sim}\; \alpha_{\mathbf{P}} - 1. \tag{7.14}$$

ϵ is usually assumed to be smaller than $(\alpha_{\mathbf{P}} - 1)$ to account for the effect of destructive interference from multiple pomeron exchanges. However, ϵ together with the normalization $\sigma_{\mathbf{P}}$ must be obtained from data. The pomeron, having the quantum numbers of the vacuum, dominates the cross-section in the high-energy limit according to the Pomeranchuk theorem [815]. It is important to understand that this fit is at odds with the perturbative, or hard, pomeron discussed in Section 5.2.5. There, the leading pomeron intercept was given by

$$A \;=\; \frac{4N_c \log 2}{\pi} \alpha_{\mathrm{s}} \approx 2.5 \cdot \alpha_{\mathrm{s}}, \tag{7.15}$$

cf. Eq. (5.226), and the total cross-section for elastic $gg \to gg$ scattering assumed the form given in Eq. (5.229), which goes beyond the simple pole of the soft pomeron idea.

However, these simple fits often are further extended by adding other **Reggeons**, which effectively sum over the exchange of full classes of particles such as ρ, ω, the corresponding f and a mesons, see Section 7.1.2. While the dominant pomeron, due to its quantum numbers, is blind to whether the colliding hadrons are particles or anti-particles, (or, indeed protons or neutrons), the sub-leading reggeons are not. This is reflected in the extension

$$\sigma_{\text{tot}}^{(pp,\,p\bar{p})} \;=\; \sigma_{\mathbf{P}} \left(\frac{s_{pp}}{\text{GeV}^2} \right)^{\epsilon} + \sigma_{\mathbf{R}} \left(\frac{s_{pp}}{\text{GeV}^2} \right)^{-\eta}, \tag{7.16}$$

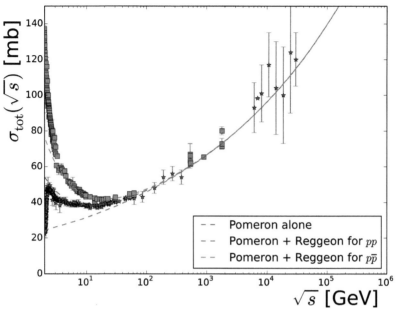

Fig. 7.4 The total pp (blue) and $p\bar{p}$ (red) cross-sections compared to the simple pomeron/reggeon fits of Eq. (7.17)

where the exponent η related to the reggeon and the normalization $\sigma_{\mathbf{R}}^{(pp,\,p\bar{p})}$ are also fitted by [485] resulting in

$$\eta = 0.4525 \quad \text{and} \quad \sigma_{\mathbf{R}} = \begin{cases} 56.08\,\text{mb for } pp \\ 98.39\,\text{mb for } p\bar{p}. \end{cases} \tag{7.17}$$

The comparison of this fit with data taken from the Review of Particle Physics by the Particle Data Group [799] is exhibited in Fig. 7.4.

Another addition that is frequently made is the inclusion of a "**hard pomeron**", which reflects the fact that the pomeron as obtained from perturbative QCD would yield a completely different — and larger — intercept $\alpha_{\mathbf{P}} - 1 \approx 0.5$ at leading order. This value is at odds with data, which would favour a much smaller intercept for the hard pomeron of about $\alpha_{\mathbf{P}} - 1 \approx 0.25$, more in line with higher-order results. However, there is some evidence of the hard pomeron in the structure function F_2 and in the interaction pattern of more exclusive processes.

Finally, note that similar fitting strategies have also been applied to different reactions such as πp or Kp–scattering. It is interesting to realize that the scaling behaviour, *i.e.* the exponents in, say, the simple fit of Eq. (7.12) are identical and that only the normalization changes by about a factor of $2/3$.[1] This is the **additive quark rule**, which was crucial in establishing the pomeron picture, see e.g. [717].

There is another important theoretical property of the total cross-section: asymptotically it cannot possibly increase faster than $\log^2 s$, with s the centre-of-mass energy squared of the incident particles. This is the **Froissart bound** [547], a manifestation of the **unitarity** requirement underpinning every reasonable field theory, supplemented with the idea that the S-matrix is analytic. Ultimately, this bound will have to kick in and thereby modify the relatively simple fits above.

Pictorially speaking, the Froissart bound ensures that the proton behaves asymptotically like a "black disk". In partonic language this means that the parton density is limited, thus limiting or counter-balancing the amount of parton creation at larger scales as driven by the DGLAP equations, Eq. (2.31). In practical terms this means that at large enough scales, there must emerge non-linear terms in the DGLAP equations, which account for parton recombination effects at large densities.

7.1.3.2 Eikonal function and cross-sections

Re-interpreting the parameterization of the **forward** or **elastic scattering amplitude** presents one way to remedy its behaviour at asymptotically high energies, thus guaranteeing that the Froissart bound is respected. This is achieved by assuming that the elastic scattering amplitude can be expressed through a Fourier transform into impact parameter space

$$\mathcal{T}(s,t) \;=\; 4s \int \mathrm{d}^2 B_\perp\, e^{i\vec{q}_\perp \cdot \vec{B}_\perp}\, a(s, \vec{B}_\perp)\,. \tag{7.18}$$

Denoting with $\vec{q}$ the three-vector of momentum transfer such that in the high-energy limit

$$t \;=\; \vec{q}^{\,2} \;=\; \vec{q}_\perp^{\,2}\,. \tag{7.19}$$

In this limit, the elastic scattering amplitude and, correspondingly, its Fourier transform are purely imaginary. This allows to rewrite it as,

$$a(s, \vec{B}_\perp) \;=\; \frac{1}{2i}\left[\exp\left(-\frac{\Omega(s, \vec{B}_\perp)}{2}\right) - 1\right], \tag{7.20}$$

where the **eikonal function** or **opacity** $\Omega(s, \vec{B}_\perp)$ has been introduced. The trick is now to identify the Regge-parameterization with the eikonal function rather than directly with the amplitude,

$$\Omega(s, \vec{B}_\perp) \;\propto\; \Omega_P \cdot \left(\frac{s}{\mathrm{GeV}^2}\right)^\epsilon + \Omega_R \cdot \left(\frac{s}{\mathrm{GeV}^2}\right)^\eta + \dots\,. \tag{7.21}$$

---

[1]In fact, [485] finds a factor of 0.63 and relates this small deviation from $2/3$ to the radius of the pion. For the Kp total cross-section the factor is even less, which may reflect an even smaller kaon radius, the fact that the pomeron couples differently to strange quarks, or merely the somewhat worse quality of the data entering the fit. Also, of course, Reggeon trajectories containing strangeness may start playing a role.

As a consequence, the cross-section will remain finite, as the exponential of Eq. (7.20) will never exceed unity — the hadronic disc will therefore never become blacker than black.

The relationship between the total cross-section and the elastic scattering amplitude gives rise to a number of further relations. For example, expressed through the eikonal, the total, elastic and inelastic cross-sections read

$$\sigma_{\text{tot}}(s) = \frac{1}{s} \, \mathfrak{Im}(\mathcal{T}(s, t = 0)) = 2 \int d^2 B_\perp \left[1 - \exp\left(-\frac{\Omega(s, \vec{B}_\perp)}{2} \right) \right]$$

$$\sigma_{\text{el}}(s) = 4 \int d^2 B_\perp \left| a(s, \vec{B}_\perp) \right|^2 = \int d^2 B_\perp \left| \exp\left(-\frac{\Omega(s, \vec{B}_\perp)}{2} \right) \right|^2$$

$$\sigma_{\text{inel}}(s) = \sigma_{\text{tot}}(s) - \sigma_{\text{el}}(s) = \int d^2 B_\perp \left[1 - \exp\left(-\Omega(s, \vec{B}_\perp) \right) \right]. \qquad (7.22)$$

Another interesting quantity is the **elastic slope** B given by

$$B(s) = \left[\frac{d}{dt} \left(\log \frac{d\sigma_{\text{el}}(s, t)}{dt} \right) \right]_{t=0} = \frac{1}{\sigma_{\text{tot}}} \int d^2 B_\perp^2 \, B_\perp^2 \left[1 - \exp\left(-\frac{\Omega(s, \vec{B}_\perp)}{2} \right) \right].$$

$$(7.23)$$

These equations manifest the relations between total and elastic scattering induced by the optical theorem. There is, however, a short-coming when comparing them to actual data, namely the conspicuous absence of any link to diffractive processes. This is related to the fact that especially **low-mass diffraction**, a process, where one or two of the incident hadrons transition to an excited state, can only be explained by the transition between scattering eigenstates. This is worked out further in the next section.

7.1.4 Diffraction

7.1.4.1 Good–Walker states and low-mass diffraction

When discussing low-mass diffractive excitations of the incident hadrons, it is sensible to introduce **diffractive eigenstates** $|\phi_i\rangle$, which are also known as **Good–Walker states** [592]. All N physical states $|\psi_j\rangle$, both the excited and the elastic ones, *i.e.* the ones, where the outgoing particles are identical to the incoming ones, can then be written as linear combinations — coherent sums — of these $|\phi_i\rangle$,

$$|\psi_j\rangle = \sum_{i=1}^{N} \alpha_{ji} |\phi_i\rangle. \qquad (7.24)$$

Assuming both the Good–Walker states and the physical ones to be orthogonal and normalized,

$$\langle \phi_i | \phi_k \rangle \;=\; \delta_{ik} \quad \text{and} \quad \sum_{i=1}^{N} |\alpha_{ji}|^2 \;=\; 1 , \tag{7.25}$$

implies that the matrix formed by the α_{ji} is unitary. The elastic amplitude of the incident particle $|\Psi\rangle = |\psi_1\rangle$ is therefore given by

$$\left\langle \Psi \left| \hat{T} \right| \Psi \right\rangle \;=\; \sum_i |\alpha_{1i}|^2 \, T_i \;=\; \left\langle \hat{T} \right\rangle , \tag{7.26}$$

the average over the diffractive eigenstates. Here it has been used that the scattering operator $\hat{T}$ is diagonal in the basis spanned by the Good–Walker states. Therefore the elastic cross-section, given by the amplitude squared, is proportional to the square,

$$\sigma_{\text{el}} \;\propto\; \langle \hat{T} \rangle^2 . \tag{7.27}$$

In contrast, the amplitude for the diffractive production of any other state $|\psi_{k\neq 1}\rangle$ reads

$$\left\langle \psi_k \left| \hat{T} \right| \Psi \right\rangle \;=\; \sum_i \alpha_{1i} \alpha_{ik}^* \, T_i , \tag{7.28}$$

which upon squaring and summing over all states k becomes

$$\sum_k \left\langle \Psi \left| \hat{T} \right| \psi_k \right\rangle \left\langle \psi_k \left| \hat{T} \right| \Psi \right\rangle \;=\; \sum_{ijk} \alpha_{1i} \alpha_{ik}^* \alpha_{j1}^* \alpha_{kj} \, T_i T_j$$

$$=\; \sum_{ij} \alpha_{1i} \alpha_{j1}^* \, T_i T_j \, \delta_{ij} \;=\; \left\langle \hat{T}^2 \right\rangle . \tag{7.29}$$

To have the diffractive component only, the elastic component must be subtracted. The interpretation of this is that the cross-section for the transition to any excited state, for diffraction, is given by the **fluctuations**

$$\sigma_{\text{diff.\,exc.}} \;\propto\; \left\langle \hat{T}^2 \right\rangle - \left\langle \hat{T} \right\rangle^2 . \tag{7.30}$$

7.1.4.2 Cross-sections with opacities and Good–Walker states

Introducing such diffractive eigenstates implies that the opacity Ω depends on the eigenstates that scatter. The cross-sections of Eq. (7.22) must be expressed in terms of the now eigenstate-dependent opacities Ω_{ik} and the expansion coefficients α_i and α_k of the incident particles:

$$\sigma_{\text{tot}}(Y) \;=\; 2 \int d^2 B_\perp \left\{ \sum_{i,k} |\alpha_i|^2 |\alpha_k|^2 \left[1 - \exp\left(-\frac{\Omega_{ik}(Y, B_\perp)}{2} \right) \right] \right\}$$

$$\sigma_{\text{el}}(Y) \;=\; \int d^2 B_\perp \left\{ \sum_{i,k} |\alpha_i|^2 |\alpha_k|^2 \left[1 - \exp\left(-\frac{\Omega_{ik}(Y, B_\perp)}{2} \right) \right] \right\}^2$$

$$\sigma_{\text{inel}}(Y) \;=\; \int \mathrm{d}^2 B_\perp \left\{ \sum_{i,k} |\alpha_i|^2 |\alpha_k|^2 \left[1 - \exp\left(-\Omega_{ik}(Y, B_\perp) \right) \right] \right\}. \quad (7.31)$$

The c.m.-energy squared s of the incident hadrons has been replaced by a rapidity

$$Y \;=\; \log \frac{s}{m_{\text{had}}^2}. \quad (7.32)$$

The differential elastic cross-section can be obtained from a Fourier back-transform,

$$\frac{\mathrm{d}\sigma_{\text{el}}(Y)}{\mathrm{d}t} \;=\; \frac{1}{4\pi} \int \mathrm{d}^2 B_\perp \left\{ e^{i\vec{q}_\perp \cdot \vec{B}_\perp} \sum_{i,k} |\alpha_i|^2 |\alpha_k|^2 \left[1 - \exp\left(-\frac{\Omega_{ik}(Y, B_\perp)}{2} \right) \right] \right\}^2.$$
$$(7.33)$$

Expressed through the opacities $\Omega_{ik}(s, \vec{B}_\perp)$, single and double diffractive differential cross-sections can be obtained similar to the elastic one above as

$$\frac{\mathrm{d}\sigma_{\text{el}+\text{SD}_1}(Y)}{\mathrm{d}t} \;=\; \frac{1}{4\pi} \sum_{i,j,k} \Bigg\{ |\alpha_i|^2 |\alpha_j|^2 |\alpha_k|^2$$
$$\times \int \mathrm{d}^2 B_\perp \exp\left(i\vec{q}_\perp \cdot \vec{B}_\perp \right) \left[1 - \exp\left(-\frac{\Omega_{ik}(Y, B_\perp)}{2} \right) \right]$$
$$\times \int \mathrm{d}^2 B'_\perp \exp\left(-i\vec{q}_\perp \cdot \vec{B}'_\perp \right) \left[1 - \exp\left(-\frac{\Omega_{jk}(Y, B'_\perp)}{2} \right) \right] \Bigg\}$$
$$(7.34)$$

for the combination of elastic scattering and diffraction of incident beam particle 1 and similar for a combination of elastic scattering and incident particle 2.

7.1.4.3 Pomerons and diffractive processes

Before discussing how pomerons and diffraction are tied together, it is worthwhile to remember that the pomeron exchange actually parameterizes the *total cross-section*. Going back to the optical theorem, Eq. (5.131), it is clear that the imaginary part of the elastic amplitude, *i.e.* the imaginary part of the amplitude for $ab \to ab$ is proportional to the total cross-section, which in turn is given by the sum over all final states n. This is schematically depicted in Fig. 7.5. A simple QCD interpretation of the pomeron follows immediately: on the level of *amplitudes*, and at lowest order in perturbation theory, one could identify the pomeron with the exchange of a single gluon. This is also known as the Low–Nussinov pomeron [735, 787],[2] see also Fig. 7.6 for a pictorial representation. In this case, the pomeron is "cut" into two single gluon exchanges, for the amplitude and its complex conjugate. In Section 5.2.5, this simplistic picture has been augmented with higher orders.

---

[2]The pomeron intercept in this model is given by $\alpha_{\mathbf{P}}(t) = 1$ leading to a constant cross-section.

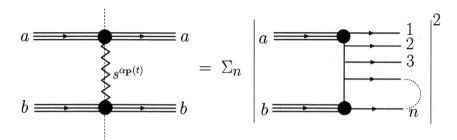

Fig. 7.5 The relationship between pomeron amplitude and total cross–section. The vertical dashed line indicates either that the imaginary part of the amplitude has to be taken, or it symbolizes a sum over all final states, n.

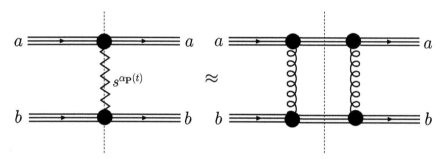

Fig. 7.6 The Low–Nussinov pomeron: in this model the pomeron is associated with the exchange of a single gluon. In the sketch, the thick blobs represent the gluons interacting with any of the valence quarks of the incoming hadrons.

It is advantageous to classify diffractive processes, in order to see their connection with pomeron exchange. First of all, there are low-mass single-diffractive or double-diffractive processes. These are essentially thosee processes that are directly captured by the transition between different Good–Walker states. They are identified with the exchange of a colour singlet between the two beam particles, *on the amplitude level*. The idea underlying this identification is that the exchange of colour, for example by a gluon, would lead to the radiation of coloured secondaries from this t-channel particle, which would result in the mostly soft production of hadrons filling the large rapidity interval between the two beams.

There are, however, also processes where the mass of the diffractively produced system is not small, or where it is not produced at forward rapidities. They thus do not manifest themselves as excitations of the beam particles, although also in such cases the absence of particles in the rapidity interval separating the diffractively produced system from the other particles is thought of as exchange of a colour singlet. In both cases, these colour singlets are readily identified with pomeron exchange on the amplitude level: in the Low–Nussinov picture this would be realized by the exchange of

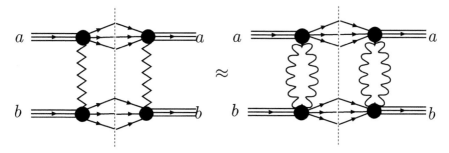

Fig. 7.7 Low-mass diffraction through (uncut) pomeron exchange.

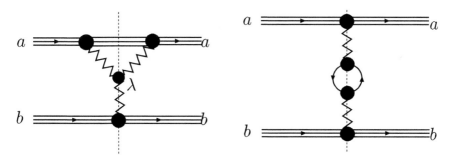

Fig. 7.8 High-mass single diffraction (left) and central exclusive production (right). Note the occurrence of a triple-pomeron vertex λ on the left plot, which plays an important role also in modelling the recombination of partons.

two gluons forming a colour singlet on the amplitude level — the pomeron thus is not cut in such cases, see Fig. 7.7 for the case of low-mass diffraction. High-mass diffraction or central exclusive production processes can thus be thought of as a combination of cut and uncut pomerons, as exhibited in Fig. 7.8. In this case, a triple-pomeron vertex appears.

It is interesting to note, however, that in some phenomenological models pomerons are thought of as objects with particle-like properties, like, e.g., an internal structure that could be resolved in a way similar to that of "usual" hadrons. In particular, in these models processes such as single diffraction or central-exclusive production are identified as the t-channel exchange of pomerons as physical particles, which are then subjected to a hard interaction with a quark or gluon or with each other. Contact to QCD is then made by defining an (equivalent) pomeron flux or similar, accompanying the incident hadrons, and a pomeron structure function, effectively parton distribution functions for these objects. While to some degree these models seem to work, they are of course essentially nothing but effective, albeit simplistic, parameterizations of a much more complicated mechanism; the basic problem with them is that the pomeron is just not a particle. In fact, in perturbative QCD the pomeron is not a simple pole, but rather a branch cut in the complex plane, which in turn renders any interpretation of the pomeron as a particle an overly naive and physically wrong idea.

7.2 Multiple parton interactions and the underlying event

The focus of this book is mostly on the determination of cross-sections and other observables at hadron colliders through perturbative methods that rest on the factorization theorem, see also Sections 2.1.1 and 2.2.1. Essentially, this theorem states that in perturbative calculations of well-defined observables, the problem can be factorized into a "hard", perturbative, and a non-perturbative part. In the former, amplitudes for the transition of incoming *partons* to the final state are calculated with perturbative methods from first principles. This has been elucidated in detail in Chapters 3–5.

In contrast, the non-perturbative part deals with the question of translating the *incoming physical hadrons into the partons*, on which the full calculation rests, and on the translation of the *outgoing partons back into the hadrons*, on which all physical observations base. For both effects, there are no qualitative methods based on first principles. The factorization theorem accounts for the former translation, incoming hadrons to partons through the parton distribution functions. The PDFs, essentially being non-perturbative in nature, have to be taken from data, with methods discussed in Chapter 6. As the PDFs basically account for finding one parton in each incoming hadron, this begs the question what happens with the rest of the hadron in an event? As the partons taking part in the hard interaction do extract quantum numbers from the hadrons, not least colour degrees of freedom, there must be some mechanism of colour and flavour compensation. This break-up of the beam particles and the formation of the **beam remnant** particles must be non-perturbative and as such is subject to heavy modelling based on a few simple principles.

But hadrons are extended objects which consist of more than one parton — if they consisted of one parton only, no PDFs would be necessary! This implies that there is a chance that more than one parton from each side enters the interaction, rendering the picture considerably more complicated. Such processes, with more than two incoming partons and more than one scatter, are called **multiple parton interactions** (**MPI**); this term specifically refers to the interaction of more than one parton pair to contribute to the final state, in contrast to **rescattering processes**, where partons entering or leaving the hard process interact with the beam remnants or each other at time scales much larger than the typical time scales for the hard scatter. Issues related to the break-up of incident hadrons and multiple scattering in the process are the subject of this section. Turning the outgoing partons into hadronic bound states is usually described by phenomenological models, discussed later in this chapter in Section 7.3.

7.2.1 Beam remnants

7.2.1.1 Flavour degrees of freedom

In this section, different aspects of the break-up of the incident hadrons in a hard scattering will be discussed, many of which can be formulated in the most convenient way in the context of full event simulations.

First of all, and most importantly, quantum numbers such as momentum, flavour, and colour must be conserved. For the latter, the limit of infinitely many colours is usually employed, which ensures that for every colour there is a unique colour partner, carrying the anti-colour. This assumption impacts on both the modelling of the **beam remnants** and the hadronization, where the latter is driven by the colour-singlet structures formed in previous stages of the event simulation.

To illustrate this while keeping things as simple as possible, imagine an event at the LHC, where a W boson decaying into a lepton–neutrino pair was created, $u\bar{d} \to W^+ \to \ell^+\nu$. Further assume that the quarks did not undergo any parton showering or that they radiated only gluons. Then, the two quarks, the u and the $\bar{d}$ must be extracted from the incident protons. Knowing their Bjorken-x at the cut-off scale of the initial-state parton shower defines how much momentum is left for the proton remnants, which will consist of other partons. These remnant partons will in turn also compensate for the flavour and colour degrees of freedom, ensuring that these quantum numbers are also conserved. In the case of the u quark extracted from a proton this is straightforward: naively, the protons have $|uud\rangle$ as valence quark flavours. Extracting a u therefore just leaves ud as the flavours of the corresponding remnant. Using the notion of diquarks as carriers of baryon number, discussed in more detail in the context of hadronization models below, *cf.* Section 7.3.1, this remnant will therefore consist of a (ud) diquark. It will be assigned the anti-colour of the u quark colour and will take the full remaining momentum, thereby fulfilling the requirement that the parton configuration replacing the proton is colour-neutral and carries all of its momentum.

A quick comment is in order here. Frequently in Monte Carlo simulations there are cases, where the three constitutent quarks have to be distributed over a diquark and a quark. The most naive way of achieving this would consist in merely giving each combination the same probability — in such a case the only small subtlety is that the diquarks come as spin-0 and some of them possibly also in spin-1 states. Alternatively, in cases where the original $|uud\rangle$ configuration of a proton is still available, one could use the proton wave-function in terms of quarks and diquarks [727],

$$|p\rangle \;=\; |uud\rangle \;=\; \frac{1}{\sqrt{2}}\,|u\rangle|(ud)_0\rangle \;+\; \frac{1}{\sqrt{6}}\,|u\rangle|(ud)_1\rangle \;+\; \frac{1}{\sqrt{3}}\,|d\rangle|(uu)_1\rangle. \qquad (7.35)$$

This is how the more intricate case of the incident $\bar{d}$ would be handled — extracting $\bar{d}$ from $|uud\rangle$ leaves $|uud + d\rangle$ as the flavour content of the proton, which must then be decomposed into one diquark, carrying the baryon number, and two quarks. One of the quarks must carry the colour matching the $\bar{d}$'s colour, while the other quark and the diquark will form another colour singlet. Assuming the most likely outcome for the flavour structure according to Eq. (7.35), a $(ud)_0$ diquark and an u quark will be formed in addition to the additional, flavour-compensating d. This leaves the task of distributing colours, answering the question if the u and the $\bar{d}$ or the d and the $\bar{d}$ form a colour-singlet. Again, different solutions present themselves, ranging from equal probabilities to a picture, where the $d\bar{d}$ stem from a gluon splitting, thus forming a colour octet. In this picture, the d and the $(ud)_0$ and the u and the $\bar{d}$ are colour singlets.

Replacing the u and $\bar{d}$ quarks in the simple example above with, say, $s\bar{c} \to W^-$ does

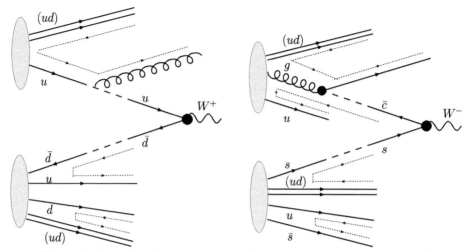

Fig. 7.9 Example colour structures of some $q\bar{q} \to W$ events: The left panel corresponds to the case of $u\bar{d} \to W^+$ with a gluon emission in initial-state parton showering, while the right panel corresponds to the case of $s\bar{c} \to W^-$, where the $\bar{c}$ is replaced by a gluon and an outgoing c quark in the initial-state showering. Colour connections are indicated by the short--dashed lines.

not alter the picture dramatically. In fact, following the initial-state parton shower to lower and lower scales, at some point the charm threshold is crossed and as a result the $\bar{c}$ quark will have become a gluon. In such a case, say an s quark and a gluon will be extracted from the incident protons. The first proton remnant will consist of the flavours $|uud + \bar{s}\rangle$, translating into one of the quarks forming a singlet with either the diquark or the $\bar{s}$, and the respective left-over $\bar{s}$ or diquark carrying the colour of the s quark. Assuming that the $s\bar{s}$-pair had emerged from the splitting of a fictitious gluon below the scale of the dissociation will fix the colours such that the s quark will form a singlet with the diquark and the remaining quark will form a singlet with the $\bar{s}$. For the proton, where the gluon is extracted, the structure is simpler. The flavour content of the proton remnant will still be $|uud\rangle$. The gluon carries a colour and an anti-colour, which will be compensated by an corresponding anti-colour assigned to the diquark and a corresponding colour assigned to the quark. The quark therefore will form a singlet with the colour-partner of the s quark in the other proton, while the diquark will be colour-connected to the c quark emerging from the $g \to c\bar{c}$ splitting that occurred in the initial-state parton shower. For a pictorial representation of the two examples discussed, *cf.* Fig. 7.9.

7.2.1.2 Momenta in hadron dissociation

Having presented some relatively naive ideas of how flavours and colours are distributed in the break-up of incident hadrons, there is one last point to be discussed, namely the momentum degrees of freedom. Up to now, the overall transverse momentum of the system produced in the hard collision — in the examples above the W bosons produced in $u\bar{d} \to W^+$ and $s\bar{c} \to W^-$ — are entirely given by the parton shower. In the first example, where the relatively unlikely case of no emission was assumed, the W^+ boson therefore would have zero transverse momentum. Events like this therefore would lead to a visible spike of the $p_\perp$ distribution of the W boson, or of Drell–Yan pairs in the case of $q\bar{q} \to \ell\bar{\ell}$ events, at $p_\perp = 0\,\text{GeV}$. This of course is not what is being seen experimentally, where the $p_\perp$ distribution apporaches zero for vanishing transverse momentum, and then increases from there until it reaches the **Sudakov peak**, which in the case of vector bosons at the LHC is located at about 5 GeV. Therefore, there must be some source of relatively small transverse momentum for such systems, beyond the parton shower.

It is quite straightforward to assume some kind of "Fermi motion" of the partons inside the incident hadrons. In the case of collisions with hadrons in the initial state, this would manifest itself as some additional **intrinsic or primordial transverse momentum** (**intrinsic** or **primordial** $k_\perp$) that the beam partons assume in the break-up of the hadron.

This presents another non-perturbative effect, similar to the **soft form factor** of Q_T resummation first encountered in Eq. (2.172) and further discussed in Chapter 5. And similar to the parameterizations used there, it is customary to employ some relatively simple form for the generation of the intrinsic $k_\perp$, like, e.g., a simple Gaussian, supplemented with a cut-off to prevent the generation of too large transverse momenta. The parameters of such functions must then be fitted to data, usually the low transverse momentum region of the lepton-pair in Drell–Yan processes.

In the dissociation of the second proton into partons, its momentum also has to be distributed. This is yet another issue where no first principles are available and simple ideas are invoked to guide the modelling. One of these ideas would be to select momenta — or, analogously, the Bjorken-x — of the quark and gluon partons emerging in the hadron break-up according to the PDF at the small parton shower cut-off scale $\mathcal{O}\,(1\,\text{GeV})$. The remaining momentum would be associated with the diquark degrees of freedom.

Even more sophisticated models can be devised for the break-up of the incident hadrons, improving the way, flavours, momenta, or colours are distributed over the outgoing quarks, see for example [855]. But any model will become more involved when the underlying event is added to the simulation.

7.2.2 The underlying event

7.2.2.1 Definition

In events with a hard interaction, described through perturbation theory in the usual framework of factorization, the **underlying event** adds further activity to the overall deposit of particles and energy in the detector. This introduces some ambiguity in the definition of what the underlying event actually is. The answers range from "everything apart from the hard scatter itself, but including parton showering in both the initial and final state plus the corresponding activity from hadronization, hadron decays, etc.", to "the additional activity after all Bremsstrahlung, hadronization, and hadron decays related to the hard signal interaction has been taken into account". The latter is the definition that will be used in this book. Correspondingly, the underlying event consists of all contributions associated with the additional scatters in multiple parton scattering, their fragmentation, and effects coming from remnants.

It is remarkable that, even with this relatively restrictive definition, the activity attributed to the underlying event is significantly higher than in soft collisions, usually associated with minimum bias events, at the same energy. In addition, fluctuations in particle multiplicity and energy flows are significantly larger in the underlying event. A simple interpretation is that with increasing scales being probed in the collision, the relative impact parameter of the two incident hadrons must decrease — in other words, by biasing the event selection towards larger momentum scales, the overlap between the colliding hadrons becomes larger, offering other partons a higher chance of interactions. This leads to an effect known as the **jet pedestal effect**: the hard objects sit on a "pedestal" of underlying activity, which becomes larger with increasing scale of the hard process.

7.2.2.2 Evidence for the pedestal effect

A traditional example for this is the increase, w.r.t. minimum bias events, of activity as a function of the transverse momentum of the hardest charged track in an event. This has been quantified for the first time by the CDF experiment at the TEVATRON [127, 526]. Looking at Drell–Yan production at the LHC, however, an even more striking example for this pedestal effect emerges. To see how this works, assume the Drell–Yan pair — effectively a Z boson decaying into leptons — to have some transverse momentum, $p_\perp^Z$. The generation of this transverse momentum is predominantly driven by the emission of a hard parton in the opposite direction, and in most events with a sufficiently sizable $p_\perp^Z$ one would thus find a jet back-to-back with the Z boson. This allows to define three regions of equal size in the transverse plane:

- a "towards" region with $|\Delta\phi| < 60°$;
- an "away" region with $|\Delta\phi| > 120°$;
- and two parts of one "transverse" region with $60° \leq |\Delta\phi| \leq 120°$,

where the orientation is such that the Z resides at $\phi = 0°$. By far and large, QCD particles from the hard scatter that produced the Z boson will be associated with the jet system, which is concentrated in the away region. In contrast, there will not be much of such primary QCD activity in the transverse region and even less in the towards region. A sizeable and even dominant fraction of particles in these regions will stem from secondary activity, the underlying event.

This has been analysed in more detail by the ATLAS collaboration at Run I of the LHC [28] with $s_{pp} = 7\,\mathrm{TeV}$, where the overall activity was measured through charged particles with $|\eta| \leq 2.5$ and $p_\perp > 500\,\mathrm{MeV}$. In Fig. 7.10, data for the number of charged tracks N_{ch} as a function of the Z transverse momentum $p_\perp^Z$ in the transverse and toward region are compared with results from the different event generators. Particle production increases with $p_\perp^Z$, and the overall activity is higher than in minimum bias events at the same energy. The same trend is also visible in the sum of the transverse momenta of the charged tracks $\sum p_\perp$ in both regions, as displayed in Fig. 7.11. It is remarkable that there are not only more particles with a

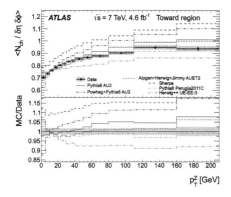
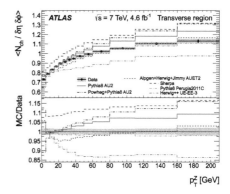

Fig. 7.10 N_{ch} vs. $p_\perp^Z$ at $s_{pp} = 7\,\mathrm{TeV}$ in the transverse (left panel) and towards (right panel) region, as measured by the ATLAS collaboration [28]. Charged tracks with $|\eta| < 2.5$ and $p_\perp > 500\,\mathrm{MeV}$ are considered. Reprinted with permission from Ref. [28].

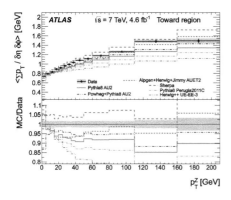
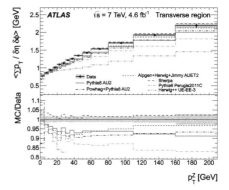

Fig. 7.11 $\sum p_\perp$ vs. $p_\perp^Z$ at $s_{pp} = 7\,\mathrm{TeV}$ in the transverse (left panel) and towards (right panel) region, as measured by the ATLAS collaboration [28]. Charged tracks with $|\eta| < 2.5$ and $p_\perp > 500\,\mathrm{MeV}$ are considered. Reprinted with permission from Ref. [28].

larger total transverse momentum, but that they also become harder — their average transverse momentum increases as well, *cf.* Fig. 7.12.

7.2.2.3 Consequences of multiple parton interactions

A phenomenologically relevant effect of such multiple interactions is that they increase the overall final-state activity of events. Despite many of these secondary scatterings happening at relatively low scales, well below the transverse momentum scales usually associated with jets, they visibly increase the total energy released in the form of

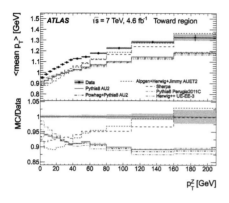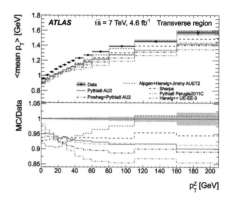

Fig. 7.12 $\langle p_\perp \rangle$ vs. $p_\perp^Z$ at $s_{pp} = 7\,\mathrm{TeV}$ in the transverse (left panel) and towards (right panel) region, as measured by the ATLAS collaboration [28]. Charged tracks with $|\eta| < 2.5$ and $p_\perp > 500\,\mathrm{MeV}$ are considered. Reprinted with permission from Ref. [28].

final-state particles through their multiplicity. One of the reasons for this increase is that they also alter the overall colour flow of the event, adding more colour sources, and thus providing many more directions along which soft emissions can proceed. This ultimately translates into many more seeds of hadron production.

Another effect is directly related to jets. From previous considerations, it is clear that QCD final-state radiation carries away energy from the primary hadrons, which may end up outside the jet. For relatively simple, cone-shaped jets with a radius of R this yields a contribution that is roughly proportional to $\log(1/R)$, coming from the integral over opening angles. Hadronization corrections, discussed later in this chapter, scale like $1/R$, and also contribute negatively to the overall jet energy, as some of the hadrons emerging may end up outside the jet. On the other hand, the underlying event adds energy back to the jets, usually in proportion to the jet area, $\propto R^2$. For more details, *cf.* [432].

7.2.2.4 Simple models for the underlying event

At the moment models for the underlying event fall, broadly speaking, into two categories: first of all, there are models based on simple parton–parton scattering, implemented in standard event generators such as PYTHIA, HERWIG, or SHERPA. Alternative models based on Regge-theory [598] and employing the notion of cut pomerons form the basis of the underlying event and minimum bias modelling in event generators such as PHOJET [514] and EPOS [889].

Concentrating on the former, more simplistic class of models first, the logic underlying their construction is that the cross-section for parton–parton scattering with transverse momentum larger than some cut-off $p_{\perp,\min}$, $\sigma_{2\to2}(p_{\perp,\min})$ becomes larger than the total proton–proton cross-section $\sigma_{pp,\mathrm{tot}}$,

$$\left[\sigma_{2 \to 2}(p_{\perp,\mathrm{min}}) \equiv \int\limits_{p_{\perp,\mathrm{min}}^2}^{s} \mathrm{d}p_{\perp}^2 \, \frac{\mathrm{d}\hat{\sigma}_{2 \to 2}}{\mathrm{d}p_{\perp}^2} \right]_{p_{\perp,\mathrm{min}} \approx 5 \, \mathrm{GeV}} \geq \sigma_{pp,\mathrm{tot}} \, . \qquad (7.36)$$

Here, $\hat{\sigma}_{2 \to 2}$ is given by Eq. (2.52), where the matrix element squared $|\mathcal{M}_{ab \to n}|^2$ is given by the sum of all partonic $2 \to 2$ QCD scatters at leading order. At the LHC the saturation of the total cross section occurs for values of $p_{\perp,\mathrm{min}}$ of the order of about 5-10 GeV, depending on the c.m.-energy of the protons, on the PDFs being used for the parton-level calculation, and on the choice of renormalization and factorization scales.

The interpretation of Eq. (7.36) is pretty straightforward: if the partonic scattering cross-section is larger than the total inelastic or, most probably the non-diffractive (ND) hadronic cross section, then there must be more than one partonic scatter per hadron interaction,

$$\sigma_{2 \to 2}(p_{\perp,\mathrm{min}}) \geq \sigma_{pp,\mathrm{ND}} \quad \longrightarrow \quad \langle N_{\mathrm{scatters}}(p_{\perp,\mathrm{min}}) \rangle \equiv \frac{\sigma_{2 \to 2}(p_{\perp,\mathrm{min}})}{\sigma_{pp,\mathrm{ND}}} \geq 1 \, . \quad (7.37)$$

This presents the starting point for a class of relatively simple models for the underlying event. In these models the underlying event emerges as the superposition of largely independent parton-parton scatters, with two partons that are oriented back-to-back in the transverse plane. The number of these scatters is distributed according to a Poissonian distribution defined by $\langle N_{\mathrm{scatters}}(p_{\perp,\mathrm{min}}) \rangle$ in Eq. (7.37), but possibly reduced by one, if the hardest partonic event — the signal event — is a QCD event. The number of actual scatters N_{scatters} is either predetermined by a Poissonian, as in JIMMY [298], the model implemented in the HERWIG family of event generators, or it is generated dynamically, as in the model [857] realized in the PYTHIA event generators. The basis of this latter model is a structure that looks like a Sudakov form factor and, analogous to the case encountered already in the construction of parton showers, yields the probability of no further scatter to happen between a higher scale Q^2 and a lower scale $t \geq p_{\perp,\mathrm{min}}^2$:

$$\Delta^{(\mathrm{UE})}(Q^2, t) = \exp\left[-\frac{1}{\sigma_{pp,\mathrm{ND}}} \int\limits_{t}^{Q^2} \mathrm{d}p_{\perp}^2 \, \frac{\mathrm{d}\hat{\sigma}_{2 \to 2}}{\mathrm{d}p_{\perp}^2} \right] \, . \qquad (7.38)$$

Equating this with a random number allows the transverse momentum squared t, at which the next scatter appears, to be determined, implying an ordering in a hardness scale given by $p_{\perp}^2$. Once the $p_{\perp}$ of the scatter is fixed, also the Bjorken–x of the incoming particles (or the rapidity of the overall system) can be selected from the differential parton-level cross-section $\mathrm{d}\hat{\sigma}_{2 \to 2}/\mathrm{d}p_{\perp}^2$.

This naive treatment has an unpleasant implication, namely a steep dependence on the value of $p_{\perp,\mathrm{min}}$, driven by the divergent structure of $\mathrm{d}\hat{\sigma}_{2 \to 2}/\mathrm{d}p_{\perp}^2 \propto 1/p_{\perp}^4$ or worse for small values of $p_{\perp}$. To cure this problem, a phenomenological *ansatz* is often used, namely replacing $p_\perp^2$ with $(p_\perp^2 + p_{\perp,0}^2)$. This also allows the elimination of $p_{\perp,\text{min}}$, as the partonic cross-section is suitably regularized. Focusing on the approximate behaviour of the differential cross-section with respect to $p_\perp^2$ and keeping only these terms, this implies a reweighting of the differential cross-section Eq. (7.38) by a factor

$$\frac{\alpha_s^2(p_\perp^2 + p_{\perp,0}^2)}{\alpha_s(p_\perp^2)} \frac{p_\perp^4}{(p_\perp^2 + p_{\perp,0}^2)^2}. \tag{7.39}$$

Comparison with data suggests that this new parameter, $p_{\perp,0}$ scales with the c.m.-energy of the colliding hadrons, in a way similar to the total cross section:

$$p_{\perp,0}(E) = \left(\frac{E}{E_{\text{ref}}}\right)^\eta p_{\perp,0}(E_{\text{ref}}). \tag{7.40}$$

Here E_{ref} is some reference scale and the exponent η is related to the pomeron intercept driving the rise of the total hadron cross section.

This relatively simple way of generating the series of $2 \to 2$ parton scatters forming the underlying event can be further modified by assuming a distribution of partons in the incoming hadron. This means that the PDFs entering the simulation become also dependent on an impact parameter b, basically the distance of the two colliding hadrons. Usually it is assumed, though, that the impact parameter dependence is on average only, and therefore there are no "positions" in transverse space associated with the individual scatters. In addition, it is assumed that the impact parameter dependence factorizes from the PDFs such that

$$f_{i/h_1}\left(x_1, \mu_F; \frac{b}{2}\right) f_{j/h_2}\left(x_2, \mu_F; \frac{b}{2}\right) = f_{i/h_1}(x_1, \mu_F) f_{j/h_2}(x_2, \mu_F) A(b), \tag{7.41}$$

where $A(b)$ denotes a **matter overlap function**. Different parameterizations are used for $A(b)$, including single and double Gaussians and forms that are inspired from the usual electromagnetic nucleon form factors. In some recent publications, these overlap functions have also been assumed to depend on the Bjorken-x of the partons [419]. What they all have in common is that the number of scatters N_{scatters} increases with the overlap $A(b)$. An important consequence of adding the impact parameter is that the b-integrated distribution of N_{scatters} becomes broader — in other words, the underlying event model supports larger fluctuations, in agreement with data.

The individual partonic sub-events undergo parton showering, which will decorrelate the two outgoing partons in the transverse plane and also contribute to the overall yield of hadrons produced. One way to impose four-momentum conservation on the event is to reduce the total energy of the incident hadrons by the amount carried away by the initial partons after the parton shower of each partonic sub-event has terminated. This is the logic underlying the models in PYTHIA and SHERPA, where it is assumed that the PDFs for the beam hadron after some partons have been extracted are just the PDFs of the same hadron, but with reduced energy. Alternatively one could also just stop the generation of additional scatters if they exceed the total energy of the hadronic system, possibly even removing the last parton scatter, which is the algorithm in HERWIG. More complicated models introduce correlations between the sub-events beyond this simplest four-momentum and flavour conservation and thereby possibly also modify the PDFs, *cf.* for instance [855].

Further details that must be taken care of in these simplest models are the colour flows between the individual partonic sub-events, which will influence the production of hadrons during hadronization. While infrared-safe observables such as energy flows etc. will remain more or less unaltered by hadronization, hadron multiplicities and the soft part of their energy and $p_\perp$-distributions are highly sensitive to this. This adds yet another dimension of model building and assumptions into the simulation, and consequently various models differ quite a lot. The answers here range from a completely random assignment of colours, with the only constraint of overall colour conservation to models, in which the overall "length" in η-ϕ space of distances between colour-connected pairs of partons is minimized.

The model presented here has to be taken with more than one pinch of salt. Due to the low scales probed — down to $p_{\perp,\min}$ — also the range of Bjorken–x extends to fairly low values of the order of 10^{-6}, which in turn introduces a sizable dependence on the parton distribution function used in the calculation. This is just one indicator of a more generic problem: any fixed-order calculation probing such scales becomes unstable, not least due to the emergence of large logarithms of the BFKL type. It therefore cannot be over-stressed that models such as the ones presented here certainly are overly simplistic and are at best only able to give some qualitative ideas about the true physics.

7.2.2.5 Including rescattering and more

In the latest PYTHIA versions, PYTHIA 6.4 and PYTHIA 8, the simple model outlined above has been enhanced by two additional ideas.

First of all, "*interleaving*" of the multiple parton scatters with the parton showering, and, in particular, initial-state parton showering, has been introduced, which basically amounts to a competition between both. To facilitate this, a combined no-emission probability has been introduced, which schematically reads

$$\frac{\mathrm{d}\mathcal{P}(Q^2, t)}{\mathrm{d}p_\perp^2} = \left(\frac{\mathrm{d}\mathcal{P}_{\mathrm{PS}}}{\mathrm{d}p_\perp^2} + \frac{\mathrm{d}\mathcal{P}_{\mathrm{MPI}}}{\mathrm{d}p_\perp^2}\right) \cdot \exp\left[-\int_t^{Q^2} \mathrm{d}p_\perp^2 \left(\frac{\mathrm{d}\mathcal{P}_{\mathrm{PS}}}{\mathrm{d}p_\perp^2} + \frac{\mathrm{d}\mathcal{P}_{\mathrm{MPI}}}{\mathrm{d}p_\perp^2}\right)\right]. \quad (7.42)$$

Naively, this seems to factorize and therefore to not alter the pattern of parton emission through the parton shower (indicated by subscript "PS") and through multiple parton interactions ("MPI"). However, the effect of flavour and especially momentum conservation will change the patterns, since both parton showering in the initial state and multiple parton scatters take energy out of the incident hadrons, and therefore the cross-talk impacts on the individual parts. As a by-product of this idea, it becomes possible that two parton scatters, that have been independent from each other when they were generated at relatively high $p_\perp$-scales, are found to point to a common "ancestor" parton in their initial-state evolution, see also the left panel of Fig. 7.13. This

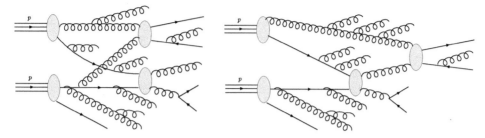

Fig. 7.13 Sketch of possible improvement of simple models for the under-lying event, available in recent versions of PYTHIA: Two scatters can have a common "ancestor" parton (left panel), or there can be rescattering effects (right panel). In both cases, parton showering effects and emissions of secondaries have been ignored.

is the manifestation of recombination effects in the DGLAP evolution of multi-parton distribution functions [679, 690, 845, 861].

As a second effect, final-state rescattering has been implemented in PYTHIA and studied in [418]. Rescattering essentially amounts to those processes, where one of the outgoing partons of one parton scatter acts as an incoming parton for another scatter, *cf.* the right panel of Fig. 7.13.

7.2.3 Double parton scattering

7.2.3.1 A simple model for double parton interactions

The simplest of multiple parton scattering processes is hard **double parton scattering**, processes with two hard scatters, where four incoming partons interact pairwise and create corresponding final states. There has been a number of proposals with varying degrees of sophistication to access such processes from a theory point of view. Most of them based on the idea that the differential parton level cross-section $\hat{\sigma}_{X+Y}$ for the production of a composite system $X+Y$ can be written as [153, 588, 637, 759, 800, 801]

$$\mathrm{d}\hat{\sigma}_{X+Y} = \mathrm{d}\hat{\sigma}_{X+Y}^{\mathrm{dir}} + \frac{m}{2}\frac{\mathrm{d}\hat{\sigma}_X^{\mathrm{dir}} \otimes \mathrm{d}\hat{\sigma}_Y^{\mathrm{dir}}}{\sigma_{\mathrm{eff}}}, \tag{7.43}$$

where the superscript ^{dir} denotes direct production in one two-parton scattering and σ_{eff} is a process-independent parameter with the dimensions of a cross-section. The factor m before the double-parton scattering contribution is used to obtain the correct symmetrization; it is $m = 1$ for $X = Y$ and $m = 2$ otherwise. The $\otimes$ symbol in the ratio and the usage of differential cross-section hints at the necessity to apply identical cuts on the final state and on possible correlation effects in the two-parton PDFs.

A number of subsequent refinements and alterations include, among others, attempts to connect σ_{eff} with total hadronic cross-sections or the geometric size of the hadrons [637, 742], a scaling of this quantity with the scale at which the hadron is being probed, and some naive inclusion of correlation effects in two-parton PDFs. The latter can be achieved, for instance, by writing the two-parton PDFs as a simple product

$$f_{p_1 p_2/h}(x_1, x_2; \mu_F^2) = (1 - x_1 - x_2) f_{p_1/h}(x_1; \mu_F^2) f_{p_2/h}(x_2; \mu_F^2) \qquad (7.44)$$

of the conventional single-parton PDFs [153, 637]. This simple picture has been refined by considerations concerning various types of correlation effects [464, 465, 748], by invoking DGLAP-type evolution equations for parameterizations of multi-parton PDFs [679, 845, 861, 903], including also different resolution scales [832, 833], and by the use of generalized two-parton distribution functions, introduced in [258–260].

With the running of the LHC a proper description of DPS has found renewed interest, as documented in a wide range of publications, which typically go beyond the simplistic picture outlined above. In any case, any models for multiple parton scattering must be compared with suitable data, like, for instance [89, 109, 113] from the TEVATRON and [17, 379, 387, 656] from the LHC. It is fair to state here that any further development of such models and probably also any serious attempt to arrive at a first-principles based theory will necessitate further measurements of such effects.

7.2.3.2 Typical processes

Regardless of the accuracy of the naive expression in Eq. (7.43), it at least roughly predicts the actual size of double-parton scattering cross-sections. Measurements to-date point at values of the phenomenological parameter σ_{eff} of the order of 10–20 mb, about 20% of the total hadronic cross-section. This immediately implies that double-scattering is observable only in those processes, where $\hat{\sigma}_X^{\text{dir}} \times \hat{\sigma}_Y^{\text{dir}} \approx \sigma_{\text{eff}} \times \hat{\sigma}_{X+Y}^{\text{dir}}$, the product of the production cross-sections for the single systems being of the order of the product of hadronic cross-section and the direct production cross-section for the combined system. This leads to combinations of individual processes, where single weak gauge bosons or pairs of jets are produced, in particular processes where $\{X\} + \{Y\}$ is for example given by $\{V\}\{V'\}$, $\{V\}\{jj\}$, $\{jj\}\{jj\}$, or $\{jj\}\{j\gamma\}$, most of which have been considered in the literature. In addition, DPS has been investigated in rare QCD processes, like for instance the production of J/ψ pairs. In order to systematize the considerations, in Table 7.1 a variety of relevant cross-section estimates at next-to-leading order are tabulated.

Cuts on the final-state particles can further improve the ratios of DPS and direct production cross-sections. One example for such a cut relies on the assumed kinematics of DPS processes, being composed of predominantly uncorrelated single parton scatter processes in one hadronic collision. The idea there is that, by far and large, systems produced in a single-parton scatter experience a very limited "kick" in transverse momentum space only, as the generation of additional transverse momentum has to be ascribed to further emissions. For example, for Drell–Yan type processes, the production of weak gauge bosons, the peak of the transverse momentum distribution and its mean are well below 10 GeV. In other words, two particles produced in a hard single scatter, such as the lepton pair in the example here, tend to be relatively well balanced in transverse momentum space,

$$\vec{p}_\perp^{\,\ell^+} + \vec{p}_\perp^{\,\ell^-} = \vec{p}_\perp^{\,(\ell\ell)} \approx 0 \,, \qquad (7.45)$$

Table 7.1 Indicative cross-sections for direct and DPS component of typical process combinations at an 8 and a 13 TeV LHC. All cross-sections are evaluated at next-to-leading order with the central set of the CT14 NLO PDF [489], and with renormalization and factorization scales equal to the scalar sum of the transverse momenta of the final-state particles, $\hat{H}_T$ The gauge bosons decay into one (different) family of leptons each, so that branching ratios have been factored in. Jets, including b jets, are defined in the anti-k_T ($D = 0.4$) algorithm with a transverse momentum of 20 (30) GeV at 8 (13) TeV; to avoid the well-known problem of diverging di-jet cross-section for identical $p_\perp$ cuts, for the jj process the leading jet must have a transverse momentum of $p_\perp^{(1)} \geq 25$ (35) GeV. Photons are subject to the same transverse momentum cuts as the jets. A value of $\sigma_{\text{eff}} = 15$ (20) mb is assumed for the two energies.

	LHC at 8 TeV		LHC at 13 TeV	
Process	$\hat{\sigma}^{\text{dir}}$ [pb]	$\hat{\sigma}^{\text{DPS}}$ [pb]	$\hat{\sigma}^{\text{dir}}$ [pb]	$\hat{\sigma}^{\text{DPS}}$ [pb]
W^+	$6.66 \cdot 10^3$	–	$11.0 \cdot 10^3$	–
W^-	$4.72 \cdot 10^3$	–	$8.20 \cdot 10^3$	–
Z	$1.06 \cdot 10^3$	–	$1.82 \cdot 10^3$	–
jj	$0.173 \cdot 10^9$	–	$0.086 \cdot 10^9$	–
$j\gamma$	$96.7 \cdot 10^3$	–	$46.9 \cdot 10^3$	–
$b\bar{b}$	$0.57 \cdot 10^6$	–	$0.27 \cdot 10^6$	–
W^+W^+	$14.1 \cdot 10^{-3}$	$2.96 \cdot 10^{-3}$	$32.6 \cdot 10^{-3}$	$6.05 \cdot 10^{-3}$
W^-W^-	$5.39 \cdot 10^{-3}$	$1.48 \cdot 10^{-3}$	$14.5 \cdot 10^{-3}$	$3.36 \cdot 10^{-3}$
W^+W^-	$602 \cdot 10^{-3}$	$2.10 \cdot 10^{-3}$	$1250 \cdot 10^{-3}$	$4.51 \cdot 10^{-3}$
ZZ	$16 \cdot 10^{-3}$	$0.074 \cdot 10^{-3}$	$34 \cdot 10^{-3}$	$0.17 \cdot 10^{-3}$
ZW^+	$47 \cdot 10^{-3}$	$0.47 \cdot 10^{-3}$	$96 \cdot 10^{-3}$	$1.00 \cdot 10^{-3}$
ZW^-	$26 \cdot 10^{-3}$	$0.33 \cdot 10^{-3}$	$60 \cdot 10^{-3}$	$0.75 \cdot 10^{-3}$
W^+jj	$0.384 \cdot 10^3$	$0.077 \cdot 10^3$	$0.410 \cdot 10^3$	$0.047 \cdot 10^3$
W^-jj	$0.259 \cdot 10^3$	$0.054 \cdot 10^3$	$0.289 \cdot 10^3$	$0.035 \cdot 10^3$
Zjj	$0.074 \cdot 10^3$	$0.012 \cdot 10^3$	$0.083 \cdot 10^3$	$0.0078 \cdot 10^3$
$W^+b\bar{b}$	1.97	0.25	1.80	0.15
$W^-b\bar{b}$	1.17	0.18	1.12	0.11
$Zb\bar{b}$	1.26	0.04	1.41	0.02
$jjjj$	$3.0 \cdot 10^6$	$2.0 \cdot 10^6$	$1.1 \cdot 10^6$	$0.370 \cdot 10^6$
$jjj\gamma$	$7.2 \cdot 10^3$	$1.12 \cdot 10^3$	$3.0 \cdot 10^3$	$0.20 \cdot 10^3$

To illustrate this idea further, consider the case of ZZ production. In the DPS contribution, there will usually be two pairs of leptons, whose transverse momenta compensate each other, as in the equation above. In contrast, in the direct contribution, the two Z bosons themselves will recoil against each other, and their individual transverse momentum usually will be larger than in the single-Z case, therefore leading

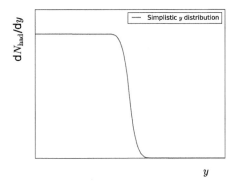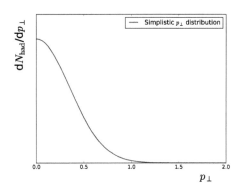

Fig. 7.14 Sketch of hadron y (left) and $p_\perp$ (right) spectra in e^-e^+ annihilations.

to the transverse momentum of the lepton pair being different from zero,

$$p_\perp^{(\ell\ell)} > 0, \qquad (7.46)$$

and the individual transverse momenta not compensating each other. Cuts like these, looking for recoiling pairs, etc., of course further enhance the DPS contribution over the direct component and allow for a better signal (DPS) to background (direct) rate.

7.3   Hadronization

7.3.1   Some qualitative statements

7.3.1.1   Motivation: Jet masses from hadronization

To understand and estimate the visible impact of hadronization, consider the case of jet production at a lepton collider. From LEP data it is known that the rapidity spectrum of hadrons is more or less flat, when rapidity is defined with respect to the main event axis, usually assumed to be the direction of the $q\bar{q}$ pair at leading order. At rapidities of the order of $\log m_{\text{had}}/E$ hadrons cannot be produced anymore, due to energy conservation effects, resulting in the hadron spectrum vanishing fast. At the same time, the transverse momentum spectrum of hadrons with respect to the same axis roughly follows a Gaussian profile, see Fig. 7.14. Defining an expectation value of this profile,

$$\langle \rho \rangle = \int_0^\infty dp_\perp \, p_\perp \rho(p_\perp) = \int_0^\infty dp_\perp \, p_\perp \exp\left(-\frac{p_\perp^2}{\sigma^2}\right) \approx \frac{1}{R_{\text{had}}} \approx m_{\text{had}} \approx 1\,\text{GeV} \quad (7.47)$$

to be approximately given by typical hadronic radii of about 1 fm or $\Lambda_{\text{QCD}} \approx 250\,\text{MeV}$, the energy and momentum of a jet is given by

$$E = \int_0^Y \mathrm{d}y \cosh y \int_0^\infty \mathrm{d}p_\perp p_\perp \rho(p_\perp) = \langle \rho \rangle \sinh Y$$

$$P = \int_0^Y \mathrm{d}y \sinh y \int_0^\infty \mathrm{d}p_\perp p_\perp \rho(p_\perp) = \langle \rho \rangle (\cosh Y - 1) \,. \tag{7.48}$$

The mass M of the jet therefore is given by

$$M = E^2 - P^2 = 2 \cosh Y \langle \rho \rangle^2 \approx 2E\langle \rho \rangle \,, \tag{7.49}$$

something of the order of about 2–3 GeV for a 100 GeV parton jet. This is an indication that in order to fully understand jet physics at the LHC, on the level of about 10%, it is important to also study non-perturbative effects such as hadronization.

7.3.1.2 Single particle fragmentation

A simple approach to describe hadronization is by considering the production of one specific hadron in a given process; consider, for example, $e^- e^+$ annihilations into one specific hadron h and an otherwise arbitrary final state, $e^- e^+ \to h + X$. In order for this to happen, pictorially speaking, one of the quarks or gluons must turn into such a hadron plus other hadronic matter, the latter in order to compensate for colour degrees of freedom. This is because a single parton is a colourful object, while a hadron is not, so some colour neutralization must take place. Schematically, and assuming a quark q to fragment, therefore a process like $q \to h + X$ must be described. Transitions like these could be called single-particle fragmentation, and they are usually described by **fragmentation functions** $D_{h/q}(z, \mu_F)$, which, *at leading order*, parameterise the probability that at a scale μ_F a hadron h emerges from a quark q, carrying a light-cone fraction z of its momentum. For the example process this implies that at leading order the cross-section for $e^- e^+ \to h + X$ is given by

$$\frac{\mathrm{d}\sigma_{e^- e^+ \to h + X}(z)}{\mathrm{d}z} = \sigma_{e^- e^+ \to q\bar{q}} \left[D_{h/q}(z, \mu_F) + D_{h/\bar{q}}(z, \mu_F) \right] \,. \tag{7.50}$$

Similar to the PDFs fragmentation functions fulfil certain sum rules; for instance, as every parton q *must* eventually hadronize into at least one hadron,

$$\sum_h \int_0^1 \mathrm{d}z D_{q/h}(z, \mu_F) = 1 \,. \tag{7.51}$$

To push the analogy a bit further, the dependence of the fragmentation functions on the factorization scale μ_F is logarithmic and given by evolution equations similar to the ones already encountered for the PDFs, see below. However, ignoring this scaling behaviour and assuming that $D_{q/h}(z, \mu_F) \equiv D_{q/h}(z)$ implies that the probability to find a 10 GeV hadron h in a jet emerging from a 20 GeV parton q is identical to finding a 20 GeV hadron h of the same kind in a jet emerging from a 40 GeV parton

q. This in turn supports the notion of **universality** of the hadronization process once it is described at the same factorization (or hadronization) scale.

7.3.1.3 Local parton–hadron duality

One idea that is somewhat related to this behaviour is that observables at the hadron level, such as momentum and energy flows and, to a lesser extent, flavour quantum numbers, roughly follow their distribution at the parton level. This concept is known as **local parton–hadron duality (LPHD)** and was introduced in [177]. This idea has already been encountered in Section 5.1.1 and has been employed there to understand the origin of the hump-backed plateau.

Related to this concept is the **preconfinement** property of parton showers. It implies that at every scale Q_0, the colour structure of the parton shower allows the decomposition into colour singlets with an asymptotically invariant mass distribution [151, 749], which depends on Q_0 and a hadronic scale Λ only, but not on the hard scale Q of the process initiating the parton showers. In fact, if Q_0 is much larger than typical hadronic scales, $Q_0 \gg \Lambda$, the mass distribution of colour-singlets can be calculated perturbatively as well as their momentum and multiplicity distributions, which, however, depend on Q: the net result is that the mass distribution exhibits a strong, power-like suppression at large masses and the asymptotic multiplicity distribution is given by a universal function of $n/\langle n \rangle$, a property known as **KNO scaling** [688].

The best way to see how this works is to use the large-N_c limit for the parton shower. In this limit, the emission pattern of the shower can be arranged as a *planar* graph, with neighbouring partons carrying colour–anti-colour quantum numbers. This idea is at the hard of cluster fragmentation models, see Section 7.3.4.

7.3.1.4 String effect

A somewhat complementary picture is based on the very property of QCD itself, which is characterized by the gluons carrying colour charges in contrast to the Abelian theory of QED, where the photons are charge-neutral. As a consequence, the Coulomb $1/r$ form familiar from QED is not realized for QCD, although the potential obtained from the one-gluon exchange approximation actually is of this form. Essentially, this implies that the form of the potential cannot be obtained from naive perturbative expansions. Based on purely gluonic dynamics, at large distances the **QCD potential** between a static quark–anti-quark pair increases linearly and thus fulfils the **confinement criterion** put forward by Wilson [894]. Such a linear form of the potential is supported by observations, exemplified by fits to the masses of heavy quarkonia.[3] It is interesting to note in passing that the linear part of the potential can be related to the linear increase of the asymptotic Regge slope parameter α', cf. Eq. (7.10).

This leads to an interesting effect, which can be best described as the QCD analogue of the **Nielsen–Olesen string** [785] between two magnetic monopoles in a

---

[3]For example, the relatively simple but hugely successful Cornell potential [496] reads

$$V(r) = -\frac{\kappa}{r} + \sigma r\,, \tag{7.52}$$

with the Coulomb parameter κ and the string tension σ fitted to quarkonia masses.

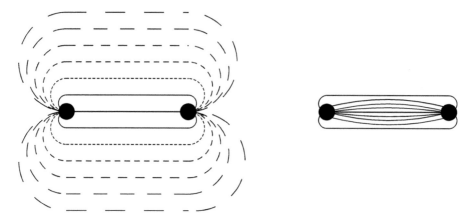

Fig. 7.15 Field lines in electromagnetic (left) and strong (right) inter-actions, spanned by two poles, like static electron–positron and quark–anti-quark pairs. While in electrodynamics the field lines occupy all space, with a strength decreasing with the distance from the poles, in strong interactions they "bunch up" in a flux tube with finite diameter between the two poles.

superconductor. In contrast to electrodynamics, where the field lines of the electric field between two monopoles extend to infinity, in QCD the charged gluon field pulls itself together into a one-dimensional string between the two poles, as illustrated in Fig. 7.15.

This effect of forming strings between colour charges is the non-perturbative continuation of the **drag effect** discussed in Section 5.1.1. As a result, hadron production during hadronization closely follows the topology defined by the strings defined by the colour connections: as a result, hadrons in **Mercedes-star** $q\bar{q}g$ and $q\bar{q}\gamma$ events in e^-e^+ collisions are produced, for the former, along the string spanned by the colour-ordered line $qg\bar{q}$ and not between the quark and the anti-quark, or, in the latter case, they are produced between the quark and the anti-quark, but away from the photon.

This very idea of hadron production along colour-lines is essentially non-local: in the string picture, hadrons are also produced in phase space regions, where no parton acts as a seed. This, however, is mainly true for relatively soft hadrons. The string picture serves as the starting point for a simple model of hadrons and for a very successful model of hadronization, as discussed below.

7.3.1.5 Dealing with baryons

Before going into any more detail concerning various fragmentation models, it is worth to briefly discuss the hadronic degrees of freedom that will be produced. While usually fragmentation functions describe only the *inclusive* production of a specfic hadron from a given parton, the more involved models encoded in event generators turn all partons into an ensemble of hadrons. This means that some underlying idea must be developed of what actually constitutes such a hadron. The answer to this, in the framework of the models that will be discussed below, is relatively simple: hadrons are bound states of quarks and antiquarks.

In the case of mesons, this is fairly obvious and straightforward to realize. They are made of quark–anti-quark pairs such that one can write $\Psi_M = |q_1 \bar{q}_2\rangle$ for their flavour part, and the only non-trivial issue is related to the flavour wave functions of neutral states such as η or η' which have some mixing in flavour and are thus a superposition of $|u\bar{u}\rangle$, $|d\bar{d}\rangle$, and $|s\bar{s}\rangle$. Practically speaking, since the parton shower implicitly also acts in the limit of infinitely many colours, $N_c \rightarrow \infty$, every colour degree of freedom will have one and only one partner with the corresponding anti-colour. In a world without gluons, the colours would be carried by quarks and the anti-colour would be carried by anti-quarks, and there would be unique pairings of both. Any of these pairs would then carry the flavour quantum numbers of mesons.

In contrast, in the large-N_c limit, the composition of baryons is not entirely straightforward. As baryons consist of three constituent quarks, for $N_c = 3$ their colour indices are completely anti-symmetric, realized by a Levi–Civita tensor in colour space. This is how they form an overall colour-singlet. For $N_c \rightarrow \infty$ such a simple reasoning of course would not work any more. Instead, usually the models underlying hadronization resort to the notion of **diquarks**, hypothetical bound states of two quarks or two anti-quarks. For large N_c, the colour state of two such combined quarks, a sextet, can be re-interpreted as an anti-triplet, which allows the formation of binary bound states of a quark and a diquark, like a meson. The flavour part of suitable baryon wave-functions, expressed as quark–diquark bound states has been discussed, for example, in [727]. In the hadronization models implemented so far, the diquarks come as spin-0 and spin-1 particles, which is what can be expected from an s-wave system made from two spin-1/2 objects. In order to maintain Fermi statistics, though, spin-0 diquarks can only exist for two different quark flavours.

Diquarks are non-perturbative objects which act as some mnemonic device to produce, carry and trace baryon number. Consequently, they are not thought to be produced in the perturbative phases of event generation, in particular the parton shower, but rather in those phases that are characterized by scales around typical hadron mass scales. This in turn means that usually only diquarks consisting of two light quarks, u, d, or s, are being produced, which have constituent masses up to the order of 1 GeV. The absence of diquarks containing one or two heavy quarks, c or b, implies that in typical simulation programs doubly or triply heavy baryons are absent. On the other hand, this restriction also implies that ordinary heavy baryons, such as Λ_c or Λ_b, consist of a heavy quark and a light diquark, a picture that is qualitatively well aligned with similar picture in heavy quark effective theory in which heavy hadrons consist of a heavy quark and some "brown muck" around them.

Various other ways to generate and trace baryon quantum numbers have been suggested, for instance by identifying Y-shaped string junctions where a "baryon" centre plays the role of a Levi–Civita tensor and is connected to three quarks through strings [854], but they will not be discussed further here.

7.3.1.6   The limitations of phenomenological hadronization models

First of all, and most obviously, phenomenological fragmentation models, like the Feynman–Field model, the string or the cluster fragmentation models, do not respect quantum mechanical principles by constructing amplitudes and adding them before squaring. Instead they model and employ probability distributions right away through a suitable ansatz. This of course means that quantum effects, like for instance correlations due to the bosonic or fermionic nature of the hadrons, cannot be accounted for in a coherent fashion.

In addition, all models to-date are more or less entirely driven by a limited set of qualitative construction paradigms but lack a more quantitative insight. As a consequence they rely, probably overly so, on the introduction of an increasing number of parameters to fit the data. This fitting to data, often called **tuning**, is exclusively performed based on LEP data from e^-e^+ annihilations at $E_{cms} = 91.2\,\mathrm{GeV}$, due to their high quality and the very large statistics. The resulting model parametrizations are then usually used verbatim not only for other c.m.–energies, but also for different processes, in particular for $p\bar{p}$ and pp collisions at the TEVATRON and the LHC experiments. While this faith probably is a consequence of the lacking ability to perform detailed checks of the assumption, e^-e^+ annihilations and hadronic collisions are different enough to raise some questions. While the former will always involve a colour dipole spanned by a $q\bar{q}$ pair plus subsequent emissions, the latter have a much more complicated colour structure and are often driven by gluons. Furthermore, the latter do have energetic additional sources of colour — the beam remnants. The possibility of the underlying event ultimately results in very complicated colour connections going back and forth between the beam remnants, including the other final-state particles, and possibly overlapping. This can and probably will mean that average distances between partons show larger fluctuations in hadronic collisions compared to the ones in e^-e^+ annihilations, which are solely driven by the self-similar structure of the parton shower.

7.3.2   Fragmentation functions

7.3.2.1   Definition and evolution equation

The most straightforward way to define fragmentation functions is by considering the cross-section for the production of a hadron h in e^-e^+ annihilations to hadrons, proceeding through a Z boson or a virtual photon for the process $e^-e^+ \to \gamma^*/Z \to h + X$. Introducing, in analogy to Eq. (5.317), the energy fraction

$$x = \frac{2E_h}{\sqrt{s}} \quad \in [0, 1] \tag{7.53}$$

of the hadron with respect to the c.m.-energy, and its angle θ with respect to the electron beam axis, allows to write the differential hadron production cross-section as

$$\frac{1}{\sigma_{e^+e^- \to q\bar{q}}} \frac{\mathrm{d}^2\sigma^h}{\mathrm{d}x \mathrm{d}\cos\theta} = \frac{3(1+\cos^2\theta)}{8} F_T^h(x) + \frac{3\sin^2\theta}{4} F_L^h(x) + \frac{3\cos\theta}{4} F_A^h(x).$$

$$\tag{7.54}$$

Here, the dependence of the fragmentation functions on the fragmentation scale μ has been suppressed. The subscripts T and L refer to the transverse and longitudinal fragmentation functions representing the contributions from the corresponding boson polarizations, while A marks the parity-violating term from the interference of vector and axial-vector parts, yielding an asymmetric fragmentation. Integrating over θ yields the total fragmentation function $F^h = F^h_T + F^h_L$, which can be related to the **parton fragmentation functions** (often called **fragmentation densities** or just fragmentation functions) $D_{h/i}$ through

$$\frac{1}{\sigma_{e^+e^- \to q\bar{q}}} \frac{d\sigma^h}{dx} = F^h(x, \mu^2) = \sum_i \int_x^1 \frac{dz}{z} C_i\left(z, \alpha_s, \frac{s}{\mu^2}\right) D_{h/i}\left(\frac{x}{z}, \mu^2\right), \quad (7.55)$$

where the coefficients are

$$C_i\left(z, \alpha_s, \frac{s}{\mu^2}\right) = g_i(s)\,\delta(1-z) + \mathcal{O}(\alpha_s). \quad (7.56)$$

In this expression, the $g_i(s)$ signify the couplings of the quarks to the γ^*/Z, and are given by linear combinations of their electrical, vector, and axial charges and the corresponding boson propagator terms. At leading order, the $C_{q,\bar{q}}$ for quarks and antiquarks are identical, and the contribution for the gluon C_g only emerges at $\mathcal{O}(\alpha_s)$, cf. [783]. Written in this form the analogy of the fragmentation functions F with the structure functions $F_{1,2}$ in DIS and of the parton densities $f_{i/h}$ with the fragmentation densities $D_{h/i}$ becomes fairly striking.

The fragmentation functions (FFs) $D_{h/i}(z, \mu^2)$ encode, at leading order, the probability that a hadron h can be found in the hadrons stemming from the fragmentation of parton i, carrying a light-cone momentum fraction z of the parton. This is in complete analogy with the parton distribution functions (PDFs). In full analogy to the PDFs again, the FFs experience logarithmic scaling violations. They depend on the fragmentation scale μ^2 as expressed through the evolution equation through

$$\frac{\partial D_{h/i}(x, \mu^2)}{\partial \log \mu^2} = \sum_j \int_x^1 \frac{dz}{z} \mathcal{P}_{ji}(z, \alpha_s) D_{h/j}\left(\frac{x}{z}, \mu^2\right), \quad (7.57)$$

in parallel to the DGLAP evolution equations for the PDFs encountered, for instance, in Eq. (2.31).

7.3.2.2 Parameterizations

Similar to the PDFs, discussed in Chapter 6 and in particular in Section 6.2.2, the evolution equations in Eq. (7.57) are used to connect data with the form of the FFs at some fixed lower scale. This lower scale μ_0 typically is of the order of a few Λ_{QCD} for light flavours like u, d, or s quarks, or gluons and of the order of the heavy quark mass in the case of c or b quarks. A typical ansatz, especially for light hadrons and flavours for the FFs at this scale is of the form [279, 624, 687, 697]

$$D_{h/i}(z, \mu_0^2) \;=\; N_i \, z^{\alpha_i^h} \, (1-z)^{\beta_i^h}, \tag{7.58}$$

but also slightly more complicated forms have been proposed, see [447, 448]. This is especially true for heavy quark fragmentation.

Typically, a number of symmetries are assumed, for instance a symmetry between particles and anti-particles

$$D_{h,\,\bar{h}/i}(z, \mu_0^2) \;=\; D_{\bar{h},\,h/\bar{i}}(z, \mu_0^2). \tag{7.59}$$

Similarly, for instance for the case of pions, the suppression of "sea-quark" formation of pions out of the "wrong" quark type is often assumed. Therefore, inequalities like the following ones are expected to hold

$$
\begin{aligned}
D_{\pi^+/d}(z, \mu_0^2) &= D_{\pi^+/s} \;<\; D_{\pi^+/u}(z, \mu_0^2) = D_{\pi^+/\bar{d}} \\
D_{K^+/\bar{u}}(z, \mu_0^2) &= D_{K^+/d,\,\bar{d}} \;<\; D_{K^+/u}(z, \mu_0^2) \;<\; D_{K^+/\bar{s}}.
\end{aligned}
\tag{7.60}
$$

Here the latter inequality in the second equation reflects the suppression of secondary $s\bar{s}$ production formation in the fragmentation process: in order to produce a K^+ out of a u quark, the strangeness quantum number of a $\bar{s}$ quark must be produced through non-perturbative $s\bar{s}$ formation. Due to the larger mass of strange quarks, this, however, is more unlikely than the corresponding non-perturbative $u\bar{u}$ formation in the case of a leading $\bar{s}$ quark fragmenting into a K^+. This reasoning will be recovered under the keyword of **strangeness suppression** later in this chapter. Further assuming "valence enhancement" of the FFs, the idea that at large z the quantum numbers of the quark flavour components would dominate the fragmentation process motivates assumptions like [697]

$$
\begin{aligned}
\beta_d^{\pi^+} &= \beta_{s,\,\bar{s}}^{\pi^+} = \beta_{u,\,\bar{d}}^{\pi^+} + 1 \\
\beta_{\bar{u}}^{K^+} &= \beta_{d,\,\bar{d}}^{K^+} = \beta_u^{K^+} + 1 = \beta_{\bar{s}}^{K^+} + 2.
\end{aligned}
\tag{7.61}
$$

Of course, another constraint on the fitting procedure of such parameterizations is the momentum sum rule,

$$\int_0^1 dz \left[z \sum_h D_{h/i}(z, \mu^2) \right] = 1, \tag{7.62}$$

which states that all hadrons stemming from the fragmentation of a single parton should carry its total energy.

Actual fits to FFs at LO and NLO, similar to the ones for PDFs, have been performed by different groups. In Fig. 7.16 results for one set of these fits for charged pions [446] and protons [447] at NLO, at the low scale of $\mu_F = 1$ GeV are displayed; the fitting function the authors use is given by

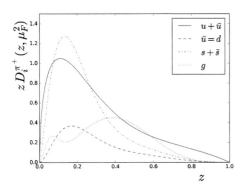 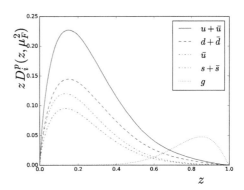

Fig. 7.16 Fragmentation functions for charged pions (left) and protons (right) at next-to-leading order and $\mu_F = 1$ GeV, with parameterizations taken from [446, 447].

$$D_{h/q}(z, \mu_F = 1\,\mathrm{GeV}) = \frac{Nz^\alpha(1-z)^\beta\left[1+\gamma(1-z)^\delta\right]}{B(2+\alpha,\,1+\beta)+\gamma\,B(2+\alpha,\,1+\beta+\delta)}, \qquad (7.63)$$

where $B(a, b)$ denotes the Euler β function.

7.3.2.3   Heavy quark fragmentation

For the fragmentation of heavy quarks, different parameterizations of the fragmentation function at low scales have been proposed, among them

$$D_{H/Q}(z, m_Q^2) \propto \begin{cases} \dfrac{1}{z}\left(1 - \dfrac{1}{z} - \dfrac{\epsilon}{1-z}\right)^{-2} & \text{(Peterson \textit{et al.} [804])} \\[2ex] z^\alpha(1-z) & \text{(Kartvelishvili \textit{et al.} [653])} \\[2ex] (1-z)^\alpha\,z^{-(1+bm_{h,\perp}^2)}\exp\left(-\dfrac{bm_{h,\perp}^2}{z}\right) & \text{(Bowler [281])} \end{cases}$$

$$(7.64)$$

but also the form in Eq. (7.58) has been suggested, for example by [407], or more complicated forms like the one by [411].

7.3.3   Feynman–Field model

7.3.3.1   Description of the model

The **Feynman–Field model** of fragmentation [527], also known as **independent fragmentation**, goes a step further than the fragmentation functions introduced in the previous section, Section 7.3.2, in its description of the hadronization process. While the fragmentation functions are typically concerned with the transition of one parton into one hadron, they are not adequate to describe the production of the full hadron ensemble in a process involving partons in the final state, due to simple colour-

and flavour-conservation arguments: turning a colourful parton into a colourless hadron without any kind of compensation will violate colour conservation. At the same time, after a hadron has been produced from a given quark, its flavour is absorbed in the hadron. Since the strong interaction does conserve flavour, this means that some other flavour quantum number may be left over. As an example, consider the case, where a u-quark fragments into a π^+. With the flavour quantum numbers of the π^+ being $u\bar{d}$, the next hadron must take care of the d-quark that would guarantee that the produced and absorbed $\bar{d}$ is compensated for. Assigning the residual colour to this d quark would actually also account for this quantum number. In a nutshell, this is the reasoning behind the Feynman–Field model.

To be more specific, consider the case of e^+e^- annihilation into a $q_1\bar{q}_1$ pair, without the emission of gluons. The two quarks move away from each other, in their rest frame exactly back-to-back, and for simplicity their momenta can be oriented along the z axis. As a consequence of their relative motion, an increasingly strong colour field along the z direction evolves, in which secondary flavours can be produced in pairs $q_2\bar{q}_2$. Now, one of the primary quarks, say q_1 will combine with the secondary $\bar{q}_2$ to form a $q_1\bar{q}_2$ meson, and the process repeats itself with the remaining $q_2\bar{q}_1$. Assuming that the "primary" $q_1\bar{q}_2$ meson carries a momentum fraction ξ of p_q away (ignoring masses), the momentum of q_2 is given by $(1-\xi)\,p_q$ and therefore the invariant mass of the remaining q_2–$\bar{q}_1$ system is reduced by the same factor. This process of "inserting" further pairs $q_i\bar{q}_i$ can continue, by forming higher-rank mesons $q_i\bar{q}_j$, taking into account the reduced invariant mass of the leftover $q\bar{q}$ pair. In the Feynman–Field model this process is completely recursive. For instance, the probability density in the momentum fraction ξ, or the probability that a specific quark transforms into a specific meson is the same for each meson production step. However, as the recursion progresses, and the invariant mass of the remaining pair becomes smaller, at some point the latter reaches typical hadronic mass scales. At this point, there are two viable choices in the Feynman–Field model: either the production of a meson is forced or this last pair is simply dropped, in both cases with momenta being adjusted by some reshuffling to ensure every meson is on a reasonable mass shell and total four-momentum is conserved.

The produced mesons will have some transverse momentum with respect to the jet axis, the original quark momentum. In the Feynman–Field model this is realised by assigning mutually compensating transverse momenta to the constituents of the secondary $q\bar{q}$-pairs with respect to the original quark–anti-quark axis (the z-direction). The transverse momenta of mesons is then obtained from the sum of the transverse momenta of their constituents. The differential distribution in these transverse momenta for the quarks is given by a Gauss-form,

$$\rho(k_\perp^2) \;=\; \mathrm{d}k_\perp^2 \, \exp\left(-\frac{\pi k_\perp^2}{\sigma}\right). \tag{7.65}$$

7.3.3.2   Shortcomings

There are a number of short–comings in the original version of this model, the most obvious one being the notable absence of any gluons in its original version. There, hadronization proceeds fully recursively and through quark degrees of freedom only, giving rise merely to mesons, but not to baryons. The introduction of diquarks would remedy the latter point, without any real addition to the model, but the former point is a bit more tricky and was never really fully addressed since more involved models such as string or cluster fragmentation took over, see the next sections. These improved models did not suffer from some of the more fundamental issues the Feynman–Field model had by construction.

There are also some issues with the way the splitting process progresses. Due to the form of the probability density $\rho(\eta)$ for the momentum fraction $\eta = 1 - \xi$ of the residual quark chosen in the original model, the model is sensitive to whether the production of mesons occurs from the q or the $\bar{q}$ end.[4]

Finally, note that the way it has been formulated, the model is not Lorentz-invariant and therefore the result has some dependence on the actual frame the hadronization is performed in.

7.3.4 Cluster fragmentation

7.3.4.1 Underlying idea

Cluster models, which are mainly based on the idea of preconfinement, have been discussed very early [899] and have been implemented in the framework of an event generator for $e^- e^+$ annihilations into hadrons in [528, 533]. They are the model of choice for the simulation of hadronization in the event generators HERWIG, HERWIG++ and SHERPA.

The key idea in this class of models was to enforce non-perturbative splittings of all gluons into quark–anti-quark pairs at the end of the parton shower. As a consequence, since the model is formulated in the $N_c \to \infty$ limit, colour-singlet clusters are formed, which consist of predominantly neighbouring quark–anti-quark pairs. The clusters will have the flavour quantum numbers given by the quark–anti-quark pairs, of mesons or, if they consist of a quark–diquark pair, of baryons. Once formed their masses are distributed in a continuous spectrum. The typical mass of these objects is relatively low and driven by the infrared cut-off Q_0 of the parton shower and by the masses of their constituents. While the bulk of low-mass clusters will be reinterpreted as actual hadrons, and enter hadron decays after some momentum reshuffling to force them onto their mass-shell, the high-mass tail of the distribution is seen as a washed out spectrum of excited hadrons. These clusters will decay into further, lighter clusters until they reach the mass scale of hadrons. This leads to a distribution of primordial hadrons that closely follows the pattern resulting from the parton shower, a manifestation of local parton–hadron duality (LPHD).

---

[4]In the original conception of the model in [527] it was noted that the energy distribution of primary mesons hints at $\rho(\eta)$ peaking close to 1, the distribution in the meson momentum fraction ξ must be peaked at low ξ to agree with data. However, the inclusion of gluon emissions certainly improves the situation, but of course it would mean that a way to deal with gluons inside this model must be defined.

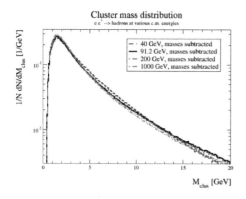
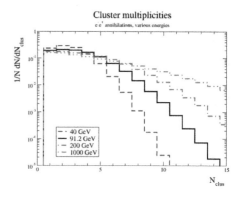

Fig. 7.17 Distribution of primary cluster masses excluding their constituent masses, $\tilde{M}_{\text{clus}} = \sqrt{m_{\text{clus}}^2 - m_1^2 - m_2^2}$ for clusters with constituents 1 and 2, in e^-e^+ annihilations at various E_{cms} (left), and the multiplicity distribution for primary clusters (right). The exemplary results shown here have been obtained from SHERPA.

7.3.4.2 Cluster formation

In a first step in cluster fragmentation, the gluons must decay into quark–anti-quark pairs or, if the phase space allows it, into the usually somewhat heavier diquark pairs. There are two ways this is achieved in practical implementations. Either, as encoded in the HERWIG and HERWIG++ realizations of the cluster fragmentation idea, the gluons acquire a non-perturbative mass, which allows a two-body decay in the rest frame of the now-massive gluon. This gluon mass (m_g) then becomes a parameter of the fragmentation model that has a direct impact on what flavours are actually produced in the transition from the perturbative to the non-perturbative regime. Usually, $m_g \approx 1\,\text{GeV}$. With quark constituent masses around Λ_{QCD}, $m_{u,d} \approx 350\,\text{MeV}$ and $m_s \approx 450\,\text{MeV}$, the gluon decays happen just above the threshold and, consequently, there is not much phase space left for the quarks. As a result, isotropic decays of the gluons will lead to the produced pairs to follow the gluon direction to good approximation, thus encoding LPHD. Alternatively, in SHERPA, the gluons are kept massless and decay by borrowing some four-momentum from a colour-connected spectator parton.

For both HERWIG++ and SHERPA, the availability or lack of phase space for the gluon decay dictates the available flavours and influences the actual flavour of the quark or diquark pair into which the gluons decay. While in HERWIG++, only light quark pairs are accessible, due to the relatively low non–perturbative gluon mass, in SHERPA all flavours can eventually be produced. To exclude the soft, non-perturbative production of heavy flavours, c and b quarks or diquarks containing them, they are explicitly disallowed. In addition, "popping" probabilities $\mathcal{P}_{\to f\bar{f}}$ modify the relative abundance of the permitted flavours f in the non-perturbatively enhanced gluon splitting process. These quantities will reoccur when the decays of clusters will be discussed. They enter the HERWIG++ as well as the SHERPA cluster fragmentation model.

After the gluons have been decayed, primary clusters are formed from the unique colour–anti-colour pairs, by combining the corresponding quarks or diquarks, which thereby also determine the cluster masses and momenta.

In the left panel of Fig. 7.17, the mass distribution of primary clusters produced in e^-e^+ annihilations is exhibited, peaking strongly at masses of around $m_{\text{clus}} \approx 1$ GeV, regardless of the c.m.-energy of the events. The near-identical shape of the mass distributions is a result of the independence from the hard scale of the process of the parton configuration at the end of the parton shower. It is interesting to note though that, trivially, mass thresholds related to the different parton species become visible, most prominently those of clusters carrying charm or bottom quantum numbers. They vanish more or less completely though when considering cluster masses with the constituent masses subtracted. In the right panel of Fig. 7.17, the cluster multiplicity distribution are shown for different c.m.-energies. The mean cluster multiplicity $\langle n \rangle$ increases with energy, but the normalized distribution $n/\langle n \rangle$ approaches a universal function for large energies.

7.3.4.3 Cluster decays to hadrons

Once primary clusters have been formed, many of them have masses close to the mass of primary hadrons with identical flavour quantum numbers. This is motivation to directly identify them with such hadrons and to compensate for the mass difference between the cluster and the hadron by reshuffling momenta in the event accordingly. This identification depends on the primary hadrons included in the simulation, with the heaviest ones introducing a scale for such transitions to be allowed.

Other clusters may be slightly too heavy for such a transition, or there is no primary hadron with a sufficiently similar mass. In cluster fragmentation models, such relatively light clusters will decay into pairs of hadrons, $C \to h_1 h_2$. The mechanism, simply put, is that a flavour pair $f\bar{f}$ (quark–anti-quark or anti-diquark–diquark) is non-perturbatively produced to form new flavour–anti-flavour pairs together with the original cluster constituents $F_1\bar{F}_2$, $F_1\bar{f}$ and $f\bar{F}_2$, which form the flavour part of the hadronic wave functions (or parts thereof, in case the flavour wave functions have more then one component, like neutral pions). The probablity of picking a hadron pair is given by considering the available spin degrees of freedom of the hadrons, n_s, the phase space for the binary decay, the overlap of the flavour configurations $F_1\bar{f}$ and $f\bar{F}_2$ with the hadron flavour wave functions, $|\langle F(f)|\Psi\rangle|^2$, and the "popping" rate for the flavour f, $\mathcal{P}_{\to f\bar{f}}$,

$$\mathcal{P}_{C \to h_1 h_2} \propto n_s^{(h_1)} n_s^{(h_2)} \frac{\sqrt{(m_{\text{clus}}^2 - m_1^2 - m_2^2)^2 + 4m_1^2 m_2^2}}{8\pi m_{\text{clus}}}$$

$$\times \left|\left\langle F_1\bar{f}|\Psi_{h_1}\right\rangle\right|^2 \left|\left\langle f\bar{F}_2|\Psi_{h_2}\right\rangle\right|^2 \mathcal{P}_{\to f\bar{f}}. \tag{7.66}$$

This is, possibly, further modified by multiplet weights reflecting the fact that for given flavour configurations there may be different viable hadrons belonging to different multiplets which are defined by their spin and excitation quantum numbers. Correspondingly there will be relative weights for the pseudoscalar, vector, *etc.*. meson multiplets. This effectively leads to an altered production rate of hadrons from heavier multiplets with respect to the lighter ones. This becomes necessary because the large available phase space for lighter hadrons is usually not correctly compensated by the increasing number of polarizations for heavier hadron species.

Having selected the hadrons, the only missing input for the cluster decay is their orientation. In the original version of cluster fragmentation models, the cluster decays have been isotropic, but this does not correctly cover leading-particle effects, in the x_P distribution of hadrons. As a consequence, also highly non-isotropic decays are used now, especially for those clusters where one of the constituents stems from the perturbative phase of the event simulation and could thus be identified as a leading parton. Naively, this means that the decay contributions must peak for vanishing transverse momentum of the outgoing hadron with respect to the cluster axis, typically with a form like $1/k_\perp^n$ or a Gaussian. This introduces a parameter determining the strength of the peak, in addition to some parameter that defines the scale up to which $C \rightarrow hh$ decays occur.

7.3.4.4 Fission of heavy clusters

Finally, there are those clusters which are too heavy to either transit into a single primary hadron or decay into two. These clusters will also decay, but into final states consisting of one or two new clusters, such as $C \rightarrow C_1 C_2$ or $C \rightarrow C'h$. Again, a new flavour pair $f\bar{f}$ needs to be created and kinematically distributed, with different algorithms.

In the HERWIG and HERWIG++ versions of the cluster model, the masses of the two new clusters are selected according to

$$M_{1,2} = m_{1,2} + (M - m_{1,2} - m_f)\#^{1/\eta},\qquad(7.67)$$

where M is the mass of the original cluster, $\#$ is an isotropic random number, $\# \in [0, 1]$, and P is a parameter which can be chosen differently for clusters containibg only light flavours, bottom, or charm. For clusters containing a beam remnant particle, the mass of the new cluster with the beam remnant parton — or possibly the masses of both clusters — are chosen with an exponential distribution. In any case, a hard veto ensures that the chosen cluster masses are larger than the sum of their constituent masses and smaller than the original cluster. The new clusters are distributed either isotropically or with preferred directions, following the original hard partons from the parton shower.

In contrast, in SHERPA the four-momenta of all four constituents are fixed by first choosing the flavour of the new pair and their transverse momenta, which directly translates into the transverse momentum of the new clusters. This choice is made according to the same Gaussian distribution as in the case of cluster decays to hadrons. Then longitudinal momenta fractions of the new clusters with respect to the two original constituents are chosen, with a function of the form

$$f(z) = z^\alpha (1 - z)^\beta .\qquad(7.68)$$

Together with the transverse momentum this fixes the masses of the new clusters which can then be formed by merely arranging the four constituents.

However, for the new clusters, again decisions of how they decay further must be made, along the lines just described.

7.3.4.5   Further thoughts

There are some interesting features of cluster fragmentation models that could pave the way for further insights into the dynamics of the hadronization process.

First of all, they thrive on the non-perturbative splitting of the gluons, predominantly produced during the parton shower, into quark or diquark pairs, which usually have an electric charge. In addition, a good fraction of the clusters produced and decaying is also charged. There are therefore various phases in cluster fragmentation where charged particles are produced or decay, processes which typically trigger QED radiation. While such effects have not been implemented in cluster fragmentation models to-date, one may speculate in how far they could account for the puzzle of the analyses [104, 105] by DELPHI, which find a significantly larger number of soft photons in e^-e^+ annihilations to hadrons than expected.

Another interesting aspect of hadronization studies at LEP experiments concerned the level of flavour-correlations in hadron production, exemplified by OPAL's measurement of the $\Lambda\bar{\Lambda}$ correlations in e^-e^+ to hadrons at the Z pole [101]. Usually the diquarks responsible for the creation of baryons are produced in late stages of hadronization, when clusters or strings decay and produce hadrons. This yields a fairly strong correlation in phase space between the diquark pairs, and, correspondingly, also between the baryons. One way of softening these correlations is to also allow diquarks to be produced in earlier stages of the fragmentation, in the non-perturbative decay of the gluons during cluster formation.

7.3.5   String fragmentation

7.3.5.1   Hadrons as strings

In contrast to cluster models, which employ some intermediate stage of clusters between the perturbative phase of hadron production and the primary hadrons, string models such as the Artru–Mennessier [172] model or the more sophisticated Lund model [162, 163, 849] are based on a direct translation of partons into hadrons. The underlying idea is relatively simple: the self-interacting low-energy confining modes of the gluon fields can be thought of as a flux tube developing between two oppositely charged poles. At large distances the confining QCD potential assumes a form where it increases linearly with distances, $V(r) \approx -\sigma r$, where the **string tension** $\sigma \approx 1\,\text{GeV/fm} \approx 1\,\text{GeV}^2$.

In this model, hadrons can be understood as strings spanned between massless quarks [162]; subjecting a classical massless quark–anti-quark pair to such a potential will result in a constant attractive force between them and an oscillatory movement. To see this, assume the two quarks are at rest and away from a common origin by a distance r_0. The overall system will then have a total energy given by its potential energy only, $E_{\text{tot}} = V(2r_0) = 2E_0 = 2\sigma r_0$, which could also be interpreted as the mass of the system, $m = 2E_0$. Assuming the quarks need no time to accelarate, they will move towards each other, with the speed of light, and on two light-cones $t \pm z$, if they have been oriented along the z axis at the start. When passing the origin at time

$t = r_0 = E_0/\sigma$, each will have energy and momentum E_0 and, therefore, light-cone momenta of $2E_0$. After another time r_0 they will have swapped positions and again be at a distance of $2r_0$. This **"yo-yo" motion** will repeat itself and after a total time $\tau = 4r_0$ the quarks are back to where they started. The total area A covered by the string during that time is given by eight right-angled triangles with legs of length r_0, thus

$$\sigma^2 A = \sigma^2 \frac{8r_0^2}{2} = 4E_0^2 = m^2 . \tag{7.69}$$

A simple calculation shows that the area is Lorentz-invariant, for details *cf.* [162]. This motivates to identify bound states — hadrons, or, more precisely, mesons – of mass m with such configurations of (massless) $q\bar{q}$ pairs connected through a linear force-field and bouncing back and forth with the speed of light.

7.3.5.2 Detour: An electromagnetic analogy

Following the logic of [162], it is interesting to note the close analogy to electromagnetic superconductors. In order to work this out, remember two of many of their interesting properties.

First of all, Cooper observed [412] that the exchange of phonons, lattice oscillations, induces a small attractive force between pairs of electrons close to the Fermi surface of a crystal. This leads to very loosely bound states, aptly named **Cooper pairs**, of size ξ, which often is macroscopic. This means that in such a case the Cooper pairs extend over many lattice spacings, and since they are bosons, they can occupy the same space[5] and "condense" into the same ground state. However, if there are enough of such pairs, they form a charged Bose gas. The formation of the pairs opens a gap in the continuous spectrum of the electron gas that normally exists in metals, translating into a minimal energy to excite the electrons out of such states. As a consequence, the small excitations that characterize electron scattering and electric resistance become forbidden; the Cooper pairs can move freely in the crystal, which in turn becomes superconducting. This sketches the **BCS theory** [201, 202] of superconductivity.

Second, applying a magnetic field to such a configuration will create a current of Cooper pairs, which in turn will counteract the original field and "expel" it from the metal. Any applied magnetic field will therefore have an exponentially falling penetration depth λ in a superconductor; this is the **Meissner effect** [758]. Increasing the temperature of the crystal will lead to the excitation of the Cooper states. When the temperature passes the critical temperature they will break up and not recombine any more, the crystal will become normally conducting again and the external magnetic field will start to permeate it.

For a type I superconductor, $\xi \gg \lambda$. There will be an interfacial layer with a thickness of about λ between the superconducting material and the external magnetic field where the latter one is practically non-existent, while the Cooper pairs, due to their finite and relatively large size, will be mostly found in the interior of the superconductor. To optimize the energy of such a system, the boundary region will thus assume a shape of minimal size with a thickness defined by a compromise between the

---

[5]It should be noted, though, that they do not form stable bosons, as the electrons will enter and leave the energetically favoured Cooper pair situation.

quickly fading external magnetic field and the density of the Cooper pairs. Conversely, for a type II superconductor, $\xi \ll \lambda$. The magnetic field will penetrate deeply into the superconductor and will have a large overlap with the region where the Cooper pairs are located. As a result, the volume and shape of the boundary region will be maximized.

7.3.5.3   String breakup: Producing $q\bar{q}$ pairs

Making the picture of hadrons consisting of $q\bar{q}$ pairs bound together by a string in a yo-yo mode more dynamical amounts to allowing them to have a larger energy, above typical hadronic scales. Then, the idea of identifying hadrons with strings can be turned into a fragmentation model. To see how this works, assume a primary pair of quarks is produced at the origin with a large c.m.-energy E. The quarks will move away from each other, back-to-back along, say, the z axis, and a string will be spanned by that movement. In the string hadronization model, hadrons are produced from such a configuration by multiple break-ups of the string through the production of new, secondary $q\bar{q}$ pairs in the strong field. These quarks, in turn, will form new endpoints of the smaller string fragments.

The idea behind the emergence of such secondary pairs is that they are spontaneously produced in one point, and then tunnel to have a separation large enough to be pulled apart by the strong field. The creation of n $(f\bar{f})$ pairs in slowly varying static strong electric fields E proceeds through the **Schwinger mechanism** [841], which yields a production rate given by

$$\Gamma = \frac{(eE)^2}{4\pi^3} \sum_{n=1}^{\infty} \frac{1}{n^2} \exp\left(-\frac{n\pi m_f^2}{eE}\right). \tag{7.70}$$

Extending this to the case at hand, of a string break-up, is straightforward, since the string actually represents a linear potential, a constant field. As a consequence, the actual flavour of the $q\bar{q}$ pairs, whose production triggers the break-up, is determined by their mass, with relative probabilities given by

$$\mathcal{P}_{q\bar{q}} \propto \exp\left(-\frac{\pi m_q^2}{\sigma}\right), \tag{7.71}$$

with σ the string tension. Assuming quark masses as above ($m_{u,d} = 350\,\text{MeV}$, $m_s = 450\,\text{MeV}$, $m_{(ud_0)} = 700\,\text{MeV}$) and hadronic scales around Λ_{QCD} for this parameter, $\sigma \approx \Lambda_{\text{QCD}}/\text{fm} \approx 0.2\,\text{GeV}^2$, leads to relative probabilties of

$$P_{u,d} : P_s : P_{(ud)} \approx 1 : 0.3 : 0.003. \tag{7.72}$$

Allowing the produced pair to have a transverse momentum, and remembering the idea of a Gaussian distribution for the bulk of the produced hadrons, *cf.* Fig. 7.14 in Section 7.3.1, the expression above is replaced by

$$\mathcal{P}_{q\bar{q}} \longrightarrow \mathcal{P}_{q\bar{q}}(p_\perp) \propto \exp\left(-\frac{\pi m_q^2}{\sigma}\right) \exp\left(-\frac{\pi p_\perp^2}{\sigma}\right) = \exp\left(-\frac{\pi m_\perp^2}{\sigma}\right). \tag{7.73}$$

This translates into a flavour–independent Gaussian distribution of transverse momenta, where the production of heavier flavours are exponentially suppressed such that charm and beauty are not being produced at all in the fragmentation. It is customary to fine-tune the distribution of flavours with additional suppression or enhancement factors for strange quarks, diquarks, and various flavour and spin configurations of the latter.

7.3.5.4   String breakup dynamics

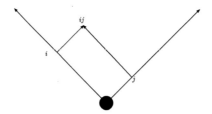

Fig. 7.18 Break-up dynamics of a string: the string breaks at two points i and j, and the resulting quarks meet at point ij

Consider the situation sketched in Fig. 7.18 where a quark q_i from the break-up point i and an anti-quark $\bar{q}_j$ from the break-up point j meet at a point ij. For simplicity only one spatial dimension x is taken into account, as this problem is practically $(1+1)$–dimensional. Since the massless quarks move with the speed of light in the linear potential $V(x) = \sigma x$, their energies and momenta are given by

$$E_{i,j} = \pm\sigma(x_{i,j} - x_{ij}) \quad \text{and} \quad p_{i,j} = \pm\sigma(t_{i,j} - t_{ij}) \tag{7.74}$$

and for the total system

$$E_{ij} = E_i + E_j = \sigma(x_i - x_j) \quad \text{and} \quad p_{ij} = p_i + p_j = \sigma(t_i - t_j). \tag{7.75}$$

The system is at rest, when its momentum is zero, or, in other words, if the two break-ups i and j happen at the same time $t_i = t_j$. In general the overall rapidity of the ij system is

$$y_{ij} = \frac{1}{2}\log\frac{(x_i - x_j) + (t_i - t_j)}{(x_i - x_j) - (t_i - t_j)}. \tag{7.76}$$

The requirement of a positive mass squared of the combined system,

$$m_{ij}^2 = E_{ij}^2 - p_{ij}^2 = \sigma^2\left[(x_i - x_j)^2 - (t_i - t_j)^2\right] \overset{!}{>} 0, \tag{7.77}$$

translates into the requirement that the two vertices i and j are separated by a space-like distance and thus causally disconnected. This implies that *all* string break-ups must be treated on an equal footing — there is just no temporal ordering like in the case of the cluster model, where clusters may decay into clusters. Every break-up partitions the original string into a left and a right-moving one (or, along the positive and negative light-cone axis). Taking into account only longitudinal degrees of freedom, the relation $m^2 = p^+ p^-$ between the mass m and the light-cone momenta $p^\pm$ implies that choosing a mass for any string fragment leaves only one of the two light-cone momenta as free degree of freedom. This hints at a *practical* way of ordering the break-ups of the string into a sequence along one of the two light-cone axes, where hadrons of mass m are being formed. Labelling the end of the string where the break-up happens with a corresponding flavour, naively there are the following possibilities for a string break-up from one of its ends:

$$\text{string}[q_1] \longrightarrow \text{meson}[q_1 \bar{q}_2] + \text{string}[q_2]$$
$$\text{string}[q_1] \longrightarrow \text{baryon}[q_1 (q_2 q_3)] + \text{string}[(\bar{q}_2 \bar{q}_3)]$$
$$\text{string}[(q_1 q_2)] \longrightarrow \text{baryon}[(q_1 q_2) q_3] + \text{string}[\bar{q}_3] \,.$$

The flavours and the transverse momenta of the hadrons are given by the Gaussian distribution in Eq. (7.73). The selection of the actual hadron is then further driven by multiplet-specific weights, corresponding to the same logic already encountered in the cluster fragmentation model. After this, the only quantity left to fix is the splitting parameter z of the longitudinal momenta between the hadron and the residual string.

The overall result for this splitting, however, should be independent of which axis is chosen for the decays in the sequence. To implicitly guarantee boost invariance along the longitudinal axes, the corresponding light-cone momentum fraction z of the hadron is used. Demanding this independence leads to the **Lund parametrization** for the fragmentation function describing the process string → string + hadron:

$$f(z) = N \frac{(1-z)^a}{z} \exp\left(-\frac{b m_\perp^2}{z}\right), \tag{7.78}$$

where a and b are free parameters of the model and m is the mass of the hadron. Typically there are two sets of parameters a and b for strings with a quark and a diquark end, but this tends not to describe the production of heavy hadrons very well. In PYTHIA, therefore, more fragmentation functions, from Eq. (7.64), for heavy hadrons are also available.

The splitting of strings into hadrons and smaller strings can be iterated until only a very light system, in the middle of the original string, is retained. For such light systems procedures similar to cluster decays into hadrons are usually employed, for more details *cf.* [853].

7.3.5.5 Kinky strings: Dealing with gluons

Up to now, only strings emerging from $q\bar{q}$ pairs have been considered. When a gluon is inserted into the game, $qg\bar{q}$, the colour flow is not a simple line connecting quark and anti-quark, but instead more complicated with q being connected to g, which in turn is also connected to $\bar{q}$. In such configurations, the gluon acts as a kink in the string, carrying energy and momentum, and the simple yo-yo motion of $q\bar{q}$ pairs is replaced by a more intricate pattern. This picture implies that the gluon experiences twice the string force, since there are two strings attached to it, rather than the single strings attached to the quark. This also reflects the ration of Casimir operators of gluons and quarks, C_A and C_F with a ratio of $C_A/C_F = 9/4 \approx 2$. As a consequence, hadron production is enhanced in the sectors of phase space spanned by the qg and the $g\bar{q}$ strings, while hadron production in the $q\bar{q}$ sector is massively suppressed. This is a direct consequence of the string picture, which can be viewed as an effective model of QCD at low energies in the limit of infinitely many colours, $N_c \to \infty$. The resulting pattern of hadron production has already been anticipated in the discussion of the drag effect in Section 5.1.2. It is interesting to note here that for relatively soft gluons the kink becomes less and less prominent — the picture of gluons as kinks in a string therefore automatically provides a smooth connection with the parton shower. The actual treatment of these kinks in string fragmentation models is relatively subtle and the reader is referred to the original literature [849, 850], the PYTHIA manual [853], or the review [290].

7.3.5.6   Advanced features of baryon production

The string fragmentation picture developed so far leads to a strong correlation of baryons in phase-space, since a diquark produced in a string break-up from which a baryon emerges leads to the corresponding anti-diquark as the new end point of the residual string. This strong correlation however is challenged by an OPAL measurement of $\Lambda\bar{\Lambda}$ correlations at LEP [101]. This re-inforced the idea of the so-called **pop-corn mechanism** [165], which relates the production of baryons to the popping of more than one $q\bar{q}$ pair. This proceeds in such a way that two quarks of two pairs conspire to form one diquark while two anti-quarks of two such pairs form an anti-diquark. This would effectively lead to configurations where one or two mesons are formed between two baryons, like $BM\bar{B}$, $BMM\bar{B}$, etc.., thereby decorrelating them. This picture is effectively realized by supplementing the potential string break-ups of Eq. (7.78) with yet another one, namely

$$\text{string}[(q_1 q_2)] \longrightarrow \text{meson}[q_1 \bar{q}_3] + \text{string}[(q_2 q_3)].$$

7.4   Hadron decays

7.4.1   General thoughts

The overwhelming majority of all objects produced in particle collisions at a hadron collider like the LHC, that interact with a detector and therefore have a sufficiently long lifetime of $c\tau \geq \mathcal{O}\,(\mathrm{mm})$, are hadrons in the lowest-lying multiplets. These include charged pions and kaons, long-lived neutral kaons — the K_L — as well as nucleons, protons and neutrons. Most of them, and also most of the photons or leptons that accompany the hadrons, actually stem from the decays of hadron resonances or hadrons containing heavy quarks. It is therefore important to have some insight into how these heavy unstable objects decay and how these decays can be described theoretically.

There are large numbers of known unstable hadrons, giving rise to a plethora of different decay modes and characteristics, which cannot be given justice by covering them in any detail in a book like this — a quick glance at the PDG [229] is testament to the incredible amount of data and knowledge gathered up to today. So, instead of trying to have an exhaustive discussion, which would be beyond the scope of this book, a brief overview supplemented with references to a number of relevant review articles must suffice. The focus will be put on those particles and their decays that are phenomenologically most relevant. Typically these are the τ-leptons, hadrons containing open heavy flavour like B and D mesons, and quarkonia states such as the J/Ψ.

7.4.2 Decays of τ–leptons

In the Standard Model the decays of the τ lepton are solely mediated by the weak interaction, giving rise to a τ-neutrino, ν_τ in the final state and either a charged lepton and its anti-neutrino or hadrons.

7.4.2.1 τ decays into leptons

For the former, the fully leptonic decay, matrix elements at leading order can be written as the product of two left-handed currents, namely

$$\mathcal{M}_{\tau \to \nu_\tau \ell \bar{\nu}_\ell} = \frac{G_F}{\sqrt{2}} \left(\bar{u}_\ell \gamma_L^\mu u_{\bar{\nu}_\ell} \right) \left(\bar{u}_{\nu_\tau} \gamma_{\mu L} u_\tau \right) = \frac{G_F}{\sqrt{2}} J_\ell^\mu L_\mu \,. \qquad (7.79)$$

This is the expression from **Fermi's theory of weak interactions**, which rests on four-fermion interactions of the type above. It is an **effective theory** which emerges from the full SM by integrating out the W boson as heavy degree of freedom. This is well justified in the decay of light objects; merely analysing the W propagator and its couplings to the fermions,

$$\frac{g_W^2}{8} \frac{g^{\mu\nu} - \frac{p^\mu p^\nu}{m_W^2}}{p^2 - m_W^2} = \frac{e^2 \, g^{\mu\nu}}{8 m_W^2 \sin^2 \theta_W} + \mathcal{O}\left(\frac{p^2}{m_W^2}\right) = \frac{G_F}{\sqrt{2}} g^{\mu\nu} + \mathcal{O}\left(\frac{p^2}{m_W^2}\right) \,. \qquad (7.80)$$

Integrated over the phase space of the three outgoing particles, Eq. (7.79) yields the well-known partial decay width for weak decays of leptons, namely

$$\Gamma_{\tau \to \nu_\tau \ell \bar{\nu}_\ell} = \frac{G_F^2 \, m_\tau^5}{192\pi^3} f\left(\frac{m_\ell^2}{m_\tau^2}\right) \quad \text{with} \quad f(x) = 1 - 8x + 8x^3 - x^4 - 12x^2 \log x \,, \qquad (7.81)$$

which gives rise of a **branching ration** of about 17 % per lepton.

7.4.2.2 τ decays into hadrons

Similar reasoning can also be applied to the decay to hadrons. There, however, the leptonic current J^μ must be replaced by some hadronic current for the production of N hadrons h_1, h_2, ... Such hadronic currents can be written as

$$J^\mu_{h_1 h_2 \dots h_N} = V_{uq} \left\langle h_1 h_2 \dots h_N \left| \bar{u}_{\bar{u}} \gamma_L^\mu u_q \right| 0 \right\rangle \,, \qquad (7.82)$$

taking into account the fact that for simple reasons of energy and charge conservation, the only allowed combinations of quarks are $\bar{u}d$ and $\bar{u}s$, reflected also in the CKM–matrix element in front of the current.

In the simplest case, when only one pseudo-scalar hadron PS is produced, like a π^- or a K^-, the current reduces to

$$J_{PS}^\mu = V_{uq} \left\langle PS \left| \bar{u}_{\bar{u}} \gamma_L^\mu u_q \right| 0 \right\rangle = -iV_{uq} f_{PS} p_{PS}^\mu \qquad (7.83)$$

with p_{PS} the four-momentum of the hadron and its **decay constant** f_{PS},

$$f_\pi = 130.2(1.7)\,\text{MeV} \quad \text{and} \quad f_K = 155.6(0.4)\,\text{MeV}. \qquad (7.84)$$

This yields the partial width

$$\Gamma_{\tau \to \nu_\tau PS^-} = \frac{G_F^2 |V_{uq}|^2 f_{PS}^2 m_\tau^3}{16\pi} \left(1 - \frac{m_{PS}^2}{m_\tau^2}\right)^2, \qquad (7.85)$$

or branching ratios of about 11 % and 0.7 % for the decay into a single charged pion or kaon.

In principle, currents similar to the one in Eq. (7.83) could also be written for the production of vector or tensor particles, but in reality these particles typically decay fairly quickly and it is thus more useful to model currents with more than one hadron in the final state. Such more complicated final states in τ decays consist of a number of pions, kaons and η-mesons, and they typically exhibit relatively rich structures resulting from a variety of different resonances such as ρ's, K^*'s, a_1's, etc.. As an example, consider the next complicated case of the production of a charged hadron h^-, such as a π^- or a K^-, accompanied by a neutral hadron h^0 like a π^0, K^0, or η. Then,

$$
\begin{aligned}
J_{h^-h^0}^\mu &= V_{uq} \left\langle h^-h^0 \left| \bar{u}_{\bar{u}} \gamma_L^\mu u_q \right| 0 \right\rangle \\
&= \sqrt{2}\, V_{uq} \left[\left(g^{\mu\nu} - \frac{q^\mu q^\nu}{q^2} \right) \left(p_{h^-,\nu} - p_{h^0,\nu} \right) F_V^{h^-h^0}(q^2) + q^\mu F_S^{h^-h^0}(q^2) \right],
\end{aligned}
$$
$$(7.86)$$

with $q = p_{h^-} + p_{h^0}$ and where two form factors, F_V and F_S have been introduced. A typical ansatz to parameterize the form factors has been provided in [703], giving rise to the phenomenological **Kühn–Santamaria model**. It has been further refined for instance in [529, 706] and forms the basis for the successful TAUOLA package [646] and other implementations of τ-decays in HERWIG++ [597] and SHERPA. In this ansatz, the form factors are composed of a number of Breit–Wigner resonances of the form such that, for instance,

$$F_V^{\pi^-\pi^0}(s) = \frac{1}{\sum_V \alpha_V} \sum_V \frac{m_V^2}{m_V^2 - s - im_V\Gamma_V(s)}, \qquad (7.87)$$

where the factor α_V define the relative weight of the contribution, the sum over V includes $V \in \{\rho, \rho', \rho''\}$ and $\Gamma_V(s)$ is the scale-dependent total width of the resonance. These terms, and in particular the relative weight α are typically fitted to experimental data; in order to further improve the agreement, sometimes additional resonances are introduced in this kind of phenomenological model, which do not necessarily correspond to a known physical state.

Alternatively, for the construction of these currents or the relevant form factors, low–energy effective theories of QCD could be invoked. The most appropriate candidate appears to be **chiral perturbation theory** (χ**PT**); for a comprehensive review see [554]. The idea underlying this effective theory is to understand chiral symmetry as one of the fundamental symmetries of the $SU(3)_F \otimes SU(3)_F$ Lagrangian with the quark mass terms breaking it [433–435, 553, 794]. This allows for an expansion in ratios of quark masses over typical momentum transfers in mesonic processes. This idea has been extended to also include resonances such as the vector mesons named above and their coupling to the pseudo-scalars in **resonance chiral perturbation theory** [494, 495], **R**χ**PT**.

7.4.2.3 Weak decays of heavy hadrons

There is some similarity when leaving the decays of the τ-leptons and moving on to deal with weak decays of heavy hadrons containing c or b quarks. There are typically three decay modes associated with them that are phenomenologically relevant, namely purely leptonic decays of mesons of the form $M \to \ell \bar{\nu}_\ell$, semi-leptonic decays of the form $H \to \ell \bar{\nu}_\ell + X$, where X denotes one or more hadrons, and fully hadronic decays $H \to h_1 h_2 \ldots h_N$.

7.4.2.4 Fully leptonic decays

At leading order, purely **leptonic decays of charged mesons** of the type $M^- \to \ell \bar{\nu}_\ell$ can be described by matrix elements based on a current–current interaction similar to the one encountered already in Eq. (7.79),

$$\mathcal{M}_{M \to \ell \bar{\nu}_\ell} = \frac{G_F}{\sqrt{2}} V_{qq'} \left\langle 0 \left| \bar{u}_{\bar{q}} \gamma_{\mu L} u_{q'} \right| M \right\rangle \left(\bar{u}_\ell \gamma_L^\mu u_{\bar{\nu}_\ell} \right) , \qquad (7.88)$$

where q and q' are the constituents of the meson M. For pseudo-scalar mesons PS, which as constituents of the lowest lying multiplet are the ones that typically decay through the weak interaction, again the hadronic part of the matrix element can be replaced with Eq. (7.83) involving the decay constants for the heavy mesons, summarized in [799] as

$$f_{D^\pm} \approx 211.9(1.1)\,\text{MeV} , \ f_{D_s} \approx 249.0(1.2)\,\text{MeV} ,$$
$$f_{B^\pm} \approx 187.1(4.2)\,\text{MeV} , \ f_{B^0} \approx 190.9(4.1)\,\text{MeV} , \quad \text{and} \quad f_{B_s} \approx 227.2(3.4)\,\text{MeV} . \qquad (7.89)$$

The corresponding branching ratios are given by

$$\Gamma_{PS \to \ell \bar{\nu}_\ell} = \frac{G_F^2 f_{PS}^2 m_{PS} m_\ell^2 |V_{qq'}|^2}{8\pi} \left(1 - \frac{m_\ell^2}{m_{PS}^2} \right)^2 . \qquad (7.90)$$

Of course such decays of the light pseudo-scalars $\pi^\pm$ and $K^\pm$ typically do not play a role for LHC physics, since the weak interaction leads to a long lifetime of these objects which in turn usually reach the detector. However, the very same decay modes of heavy mesons are relevant for a number of reasons. First of all, on their own they provide an interesting laboratory testing the interface of perturbative QCD and derived effective theories with lattice QCD and data and thus allow for the measurement of important quantities necessary for the description of the phenomenologically relevant rare decays. In addition, due to the presence of heavy quarks the weak decays would be potentially susceptible to interactions that are sensitive to the quark mass. They thereby allow for a relatively clean way to study, for instance, the effects of interactions mediated by charged Higgs bosons. The case of a B meson decay illustrates this, where the inclusion of charged Higgs bosons would modify the above partial width by a factor [635]

$$r = \left(1 - \tan^2\beta \, \frac{m_{PS}^2}{m_{H^\pm}^2}\right)^2. \tag{7.91}$$

7.4.2.5 Semi-leptonic decays

The analogy with the treatment through current–current interactions similar to the decays of τ decays is also manifest for semi-leptonic decays of the type $H \to \ell\bar{\nu}_\ell + X$. These decays are often used to **tag** heavy flavours produced at relevant hard processes at larger scales, and they also allow for the determination of form factors relevant for rare decays, see below. Depending on the hadronic final state X, different considerations for their actual evaluation will apply. Consider first the case of **exclusive semi–leptonic weak decays**, where X is a specific hadron, $X = h$, like for instance in the B-meson decay $B \to \ell\bar{\nu}_\ell D$. In such cases the matrix element reads

$$\mathcal{M}_{H \to \ell\bar{\nu}_\ell h} = \frac{G_F}{\sqrt{2}} V_{qq'} \left\langle h \left| \bar{u}_{\bar{q}}\, \gamma_{\mu L}\, u_{q'} \right| H \right\rangle \left(\bar{u}_\ell\, \gamma_L^\mu\, u_{\bar{\nu}_\ell} \right) = \frac{G_F}{\sqrt{2}} J_\mu^{Hh}\, L^\mu, \tag{7.92}$$

and, once again, one is left with the determination of the hadronic current. Similar to the case of fully leptonic decays, though, also semi-leptonic decays of most light hadrons are not relevant for LHC phenomenology. This is due to the relatively long lifetime of the involved hadrons, which, again, would consequently arrive at the detector to decay in it. For the few relevant decays of, e.g., K_S, χPT and other effective theories or sum rules help to constrain or evaluate the hadronic decay currents.

7.4.2.6 Heavy quark symmetry

In contrast, in the case of transitions from one heavy quark to another, like in the previous example, an interesting symmetry of QCD comes to help: the heavy meson in question can be seen as the quark surrounded by some light "brown muck" dragged along with it. Since the weak decays happen on time-scales shorter than the scales dominating the bound state, and as long as it continues to be dragged along at about the same velocity, this muck will, at leading order, not resolve the decay or the difference between heavy quarks of different flavour and with different mass. This idea can be formalized as **heavy quark symmetry** [784]. There it gives rise to an expansion in $\Lambda_{\mathrm{QCD}}/m_Q$, aptly known as **heavy mass expansion**, with m_Q the mass of the heavy quark or meson.[6] Following these ideas it is possible to formulate an effective theory, **heavy quark effective theory (HQET)** [497, 567, 736], which is particularly powerful for analysing form factors for hadronic currents like the heavy-to-heavy current in Eq. (7.92). Following ideas introduced in [847, 848], for pseudo-scalar mesons h and H,

$$\left\langle h \left| \bar{u}_{\bar{q}} \gamma_\mu u_{q'} \right| H \right\rangle = \xi(v_h \cdot v_H)(v_h + v_H)_\mu \qquad (7.93)$$

where $\xi(v_h \cdot v_H)$ denotes the Isgur–Wise function [640, 642], which depends on the velocities v_h and v_H of the two mesons h and H. $\xi(vv')$ is normalized such that

$$\xi(1) = 1 + \mathcal{O}\left(1/m_Q^2\right). \qquad (7.94)$$

Similar equations also hold true for cases with a more complicated spin structure, *cf.* [784], where due to the heavy quark symmetry the form factors enjoy similar properties and some remarkable relations among each other.

It should be noted here that such an analysis can also be extended to decays of heavy baryons [568, 569, 643, 745] and to heavy-to-light transitions, to cases where h is a light meson, see [293, 519, 641] for some early work. In contrast, the hadronic state X in semi-leptonic decays can also be composed of a variety of hadrons, leading to **inclusive semi-leptonic decays** when all configurations are summed over. From a simple physical point of view such configurations stem from the fragmentation of the outgoing quark produced in the weak decay and the spectator quark or diquark which together with the decaying quark have formed the hadron. In this simplistic picture, the fragmentation does not change the partial widths on the parton level and the inclusive partial widths are thus given by the quark-level expression only, the quark current is not replaced by hadronic matrix elements. In principle this would allow for a fairly straightforward determination of relevant quantities describing the decay, like for instance the CKM element $V_{qq'}$ present in the transition amplitude of Eq. (7.92). In view of the larger number of form factors in heavy-to-light transitions this was thought to be particularly relevant for the case of V_{ub} [639, 719]. The experimental difficulty in such a measurement, however, is to collect all relevant final-state particles and to cover all possible phase space for them. This poses a challenging problem at hadron colliders such as the LHC.

---

[6]As a first result of this symmetry consider for example the mass splitting between pseudo-scalar and vector mesons with the same flavour quantum numbers such as B^* and B:

$$m_{B^*} - m_B \propto 1/m_B,$$

and as a result the quadratic mass differences are nearly constant

$$m_{D^*}^2 - m_D^2 \approx m_{D^*}^2 - m_D^2 \approx m_{D_s^*}^2 - m_{D_s}^2 \approx 0.5\,\mathrm{GeV}^2.$$

7.4.2.7 Non-leptonic decays

In principle, similar operators also account for the weak non-leptonic decays of hadrons on the parton level: looking at the current–current interaction in Eq. (7.79) one would only have to replace the τ-lepton with the corresponding decaying quark and the outgoing leptons and neutrinos with the quarks that are produced in the decay. However, now the hadronic final state originates from the fragmentation of three outgoing quarks from the Fermi interaction plus the spectator. This renders the calculation of partial widths related to such decays a hard nut to crack theoretically. There are some simplifications, most notably in cases where the decaying quark is a heavy quark (c or, even better, a b quark). Some progress is possible by again using technologies deeply rooted in the fact that the heavy quark mass sets a scale m_Q allowing for an expansion in $\Lambda_{\rm QCD}/m_Q$ [247, 248, 261, 388]. In the spirit of the treatment of exclusive processes with large momentum transfers in QCD through light-cone methods [726] and first worked out for the case of $B \to \pi\pi$ decays, the current–current interaction can be decomposed — factorized — such that

$$\mathcal{M}_{B\to\pi\pi} = \left\langle \pi \left| J_\mu^{b\to q} \right| B \right\rangle \left\langle \pi \left| J_{\mu,q\bar{q}} \right| 0 \right\rangle \left[1 + \sum r_n \alpha_{\rm s}^n + \mathcal{O}\left(\frac{\Lambda_{\rm QCD}}{m_B} \right) \right] \qquad (7.95)$$

with perturbatively calculable coefficients r_n [212]. The picture underlying this factorization is as follows: due to the large mass difference between the incident B meson and the pions the $q\bar{q}$ pair forming one of them will be fairly collimated. As it is a colour singlet, soft gluons with momentum of the order of $\Lambda_{\rm QCD}$ will not see it and thus decouple at leading order in $\Lambda_{\rm QCD}/m_B$. Similar reasoning also holds true for the spectator quark, which in the B rest frame typically also carries momentum of order $\Lambda_{\rm QCD}$. If it does not take part in any way in the hard interaction, it merely contributes the quantum number to the pion formed on the b-quark side of the current. This is then absorbed into a corresponding $B \to \pi$ form factor already encountered in the treatment of semi-leptonic decays. Doe to the size of this colour singlet and its composition, therefore the two pions at leading order factorize nicely. This reasoning could be extended to some degree to other final states, and in particular also to those containing a heavy meson instead of the pion [213].

7.4.2.8 Rare decays

A specific class of weak decays is represented by rare decays which, in contrast to the decays discussed up to now, cannot directly be identified with the tree-level exchange of a W-boson. In contrast to the this case, in which in the low-energy limit therefore becomes equivalent to the Fermi four-fermion interaction, the rare decays typically are mainly driven by loop-induced interactions. Fig. 7.19 shows some examples, where W bosons and additional quarks enter as virtual particles. These decays on the quark level quite often have the form of **flavour-changing neutral current** (**FCNC**) processes, where the overall charge transmitted from the heavy quark line to other final-state objects is zero, but the flavour of this heavy quark line changes. They may manifest themselves as purely leptonic decays, like for example $B_s \to \mu^- \mu^+$, the most prominent one at the LHC. In addition, there are also semi-leptonic decays, related to, e.g., $b \to s\gamma$,

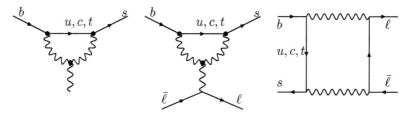

Fig. 7.19 Some example Feynman diagrams for rare decays of heavy quarks: $b \to s\gamma$ (left) and $b \to s\ell\bar{\ell}$ (centre and right). The graph in the middle motivated the slightly unintuitive name of "penguin-diagrams".

$b \to s\ell\bar{\ell}$, $b \to ss\bar{s}$ etc.. Quite often these processes are also known as "**penguin**" decays. The most prominent one at the LHC to date are the decays $B \to K^*\mu^-\mu^+$ and $B_s \to \phi\mu^-\mu^+$, triggered by the quark-level $b \to s\ell\bar{\ell}$ transition. In addition, although not a decay, also **mixing in systems of neutral mesons** such as $B^0\bar{B}^0$-mixing, falls into the category of such processes.

In FCNC processes, the change of flavour quantum numbers along the quark lines typically stems from a loop consisting of W bosons and a combination of quarks — u, c, and t in the examples depicted in Fig. 7.19. By invoking the **unitarity of the CKM matrix** and by assuming massless charm quarks, the contribution of the c and u quarks can be summed and combined with the t contribution using the same combination of CKM elements $V_{bt}V_{ts}^*$.[7] This is the celebrated **GIM mechanism** [581]. As a consequence, rare processes such as $b \to s\gamma$ are driven by the large mass hierarchy between the up-type quarks; conversely, similar processes such as $c \to u\gamma$ are GIM-suppressed.

The interest in rare processes is fed by the observation that typically the particles running in the loops are fairly heavy — W bosons and t quarks — and that similar loops could also originate from hitherto unknown heavy particles from new physics scenarios. Therefore, such rare processes allow indirect probes of physics beyond the SM. In order to systematically study such processes, **operator product expansion**

---

[7]To see how this works assume that the loop contributions to a process such as $b \to s\gamma$ can be written as a linear combination of functions that depend on the ratio of quark and W-mass multiplied with the corresponding CKM elements. Then

$$\mathcal{M}_{b \to s\gamma} \propto f\left(\frac{m_t^2}{m_W^2}\right) V_{bt}V_{ts}^* + f\left(\frac{m_c^2}{m_W^2}\right) V_{bc}V_{cs}^* + f\left(\frac{m_u^2}{m_W^2}\right) V_{bu}V_{us}^*$$

$$\xrightarrow{m_c, m_u \to 0} f\left(\frac{m_t^2}{m_W^2}\right) V_{bt}V_{ts}^* + f(0)\left(V_{bc}V_{cs}^* + V_{bu}V_{us}^*\right)$$

$$= \left[f\left(\frac{m_t^2}{m_W^2}\right) - f(0)\right] V_{bt}V_{ts}^*,$$

where the unitarity of the CKM matrix has been used in the form

$$0 = \sum_q \left(V_{bq}V_{qs}^*\right) = V_{bt}V_{ts}^* + V_{bc}V_{cs}^* + V_{bu}V_{us}^* \quad \longrightarrow \quad V_{bt}V_{ts}^* = -\left(V_{bc}V_{cs}^* + V_{bu}V_{us}^*\right)$$

(**OPE**) is frequently used [893], which rests on a factorization of short- and long-distance contributions to a given process. This results in operators describing the latter, multiplied by **Wilson coefficients** taking account of the former. The result of the loop and possibly higher-order corrections drives the coefficients, which are therefore the relevant quantities sensitive to potential new physics effects. The construction of this operator basis has been intensively discussed in the literature, for a comprehensive review *cf.* [289].

7.4.2.9 Aside: CP violation

At this point it is interesting to point out that the **complex phase in the CKM matrix** induces **CP violation** in the SM. This manifests itself as differences between the branching ratios or kinematic distributions of decays $I \to F$ of an initial state I into a final state F and the corresponding properties of the conjugate decay $\bar{I} \to \bar{F}$. As **direct CP violation** this phenomenon may occur in decays of charged and neutral particles; it is triggered by having different amplitudes — quite often tree-level amplitudes and loop-induced ones — contributing to the same decay. Then different combinations of CKM elements may be invoked, some of which are purely real and some others exhibiting the weak phase. Since the weak phase flips sign when anti-particles are involved, different contributions may exhibit different interference patterns, thus yielding different total amplitudes and therefore different decay characteristics. Additionally, in neutral mesons, the phenomenon of $M\bar{M}$-mixing introduces another source of CP violation. The textbook example is CP-violation in kaon mixing, parameterized by ϵ_K, which manifests itself by a small branching probability of K_S (K_L) mesons to decay into three (two) pions. Similar mixing also occurs in the neutral $D^0\bar{D}^0$ and $B_{d,s}\bar{B}_{d,s}$ meson systems. Finally of course, CP violation can also be triggered through the interference of decay and mixing amplitudes. For a pedagogical introduction into this subject, *cf.* for instance the BABAR physics book [616].

7.4.3 Strong decays

In addition to hadron decays mediated by the weak interaction there are also decays that proceed through the strong interaction, such as $\eta \to \pi\pi\pi$, $\phi \to KK$, or the decay of quarkonia like J/Ψ to lighter hadrons. They all have in common that flavour quantum numbers are conserved, in contrast to the weak decays discussed up to now. Consequently, on the parton level, such decays could possibly be seen as the annihilation of the quark–anti-quark pair constituting the decaying hadron, mainly into gluons, which in turn would fragment into the observed final-state hadrons. For such decays of the lighter hadrons in particular, effective theories like chiral perturbation theory could be invoked again.

Alternatively, for quarkonia — bound states of two heavy quarks — using perturbative QCD is a viable option as well. Depending on the spin of the quarkonium in question, different decay routes would apply. For (pseudo-)scalar particles such as the η_c decays typically would yield two gluons or photons in the final state while for (axial-)vector particles there are more parton level channels available. For instance, for the case of J/Ψ this would include the annihilation of the $c\bar{c}$–pair forming the J/Ψ into a virtual photon, which subsequently would decay into a pair of leptons or quarks or

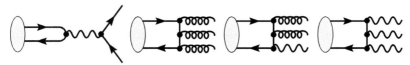

Fig. 7.20 Parton-level decay channels of J/Ψ: annihilation into fermion pairs, annihilation into three gluons, two gluons and a photon, and into three photons (from left to right).

the annihilation into three gluons, two gluons and a photon, or three photons, see also Fig. 7.20. By far and large, the rates calculated for leptonic, photonic and hadronic decays follow the pattern suggested by these parton–level considerations. A simple back-of-the-envelope estimate suggests that the branching ratio to a single lepton is given by

$$\mathcal{BR}_{j/\Psi\to\ell\bar{\ell}} \approx \frac{e_c^2\alpha}{C\alpha_s^3(m_c) + e_c^2\,\alpha\sum\limits_{f} e_f^2} \approx \frac{e_c^2\alpha}{5\alpha_s^3(m_c) + 4\alpha e_c^2} \approx 5\%\,, \qquad (7.96)$$

where $e_c = 2/3$ is the charge of the charm quark, and the factor $C = 5$ stems from the colour factor related to the transition of a singlet to three gluons including their symmetrization. The sum over accessible fermions includes e, μ, u, d, and s for the case of charmonia, yielding a factor of four. This crude estimate of a branching ratio of 5% has to be compared with the measured branching ratio of about $\mathcal{BR}_{j/\Psi\to\ell\bar{\ell}} = 5.94\%$ [799] for the decay of a J/Ψ into one lepton flavour. It has to be noted, though, that this result of course depends crucially on the choice of scale in α_s.

Decays of this kind are important for two reasons. First of all, they provide a test of perturbative QCD at relatively low scales, which is interesting in its own right. Furthermore, and probably far more importantly, at hadron colliders such as the TEVATRON or the LHC, the lepton pairs produced in decays of J/Ψ, Ψ', or Υ mesons, at well-defined and relatively sharp invariant masses of about 3.097 GeV, 3.686 GeV, and 9.46 GeV, respectively, allow, together with leptons from the Z boson, a calibration of the muon chambers and electromagnetic calorimeters. They therefore directly contribute to the overall success of every measurement with leptons in the final state.

8

Data at the TEVATRON

In this chapter, many of the ideas developed in the book up to now will be put
to the test, focusing on the comparison of QCD theory predictions from fixed–order
calculations, analytic resummation calculations, and full hadron–level simulations with
data from the TEVATRON era. Not only did the experiments at the TEVATRON test the
theory of QCD over a wide range of scales, but in addition they also probed interesting
non–perturbative aspects such as soft QCD interactions and multiple parton–parton
scattering. While such an environment is very close to the conditions encountered at
the LHC, the experimental data discussed here have been taken at much lower energies
and in a comparably-reduced phase space.

Nevertheless, the goal of this chapter is to appreciate the precision of theory and
experiment achieved at the TEVATRON, that was instrumental in inspiring a profound
faith in the technology that is now employed at the LHC. We will review some of
the salient analyses at the TEVATRON, dealing with many of the same processes that
will also be discussed for the LHC. The aim is not to present a complete review of
TEVATRON physics, including all of the most up-to-date results, but to discuss those
physics topics which will serve as a good pedagogical introduction to QCD physics at
the LHC. This will help to set the scene for Chapter 9, where results from Run I of
LHC will be confronted with theoretical predictions in the same spirit.

8.1 Minimum bias and underlying event physics

8.1.1 Minimum bias events

The overwhelming majority of the events produced at the TEVATRON are due to pe-
ripheral collisions of the proton and anti-proton, the so-called **minimum bias events**.
These events occur at such a high rate (50 mb) that it would be impossible to record
all of them. However, through the use of a highly pre-scaled trigger, a small fraction
of these events are saved. The events contain interesting physics in their own right, as
well as providing a means of monitoring the performance of the detector. In the high
luminosity running typical of Run II at the TEVATRON, there are many such minimum
bias events in every beam crossing, and the effects of such **pileup** must be dealt with
for the precision determination of physics at higher scales.

The Black Book of Quantum Chromodynamics. John Campbell, Joey Huston and Frank Krauss, Oxford University Press. © John Campbell,
Joey Huston, and Frank Krauss 2018, First published in paperback 2022. DOI: 10.1093/oso/9780192871961.001.0008

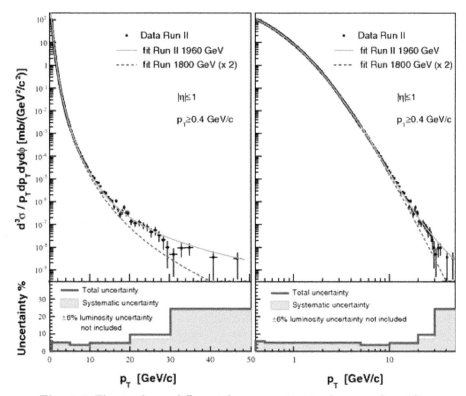

Fig. 8.1 The track p_T differential cross-section in the central rapidity region for CDF in Run I and Run II. Reprinted with permission from Refs. [74, 75].

The transverse momentum distribution for charged tracks with pseudo-rapidity less than 1 is shown in Fig. 8.1 for CDF in Run II, compared to the similar distribution in Run I [74, 75]. As expected, the majority of tracks are at very low transverse momentum. On average, there are slightly more than 2 tracks per unit pseudo-rapidity for events in which at least one track has a transverse momentum larger than 0.5 GeV. Tracks in the higher p_T range are constituents of jets; as can be observed in the figure there is a larger cross-section for the production of these tracks at 1.96 TeV than at 1.8 TeV. Fig. 8.2 shows that the track–p_T distribution is reasonably well-described by PYTHIA using Tune A.[1]

Late in its career, the TEVATRON carried out an energy scan, running at energies of 300 and 900 GeV, in addition to its normal Run II energy of 1.96 TeV. The data obtained from the relatively short runs proved to be very useful in tuning models for minimum bias production. The average particle density $(dn/d\eta)$ for charged particles

---

[1] Tune A refers to the values of the parameters describing multiple-parton interactions and initial state radiation which have been adjusted to reproduce the energy observed in the region transverse to the jet.

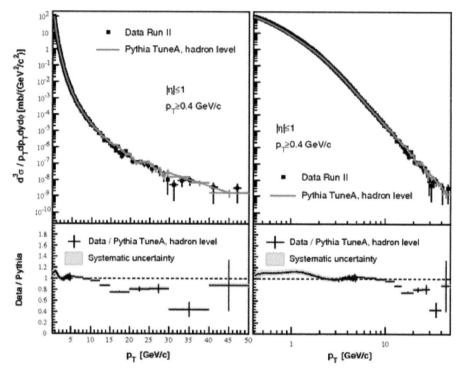

Fig. 8.2 The track p_T differential cross-section in the central rapidity region for CDF in Run II compared to the prediction from PYTHIA Tune A. Reprinted with permission from Refs. [74, 75].

with $|\eta| \leq 0.8$ and $p_T \geq 0.5$ GeV is shown in Fig. 8.3 [72]. An extrapolation predicts a charged particle density of 3.1 at 7 TeV and 3.3 at 8 TeV.

8.1.2   Underlying event

Any event at a hadron-hadron collider consists of a hard collision of two incoming partons, with possible QCD radiation from both the incoming and outgoing legs, along with softer interactions from the remaining partons in the colliding hadrons. Such interactions represent the **underlying event energy** discussed in Chapter 2. A schematic depiction of this situation is shown in Fig. 8.4 for the case of a dijet event.

The underlying event energy is due to the interactions of the spectator partons in the colliding hadrons. It results in an energy deposit of approximately 0.5 GeV for a cone of radius 0.7 and is similar to the amount of energy observed in minimum bias events with a high track multiplicity. The rule-of-thumb has always been that the underlying event energy in a jet event looks very much like that observed in minimum bias events, *i.e.* that there is a rough factorization of the event into a hard scattering part and a soft physics part [121].

Studies have been carried out with inclusive jet production in CDF, examining the transverse momentum carried by charged particles inside and outside of jets [121, 127].

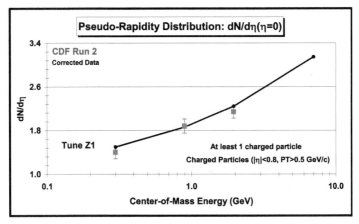

Fig. 8.3 The average track multiplicity distribution for CDF in the central rapidity region ($|\eta| \leq 0.8$), for tracks with $p_T \geq 0.5$ GeV, as a function of the centre-of-mass energy. Reprinted with permission from Ref. [72].

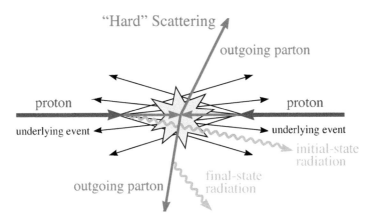

Fig. 8.4 Schematic cartoon of a $2 \rightarrow 2$ hard scattering event.

For example, the geometry for one study is shown in Fig. 8.5, where the **towards** and **away** regions have been defined with respect to the direction of the leading jet.

Of the two transverse regions indicated in Fig. 8.5, the one with the largest transverse momentum is designated the TransMAX region and the one with the lowest, the TransMIN region.[2] The transverse momenta in these two regions is shown in Fig. 8.6. As the lead jet transverse momentum increases, the momentum in the TransMAX region increases; the momentum in the TransMIN region does not. The amount of transverse momentum in the TransMIN region is consistent with that observed in high multiplicity minimum bias events at the TEVATRON. At the parton level, the TransMAX region can receive contributions from the extra parton present in NLO

---

[2]A similar analysis was carried out in Run I, using cones of radius 0.7, at the same η as the lead jet in the event and $\pm 90^o$ in ϕ, to define the MIN and MAX regions [127].

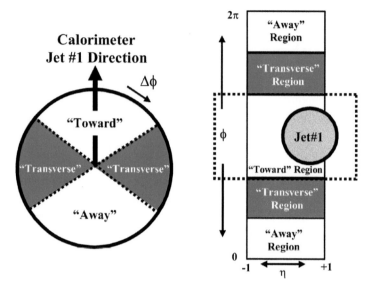

Fig. 8.5 Definition of the *toward*, *away* and *transverse* regions.

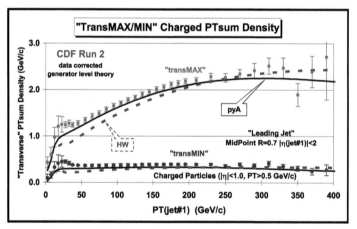

Fig. 8.6 The sum of the transverse momenta of charged particles inside the TransMAX and TransMIN regions, as a function of the transverse momentum of the leading jet. from Ref. [133]. The solid curves are the predictions from PYTHIA and the dashed curves are the predictions from HERWIG. Reprinted with permission from Ref. [133].

inclusive jet calculations. The TransMIN region can not. There is good agreement between the TEVATRON data and the PYTHIA tunes, not surprising since the data was used in the creation of the tunes.

8.2 Drell-Yan production

As discussed in Chapter 2, W/Z production at hadron-hadron colliders serves as a precision benchmark for Standard Model physics. This is true especially for the TEVATRON where, on the experimental side, the systematic errors are small. The decay leptons are easy to trigger on and the backgrounds are under good control. An electron cluster is formed from all towers in the electromagnetic calorimeter containing energy from the electron shower. The inclusion of such **towers**[3] in the electron energy will also effectively add to the electron 4-vector the energy from any collinear photons radiated by the electron. Thus, in the language of Section 2.1.6, these would be referred to as **dressed leptons**. This is not true for muon candidates, where no calorimeter energy is added to the muon 4-vector. Thus, these would be referred to as **bare leptons**.

Identification cuts, using both tracking and calorimetric information, are applied to the lepton candidates to improve their purity. In addition, for electrons, an **isolation cut** is applied which requires that the energy around the electron candidate in a cone of radius $R = 0.4$ be less than a fixed amount (typically 4 GeV). As an alternative, it is sometimes required that the isolation energy is less than a fixed fraction of the electron's transverse momentum. The isolation cut serves to reduce the rate for jets faking electrons. The isolation energy sum excludes the towers already included in the electron cluster. No such isolation cut is applied for muons. However, the energy deposited by the muon in the calorimeter is required to be consistent with that expected from a minimum ionizing particle.

For some of the theoretical predictions used for W/Z production at the TEVATRON PHOTOS [591] or other QED FSR simulations can be used to generate QED radiation for the leptons. This is true for PYTHIA as well as for RESBOS [300], a generator that accounts for the effects of resummation at small W/Z transverse momenta (as described in Section 5.3). Alternatively, the W and Z boson cross-sections and differential distributions can be compared to the known NNLO QCD predictions, such as those provided by FEWZ2 [556].

The W and Z cross-sections measured at the TEVATRON are shown in Fig. 8.7 for both Run I and Run II [122]. The experimental cross-sections agree well with the theory predictions at NNLO and the Run II cross-sections show the rise expected from the increase in centre-of-mass energy over Run I.

The Z rapidity distribution measured by DØ in Run II is shown in Fig. 8.8 [94], with the measurement agreeing with the NNLO prediction over the entire rapidity range. This agreement is non-trivial since there is a shape change in the rapidity distribution between LO to NLO results, which is primarily driven by the differences between LO and NLO PDFs discussed in Chapter 6, as well as an increase in normalization. On the other hand, the transition from NLO to NNLO is essentially just a small K-factor with little change in shape (*cf.* Fig. 3.10). It is noteworthy that the largest experimental uncertainty for the W and Z cross-sections is the uncertainty in the luminosity, typically on the order of 5% at the TEVATRON (it is lower at the LHC).

---

[3]A tower refers to the lateral segment of a calorimeter reading out a specific region in $\Delta\eta$ and $\Delta\phi$. There may be more than one longitudinal division of the calorimeter within the tower, in which case the energies of all of the longitudinal divisions are added together to form the tower energy.

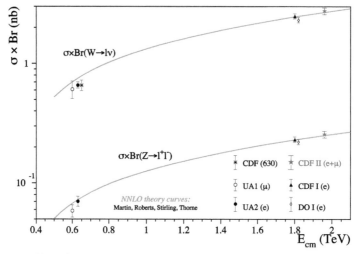

Fig. 8.7 *W* and *Z* cross-sections as a function of the centre-of-mass energy at the TEVATRON, from Ref. [122]. Copyright IOP Publishing. Reproduced with permission. All rights reserved.

The theoretical systematic errors are primarily from the PDF uncertainty. Such cross-sections have thus proven to be useful as inputs to global PDF fits.

The transverse momentum distribution for *Z* bosons, measured at CDF in Run II of the TEVATRON [65], is shown in Fig. 8.9 along with comparisons to predictions from RESBOS and FEWZ2. The FEWZ2 and RESBOS predictions agree well with the data at high transverse momenta. The inset shows the low transverse momentum region, where the RESBOS prediction also matches the data, including the turn-over of the cross-section for $p_T < 5$ GeV, as expected from a resummation calculation. This very low p_T region is sensitive to non-perturbative effects. FEWZ2 is not shown for the low-p_T region as, being a fixed-order calculation, it will not provide sensible predictions. It is worth noting that the CDF p_T data have been plotted with a very fine binning, made possible by the excellent tracking resolution of the CDF detector. Such binning allows a better determination of the low p_T physics.

For Drell-Yan production, the average transverse momentum of the lepton pair has been measured as a function of their invariant mass. Fig. 8.10 [65] shows the measurement made by CDF in Run II. The average transverse momentum increases roughly logarithmically with the square of the Drell-Yan mass, as expected from the discussion in Section 2.3.2. The data agree well with the default PYTHIA 6.2 prediction using Tune A. Also shown are two predictions involving tunes of PYTHIA that give larger or smaller values for the average Drell-Yan transverse momentum as a function of the Drell-Yan mass. The **Plus/Minus** tunes were used to estimate the initial-state-radiation uncertainty for the determination of the top mass in CDF. Most of the $t\bar{t}$ cross-section at the TEVATRON arises from $q\bar{q}$ initial states, so the Drell-Yan measurements serve as a good model. This will not be true at the LHC, where the dominant initial state for $t\bar{t}$ production is gg.

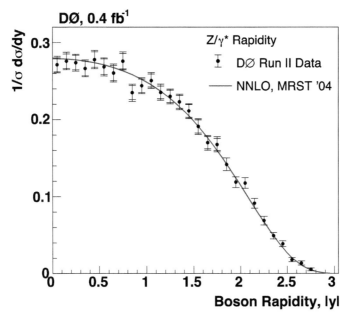

Fig. 8.8 The Z rapidity distribution from DØ in Run II. Reproduced with permission from Ref. [94].

8.3 Inclusive jet production

Inclusive jet production at the TEVATRON plays an important role in this book for a number of reasons. First, it probes the highest transverse momentum range accessible at the TEVATRON. Second, it has a large impact on global PDF analyses. Finally, many of the subtleties regarding measurements with jets in the final state and the use of jet algorithms come into play, providing an introduction to some of the issues which will also be encountered at the LHC.

In many events, the assignment of individual calorimeter towers into jets is fairly unambiguous and the jet structure in such final states is relatively clear. However, in some events, the complexity of the energy depositions means that different algorithms will result in different assignments of towers to the various jets. This is no problem to the extent that a similar complexity can be matched by the theoretical calculation to which it is being compared. This is the case, for example, for events simulated with parton shower Monte Carlos. However, a NLO calculation for inclusive jet production (or, indeed, any other process) can place at most two partons in a single jet.

Before proceeding with a discussion of the TEVATRON jet analyses, it is worthwhile to step back and consider a few practical experimental implications. At the TEVATRON, IR-unsafe algorithms (*cf.* Section 2.1.6) were universally used. These included JetClu in CDF [108] and a comparable cone jet algorithm in DØ in Run I, and then the Midpoint cone algorithm in Run II (although JetClu continued to be used in CDF for many analyses). These algorithms required the presence of seeds in order for the jet clustering to begin, introducing the IR-safety problem that was discussed in Section 2.1.6. For

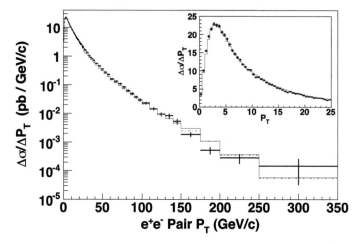

Fig. 8.9 The transverse momentum distribution for $Z \rightarrow e^+e^-$ from CDF in Run II (black crosses), along with comparisons to predictions from FEWZ2 (dot-dash histogram) and RESBOS (solid histogram). The inset shows a closeup of the low transverse momentum region. Reproduced with permission from Ref. [65].

Run II, the Midpoint jet algorithm was developed, which pushed the IR-safety problem to a higher order (NNLO).[4] The SISCone algorithm, which is IR-safe to all orders, was developed late in Run I, too late for any analyses,[5] but Monte Carlo comparisons using this algorithm and the Midpoint algorithm were carried out for many analyses. A few analyses were carried out with the (IR-safe) k_T algorithm. The anti-k_T clustering algorithm was developed too late for any TEVATRON analyses.

The IR-safety problem applies only to fixed-order calculations, *i.e.* any of the jet algorithms mentioned above are IR-safe when applied to data/Monte Carlo, simply, because for hadrons there is no infrared problem with the minimal hadron mass being $m_\pi \approx 135\,\text{MeV}$. The differences between the Midpoint and SISCone algorithm are finite and typically of the order of a few percent at most, so it is perfectly acceptable to use a SISCone jet algorithm in a fixed-order prediction to compare to data taken with the Midpoint algorithm. The stochastic instabilities introduced in Monte Carlo simulations, and inherent in the data itself, tend to *level the playing field* for many of the jet algorithms, since the extra effective seed in the Midpoint algorithm and the extra seeds in the SISCone algorithm have little impact [508].

---

[4]The Midpoint algorithm also introduced the use of the kinematic variables p_T and y, in contrast to the variables E_T and η used for the earlier algorithms.

[5]Tradition and inertia can make it difficult for experiments to give up old jet clustering algorithms, even if better algorithms are available. Thus, it is very good that both ATLAS and CMS have used the anti-k_T algorithm from the start.

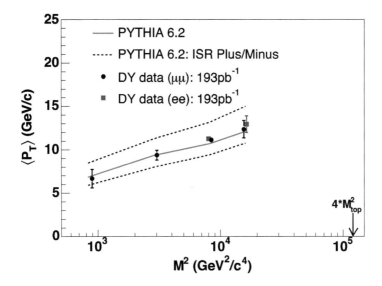

Fig. 8.10 The average transverse momentum for Drell-Yan pairs from CDF in Run II, along with comparisons to predictions from PYTHIA. Reproduced with permission from Ref. [65].

8.3.1 Corrections

For comparison of data to theory, the calorimeter tower energies clustered into a jet must first be corrected for the detector response. The calorimeters in the CDF experiment (or basically any experiment) respond differently to electromagnetic showers than to hadronic showers, and the difference varies as a function of the transverse momentum of the jet. The detector response corrections are determined using a detector simulation in which the parameters have been tuned to test-beam and in-situ calorimeter data. PYTHIA, with Tune A, is used for the production and fragmentation of jets. The same clustering procedure is applied to the final state particles in PYTHIA as is done for the data. The correction is determined by matching the calorimeter jet to the corresponding particle jet. An additional correction accounts for the smearing effects due to the finite energy resolution of the calorimeter. At this point, the jet is said to be determined at the **hadron level**.

One of the observables that is crucial to be able to described is the **jet shape**. A study of this quantity is shown in Fig. 8.11 [122], where the jet energy away from the core of the jet (*i.e.* in the annulus $0.3 < R < 0.7$) is plotted as a function of the transverse momentum of the jet. The general feature of these curves, that jets become more collimated as the jet transverse momentum increases, can be understood as due to three effects: First, power corrections that tend to broaden the jet decrease as $1/p_T$ or $1/p_T^2$; second, a larger fraction of jets are quark jets rather than gluon jets; third, the probability of a hard gluon to be radiated (the dominant factor in the jet shape)

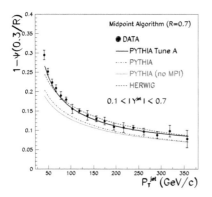
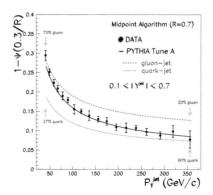

Fig. 8.11 The fraction of the transverse momentum in a cone jet of radius 0.7 that lies in the annulus from 0.3 to 0.7, as a function of the transverse momentum of the jet. Comparisons are made to several tunes of PYTHIA (left) and to the separate predictions for quark and gluon jets (right). Reprinted with permission from Ref. [122].

decreases as $\alpha_S(p_T^2)$. As can be seen in Fig. 8.11, the PYTHIA predictions using Tune A describe the data well, even better than with the default PYTHIA prediction. In fact a reasonable description of the jet shape can also be provided by the pure parton-level NLO prediction [510], perhaps supplemented by non-perturbative corrections, as discussed in Section 4.1, and in Ref. [680].

For data to be compared to a parton level calculation, the theory must be corrected to the hadron level.[6] In general, the data should be presented at the hadron level, and the corrections between hadron and parton level should be clearly stated. In retrospect this seems obvious, but the TEVATRON jet measurements were one of the first analyses where this was true.

The hadronization corrections consist of two components: the subtraction from the jet of the underlying event energy discussed in Section 8.1 and the correction for a loss of energy outside a jet due to the fragmentation process. The hadronization corrections can be calculated by comparing the results obtained from PYTHIA at the hadron level to the results from PYTHIA when the underlying event and the parton fragmentation into hadrons has been turned off. The underlying event energy is due to the interactions of the spectator partons in the colliding hadrons and the size of the correction depends on the size of the jet cone. As discussed earlier in this chapter, the rule-of-thumb has always been that the underlying event energy in a jet event looks very much like that observed in minimum bias events.

---

[6]In some analyses at the TEVATRON, the data was corrected to the parton level using the inverse of the parton to data corrections.

The fragmentation correction accounts for the daughter hadrons ending up outside the jet cone from mother partons whose trajectories lie inside the cone (also known as **splash-out**); it does not correct for any out-of-cone energy arising from perturbative effects as these should be correctly accounted for in a NLO calculation. It is purely a power correction to the cross section. The numerical value of the splash-out energy is roughly constant at 1 GeV for a cone of radius 0.7, independent of the jet transverse momentum. This constancy may seem surprising. But, as just discussed, as the jet transverse momentum increases the jet becomes more collimated. The result is that the energy in the outermost annulus, the region responsible for the splash-out energy, is roughly constant. The correction for splash-out derived using parton shower Monte Carlos can be applied to a NLO parton level calculation to the extent to which both the parton shower and the two partons in a NLO jet correctly describe the jet shape.

The two effects of underlying event and splash-out produce corrections which go in opposite directions. Therefore they partially cancel when computing the total correction for parton level predictions. For a jet of radius 0.7, the underlying event correction is larger, so the correction for the parton level prediction is positive. The total correction is of the order of 7% for the lowest transverse momentum values in the inclusive jet cross-section measurement, decreasing rapidly to less than 1% at higher p_T values (falling roughly as $1/p_T^2$, as would be expected for such power corrections). The correction is roughly independent of rapidity. For a jet cone radius of 0.4, the fragmentation correction is somewhat larger (increasing as $1/R$ for small R) but the underlying event correction scales by the ratio of the cone areas (R^2) [432]; as a result the two effects basically cancel each other out over the full transverse momentum range at the TEVATRON.

Note that these two corrections deal with non-perturbative physics only. The assumption for the comparison to a NLO parton-only prediction is that the perturbative aspect of the jet shape is reasonably well-described by one gluon (in the NLO calculation) as with the parton shower (in the Monte Carlo). Thus, the fragmentation corrections determined for the latter can be applied to the former. Studies of the jet shape at NNLO should prove useful in testing this assumption.

8.3.2 CDF inclusive jet results

The inclusive jet cross-section measured by the CDF Collaboration in Run II in the central rapidity region using the Midpoint cone algorithm is shown in Fig. 8.12, as a function of the jet transverse momentum [116]. Due to the higher statistics compared to Run I, and the higher centre-of-mass energy, the reach in transverse momentum for the inclusive jet cross-section increased by approximately 150 GeV. The CDF measurement used the midpoint cone algorithm with a cone radius of 0.7. As discussed earlier in this section, the Midpoint algorithm places additional seeds (directions for jet cones) between stable cones having a separation of less than twice the size of the clustering cones. The Midpoint algorithm uses four-vector kinematics for clustering individual partons, particles or energies in calorimeter towers, and jets are described using rapidity (y) and transverse momentum (p_T).

A comparison of the inclusive jet cross-section measured by CDF in Run II with the Midpoint cone algorithm, to NLO QCD predictions using the EKS [509] program

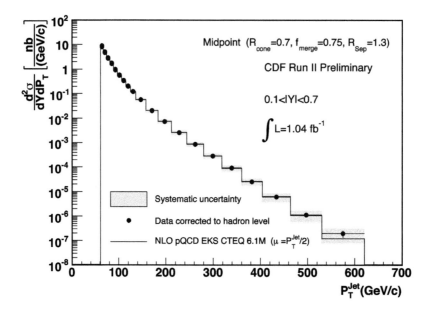

Fig. 8.12 The inclusive jet cross-section from CDF in Run II. Reprinted with permission from Ref. [116].

with the CTEQ6.1 and MRST2004 PDFs, is shown in Fig. 8.13 for the five rapidity regions in the analysis [61].

A renormalization and factorization scale of $p_T^{jet}/2$ has been used in the calculation. Typically, this leads to the highest predictions for inclusive jet cross-sections at the TEVATRON (for $R{=}0.7$), as discussed in Section 4.1.[7] There is good agreement with the CTEQ6.1 predictions over the transverse momentum range of the prediction, in all rapidity regions. The MRST2004 predictions are slightly higher at lower p_T and slightly lower at higher p_T, but still in good overall agreement.

As noted before, the CTEQ6.1 and MRST2004 PDFs have a higher gluon at large x as compared to previous PDFs, due to the influence of the Run I jet data from CDF and DØ. This enhanced gluon provides a good agreement with the high p_T CDF Run I measurement as well. The DØ inclusive jet data taken in Run II, however, do not favour this higher gluon at large x, but instead prefer a weaker gluon. As will be observed in the next chapter, the inclusive jet data from the LHC do not yet provide a definitive answer. The curves indicate the PDF uncertainty for the prediction using the CTEQ6.1 PDF error set. The shaded band indicates the experimental systematic uncertainty, which is dominated by the uncertainty in the jet energy scale (on the order of 3%). It is important to note that for much of the kinematic range, the experimental

---

[7]For smaller jet sizes, a central scale of p_T^{jet} is perhaps more appropriate, and that is what is generally used in PDF fits currently.

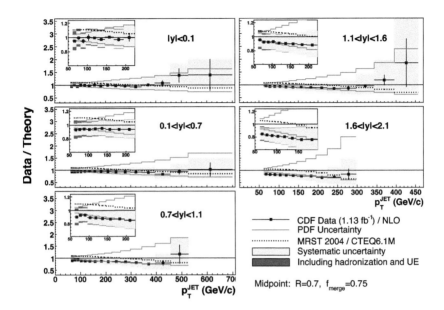

Fig. 8.13 The inclusive jet cross-section from CDF in Run II, for several rapidity intervals using the Midpoint cone algorithm, compared on a linear scale to NLO theoretical predictions using CTEQ6.1 PDFs. Reprinted with permission from Ref. [61].

systematic errors are less than PDF uncertainties; thus, the use of this data has proven to be useful in global PDF fits.

8.3.3 Jet algorithms and data

The experimental jet cross-sections have also been measured in CDF using the k_T algorithm with the same data sample and kinematic region as for the Midpoint analysis [119]. See Fig. 8.14 for a comparison of the ratios of the two algorithms in data and in theory. Similarly good agreement to that obtained for the Midpoint cone algorithm is observed. This is an important observation. The two different jet algorithms have different strengths and weaknesses and it is useful to have data comparisons to both. However, this is one of the few cross-sections at the TEVATRON for which this comparison was performed. As will be observed at the LHC, the use of the anti-k_T cross-section alone has become fairly universal, and there are few comparisons to other jet algorithms. The former point is good, as the anti-k_T jet algorithm is arguably the best on the market. The latter is bad, as different jet algorithms (as well as different jet sizes) can spotlight different aspects of the underlying physics.

It was noted in Section 2.1.6 that on general principles, for NLO parton level predictions, the cone jet cross-section is larger than the k_T jet cross-section when $R_{cone} = D$. At the hadron level, this is no longer true; the cone jet loses energy by the

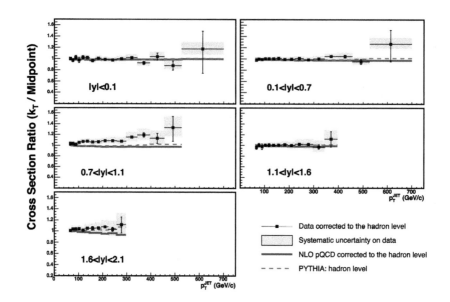

Fig. 8.14 The ratios of the inclusive jet cross-sections measured with the k_T algorithm (with $D = 0.7$) to those measured with the Midpoint algorithm (with $R = 0.7$) from CDF in Run II, for several rapidity intervals, with comparisons to the predictions of a NLO fixed-order QCD calculation and from Pythia. The results are from Ref. [61].

splash-out effect while the k_T algorithm has a tendency to **vacuum up** contributions from the underlying event. This will be corrected at least partially by the hadron to parton level corrections for each algorithm.

A particular complexity with the cone algorithm occurs when two jets overlap; a decision must be made whether to merge the two jets into one, or to separate them. This is an experimental decision; in CDF, the two overlapping jets are merged when more than 75% of the smaller jet energy overlaps with the larger jet. When the overlap is less, the towers are assigned to the nearest jet. DØ uses a criterion of a 50% fraction. NLO theory is agnostic on the subject as there is no overlap between the two partons that can comprise a jet. This point has become moot at the LHC in that cone algorithms are rarely used.

Another problem that can arise on the particle or calorimeter level for a cone jet, but not on the NLO parton level, occurs when particles or calorimeter towers remain unclustered in any jet, due to the strong attraction of a nearby larger jet peak that will attract away any trial jet cone placed at the location of the original particles/calorimeter towers. The result will be what [507] calls **dark towers**, *i.e.* clusters that have a transverse momentum large enough to be designated either a separate jet or to be included in an existing nearby jet, but which are not clustered into either. This is a feature endemic to any cone algorithm (including SISCone), but not to the k_T family of jet algorithms. Thus, this is another advantage for the use of the anti-k_T algorithm at the LHC.

The TeV4LHC workshop writeup [133] recommended the following solution to the problem of unclustered energy with cone jet algorithms. The standard midpoint algorithm should be applied to the list of calorimeter towers/particle/partons, including the full split/merge procedure. The resulting identified jets are then referred to as first pass jets and their towers/particles/partons are removed from the list. The same algorithm is then applied to the remaining unclustered energy and any jets that result are referred to as second pass jets. There are various possibilities for making use of the second pass jets. They can be kept as separate jets, in addition to the first pass jets, or they can be merged with the nearest first pass jets. The simplest solution, until further study, is to keep the second pass jets as separate jets.

It was originally thought that with the addition of a midpoint seed, the value of R_{sep} used with the NLO theory could be returned to its **natural** value of 2.0 (*cf.* Section 2.1.6). Now it is realized that the effects of parton showering/hadronization result in the midpoint solution virtually always being lost. Thus, a value of R_{sep} of 1.3 (for split/merge fraction f=0.75) is required for the NLO jet algorithm to best model the experimental one. The inclusive jet theory cross-section with $R_{sep} = 1.3$ is approximately $3 - 5\%$ smaller than with $R_{sep} = 2.0$, decreasing slowly with the jet transverse momentum.

8.3.4   Inclusive jet production at the TEVATRON and global PDF fits

Inclusive jet production receives contributions from gg, gq, and $qq(q\bar{q})$ initial states as shown in Fig. 4.1. Thus, in principle, this process is sensitive to the nature of all the PDFs. The experimental precision of the measurement, along with the remaining theoretical uncertainties, means that the cross-sections do not serve as a meaningful constraint on the quark or antiquark distributions. However, they do serve as an important source of information on the gluon distribution, especially at high x. The addition of the jet data from CDF and DØ resulted in a larger gluon distribution at high x than present in PDFs determined without the TEVATRON jet data. The influence of the high E_T Run I jet cross-section on the high x gluon is evident. There is always the danger of sweeping new physics under the rug of PDF uncertainties. Thus, it is important to measure the inclusive jet cross-section over as wide a kinematic range as possible, as was done by DØ in Run I [103] and by CDF [61] and DØ [86] in Run II. The generic expectation is that most signals of new physics would tend to be central while a PDF explanation should be universal, *i.e.* fit the data in all regions.

As inclusive jet production probes high scales, it serves as a useful observable to search for the presence of quark compositeness. Fig. 8.15 compares the DØ jet cross-sections measured in Run I to the NLO QCD predictions using the CTEQ6.1 PDFs, along with the cross-sections for jet production including a four-Fermi contact interaction (as discussed in Section 4.1). The mass scale of the contact interaction, Λ (*cf.* Eq. (4.18)), is probed at three values (1.6, 2.0 and 2.4 TeV), assuming constructive interference [867]. The cross-section is plotted as a ratio to the pure QCD prediction. The effect of the contact term is limited to the central rapidity regions (with of course

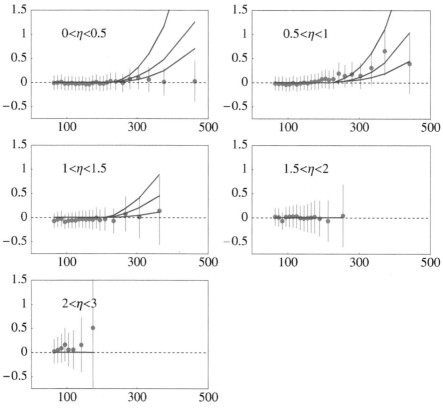

Fig. 8.15 Comparison of the DØ inclusive jet data from Run I to the pure NLO QCD prediction using the CTEQ6.1 PDFs, and to the addition of a contact interaction term with Λ values of 1.6, 2.0 and 2.4 TeV (solid curves, from top to bottom). Reprinted with permission from Ref. [867].

the size of the effect decreasing with increasing mass scale). This DØ data was used in the determination of the CTEQ6.1 PDFs. If there were a contact interaction, then the PDFs would need to be refit, comparing the data to the theory with compositeness included.

8.4 Inclusive photon and diphoton production

Measurements of single and double photon production also serve as precision tests of perturbative QCD, while avoiding the subtleties of jet definition, and were extensively studied at the TEVATRON [63, 67, 82, 90, 91, 93, 102, 110, 120]. Due to the presence of the electromagnetic coupling α, and the more limited number of subprocesses, the rate is suppressed with respect to jet production. In fact, photon measurements suffer from backgrounds due to the rare jets in which a large fraction of the momentum of the jet is taken by one or more π^0 (or η) mesons. Each meson decays into two photons that typically can not be resolved due to the finite granularity of the calorimeter. Such

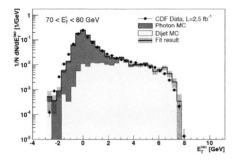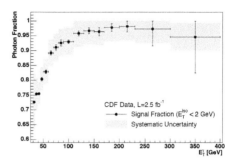

Fig. 8.16 (left) The energy distribution in an isolation cone about the photon direction. Shown are the contributions from true photons and from backgrounds. (right) The resultant photon signal fraction as a function of photon transverse energy. Reprinted with permission from Ref. [58].

events fall on the tail of the (rapidly falling) jet fragmentation function, but the large rate of jet production means that the background can not be ignored.

To reduce the jet backgrounds, the photons are typically required to be isolated, loosely at the trigger level, and more tightly off-line, similar to what is done for electrons. For example, in the CDF measurements in Run II, a requirement is made that the additional energy in a cone of radius $R = 0.4$ about the photon direction is less than 2 GeV, once the pileup energy from additional minimum bias events has been subtracted. An isolation requirement reduces not only jet backgrounds, but also contributions from photon fragmentation functions. This isolation is tighter than the typical isolation cuts applied at the LHC.

An example of a photon isolation distribution is shown in Fig. 8.16 (left) [58]. The negative energy tail is a result of the pileup subtraction. The photon fraction in each kinematic bin can be determined using templates of the isolation energy for photon candidates and backgrounds for each kinematic bin. The resulting true photon fraction is shown in Fig. 8.16 (right). The photon fraction of the sample rises rapidly towards one as the photon candidate transverse momentum increases, as expected, since the fixed isolation cut requires the fraction z of the jet momentum taken by the leading π^0 also increases. This is a general rule-of-thumb for both the TEVATRON and the LHC: an isolated *photon-like* object is almost always a real photon.

The resulting cross-section is shown in Fig. 8.17 [58]. Good agreement is observed with the NLO prediction from JETPHOX, except perhaps at low E_T, where the data is higher than theory. This has been observed in several other TEVATRON photon measurements, and has been attributed to the effects of soft gluon radiation. However, as will be seen in the next chapter, this has not been seen for similar measurements at the LHC. Note that the cone isolation cut greatly suppresses the fragmentation contribution to photon production, but does not explicitly remove it.

The photon + jet cross-section is interesting in its own right and, as discussed already in Section 4.2, as an input to PDF fits. It can also be used for a calibration of the jet energy scale. The electron energy scale is known very precisely from mea-

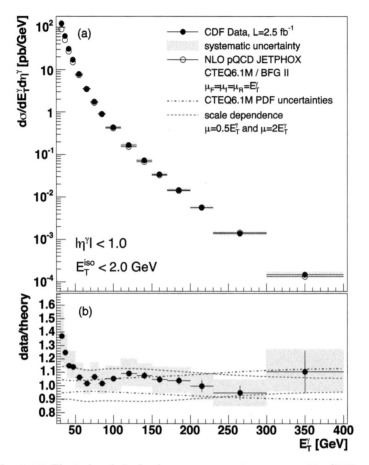

Fig. 8.17 The isolated single photon cross-section measured in CDF compared to NLO QCD predictions from JETPHOX [348]. Reprinted with permission from Ref. [58].

surements of $Z \to e^+ e^-$ [95]. This energy scale can be transferred to photons, taking into account the differences in electromagnetic showers between photons and electrons, and is limited primarily by an incomplete knowledge of the material in front of the calorimeter. In DØ, for example, a tight selection is made on photon candidates to reject as much background from jet fragmentation as possible [98]. A requirement is made that no other jets be present in the event, that no additional minimum bias events (pileup) be present, and that the photon and the jet be back-to-back ($\Delta\phi > 2.9$ radians). The jet energy response is then corrected by looking at the balance in p_T between the photon and the jet using the missing transverse energy projection fraction method [98]. Corrections are also made for the presence of underlying event energy, and for gluon radiation outside of the jet cone. The resulting uncertainty on the jet

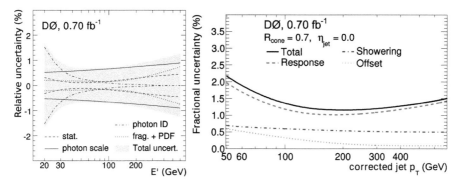

Fig. 8.18 (left) The sources for uncertainty of the jet p_T response in the central rapidity region. Here, the variable $E = p_{T,\gamma}\cosh\eta_{jet}$ is used on the horizontal axis, as the resolution is better than for jet energy. (right) The resultant jet energy scale uncertainty, as a function of jet transverse momentum, for the central rapidity region. Reprinted with permission from Ref. [98].

p_T response is shown in Fig. 8.18 (left) and is dominated by the photon energy scale error. The method is restricted to the central rapidity region and is limited by the statistics of the photon + jet sample to jets below 350 GeV. The energy scale can be transferred to non-central rapidities using transverse momentum balancing in dijet events. The energy scale for jets above that transverse momentum must be extrapolated. The final fractional uncertainty on the jet energy scale is shown in Fig. 8.18 (right) for the central rapidity region.

Diphoton production has also been extensively studied at the TEVATRON. In CDF, for example, a study was carried out using the full Run II data sample, using the same cuts and analysis techniques as for single photon production [67].[8] In general, the phenomenology is *richer* than for single photons. For example, the transverse momentum cuts on the photons sculpt the diphoton p_T spectrum, giving rise to a bump in the spectrum often referred to as the *Guillet shoulder* [254, 255]. This can be seen in Fig. 8.19, which shows a CDF measurement of the diphoton p_T distribution [67], with the shoulder around 35 GeV. This feature is due to configurations in which the photon pair is accompanied by significant additional hard radiation.

As pointed out in Section 4.2.3, in particular in Table 4.1, this distribution is difficult to describe with the first few orders of a fixed-order calculation. An NLO calculation of the total diphoton cross-section, corresponding to the MCFM prediction in Fig. 8.19, gives the first non-trivial prediction. Since the diphoton pair recoils against a single parton, this calculation is insufficient to capture the shoulder and the description of the data is relatively poor. In the NNLO calculation, which accesses configurations with two recoiling partons for the first time, the shoulder begins to be reproduced and the theoretical description is much improved. The SHERPA prediction, which here

---

[8]Diphoton production has backgrounds from either one or both photons being faked by jets. Again, as the E_T of the photons increases, the purity fraction increases.

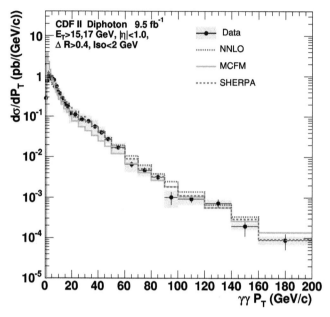

Fig. 8.19 The diphoton p_T distribution compared to theoretical predictions that are accurate for the total cross-section at NNLO QCD [340], at NLO QCD using MCFM [311], and to SHERPA [583]. Reprinted with permission from Ref. [67].

contains LO matrix elements for multiple parton emission, is similarly well-suited to describe the data. The same events are also responsible for an excess of events, with respect to pure NLO predictions, at low diphoton mass, and at low $\Delta\phi$.

8.5 Vector boson plus jet physics

As discussed in Section 4.3, measurements of vector boson production in association with jets serve as a precision test of perturbative QCD over a wide dynamic range, in the presence of a large mass scale $(m_{W/Z})$. The process also serves as a background to $t\bar{t}$ production (for the $W \to l\nu$ + jets final state), as well as to possible new physics signals. Typically, a smaller jet size is preferred for final states that may be complicated by the presence of a large number of jets, so in most cases at the TEVATRON a jet size of $R = 0.4$ was used for measurements of this process. This process was well-studied at the TEVATRON by both CDF and DØ [59, 60, 85, 88, 96, 99].

For example in Fig. 8.20 (left), the W boson p_T is shown for several different jet multiplicities and compared to predictions at LO and NLO in QCD. Good agreement is observed for the BLACKHAT+SHERPA NLO prediction. Here, the jet transverse momentum cut is 20 GeV.[9] Below a W boson p_T of 20 GeV the W boson recoils against more than one hard jet, and the effects of soft gluon emission also become important.

The H_T distribution (the sum of the transverse momenta of all jets and leptons (including neutrinos) is plotted in Fig. 8.20 (right) as a function of the jet multiplicity. The H_T variable is particularly sensitive to higher order QCD effects, but is also often chosen as a variable in which to look for the presence of BSM physics. There is a significant variation observed in the level of agreement between the data and the predictions evident in the figure, possibly allowing for the possibility of improvements to these predictions. Note in particular the tendency of the NLO BLACKHAT+SHERPA predictions to lie below the data at high H_T for the ≥ 1 jet bin. We will return to this observation in Chapter 9.

Note that the DØ measurements were carried out with the midpoint Cone jet algorithm, but the results were corrected to the (infra-red safe) SISCone jet algorithm using SHERPA. For the leading jet transverse momentum, this correction is very small as the two algorithms are very close in behaviour.

A measurement of Z+jets production was carried out at CDF using the full data sample [76]. The jet multiplicity distribution is shown in Fig. 8.21 for data compared to BLACKHAT+SHERPA (left), and ALPGEN +PYTHIA, POWHEG +PYTHIA and LoopSim+MCFM (right). Recall that this last prediction is obtained by ameliorating an exact NLO calculation with an approximate treatment of NNLO effects, according to the procedure described in Section 3.4.2. The midpoint jet algorithm with $R = 0.7$ was used (in contrast to the other V+jets measurements which used $R = 0.4$). It is noticeable that the measured cross-section for $Z+ \geq 3$ jets is approximately 30% larger than the BLACKHAT+SHERPA prediction, contrary to what has been observed at the LHC using the anti-k_T jet algorithm with $R = 0.4$[10] and with $W+ \geq 3$ jets (using $R = 0.5$) at the TEVATRON [96]. The Monte Carlo predictions are in better agreement with the high jet multiplicity data, albeit with larger uncertainties. Looking back at Fig. 4.22, the cross-section predictions for the SISCone jet algorithm (for ≥ 4 jets, but the situation is roughly similar for ≥ 3 jets) tend to peak at smaller scales than do the predictions for the anti-k_T jet algorithm. The differences increase as the jet multiplicity and the jet size increases. The peak cross-section values for the TEVATRON measurement are actually quite similar for the two algorithms, as discussed in Section 4.3 [227]. The $Z+ \geq 3$ jet cross-section was not determined using the anti-k_T jet algorithm in the CDF measurement, but a comparison was carried out using simulated data. The resulting cross-sections, for the two jet algorithms, are much closer than implied by the BLACKHAT+SHERPA predictions using a scale of $H_T/2$.

8.6 *t̄t̄* production at the TEVATRON

The greatest discovery at the TEVATRON was that of the top quark [81, 112]. The production mechanism was through $t\bar{t}$ pair production, dominated by a $q\bar{q}$ initial state. The top quark decays essentially 100% into a W and a b quark; thus the final states

---

[9]As we will see in Chapter LHC, the jet cuts at the LHC are typically higher than 20 GeV, due to increased backgrounds from pileup and the underlying event.

[10]Although the cross-section for $Z+ \geq 4$ jets has been calculated for the LHC, it has not been calculated for the TEVATRON. Thus, there are no BLACKHAT+SHERPA predictions for that final state.

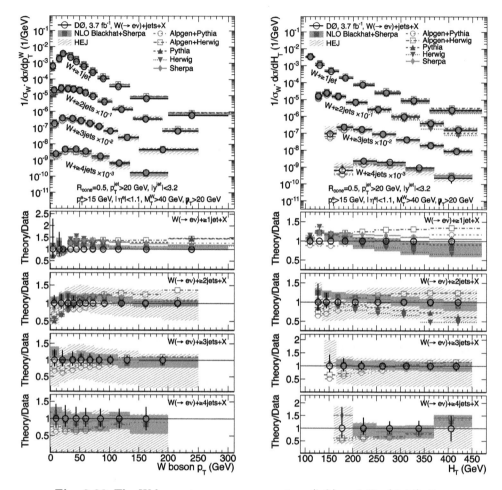

Fig. 8.20 The W boson transverse momentum (left) and H_T (right) distributions in inclusive $W + n$-jet events, for $1 \leq n \leq 4$, measured by DØ. Reprinted with permission from Ref. [99]. The measurements are compared to several theoretical predictions.

being investigated depend on the decays of the two W's. The most useful (combination of rate and background) final state occurs when one of the W's decays into a lepton and neutrino and the other decays into two quarks. Thus, the final state consists of a lepton, missing transverse energy and of the order of four jets. The number of jets may be less than 4 due to one or more of the jets not satisfying the kinematic cuts, or more than 4 due to additional jets being created by gluon radiation off the initial or final state. Because of the relatively large number of jets, a smaller cone size ($R = 0.4$) has been used for jet reconstruction, with the CDF analyses in Run 2 using the JetClu cone algorithm and the DØ analyses the Midpoint cone algorithm. No top analysis has been performed using the k_T jet algorithm. There is a sizeable background for

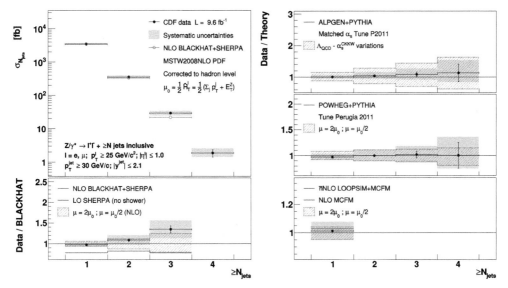

Fig. 8.21 The jet multiplicity distribution for data compared to BLACKHAT+SHERPA (left), and ALPGEN +PYTHIA, POWHEG +PYTHIA and LoopSim+MCFM (right). Reprinted with permission from Ref. [76].

this final state through QCD production of $W+$ jets. Two of the jets in $t\bar{t}$ events are created by b quarks; thus there is the additional possibility of an improvement of signal purity by the requirement of one or two b-tags.

Top pair events also have a harder H_T (sum of the transverse energies of all jets, leptons and missing transverse energy in the event) than does the W + jets background. This is due to the harder spectrum of the jets from $t\bar{t}$ decays (compared to the background), resulting from the large mass scales inherent in top production. A requirement of large H_T thus improves the $t\bar{t}$ purity.

The jet multiplicity distribution for the top candidate sample from CDF in Run II is shown in Fig. 8.22 for the case of one of the jets being tagged as a b-jet (left) and two of the jets being tagged (right) [118]. The requirement of one or more b-tags greatly reduces the $W+$ jets background in the 3 and 4 jet bins, albeit with a reduction in the number of events due to the tagging efficiency. The b-tagging efficiency at the TEVATRON was typically of the order of 40%. As will be seen in the next chapter, the b-tagging efficiency is higher at the LHC, primarily due to a larger rapidity coverage for the silicon detectors. The high jet multiplicity double-b-tagged events are almost exclusively $t\bar{t}$ events.

The top pair cross-section has been measured in a variety of final states (depending on the W boson decays) with a variety of techniques. A compilation of measurement results from CDF and DØ is shown in Fig. 8.23 [70]. The cross-sections from the two experiments agree with each other and with the theoretical predictions.

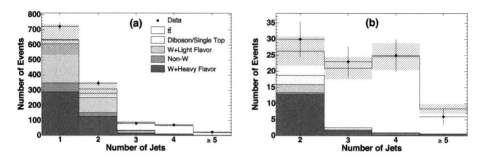

Fig. 8.22 The expected number of $W+$ jets events that are b-jet tagged (left) and double-b-tagged (right), indicated by source. Reprinted with permission from Ref. [118].

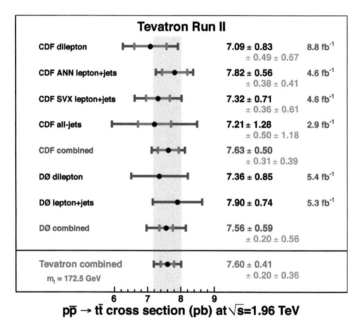

Fig. 8.23 A compilation of cross-section measurements for $t\bar{t}$ final states from CDF and DØ at the TEVATRON. Reprinted with permission from Ref. [70].

8.6.1 Measuring the top mass

The lepton + jets final state (with one or more b-tags) is also the most useful one for the determination of the top mass (although there is top mass information in all of the final states). This final state optimizes the most information as to the top mass, with the least background from non-top processes. For example, the final state with two high–p_T leptons, large missing transverse energy, and two (b) jets has the least background, but the presence of two neutrinos makes the reconstruction of the final-state kinematics difficult. For the final state with 6 jets (2 b-jets), complete kinematic information is available; however, the QCD background from six-jet production is very high.

The high statistics for top production accumulated at the TEVATRON has allowed cross-checks of the calibration techniques for the top mass reconstruction, as for example the calibration of the light quark jet response using the decay of the hadronically decaying W boson (in $t\bar{t}$ events) into light quarks.

There are two main techniques at the TEVATRON for top mass determination: the template method and the matrix element method. In the template method, the information from the lepton + jets final state is input to a χ^2 determination, where the reconstructed top mass is a free parameter. The χ^2 is minimized for each possible way of assigning the 4-vector information for each of the four leading jets to the top decay products (if any jets are b-tagged, then they are required to be from the top decay and not from the W boson decay). The χ^2 expression has terms for the uncertainty on the measurements of the 4-vectors of the decay products. There are two possible solutions for the longitudinal momentum of the neutrino, so both are used. The minimum chi-square solution (for jet assignment, neutrino solution and m_{top}) is chosen for each event.

The matrix element method has the ability to use theoretical information from the matrix element, retaining all of the hard scattering correlations, in the top mass determination. A likelihood is determined for each event that the theoretical model from the matrix element describes the kinematics of the event. The technique is very CPU-intensive, and until recently was restricted to the use of leading order matrix elements. In [318], the matrix element method was extended to allow the calculation of next-to-leading order weights on an event-by-event basis.

In either method, the determination of the top mass is obtained by comparing data with Monte Carlo predictions. Thus, the top mass can not be strictly identified with any precise theoretical definition, such as the $\bar{M}S$ mass or the pole mass discussed in Section 4.5. However, the differences should be smaller than the current uncertainties on the top mass from the TEVATRON measurements, but may be an issue for future more precise determinations at the LHC.

A compilation of top mass determinations from the TEVATRON, with the global mass fit dominated by lepton+jets final states, is shown in Fig. 8.24 [599]. The top mass has been determined from the TEVATRON measurements to 0.64 GeV, a precision of about 0.4%. All of the individual determinations of the top mass are consistent with each other. These determinations of the top quark mass, together with the measured W-boson mass, provide an indirect constraint on the mass of the Higgs boson. This is shown in Fig. 8.25 [599].

The precision of the top mass determination at TEVATRON has reached the point where some of the systematics due to QCD effects must be considered with greater care. One of the potentially important systematics is that due to the effects of initial-state radiation. Jets created by initial state radiation may replace one or more of the jets from the top quark decays, affecting the reconstructed top mass. In the past, the initial state radiation (ISR) systematics was determined by turning the radiation off/on, leading to a relatively large impact. A more sophisticated and more correct

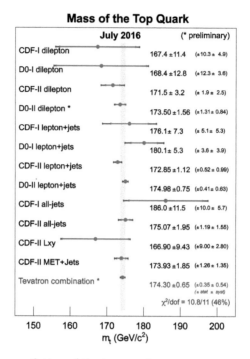

Fig. 8.24 A compilation of the top quark mass measurements from CDF and DØ, from arXiv:1608.01881.

treatment was adopted in Run II, where the tunings for the parton shower Monte Carlos were modified leading to more/less initial-state radiation, in keeping with the uncertainties associated with Drell-Yan measurements as discussed in Section 8.5. The resultant $t\bar{t}$ pair transverse momentum distributions are shown above in Fig. 8.26. The changes to the $t\bar{t}$ transverse momentum distribution created by the tunes are relatively modest, as is the resultant systematic error on the top mass determination.

Note that the peak of the $t\bar{t}$ transverse momentum spectrum is somewhat larger than that for Z production at the TEVATRON, due to the larger mass of the $t\bar{t}$ system. As both are produced primarily by $q\bar{q}$ initial states, the difference is not as large as it is at the LHC, where the primary $t\bar{t}$ production mechanism is through gg fusion.

It is also interesting to look at the mass distribution of the $t\bar{t}$ system, as new physics (such as a Z' [622]) might couple preferentially to top quarks. Such a comparison for Run II is shown in Fig. 8.27 without any signs for a high mass resonance [69]. The simulation of the Standard Model $t\bar{t}$ signal for this analysis was carried out with POWHEG using MSTW2008 NLO PDFs. Often, in previous TEVATRON studies, a LO Monte Carlo such as PYTHIA was used, along with a LO PDF. Note that if we compare the predictions for the $t\bar{t}$ mass distribution at LO and NLO, we see that the NLO cross-section is substantially less than the LO one at high mass. Further investigation shows that the decrease of NLO compared to LO at high mass is found only in the $q\bar{q}$ initial state and not in the gg initial state. In fact, at the TEVATRON, the ratio of NLO to

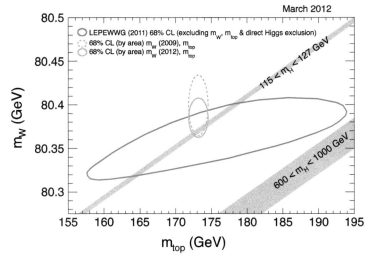

Fig. 8.25 Implications of the measured top quark and W-boson masses for the mass of the Higgs boson, from ref. [599].

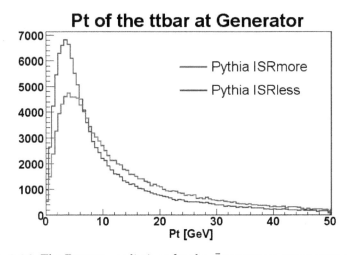

Fig. 8.26 The PYTHIA predictions for the $t\bar{t}$ transverse momentum using the **Plus/Minus** tunes. We thank Prof. Un-Ki Yang for this figure.

LO for gg initial states grows dramatically with increasing top pair invariant mass. This effect is largely due to the increase in the gluon distribution when going from CTEQ6L1 in the LO calculation to CTEQ6M at NLO. For instance, at $x \sim 0.4$ (and hence an invariant mass of about 800 GeV) the gluon distribution is about a factor two larger in CTEQ6M than in CTEQ6L1, giving a factor four increase in the cross-section. Conversely, the quark distribution is slightly decreased at such large x. If NLO PDFs were used for both the numerator (NLO) and denominator (LO), this dramatic effect

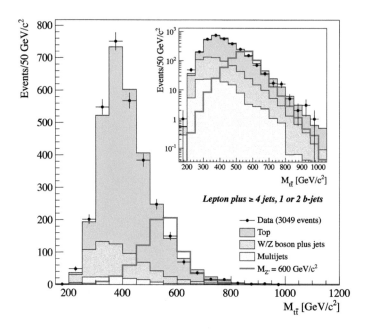

Fig. 8.27 The $t\bar{t}$ mass distribution measured by CDF in Run II. Reprinted with permission from Ref. [69].

would not exist. This is an example of the danger if LO PDFs are used, especially in discovery physics regions.

In any case, the absolute contribution of the $t\bar{t}$ cross at high masses from gg initial states at the TEVATRON is small, due to the rapidly falling gluon distribution at high x. The dominant $t\bar{t}$ production mechanism at the LHC in all mass regions is through gg fusion.

8.6.2 The forward-backward asymmetry

Given the symmetric initial and final states for the reaction $p\bar{p} \to t\bar{t}$, it would be natural to expect that the angular distribution of the t and the $\bar{t}$ would be symmetric around rapidity $y = 0$. This is certainly true in a leading order QCD matrix element calculation. However, as discussed at length in Section 4.5.3, this is no longer true at NLO. The net production asymmetry at NLO in QCD, including EW effects which increase the asymmetry by a factor of 1.26 [702, 747], is of the order of 7% at the TEVATRON. The larger values for the asymmetry (compared to theory) measured by CDF [68, 71], and DØ [83, 100], have resulted in a number of new calculations of both SM effects as well as possible BSM contributions to the (larger than expected) asymmetry. Given the large mass of the top quark, it seems a likely place to look for the presence of new physics. The recently calculated NNLO QCD asymmetry increases the inclusive asymmetry prediction by a factor of 1.3 [428], in better agreement with

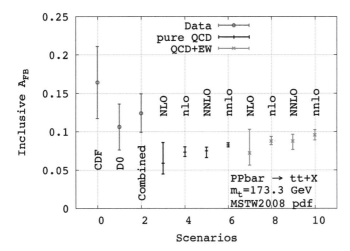

Fig. 8.28 The inclusive $t\bar{t}$ forward-backward asymmetry at the TEVATRON is shown for CDF and DØ, compared to theoretical predictions (QCD and QCD+EW) using two different definitions for the ratio, as discussed in the text. Reprinted with permission from Ref. [428].

the measurements. The inclusive asymmetry values from experiment and the NNLO calculation are shown in Fig. 8.28. There is an ambiguity in the manner in which the asymmetry can be calculated, *i.e.* whether the exact results are used for the numerator and denominator, or whether the asymmetry ratio is expanded in terms of powers of $\alpha_s(m_Z)$. In the figure, the exact results at each order are shown in capital letters, while the small letters refer to the expanded version of the calculation. The first four theory predictions in the figure are QCD-only while the second four are QCD+EW. The expanded version results in a larger value for the asymmetry with smaller errors. The authors of [428] prefer the exact result, so that is used in the figures below.

The asymmetry typically is measured as a function of the three dynamical variables: $\Delta Y_{t\bar{t}}$, $m_{t\bar{t}}$, and $p_{T,t\bar{t}}$. The expectation is that at NLO in QCD the asymmetry should increase linearly with the variables $\Delta Y_{t\bar{t}}$ and $m_{t\bar{t}}$,[11] while it should switch signs at non-zero p_T (the ISR/FSR radiation is the only contribution to the asymmetry away from $p_{T,t\bar{t}} = 0$).

At the TEVATRON, the asymmetry is not a subtle effect [68]. It is observable even at the raw data level, as shown in Fig. 8.29 (left). The effects of the background and of the non-perturbative effects are to dilute the net asymmetry, so after corrections for these effects, the asymmetry increases, as shown in Fig. 8.29 (right). Note that the parton level asymmetry is larger than the background-subtracted asymmetry, which is larger than the asymmetry at the reconstructed level.)

Comparisons of the parton-level NLO and NNLO QCD (no EW) asymmetry predictions to the CDF and DØ measurements for $\Delta Y_{t\bar{t}}$ (left) and $m_{t\bar{t}}$ (right) are shown

---

[11]The asymmetry is primarily proportional to β, the velocity of the top (or anti-top) in the $t\bar{t}$ centre-of-mass frame [141].

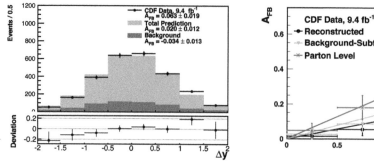

Fig. 8.29 (left) The Δy distribution at the reconstruction level, before any background subtraction or unfolding, compared to predictions from signal and background models. (right) A comparison of the asymmetry as a function of the rapidity separation between the t and $\bar{t}$ at the reconstructed, background-subtracted and parton levels. Reprinted with permission from Ref. [68].

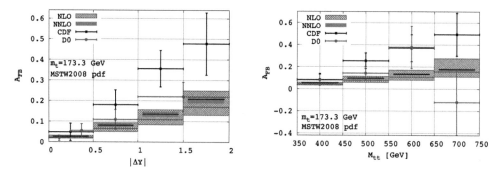

Fig. 8.30 A comparison of the pure QCD (NLO and NNLO) $t\bar{t}$ asymmetry predictions to the data from CDF and DØ, as a function of (left) the rapidity separation between the t and the $\bar{t}$, and (right) the $t\bar{t}$ mass distribution. Reprinted with permission from Ref. [428].

in Fig. 8.30, from [428]. The CDF results tend to be higher than those from DØ, but both are in statistical agreement. Note that many physics models that could increase the asymmetry at high $m_{t\bar{t}}$ should also cause a change in the observed $t\bar{t}$ mass spectrum. However, as shown in Fig. 8.27, no such deviation is observed, and the mass distribution agrees with NLO predictions.

Predictions at NLO and NNLO (pure) QCD for the asymmetry as a function of the transverse momentum of the $t\bar{t}$ pair are shown in Fig. 8.31 (left). The NNLO corrections greatly decrease the size of the (negative) asymmetry for non-zero $p_T^{t\bar{t}}$ values, leading to the net increase in the inclusive asymmetry discussed previously. The CDF data are shown in Fig. 8.31 (right), compared to predictions from POWHEG and

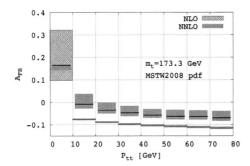 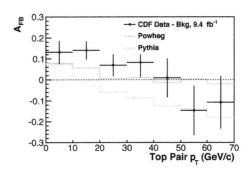

Fig. 8.31 (left) Predictions for the QCD asymmetry as a function of the *t̄t* transverse momentum at NLO and NNLO. Reprinted with permission from Ref. [428]; (right) the CDF *t̄t* asymmetry distribution as a function of the transverse momentum of the *t̄t* pair, compared to predictions from POWHEG and PYTHIA. Reprinted with permission from Ref. [68].

PYTHIA. The asymmetry does decrease with increasing transverse momentum, as do the predictions, but there is not good agreement. It is interesting that the slope of the PYTHIA prediction seems similar to that observed in the data. It might at first glance seem surprising that PYTHIA, using LO QCD matrix elements, predicts any inclusive asymmetry at all. The parton shower will create additional jets, but as they result from the shower, and not from a matrix element, there is no net ISR/FSR interference as discussed above. However, as discussed in Section 4.5.3, there is a greater likelihood for gluons to be emitted if the colour charges are accelerated. As the production of the jet is more likely if the top (*t̄*) is produced in the backward (forward) direction, the net asymmetry for the *t̄tj* final state is negative. Also, the result of the gluon emission is that the *t̄t* system moves from $p_T = 0$ to a higher transverse momentum, leaving the events where no radiation has taken place (near or at $p_T = 0$), with a positive asymmetry, since the sum of the two asymmetries for the leading order prediction must be zero. Detailed investigations of colour coherence effects and of the treatment of recoils in the parton showering have shown that the asymmetry as a function of the transverse momentum of the *t̄t* pair can in addition be substantially affected by the details of the treatment of those effects [627, 859].

Unfortunately, the study of the *t̄t* asymmetry at the TEVATRON has ended, and similar investigations at the LHC are much more difficult due to the dilution of the effect (the *gg* initial state dominates, and the initial hadrons are both protons). There is no evidence for the presence of any BSM physics at either the TEVATRON or the LHC and improved calculations of QCD and EW effects have lead to better improvements of the data with the theory. A complete understanding, though, may require the combination of investigations at both the NNLO and the parton shower levels. This observable seemed to be a clear sign of possible new physics, but instead showed the complexity and subtlety of QCD+EW predictions in a hadron-hadron environment. The Standard Model once again appears to rule.

8.6.3 Single top production

Single top production was first observed at the TEVATRON [73]. As discussed in Section 4.6, the measurement of single top production provides a sensitivity to the value of the CKM matrix element V_{tb}, and thus a sensitivity to a variety of possible new physics [149, 357, 870]. There are three production channels, s-channel, t-channel, and tW production, with the first two dominating at the TEVATRON. Closer inspection reveals that the latter channel tW-associated production is the part of a larger set of Feynman diagrams at leading order, which lead to a $b\bar{b}W^+W^-$ final state. Clearly, one could try to differentiate three kinematic regions, by allowing zero, one, or two top quarks becoming resonant, with a corresponding peak in the bW mass distribution – their decay products. In addition to this complication, there are also ambiguities in the definition of s-channel and t-channel production; indeed as discussed in Section 4.6, there is no distinction between them at higher orders in QCD. However, t-channel events tend to produce light flavour jets at high rapidity, while s-channel events are more likely to contain two b-jets in the central rapidity region. The jet–η distribution in t-channel production is strongly peaked in the forward direction due to the larger momentum that the incident valence quark participating in the t-channel process has compared to the incident gluon (the jet η direction for the spectator jet accompanying single $\bar{t}$ production would be peaked in the backward direction). The order α_s corrections shift the spectator jet to even more forward η, due to the impact of gluon radiation.

There are significant backgrounds to both channels, from W/Z+jets, dibosons, $t\bar{t}$, and even Higgs boson production. The W boson + jets normalization is taken from data (in regions which are not likely to contain single-top events; the shape is taken from the NLO prediction), while the cross-sections for the other background processes are taken from their theoretical predictions.

There is no kinematic variable that allows for a distinct separation between single top production and its backgrounds. Thus, multivariate discriminants are used to optimize the separation. The distribution of the discriminant is shown in Fig. 8.32 (left), for the combined CDF +DØ combination, with a t-channel signal being evident at large negative values of the discriminant and an s-channel signal being evident at large positive values [73]. The measured single top cross sections for CDF, DØ and for the combination are shown in Fig. 8.32 (right). The results are in good agreement with Standard Model predictions [73], and allow a determination of $|V_{tb}| = 1.02^{+0.06}_{-0.05}$. At 95% CL, $|V_{tb}| > 0.92$.

8.7 Higgs boson searches

The most challenging analysis of the TEVATRON involved the search for the Higgs boson. The relatively low signal cross section and the large backgrounds required (1) a large integrated luminosity, (2) the use of multivariate analysis techniques, (3) the combination of a large number of Higgs boson production and decay channels, and (4) the combination of results from the CDF and DØ experiments. The final results [66] involved the complete TEVATRON data sample (approximately 10 fb^{-1}), the gg fusion, associated production (VH), vector boson fusion and $t\bar{t}H$ production channels, and the $b\bar{b}$, W^+W^-, $\tau^+\tau^-$, $\gamma\gamma$ and ZZ decay modes. With the multivariate analyses, it

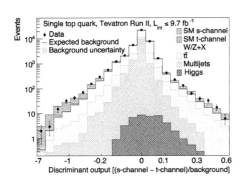
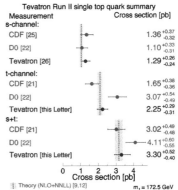

Fig. 8.32 (left) The distribution of the discriminant used for the separation of single top production and its backgrounds. The black solid line shows the total background. (right) The measured single top production cross-sections from CDF and DØ, and the TEVATRON combination, compared to a prediction at NLO+NNLL. Reprinted with permission from Ref. [73].

was important to separate the event selections into orthogonal search regions, so that the multivariate analyses could be optimized for each region. For example, for the $H \to W^+W^-$ decay mode, where both W bosons decay leptonically, it was useful for the searches to be broken into 0,1 and ≥ 2 jet final states.

One of the backgrounds for the $VH(\to b\bar{b})$ searches is $VZ(\to b\bar{b})$ production; however, this also serves as a useful calibration tool. The background-subtracted distribution for the reconstructed dijet mass in VZ final states is shown in Fig. 8.33. The reconstructed cross section agrees well with the Standard Model prediction.

The best-fit cross Higgs boson signal cross section as a function of Higgs boson mass for the final TEVATRON result is shown in Fig. 8.34 (left). A broad excess in the range of 120-140 GeV can be observed; also shown are the expectations for the production of a Higgs boson with either the Standard Model cross section, or the Standard Model cross section multiplied by a factor of 1.5. This excess has a significance of 3.0 standard deviations and can be associated with the production of a Higgs boson with a cross section that is a factor $1.44^{+0.59}_{-0.54}$ times the SM prediction. The best-fit values for the cross sections from the four decay modes shown in Fig. 8.34 (right) are all consistent with each other, and with the SM predictions.

8.8 Summary

The TEVATRON was the first hadron-hadron machine in which *modern* techniques could be used for event reconstruction and analysis, and for comparison of data to theory. Jet algorithms were developed that allowed more precise theoretical comparisons, although in most cases not with the full all-orders infrared-safety desired. Some, but not all, measurements were presented at the hadron level, with complete information about the parton-to-hadron corrections. Results at the hadron level allow theorists

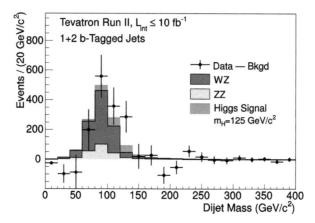

Fig. 8.33 The background-subtracted data for the reconstructed dijet mass for the combined CDF and DØ measurements of VZ production. Reprinted with permission from Ref. [66].

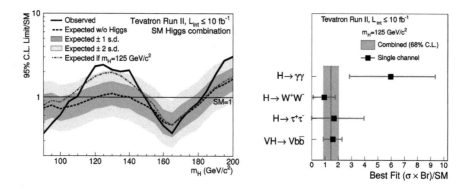

Fig. 8.34 (left)The best-fit signal cross sections, expressed as a ratio to the SM prediction, as a function of the Higgs boson mass, using the combined CDF and DØ data samples. (right) The best fit values ($\sigma \times$ Br) for the combined CDF and DØ Higgs boson search channels, for the $\gamma\gamma$, W^+W^-, $\tau^+\tau^-$ and $Vb\bar{b}$ final states. Reprinted with permission from Ref. [66].

(or other experimentalists) to better compare their predictions against experimental data. An alternative commonly used in the experimental community was to compare data and theory at some intermediate level of reconstruction, accessible only to the experimentalists who carried out the measurement. Many measurements were carried out in **fiducial** regions, allowing the comparison of data to theory without extrapolations which may have model-dependence. As will be seen in the next chapter, this extrapolation can still cause problems at the LHC.

Advances in theoretical techniques, and in computing power, allowed the calculation of processes with multi-jet final states at LO and NLO, often in the context of a parton shower Monte Carlo. With the new theoretical predictions, and the access to larger scales at the TEVATRON than previously accessible, the Standard Model was tested with great precision. Alas, although the TEVATRON was capable of discovering and of measuring the properties of the top quark, and of finding evidence for the existence of the Higgs boson, it completed its run without the discovery of new physics. Searches for new physics, and for further precision measurements of the standard model, had to await the LHC.

9
Data at the LHC

This chapter on LHC results presents the culmination of the theoretical techniques developed in the earlier chapters, along with the data analysis experiences from TEVATRON. Here, a wide range of exemplary data from Run I at 7 and 8 TeV will be discussed.[1] For the cross-sections discussed in this chapter, the data have been corrected for all experimental effects, so that effectively they are at the hadron level. Corrections for detector inefficiencies and resolution effects have been taken into account by an unfolding procedure. The theoretical predictions have been corrected for non-perturbative effects, either in a Monte Carlo framework, or by the addition of non-perturbative corrections to parton-level predictions.

Note that the latter approach is often ignored by experimenters at the LHC, outside of the Standard Model groups, even though in many cases it offers the highest precision comparison. This is especially true if the cross section has been calculated to NNLO. If the cross-section is suitably inclusive, as for example the transverse momentum distribution for the leading jet in Higgs+≥ 1 jet events, resummation effects should be small, and a comparison of the data to a NLO or NNLO fixed order prediction should certainly be carried out.[2]

Cross-sections for Standard Model processes at the LHC have been measured over 14 orders of magnitude, as shown in Fig. 9.1 for the ATLAS experiment. In general, SM predictions are in remarkable agreement with data. It is a hallmark of the abilities of the human mind that the technology developed and the calculations made in the past decades prove to be so amazingly powerful and accurate.

The only grain of salt in this success story is that up to now no clear sign of any physics Beyond the Standard Model (BSM) has shown up, despite a plethora of searches.[3] The higher energy (and higher integrated luminosity) for Run II holds the promise that a threshold for new physics may be reached. However, it is clear that the signatures for new physics may be subtle and a thorough understanding of pQCD is

---

[1] At the time of the writing of this book, first measurements at 13 TeV were available. However, comparisons have been limited to data from the 7 and 8 TeV running.

[2] For this example, the cross-section is not totally inclusive, in the sense that a p_T requirement has been imposed on the jets, typically 30 GeV. But, the resulting restriction on the phase space for gluon emission is minimal and the effects of resummation thus are suitably small [198].

[3] Unfortunately, the 750 GeV bump in the diphoton mass spectrum appears to have gone away in the most recent 13 TeV data.

The Black Book of Quantum Chromodynamics. John Campbell, Joey Huston and Frank Krauss, Oxford University Press. © John Campbell, Joey Huston, and Frank Krauss 2018, First published in paperback 2022. DOI: 10.1093/oso/9780192871961.001.0009

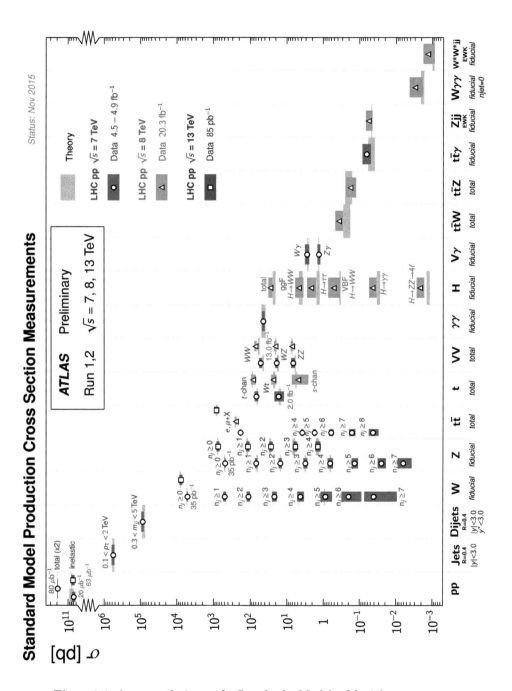

Fig. 9.1 A compilation of Standard Model fiducial cross-sections measured by the ATLAS experiment in Run I, taken from `atlas.web.cern.ch/Atlas/GROUPS/PHYSICS/CombinedSummaryPlots`. Reprinted with permission from CERN.

necessary for such discoveries to take place.

In this chapter, data for relatively inclusive strong interaction observables are presented, including total cross-sections, particle spectra in Minimum Bias events, and the Underlying Event, which constitutes an important and non–negligible nuisance for nearly every LHC measurement, *cf.* Section 9.1. This is followed in Section 9.2 by a discussion of pure QCD production of jets, ranging from inclusive jet spectra to multijet topologies, which also play a role in some of the searches for BSM physics. Next, a benchmark signature, namely Drell-Yan production, will be discussed in Section 9.3, including total and differential production rates of the heavy gauge bosons. Topologies of single electroweak gauge bosons accompanied with jets are covered in Section 9.3.4. This is followed by some data for di– or multi-boson production, including diphotons, in Section 9.4. These processes were an important background for the discovery of the Higgs boson, and, in general, are also important for BSM signatures involving leptons. One of the most dominant particle production processes at the LHC is related to the production of top-quarks, either in pairs or as single-top. Its large cross-sections render the LHC a prime laboratory for precision studies of this heaviest quark, which in turn also means that tops play an important role as background for nearly every important signal process at the LHC. Data for this class of processes, including a discussion of measurement of the top quark mass and of the QCD radiation pattern is presented in Section 9.5. Having discussed all relevant Standard Model candles, finally signals for and signatures involving precision studies of the Higgs boson will be presented, Section 9.6. One caveat is that unlike the TEVATRON or HERA, results are still coming out of the LHC at a furious rate. So the results shown in this chapter are a snapshot taken at the time of writing of this book and do not represent the full on-going LHC story.

Of course, the data presented here constitute only one aspect of the overall LHC programme, which also encompasses *B* physics and the study of highly energetic collisions of heavy ions. While these indeed are intriguing subjects, they somewhat fall outside the remit of this book and the authors apologize for this shortcoming.

9.1   Total cross-sections, minimum bias and the underlying event

The most inclusive observables in hadronic collisions are the total and elastic cross-sections, σ_{tot} and σ_{el}, the latter possibly given in the form of a distribution with respect to the momentum transfer t between the hadrons. However, quite often these observables are not trivial to measure, and instead inclusive, predominantly soft, particle production studies constitute the material for the first physics publication related to collisions. This indeed was the case at the LHC, where the Minimum Bias data were the first to appear. In this section, however, the presentation will follow an order from the more inclusive to the more exclusive processes, starting with total hadronic cross-sections, touching on Minimum Bias data,then discussing the Underlying Event in different settings before, finally, some first results for double-parton scattering measurements will be presented.

Table 9.1 Results of measurements of total cross-sections $\sigma_{\rm tot}$, elastic cross-sections $\sigma_{\rm el}$, and the elastic slope B at the LHC.

Collab.	$E_{\rm c.m.}$	Result	
ATLAS	7 TeV	$\sigma_{\rm tot} = 95.35 \pm 0.38$ (stat.)± 1.25 (exp.)± 0.37 (extr.) mb	[33]
		$B = 19.73 \pm 0.14$ (stat.)± 0.26 (syst.) GeV^{-2}	
ATLAS	8 TeV	$\sigma_{\rm tot} = 96.07 \pm 0.18$ (stat.)± 0.85 (exp.)± 0.31 (extr.) mb	[1]
		$B = 19.74 \pm 0.05$ (stat.)± 0.25 (syst.) GeV^{-2}	
TOTEM	7 TeV	$\sigma_{\rm tot} = 98.0 \pm 2.5$ mb	[167]
		$\sigma_{\rm el} = 25.1 \pm 1.1$ mb	
		$\sigma_{\rm inel} = 72.9 \pm 1.5$ mb	
	8 TeV	$\sigma_{\rm tot} = 101.7 \pm 2.9$ mb	[166]
		$\sigma_{\rm el} = 27.1 \pm 1.4$ mb	
		$\sigma_{\rm inel} = 74.7 \pm 1.7$ mb	

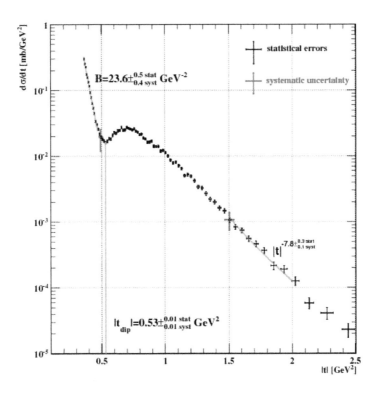

Fig. 9.2 The differential elastic cross-section $\mathrm{d}\sigma_{\rm el}/\mathrm{d}|t|$ at the 7 TeV LHC. Reprinted with permission from Ref. [167].

9.1.1 Total and differential pp cross-sections

At the LHC, there are a few measurements of the total cross-section, typically based on the **optical theorem**, which are summarized in Table 9.1. The elastic cross-section σ_{el}, or the elastic slope B, serve as important inputs to this kind of determination of σ_{tot} and they are therefore quoted here as well. The crux of this method is that the total and elastic cross-sections are related through

$$\sigma_{\text{tot}}^2 = \frac{16\pi}{1+\rho^2} \left.\frac{\mathrm{d}\sigma_{\text{el}}}{\mathrm{d}|t|}\right|_{t=0} = \frac{16\pi}{1+\rho^2} \frac{1}{\mathcal{L}} \left.\frac{\mathrm{d}N_{\text{el}}}{\mathrm{d}|t|}\right|_{t=0}, \tag{9.1}$$

where t is the momentum transfer between the protons, and the elastic cross-section is obtained from the number of elastic events N_{el} divided by the integrated luminosity $\mathcal{L}$. Here $\rho \approx 0.15$ captures the effect of the small real parts of the elastic amplitude. A similar expression relates the elastic slope B (the slope of the elastic cross section at $|t| \to 0$) to the total cross-section. Obviously, in this type of determination of σ_{tot}, one must extrapolate the differential elastic cross-section $\mathrm{d}\sigma_{\text{el}}/\mathrm{d}|t|$ and it is important extend the measurement to the smallest possible t. Results from the TOTEM collaboration [167], including parameters of simple fits to different features, on this observable are exhibited in Fig. 9.2, where the peak as $|t| \to 0$ is evident.[4]

It is worth noting that in contrast to the TEVATRON with its $p\bar{p}$ collisions,[5] it is possible at the LHC to use the **van-der Meer** method [877], which allows the direct determination of the luminosity in collisions of like-sign charged particles. Practically eliminating the large systematic uncertainties due to the luminosity in turn translates to significantly reduced uncertainties on the cross-section measurement. The results of the measurements quoted here put the cross section fits and the models and assumptions underlying them, *cf.* Section 7.1.3, to a stringent test; many of the models predicted σ_{tot} and related quantities to be somewhat larger than observed.

Moving on to measurements of inelastic cross-sections, it is fairly obvious that one could define the inelastic cross-section as the difference of total and elastic cross-section,

$$\sigma_{\text{inel}} = \sigma_{\text{tot}} - \sigma_{\text{el}}. \tag{9.2}$$

This is precisely what the TOTEM collaboration uses in their determination of the inelastic cross-sections, which for this reason are also quoted in Table 9.1. However, quite often a distinction is being made between **low-mass diffractive** and "truly" inelastic events. Defining the scaled **diffractive mass** of the dissociating proton ζ through

$$\zeta = \frac{M_X^2}{E_{\text{c.m.}}^2}, \tag{9.3}$$

inelastic events are defined as those events where the larger of the two diffractive masses M_X – or, correspondingly ζ – is larger than some critical value. At the LHC

---

[4] The β parameter defines the beam envelope. Its value at the collision point is termed $\beta*$. For the highest luminosities, it is desirable to have a small value of $\beta*$. Measurements of the elastic scattering cross-section, however, require special runs with large values of $\beta*$.

[5] As a reminder, the luminosity uncertainty at the TEVATRON was on the order of 6%.

Table 9.2 Results of measurements of inelastic cross-sections σ_{inel} in different definitions at the Lhc at 7 TeV. For the Alice result, the relatively tight diffractive mass cut was extrapolated down to zero, introducing a modeling uncertainty. ζ is defined in Eq. (9.3). For the Cms measurements, the tracks must be in the pseudorapidity region of $|\eta| < 2.4$, with $p_\perp > 200$ MeV.

Collab.	Def.	σ_{inel} [mb]	
Alice	$M_X > 200$ GeV	73.2 ± 2.6 (lumi)$^{+2.0}_{-4.6}$ (model) mb	[114]
Atlas	$\zeta > 5 \cdot 10^{-6}$	60.3 ± 0.05 (stat.)± 0.5 (syst.)± 2.1 (lumi) mb	[5]
Cms	$\zeta > 5 \cdot 10^{-6}$	60.2 ± 0.2 (stat.)± 1.1 (syst.)± 2.4 (lumi) mb	[374]
	>1 track	58.7 ± 2.0 (syst.)± 2.4 (lumi) mb	
	>2 tracks	57.2 ± 2.0 (syst.)± 2.4 (lumi) mb	
	>3 tracks	55.4 ± 2.0 (syst.)± 2.4 (lumi) mb	

$M_X > 15.7\,\text{GeV}$ or $\zeta > 5 \cdot 10^{-6}$ is often used.[6] Events that do not satisfy this condition are then dubbed single- or double-diffractive, and their cross-section is often given w.r.t. the inelastic one. The corresponding results at c.m.-energies of $E_{\text{c.m.}} = 7\,\text{TeV}$ are exhibited in Table 9.2.

In Fig. 9.3 the results for the total and elastic cross-sections in pp and $p\bar{p}$ collisions are compared to data from lower energy measurements and from (higher energy) cosmic ray measurements. It is clear from the collider data that the cross-sections continue to increase logarithmically with the centre-of-mass energy of the reactions. Looking at the data from the highest energies, obtained by astroparticle experiments, it also appears that the data continue to rise with energy above the current collider reach. The **unitarization of the total hadronic cross-section** has not set in, and the **Froissart bound** is still ahead of us.

9.1.2   Minimum Bias physics and inclusive hadron spectra

By definition, Minimum Bias (MB) data provide a very inclusive picture of particle collisions. In the case of pp collisions, the MB data encompass a plethora of physics processes, with the bulk of events dominated by the production of relatively few and relatively soft particles. This is indeed identical to the case of $p\bar{p}$ collisions, as studied at the Tevatron, with some results presented in the previous chapter in Section 8.1.

As in the Tevatron case, the production mechanisms can be classified as **elastic hadron scattering**, **diffractive scattering** and **inelastic particle production**. But of course, the boundaries are somewhat blurred, in particular between diffractive and inelastic events. Common lore has it that the former are characterized by the emergence of **rapidity gaps**. These are empty regions of rapidity in the particle production phase space, with no particles in them. Typically, diffractive events are thought to have such rapidity gaps with sizes of the order of a few units, typically about 3 or more. Conversely, from a theoretical perspective, such events are thought to emerge when **pomerons** are involved in the process. In fact, in most event generators,

---

[6] M_X is the (diffractive) mass of the system emerging from the proton breakup, and $E_{\text{c.m.}}$ is the hadronic centre-of-mass energy.

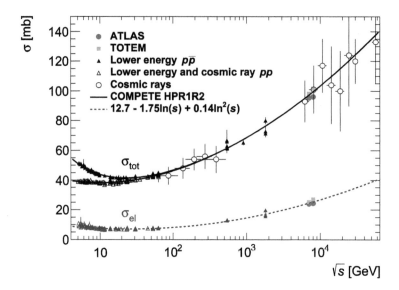

Fig. 9.3 Total and elastic cross-sections for pp and $p\bar{p}$ scattering vs. centre-of-mass energies ranging from a few GeV to 60 TeV, along with the results of a global fit. Reprinted with permission from Ref. [1].

such a distinction of the different event categories is made, and in the simulation of MB events, quite often an admixture of inelastic and single and double diffractive event samples must be used. Historically, this has led to correcting the MB data for diffractive events, which are effectively subtracted by the use of event generators, typically PYTHIA. Quite often the data have been extrapolated to the full phase space, *i.e.* to all pseudorapidities and to zero transverse momentum, again by invoking the simulation tools. This can be risky.

MB measurements at the LHC include only those events which satisfy relatively inclusive requirements on visible particles, such as the requirement of a certain, small number of (charged) particles with a minimal, usually low, transverse momentum inside the acceptance region of the detector. Usually, this boils down to the requirement of something like one to six charged particles with a minimal transverse momentum between about 100 and 500 MeV inside a pseudo-rapidity regime given by $|\eta| < 2.5$ or similar.

In Fig. 9.4 such data, taken by the ATLAS collaboration [2] are shown, based on events with at least one charged particle, where all charged particles are inside the interval $|\eta| < 2.4$ and have transverse momenta $p_\perp > 500\,\text{MeV}$. The particles are distributed relatively evenly in pseudo-rapidity, forming a plateau. (If a smaller cut on the transverse momentum (100 MeV) is instead used, the distribution has a slight peaking around $|\eta|=2$.) All MC distributions predict a flat η distribution, but there are some differences in the normalization. The PYTHIA 6 ATLAS MC09 tune was fit to data from 200 GeV to 1.96 TeV. The PYTHIA 6 AMBT1 tune derived from the MC09 tune, but was also fit to early LHC minimum bias data from 0.8 and 7 TeV. It

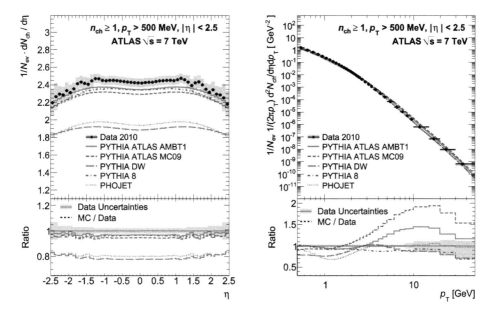

Fig. 9.4 Single charged particle spectra in Minimum Bias events at the LHC at 7 TeV: pseudorapity left), transverse momentum right). All results refer to events with at least one charged particle in the interval $|\eta| < 2.4$ with a transverse momentum of $p_\perp > 0.5$GeV. Reprinted with permission from Ref. [2].

is not surprising then that the best agreement is with the AMBT1 tune, although the MC09 tune also works well. In any case, the charged particle density for MB events at the LHC is larger than that observed at the TEVATRON (Section 8.1). The transverse momentum distribution falls seven orders of magnitude between 500 MeV and 10 GeV. The high transverse momentum range is hard to fit; here, the AMBT1 and MC09 tunes actually perform the worst.

In Fig. 9.5 (left), the charged particle multiplicity distribution is shown for MB events. The various predictions agree reasonably well with the data and each other at low n_{ch}, but there are clear deviations at large values of n_{ch}, and no prediction describes the data well. The mean transverse momentum is plotted versus the track multiplicity in Fig. 9.5 (right). The figure indicates that the greater the charged track multiplicity, the larger the mean track multiplicity is. This is not surprising in that the larger the charged particle multiplicity is in an event, the more likely it is that the protons have suffered a violent collision. One possible surprise is the degree of linearity of the correlation for charged particle multiplicities above 20. The ATLAS AMBT1 tune describes this correlation the best.

It is clear that such inclusive distributions present a formidable challenge to our understanding of the data and of the strong interaction. Typically, the level of agreement is in the range of a few to about 30%, even for the steepest distributions. That

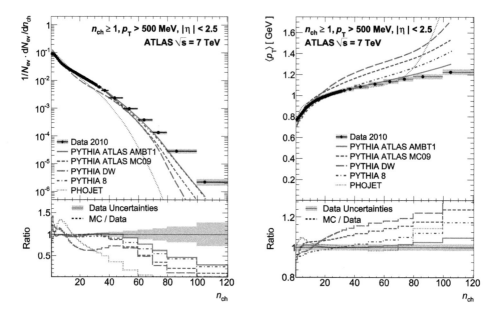

Fig. 9.5 Single charged particle spectra in Minimum Bias events at the LHC at 7 TeV:charged particle multiplicity left), mean transverse momentum versus track multiplicity (right). All results refer to events with at least one charged particle in the interval $|\eta| < 2.4$ with a transverse momentum of $p_\perp > 0.5$GeV. Reprinted with permission from Ref. [2].

the event generators are able to describe the data to this level is somewhat remarkable, even acknowledging that the MC parameters were often tuned to this data, in addition to lower energy data. This agreement could not necessarily have been expected from the beginning, especially keeping in mind that the simulation of MB events is based on relatively primitive paradigms, namely **multiple parton–parton scattering** in collinear factorization (see also Sections 7.1 and 7.2). While this provides some confidence that inclusive features of MB physics are under control well enough to allow for more complex measurements, it should not be forgotten that the agreement is less than perfect.

One of the places where such cracks in the otherwise relatively good description of MB data shows up, is in the production of individual hadron species, something that may be dubbed "**hadro-chemistry**, and in particular in the production of hadrons with multiple strange quarks or of baryons. As an example, consider the case of hyperon and cascade production (Λ and Ξ production), studied by the CMS collaboration [655], with some characteristic distributions shown in Fig. 9.6. Not surprisingly, the rapidity distributions for both the Λ and Ξ^- are fairly flat, as in the inclusive case above. However, the normalizations of the predictions are off by sizable factors at both 0.9 and 7 TeV for all predictions. The ratio of Ξ^-/Λ production is also more or less flat in rapidity, and the normalization of the predictions is again off. The ratio of MC to

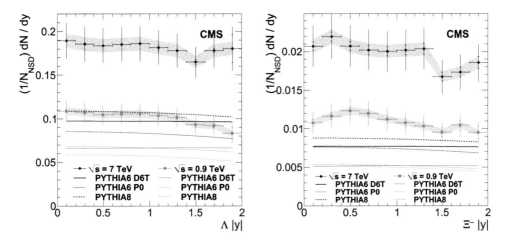

Fig. 9.6 Strange particle spectra in Minimum Bias events at the LHC at 7 TeV: The rapidity distribution of Λ baryons is shown on the left and the rapidity distribution of Cascade baryons Ξ is shown on the right, compared to several MC predictions. Reprinted with permission from Ref. [655].

data, for the transverse momentum distributions of several strange particles, is shown in Fig. 9.7, again at both 0.9 and 7 TeV. There is a sizable model dependence to the description of these data, and no MC prediction describes the data well.

When comparing these data with various MC predictions, it is obvious that the event generators struggle to achieve an agreement for baryons which is of similar quality as in the inclusive case in Fig. 9.4 and Fig. 9.5. This is a testament to the fact that the parameters of hadronization have been tuned to e^-e^+ annihilation data, typically from LEP 1, involving no incoming hadrons and therefore no highly energetic sources of additional colour such as the beam remnants in hadronic collisions. This is a clear hint at some deficiencies in our current understanding of some of the aspects of particle production at the LHC, while predictions for the bulk of particle production are under reasonably good control.

Another region where the description of data is less than perfect has been studied by the ATLAS collaboration in [16], namely the emergence of rapidity gaps. These are regions in pseudorapidity where no particle above a certain minimal transverse momentum is observed. In this study, rapidity gaps with respect to the edges of the detector at $\eta = \pm 4.9$, and in different bins of the minimal $p_\perp$, are analysed. The largest rapidity gap, $\Delta\eta_F$, is reported. Diffractive events are responsible for the region of large rapidity gaps. Regions of relatively small gap sizes, of up to $\Delta\eta_F \approx 3$, are dominated by fluctuations in standard QCD events, which emerge from the absence of parton radiation into that region (and its interplay with the hadronisation model). It is clear, therefore, that for $p_\perp$ of up to $\mathcal{O}(1\,\text{GeV})$, this is a highly model-dependent statement, with non–perturbative effects such as colour reconnections or so having a strong impact. This in turn depends on the admixture of non–diffractive and diffractive events, rendering this type of observable an interesting testbed for various reaction

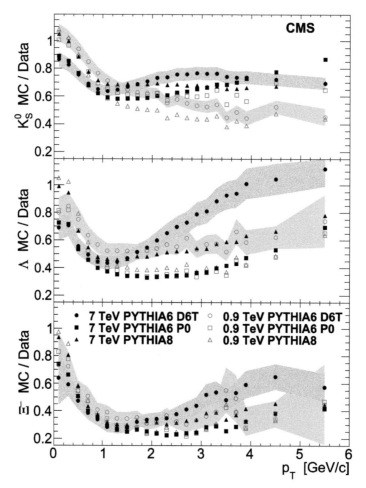

Fig. 9.7 The ratio of MC predictions to data for K_s^o, Λ and Ξ transverse momentum distributions. Reprinted with permission from Ref. [655].

mechanisms.

In Fig. 9.8 (left), the cross-section as a function of $\Delta\eta^F$ is shown for particles with transverse momenta above 200 MeV (the lowest limit for this measurement), along with several MC predictions. The PYTHIA tunes have been fit to ATLAS data. In Fig. 9.8 (right), the same data is shown, now compared to separate predictions, from PYTHIA 8, of the non-diffractive, single diffractive and double diffractive components. Note the exponentially falling non-diffractive component at small gap sizes. At larger gap sizes, there is a plateau, corresponding to a combination of single-diffractive and double-diffractive processes. All MC models are able to reproduce the general trends of the data, but none provide a perfect description.

It is noteworthy that the observables are defined entirely on the basis of visible, and therefore physical final states, allowing for easier (and unbiased) interpretation of

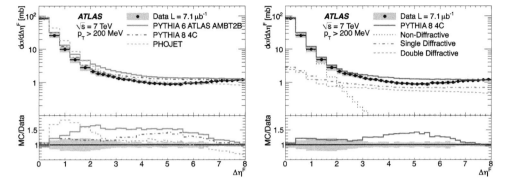

Fig. 9.8 Pseudorapidity gap distributions at 7 TeV. The cross-sections are measured differentially in terms of $\Delta\eta_F$, the largest of the pseudo-rapidity regions extending to the limits of the ATLAS sensitivity, at $\eta = \pm 4.9$, in which no final state particles are produced above a transverse momentum threshold $p_\perp$ cut. The ATLAS data are compared to several MC predictions on the left, and to a decomposition into contributions from non-diffractive, single-diffractive and double-diffractive on the right. Reprinted with permission from Ref. [16].

the data in the context of MB physics models.

9.1.3   The Underlying Event

The **Underlying Event** (UE) is to some degree related to Minimum Bias (MB) physics, since in most simulations both aspects of relatively soft QCD are described by **multiple parton–parton scattering.** As in the first UE analyses at the TEVATRON, it has become customary to divide the azimuthal phase space into a *"towards"*, an *"away"*, and a two–part *"transverse"* region such that each of the regions covers an identical size of $2\pi/3$ or $120°$, *cf.* Fig. 8.5. In some analyses, the two parts of the transverse region, are further decomposed into the *"trans-min"* and the *"trans-max"* part, depending on their occupation with particles or energy. Usually the regions are oriented in such a way that the signal object – the hardest charged track, the jet–axis or the lepton pair – is at the centre at $0°$ of the towards region, which in turn ranges from $300°$ to $60°$. The away region then covers the interval $[120°, 240°]$ and the two parts of the transverse region are in $[60°, 120°]$ and $[2400°, 300°]$.

The first published UE analysis at the LHC, undertaken by the ATLAS collaboration [6], relied on the leading charged track to orient the regions; its results are depicted in Fig. 9.9. The observables that are most sensitive to UE are the density of charged particles in the transverse region and the sum of their $p_\perp$; in addition, ATLAS reported the standard deviations of these observables and the average $p_\perp$, all as functions of the leading track $p_\perp$ and applying a cut of $p_\perp > 500\,\text{MeV}$ on all particles.

The qualitative features of the distributions are the same as at the TEVATRON; there is a rapid rise with the leading track $p_\perp$, then a plateau at higher leading track $p_\perp$. The predictions shown in Fig. 9.9 are all derived from MC tunes based on TEVATRON

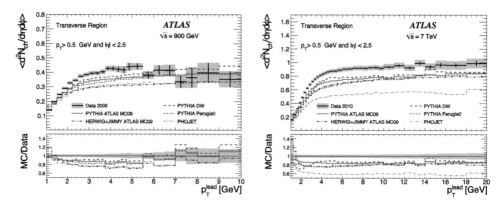

Fig. 9.9 Underlying Event observables with respect to the leading charged track transverse momentum in the transverse region at LHC at 7 TeV: the charged particle density N_{chg} (left) and the mean of the charged particle transverse momentum, $\langle p_\perp \rangle$ (right). Only charged particles with $|\eta| < 2.4$ and $p_\perp > 500\,\mathrm{MeV}$ have been considered. Reprinted with permission from Ref. [6].

data. It is interesting to note that all of the predictions fall short of the ATLAS data at 7 TeV. The agreement is better, typically within 10% or less, for predictions with subsequent tunes that have incorporated this data, as might be expected.

A similar analysis based on jets rather than leading tracks has been presented by CMS [362] and several of the results are displayed in Fig. 9.10. By the use of jets, the range of the measurement is greatly extended from that obtained using only the leading track. The mean charge density and the mean summed transverse momentum are shown in the figure, compared to several MC predictions. The best agreement is with the PYTHIA predictions using the Z1 tune. However, data from CMS at 7 TeV was used in defining this tune, so the level of agreement is not surprising. The other tunes, developed using lower energy data, also provide good agreement for the mean charge density distribution, but not for the mean summed transverse momentum distribution.

Fig. 9.11 provides a more differential look at the UE data, using the same observables as in the previous plot. Good agreement between the data and the MC predictions is found, especially for PYTHIA using the Z1 tune.

The event generators generally work well at the LHC for the description of the UE. This finding further strengthens the statement made in the discussion of MB data, that the event generators are adept in describing the bulk of LHC events. This indicates that the ideas underlying the construction of the non–perturbative models responsible for the simulation of MB and UE physics cannot be completely off the mark. However, similar to the case of MB, some deficiencies in the UE simulations start to appear when going to more taxing regions of phase space.

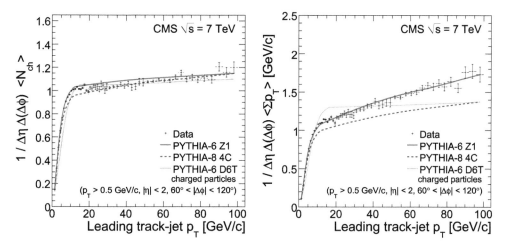

Fig. 9.10 Underlying Event observables with respect to the leading jet transverse momentum in the transverse region at LHC at 7 TeV: the charged particle density N_{chg} (top left), the sum of the charged particle transverse momenta $\sum p_{\perp}$ (top right), and their distributions (bottom left and right), all with respect to the leading jet transverse momentum. Only charged particles with $|\eta| < 2$ and $p_{\perp} > 500$ MeV have been considered. Reprinted with permission from [362].

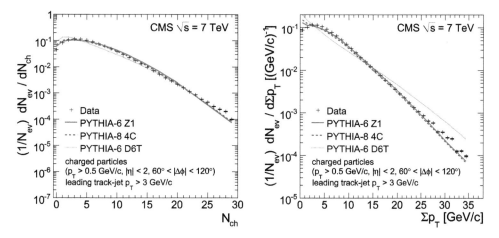

Fig. 9.11 Underlying Event observables with respect to the leading jet transverse momentum in the transverse region at the LHC at 7 TeV: the charged particle density N_{chg} distribution (left), the sum of the charged particle transverse momenta $\sum p_{\perp}$ distribution (right). Only charged particles with $|\eta| < 2$ and $p_{\perp} > 500$ MeV have been considered. Reprinted with permission from Ref. [362].

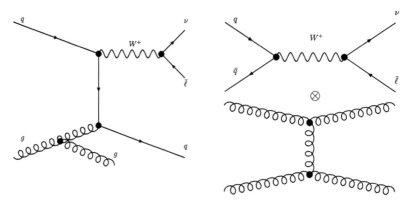

Fig. 9.12 Some example diagrams illustrating single (left) vs. double-parton (right) scattering production of Wjj

9.1.4   Multiple parton scattering

An interesting feature of the Underlying Event (UE) is that the secondary interactions of the hadron constituents that give rise to most of the UE can themselves become hard and give rise to physical objects such as additional jets, or even gauge bosons.

To illustrate the latter consider the production of same–sign W pairs, such as W^+W^+ pair production. Clearly, parton–level processes such as $uu \to W^+W^+dd$ must be invoked already at the lowest order in the Standard Model. The cross-section for this process at the LHC is of the order of a few (about five) femtobarns, and, using the simplifed model for double-parton scattering (DPS), Eq. (7.43), would yield a similar size for the DPS production cross-section of W^+W^+ pairs. This renders same-sign W pairs a smoking gun for double-parton scatttering.

However, the first measurement of DPS at the LHC was achieved with the associated production of a W boson with two jets, Wjj, by the ATLAS collaboration in [17]. The defining feature of DPS in contrast to the production of the same final state in one single partonic interaction, *cf.* Fig. 9.12 for an illustation of the two production modes, is that in the former the W and the di-jet systems are kinematically decoupled, and each of them typically has a relatively small total transverse momentum. This motivated ATLAS to use the total transverse momentum of the di-jet system or the total transverse momentum of the two jets divided by the sum of their individual transverse momenta,

$$\Delta_{jets}^n = \frac{|\vec{p}_\perp^{j_1} + \vec{p}_\perp^{j_2}|}{|\vec{p}_\perp^{j_1}| + |\vec{p}_\perp^{j_2}|} \tag{9.4}$$

as sensitive observables. The latter, Δ_{jets}^n yields numbers between 0 – when the two jets balance each other exactly, with the same transverse momentum oriented back-to-back – and 1 – when the jets point into the same direction. In this observable, the DPS region is clearly related to relatively low values of this quantity.

This is shown in the left panel of Fig. 9.13, where ATLAS data are compared with a combination of simulation results and a DPS sample constructed from data. The for-

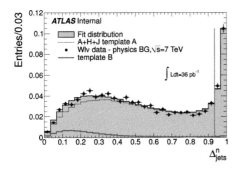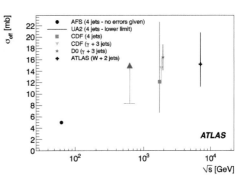

Fig. 9.13 (left) The distribution of Δ_{jets}^{n} is shown for Wjj events, along with the two templates described in the text. (right) The extracted value of σ_{eff} from this measurement is plotted along with other measurements of this parameter at the LHC and at lower energies. Reprinted with permission from Ref. [17].

mer, labelled as A+H+J is derived from a single-parton scattering simulation of $W+$ jets, using a combination of leading order matrix elements from ALPGEN [743] supplemented with the parton shower of HERWIG [415] and its underlying event, obtained from JIMMY [298]. The latter is actually constrained to produce parton-parton scatters only with a transverse momentum below $p_\perp < 15$ GeV, well below the jet cut in the data of $p_\perp > 20$ GeV. The latter, labelled as template B, is taken from data, where inclusive W production events have been overlayed with di-jet events, both taken in the early stages of Run I at the LHC. Extracting the overall normalisation of template B from a fit corresponds to fixing the value of σ_eff in the simple model of Eq. (7.43). In the right panel of the same figure, Fig. 9.13 the result of this fit and one obtained from a similar analysis by the CMS collaboration [387] are compared with determinations of the cross-section from lower energy data. The values of σ_eff are all consistent, except the result from the AFS collaboration.

Studies of DPS scattering with four-jet topologies [379] or involving the prompt production of J/ψ pairs [656] will further contribute to a more detailed understanding of the underlying mechansim and will ultimately help to improve on the simple model with a single parameter σ_eff.

9.2 Jets

9.2.1 Inclusive jet production

At the LHC, in contrast to the TEVATRON, IR-safe jet clustering algorithms, in particular the anti-k_T algorithm, are universally used. As mentioned previously, the anti-k_T jet algorithm was developed only near the end of the activity of the TEVATRON. In addition to its ease of use with fixed-order predictions, the anti-k_T algorithm provides jets that are very close to perfect cone-shaped, allowing an easy determination of the effective jet area. ATLAS typically uses jet sizes of 0.4 and 0.6 [31], while CMS uses jet sizes of 0.5 and 0.7 [383, 405]. Both experiments will expand the range of jet sizes used, in particular to be able to compare directly to the other experiment's results, for example with a common jet size of 0.4. Here, we discuss CMS results for inclusive jet cross-section measurements with the anti-k_T ($D = 0.5$) and anti-k_T ($D = 0.7$) jet algorithms, at a centre-of-mass energy of 8 TeV. Similar results exist for ATLAS with the anti-k_T ($D = 0.4$) and anti-k_T ($D = 0.6$) algorithms.

The calorimeter coverage for both ATLAS and CMS goes out to rapidities of the order of 4.5. However, many of the analyses involving jets restrict themselves to jets in more central rapidity regions (typically $y_{jet} \leq 2.5 - 3$), where more tracking information is available. The tracking information serves both to improve the energy resolution of the measured jet, and as a way of discriminating between jets from the event of interest and jets produced by pileup. Fewer tools are available at higher rapidities, but this region is still important for many physics results. Jet measurements at high y provide useful information in PDF fits and for discriminating between the VBF and gluon-gluon fusion mechanisms for Higgs boson production, for example. A number of analyses at both ATLAS and CMS have measured jets out to the full rapidity coverage, as will be shown in this chapter.

In CMS, jet measurements are conducted primarily with the particle-flow event reconstruction algorithm [400], in which tracking and calorimetry information is used in a framework optimized to provide the best jet energy resolution. An offset correction is used to remove the energy contributed by additional proton-proton interactions [303]. The offset method calculates a rapidity-dependent energy density ρ, which when multiplied by the jet area, provides an indication of the energy to be subtracted from the jet. Most of the pileup is due to collisions in the same bunch-crossing, with smaller contributions from out-of-time pileup. Note, though, that unlike at the TEVATRON, the underlying event energy is subtracted by the offset correction along with the pileup energy. The non-perturbative corrections for the underlying event can effectively be added back to the data for easier comparison to Monte Carlo predictions. This offset method is becoming the standard for both experiments.

Results for the measurement of the inclusive jet cross-section for the anti-k_T ($D = 0.7$) jet algorithm for all rapidity intervals are shown in Fig. 9.14 compared to NLO QCD predictions using CT10 PDFs, and modified by non-perturbative corrections [405]. A linear comparison of the data to NLO jet cross-section predictions with various PDFs, for the central rapidity region, is shown in Fig. 9.15. The jet measurement reaches transverse momenta greater than 2 TeV, with the integrated data sample of 19.7 fb^{-1}. Better agreement is observed for the anti-k_T ($D = 0.7$) results than for the results using the anti-k_T ($D = 0.5$) algorithm (not shown), perhaps indicating that while the NLO prediction (where at most two partons can be in a jet) describes the jet shape reasonably well, it does not describe the jet shape completely. See also the discussion below.

Over most of the kinematic range, the experimental uncertainties are smaller than the theoretical uncertainties (both PDF and scale). The scale uncertainties have greatly improved upon completion of the NNLO jet calculation. Given the spread of PDF predictions, this data should be useful in parton distribution function fits.

Note that NLO electroweak corrections are already on the order of approximately 5% at 1 TeV and increase fairly rapidly with jet transverse momentum, *cf.* the simple estimate provided in Eq. (3.232) and Fig. 3.12. This estimate is confirmed by an exact

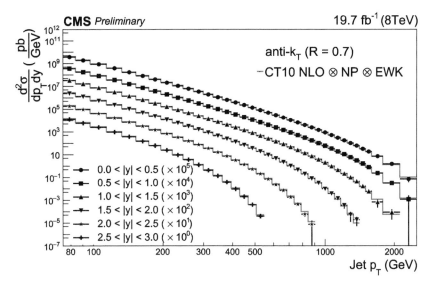

Fig. 9.14 A comparison of the CMS inclusive jet cross-section measured using the anti-k_T ($D = 0.7$) jet algorithm to NLO predictions using the CT10 PDFs. Reprinted with permission from Ref. [405].

calculation of Ref. [469].

Many of the systematic uncertainties, both experimental and theoretical, cancel in the ratio of measurements of jet cross-sections for two different jet sizes. The ratio for anti-k_T ($D = 0.5$) to anti-k_T ($D = 0.7$), using CMS data from 7 TeV [383], is shown in Fig. 9.16 The ratio of the two cross-sections starts at approximately 0.7 and rises as the jet transverse momentum increases, as expected given the increasing collimation of the jet. Fixed-order predictions, either at LO or NLO, do not describe the shape of the measured ratio. Better agreement is provided by the incorporation of non-perturbative corrections to the fixed-order predictions, but the best agreement is achieved by the POWHEG +PYTHIA 6 prediction, which combines a NLO matrix element calculation within a parton shower Monte Carlo framework. It will be interesting to compare the calculated NNLO (with non-perturbative corrections) ratio to this data, to see to what extent the higher order prediction can describe the jet shape. Similar results have been reported by ATLAS [31].

At the time of writing of this book,the 7 TeV data has been incorporated into global PDF fits [194, 489, 614], using the larger jet size where the NLO predictions may be more applicable.

9.2.2 Dijet production

Measurements of the inclusive dijet cross-section provide a means to test precision predictions of perturbative QCD at the highest mass scales achievable at the LHC. Such comparisons can help to constrain the high-x gluon distribution, as well as to search

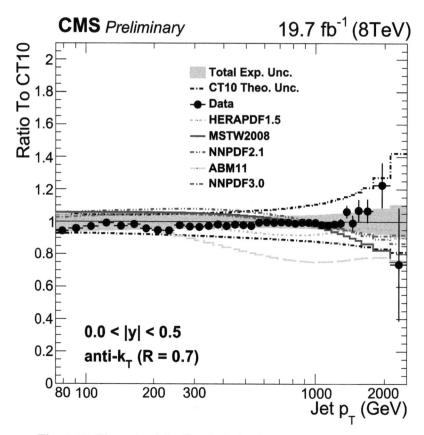

Fig. 9.15 The ratio of the CMS inclusive jet cross-section measured using the anti-k_T ($D = 0.7$) jet algorithm to NLO predictions using the CT10 PDFs, as well as the ratios of NLO predictions using other PDFs to predictions using CT10. Reprinted with permission from Ref. [405].

for the presence of new contact interactions, for example due to quark compositeness. At the LHC, the dijet cross-section has been measured out to dijet masses of 5 TeV.

Typically, the measurement is divided into bins of y^*, with y^* defined as half of the absolute value of the rapidity difference between the two jets. Using the kinematic relations of Section 4.1 it is straightforward to show that, at leading order, this quantity is related to the centre-of-mass scattering angle θ^* by $|\cos\theta^*| = \tanh y^*$. Thus high values of y^* correspond to larger values of $\cos\theta^*$. A measurement of the dijet mass cross section, in bins of y^*, carried out by ATLAS [27] is shown in Fig. 9.17, compared to NLO parton level predictions from NLOJET++. The measurement in the plots has been carried out with the anti-k_T ($D = 0.6$) jet clustering algorithm (similar measurements are available with the anti-k_T ($D = 0.4$) algorithm). Jets are reconstructed using topological cell clusters [39]. These clusters are determined from calorimeter cells and local hadronic calibration weighting. The latter depends on the good 3-D

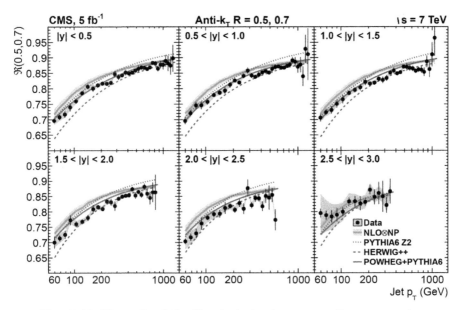

Fig. 9.16 The ratio of the CMS inclusive jet cross-section measured using the anti-k_T $(D = 0.5)$ jet algorithm to the jet cross-section using $D = 0.7$. Comparisons are made to LO and NLO theoretical predictions, with and without non-perturbative corrections. Reprinted with permission from Ref. [383].

(lateral and longitudinal) segmentation of the ATLAS calorimeters. Pileup corrections are determined from Monte Carlo calculations, as a function of the number of track vertices in the event (for in-time pileup) and the average instantaneous luminosity at the time of the event (for out-of-time pileup), in bins of jet p_T and rapidity. Note the differences with respect to the technique described in the previous section for the CMS inclusive jet analysis. ATLAS is switching to an offset method similar to that described for CMS.

The NLOJET++ predictions have been corrected for non-perturbative effects. In the calculations, a common renormalization and factorization scale of $p_T^{max} \exp(0.3y^*)$ has been used, where p_T^{max} refers to the transverse momentum of the largest jet. For values of y^* near zero, and for $2 \rightarrow 2$ scattering, the scale reverts to just the jet transverse momentum. As noted in Section 4.1, the peak inclusive jet cross-section at NLO tends to move to higher scales as the jet rapidity increases. The same is true for dijet cross-sections as a function of y^*, and the scale choice given above tends to be near the peak value of the dijet cross section at high y^*. In fact, at very high dijet mass, and at large values of y^*, a scale choice close to p_T^{max} can actually lead to negative cross-sections at NLO.

Electroweak corrections have also been taken into account in the data comparisons. These corrections are typically less than 1% for $y^* \geq 0.5$, but can be larger than 9% for high dijet mass (> 3 TeV) for $y^* < 0.5$, as shown in Fig. 9.18 . The corrections include

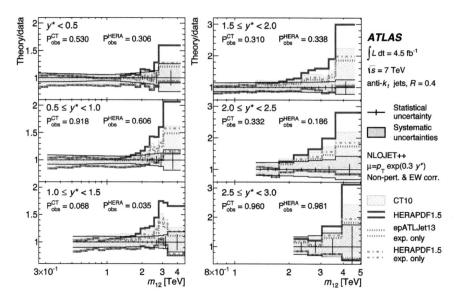

Fig. 9.17 The dijet mass cross-section from ATLAS as a function of y^* compared to predictions from NLOJET++ using several PDFs. Reprinted with permission from Ref. [27].

both tree-level effects of order $(\alpha\alpha_s, \alpha^2)$ and weak loop effects of order $(\alpha\alpha_s^2)$ [469].

The probability for the NLO predictions to describe the data, taking into account the experimental uncertainties, is indicated in the plots for each y^* bin. Both the CT10 and HERAPDF1.5 PDFs describe the data well, except perhaps in the y^* interval from 1.0 to 1.5. Contact interactions preferentially produce events at low y^* compared to QCD. Thus, the most sensitive region to search for the effects of contact interactions is at low y^* (< 0.5) and high dijet mass (> 1.31 TeV). Using a model of QCD+contact interactions with left-left coupling and destructive interference, the ATLAS data is sufficient to exclude contact interactions at a scale Λ less than 7.1 TeV (using the CT10 PDFs). Similar results are also available from CMS [376].

Dijet production is not expected to be well-described by fixed-order predictions in kinematic configurations where either a large rapidity interval exists between the two jets, and/or there is a veto on the existence of a third jet in the rapidity interval bounded by the dijet system. In these situations, higher order corrections can become important, and logarithmic terms depending on the rapidity separation between the two leading jets[7] or on the average transverse momentum of the dijets may need to be resummed in order to achieve a good description of the data.

The region where two jets, with a fixed p_T threshold, are separated by a large rapidity interval corresponds to a large value of $\hat{s}$ and a small value of $\hat{t}$. Such regions

---

[7]Technically, the logarithmic terms depend on the dijet mass, but in situations where the jets are separated by a large rapidity interval and the transverse momenta of the jets are similar, the argument of the logarithm reduces to the rapidity separation.

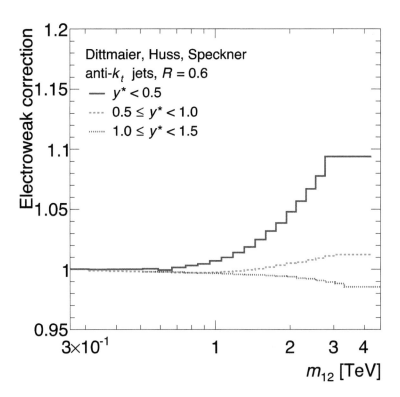

Fig. 9.18 Electroweak corrections for the dijet mass cross-section from ATLAS for several different y^* bins. Reprinted with permission from Ref. [27].

are dominated by t-channel gluon exchange and one expects a linear growth of jet multiplicity with increasing Δy_{jj}. This can be observed in Fig. 9.19, which shows the mean number of jets (defined by $p_T > 20$ GeV) in the rapidity interval bounded by the dijet system, for different average transverse momenta for the tagging jets. The ATLAS data is compared to predictions from HEJ and POWHEG [3]. Here, the Δy_{jj} interval is defined by the two jets in the event that have the greatest rapidity separation. If instead, the two highest p_T jets were used, the growth with Δy_{jj} would be reduced by about a factor of two [3]. Thus, rapidity ordering may be more efficient in rejecting gluon-gluon fusion production of a Higgs boson, in order to measure VBF Higgs boson production, than using the two highest p_T jets, as is currently done. The reason for the faster increase with rapidity ordering can be easily understood. If the bounding jets are the two highest p_T jets, then any additional jets produced in the gap are required to be in the transverse momentum range determined by the jet cutoff and the second highest p_T jet in the event. For rapidity ordering, there is no such bound, and there is effectively a larger phase space. There are practical considerations, however, in the experimental difficulties with dealing with jets at very forward rapidities, especially in high pileup conditions.

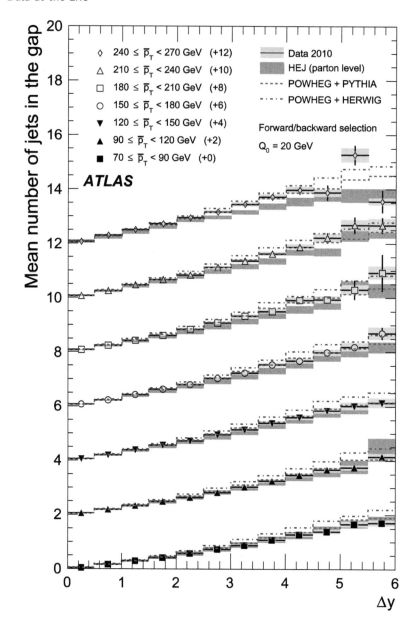

Fig. 9.19 The average jet multiplicity in the rapidity interval between two jets separated by the largest rapidity interval, as a function of the dijet rapidity interval. Results are shown for different values of the average transverse momenta of the tagging jets. Reprinted with permission from Ref. [3].

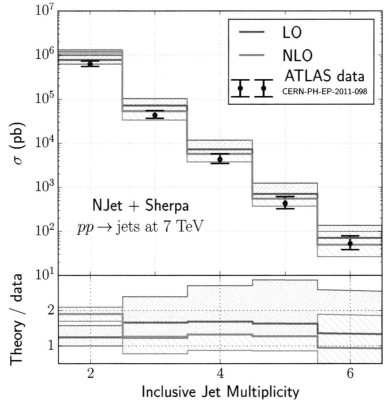

Fig. 9.20 The ATLAS inclusive jet multiplicity distribution at 7 TeV compared to LO and NLO predictions from NJET. The jets are measured using the anti-k_T ($D = 0.4$) jet algorithm. The plot is taken from Ref. [183] using data from Ref. [4]. Reprinted with permission from Ref. [183].

In the figure, comparisons are made to POWHEG, coupled either with the PYTHIA or HERWIG parton shower and to HEJ, at the partonic level. The POWHEG predictions include a full NLO partonic description of the dijet system and the PYTHIA and HERWIG parton showers provide a resummation of soft and collinear gluon radiation. The HEJ formalism provides a leading logarithmic resummation of terms proportional to the rapidity separation of the two jets, embedded in a framework that includes fixed-order corrections from multi-jet matrix elements.

9.2.3 Multijet production

Multijet production at the LHC is an interesting and important process, in that it allows for precision tests of the perturbative QCD framework, serves as a platform for measuring the running of α_s, and also forms a background for many types of new physics. Final states with 7 or more jets have been measured at the LHC [4, 31, 404, 663], and NLO predictions are available for states with up to 5 jets [183].

ATLAS, for example, has measured final states with up to 6 jets at 7 TeV, with the requirement that the leading jet have a transverse momentum greater than 80 GeV and additional jets have transverse momentum greater than 60 GeV. The anti-k_T jet algorithm with $D = 0.4$ and $D = 0.6$ was used. A comparison of the ATLAS data for the jet multiplicity distribution is shown in Fig. 9.20, along with LO and NLO predictions from NJET [183]. The NLO calculation significantly decreases the scale uncertainty from that obtained at LO, and in general is in better agreement with the data, with the exception of the 2-jet bin, where there are large negative NLO corrections. For the higher jet multiplicities (for the bins where NLO predictions are available), the ratio between theory and data is in the range of 1.2–1.3. As Ref. [183] notes, the main driver for the difference between the LO and NLO predictions is the use of LO PDFs for the former. If NLO PDFs are used for both predictions, the results at the two orders are very close. In particular, if a scale of $\hat{H}_T/2$ is used, the ratio of the NLO to LO predictions tends to be very flat as a function of the relevant kinematic variables. This same behaviour has been observed for W/Z+jets at the TEVATRON, as discussed earlier, and will be encountered again in the context of W/Z+jets at the LHC.

In Figure 9.21 (left), the ratio of the cross-section for the production of $(n + 1)$ and n jets for the ATLAS data is compared to NLO predictions using several PDFs. Within uncertainties, the predictions agree with the ATLAS data for σ_4/σ_3 and σ_5/σ_4. In this case, the LO and NLO predictions for the ratios are within 10% of each other due to cancellations of PDF effects. In Figure 9.21 (right) is shown the σ_3/σ_2 ratio (for $D = 0.6$) as a function of the lead jet transverse momentum. Good agreement is obtained for predictions from all PDFs at large leading jet transverse momentum.

9.2.4 Jet substructure

Knowledge of the jet four-vector allows for calculation not only of the transverse momentum and rapidity of the jet, but also of an additional degree of freedom, the jet mass. Jets acquire mass dynamically through perturbative gluon radiation, and to some extent through the non-perturbative fragmentation process. As with the jet shape, the one hard gluon present in a NLO jet calculation describes the perturbative contribution to the jet mass reasonably (but not perfectly) well, i.e. a parton shower is not required for tolerable agreement.

The general form for a typical jet mass can be determined through dimensional analysis: the dominant contribution to the jet mass squared (at NLO parton level) should scale with the square of the transverse momentum of the jet, with the square of the size of the jet R, and be proportional to one factor of α_s, where the argument of α_s should be related to the transverse momentum of the jet: $m^2 \propto \alpha_s(p_T)\, p_T^2 R^2$. Thus, there is a roughly linear dependence of the jet mass on both the p_T and the size of the jet. Of course, in any particular event, the gluon radiation process is stochastic, and thus there will be a great variation of jet masses in practice, as will be shown later. In general, the jet mass distribution will be strongly suppressed for low jet masses (this corresponds to little or no gluon radiation), will rise roughly linearly to a peak value, and then fall off slowly with jet mass, with a slope between $1/m$ and $1/m^2$. The peak of the jet mass distribution occurs for $m \sim (0.1 - 0.2)p_T R$, and is dominated by multiple soft wide-angle gluon emissions. There is also a shoulder at

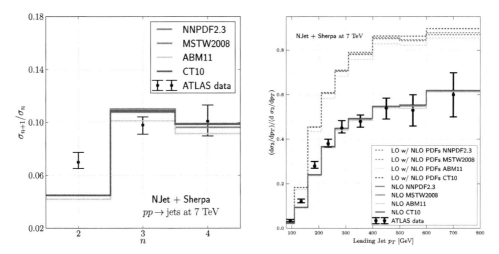

Fig. 9.21 (left) The ratio of the cross-sections for $(n+1)$ and n-jet production, for $n = 2$, 3 and 4, in ATLAS data [4] and from the theoretical predictions of NJET +SHERPA. The jet clustering uses the anti-k_T $(D = 0.4)$ algorithm. (right) The 3-jet to 2-jet ratio as a function of the leading jet transverse momentum using anti-k_T $(D = 0.6)$ jet clustering. Predictions are shown at LO and NLO for several PDFs. Reprinted with permission from Ref. [183].

larger masses $(0.3 \leq m/(p_T R) \leq 0.5)$ dominated by a single hard gluon emission. This shape means that the average jet mass is above the peak value. A simple rule-of-thumb, that describes the result of an exact NLO calculation of the jet mass to reasonable accuracy, is that the average mass of a jet at the LHC (13 TeV) measured with the anti-k_T jet algorithm is approximately given by $\langle m \rangle = 0.16\, p_T R$ [508]. Since the factor of α_s should be accompanied by the colour charge of the hardest parton in the jet, gluon jets should have an average mass a factor of $\sqrt{C_A/C_F}$ greater than quark jets. The general formula is for an average of gluon and quark jets. The jet mass distribution does not depend strongly on the centre-of-mass energy, but will depend on the jet algorithm (as well as the jet size). At the particle level, jet masses will be larger than at the parton level, due to non-perturbative contributions but with the difference growing smaller with increasing jet p_T.

The jet mass is an interesting variable to measure, not just because of the perturbative QCD aspects, but because a jet may be massive because it's a (boosted) W boson, or a top quark, or even a Higgs boson [106]. For example, it was shown that the discrimination of the signal for a Higgs boson decaying into a $b\bar{b}$ pair can be improved in the kinematic region where the Higgs is at high transverse momentum, and the $b\bar{b}$ final state is reconstructed within a single (fat) jet [297].

With a few exceptions [64], there was little investigation into jet masses at the TEVATRON, but this information has become an integral part of many analyses at the LHC [8], especially in the context of jet grooming. In general, all jet grooming techniques are designed to provide a separation between the decays of heavy objects and the QCD branchings that are a normal aspect of parton evolution inside jets. In addition, the grooming techniques try to remove soft energy depositions inside the jets, which can arise both from the underlying event in a hard collision, as well as from the multiple minimum bias interactions that are present in high luminosity LHC running conditions. Three of the major grooming tools are filtering [297], trimming [699], and pruning [512, 513], which are described below in the context of a CMS analysis.

CMS has measured jet mass distributions in both V+jets and inclusive dijet events, and has examined the impact of filtering, trimming, and pruning on the jet mass distributions [377]. In filtering, a jet determined through a regular jet clustering algorithm, most typically the anti-k_T algorithm for the LHC, is re-clustered using the Cambridge-Aachen algorithm with a smaller jet size ($R = 0.3$ for the CMS analysis). The resulting new sub-jets are ordered in transverse momentum and the jet is redefined using only the three hardest sub-jets. In the trimming algorithm, jets are again re-clustered using a smaller jet size (using the k_T-clustering algorithm), and sub-jets are only kept if they pass the requirement that $p_{Tsub} > f_{cut}\lambda_{hard}$. Typically, λ_{hard} is chosen to be equal to the transverse momentum of the original jet. For this CMS analysis, R_{sub} has been chosen to be 0.2 and f_{cut} to be 0.03. The pruning algorithm re-clusters the constituents of a jet with the Cambridge-Aachen algorithm, using the original parameters, but requiring that for two sub-constituents i and j, the softer of the two constituents is removed when the following conditions are satisfied:

$$z_{ij} = \frac{\min(p_{\perp i}, p_{\perp j})}{p_{\perp i} + p_{\perp j}} < z_{\text{cut}} \tag{9.5}$$

$$\Delta R_{ij} > D_{\text{cut}} \equiv \alpha \cdot \frac{m_J}{p_\perp}, \tag{9.6}$$

where m_J and p_T are the mass and transverse momentum of the (original) jet, and the parameters z_{cut} and α have been chosen to be 0.1 and 0.5.

Fig. 9.22 shows distributions of the ratios of jet mass distributions after grooming (using either the filtering, trimming, or pruning techniques), for reconstructed data, for reconstructed simulated PYTHIA 6 events, and for generator-level PYTHIA 6 events. The events are from inclusive dijet production and the jets have been reconstructed with the anti-k_T ($D = 0.7$) jet algorithm. All distributions have been corrected back to the particle level, to allow for direct comparison to theoretical predictions. In general, filtering results in the smallest changes to the original mass distributions, followed by trimming and then pruning (with the parameters chosen in the CMS analysis).

Fig. 9.23 and Fig. 9.24 show the unfolded jet mass distributions for anti-k_T ($D = 0.7$) jets from $Z \to ll$+jet events for ungroomed jets and pruned jets respectively, with the parameters for the grooming as described above.[8] The data are shown for four different jet transverse momentum bins and are compared to predictions from PYTHIA 6 and HERWIG ++, with the tunes indicated in the legends. The Sudakov suppression at low jet mass, peaking and then slow fall-off with increasing jet mass described earlier, can be seen for the ungroomed jet distributions in Fig. 9.23. The

---

[8]Results for filtered and trimmed jets can be found in Ref. [377].

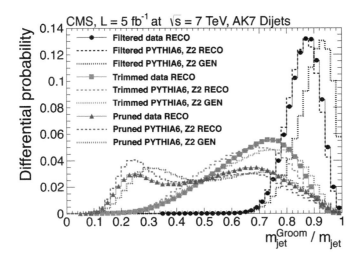

Fig. 9.22 The differential probability distributions for jet mass ratios for groomed jets to ungroomed jets for three different grooming techniques. The data are CMS dijet events,and the Monte Carlo predictions use PYTHIA 6. The Monte Carlo predictions are given both at the generated and reconstructed levels. Reprinted with permission from Ref. [377].

dip for low jet mass fills in with the use of an aggressive grooming procedure such as pruning. The peak region of the jet mass distribution receives substantial contributions from soft gluon emissions at wide angle. After pruning, these emissions are largely removed, resulting in events in the peak region migrating to lower masses [506].

In general, the data are in good agreement with the Monte Carlo predictions, except perhaps for smaller jet masses. Ref. [377] notes that an aggressive grooming procedure (like pruning) tends to lead to better agreement between data and the Monte Carlo simulation, and that the data/Monte Carlo agreement is better in general for the V+jets analysis than for the dijet analysis, perhaps indicating that quark jets (more typical in V+jet final states) are better modelled in Monte Carlo than gluon jets.

Formerly, it was thought difficult to provide analytic calculations that describe the impact of the jet grooming techniques on distributions, such as jet masses, but great progress has been made in recent years [431]. These calculations not only provide a better understanding of the different jet grooming techniques, but have also allowed for the development of new tools, such as the mass-drop tagger [431] and the soft drop tagger [722].

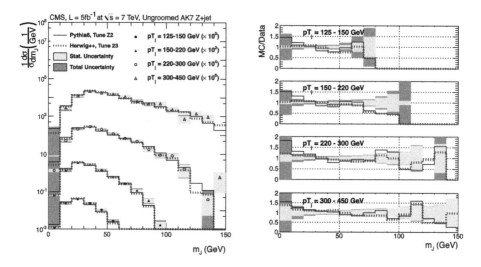

Fig. 9.23 The unfolded ungroomed jet mass distributions for $Z \to ll +$ jet events. Data from CMS is compared to Monte Carlo predictions from MADGRAPH + either PYTHIA or HERWIG. The ratio of Monte Carlo to data is given on the right, with the statistical uncertainty shown in light shading and the total uncertainty in dark shading. Reprinted with permission from Ref. [377].

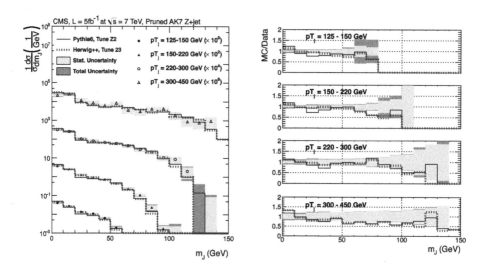

Fig. 9.24 The same comparison as in Fig. 9.23, for the unfolded pruned–jet mass distributions for $Z \to ll +$ jet events. Reprinted with permission from Ref. [377].

9.3 Drell-Yan type production

9.3.1 Inclusive spectra

One of the primary benchmarks, and often one of the first cross-sections to be measured, is that for W and Z production. ATLAS and CMS have both measured the W and Z cross-section at 7 TeV (during the 2010 running) [12, 361] and in addition CMS has measured the cross-sections at 8 TeV (in special low luminosity, and thus low pile-up, running conditions) [380]. The cross-sections are measured in both the electron and muon channels, with similar requirements for the transverse momenta and rapidities for the electron and the muon. The data are corrected for the effects of final-state QED radiation and an isolation cut is placed on the lepton candidates. The differential cross-sections are then combined after extrapolating each measurement to a common fiducial kinematic region. For ATLAS, the missing transverse energy for W boson production is required to be greater than 25 GeV and the transverse mass is required to be greater than 40 GeV. For CMS, no explicit cut is placed on the missing transverse energy, but the missing transverse energy distribution is used to determine the background. In CMS, the Z boson candidates are required to have a mass between 60 and 120 GeV, while for ATLAS the range is 66 to 116 GeV.

The cross-sections are measured in the fiducial regions as well as being extrapolated to the full phase space. The latter involves a calculation for the geometrical and kinematic acceptances for the measurement, and thus the introduction of possible model dependence. Typically, POWHEG +PYTHIA is used for the extrapolation. The theoretical systematic uncertainties for the acceptance calculation can be evaluated by varying the PDFs (using the PDF4LHC prescription for the three global PDF families), examining the impact of NNLL soft gluon resummation using the program RESBOS, and examining the impact of higher-order corrections by varying the renormalization and factorization scales in the program FEWZ within a factor of two. The effects of higher-order EW corrections can also be simulated by the use of the program HORACE. The effect of extrapolating from the fiducial to the full phase space is typically to increase the cross section by a factor of about 2(2.5) for $W(Z)$ production. The total uncertainties for the extrapolation corrections for W and Z production range between approximately 1 and 1.5%, with each of the sources mentioned above contributing.

The ratio of the W to Z cross-section can be especially interesting, as many of the systematic errors, from both experiment and theory, cancel out. The total W and Z cross-sections measured by CMS at 8 TeV are shown in Fig. 9.25, along with the NNLO predictions from the three global PDF sets. The (solid) ellipse indicates the 68% CL region for the total experimental uncertainty. The (open) ellipses for the theory predictions indicate the 68% CL PDF uncertainties from each group. The three predictions are all consistent with the CMS data, so there is no great discrimination among them provided by the total cross section measurements. All of the PDFs provide a somewhat lower prediction for the Z boson cross-section than observed in the data, though. The ATLAS cross-sections are consistent with those from CMS. The ratios of various W and Z cross-sections from CMS at 8 TeV with NNLO predictions are plotted in Fig. 9.26.

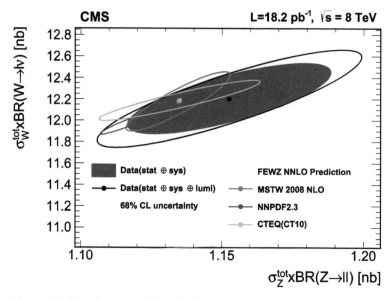

Fig. 9.25 The CMS total W and Z cross-sections at 8 TeV (extrapolated from the fiducial cross-sections) compared to NNLO predictions from the CT10, MSTW2008, and NNPDF2.3 PDFs. Reprinted with permission from Ref. [380].

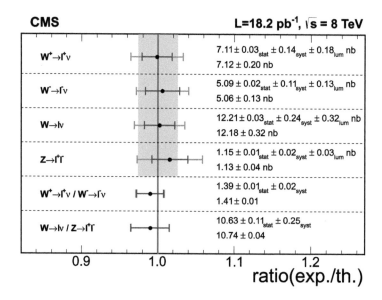

Fig. 9.26 The ratios of CMS W and Z cross-sections (extrapolated from the fiducial cross-sections) to NNLO predictions. Reprinted with permission from Ref. [380].

9.3.2 Differential cross-sections in mass and rapidity

More information can be obtained by measuring double differential Drell-Yan cross-sections. CMS, for example, has measured the cross-section for electron and muon pairs as a function of the dilepton mass and the cross section for muon pairs as a function of mass and rapidity, at centre-of-mass energies of 7 TeV [373] and 8 TeV [667]. The data cover a wide kinematic range, from 20 GeV to 2000 GeV. The data are corrected for the effects of final-state QED radiation and are extrapolated from the fiducial measurement to the full phase space. The consistency of the electron and muon channels allows them to be combined. All of the differential cross-sections have been normalized to the cross-section for the Z-peak region ($60 < m < 120$ GeV).

The combined (electron and muon) Drell-Yan differential cross-section at 7 TeV as a function of the dilepton mass is shown in Fig. 9.27, compared to theoretical predictions at NNLO QCD using FEWZ 3.1 [761] and the CT10 PDFs, and with electroweak corrections at LO and at NLO. The contributions of lepton pair production from $\gamma\gamma$ initial states have been taken into account. These contributions increase with dilepton mass, reaching up to 10% for the highest mass bin. The higher mass reach at Run II will result in even larger contributions from $\gamma\gamma$ initial states, necessitating a better determination of the photon PDF in the proton. (See, however, the discussion in Section 6.2.2 on the photon PDF and recent advances in its determination.) The data are in good agreement with the theoretical predictions using CT10. The measurement does not provide, however, sufficient sensitivity to distinguish CT10 from other PDFs.

The Drell-Yan cross-section at 8 TeV as a function of rapidity is shown for the Z boson mass region (60-120 GeV) and the highest Drell-Yan mass bin (200-1500 GeV) in Fig. 9.28 (from a total of 6 mass bins in the CMS measurement) [667]. NNLO predictions from FEWZ 3.1 are shown for two NNLO PDFs. Good agreement with the data is observed for both PDFs. Here, in the double-differential distributions, the sensitivity is sufficient to distinguish among the various PDF families, and this information will be useful in PDF fits utilizing LHC data.

The ATLAS and CMS measurements of W and Z production are predominantly in the central rapidity region, $|y| \leq 2.5$. The LHCB experiment, having been designed primarily for the study of forward production of particles containing b- and c-quarks, extends the kinematic reach for LHC vector boson measurements up to a rapidity of 4.5. This is very useful, for example, for better constraints on PDFs at both low and high x. LHCB has measured W production at a centre-of-mass energy of 7 TeV in the muon channel [408]. The muons in this measurement are required to have a transverse momentum greater than 20 GeV and a rapidity between 2 and 4.5. Given the non-hermiticity of the detector, there is no requirement on missing transverse energy. To reduce backgrounds from heavy flavour decays, the muons are required to be isolated, with the sum of the transverse momenta of all charged tracks within a radius of 0.5 of the muon direction (as well as a related quantity involving calorimeter information) required to be less than 2 GeV. A final state QED radiation correction is applied to account for the events lost due to the photon radiation resulting in the muon failing the kinematic cuts.

The resultant cross-sections for W^+ and W^- production, and the ratio between the two, are shown in Fig. 9.29 and compared to NNLO predictions using six different

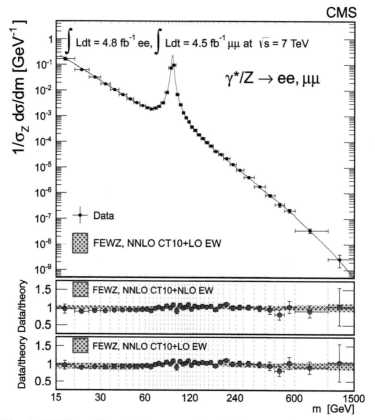

Fig. 9.27 The differential cross-section for lepton pair production as a function of the dilepton mass. The cross-section has been normalized to the Z-peak region and the electron and muon channels have been added. Comparisons are made to the NNLO predictions from FEWZ using the CT10 NNLO PDFs, with EW corrections at LO and NLO added. The shaded bands in the lower two plots represent the statistical uncertainty from the FEWZ calculation added in quadrature with the 68% CL PDF errors from CT10. Reprinted with permission from Ref. [373].

PDFs. It is interesting to note that the W^- cross-section is larger than the W^+ cross-section at high muon pseudorapidity, perhaps counter to expectations. The W^+ (W^-) boson at high rapidity is produced primarily from the collision of an up quark (down quark) with large momentum fraction x, and a $\bar{d}$ ($\bar{u}$) anti-quark with low momentum fraction x. As the high-x up quark distribution is larger than that of the down quark (and the low-x anti-quark distributions are essentially equal), one would expect that the W^+ cross-section would be higher than that of the W^- boson. This would be true if the cross-sections were plotted against the rapidity of the W boson, *cf.* Fig. 2.16. However, the muon in the decay of the W^- tends to travel in the direction of the boson

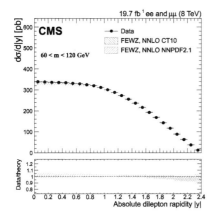 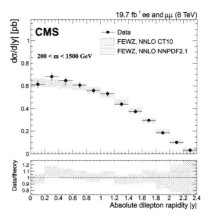

Fig. 9.28 The Drell-Yan differential cross-sections with respect to rapidity, in 2 different mass bins. Comparisons are made to predictions obtained from FEWZ 3.1 using two NNLO PDFs. Reprinted with permission from Ref. [667]

in the boson centre-of-mass frame (and the muon in the decay of the W^+ opposite the direction). Thus, the μ^- leptons tend to have higher transverse momenta than the μ^+ leptons, and binning versus the muon pseudorapidity results in a greater proportion of μ^- at high pseudorapidity (*cf.* the discussion at the end of Section 2.2.3).

9.3.3 Differential cross-sections in transverse momentum

ATLAS [34, 52], CMS [366, 666] and LHCB [57] have extensively studied the transverse momentum distribution of lepton pairs at centre-of-mass energies of both 7 and 8 TeV. The large integrated luminosities allow for differential measurements of the transverse momentum to be carried out both as a function of the lepton pair rapidity and of its mass, allowing for tests of the resummation and parton showering formalisms in different kinematic regions.

In this section, the focus will be on the ATLAS results at 8 TeV. Cross-sections were measured in both the di-muon and di-electron channels. Relatively broad mass ranges were chosen to minimize the effect of QED FSR on the signal acceptance. The transverse momentum distribution in the Z boson mass range for the di-electron and di-muon channels (and the combined result) is shown in Fig. 9.30 (left). The leptons were required to have opposite sign, transverse momenta greater than 20 GeV, and to be within a rapidity (absolute value) of less than 2.4 (2.47 for the electron channel). In addition, an isolation cut was applied in the muon channel to reduce the backgrounds from heavy flavour decays. The isolation requirement induces a sizable p_T dependence in the muon selection efficiency and must be accounted for in each bin. The cross-sections in the two channels were then corrected back to the Born level and combined, with the relevant correction factors between Born, bare and dressed levels specified for the two channels. This is a process which can be measured very accurately. The total uncertainty for the normalized cross-section is less than 1% for transverse momenta up

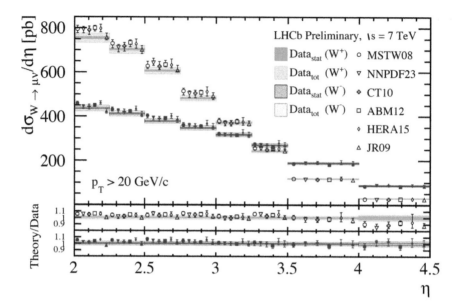

Fig. 9.29 The differential cross-sections for W^+ and W^- production as a function of the muon pseudorapidity. Comparisons are made to NNLO predictions using six PDFs. Reprinted with permission from Ref. [408].

to about 200 GeV. The range at 8 TeV extends from a transverse momentum at 1 GeV to approximately 900 GeV. There is an appreciable broadening of the p_T distribution in going from 1.96 TeV to 8 TeV. This is expected as there is more phase space for gluon radiation at the higher energy, due to the lower average x values for the colliding partons.

The data for all three experiments have been compared to a number of theoretical predictions, varying from NNLO fixed-order, to NNLO, but also including the effects of soft gluon resummation at NNLL, to parton shower Monte Carlos, in some cases including fixed-order information at NLO and with various tunes. A comparison of the normalized cross-section to the resummed predictions from the RESBOS program (NNLO+NNL) is shown in Fig. 9.30 (right). Good agreement with the data is observed below 20 GeV. There is a dip around 40 GeV, and then a rise in the theory prediction with respect to the data at high transverse momentum. The dip is in the transition region where there is a matching between the resummed part of the RESBOS prediction and the fixed-order part. Improvements in this matching should reduce this dip. The rise at high p_T is just an artefact of the scale choice which does not take into account the transverse momentum of the Z boson (and thus results in too small of a scale).

The absolute (un-normalized) high p_T (≥ 20GeV) Z boson cross-section is shown in Fig. 9.31 (left), compared to the NNLO predictions of NNLOJET [565]. The NNLO corrections are relatively small, but result in a significant reduction of the scale uncertainty. The data are above the absolute prediction for most of the Z boson p_T

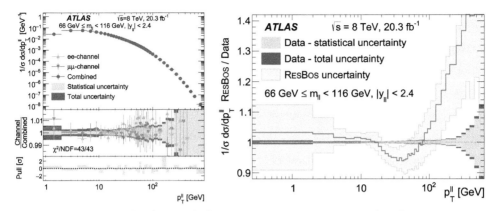

Fig. 9.30 The ATLAS Z boson transverse momentum distribution at 8 TeV is shown for the two Z boson decay channels and their combination (left). On the right, the data are compared to the prediction from RESBOS. Reprinted with permission from Ref. [52].

range. (There is also a 2.8% luminosity uncertainty which is not shown.) If instead, the data and theory are normalized to the total Z boson cross-section in this mass region, the result is a significant improvement in the agreement, as shown in Fig. 9.31 (right). There is a tension between the data cross-section and the theory prediction, also observed to some extent in Fig. 9.25 and Fig. 9.26.

The transverse momentum distributions have also been measured, and compared to NNLOJET, in various mass and rapidity bins. More details can be found in Ref. [52] and Ref. [565].

9.3.4 Vector bosons plus jets

The kinematic reach of the LHC allows for the measurement of W and Z bosons with 7 or more jets. Perturbative QCD theory has reached the point where NLO predictions are available for the production of $W + 5$ jets and $Z + 4$ jets [232, 644], and the cross-sections for both W and $Z+ \geq 1$ jet have been calculated to NNLO [269, 274, 276–278, 564]. Jet transverse momenta of 1 TeV or greater have been measured so far, a region where both QCD and EW higher-order effects become important.

ATLAS has measured W and Z boson plus jets production using the anti-k_T jet algorithm, with a jet size of 0.4 and a jet transverse momentum threshold of 30 GeV [20, 44]. Comparisons to the data are performed using dressed leptons and with parton-level predictions being corrected for the effects of fragmentation and underlying event energy. The inclusive jet multiplicity distribution for W+jets production in ATLAS is shown in Fig. 9.32 compared to predictions at LO and NLO. BLACKHAT +SHERPA, which provides NLO predictions for all jet multiplicities up to 5 (for W+jets), is in good agreement with the data, along with MEPS@NLO, which has NLO information for up to 2 jets. HEJ, which is based on an all-orders summation of terms describing at least two well-separated jets, provides good agreement for jet multiplicities of 2, 3 and

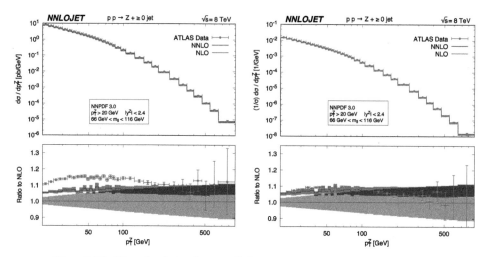

Fig. 9.31 The absolute ATLAS Z boson transverse momentum distribution at 8 TeV is compared to the predictions of NNLOJET. On the right, the normalized data is compared to the normalized prediction from NNLOJET. The data are from Ref. [52]. Reprinted with permission from Ref. [565].

4. ALPGEN +PYTHIA and SHERPA provide good agreement for up to 4 jets in the final state, but the predictions from the two programs diverge for higher jet multiplicities.

The measured transverse momentum distribution for the lead jet for $W+ \geq 1$ jet events is shown in Fig. 9.33, compared to a number of theoretical predictions. It is noticeable that the BLACKHAT +SHERPA predictions undershoot the data at high transverse momentum. In this region, significant contributions are expected from processes such as $qq \to qqW$, basically dijet production with a W boson emitted from a quark line. This process grows with the transverse momentum of the jets primarily because W emission becomes competitive with hard gluon emission when the jet p_T is much larger than the W boson mass.

The LoopSim [757] (*cf.* Section 3.4.2) and BLACKHAT +SHERPA exclusive sums [134] predictions include more contributions from such final states, but these can be seen to have little impact. Note that EW corrections at 1 TeV are negative, which would increase the size of the discrepancy. SHERPA and ALPGEN +PYTHIA each provide better agreement with the higher p_T range, but with the larger theoretical uncertainties (not shown) inherent with LO predictions. The prediction from MEPS@NLO, which includes NLO information for $W + 1,2$ jets, is still below the data at high transverse momentum, but closer than BLACKHAT +SHERPA. A similar effect, albeit limited to somewhat smaller p_T can be seen with Z+jets in ATLAS [20]. However, the situation is not as clear for W/Z+jets measurements in CMS [660, 668].

The agreement for the inclusive lead jet transverse momentum distribution is better when compared to the NNLO theory prediction from Ref. [276], as shown in Fig. 9.34. It is interesting as well that better agreement at high p_T is observed with the NLO prediction from this paper, albeit with a large scale uncertainty. The discrepancy

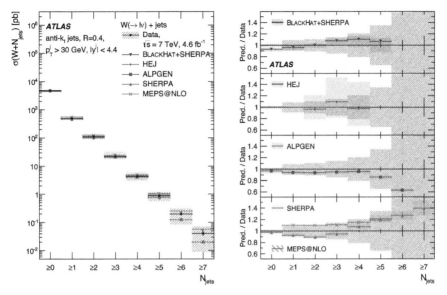

Fig. 9.32 The cross-section for the production of a W boson plus jets as a function of the inclusive jet multiplicity. The statistical uncertainties for the data are shown by the vertical bars, and the combined statistical and systematic uncertainties are shown by the black-hashed regions. The data are compared to predictions from BLACKHAT +SHERPA, ALPGEN +PYTHIA, SHERPA, and MEPS@NLO. The left-hand plot shows the differential cross–sections and the right-hand plot show the ratios of the predictions to the data. Reprinted with permission from Ref. [44].

between the two NLO predictions may be due to different forms for the central scale used in the two theory calculations, an indication that the optimal choice of scale can often be difficult.

Similar comparisons are shown in Fig. 9.35 for the exclusive final state in which only one jet is present (above the jet p_T threshold of 30 GeV). The transverse momentum range is more limited than for the inclusive case as the production of a very high p_T jet, and no other jets, is very strongly Sudakov-suppressed. In contrast to the inclusive case, the BLACKHAT +SHERPA prediction is in very good agreement with the data. This is somewhat of a surprise, as the presence of two disparate scales (the p_T of the lead jet compared to the jet p_T threshold of 30 GeV) should lead to the presence of large logarithms, which should spoil any agreement with the fixed-order prediction. Here again, though, note that the EW corrections are even larger (and negative) for the exclusive case than for the inclusive case. A similar level of agreement was observed for the ATLAS $Z+1$ jet exclusive jet p_T distribution. This mystery was solved in Ref. [273], where it was shown that the ATLAS analysis removed only the jet (and not the event) in the situation where there is an overlapping jet and a lepton (within $\Delta R < 0.4$). Thus, an event classified as a W/Z plus exactly one jet event may also have a second

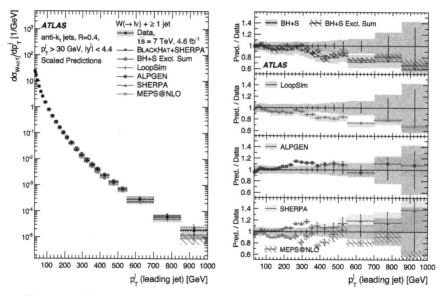

Fig. 9.33 The cross-section for the lead jet transverse momentum for $W+ \geq 1$ jet events, with theory comparisons as in Fig. 9.32. The theoretical predictions have been scaled to the data to allow for easier comparisons of the shapes. Reprinted with permission from Ref. [44].

jet, in close proximity to a lepton, a configuration which has a collinear enhancement. This feature of the analysis (and of similar analyses in ATLAS) has since been removed.

The cross-section for the fifth leading jet transverse momentum is shown in Fig. 9.36. Of course, the dynamic range is smaller than for the leading jet p_T, but it is evident that the NLO predictions describe the data well even for a $2 \rightarrow 6$ process. Finally, the H_T distribution for events with $W+ \geq 1$ jet is shown in Fig. 9.37. As a reminder, H_T is the scalar sum of the transverse momenta of all jets and leptons (including the missing transverse momentum from the neutrino) in the event. Here the NLO predictions from BLACKHAT +SHERPA are a factor of two below the data for H_T values of the order of 2 TeV. For high H_T, as for high p_T^{jet1}, the dominant sub-process becomes $qq \rightarrow qqW$, where a W boson is emitted off of one of the quark lines. This subprocess is present as a real correction to $W+ \geq 1$ jets at NLO. Formalisms that include the virtual corrections for $W+ \geq 2$ jets, such as BLACKHAT +SHERPA exclusive sums, or LoopSim, reduce, but do not eliminate the discrepancy. MEPS@NLO, which has the one and two jet matrix elements at NLO included, provides a good description over the full dynamic range. Better agreement with the data is also seen using the NNLO calculation from Ref. [276], as observed in Fig. 9.38. The NNLO $W+ \geq 1$ jet prediction naturally includes the $W+ \geq 2$ jet cross-section at NLO.

The exclusive jet multiplicity distribution is shown for $Z+$jets in ATLAS in Fig. 9.39 (left) [20]. Similar to the $W+$jets analysis, a transverse momentum cut of 30 GeV and an absolute rapidity cut of 4.4 have been applied. The exclusive jet multiplicity

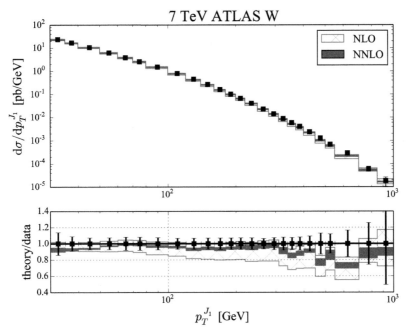

Fig. 9.34 The cross-section for the lead jet transverse momentum for $W+ \geq 1$ jet events, with theory comparisons from Ref. [276]. Reprinted with permission from Ref. [276].

distribution for Z+jets (as for W+jets) follows a *staircase* pattern indicative of Berends scaling, as discussed in Sections 4.1 and 4.3. This scaling is a result of the same p_T cut being applied to every jet, and to the non-Abelian nature of the gluon branching process, *i.e.* each final state gluon itself carries a colour charge and can itself radiate. If instead, there is a difference in the transverse momentum threshold between the leading jet in the event, and any additional jets, a Poisson-type behaviour will instead be evident. In Fig. 9.39 (right),the lead jet p_T is required to be greater than 150 GeV, while still retaining a cut of 30 GeV on the other jets. The result is described well by a Poisson scaling,

$$\frac{\sigma\left(Z + (n+1) \text{ jets}\right)}{\sigma\left(Z + n \text{ jets}\right)} = \frac{\bar{n}}{n}, \tag{9.7}$$

with an expectation value $\bar{n} = 1.04 \pm 0.04$. Note that there is basically no suppression for the second jet emission, given the core process of Z+jet, with the jet having a transverse momentum greater than 150 GeV. For both situations, the data are well-described by the theoretical predictions.

9.3.5   Vector bosons plus heavy flavours

The case where a vector boson is produced in association with one or more jets that originates from a heavy quark, either a *b*- or a *c*-quark, is especially interesting. Such

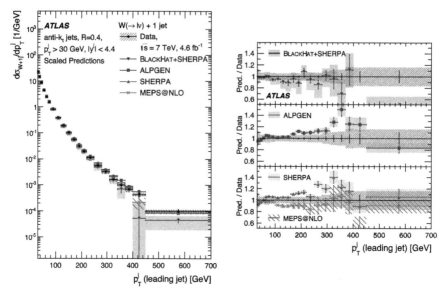

Fig. 9.35 As for Fig. 9.33, but for the lead jet transverse momentum in $W+$ exactly one jet events. Reprinted with permission from Ref. [44].

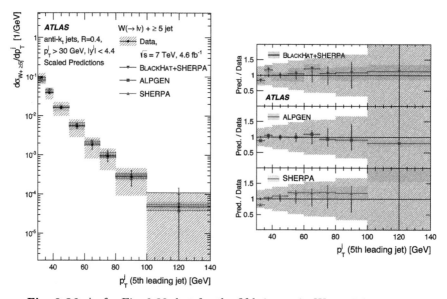

Fig. 9.36 As for Fig. 9.33, but for the fifth jet p_T in $W+ \geq 5$ jet events. Reprinted with permission from Ref. [44].

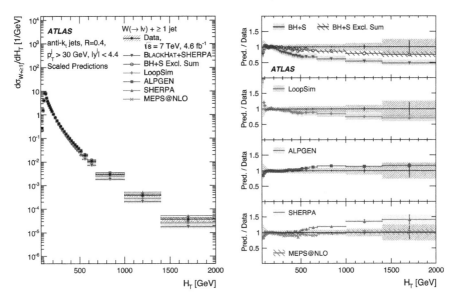

Fig. 9.37 As for Fig. 9.33, but for the H_T distribution in $W + \geq 1$ jet events. Reprinted with permission from Ref. [44].

processes are important as backgrounds for other physics studies, such as associated Higgs boson production (VH), where the Higgs boson decays into a $b\bar{b}$ final state, for single top production [14] or for searches for physics beyond the Standard Model [23]. However, they are also interesting from the standpoint of perturbative QCD, since the presence of a heavy quark mass scale introduces additional complications into the calculation. As in the case of single top production, *cf.* Section 4.6, the calculation may either be performed in a 4-flavour scheme (in which the only active quark flavours are u, d, c, s) or the 5-flavour scheme, in which the b parton is also present in the initial state. As an example, consider the Born-level predictions for the production of a final state containing a W-boson and at least one b-jet. In the 4-flavour scheme this can be described by the processes $q\bar{q} \rightarrow Wb\bar{b}$ and $gq \rightarrow Wb\bar{b}q$, where the presence of the initial-state gluon in the latter is important at the LHC. In the 5-flavour scheme this second process is replaced by $bq \rightarrow Wbq$. The advantages and drawbacks of each scheme have been summarized in Section 4.6.

Such processes have been measured at the TEVATRON by both the CDF [62] and DØ [92] collaborations. The CDF collaboration found a result for the $W + b$ jet cross-section larger than the SM prediction, while DØ measured a cross-section smaller than the SM prediction, although both were consistent within the quoted theoretical uncertainties. Such measurements have also been carried out at the LHC [11, 381].[9] ATLAS, for example, has measured the fiducial $W + b$ jet cross-section, as a function of the jet multiplicity and as a function of the $b - jet$ transverse momentum [19].

---

[9]Similar results are also available for $Z + b$ jet cross-sections [26, 367, 372, 382].

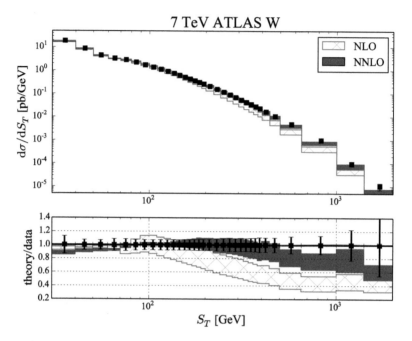

Fig. 9.38 The cross-section for the H_T (also known as S_T) distribution for $W+ \geq 1$ jet events, with the theory comparisons from Ref. [276]. Reprinted with permission from Ref. [276].

The results are reported either with the single-top contribution subtracted, or not-subtracted. They have the same (inclusive) final state, but single top production has different kinematics, so can be separated.

As for other measurements involving W boson decay, a high p_T isolated lepton (either a muon or electron) is required, along with a requirement on a minimum missing transverse energy. The analysis requires one or more jets, with one (and only one) jet being tagged as a b-quark jet. It is necessary to veto on events with two or more b-tagged jets to reduce the background from the (sizeable) $t\bar{t}$ cross-section. Jets are reconstructed with the anti-k_T ($D = 0.4$) jet clustering algorithm and a jet threshold of 25 GeV. The jets are required to have a rapidity $|y| < 2.1$, so that the jet lies within the tracking (and thus b-tagging) region, and any jets within a distance $\Delta R = 0.5$ of the lepton candidate are removed. Jets are tagged as originating from b-quarks using a combination of two tagging algorithms. The first algorithm involves either explicit reconstruction of a secondary vertex consistent with originating from a b-quark decay. The second calculates the impact parameter significance of each track within the jet to determine the probability of the jet being a b-quark jet.

The measurement of this process has backgrounds, both from processes where a real b is present in the final state (single-top, $t\bar{t}$, and multi-jet), and from processes where a jet has been mis-tagged as a b-jet (such as $W + c - jet$ and $W +$ light

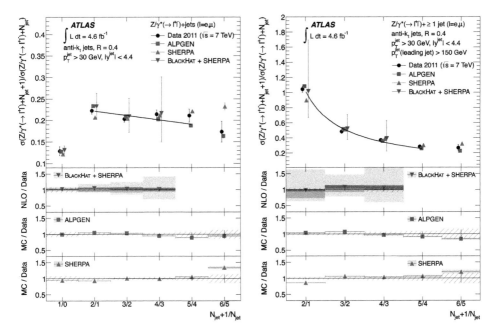

Fig. 9.39 Ratios of exclusive jet multiplicity distributions for Z+jets in ATLAS with (left) the standard selection cuts, and (right) requiring at least one jet with transverse momentum above 150 GeV. Reprinted with permission from Ref. [20].

jets). The backgrounds can be largely determined from the data itself, either by the presence of different kinematics (for the first category of backgrounds), or from different characteristics of the b-tagged jets (for the second category). For example, in the 2-jet bin, the contributions from $W + b$ and single-top processes are comparable. However, since most of the single top events have a relatively narrow $(W, b-\text{jet})$ mass distribution around the top quark mass, this distribution can be used to discriminate the two processes.

The measured (unfolded) cross-section for the production of a W boson and a b-jet is shown in Fig. 9.40 as a function of the jet multiplicity. Results are shown for the electron, muon and combined (electron+muon) channels, and comparisons are made to theoretical predictions from fixed order (MCFM) and parton shower Monte Carlo predictions (POWHEG +PYTHIA and ALPGEN +HERWIG). The Monte Carlo predictions use the 4-flavour scheme, while the MCFM prediction includes higher-order corrections from the 5-flavour scheme, *i.e.* allowing b-partons in the initial state. The fixed-order predictions have been corrected for non-perturbative effects and for double-parton scattering, where a W boson, and a $b\bar{b}$ final state, are produced in two separate proton-proton collisions. The latter cannot be ignored for this measurement at the LHC, and amounts to a 25% correction to the total cross-section, concentrated in the lowest p_T bins. The kinematics of this process are complex and are reflected in the choice of the central scale for the theoretical calculations,

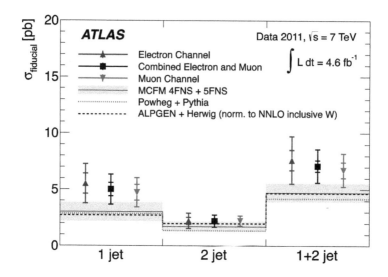

Fig. 9.40 The measured $W + b$-jet cross-section is compared to several theoretical predictions, in the one-jet, two-jet and one-jet+two-jet bins. Reprinted with permission from Ref. [19].

$$\mu_F^2 = \mu_R^2 = m_{\ell\nu}^2 + p_\perp^2(\ell\nu) + \frac{m_b^2 + p_\perp^2(b)}{2} + \frac{m_{\bar{b}}^2 + p_\perp^2(\bar{b})}{2}. \qquad (9.8)$$

Somewhat conservatively, the scale is varied by a factor of four about the central scale, rather than the traditional factor of two.

The results for the 1-jet bin are slightly higher than the theoretical predictions (but within the combined experimental and theoretical uncertainty bands) while the results for the 2-jet bin are in good agreement with the theory. The differential cross-sections are shown in Fig. 9.41, for the 1-jet (left) and 2-jet (right) bins. In both cases, the agreement of data with theory worsens as the b-jet transverse momentum increases.

CMS has measured the $Wb\bar{b}$ final state in which both jets are required to be tagged as b-jets [381]. There the results, over a similar phase space as the ATLAS measurement (except for the requirement of 2 b-tags rather than a restriction to one and only 1), are in good agreement with the SM prediction. The ΔR distribution between two b-jets, for the process $Z+ \geq 2\, b$ jets, has been measured in Ref. [26], and is shown in Fig. 9.42, compared to a variety of predictions. Good agreement is observed with the fixed-order prediction from MCFM, with SHERPA and with aMC@NLO, in the 4-flavour scheme. It is noteworthy, however, that no prediction describes the first bin well, when the two b-jets approach collinearity,indicating perhaps a difficulty in describing the collinear splitting of a gluon into a $b\bar{b}$ final state. This was also a difficulty at the TEVATRON, and could have implications for other final states involving $b\bar{b}$ pairs at the LHC, such as for associated Higgs boson production, especially in the boosted region.

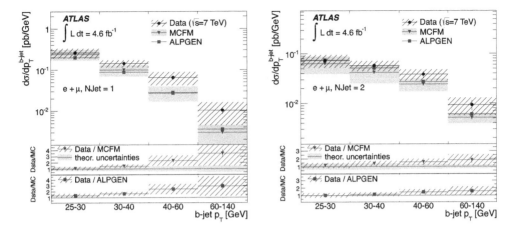

Fig. 9.41 The measured $W + b$-jet differential jet cross-section, compared to theoretical predictions from ALPGEN and MCFM, in the one-jet (left) and two-jet (right) bins. The MCFM cross-sections have been corrected for non-perturbative effects. The ALPGEN prediction has been interfaced to HERWIG and JIMMY and has been scaled to the inclusive NNLO W cross-section. The ratios between the measured and predicted cross-sections are also shown. Reprinted with permission from Ref. [19].

9.3.6 Single photons

At the LHC, in contrast to the TEVATRON, the single inclusive photon production process is dominated by the gq initial state for moderate to high E_T photons. Thus, a precision measurement of the process provides information on the gluon distribution complementary to that provided by the inclusive jet cross-section. The energy and the direction of the photon can be measured very well (better than that for a comparable jet); thus, in principle the photon cross-section can be measured with greater precision than the jet cross-section. Photon production does suffer, however, from backgrounds resulting from jets that fragment into one or more mesons (such as neutral pions) carrying a large fraction of the parent jet's momentum, as discussed in Section 8.4. As at the TEVATRON, this background can be reduced by imposing an isolation cut, and quality identification cuts, on the photon candidates. The impact of the isolation cut is not only to reduce the background due to jets, but also to reduce the photon production cross-section due to fragmentation processes.

In ATLAS [30, 51], an isolated photon is defined as one in which there is a restricted amount of additional energy around the photon candidate in a cone of radius $R = 0.4$.[10] The energy in the isolation cone has already been corrected for both the underlying event and the effects of multiple interactions. In Fig. 9.43 is shown (left) the energy distribution in the isolation cone for *tight* and *non − tight* photon candidates in the

---

[10]For example, for the 8 TeV analysis, the requirement on the isolation energy is: $E_T^{iso} < 4.8$ GeV $+$ $4.2 \times 10^{-3}\, E_T^{\gamma}$.

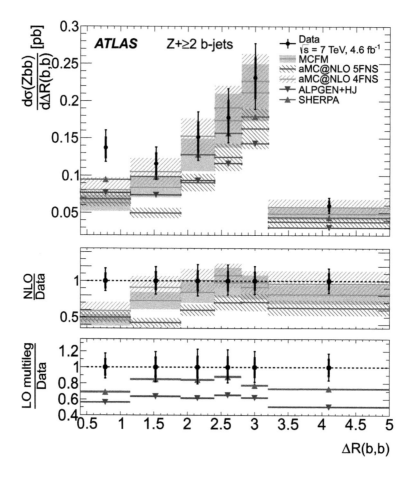

Fig. 9.42 The measured ΔR distribution between 2 b jets, in $Z+ \geq 2\, b$ jet events, compared to a variety of predictions. Reprinted with permission from Ref. [26].

7 TeV data from Run I. The non-tight photons fail the tight identification criteria in one or more categories, and thus are likely to be produced as a result of jet fragmentation. In general, the non-tight photon candidates are less isolated than the tight ones. By determining the fraction of loose photons present in the tight isolation region, the photon backgrounds can be determined for each kinematic bin. The resulting photon purity is shown in Figure 9.43 (right) for two rapidity intervals in the 7 TeV analysis. The imposition of the isolation cut results in a high photon purity, which approaches 100% for high photon transverse energy. As stated in Section 8.4, an isolated high transverse energy photon candidate is much more likely to be a true photon, than to be a product of jet fragmentation.

The resulting inclusive photon cross-sections for a centre-of-mass energy of 8 TeV

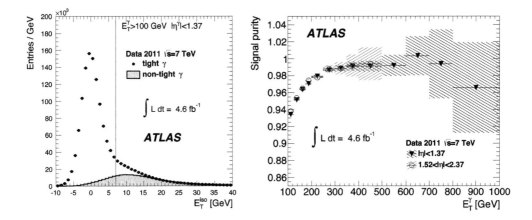

Fig. 9.43 Left: the transverse energy in the isolation cone for photon candidates (both tight and non-tight) for photon transverse energies above 100 GeV. The non-tight distribution has been normalized to the tight distribution in the (background-rich) region above 15 GeV. Right: the resulting photon purity is shown as a function of the photon transverse energy in the ATLAS barrel and end-cap regions. The shaded bands indicate the statistical uncertainties. Reprinted with permission from Ref. [30].

are shown in Fig. 9.44 for four rapidity intervals. A large dynamic range is represented in this plot, from 30 GeV to over 1 TeV. There was a considerable reduction in the size of the experimental systematic errors from the 7 TeV analysis to the 8 TeV analysis. The data is systematically below the NLO predictions from JETPHOX [348] for transverse momenta below 500 GeV. A somewhat better agreement with the data is achieved with the PeTeR prediction [208]. Here, the calculation is again carried out at NLO, but in addition threshold logarithms are resummed at next-to-next-to-leading accuracy. Note that since the calculations are at NLO, the scale uncertainty is still sizable. The uncertainty will be reduced once a NNLO calculation for inclusive photon production is completed, which will add to the attractiveness of the process as an input into global PDF fits.[11]

Similar methods have been used by CMS [359, 384] for measuring the isolated inclusive photon cross-section, with similar results obtained. More potential information on the PDFs of the colliding protons can be obtained by measuring the distribution of the accompanying jet, in addition to the photon, at the cost of a somewhat less-inclusive theoretical prediction [13, 384]. As at the TEVATRON, the inclusive photon+jet data are very useful for the calibration of the jet energy scale.

---

[11]Late in the editing of this book, the NNLO calculation has in fact been completed, in Ref. [316].

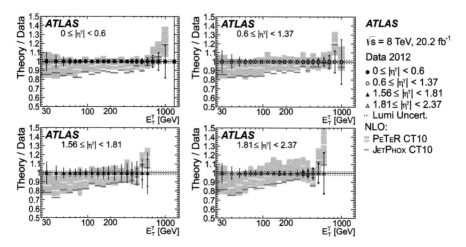

Fig. 9.44 The ATLAS isolated photon cross-section, for four rapidity region. The inner error bars on the data points indicate statistical errors only, while the outer error bars show the statistical and systematic errors added in quadrature. The error band given is that from the PeTeR calculation, and corresponds to the combination of the scale, PDF and electroweak undertainties. Reprinted with permission from Ref. [51].

9.4 Vector boson pairs

9.4.1 Diphotons

One of the key Higgs boson final states at the LHC is the decay into two photons. The Higgs boson diphoton signal is typically swamped by the much larger QCD diphoton production rate, but with fine enough diphoton mass resolution, the presence of a narrow Higgs boson resonance can be detected (and indeed has been). Nevertheless, it is important to understand QCD production of diphotons, especially as approximately half of the production proceeds through gg scattering, the same initial state that dominates Higgs boson production. A cut is placed on the transverse energy of the leading (second leading) photon of 25 GeV (22 GeV). The slight asymmetry helps to reduce any instability in the higher-order calculations. As for single photon production, backgrounds due to jet fragmentation are suppressed by the imposition of an isolation cut, similar to the one used for the inclusive photon measurement. This isolation cut also suppresses production mechanisms where one or both photons results from photon fragmentation from a quark line. Results from an ATLAS measurement of diphoton production at 7 TeV [18] are shown in Fig. 9.45 for the diphoton transverse momentum. As at the TEVATRON, there is a shoulder (the *Guillet shoulder*) at a transverse momentum of approximately 50 GeV, corresponding to events that also have a low azimuthal separation. The agreement with the DIPHOX +GAMMA2MC prediction is poor for these two variables, but much better for the 2γNNLO and SHERPA predictions. The key aspect for both of the latter calculations is the presence of tree-level $2 \rightarrow 4$ processes, needed to describe the two observables.

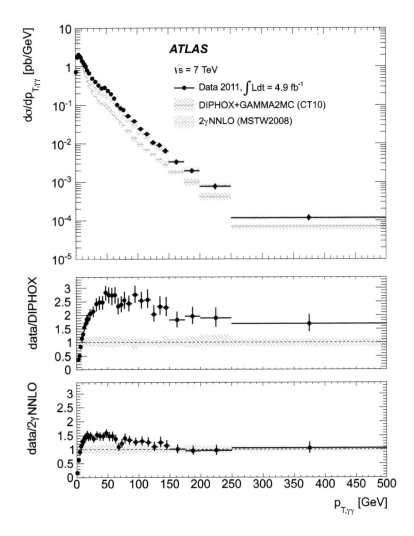

Fig. 9.45 The ATLAS diphoton transverse momentum data compared to predictions from DIPHOX +GAMMA2MC, 2γNNLO and SHERPA. The error bars correspond to the total experimental uncertainties, which are dominated by the systematic errors. Reprinted with permission from from Ref. [18].

The diphoton mass distribution is shown in Fig. 9.46. The agreement with the 2γNNLO prediction is good for the entire mass range; there is a disagreement with the DIPHOX prediction at low diphoton masses, where the $2 \to 4$ subprocesses are important, and to a lesser extent for masses of a few hundred GeV, where NNLO corrections may be important.

Similar results have been obtained by CMS [378].

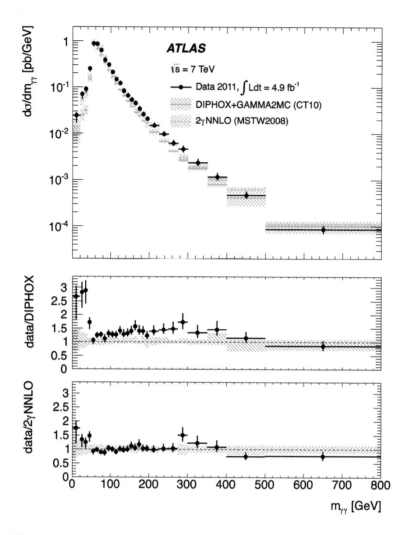

Fig. 9.46 The ATLAS diphoton data compared to predictions from DIPHOX +GAMMA2MC, 2γNNLO and SHERPA. The error bars correspond to the total experimental uncertainties, which are dominated by the systematic errors. Reprinted with permission from Ref. [18].

9.4.2 Dibosons

The measurement of diboson final states provides an important test of the non-Abelian nature of the Standard Model, and a sensitivity to anomalous triple gauge boson couplings. In addition, the WW and ZZ final states are important for the measurement of the Higgs boson decays into those two channels. There is a large background for the measurement of WW production from $t\bar{t}$ production; thus, commonly a jet veto requirement is applied to reduce the latter. Cross-sections then are commonly corrected for the geometric and kinematic acceptances as well as for the impact of the jet veto to obtain a fully inclusive cross-section [21, 53, 375, 673], more easily compared to theoretical predictions. The imposition of a jet veto restricts the phase space for gluon emission and thus results in an increased uncertainty in the predicted cross-section. However, in the case of diboson production, the scale dependence at NLO is inherently small, and thus the increased uncertainty for the vetoed cross-section is still less than the experimental systematic uncertainty, and may not represent the full theoretical uncertainty.

WW production is typically measured in the final state where both W bosons decay into a lepton (electron or muon) and a neutrino. Thus, the signature consists of two high p_T leptons and a substantial amount of missing transverse energy. The dominant sub-process is $q\bar{q} \to WW$, with much smaller contributions (approximately 10% total) from $gg \to WW$ and $gg \to H \to WW$. The main backgrounds come from Drell-Yan, top ($t\bar{t}$ and single top) production, W+jets, and diboson (WZ, ZZ, $W\gamma$) production. The top background is suppressed by the requirement that there can be no jets with transverse momentum above 25 GeV within a rapidity interval of ± 4.5.[12]

The fiducial cross-section is corrected for identification and isolation requirements and detector resolution effects. The total WW cross-section is calculated by correcting the fiducial cross-section for the extrapolation to the full WW phase space. A small excess with respect to the NLO predictions is observed by ATLAS, with similar results obtained in CMS. There has been speculation that this excess might be the result of new physics [423, 424, 522, 647, 828]. However, there are a few caveats. The WW cross-section has recently been calculated to NNLO, resulting in an increase over the NLO result of the order of 10% [557], thus significantly decreasing the excess. (A further 2% increase in the theoretical prediction results from a 2-loop calculation of the subprocess $gg \to WW$ [324].) In addition, Ref. [768] points out that the ATLAS fiducial cross-section is in agreement with the theoretical prediction for the same, and the disagreement for the total cross section results from the extrapolation to the full phase space using the POWHEG box. The Monte Carlo result for the jet-veto efficiency overestimates Sudakov suppression effects with respect to a calculation using analytic resummation. This is one of the perils of comparisons at the fully inclusive (corrected) level, compared to fiducial comparisons.

A comparison of the NLO and NNLO predictions, and of ATLAS and CMS data, for WW production from Ref. [557] is shown in Fig. 9.47.

Other diboson final states (for example ZZ, WZ, $W\gamma$, $Z\gamma$) have also been measured by both ATLAS and CMS [22, 41, 375, 665]. A summary of CMS results is shown in Fig. 9.48, where good agreement with NLO and NNLO Standard Model predictions is observed. Differential distributions can be used to place limits on anomalous couplings. For example, in Fig. 9.49 are shown (left) the unfolded transverse momentum distribution for the leading Z boson and (right) the four-lepton reconstructed mass distribution, both from CMS [664]. The presence of anomalous triple gauge couplings would manifest itself as deviations from the Standard Model predictions at high Z p_T/ high four-lepton mass. In both cases, good agreement with the Standard Model predictions is observed.

---

[12]These specific cuts are for ATLAS but are similar for CMS.

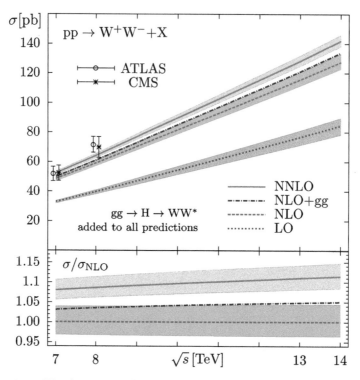

Fig. 9.47 The ATLAS and CMS WW cross-sections compared to the predictions at NLO and NNLO from Ref. [557]. Reprinted with permission from Ref. [557].

9.4.3 Vector boson backgrounds for BSM searches

The Standard Model has been extremely successful at LEP, HERA, the TEVATRON and now at the LHC. However, as discussed earlier, the SM is incomplete, and one of the main goals of the LHC is the discovery of new physics beyond the current paradigm. Such Beyond-the-Standard Model (BSM) physics is mostly expected at high mass scales, where the energies of previous colliders would not have been sufficient to discover it already. The signatures of BSM physics are mostly comprised of the same observables considered in this chapter: photons, leptons, jets (with or without b-tags) and missing transverse energy, but with cuts appropriate to the expected higher mass scale. Often the dominant contribution to these final states comes from the production of vector bosons, either singly or in pairs, plus jets. In that case the measured SM cross-sections can serve to determine the backgrounds to new physics processes, for instance by extrapolating to new kinematic regions. The possibility that new physics (such as stop pair production) could be *hiding* in the WW cross-section measurement, for final states involving two leptons and large missing transverse energy, was already mentioned in Section 9.4.2.

As an example, consider a BSM search performed by CMS using the full 2012 data

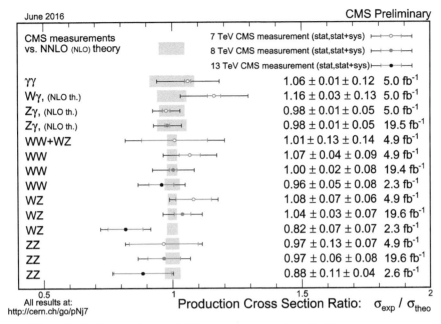

Fig. 9.48 Comparison of CMS measurements of diboson cross-sections at 7, 8 and 13 TeV with NLO and NNLO predictions, from `twiki.cern.ch/twiki/pub/CMSPublic/PhysicsResultsCombined/`. Reprinted with permission from CERN.

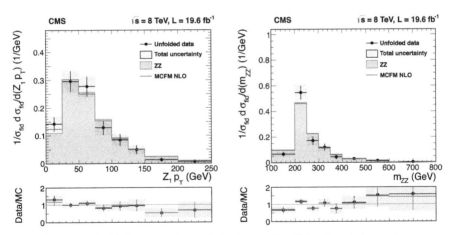

Fig. 9.49 (left) A comparison of the (normalized) leading Z boson transverse momentum distribution from CMS to a NLO prediction from MCFM. (right) A comparison of the (normalized) four-lepton mass distribution from CMS to a NLO prediction from MCFM. Reprinted with permission from Ref. [664].

sample at 8 TeV, targeting new physics in multi-jet final states [386]. To focus on the region most sensitive to BSM signals, the search requires both large amounts of total transverse energy of the jets (H_T) and missing transverse energy $\slashed{H}_T$. Specifically, the search region is defined by the requirements that there be three or more jets, with $p_T \geq 50$ GeV and $|\eta| \leq 2.5$, a total transverse energy sum of the jets greater than 500 GeV, and a missing transverse energy greater than 200 GeV. This final state is sensitive to the production of pairs of squarks and gluinos, where the squarks (gluinos) each decay into one (two) jets and a lightest supersymmetric partner (LSP). There are substantial contributions to this final state from $Z(\to \nu\bar{\nu})$ + jets and from $W(\to \ell\nu)$+jets (and $t\bar{t}$ production), when an electron or muon is lost or when a τ lepton decays hadronically. The cross-section for $Z(\to \nu\bar{\nu})$ + jets is estimated using the larger measured cross-section for γ+jets, correcting for the electroweak coupling differences, and making use of the similar kinematic properties exhibited by the two processes. To reduce the backgrounds from W+jets and $t\bar{t}$ processes, events with isolated electrons or muons with transverse momentum greater than 10 GeV are vetoed. The surviving events are those in which any leptons escape detection, and thus a good knowledge of the lepton reconstruction efficiency is necessary to accurately predict this background.

In addition the analysis is subdivided into bins that correspond to the number of jets in the final states, 3–5 jets, 6–7 jets and 8 or more jets, in order to retain sensitivity to possible longer cascades of squark and gluino decays. The CMS data is shown as a function of $\slashed{H}_T$ for one of the jet bins in Figure 9.50, and compared to the predicted backgrounds and possible signals for several squark and gluino production and decay modes. The number of events observed in the data is consistent with the number expected from SM background processes. For low jet multiplicities the primary background for high values of $\slashed{H}_T$ is from $Z(\to \nu\bar{\nu})$ + jets events, whereas in the highest jet multiplicity bin the largest background comes from W+jets and $t\bar{t}$ events. Smaller backgrounds result from QCD multi-jet production, where the large $\slashed{H}_T$ is produced primarily from heavy flavour decays inside the jets or from jet energy mis-measurement. This background is more significant (but still sub-leading) for the higher jet multiplicity bins. The larger the number of jets, the greater is the chance that one or more jets will contribute significant $\slashed{H}_T$ due to the causes mentioned above.

9.5 Tops

9.5.1 Distributions: top-pairs (plus jets)

Measurement of the top pair cross-section at the LHC allows for precision tests of QCD, particularly with the theoretical prediction now known to NNLO. The top pair cross-section is significantly larger at the LHC than at the TEVATRON, partially because of the dominance of the gg initial state for top production at the LHC, and the rapid increase of the gg PDF luminosity with energy, as discussed in Chapter 6. A comparison of the cross-section measurements at the TEVATRON and at the LHC (7 and 8 TeV) is shown in Fig. 9.51. Top pair production can be measured in a number of final states, depending on the decay modes of the W bosons that are produced. The most useful of these states have at least one leptonic W-decay, with any leptons produced at high-p_T and well-isolated. In the dilepton mode the two leptons are of opposite charge and

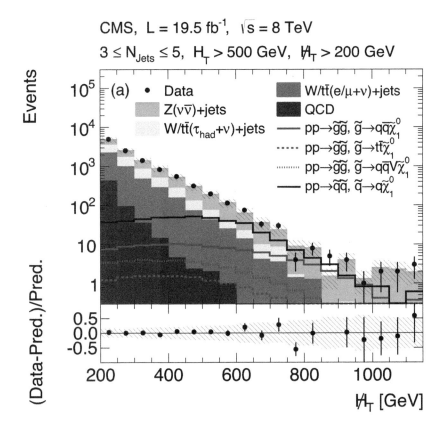

Fig. 9.50 The observed $\slashed{H}_T$ distributions from CMS for events with $H_T \geq 500$ GeV and jet multiplicities from 35. The data are compared to the SM backgrounds and the predictions for several different SUSY scenarios. Reprinted with permission from Ref. [386].

are accompanied by two jets. In the case where one W-boson decays hadronically, the final state corresponds to a single lepton and three or four jets. The jets should have transverse momenta above $25-30$ GeV and at least one of them should be tagged as a b-jet. The jet threshold is higher at the LHC than the TEVATRON because of the larger underlying event, as well as the greater pileup through much of the running at 7 and 8 TeV. The better tracking detectors in ATLAS and CMS than at the TEVATRON have resulted in higher b-tagging efficiencies, typically of the order of 80%, compared to the 50-60% efficiencies for CDF and DØ. ATLAS and CMS both use the anti-k_T jet algorithm for jet reconstruction, with jet sizes of 0.4 and 0.5 respectively used for the two experiments (at 7 and 8 TeV). The cross-sections determined from the different final states agree with each other, as do the results from the two experiments. The experimental results also agree well with the NNLO+NNLL predictions of Ref. [427].

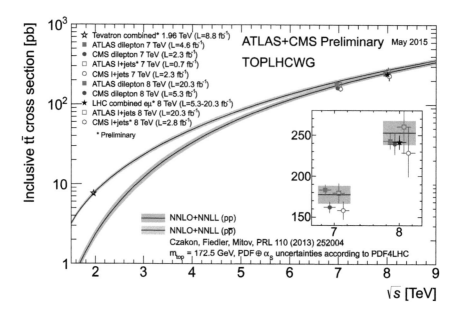

Fig. 9.51 A compilation of top pair production cross-sections as a function of centre-of-mass energy from the LHCTopWG (`twiki.cern.ch/twiki/bin/view/LHCPhysics/LHCTopWG`) compared to the NNLO+NNLL predictions of Ref. [427]. Reprinted with permission from CERN.

It is notable that, despite the impressive precision of the theoretical prediction, the experimental errors are smaller still.

The top mass can be measured as well from the same final states used in the cross-section measurements. A compilation of the top mass measurements at ATLAS and CMS compared to the TEVATRON average, the LHC average, and the world average, is shown in Fig. 9.52. The precision at the LHC is still not at the same level as the final measurements from the TEVATRON, but it should surpass the latter in Run 2. Theoretical issues (such as recombination effects and uncertainties as to what exactly the measured top mass represents), discussed in Section 4.5, will now have to be addressed.

Differential measurements of $t\bar{t}$ final states allow for additional precision tests of perturbative QCD, probes of high mass regions sensitive to new physics, and more detailed information on $t\bar{t}$ kinematics useful for PDF determination. Differential predictions for $t\bar{t}$ production at NNLO have been recently calculated [426, 429]. Differential measurements have been performed for such variables as the $t\bar{t}$ mass, the $t\bar{t}$ rapidity distribution, the transverse momentum of the top quark, and the $t\bar{t}$ transverse momentum distribution (as well as others). In Ref. [56], the experimental results have been unfolded both to a fiducial particle-level phase space and to a fully-corrected phase space. The former has less of a model dependence, and thus smaller uncertainties, and the latter is often more appropriate for comparison to higher-level predictions.

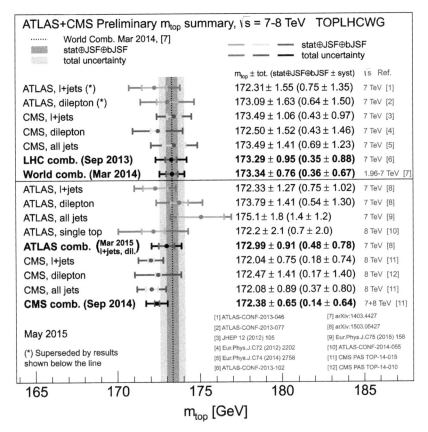

Fig. 9.52 A compilation of top mass determinations from ATLAS and CMS compared to the LHC and world averages, from `atlas.web.cern.ch/Atlas/GROUPS/PHYSICS/CombinedSummaryPlots/TOP/`. Reprinted with permission from CERN.

The $t\bar{t}$ mass and rapidity distributions, corrected to the full phase space, are shown in Fig. 9.53 for the ATLAS 8 TeV measurements. (Results at 7 TeV for ATLAS can be found at [40] and 7 and 8 TeV results for CMS can be found at [371] and [662]. It is noteworthy that ATLAS and CMS have adopted the same binning for the fully-corrected distributions, allowing for easier future combinations.) The data are compared both to a POWHEG +PYTHIA 6 prediction and to the NNLO differential (fixed-order) prediction, both using the MSTW2008 NNLO PDFs. Good agreement with the data is observed for both predictions for the $m_{t\bar{t}}$ distribution, while the NNLO prediction agrees better with the $y_{t\bar{t}}$ data. In general, the agreement seems to be better for NNLO comparisons than for NLO comparisons, and better with the more recent PDFs than with previous generations. There is still a sizable PDF sensitivity, especially at high $m_{t\bar{t}}$ and $y_{t\bar{t}}$, paving the way for the inclusion of this data into global PDF fits. One

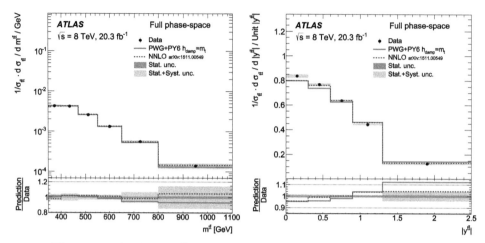

Fig. 9.53 The ATLAS $t\bar{t}$ normalized, differential cross-sections for the $t\bar{t}$ mass (left) and $t\bar{t}$ rapidity (right), compared to theoretical predictions at NNLO using the MSTW2008 PDFs. Reprinted with permission from Ref. [56].

caveat is that electroweak effects, which can be sizable, are also not yet available for all observables [705].

Measurements of the $t\bar{t}$ asymmetry at the LHC are much more difficult than at the TEVATRON due both to the symmetric nature of the colliding beams, and the dominance of the (symmetric) gg subprocess for $t\bar{t}$ production. Also, as a result of this symmetry, the forward-backward asymmetry measured at the TEVATRON ($\Delta y = y_t - y_{\bar{t}}$) is no longer a useful variable, and the charge asymmetry instead ($\Delta y = |y_t| - |y_{\bar{t}}|$) must be measured. Perturbative QCD predicts top anti-quarks to be produced more centrally than top quarks. The expected value for this asymmetry is smaller than the forward-backward asymmetry measured at the TEVATRON. A number of measurements have been carried out by both ATLAS and CMS of the inclusive charge asymmetry [10, 32, 50, 363, 365, 670, 672, 674], and the results are in agreement with the NLO+EW prediction, as observed in Fig. 9.54. Measurements have also been made for high $t\bar{t}$ mass ($m_{t\bar{t}} > 600$ GeV) and for boosted $t\bar{t}$ systems ($\beta_{t\bar{t}} > 0.6$) [32], where the effects of any new physics may be expected to be magnified [128]. No deviation from the SM predictions is observed for these special kinematic regions. Note that unlike the TEVATRON, NNLO predictions for this observable are not yet available for the LHC (at the time of this book).

9.5.2 Single top production

Single top production, measured at the TEVATRON in the s and t channels (see Section 8.6), is dominated by $t-$channel production at the LHC, as shown in Figure 9.55. There is a large growth in the single top cross-section from the TEVATRON to the LHC (8 TeV), of about a factor of 60. ATLAS and CMS have measured single top cross sections for the $t-$channel [24, 657] and for the Wt final state [7, 385] while setting upper

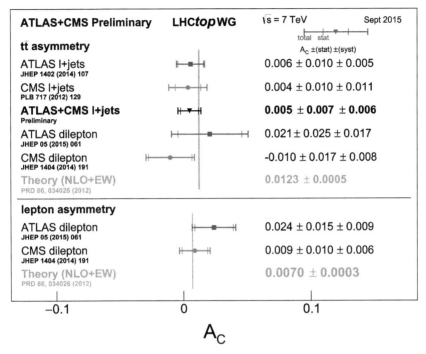

Fig. 9.54 A compilation of top pair charge asymmetry measurements at the LHC compared to SM predictions, from `atlas.web.cern.ch/Atlas/GROUPS/PHYSICS/CombinedSummaryPlots/TOP/`. Reprinted with permission from CERN.

limits for the $s-$ channel contribution, which has of course already been measured at the TEVATRON [47]. Since the tW final state can be distinguished from the other single top final states, its analysis is usually conducted separately from the other two.

As at the TEVATRON, the signature for t channel production involves the presence of an isolated high p_T lepton, substantial missing transverse energy, and two or three jets, with at least one jet tagged as a b-jet and with one jet at high rapidity. The rapidity distribution for the (untagged) forward jet serves as a good discriminant for separating t-channel single top production from the s-channel mode and other SM backgrounds. Measurements of the t-channel cross-section at the LHC, and the theoretical prediction of NLO QCD, are shown in Fig. 9.56. The cross-sections measured by ATLAS and CMS are in good agreement both with each other and with the theoretical prediction at this order. Although the NNLO prediction for this cross-section [288] has not been compared in this figure, it also agrees very well. For instance, the comparison with the combined results of ATLAS and CMS for the $(t + \bar{t})$ single top t-channel process is,

$$\sigma^{8\ TeV}_{ATLAS+CMS} = 85 \pm 12 \text{ pb}, \qquad \sigma^{8\ TeV}_{NNLO} = 83.9^{+0.8}_{-0.3} \text{ pb}. \qquad (9.9)$$

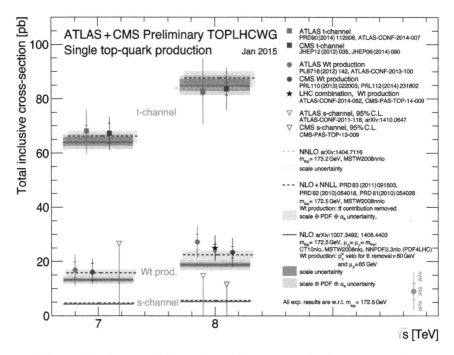

Fig. 9.55 A compilation of single top production cross-sections from ATLAS and CMS as a function of the centre-of-mass energy, from `atlas.web.cern.ch/Atlas/GROUPS/PHYSICS/CombinedSummaryPlots/TOP/`. Reprinted with permission from CERN.

The difference between the two uncertainties underscores both the difficulty of measuring this process at the LHC and the precision of the NNLO calculation.

Due to the LHC being a pp collider, the production of single *top quarks* is larger than that of single *anti-top quarks*. The NNLO prediction for the ratio of $t/\bar{t}$ production is 1.825 ± 0.001. This ratio is sensitive to the distribution of up and down type quarks in the proton, as well as possible new physics that may couple to the Wtb vertex. The ratios measured by ATLAS and CMS are consistent with the Standard Model predictions, albeit with relatively large statistical and systematic uncertainties, as observed in Fig. 9.57. If one assume that no anomalous form factors are present at the Wtb vertex it is possible to extract the size of the (t, b) CKM matrix element, $|V_{tb}|$. ATLAS and CMS measure 1.02 ± 0.07 and 0.998 ± 0.041 respectively, where the total error includes both the experimental and theoretical errors added in quadrature.

9.6   Higgs boson

9.6.1   Introduction

The discovery of the Higgs boson at the LHC by the ATLAS [15] and CMS [370] experiments can be considered as the culmination of the Standard Model. Because of its importance, a great deal of theoretical effort has been devoted towards precision

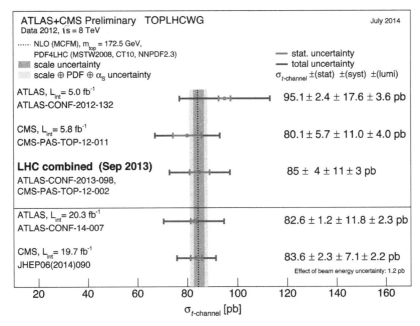

Fig. 9.56 A compilation of t-channel single top cross–section measurements from ATLAS and CMS, taken from atlas.web.cern.ch/Atlas/GROUPS/PHYSICS/CombinedSummaryPlots/TOP/. Reprinted with permission from CERN.

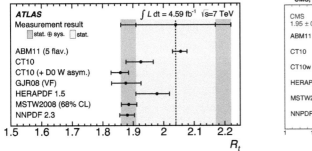
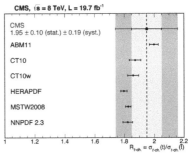

Fig. 9.57 A comparison of the measured ATLAS (left) [24] and CMS (right) [657] ratios of single top to single anti-top production with NLO predictions using various PDFs. Reprinted with permission from Refs. [24] and [657].

calculations of both the Higgs boson production cross-sections and the decay branching ratios [161, 296]. Higgs boson final states involve the measurements of photons, leptons, jets (including b-tagged and c-tagged jets), τ leptons, and missing transverse energy, i.e. the building blocks of the LHC SM measurements discussed in this chapter. The tools developed for SM measurements, both theoretical and experimental, can be directly adapted for measurements of the Higgs boson production and decay rates, and its properties. For example, for some of the Higgs boson measurements, a better signal-to-background ratio can be gained by requiring the Higgs boson system to be boosted. Boosted systems have received a great deal of attention at the LHC, as discussed in Section 9.2.4 and, for example, Ref. [431]. The knowledge of the SM processes also serves to improve the determination of the backgrounds to measurements of Higgs boson final states. Some of the backgrounds can be determined from the data; for others some dependence on theoretical predictions is necessary.

The relative rates for the production of a 125 GeV SM Higgs were shown in Fig. 4.53. The dominant production mode is gg fusion for all centre-of-mass energies. The other modes added to the discovery potential but are also important for a complete understanding of the Higgs boson properties. For example, the VBF process probes the couplings to W/Z bosons, while $t\bar{t}H$ probes the coupling to the top quark. The decay branching ratios have been previously shown in Fig. 4.55. For a Higgs boson mass of 125 GeV, the dominant decay is into a $b\bar{b}$ pair, followed closely by a decay into WW^*. As the Higgs mass is below the threshold for WW production, one of the W bosons has to be off mass-shell. The decay into two photons has one of the smallest branching ratios, but was still important for the discovery because of the precision in which the 4-vectors of the two photons could be measured. With sufficiently precise resolution, the two photon mass peak for the Higgs boson can be discerned from the copious backgrounds for QCD diphoton production. To some extent, the ATLAS and CMS detectors were designed to optimize the search for the Higgs boson. For example, both experiments chose solutions for their electromagnetic calorimetry that prioritized precise energy resolution so as to be able to reduce the observed width of the Higgs boson in its two photon final state.

The total number of inelastic pp collisions in Run 1 at the LHC was of the order of 1.5×10^{15}. In total, over 500,000 Higgs bosons were produced per experiment in these final states (before acceptance and reconstruction).

The discovery of the Higgs boson (or rather of a new particle with a signature consistent with the Higgs boson, as noted in the discovery) occurred on July 4, 2012, with both ATLAS and CMS observing a significance for a signal at 125 GeV of approximately 5 sigma. The discovery resulted from approximately 5 fb^{-1} of data at a centre-of-mass energy of 7 TeV and approximately 6 fb^{-1} at 8 TeV. Further data-taking increased the integrated luminosity at 8 TeV to over 20 fb^{-1}, allowing not only 10 standard deviation evidence for the Higgs boson, but also detailed investigations of its couplings. For some of the final states, differential distributions were also measured.

The discovery and first measurements of the Higgs boson in Run 1 required the development of sophisticated analysis techniques designed to optimize the signal-to-background discrimination in the various analysis channels. This optimization could be applied directly to cut-based analyses, or as input to multivariate analysis techniques.

Since the signal-to-background discrimination was based on SM theory, in some sense this was the discovery of the Standard Model Higgs boson. With the increased statistics expected in Run 2, some of this theory dependence in the analyses can be relaxed.

The rest of this section will be as follows. In Section 9.6.2, the analyses of Higgs boson production in several specific channels will be discussed (diphoton, WW^*, ZZ^*), followed in Section 9.6.3 by a summary of the signal strengths for the different modes. In Section 9.6.4, the determination of the Higgs mass and width is discussed. Finally, in Section 9.6.5, differential Higgs measurements will be discussed. In most cases, the details of the analyses are taken from the ATLAS measurements, but similar techniques are used by the CMS experiment.

9.6.2   Decay channels

9.6.2.1   Diphoton final states

The Higgs boson discovery potential in the diphoton channel depends critically on the precision with which the Higgs 4-vector can be measured. This precision depends both on the resolution for the photon energy determination and on the resolution for the determination of its direction. If the only interaction per crossing were that of the collision that produced the Higgs boson, then there would be no ambiguity for the latter, as the charged particle tracking for both ATLAS and CMS allows a precision determination of the interaction vertex. However, this becomes more problematic under the high pileup conditions with which the Higgs boson was discovered, as there can be on the order of 20 or more additional interaction vertices. The ATLAS detector was designed with a pre-shower detector (and fine lateral and longitudinal segmentation) that allows some degree of pointing of the photon back to the correct interaction vertex. The correct interaction vertex can be identified over 90% of the time for low pileup conditions, decreasing to the order of 70% for high pileup conditions (25 interactions per crossing) [29]. The CMS detector has fine lateral segmentation but only one depth segment (and no pre-shower detector in the central rapidity region), so the interaction vertex chosen is chosen using a boosted decision tree using kinematic information on the charged tracks from each vertex and on the diphotons in order to provide the best match [658]. The efficiency to determine the correct interaction vertex increases as the transverse momentum of the diphoton system increases, reaching above 90% for $p_T^{\gamma\gamma} > 50$ GeV. Both experiments have a significant amount of material in front of the electromagnetic calorimeter (tracking, services, the solenoid coil in the case of ATLAS), so a significant fraction of the photons will have converted into electron-positron pairs. Algorithms then have to be used to identify those conversions and to correct the photon energy accordingly. In general, the same reconstruction algorithms are applied for photons from Higgs candidates as for photons in other SM analyses.

As for other SM measurements, the two photons from the Higgs boson decay have an isolation cut imposed; in the case of ATLAS, this involved a combination of the requirement of less than 6 GeV of energy in the calorimeters in a cone of $R = 0.4$ about the photon direction and a requirement that the sum of all charged track momenta within a cone of 0.2 about the photon direction be less than 2.6 GeV. The isolation cut is applied after the event-by-event subtraction of the underlying event and pileup energy, similar to what is done in the QCD diphoton measurement. The $H \to \gamma\gamma$ search

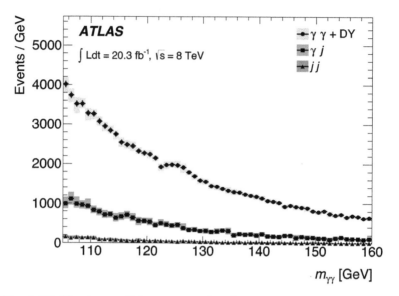

Fig. 9.58 The ATLAS diphoton mass distribution in the Higgs search region. The fractions of the events resulting from real diphotons, from photon+jet events and from jet-jet events have been estimated using a double two-dimensional sideband method as discussed in the text. Reprinted with permission from from Ref. [29].

in ATLAS requires (at least) two photons in the (absolute) rapidity range less than 2.37 (excluding the crack region $1.37 < |\eta| < 1.56$ where the energy resolution is degraded). A requirement is made that the ratio of the photon transverse energy to the diphoton mass ($E_T/m_{\gamma\gamma}$) is less than 0.35 (0.25) for the leading (2nd leading) photon. The photon transverse energy cuts are larger than for the diphoton measurement discussed in Section 9.4.1 and more asymmetric. Such a large (asymmetric) cut emphasizes the Higgs boson signal over the continuum diphoton background.

The mass spectrum for Higgs diphoton candidates with these cuts at 8 TeV is shown in Fig. 9.58. Notice that the real QCD diphoton signal dominates over photon-jet and jet-jet backgrounds, where one or more jets mimics a real photon (for example with a π^o which takes most of the momentum of the jet). The isolation cut greatly reduces the background rate. The real diphoton and jet background fractions are determined by the use of a double two-dimensional sideband method involving (1) loose and tight photon identification criteria and (2) loose and tight photon isolation criteria [29].

As for the inclusive Higgs boson cross-section, the dominant subprocess involving a diphoton final state is gg fusion (87%), followed by VBF, VH (5%), and $t\bar{t}H$ (1%). As the signal to background ratio is small, and varies according to the Higgs subprocess, the diphoton events are assigned to 12 exclusive categories, with each category optimized to maximize the expected signal strength of the subprocess (for example, the presence of leptons for VH, two widely separated jets for VBF, etc). The expected

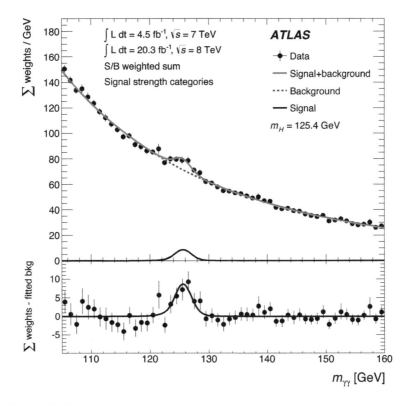

Fig. 9.59 The diphoton mass spectrum measured in ATLAS in the 7 and 8 TeV data. Each event has been weighted by the signal to background ratio for the category to which it belongs. The solid red curve shows the sum of the signal (for a Higgs of mass 125.4 GeV) and background fits. Reprinted with permission from Ref. [29].

diphoton mass resolution also differs among the various categories. These exclusive categories can be combined into the four channels discussed above (gg fusion, VBF (7%), VH and $t\bar{t}H$).

For each category of event, a weight is determined based on the expected signal to background for that category, where the S/B ratio is determined from SM theory. The (S/B) weighted distribution is shown in Fig. 9.59, where a clear bump is evident at about a mass of 125 GeV. The width of the bump is entirely determined by the resolution of the photon measurements, as the intrinsic width for a 125 GeV Higgs boson is on the order of 4 MeV. The signal strengths for the diphoton channels are shown in Fig. 9.64.

9.6.2.2 $ZZ^* \rightarrow 4l$ final states

Higgs boson candidates decaying into ZZ^* final states are formed by selecting two same-flavour, opposite-sign lepton pairs, with the dilepton mass combination closest

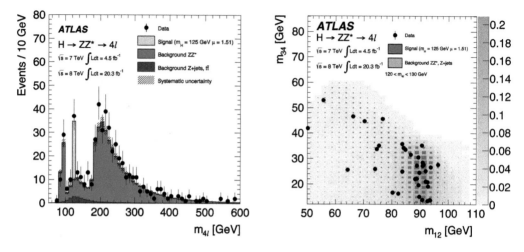

Fig. 9.60 (left) The 4 lepton mass spectrum measured in ATLAS in the 7 and 8 TeV data. (right) A plot of the sub-leading vs leading dilepton pair masses, where the 4 lepton mass was required to be between 120 and 130 GeV. The dominant probability for the leading pair to be at the Z pole mass can be observed. Reprinted with permission from Ref. [42].

to the Z mass being termed the leading dilepton pair with the second dilepton pair being formed from the remaining two leptons. The leading pair is required to have a mass between 50 and 106 GeV. Electromagnetic radiation from the leptons can often be measured in the electromagnetic calorimeters and used to correct the lepton momentum. Collinear photons are associated with muons and non-collinear photons are associated with either electrons or muons. Both track and calorimeter isolation requirements are applied to the leptons, after the event-by-event subtraction of the underlying event and pileup energy.

As for the diphoton final state, the ZZ^* event candidates are assigned to categories (4 in this case; high mass 2 jets (VBF-enriched), low mass 2 jets (VH-enriched), additional lepton (VH-enriched), and gg fusion) in order to optimize the Higgs cross-section determination. For the VBF category the dijet mass is required to be above 140 GeV; for the low mass 2 jets VH category, the dijet mass is required to be between 40 and 130 GeV.

The resultant 4-lepton mass distribution is shown in Fig. 9.60 (left), where a clear (but statistically limited) peak is observed at about 125 GeV [42]. Note also the presence of the 4-lepton decay of the Z boson at 90 GeV (useful for calibration) and the large increase of the 4-lepton cross-section once both Z's can be on mass-shell, above 200 GeV. The mass distribution for the two dilepton pairs is shown in Fig. 9.60 (right), where the dominance of the leading pair to be at the Z-pole mass is evident. The signal strengths for the ZZ^* channels are shown in Fig. 9.64.

9.6.2.3 $WW^* \to l\nu l\nu$ final states

The $WW^* \to l\nu l\nu$ final state is perhaps the most challenging from the analysis perspective and from the viewpoint of perturbative QCD predictions. There is less information available due to the presence of 2 neutrinos in the final state, and there are significant backgrounds which depend on the jet multiplicity, necessitating separate analyses depending on the number of jets present. The latter requires the use of exclusive final states (exactly 0 jets, exactly 1 jet) which have intrinsically larger theoretical uncertainties than inclusive final states.

The most sensitive final state is with $e\mu + 0$ jet. The dominant background for this final state is from WW production. WW backgrounds can be suppressed by exploiting the properties of W boson decays and of the (expected) spin 0 nature of the Higgs boson. The latter results in the two leptons being produced relatively close to each other, resulting in a small dilepton mass ($< m_{Higgs}/2$). The dilepton invariant mass is used to select signal events and a signal likelihood fit is performed in two ranges of m_{ll} in $e\mu$ final states with 0 or 1 jet. The final states are also separated according to the value of the transverse momentum of the sub-leading lepton (W^* leptons will have on average lepton momenta smaller than from on-shell W decays). It is also useful to calculate the transverse mass:

$$m_T = \sqrt{\left(E_T^{ll} + p_T^{\nu\nu}\right)^2 - \left|\, \mathbf{p_T^{ll}} + \mathbf{p_T^{\nu\nu}}\, \right|^2}, \tag{9.10}$$

where $E_T^{ll} = \sqrt{(p_T^{ll})^2 + (m_{ll})^2}$, $\mathbf{p_T^{\nu\nu}}$ ($\mathbf{p_T^{ll}}$) is the vector sum of the neutrino (lepton) transverse momenta.

The distribution has a kinematic upper bound at the Higgs boson mass, effectively separating Higgs boson production from the WW and top quark backgrounds. The transverse mass distributions are shown for ATLAS for two different analysis categories in Fig. 9.61 [45]. The dominant background is WW production for low jet multiplicities and $t\bar{t}$ for the higher jet multiplicity.

The exclusive nature of the 0 and 1 jet bins (basically the large restriction in phase space that leads to the creation of large logarithms that need to be resummed) adds to the uncertainty for the determination of the Higgs boson cross-sections in this decay channel and to the extraction of the Higgs couplings. A great deal of theoretical effort has been devoted towards understanding those increased uncertainties, and in trying to reduce them through the use of resummation techniques. The uncertainty on the jet multiplicity distributions is calculated using the jet-veto-efficiency (JVE) method [199] for the gluon-gluon fusion categories and with the Stewart-Tackmann method (ST) [865] for the VBF category.

The signal strengths for the WW^* channels are shown in Fig. 9.64.

9.6.2.4 Associated production

As seen in Fig. 4.55, the largest branching ratio for a 125 GeV Higgs boson is into a $b\bar{b}$ final state. Production through gg fusion is not measurable due to the overwhelming background from QCD $b\bar{b}$ production. However, measurement of the $b\bar{b}$ final state is possible when the Higgs boson is produced in association with a vector boson. As

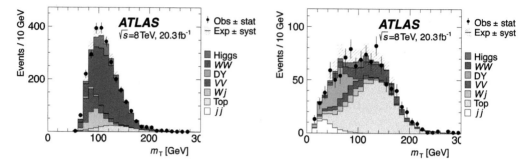

Fig. 9.61 The transverse mass distributions measured in ATLAS in the 8 TeV data in the electron-muon channel, for the gluon-gluon fusion enriched region. The transverse mass distribution on the left is for the 0 jet bin, while the transverse mass distribution on the right corresponds to the ≥ 2 jet bin. The Higgs boson signal is given by the uppermost filled portion of the histograms. Reprinted with permission from Ref. [45].

discussed in Section 8.7, this was the primary search channel for the Higgs boson at the TEVATRON. The measurement of the associated production of a vector boson (W/Z) with a Higgs boson decaying into a $b\bar{b}$ pair allows for the direct measurement of the coupling of the Higgs boson to b-quarks. There is still a significant background from $Vb\bar{b}$ production, itself not perfectly understood, as discussed in Section 9.3.5.

In the Higgs boson analysis in this channel, the events are first categorized according to the number of leptons (0,1 and 2), jets (2 or 3 with traverse momenta great than 20 GeV and (absolute) rapidity less than 2.5 (inside the b-tagging range) and b-tagged jets. Events are rejected if any additional jets with transverse momenta great than 30 GeV are found with rapidity greater than 2.5 (in order to reduce $t\bar{t}$ backgrounds). Dedicated boosted decision trees are then constructed for each channel, with the boosted decision trees trained to separate the associated production signal from the backgrounds. The weighted event distribution is shown in Fig. 9.62. The signal strengths for the $VH(\rightarrow b\bar{b})$ channels are shown in Fig. 9.64 [48]. There is a significance of 1.4 with an expected significance of 2.6.

9.6.2.5 $\tau\tau$ final states

The ATLAS analysis channels require either the presence of 2 isolated opposite-sign leptons above the transverse momentum threshold, exactly one isolated lepton and one hadronic candidate with opposite sign charges, above threshold, or two hadronic candidates above threshold. The events are divided into two categories: VBF Higgs boson production, with a requirement of two high transverse momentum jets separated in rapidity, or boosted, requiring a transverse momentum for the Higgs boson candidate of above 100 GeV. The signal-to-background ratio is improved by going to the boosted regime. The weighted mass distribution is shown for the Higgs boson search into the $\tau\tau$ final state in Fig. 9.63 (left), while the signal strength is shown in Fig. 9.63 (right). The observed significance is 4.5 with an expectation of 3.4.

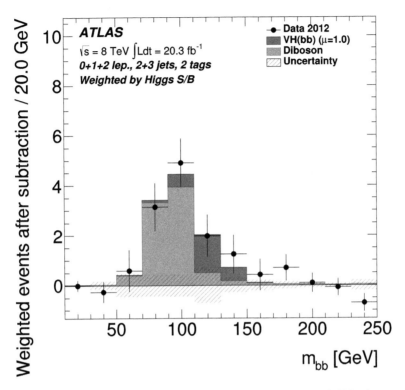

Fig. 9.62 The distribution of $m_{b\bar{b}}$ after the subtraction of all backgrounds (except for diboson) in ATLAS in the 8 TeV data. The contributions have been summed weighted by their values of expected Higgs signal to background. Reprinted with permission from Ref. [48].

9.6.3 Signal strengths

The measured signal strengths for the Higgs boson analysis channels for the individual ATLAS and CMS experiments in Run I are reviewed in references ATLAS [55] and CMS [669]. A joint determination by the two experiments of the signal strengths for the various Higgs boson production and decay channels is shown in Fig. 9.64 [54]. All signal strengths are consistent with the SM predictions within the relatively large Run 1 uncertainties. In most cases, the precision of the joint results improves by the expected factor of $1/\sqrt{2}$ from those of the individual experiments. The combined signal yield (with respect to the SM prediction) is $1.07\pm0.07(stat)\pm0.08(syst)$. The dominant systematic uncertainty is due to the theoretical uncertainties for the inclusive cross section predictions.

9.6.4 Higgs boson mass and width

Precision measurements of the photon and lepton 4-vectors allow for the best determination of the Higgs boson mass in the diphoton and 4-lepton final states. The

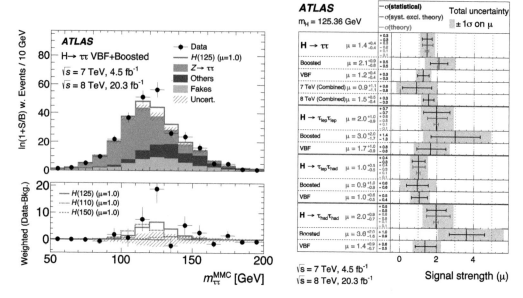

Fig. 9.63 (left) The reconstructed weighted $m_{\tau\tau}$ distribution, where the weights are determined by the signal-to-background predictions for the different final states. The data are from ATLAS in the 7 and 8 TeV data samples. (right) The fitted values of the Higgs boson signal strength for the different $\tau\tau$ final states, and their combination, from the combined 7 and 8 TeV data. Reprinted with permission from Ref. [38].

measurements by ATLAS and CMS for these 2 final states are shown in Fig. 9.65 [36]. It would be potentially interesting if the 4-lepton and diphoton final states had different masses, but the hierarchy between the 2 states is opposite for ATLAS and CMS, pointing to statistical fluctuations as the root cause. The measurements are all consistent with each other, allowing a combined determination of the Higgs boson mass from Run 1 of $m_{Higgs} = 125.09 \pm 0.21(stat) \pm 0.11(syst)$. Note that the dominant error is statistical; both the statistical and systematic errors should significantly improve in Run 2 of the LHC. This particular value for the Higgs boson mass has interesting implications for the (meta)stability of the vacuum.

As mentioned previously, the width of a 125 GeV Higgs boson is too small to be measurable, on the order of 4 MeV. However, as discussed in Section 4.8, the high mass ZZ^* region is sensitive to Higgs boson production through off-shell and background interference effects. Amazingly enough, approximately 15% of Higgs boson production in the ZZ^* channel occurs above the ZZ threshold. The cross-section for $H \to ZZ^*$ is comparable to the cross-section for continuum production of $gg \to ZZ$ (with which it destructively interferes) above this threshold. The dominant sub-process for high mass ZZ final states is through $q\bar{q} \to ZZ$. The leading order cross-section for $gg \to ZZ$ is through a box diagram as shown in Fig. 4.57. At the time that the 8 TeV analyses were carried out, the NLO (2-loop) calculation for this process was beyond

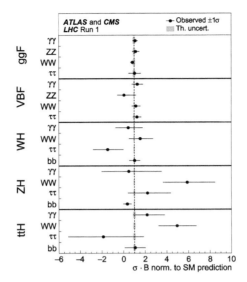

Fig. 9.64 The joint ATLAS +CMS Higgs boson signal strengths, by final state and production mode. Reprinted with permission from Ref. [54].

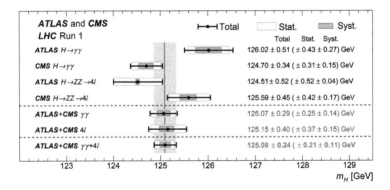

Fig. 9.65 The measured ATLAS and CMS Higgs boson masses separated by final state. Reprinted with permission from Ref. [36].

the current technology. Given the progress in calculating two-loop integrals with two massless and two massive external lines, this calculation has now been carried out. The resulting QCD corrections increase the $gg \rightarrow ZZ$ cross-section by the order of 50-100%, depending on the exact scale choice [323].

The ATLAS analysis varies the possible K-factors for this process as part of the systematic uncertainties (CMS assumes the same K-factors for both resonant and non-resonant $gg \rightarrow ZZ$ production). The ratio of the off-shell to on-shell signal strengths is directly proportional to the Higgs width. The 4-lepton mass distribution in the ATLAS Higgs search is shown in Fig. 9.66 in the mass range from 220-1000 GeV, along with

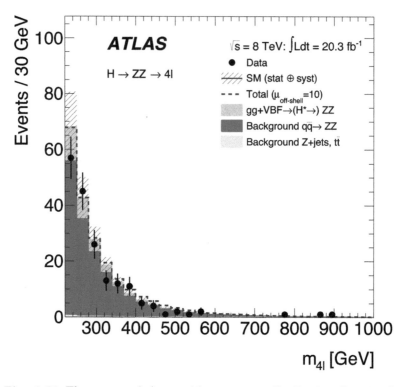

Fig. 9.66 The measured ATLAS 4-lepton mass distribution. Reprinted with permission from Ref. [37].

the contributions from the Standard Model, including that of the Higgs boson [37]. The dashed line indicates the impact of an off-shell coupling 10 times the SM value. Assuming that the relevant Higgs boson couplings are independent of the energy scale of the Higgs production, the combination of the ZZ and WW results yields 95% confidence level upper limits for Γ_H/Γ_H^{SM} in the range between 4.5-7.5 (with ATLAS and CMS having similar results).

9.6.5   Higgs differential distributions

More information on the properties of the Higgs boson can be gained by studies of differential distributions of the Higgs boson itself and of accompanying jets. This is easiest done in two decay modes: the diphoton mode, because of its relatively large number of signal events, and the 4-lepton mode, because of its large signal to background ratio. The diphoton final state is described first.

Whether jet measurements can be conducted with the diphoton final states can be easily determined from the diphoton mass distributions for different jet multiplicities observed in Fig. 9.67 and Fig. 9.68 [35]. The diphoton mass bump becomes more prominent as the number of jets increases. This implies that events with a Higgs

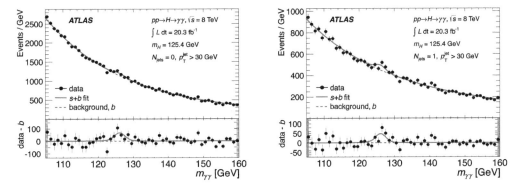

Fig. 9.67 The diphoton mass distributions measured in ATLAS in the 8 TeV data for the 0 jet channel (left) and the 1 jet channel (right). Reprinted with permission from Ref. [35].

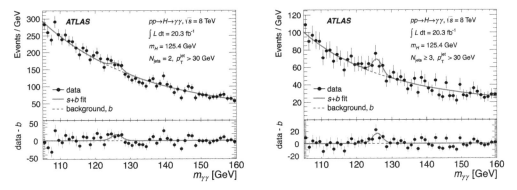

Fig. 9.68 The diphoton mass distributions measured in ATLAS in the 8 TeV data for the 2 jet channel (left) and the ≥ 3 jet channel (right). Reprinted with permission from Ref. [35].

boson are jettier than events produced from the QCD continuum background. This is expected as the bulk of Higgs boson production occurs through gg fusion into a colour-singlet (Higgs boson) final state, a situation that leads to a large probability for the production of additional jets, as discussed in Section 4.8.4.

The number of events is much more limited for the case of the 4-lepton final state, but it is still useful to combine the two measurements within a common fiducial volume, especially as they are consistent with each other. The Higgs boson transverse momentum distribution for the diphoton, the 4-lepton and the combined final states is shown in Fig. 9.69 [43]. The two final states produce similar results. The Higgs boson transverse momentum distribution in the data is observed as somewhat shifted towards higher p_T compared to the theoretical predictions.

The jet multiplicity distributions for the combined diphoton and 4-lepton final states are shown in Fig. 9.70 for the inclusive (left) and exclusive (right) cases [43].

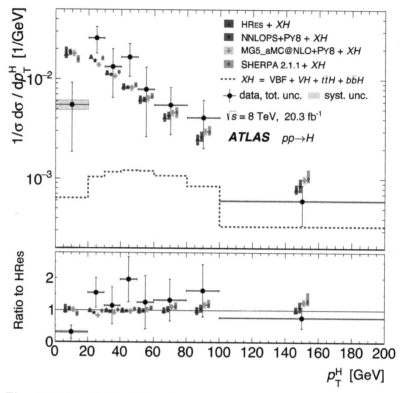

Fig. 9.69 The ATLAS Higgs boson transverse momentum distribution, combining the diphoton and ZZ^* channels. Reprinted with permission from Ref. [43].

The measured cross-sections show somewhat jettier final states than predicted by the various theory predictions for the gg fusion process shown.[13] Also shown in the figures (and added to the gg fusion results) are the contributions from VBF, VH, $t\bar{t}H$ and $b\bar{b}H$ production. These contributions form a more significant fraction of the total production as the jet multiplicity increases.

The transverse momentum distribution for the lead jet is shown in Fig. 9.71 (left), compared to several theoretical predictions [43]. The Δy distribution between the two leading jets is shown in Fig. 9.71 (right) [35]. Due to limited 4-lepton statistics for this observable, the data for this plot is only from the diphoton final state. The excess of data over theory occurs more for small jet rapidity separations. Note that larger jet separations are dominated by VBF production.

A useful summary of the diphoton channel differential cross-sections, compared to a number of theoretical predictions, is shown in Fig. 9.72, from Ref. [35]. For reference,

---

[13]The differential distributions reported by the CMS collaboration in [671] are closer to the SM predictions.

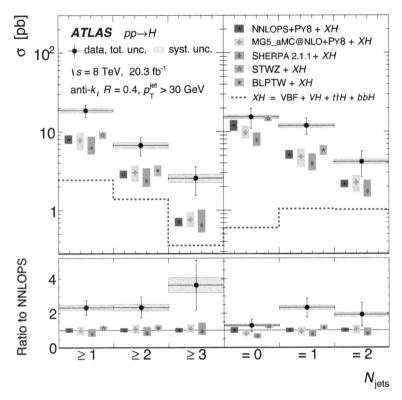

Fig. 9.70 The ATLAS Higgs boson jet multiplicity distribution, combining the diphoton and ZZ^* channels. Reprinted with permission from Ref. [43].

the NNLO Higgs+$\geq$ 1 jet fiducial cross-section (not shown on the plot) is 10 fb^{-1} (including top quark mass effects) [325].

9.6.6 Short aside/rant

The highest precision for inclusive jet observables associated with the production of a Higgs boson, or for any other suitable final state, is typically achieved with fixed order predictions. This is especially true given the number of such final states for which NNLO predictions are available. Resummed/parton shower calculations provide a better description for non-inclusive observables where Sudakov effects are important, such as the transverse momentum distribution of the Higgs boson. It is notable that, unlike comparisons shown for the W/Z + jets finals states earlier in this chapter, there are no similar comparisons to fixed order predictions for Higgs boson + jets at the LHC. The 2015 Les Houches Standard Model working group report [161] provided a detailed comparison of a wide variety of predictions for both inclusive and exclusive final states for Higgs boson (+ jets) production at the LHC. In order to control extraneous differences, the same PDF was used for all predictions, as well as the same scale (to the

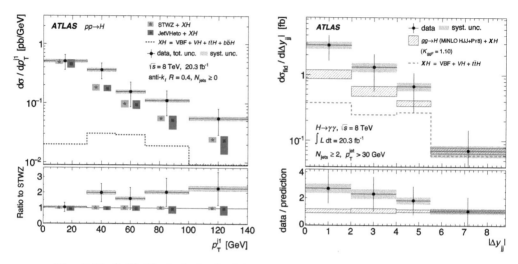

Fig. 9.71 (left) The inclusive lead jet transverse momentum distribution measured in ATLAS in the 8 TeV data from Ref. [43]. The dijet rapidity separation between the two leading jets for Higgs $+ \geq 2$ jets. Reprinted with permission from Ref. [35].

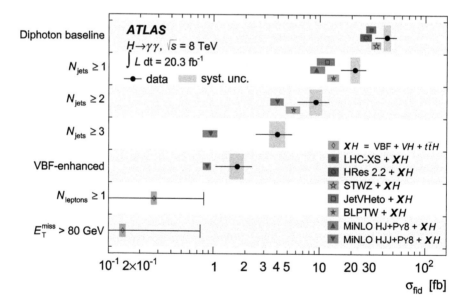

Fig. 9.72 A compilation of the Higgs differential fiducial cross-section measurements in the diphoton channel, compared to a number of theoretical predictions. Reprinted with permission from Ref. [35].

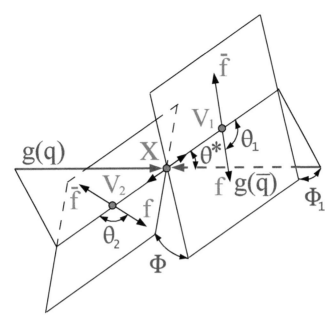

Fig. 9.73 The angles in the $H \to ZZ^* \to 4l$ decay channel useful for spin-parity determination of the the Higgs boson. Reprinted with permission from Ref. [659].

degree for which this was possible). Good agreement was observed between fixed order predictions and predictions involving parton showering/resummation for inclusive observables, such as the lead jet p_T distribution for $H+ \geq 1$ jet. That is, contrary to some conventional wisdom, parton showers and/or resummation do not affect fixed order results, for suitably inclusive observables.

9.6.7   Spin-parity

As discussed in Chapter 4, the Higgs boson is predicted to be a pure scalar particle ($J^P = 0^+$). With the addition of beyond Standard Model physics, resulting in different allowed interactions in the Lagrangian, the Higgs boson(s) may have a different spin or CP state, or even states of mixed CP. The different CP/spin structures may alter the kinematic distributions, especially angular distributions, of the Higgs boson decay particles. The most useful modes are those involving $H \to ZZ^* \to 4l, H \to \gamma\gamma$, and $H \to WW^* \to l\nu l\nu$ final states. See for example the angular observables sensitive to the spin and parity of the Higgs boson in the $H \to ZZ^* \to 4l$ decay channel in Fig. 9.73. Both the ATLAS [49] and CMS [659] results strongly disfavour the spin 1 and spin 2 hypotheses, as well as the 0^- state. In studies of potential mixed states, for example having the Higgs boson couple via a CP-odd term (thus implying CP violation), ATLAS and CMS have found no evidence of non-Standard Model (0^+) behaviour.

9.6.8 Vector boson scattering

Vector boson scattering is a key process to fully understand the nature of electroweak symmetry breaking. Without the presence of a Higgs boson, the cross section for the process would diverge at high diboson masses. Even given the discovery of a Higgs boson with a mass of 125 GeV, the process may still not respect unitarity, and a detailed study of the process is necessary.

First evidence for vector boson scattering (VBS) has been found by the ATLAS experiment in a measurement of same sign W boson pair production, accompanied by two or more jets ($W^{\pm}W^{\pm}jj$), at 8 TeV [25]. The W bosons are required to decay into either electrons or muons, with both leptons having the same sign. This suppresses the background from normal diboson production. The leptons are required to have a transverse momentum greater than 25 GeV, an absolute rapidity less than 2.5, a dilepton mass greater than 20 GeV, and a dilepton separation (ΔR) greater than 0.3. The jets are formed using the anti-k_T algorithm with radius 0.4 and are required to have a transverse momentum greater than 30 GeV, and an absolute rapidity less than 4.4. Each jet and lepton pair are required to have a separation (ΔR) greater than 0.3. The invariant mass of the two jets with the highest transverse momentum is required to be greater than 500 GeV, and the missing transverse energy is required to be greater than 40 GeV.

The VBS process under investigation is of order α_{EW}^4. The same final state can also be produced with a mixed strong-electroweak process of order $\alpha_{EW}^2\alpha_s^2$. The VBS process is similar to that of vector boson fusion production of the Higgs boson, in that the initial state W bosons are radiated off of incoming quarks, as shown in Figure 9.74(left). Those quarks receive a transverse momentum kick and appear as jets. Given their longitudinal boost, the jets in the VBS process tend to have a wider separation in rapidity than for the mixed strong-electroweak process. To enhance the VBS purity, an additional requirement is made that two jets be separated by a rapidity interval of 2.4 or more.

The measured Δy_{jj} distribution is shown in Figure 9.74(right), along with the predicted signal and background components. The $|\Delta y_{jj}|$ cut of 2.4 for the VBS region is indicated. There is both strong evidence for inclusive production of $W^{\pm}W^{\pm}jj$ (4.5 standard deviations) and of pure electroweak production (3.6 standard deviations). The measured fiducial cross sections of $2.1 \pm 0.5(stat) \pm 0.3(syst)$ for the inclusive region and $1.3 \pm 0.4(stat) \pm 0.2(syst)$ fb in the VBS region, are consistent with the respective Standard Model predictions of 1.52 ± 0.11 fb and 0.95 ± 0.06 fb. More detailed investigations will be possible with the higher statistics expected in Run 2.

9.7 Outlook

9.7.1 Standard Model physics at the LHC

Two key aspects of Run II (and beyond) physics at the LHC relate to higher precision and extended kinematic reach. No clear signs of beyond Standard Model physics have been discovered (to date) at the LHC. Searches for new physics will necessarily require precision measurements of SM processes (especially of Higgs boson production) seeking deviations that may indicate the presence of new physics. The higher running energy

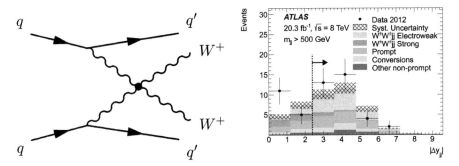

Fig. 9.74 (left) One of the diagrams for VBS. (right) The $|\Delta y_{jj}|$ distribution for events passing the cuts for the inclusive region. The cut in $|\Delta y_{jj}|$ denoting the VBS region is indicated. The $W^\pm W^\pm jj$ prediction has been normalized to the Standard Model value. Reprinted with permission from Ref. [25].

and increased luminosity also result in greater access to the TeV range, in which signs of new physics may be more obvious.

In most cases, increased precision means the calculation of a process to NNLO in QCD, and often NLO in EW. There has been great progress in the calculation of LHC processes to NNLO and beyond, as detailed both in this book and in workshops such as Les Houches [161]. The technology for $2 \to 2$ processes at NNLO is already relatively mature, and the reasonably near future should see NNLO being extended to $2 \to 3$ processes. Higgs boson production through gg fusion has already been calculated at N^3LO and the next obvious extension is to carry out the calculation of Drell-Yan production to this order. So far, only PDFs at NNLO are available, but for ultimate precision a determination of PDFs at N^3LO may be necessary. Of course, this also requires the calculation of the processes in global PDF fits at this order. At this level of precision, NLO EW corrections can become equally important as those from NNLO (and above) QCD. Above the TeV scale, EW effects are often not subtle, and the radiation of W and Z bosons will compete with QCD gluon radiation. In most cases, the NNLO QCD and NLO EW calculations may factorize, but in some instances mixed corrections may be required, especially if the QCD corrections are large. In general, the higher the order of calculation, the smaller the scale dependence will be. However, some care will still have to be taken with regards to the choice of a physical scale for the process, especially in the presence of a complex final state.

In the TeV range, photon-initiated processes will become increasingly important, for example for high-mass W boson pair production. Most of this book has concentrated on fixed-order calculations, but the TeV range also means that high-x effects will become important and threshold resummation corrections will be crucial to calculate.

From the experimental side, larger data samples by definition mean smaller statistical errors, and often smaller systematic errors due to an improved knowledge of the experimental measurement. A high integrated luminosity necessarily requires a high instantaneous luminosity and the presence of many pileup events. This will in many cases degrade the quality of the experimental measurements, and require the increase of trigger/analysis thresholds. It will still be crucial, however, to have the ability to trigger on standard benchmarks such as W and Z boson production. Even though the experimental environment may be daunting, the ATLAS and CMS experiments have been designed to operate in those conditions.

9.7.2 Higgs boson physics at the LHC

The culmination of Run 1 at the LHC was the discovery of the Higgs boson by ATLAS and CMS. A precision determination of its properties, though, requires the higher energy and integrated luminosity of Run 2 (and beyond). Data samples of 300 fb^{-1} are projected for each experiment in Run 2 at a centre-of-mass energy of 13-14 TeV. Through the full running of the LHC, an integrated luminosity of 3000 fb^{-1} can be expected. Cross-sections for Higgs boson production subprocesses are a factor of 2-4 times higher than at 8 TeV, as seen in Fig. 4.53. In this section, we briefly describe the experimental improvements in precision expected for the full Run 2 (and beyond) data sample, and the needed theory improvements to best match those experimental improvements. The discussion roughly follows that in Refs. [161, 439].

The largest production cross-section at 13 TeV remains that of the gg fusion sub-process. As discussed in Section 9.6, the current experimental uncertainties for this subprocess are on the order of 20-40%. Theoretically, the uncertainty was formerly on the order of 15%, with the scale and PDF+$\alpha_s(m_Z)$ uncertainties both having roughly equal values. The calculation of the gg Higgs boson production to NNNLO has reduced the scale uncertainty to 2-3%, while the recent PDF4LHC combination has resulted in a PDF uncertainty of the same order. The recommendation for the $\alpha_s(m_Z)$ variation from the PDF4LHC combination results in a similar level of uncertainty on the gg fusion cross-section.

The experimental uncertainty is expected to decrease to less than 10% (4%) in the 300 fb^{-1} (3000 fb^{-1}) data sample. This may require an improvement of the theoretical accuracy for the production cross-section, with a knowledge of the combined NNLO QCD+EW contributions retaining the top quark mass effects [161].

With a data sample of 300 fb^{-1}, a very rich program of measurements of Higgs boson + jets final states is possible. There is a comparable (or larger) increase in the Higgs boson + jet cross-section as observed for inclusive Higgs production. As the production proceeds primarily through a top quark loop, it is important to probe inside that loop to understand the dynamics of production, and in particular to determine if any BSM particles may contribute to the loop. Each experiment will have on the order of 3000 events (in the diphoton channel) with a jet with transverse momentum above the top quark mass. With 3000 fb^{-1}, the reach in jet transverse momentum is over 700 GeV. At jet transverse momenta of this order, there is a very large suppression (over a factor of 5) of the Higgs+jet cross-section due to finite top-mass effects (over the effective theory). Even without the presence of new physics, there may be new dynamics present at these scales. To properly understand the physics of Higgs boson + jet production, it is necessary to calculate the finite top quark mass effects at NLO QCD+NLO EW.

Higgs boson final states with at least 2 jets are crucial to understand Higgs boson couplings, especially the coupling to vector bosons through the vector boson fusion process. With a 300 fb^{-1} (3000 fb^{-1}) data sample, this coupling can be determined to the order of 5% (2-3%). On the theoretical side, this may require the calculation of both vector boson fusion and gluon gluon fusion production of the Higgs boson + 2 or more jets to be known to NNLO QCD and for the finite top mass effects to be known to NLO QCD+NLO EW.

Higgs boson couplings to b-quarks are known primarily through associated production (VH). Currently, this coupling is known to the order of 50%. With 300 fb^{-1} (3000 fb^{-1}), this can be improved to 10-15% (7-10%). On the theoretical side, one bottleneck has been the knowledge of the $gg \to HZ$ process, currently known only to LO. This process has a sizable contribution to the total rate, and contributes significantly to the total theoretical uncertainty. It is desirable to combine Higgs production and decay to the same order, NNLO in QCD and NLO in EW.

With 300 fb^{-1} (3000 fb^{-1}), the top quark Yukawa coupling should be measured to the order of 15% (5-10%) (through the tH and $t\bar{t}H$ subprocesses). Current $t(\bar{t}H$ is only known to LO in QCD and $t\bar{t}H$ is known to NLO. For a full understanding of this coupling, it is desireable to know both cross-sections (with top quark decays) to NLO QCD including NLO EW effects.

Fiducial cross-sections have been measured for several of the Higgs boson (+jets) channels in Run I. With the higher statistics of Run II, this will happen for more final states. For most channels in Run I, however, the end result has been measured signal strengths and multiplicative coupling modifiers. In Run II there will be a transition to simplified template cross-sections (STCS), as discussed in [161]. The primary goals of the STCS method are to maximize the sensitivity while minimizing the theory dependence of the measurement. This entails: the combination of decay channels, the measurement of cross-sections rather than signal strengths, and the determination of cross-sections for specific production modes. The physics interpretation (and model-dependence) is left for the final stage of the analysis.

10
Summary

10.1 Successes and failures at the LHC

Perhaps the greatest success at the LHC (besides the discovery of the Higgs boson) is the non-discovery of new physics. This statement may seem counter-intuitive. Of course, the discovery of new physics would have been desirable, but the experimental analysis techniques and the comparisons to theoretical predictions have worked well enough that Standard Model physics has not been confused with BSM physics.[1] Another seemingly counter-intuitive statement is that the LHC benefitted by turning on at a lower energy, with a reduced luminosity, in 2010. The lower energy and smaller data sample precluded most beyond-the-Standard Model searches. This forced more physicists to work on SM physics measurements (leading to the re-discovery of the Standard Model), thus forming benchmarks and tools that were useful with higher luminosity samples where discovery potential was present.

The resolution of the LHC detectors, both calorimetry and tracking, is superior to that of the CDF and DØ detectors. Tracking, in particular, has higher precision and extends to a higher rapidity than possible at the TEVATRON. The improvement in computing power has meant that detailed event simulations, tracing the electromagnetic and hadronic showers, are possible for a variety of physics processes, allowing a better understanding of the detector response.

The theoretical tools and analysis techniques available to LHC physicists are for the most part more sophisticated than those available at the TEVATRON. Fixed-order predictions at NLO (interfaced to parton shower programs) are available for basically any reasonable process, and NNLO calculations for $2 \to 2$ processes have reached a degree of maturity, with calculations of $2 \to 3$ processes to be expected. The $gg \to H$ process has been calculated to NNNLO and similar calculations for Drell-Yan production are not far off. The higher order calculations have resulted in smaller theoretical uncertainties from scale variations. Since it is possible that new physics may not show up as a clear peak on a distribution, but rather in subtle variations from SM predictions, precision comparisons are crucial for discovery/exclusion of BSM physics.

In the precision physics region (50–500 GeV), PDF uncertainties are small for most

---

[1] The authors hold out hope that new physics will indeed be discovered at the LHC.

The Black Book of Quantum Chromodynamics. John Campbell, Joey Huston and Frank Krauss, Oxford University Press. © John Campbell, Joey Huston, and Frank Krauss 2018, First published in paperback 2022. DOI: 10.1093/oso/9780192871961.001.0010

parton-parton luminosities, but are relatively unconstrained at high mass, especially for initial states involving gluons. Further reduction in PDF uncertainties, especially in the high mass region, can come only from data at the LHC. However, in order to provide constraining information on PDFs at high x, the data must be consistent: different distributions in the same measurement that provide overlapping PDF information (for example $y_{t\bar{t}}$ and $m_{t\bar{t}}$) must be consistent, results from different measurements in the same experiment must be consistent (for example inclusive jet production and $t\bar{t}$ production both provide information on the high x gluon), and results from the LHC experiments must be consistent. Otherwise, the measurements may change the central PDFs, but the tension will result in the uncertainty not changing (or even growing larger).

Theoretical predictions are most powerful when they relate to fiducial cross-sections; extrapolating to the full phase space most often introduces an extra layer of uncertainty, as witnessed for example in the ATLAS measurement of the WW cross-section, discussed in Section 9.4.2. Fiducial measurements are more common at the LHC than at the TEVATRON, and hopefully the trend towards more fiducial measurements will continue. This requires that the theoretical calculations also provide predictions at the fiducial level, incorporating for example decays for all unstable particles.

There are still issues with predictions at the very highest masses; in addition to the larger PDF uncertainties in this region, projections of cross-sections/backgrounds are often made using parton shower Monte Carlo programs, where parameter variations in the Monte Carlo can lead to sizeable uncertainties. In some cases, this increased uncertainty is not warranted, especially if the parameter variations can be constrained by (higher precision) fixed-order calculations.

In order to reach the sensitivity needed for new physics searches, the LHC must be run at as high a luminosity as possible. This necessarily results in a large number of additional interactions in each bunch crossing (pileup), creating problems with particle identification and with precision measurements of the particle/jet energies. Techniques have been developed for dealing with pileup, in particular the jet area subtraction technique discussed in Section 9.2.1. Topology dependences of the pileup energy density can limit the ultimate efficacy of the subtraction method.

By necessity, the jet area subtraction technique removes not only the pileup energy, but also the energy associated with the underlying event. Previous measurements at the TEVATRON and LHC have included the underlying event in the physical observables.[2] Since the underlying event information has been removed by the subtraction technique, the choice of the LHC experiments has been to add it back in by including a Monte Carlo prediction for that energy. In some sense, this, although necessary, is a step backwards from the trend towards removing as much Monte Carlo extrapolation on an observable as possible.

Tracking at the LHC is better than that at the TEVATRON, and in particular it is most often possible to distinguish the interaction vertex for the interaction of interest from those of pileup events. Thus, one can distinguish hard scatter jets from pileup

---

[2]A prediction for the underlying event is present in every parton shower Monte Carlo program, but not in fixed-order calculations. For these, non-perturbative corrections must be calculated by the experimenters to allow comparison of parton-level predictions to hadron level observables.

jets using the jet tracking information, and reject jets if too much of the jet energy arises from pileup contributions. Alas, this is possible only for jets produced in the precision tracking region ($|y| \leq 2.5$) and pileup jets are much more of a problem at more forward rapidities. Unfortunately, this is a region where jet identification can be crucial, as for example in measuring the tagging jets in VBF Higgs production. The problem will only get worse as the instantaneous luminosity increases. The solution is to provide more information to discriminate between pileup and hard scatter jets (such as timing for the forward calorimetry), or to simply raise the jet transverse momentum cutoff for forward jets.

10.2 Lessons for future colliders

10.2.1 Standard Model cross-sections beyond 14 TeV

To understand the physics potential of future proton-proton colliders, it is imperative to understand the centre-of-mass energy dependence of notable cross-sections at such machines. Fig. 10.1 shows the predicted cross-sections for a selection of basic processes, ranging over twelve orders of magnitude from the total inelastic proton-proton cross-section to Higgs boson pair-production. For inclusive jet and direct photon production, 50 GeV transverse momentum cuts are applied to the jet and the photon respectively.

The growth of the cross-sections with $\sqrt{s}$ largely reflects the behaviour of the underlying partonic luminosities, *cf.* Section 6.5. For instance, the top pair cross-section is dominated by the partonic process $gg \to t\bar{t}$ and the gluon-gluon luminosity rises significantly at higher values of $\sqrt{s}$. The same holds true for the Higgs production channel $t\bar{t}H$ but, in contrast, the associated production channels are dominated by quark-antiquark contributions and rise much more slowly. The different behaviour means that, unlike at current LHC operating energies, the $t\bar{t}H$ channel becomes the third-largest Higgs production cross-section at 33 TeV and above. As a figure of merit for estimating the difficulty of observing the Higgs pair production process it is not unreasonable to consider the ratio of its cross-section to the top pair cross-section. In many of the possible Higgs boson decays the final states receive significant background contributions from the top pair process. The fact that both processes are predominantly gluon-gluon induced means that this measure is approximately constant across the range of energies considered. From a consideration of total cross-sections alone, it is therefore not clear that the prospects for extracting essential information from the Higgs-pair process are significantly better at a higher-energy hadron-collider, even though the rates increase dramatically.

A different sort of contribution to event rates can also be estimated from this figure. The contribution of double parton scattering events, of the type discussed in Section 7.2.3, can be crudely estimated from Eq. (7.43). The value of σ_{eff} can be considered to be approximately energy-independent and around 20mb. Although this is not exactly true, the uncertainty on this parameter, and indeed on the accuracy of Eq. (7.43) itself, is such that this should be considered sufficient for an order-of-magnitude estimate only. A particularly simple application of this is the estimation of the fraction of events for a given final state in which there is an additional DPS contribution containing a pair of b-quarks. This fraction is clearly given by the ratio, $\sigma_{b\bar{b}}/(20 \text{ mb})$. From the figure this fraction ranges from a manageably-small 2% effect

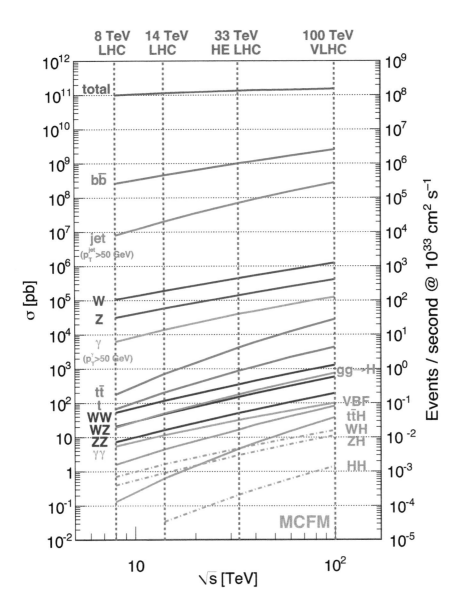

Fig. 10.1 Cross-sections for select hadron collider processes as a function of the operating energy, $\sqrt{s}$. The cross sections presented in this figure have been calculated at next-to-leading order in QCD using the MCFM program [311, 314].

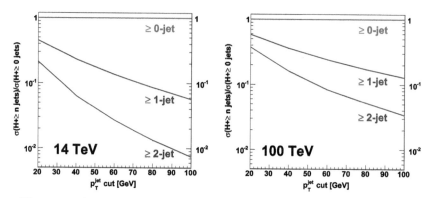

Fig. 10.2 Cross-sections for the production of a Higgs boson in association with n or more jets, for $n = 0, 1, 2$, normalized to the inclusive Higgs cross-section ($n = 0$). Cross-sections are shown as a function of the minimum jet p_T and are displayed for a proton-proton collider operating at 14 TeV (left) and 100 TeV (right).

at 8 TeV to a much more significant 15% at 100 TeV. More study would clearly be required in order to obtain a true estimate of the impact of such events on the physics that could be studied at higher energies, but these simplified arguments can at least give some idea of the potentially troublesome issues.

As an example of the behaviour of less-inclusive cross-sections at higher energies, Fig. 10.2 shows predictions for $H + n$ jets $+ X$ cross-sections at various values of $\sqrt{s}$ and as a function of the minimum jet transverse momentum. The cross-sections are all normalized to the inclusive Higgs production cross-section, so that the plots indicate the fraction of Higgs events that contain at least the given number of jets. The inclusive Higgs cross-section includes NNLO QCD corrections, while the 1- and 2-jet rates are computed at NLO in QCD. All are computed in the effective theory with $m_t \to \infty$.

The extent to which additional jets are expected in Higgs events is strongly dependent on how the jet cuts must scale with the machine operating energy. For instance, consider a jet cut of 40 GeV at 14 TeV, a value in line with current analysis projections. For this cut, approximately 20% of all Higgs boson events produced through gluon fusion should contain at least one jet. The fraction with two or more jets is expected to be around 5%. To retain approximately the same jet compositions at 100 TeV requires only a modest increase in the jet cut to 80 GeV.

However, this analysis is not the full story, due to effects induced by a finite top-mass that are neglected in the effective theory. This is illustrated in Fig. 10.3, which shows the rates for Higgs production in association with up to three jets, taking proper account of the top-mass, as a function of the minimum jet p_T. As shown in the lower panel, a comparison of these results with those obtained in the effective theory reveals significant differences. Even for moderate jet cuts of around 50 GeV a finite top-mass results in differences in the $H + 3$ jet rate of approximately 30%. For significantly harder jet cuts the effective theory description clearly fails spectacularly. Although

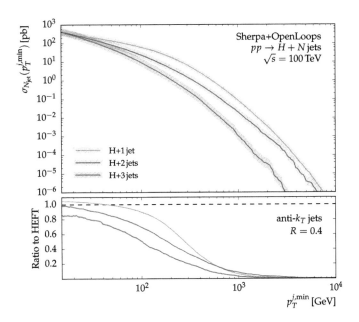

Fig. 10.3 Cross-sections for the production of a Higgs boson in association with 1, 2 or 3 jets, taking into account finite top-mass effects. Cross-sections are shown as a function of the minimum jet p_T for a proton-proton collider operating at 100 TeV. The lower panel shows the ratio of these results to the ones obtained in the effective theory. Reprinted with permission from Ref. [267].

this should not be a great surprise, given the energy scales being accessed, it is a useful reminder of the limitations of approximations that are commonly used at the LHC. Such approximations must clearly be left behind in order to obtain meaningful predictions for relatively common kinematic configurations at a 100 TeV collider.

Of course, the differences that exist between the theoretical predictions at 14 TeV and 100 TeV offer significant opportunities that are only beginning to be explored. The event rates will be sufficiently high that analysis cuts can be devised to take advantage of the unique kinematics at a 100 TeV collider, rather than simply "scaling up" the types of analyses currently in use at the LHC. For instance, substantially harder cuts on the transverse momenta of jets will lead to a predominance of boosted topologies, which can be analysed with the types of jet substructure techniques that are still relatively new at the LHC, *cf.* Section 9.2.4.

10.2.2 Necessary theory developments

Improvements to the theoretical description of hadronic collisions are of course driven by the accuracy of the experimental measurements that can be made. The outstanding level of detail that the LHC detectors have been able to provide, from particle identification to jet tracking, has enabled experimental uncertainties to be controlled at the few-percent level for quantities such as the transverse momentum of single photon or Z-bosons. Such exquisite measurements have thrown down the gauntlet to the theoretical community.

Of course, some of these challenges have been foreseen. Going from the early days of the TEVATRON to the build-up to the LHC saw a sea-change in the quality of perturbative predictions. Rather than being limited to LO predictions for $2 \rightarrow 2$ processes, by the advent of the LHC NLO predictions were available for almost all final states of immediate interest. At the beginning of Run II of the LHC, even NNLO calculations have matured to the level of providing differential predictions for events containing jets. The pace of these developments has been so fast that it is easy to take for granted a level of sophistication that many never believed would have been achieved by now. The availability of N^3LO predictions for Higgs production, multiple examples of NNLO calculations matched to a parton shower, and the ability to go from a Lagrangian to NLO-accurate showered events, are just a few such examples. Such progress, to a level of precision that in some cases borders the ridiculous, may leave the reader wondering if challenges remain. Yet, undeniably, much work lies ahead.

In terms of fixed-order descriptions, the march to higher orders is not yet over. It is not clear whether existing techniques for performing NNLO calculations will be able to be applied to more complex final states. While continued improvements in computer processing power will certainly help, it is almost certain that alternative, superior approaches have yet to be devised. Similar arguments apply to the case of N^3LO predictions, where extensions to the method that could provide more differential information, or perhaps be suitable for more general processes, are far from obvious. As highlighted in earlier chapters, the presence of substantial electroweak corrections at high energies is just beginning to be probed. As the LHC becomes more sensitive to even higher energies, the inclusion of higher-order electroweak effects will become mandatory in order to retain theoretical predictions of sufficient precision. A simultaneous expansion in both parameters, *i.e.* correctly including corrections that contain a mix of strong and electroweak couplings, will also become important. At present no complete calculation of such effects exists, even for a single process. In addition, a number of approximations are routinely used to simplify existing calculations. Examples include neglecting quark masses, working in the limit $m_t \rightarrow \infty$, and considering production and decay stages of resonance production separately. These will all need to be revisited, for various physics processes, in the coming years.

As improved fixed-order predictions become available it will be important that their effects are included in parton shower predictions. This will enable the improved modelling of the computed processes to be properly taken into account across a wide range of experimental analyses. The parton showers themselves will be the subject of greater scrutiny as they are held up to the light of experimental data that is ever more precise. This may reveal deficiencies in our modelling, either related to an incomplete treatment of towers of logarithms, or simply from an unavoidable choice in how the shower is constructed. Further subtleties, related to non-perturbative effects such as hadronization, fragmentation, and even the quality of the factorization picture itself, will eventually require new theoretical understanding as they become the dominant sources of theoretical uncertainty.

Finally – and, perhaps, most critically – it is important to not lose sight of the fact that the ultimate goal of this program is to extract the most possible information from the data that the LHC provides. To this end it is imperative to also continually develop new tools and novel approaches for doing just that. An excellent example of this is the development of jet substructure techniques, that have already had applications to top-tagging, jet discrimination, and a host of other analysis methods besides. No doubt there are many more insightful theoretical observations of this nature waiting to be made in the years ahead.

Appendix A
Mathematical background

A.1 Special functions

This section outlines the definitions of the special functions that appear in this book, together with some simple properties that are germane to the discussion here.

A.1.1 Gamma function

The Gamma function is defined by the integral,

$$\Gamma(\alpha) = \int_0^\infty \mathrm{d}x \, x^{\alpha-1} e^{-x} \tag{A.1}$$

and satisfies the relation,

$$\Gamma(\alpha+1) = \alpha \, \Gamma(\alpha). \tag{A.2}$$

If α is a positive integer then the Gamma function reduces to a factorial,

$$\Gamma(\alpha) = (\alpha-1)! \tag{A.3}$$

In calculations performed in dimensional regularization a useful representation of the Gamma function is,

$$\Gamma(1+\varepsilon) = \exp\left(-\gamma_E \varepsilon + \frac{\pi^2}{12} \varepsilon^2\right) + \mathcal{O}(\varepsilon^3), \tag{A.4}$$

where $\gamma_E \approx 0.57221$ is the Euler-Mascheroni constant. Using this representation it is easy to see that the following relation holds,

$$\frac{\Gamma^2(1-\varepsilon)}{\Gamma(1-2\varepsilon)} = 1 - \frac{\varepsilon^2 \pi^2}{6} + \mathcal{O}(\varepsilon^3). \tag{A.5}$$

The beta function $B(\alpha, \beta)$ is defined by the integral,

$$B(\alpha, \beta) = \int_0^1 \mathrm{d}x \, x^{\alpha-1}(1-x)^{\beta-1} = \frac{\Gamma(\alpha)\Gamma(\beta)}{\Gamma(\alpha+\beta)}, \tag{A.6}$$

and, as shown, can immediately be re-expressed in terms of Gamma functions.

A.1.2 "+" functions

The notation of "+" functions appears, for example, in the definition of the splitting kernels, *cf.* Eq. (2.10). The quantity $[g(x)]_+$ is defined through its integral together with a test function $f(x)$ such that

$$
\int_0^1 dx\, f(x)g(x) = \int_0^1 dx\, \Big(f(x) - f(1) \Big) g(x) + f(1) \int_0^1 dx\, g(x)
$$
$$
= \int_0^1 dx\, f(x) \left([g(x)]_+ + \delta(1-x) \int_0^1 dy\, g(y) \right). \qquad (A.7)
$$

In other words,

$$
\int_0^1 dx\, f(x)[g(x)]_+ = \int_0^1 dx\, [f(x) - f(1)]\, g(x). \qquad (A.8)
$$

A.1.3 Dilogarithm

The dilogarithm (or Spence's) function $\mathrm{Li}_2(x)$ is defined by,

$$
\mathrm{Li}_2(x) = - \int_0^x dy\, \frac{\ln(1-y)}{y}, \qquad (A.9)
$$

and has the series expansion,

$$
\mathrm{Li}_2(x) = \sum_k^\infty \frac{x^k}{k^2}. \qquad (A.10)
$$

Useful values for particular arguments are,

$$
\mathrm{Li}_2(0) = 0, \qquad (A.11)
$$
$$
\mathrm{Li}_2(1) = \frac{\pi^2}{6}. \qquad (A.12)
$$

There are also a number of identities that combine values of the function for related arguments such as,

$$
\mathrm{Li}_2(x) + \mathrm{Li}_2(1-x) = \frac{\pi^2}{6} - \log x \log(1-x). \qquad (A.13)
$$

A.1.4 Mellin transforms

The Mellin transform $\mathbf{M}_N\,[f(x)]$ of a function $f(x)$ is given by[1]

---

[1] In mathematical sciences, tpyically the integration limits are 0 and ∞, in which case the back-transformation reads

$$
f(x) = \frac{1}{2\pi i} \int_{-i\infty}^{i\infty} dN\, x^{-N-1} f(N). \qquad (A.14)
$$

However, in what follows the back-transformation is often performed by merely identifiying known expressions for Mellin transforms.

$$\mathbf{M}_N\left[f(x)\right] = \int\limits_0^1 \mathrm{d}x\, x^N f(x). \tag{A.15}$$

There are two reasons why the technology of this transform is interesting. First of all, it can be seen very quickly that convolutions of the type encountered in the cross-section calculation involving PDFs factorize to trivial products of the Mellin transforms. In order to see how this works, consider the convolution of $f \otimes \hat{\sigma}$ which often occurs in the calculation of cross-sections,

$$
\begin{aligned}
\sigma &= \int\limits_0^1 \mathrm{d}x (f \otimes \hat{\sigma})(x) = \int\limits_0^1 \mathrm{d}x \int\limits_x^1 \frac{\mathrm{d}y}{y}\, f(y)\, \hat{\sigma}(x/y) \\
&= \int\limits_0^1 \mathrm{d}x\mathrm{d}y\mathrm{d}z\, \delta(x - yz)\, f(y)\hat{\sigma}(z).
\end{aligned} \tag{A.16}
$$

In Mellin space then

$$
\begin{aligned}
\mathbf{M}_N\left[(f \otimes \hat{\sigma})(x)\right] &= \int\limits_0^1 \mathrm{d}x\, x^N\, (f \otimes \hat{\sigma})(x) = \int\limits_0^1 \mathrm{d}x\mathrm{d}y\mathrm{d}z\, x^N\, \delta(x - yz)\, f(y)\, \hat{\sigma}(z) \\
&= \int\limits_0^1 \mathrm{d}y\mathrm{d}z\, (yz)^N\, f(y)\, \hat{\sigma}(z) = \mathbf{M}_N\left[f(x)\right] \cdot \mathbf{M}_N\left[\hat{\sigma}(x)\right].
\end{aligned} \tag{A.17}
$$

Consider, as a useful example, the PDFs transformed to Mellin space, $\mathbf{M}_N\left[f_{i/A}(x,\,\mu_F)\right]$, which, in parallel to the original PDFs, fulfil the DGLAP equation

$$\frac{\mathrm{d}}{\mathrm{d}\log\mu^2} \mathbf{M}_N\left[f_{i/A}(x,\,\mu)\right] = \gamma(N,\,\alpha_s(\mu^2))\, \mathbf{M}_N\left[f_{i/A}(x,\,\mu)\right] \tag{A.18}$$

with the solution

$$\mathbf{M}_N\left[f_{i/A}(x,\,\mu)\right] = \exp\left[-\int\limits_{\mu^2}^{Q^2} \frac{\mathrm{d}q^2}{q^2}\, \gamma(N,\,\alpha_s(q^2))\right] \mathbf{M}_N\left[f_{i/A}(x,\,Q)\right]. \tag{A.19}$$

Here, the $\gamma(N,\,\alpha_s(\mu^2))$ are the anomalous dimensions and depend on the strong coupling. Similar to all other quantities encountered so far they can be expanded as a power series in α_s as

$$\gamma(N,\,\alpha_s(\mu^2)) = \sum_{i=1}^{\infty} \left(\frac{\alpha_s(\mu^2)}{2\pi}\right)^i \gamma^{(i)}(N). \tag{A.20}$$

Concentrating on the first order term only, $\gamma^{(1)}(N)$, and by direct comparison with the DGLAP equation, *cf.* Eq. (2.31), it is clear that the moment related to the $q \to qg$ splitting is given by

$$
\gamma_{qq}^{(1)}(N) = \mathbf{M}_N \left[P_{qq}^{(1)}(x) \right] = \int_0^1 dx\, x^N P_{qq}^{(1)}(x) = C_F \int_0^1 dx\, x^N \left(\frac{1+x^2}{1-x} \right)_+
$$

$$
= C_F \int_0^1 dx\, \frac{(1+x^2)(x^N - 1)}{1-x} = C_F \xi(N),
$$

(A.21)

a finite number.

Furthermore, logarithms can be directly identified by analysing the analytical structure in the Mellin parameter N, by identifiying poles in $1/(N - N_0)$. A straightforward way to see this is by realising that

$$
\frac{d^L}{dN^L} \mathbf{M}_N \left[f(x) \right] = \mathbf{M}_N \left[\log^L(x)\, f(x) \right],
$$

(A.22)

which follows directly from the definition of the Mellin transform when realising that $x^N = \exp(N \log x)$. In a similar way,

$$
\mathbf{M}_N \left[x^k f(x) \right] = \mathbf{M}_{N+k} f(x),
$$

(A.23)

and, if $f(x)$ is regular in the limit $x \to 1$,

$$
\mathbf{M}_N \left[\frac{df(x)}{dx} \right] = x^N f(x) \big|_0^1 - N \mathbf{M}_{N-1} f(x).
$$

(A.24)

For further reference, in Eq. (A.25) Mellin transforms of different relevant functions are listed.

$$
\mathbf{M}_N [1] = \frac{1}{N+1}
$$

$$
\mathbf{M}_N \left[\left[\frac{1}{1-x} \right]_+ \right] = - \sum_{k=1}^N \frac{1}{k}
$$

(A.25)

$$
\mathbf{M}_N \left[\left[\frac{\log(1-x)}{1-x} \right]_+ \right] = \frac{1}{2} \left\{ \psi'(N) + \zeta(2) + \left[\psi(N) + \gamma_E \right]^2 \right\}
$$

Here, the ψ function is related to derivatives of the Γ function through

$$
\psi(x) = \frac{d \log \Gamma(x)}{dx} \quad \text{and} \quad \psi'(x) = \frac{d\psi(x)}{dx} = \frac{d^2 \log \Gamma(x)}{dx^2},
$$

(A.26)

and ζ denotes, as usual, Riemann's ζ function with $\zeta(2) = \pi^2/6$ and $\zeta(3) \approx 1.2021$.

A.1.5 Feynman parameterization

The generic identity for combining propagators into a single denominator through the use of Feynman parameters is,

$$\frac{1}{A_1^{\nu_1}\dots A_n^{\nu_n}} = \frac{\Gamma(\nu)}{\prod_i \Gamma(\nu_i)} \int_0^1 d^n x_i\, \delta\left(\sum_i x_i - 1\right) \frac{\prod_i x_i^{\nu_i-1}}{\left(\sum_i x_i A_i\right)^{\sum_i \nu_i}}. \tag{A.27}$$

A.1.6 D dimensions and the Gauss integral

The use of the Feynman parameterization of loop integrals given in Eq. (A.27) leads directly to integrals that take the form,

$$\int \frac{d^D\ell}{(2\pi)^D} \frac{(\ell^2)^k}{(\ell^2 - \Delta + i\varepsilon)^n}. \tag{A.28}$$

In order to evaluate this integral it is easiest to perform a **Wick rotation** from Minkowski to Euclidean space. This is accomplished through the transformation, $\ell_0 \to i\ell_0$, with the space-like components of ℓ unchanged. This is permissible because of the location of the poles in the original integrand in Eq. (A.28), which are at $\ell_0 = \pm(|\ell|^2 + \Delta - i\varepsilon)$ where $\ell^2 = \ell_0^2 - |\ell|^2$. These are not encountered when rotating the real axis by an angle $\pi/2$ (counter-clockwise) to the imaginary one. The Euclidean integral can then be parameterized by a generalization of spherical coordinates to D dimensions so that,

$$\int d^D\ell = \int d\ell_E\, \ell_E^{D-1} \sin^{D-2}\theta_{D-1} \sin^{D-3}\theta_{D-2}\dots \sin\theta_2\, d\theta_{D-1}\, d\theta_{D-2}\dots d\theta_1, \tag{A.29}$$

where $\ell_E = \sqrt{\ell_0^2 + |\ell|^2}$. Since the integrand is only a function of ℓ_E the angular integrations can be performed immediately by using

$$\int_0^\pi d\theta \sin^n\theta = \sqrt{\pi}\,\frac{\Gamma\left(\frac{n+1}{2}\right)}{\Gamma\left(\frac{n+2}{2}\right)}. \tag{A.30}$$

This result leads to

$$\int \frac{d^D\ell}{(2\pi)^D} = \frac{2i\pi^{D/2}}{(2\pi)^D\Gamma(D/2)} \int d\ell_E\, \ell_E^{D-1}. \tag{A.31}$$

The final integral over ℓ_E can be cast into the form of a beta-function integral,

$$\int \frac{d\ell_E\, \ell_E^{D-1}(-\ell_E^2)^k}{(-\ell_E^2 - \Delta + i\varepsilon)^n} = \frac{(-1)^{n-k}}{2}(\Delta - i\varepsilon)^{D/2-n+k} \int_0^1 dx\, x^{n-k-D/2-1}(1-x)^{D/2+k-1} \tag{A.32}$$

Using the result for such an integral given in Eq. (A.6), together with Eq. (A.31), one arrives at the identity

$$\int \frac{\mathrm{d}^D \ell}{(2\pi)^D} \frac{(\ell^2)^k}{(\ell^2 - \Delta)^n} = \frac{i\,(-1)^{n-k}}{(4\pi)^{D/2}} \frac{\Gamma(D/2+k)}{\Gamma(D/2)} \frac{\Gamma(n-k-D/2)}{\Gamma(n)} (\Delta - i\varepsilon)^{D/2-n+k}.$$

$$(A.33)$$

A.2 Spinors and spinor products

A.2.1 A first representation

There are different related techniques to evaluate spinor products. Here a fairly old approach [197, 684] will be briefly presented, which is employed in the HELAS library [604] forming the basis of MADGRAPH [150] and in AMEGIC++ [696]. The starting point of all constructions lies in the fact that spinors such as $u(p, \lambda)$ and $v(p, \lambda)$, related to fermions or anti-fermions with mass m, momentum p and with definite helicity λ obey Dirac's equation of motion

$$(\slashed{p} - m)u(p, \lambda) = 0, \qquad (\slashed{p} + m)v(p, \lambda) = 0, \qquad (A.34)$$

where $\slashed{p}$ is not necessarily Hermitean and m does not need to be real, while in any case $p^2 = m^2$ must be fulfilled. Additionally, the spinors fulfil the spin projection identity

$$(1 \mp \gamma^5 \slashed{s})u(p, \pm) = 0, \qquad (1 \mp \gamma^5 \slashed{s})v(p, \pm) = 0 \qquad (A.35)$$

for all polarization vectors s obeying $s \cdot p = 0$ and $s^2 = 1$. This allows to construct *massless chiral spinors* w, spinors satisfying

$$w(k_0, \lambda)\bar{w}(k_0, \lambda) = \frac{1 + \lambda\gamma_5}{2} \slashed{k}_0, \qquad (A.36)$$

for an arbitrary light-like four-vector k_0, which will act as some kind of a "spinor gauge vector". Spinors of opposite chirality $w(k_0, -\lambda)$ can be constructed through

$$w(k_0, \lambda) = \lambda \slashed{k}_1 w(k_0, -\lambda) \qquad (A.37)$$

with the vectors $k_{0,1}$ satisfying

$$k_0^2 = 0, \qquad k_0 \cdot k_1 = 0 \quad \text{and} \quad k_1^2 = -1. \qquad (A.38)$$

Arbitrary, and potentially massive, spinors can be expressed in terms of these chiral spinors as

$$\begin{aligned} u(p, \lambda) &= \frac{\slashed{p} + m}{\sqrt{2p \cdot k_0}} w(k_0, -\lambda) \\ v(p, \lambda) &= \frac{\slashed{p} - m}{\sqrt{2p \cdot k_0}} w(k_0, -\lambda). \end{aligned} \qquad (A.39)$$

The relations above also hold true for $p^2 < 0$ and imaginary m.

For the construction of conjugate spinors the proper definitions

$$\bar{u} = u^\dagger \gamma^0 \quad \text{and} \quad \bar{v} = v^\dagger \gamma^0, \qquad (A.40)$$

applied on the spinors obtained so far do not lead to the correct E.o.M.

$$\bar{u}(p,\lambda)(\not{p} - m) = 0 \quad \text{and} \quad \bar{v}(p,\lambda)(\not{p} + m) = 0, \tag{A.41}$$

if the mass is imaginary. Using instead

$$\bar{u}(p,\pm)(1 \mp \gamma^5\not{s}) = 0 \quad \text{and} \quad \bar{v}(p,\pm)(1 \mp \gamma^5\not{s}) = 0 \tag{A.42}$$

and choosing the normalisation conditions

$$\bar{u}(p,\lambda)u(p,\lambda) = 2m \quad \text{and} \quad \bar{v}(p,\lambda)v(p,\lambda) = -2m, \tag{A.43}$$

the conjugate spinors are constructed as

$$\begin{aligned}
\bar{u}(p,\lambda) &= \bar{w}(k_0, -\lambda)\frac{\not{p} + m}{\sqrt{2p \cdot k_0}} \\
\bar{v}(p,\lambda) &= \bar{w}(k_0, -\lambda)\frac{\not{p} - m}{\sqrt{2p \cdot k_0}}.
\end{aligned} \tag{A.44}$$

This yields the following products for massless spinors

$$\bar{u}(p_1, +)u(p_2, p_2) = \frac{(p_1k_0)(p_2k_1) - (p_1k_1)(p_2k_0) + i\epsilon_{\mu\nu\rho\sigma}p_1^\mu p_2^\nu k_0^\rho k_1^\sigma}{\sqrt{(p_1k_0)(p_2k_0)}} \tag{A.45}$$

$$\bar{u}(p_1, -)u(p_2, +) = [\bar{u}(p_1, +)u(p_2, -)]^*.$$

Terms of the form $\bar{u}(+, p_1)u(+, p_2)$ are proportional to the masses of the spinors. The such-defined spinors fulfil the completeness relation

$$1 = \sum_\lambda \frac{\bar{u}(p,\lambda)u(p,\lambda) - \bar{v}(p,\lambda)v(p,\lambda)}{2m}. \tag{A.46}$$

This ensures that the identity

$$\not{p} + m = \frac{1}{2}\sum_\lambda \left[\left(1 + \frac{m}{\sqrt{p^2}}\right)u(p,\lambda)\bar{u}(p,\lambda) + \left(1 - \frac{m}{\sqrt{p^2}}\right)v(p,\lambda)\bar{v}(p,\lambda)\right] \tag{A.47}$$

holds true, which allows to rewrite propagator numerators as spinor products. This allows terms of the form $\bar{u}(p_1)\not{k}u(p_2)$ to be rewritten by decomposing $\not{k}$ into spinors, resulting in the spinor products of Eq. (A.45). In addition, terms of the form $(\bar{u}\gamma^\mu u) \times (\bar{u}\gamma_\mu u)$ are dealt with by employing Chisholm identities, yielding, again, spinor products of the type $\bar{u}u$ from Eq. (A.45). Furthermore, a spinor representation of polarization vector for external particles may become necessary, unless they are explicitly constructed and contracted into Lorentz-invariant scalar products or similar. To achieve this, first for massless vector particles like gluons or photons, it is clear that any representation must satisfy the identities

$$\epsilon_\mu(p, \lambda)\, p^\mu = 0$$

$$\sum_{\lambda=\pm} \epsilon_\mu(p, \lambda)\epsilon_\nu^*(p, \lambda) = -g_{\mu\nu} + \frac{q_\mu p_\nu + q_\nu p_\mu}{pq}, \tag{A.48}$$

where q denotes an arbitrary, light-like four-vector not parallel to p. This can be achieved by

$$\epsilon_\mu(p, \lambda) = \frac{1}{2\sqrt{pq}}\, \bar{u}(q, \lambda)\, \gamma_\mu\, u(p, \lambda). \tag{A.49}$$

Similarly, for massive vector bosons, the polarization vectors must satisfy the completeness relation

$$\sum_{\lambda=\pm,0} \epsilon_\mu(p, \lambda)\epsilon_\nu^*(p, \lambda) = -g_{\mu\nu} + \frac{p_\mu p_\nu}{p^2}, \tag{A.50}$$

It can be shown that this relation can be obtained by writing

$$\epsilon_\mu(p, \lambda) \longrightarrow \sqrt{\frac{3}{8\pi p^2}}\, \bar{u}(q_1, \lambda)\, \gamma_\mu\, u(q_2, \lambda)\Big|_{p^\mu = q_1^\mu + q_2^\mu} \tag{A.51}$$

and integrating over the solid angle of q_1 in the rest-frame of p. The apparent problem is that this does allow only for unpolarized cross section calculations and the additional integration renders the direct construction of polarization vectors potentially advantageous in terms of computing speed.

A.2.2   Weyl–van der Waerden spinors

A more convenient way to deal with spinors is encoded in the Weyl–van der Waerden formalism [878, 890]. Right- and left-handed chiral spinors in the $\mathcal{D}(\frac{1}{2}, 0)$ and $\mathcal{D}(0, \frac{1}{2})$ representations of the Lorentz group are denoted, as usual, by undotted and dotted spinor indices, with complex conjugation connecting them:

$$\psi_{\dot{a}} = (\psi_a)^* \quad \text{and} \quad \psi^a = (\psi^{\dot{a}})^* \tag{A.52}$$

Raising and lowering these indices is achieved by applying a tensor ϵ given by

$$\epsilon_{ab} = \epsilon^{ab} = \epsilon_{\dot{a}\dot{b}} = \epsilon^{\dot{a}\dot{b}} = \begin{pmatrix} 0 & 1 \\ -1 & 0 \end{pmatrix}. \tag{A.53}$$

There are two inner products in spinor space, for undotted and for dotted indices, namely

$$\begin{aligned} \langle \zeta\eta \rangle &= \zeta_a \eta^a \\ [\zeta\eta] &= \zeta_{\dot{a}} \eta^{\dot{a}} = \langle \zeta\eta \rangle^*. \end{aligned} \tag{A.54}$$

In the literature it has become customary to replace the spinors by their momentum argument or its label; for example $\zeta_a(k) = |k\rangle$ and $\zeta_{\dot{a}}(k) = |k]$.

With four-vectors residing in the $\mathcal{D}(\frac{1}{2}, \frac{1}{2})$ representation, they are constructed using two spinors and the four-vectors of Pauli matrices

$$\sigma^{\mu\dot{a}b} = \left(\sigma^0, \vec{\sigma}\right) \quad \text{and} \quad \sigma^\mu_{a\dot{b}} = \left(\sigma^0, -\vec{\sigma}\right) \tag{A.55}$$

such that any four-vector k^μ can be written as

$$k_{a\dot{b}} = \sigma^\mu_{a\dot{b}} k_\mu = \begin{pmatrix} k^+ & k_\perp \\ k^*_\perp & k^- \end{pmatrix}, \quad \text{where} \quad \begin{aligned} k^\pm &= k^0 \pm k^3 \\ k_\perp &= k^1 + ik^2, \end{aligned} \tag{A.56}$$

such that for massless vectors $\vec{k}_\perp^2 = k^+k^-$. This allows to define spinors $\zeta(k)$ such that

$$k_{a\dot{b}} = \zeta_{\dot{a}}(k)\zeta_b(k), \quad \text{with} \quad \zeta_a(k) = \begin{pmatrix} \sqrt{k^+} \\ \sqrt{k^-}\,e^{i\phi_k} \end{pmatrix}, \tag{A.57}$$

and where $\phi_k = \arg k_\perp$. There are of course a few choices that can be made: first of all, the spinor representation above can be multiplied with a freely chosen total phase $\exp(i\theta)$. In addition, of course, the choice of which axis defines the $\pm$-direction, is arbitrary; instead of using the z-axis it would be possible to choose another axis, corresponding to a rotation of the Pauli matrices. Irrespective of such details in the spinor definition, four-vectors are given by

$$k^\mu = \sigma^\mu_{\dot{a}b}\zeta^{\dot{a}}(k)\zeta^b(k). \tag{A.58}$$

Massive four-vectors can be constructed by decomposing them into two massless ones, introducing yet another gauge degree of freedom. As a by-product of this, for massless vectors k_i and k_j

$$2k_i k_j = \langle ij \rangle\, [ij]\,, \tag{A.59}$$

so that a Lorentz product can be cast as a Dirac product. The fact that squared scattering amplitudes can always be expressed as Lorentz invariants translates into an independence on all choices made in constructing the underlying spinors, thus providing a welcome check of any calculation.

External particles can then be represented in the following way:

- Fermions are decomposed into left- and right-handed components, $u_\pm = P_\pm u$ with the projection operators

$$P_{R,L} = P_\pm = \frac{1 \pm \gamma_5}{2}. \tag{A.60}$$

Such chiral fermions with momentum $p = (p_0, \vec{p})$, with signed three-momentum $\bar{p} = \text{sgn}(p_0)|\vec{p}|$ and with the light-like four-vector $\hat{p} = (\bar{p}, \vec{p})$ are represented by

$$\begin{aligned} u_+(p, m) &= \frac{1}{\sqrt{2|\bar{p}|}} \begin{pmatrix} \sqrt{p_0 - \bar{p}}\,\chi_+(\hat{p}) \\ \sqrt{p_0 + \bar{p}}\,\chi_+(\hat{p}) \end{pmatrix} \\ u_-(p, m) &= \frac{1}{\sqrt{2|\bar{p}|}} \begin{pmatrix} \sqrt{p_0 + \bar{p}}\,\chi_-(\hat{p}) \\ \sqrt{p_0 - \bar{p}}\,\chi_-(\hat{p}) \end{pmatrix} \end{aligned} \tag{A.61}$$

$$v_+(p, m) = \frac{1}{\sqrt{2|\vec{p}|}} \begin{pmatrix} \sqrt{p_0 - \bar{p}} \, \chi_-(\hat{p}) \\ -\sqrt{p_0 + \bar{p}} \, \chi_-(\hat{p}) \end{pmatrix}$$

$$v_-(p, m) = \frac{1}{\sqrt{2|\vec{p}|}} \begin{pmatrix} -\sqrt{p_0 - \bar{p}} \, \chi_+(\hat{p}) \\ \sqrt{p_0 + \bar{p}} \, \chi_+(\hat{p}) \end{pmatrix},$$

(A.62)

where the Weyl spinors $\chi_\pm$ read

$$\chi_+(\hat{p}) = \frac{1}{\sqrt{\hat{p}^+}} \begin{pmatrix} \hat{p}^+ \\ \hat{p}_\perp \end{pmatrix} = \begin{pmatrix} \sqrt{\hat{p}^+} \\ \sqrt{\hat{p}^-} \, e^{i\phi_p} \end{pmatrix}$$

$$\chi_-(\hat{p}) = \frac{e^{i\pi}}{\sqrt{\hat{p}^+}} \begin{pmatrix} -\hat{p}_\perp^* \\ \hat{p}^+ \end{pmatrix} = \begin{pmatrix} \sqrt{\hat{p}^-} \, e^{-i\phi_p} \\ -\sqrt{\hat{p}^+} \end{pmatrix}.$$

(A.63)

Since they are orthogonal and normalised to $2\hat{p}_0$, the resulting Dirac spinors above are normalised to $\pm 2m$. In most cases, massless particles are considered, which leads to a massive simplification of the Dirac spinors, since only the upper or lower components survive. In such a case, the short-hand notation

$$u_\pm(k) = v_\mp(k) = |k^\pm\rangle \quad \text{and} \quad \bar{u}_\pm(k) = \bar{v}_\mp(k) = \langle k^\pm| \tag{A.64}$$

has become customary.

- Polarization vectors for massless particles with momentum p are obtained in a way similar to the one already encountered by introducing a light-like gauge vector q and read

$$\epsilon_\pm^\mu(p, q) = \pm \frac{\langle q^\mp| \gamma^\mu |p^\mp\rangle}{\sqrt{2} \, \langle q^\mp| p^\pm\rangle}. \tag{A.65}$$

Again, this identification yields the properties of polarization vectors and their completeness relation in axial gauge, *cf.* Eq. (A.48).

- Polarization vectors for massive particles will have a slightly different representation; there the light-like gauge vector q is used to construct a vector $\tilde{p}^\mu = p^\mu - p^2/(2pq) \, q^\mu$ and the regular transverse polarizations

$$\epsilon_\pm^\mu(p, q) = \pm \frac{\langle q^\mp| \gamma^\mu |\tilde{p}^\mp\rangle}{\sqrt{2} \, \langle q^\mp| \tilde{p}^\pm\rangle} \tag{A.66}$$

are augmented by a longitudinal one,

$$\epsilon_0^\mu(p, q) = \frac{1}{\sqrt{p^2}} \left[\langle \tilde{q}^-| \gamma^\mu |\tilde{q}^-\rangle - \frac{p^2}{2pq} \langle q^-| \gamma^\mu |q-\rangle \right], \tag{A.67}$$

which then yield the right polarization sum of Eq. (A.50).

The connection to the spinor formalism that has been introduced can be made transparent in a very straightforward way by realizing that

$$|k^\pm\rangle\langle k^p m| = \frac{1 \pm \gamma_5}{2} \, k\!\!\!/, \tag{A.68}$$

which is exactly the definition defining the massless spinors there; thus, up to a potentially different phase convention, the objects $|k\rangle$ and $\langle k|$ are identical to the basic spinors $\bar{u}$ and u. Therefore, the spinor products $\bar{u}u$ of Eq. (A.45) become

$$
\begin{aligned}
\bar{u}(+, p_1)u(-, p_2) &= \frac{\langle p_1 p_2 \rangle \langle k_0 k_1 \rangle [p_2 k_0][k_1 p_1]}{\sqrt{4(p_1 k_0)(p_2 k_0)}} \\
\bar{u}(-, p_1)u(+, p_2) &= \frac{\langle p_2 k_0 \rangle \langle k_1 p_1 \rangle [p_1 p_2][k_0 k_1]}{\sqrt{4(p_1 k_0)(p_2 k_0)}}.
\end{aligned}
\tag{A.69}
$$

A.3   Kinematics

In most cases, there is a special axis defined through the geometry of particle physics experiments, namely the beam axis — the axis parallel to the incoming beams. In most experiments (BaBar is a famous exception), this axis is uniquely defined.[2] Usually, this beam axis is chosen to be the z-axis. In most cases, the position, where the beams are brought to collision, is pretty well known; this knowledge is used to fix an "origin" of the coordinate system. As long as the incoming beams are not polarized there is thus only one particular axis, and in such cases events exhibit cylindrical symmetry w.r.t. the z axis. Usually the related azimuthal angle is denoted by ϕ. Naively, then, other meaningful variables to determine momenta are the polar angle, typically denoted by θ, and either the energy or the absolute value of the three-momentum of the particle, where the ladder two are connected by the on-shell condition above. This set of parameters, say $\{p, \theta, \phi\}$ is particularly useful for lepton-lepton colliders where the longitudinal (w.r.t. the beam axis) momenta of the colliding partons — the leptons — are well known. However, for collisions involving hadrons this ceases to be true. since usually only some constituents of them, the partons, interact. In such cases, the energies and therefore the momenta of the incoming hadrons are known, but the energies and momentum fractions of the respective constituents that interact are not known a priori. Assuming that the initial partons of the process, the colliding hadron constituents, move in parallel to the incoming hadrons, this implies that the overall momentum of the colliding constituents along the beam axis is essentially unknown. One could then characterise their collision by their centre-of-mass energy and by the relative motion of their centre-of-mass system in the lab system. This relative motion can be understood as a boost of the constituent system with respect to the lab or beam system. Therefore, instead of using the polar angle θ in this cases it is more useful to have a quantity with better properties under boosts along the beam axis. Such a quantity is the *rapidity*, usually denoted by y. For a given four-momentum p, it is defined as

$$
y = \frac{1}{2} \log \frac{E + p_z}{E - p_z}.
\tag{A.70}
$$

It is simple to show that rapidity differences remain invariant under boosts along the z axis. To do so, it is enough to prove that rapidities change additively under boosts. Any boost is parameterized by a boost parameter γ and by defining an axis. Energy

---

[2]Even in BaBar, where the beams cross under an angle in the laboratory system, a boost (Lorentz transformation) can be applied, to find a system, where the beams collide "head on".

and three-momentum along this axis (here for obvious reasons the z axis) then change according to

$$E' = E \cosh \gamma - p_z \sinh \gamma$$
$$p'_z = p_z \cosh \gamma - E \sinh \gamma \qquad \text{(A.71)}$$

and thus

$$y' = y - \gamma. \qquad \text{(A.72)}$$

Unfortunately the rapidity does not provide a very intuitive interpretation, based on geometry. Therefore, another quantity has been introduced, called the *pseudo-rapidity*, commonly denoted by η. Employing the polar angle θ, it is defined through

$$\eta = \log \tan \frac{\theta}{2}. \qquad \text{(A.73)}$$

It is worth stressing here that in the limit of massless particles, their rapidities and pseudorapidities coincide. On the other hand, for massive particles, a finite rapidity y may be achieved my a mere boost along the beam axis, leading to an angle $\theta = 0$ w.r.t. this axis and hence an infinite pseudo-rapidity.

Having thus characterized the longitudinal component of the momentum, only the transverse component needs to be described — which is typically achieved by quoting its absolute value $p_\perp$ and the azimuthal angle ϕ. For massless particles therefore the four-momenta can be written as

$$p^\mu = p_\perp (\cosh \eta, \cos \phi, \sin \phi, \sinh \eta), \qquad \text{(A.74)}$$

while for massive particles

$$p^\mu = (m_\perp \cosh y, p_\perp \cos \phi, p_\perp \sin \phi, m_\perp \sinh y), \qquad \text{(A.75)}$$

in terms of the *transverse mass*

$$m_\perp^2 = p_\perp^2 + m^2. \qquad \text{(A.76)}$$

This also allows to rewrite the Lorentz-invariant phase-space element of one particle as follows:

$$\frac{\mathrm{d}^4 p}{(2\pi)^4} (2\pi) \delta(p^2 - m^2) \Theta(p_0) = \frac{\mathrm{d}^3 p}{2E(2\pi)^3} = \frac{p_\perp \mathrm{d}p_\perp \mathrm{d}y \mathrm{d}\phi}{2(2\pi)^3}. \qquad \text{(A.77)}$$

A.3.1 Light-cone decomposition

From here, the transition to light-cone variables is fairly straightforward. They are constructed by defining two momenta P_+ and P_- — in hadronic collisions they are typically given by the incoming beams. Orienting them along the z axis, and assuming symmetric collisions, they read

$$P_\pm = (E, 0, 0, \pm E), \qquad \text{(A.78)}$$

where the projectiles' masses have been neglected. Then the total hadronic centre-of-mass energy squared can be expressed as

$$S = 2P_+P_-.$$ (A.79)

Then, any momentum p^μ can be decomposed as

$$p^\mu = \alpha P_+^\mu + \beta P_-^\mu + \vec{p}_\perp^\mu,$$ (A.80)

where α and β are the plus and minus components of the momentum, respectively. The rapidity of p is given by

$$y = \frac{1}{2} \log \frac{E + p_z}{E - p_z} = \frac{1}{2} \log \frac{p_+}{p_-} = \frac{1}{2} \log \frac{\alpha}{\beta}.$$ (A.81)

In addition,

$$p^2 = \alpha\beta S - p_\perp^2,$$ (A.82)

which, together with $p^2 = m^2$ allows α or β to be eliminated through

$$\alpha = \frac{m^2 + p_\perp^2}{\beta S} = \frac{m^2 + p_\perp^2}{S} e^{+y} \quad \text{or} \quad \beta = \frac{m^2 + p_\perp^2}{\alpha S} = \frac{m^2 + p_\perp^2}{S} e^{-y}.$$ (A.83)

In a scattering process $p_1 + p_2 \to p_3 + \cdots + p_n$, four-momentum conservation then translates into

$$\alpha_1 + \alpha_2 = \alpha_1 = \sum_{i=3}^{n} \alpha_i$$

$$\beta_1 + \beta_2 = \beta_2 = \sum_{i=3}^{n} \beta_i$$ (A.84)

$$\vec{p}_{\perp,1} + \vec{p}_{\perp,2} = 0 = \sum_{i=3}^{n} \vec{p}_{\perp,i},$$

where it has been assumed that the two incident partons $p_{1,2}$ move along the positive and negative z axis respectively, implying that they have zero transverse momentum and that $\alpha_2 = \beta_1 = 0$. This also allows to identify α_1 and β_2 as the light-cone momentum fractions the partons carry with respect to the incoming hadrons. It is customary to identify these with the respective *Bjorken-x*,

$$x_1 \equiv \alpha_1 \quad \text{and} \quad x_2 \equiv \beta_2.$$ (A.85)

Appendix B
The Standard Model

B.1 Standard Model Lagrangian

B.1.1 Constructing the Standard Model

B.1.1.1 Gauge invariance: $U(1)$

The Standard Model (SM) of particle physics is arguably the most successful model in physics to date, explaining practically all known phenomena on sub-nuclear-length scales with only 19 parameters, which are being determined at ever-increasing precision. The construction of the model rests on one paradigm, namely **gauge invariance**.

The idea is the following: **global phase invariance** is the invariance of a Lagrangian under phase transformations of its fields. As an example consider the Lagrangian for a simple massive Dirac fermion ψ without interactions,

$$\mathcal{L} = \bar{\psi}\left(i\slashed{\partial} - m\right)\psi, \tag{B.1}$$

which is invariant under transformations parameterized by a single real phase θ,

$$\psi \longrightarrow \psi' = e^{i\theta}\psi \quad \text{and} \quad \bar{\psi} \longrightarrow \bar{\psi}' = \bar{\psi}e^{-i\theta}. \tag{B.2}$$

The Dirac fields above play the role of matter and are therefore usually also called the **matter fields**. Their invariance under the transformation actually guarantees that they have associated, **conserved charges**.

The gauge principle introduces interactions to this Lagrangian by *postulating* that the Lagrangian remains invariant even if the phase depends on space-time x^μ, $\theta = \theta(x)$, or, in other words, that the Lagrangian is also **local phase** or **gauge-invariant**. Naively, however, this is not the case, since

$$\slashed{\partial}\psi' = e^{i\theta(x)}\slashed{\partial}\psi + e^{i\theta(x)}\left(i\slashed{\partial}\theta(x)\right)\psi \neq e^{i\theta(x)}\slashed{\partial}\psi. \tag{B.3}$$

and the second, additional term must be compensated for. This is achieved by defining a self-compensating **gauge-invariant derivative**,

$$D_\mu = \partial_\mu - ieA_\mu(x) \tag{B.4}$$

where $A_\mu(x)$ is a new, additional field that transforms as

$$A_\mu(x) \longrightarrow A'_\mu(x) = A_\mu(x) + \frac{1}{e}\partial_\mu\theta(x). \tag{B.5}$$

As a consequence

$$(\not{D}\psi)' = \not{D}\psi \tag{B.6}$$

and similar for the $\bar{\psi}$. The new field(s) A_μ are the gauge fields. Through their introduction enforced by the gauge postulate of local phase invariance the Lagrangian is modified and reads

$$\mathcal{L} = \bar{\psi}\left(i\not{D} - m\right)\psi = \bar{\psi}\left(i\not{\partial} + e\not{A} - m\right)\psi. \tag{B.7}$$

As a consequence, an interaction emerged in the previously free Lagrangian, namely between two of the Dirac spinors and one of the gauge fields. Dynamics of the new gauge field A_μ is generated through adding a kinematic term, which in general has the form

$$\mathcal{L}_{\text{gauge}} = [D_\mu, D_\nu] = (\partial_\mu A_\nu - \partial_\nu A_\mu)(\partial^\mu A^\nu - \partial^\nu A^\mu) \tag{B.8}$$

where the subscript "g" refers to gauge. Simple inspection shows that on this level, without any further fields introduced, mass terms such as

$$\mathcal{L}_{gm} = \frac{m^2}{2} A_\mu A^\mu \tag{B.9}$$

violate gauge invariance and are thus forbidden.

B.1.1.2 Non-Abelian groups: $SU(N)$

In the previous example, the gauge transformations of Eq. (B.2) are effected by real numbers θ, the phases. These transformations form a group with elements labelled by a continuos index, the phase. This structure is known as a **Lie group**; the **algebra of its generators** consists of exactly one element, 1, and in turn the gauge group from the example above is known as $U(1)$.

Of course more complicated gauge transformations are also possible by arranging the **matter fields** in multiplet structures Ψ, and labelling the ψ with some index i:

$$\Psi = (\psi_1, \psi_2, \dots \psi_n)^T. \tag{B.10}$$

Global and local phase transformations then mix the various components; this is achieved through

$$\Psi \longrightarrow \Psi' = \exp\left(i\theta^a \tau^a\right) \Psi \tag{B.11}$$

or, in component notation,

$$\psi_i \longrightarrow \psi_i' = [\exp\left(i\theta^a \tau^a\right)]_{ij}\, \psi_j = U_{ij}\, \psi_j, \tag{B.12}$$

exhibiting the fact that the generators τ_a in fact are $n \times n$ matrices in the space of the indices i, as is their exponential. The phases θ_a may or may not depend on x, depending on whether the transformations above are local or global.

As a consequence, the gauge-invariant derivative reads

$$D_\mu = \partial_\mu - ig\tau^a A_\mu^a(x) = \partial_\mu - igA_\mu. \tag{B.13}$$

In order to ensure gauge invariance, the gauge fields now transform as

$$A_\mu(x) \longrightarrow A'_\mu(x) = U(x)\,A_\mu(x)\,U^\dagger(x) + \frac{i}{g}\,[\partial_\mu U(x)]\,U^\dagger(x) \tag{B.14}$$

and the gauge-invariant derivative transforms as

$$D_\mu(x) \longrightarrow D'_\mu(x) = U(x)\,D_\mu(x)\,U^\dagger(x). \tag{B.15}$$

In all cases the gauge fields must be massless in order to guarantee gauge invariance.

The only thing necessary to fix now is the **gauge group**. In the case of the SM it is given by a direct product of three groups, namely $SU(3)_c \otimes SU(2)_L \otimes_U (1)_Y$.

The subscript c of the first group $SU(3)_c$ stands for "colour", and it is the strong interactions which are susceptible to colour charges. These interactions are enjoyed by the quarks and mediated by the gluons. The most **fundamental representation** of this group is by arranging the individual quarks in triplets with a **colour index** running from 1 to 3, $i \in [1, 3]$ such that a quark field q is given by three spinors $\psi_{q,i}$

$$\Psi_q = (\psi_{q,1}, \psi_{q,2}, \psi_{q,3})^T. \tag{B.16}$$

The three "colours" i that the quark fields can carry are often also denoted as "red", "blue", and "green", a reminiscence to the first day of colour television. The anti-quarks of course carry anti-colour quantum numbers. The interactions mediated by the gluons are able to change the colour of a quark. They are related to the eight **Gell-Mann matrices** λ^a so that in the case of $SU(3)$, $\tau^a \equiv \lambda^a/2$ where the latter are given by

$$\lambda_1 = \begin{pmatrix} 0 & 1 & 0 \\ 1 & 0 & 0 \\ 0 & 0 & 0 \end{pmatrix}, \quad \lambda_2 = \begin{pmatrix} 0 & -i & 0 \\ i & 0 & 0 \\ 0 & 0 & 0 \end{pmatrix}, \quad \lambda_3 = \begin{pmatrix} 1 & 0 & 0 \\ 0 & -1 & 0 \\ 0 & 0 & 0 \end{pmatrix},$$

$$\lambda_4 = \begin{pmatrix} 0 & 0 & 1 \\ 0 & 0 & 0 \\ 1 & 0 & 0 \end{pmatrix}, \quad \lambda_5 = \begin{pmatrix} 0 & 0 & -i \\ 0 & 0 & 0 \\ i & 0 & 0 \end{pmatrix}, \quad \lambda_6 = \begin{pmatrix} 0 & 0 & 0 \\ 0 & 0 & 1 \\ 0 & 1 & 0 \end{pmatrix}, \tag{B.17}$$

$$\lambda_7 = \begin{pmatrix} 0 & 0 & 0 \\ 1 & 0 & -i \\ 0 & i & 0 \end{pmatrix}, \quad \lambda_8 = \frac{1}{\sqrt{3}}\begin{pmatrix} 1 & 0 & 0 \\ 0 & 1 & 0 \\ 0 & 0 & -2 \end{pmatrix}$$

and satisfy the commutator relation

$$[\lambda_a, \lambda_b] = i f_{abc}\lambda_c. \tag{B.18}$$

Furthermore, the λ are Hermitian and traceless, a property that usually is shared by generators of a group. One of the invariants of this group is given by the **Casimir operator**,

$$C_F = \sum_a \tau_a^2 = \frac{1}{4}\sum_{a=1}^{8} \lambda_a^2 = \frac{4}{3}. \tag{B.19}$$

This finishes the quick summary of the properties of the generators of $SU(3)$ in its **fundamental representation**.

The f_{abc} are the **structure constants** of $SU(3)$; they are completely anti-symmetric in the three indices and are given by

$$f_{123} = 1$$

$$f_{147} = f_{165} = f_{246} = f_{257} = f_{345} = f_{376} = \frac{1}{2}$$

$$f_{458} = f_{678} = \frac{\sqrt{3}}{2}. \qquad (B.20)$$

The structure constants form the **adjoint representation** of the group — in this repesentation the generators T_{ik}^a are matrices of dimension 8×8 given by

$$T_{ik}^a = if_{aik}. \qquad (B.21)$$

The corresponding Casimir operator is given by

$$C_A = \sum_a T^a T^a = 3. \qquad (B.22)$$

In general for $SU(N)$ the generators in the fundamental repesentation are $(n^2 - 1)$ matrices of dimension $n \times n$, and in the adjoint representation there are $(n^2 - 1)$ generators of dimension $(n^2 - 1) \times (n^2 - 1)$.

The other group structure relevant for the SM is $SU(2)_L$, acting on the left-handed spinor fields. For this group the fundamental representation puts the left–handed spinor fields in corresponding doublets, with weak isospin charges of $\pm 1/2$ for the upper and lower field. The generators are given by $\tau_a = \sigma_a/2$, where the Pauli matrices σ_a read

$$\lambda_1 = \begin{pmatrix} 0 & 1 \\ 1 & 0 \end{pmatrix}, \ \lambda_2 = \begin{pmatrix} 0 & -i \\ i & 0 \end{pmatrix}, \ \lambda_3 = \begin{pmatrix} 1 & 0 \\ 0 & -1 \end{pmatrix}, \qquad (B.23)$$

and enjoy the commutation relation

$$[\sigma_a, \sigma_b] = i\epsilon_{abc}\,\sigma_c. \qquad (B.24)$$

Here, the structure constants are the completely anti-symmetric **Levi-Civita** symbols.

B.1.1.3 Standard Model before electroweak symmetry breaking

As indicated earlier, the SM of particle physics is built from fermionic matter fields that are subjected to gauge invariance with respect to the gauge group $SU(3)_c \times SU(2)_L \times U(1)_Y$. The matter fields come in **three generations**, labelled with a generation index I. Fermions are **chiral**, they are **left- or right-handed**; the decomposition of Dirac fermions into chiral ones is achieved with projectors such that

$$\psi = \psi_L + \psi_R = P_L\,\psi + P_R\,\psi = \frac{1 - \gamma_5}{2}\,\psi + \frac{1 + \gamma_5}{2}\,\psi, \qquad (B.25)$$

Table B.1 The matter fermions of the Standard Model, with the corresponding charge assignments. The charges fulfil $Q = T_3 + Y_W/2$.

Fields	$SU(3)_c$	$SU(2)_L\colon T_3$	$U(1)_Y\colon Y_W$	Q
$Q_{L,i}^{(I)} = \begin{pmatrix} u_{L,i}^{(I)} \\ d_{L,i}^{(I)} \end{pmatrix}$	C_F	$+\frac{1}{2}$ $-\frac{1}{2}$	$+\frac{1}{3}$	$+\frac{2}{3}$ $-\frac{1}{3}$
$u_{R,i}^{(I)}$	C_F	0	$+\frac{4}{3}$	$+\frac{2}{3}$
$d_{R,i}^{(I)}$	C_F	0	$-\frac{2}{3}$	$-\frac{1}{3}$
$L_{L,i}^{(I)} = \begin{pmatrix} \nu_{L,i}^{(I)} \\ \ell_{L,i}^{(I)} \end{pmatrix}$	0	$+\frac{1}{2}$ $-\frac{1}{2}$	-1	0 -1
$\ell_{R,i}^{(I)}$	0	0	-2	-1

where $P_L^2 = P_L$ and $P_R^2 = P_R$ as well as $P_L + P_R = 1$. Each generation contains a left-handed doublet of quarks, with a colour charge i in the fundamental representation, $Q_{L,i}^{(I)} = (u_{L,i}^{(I)}, d_{L,i}^{(I)})^T)$. Here, $u_{L,i}^{(I)}$ and $d_{L,i}^{(I)}$ denote left-handed up-type and down-type quarks, respectively. Each generation also includes a similar left-handed doublet of leptons, containing a neutrino $\nu_L^{(I)}$ and a charged lepton $\ell_L^{(I)}$, $L^{(I)} = (\nu_L^{(I)}, \ell_L^{(I)})^T$. The presence or absence of colour charges indicates that the quarks enjoy the $SU(3)$ interaction, while the leptons don't. In an equivalent way there are also fields that do not take part in the left-handed interaction: in each generation there are right-handed colour-charged quark fields $u_{R,i}^{(I)}$ and $d_{R,i}^{(I)}$ and a right-handed lepton field $\ell_R^{(I)}$. There are no right-handed neutrinos in the SM. In Table B.1 all matter fields of the SM are listed, including their charges: the third component of the weak isospin T_3, related to $SU(2)_L$, and the weak hypercharge Y_W, related to the $U(1)_Y$, as well as the electrical charge Q, which only emerges after breaking the symmetry $SU(2)_L \times U(1)_Y$ down to $U(1)_Q$. The charges are related through

$$Q = T_3 + \frac{Y_W}{2}. \tag{B.26}$$

Defining the gauge-invariant derivatives through their action on the various fermion fields as

$$D_\mu Q_{L,i,\alpha}^{(I)} = \left(\partial_\mu + ig_3 \frac{\lambda_{ij}^a}{2} G_\mu^a \delta_{\alpha\beta} + ig_2 \frac{\sigma_{\alpha\beta}^a}{2} W_\mu^a \delta_{ij} + ig_1 \frac{Y_W}{2} B_\mu \delta_{ij}\delta_{\alpha\beta} \right) Q_{L,j,\beta}^{(I)}$$

$$D_\mu u_{R,i}^{(I)} = \left(\partial_\mu + ig_3 \frac{\lambda_{ij}^a}{2} G_\mu^a + ig_1 \frac{Y_W}{2} B_\mu \delta_{ij} \right) u_{R,j}^{(I)}$$

$$D_\mu d_{R,i}^{(I)} = \left(\partial_\mu + ig_3 \frac{\lambda_{ij}^a}{2} G_\mu^a + ig_1 \frac{Y_W}{2} B_\mu \delta_{ij} \right) u_{R,j}^{(I)}$$

$$D_\mu L_{L,\alpha}^{(I)} = \left(\partial_\mu + ig_2 \frac{\sigma_{\alpha\beta}^a}{2} W_\mu^a + ig_1 \frac{Y_W}{2} B_\mu \delta_{\alpha\beta} \right) L_{L,\beta}^{(I)}$$

$$D_\mu \ell_R^{(I)} = \left(\partial_\mu + ig_1 \frac{Y_W}{2} B_\mu \right) \ell_R^{(I)},$$

$$(\text{B.27})$$

where the λ and σ are the Gell–Mann and Pauli matrices, respectively, labelled by $a \in [1, 8]$ and $a \in [1, 3]$. The colour and weak isospin indices i and j and α and β have been made explicit here. The gauge fields are the eight gluons G_μ^a, the three weak isospin bosons W_μ^a, and the weak hypercharge field B_μ. The gauge-invariant derivatives enter the Lagrangian of the SM before electroweak symmetry breaking (EWSB) as

$$\mathcal{L}_{\text{SM}} = \mathcal{L}_{\text{matter}} + \mathcal{L}_{\text{gauge}}$$

$$\mathcal{L}_{\text{matter}} = \sum_{I=1}^{3} \left[\bar{Q}_L^{(I)} \, \slashed{D} Q_L^{(I)} + \bar{u}_R^{(I)} \, \slashed{D} u_R^{(I)} + \bar{d}_R^{(I)} \, \slashed{D} d_R^{(I)} + \bar{L}_L^{(I)} \, \slashed{D} L_L^{(I)} + \bar{\ell}_R^{(I)} \, \slashed{D} \ell_R^{(I)} \right]$$

$$\mathcal{L}_{\text{gauge}} = -\frac{1}{4} G_{\mu\nu}^a G^{a,\mu\nu} - \frac{1}{4} W_{\mu\nu}^a W^{a,\mu\nu} - \frac{1}{4} B_{\mu\nu} B^{\mu\nu},$$

$$(\text{B.28})$$

where the summation over the colour or weak isospin labels a is understood and where the non-Abelian generalization of Eq. (B.8) yields for the field strength tensors

$$G_{\mu\nu}^a = \partial_\mu G_\nu^a - \partial_\nu G_\mu^a + ig_1 f^{abc} G_\mu^b G_\nu^c, \tag{B.29}$$

introducing self-interactions of the non-Abelian gauge fields G_μ^a and W_μ^a into the gauge part of the Lagrangian above.

Note that in principle also gauge-fixing terms $\mathcal{L}_{\text{g.f.}}$ would have to be added, which could further neccessitate the introduction of **Fadeev–Popov ghosts**. These unphysical degrees of freedom manifest themselves as Grassman scalars, scalars with fermionic behaviour, which carry the gauge quantum numbers of the gauge fields.

B.1.1.4 The need for electroweak symmetry breaking

Analysing the structure of the Lagrangian in Eq. (B.28) in some detail exposes a phenomenological shortcoming of the SM: all particles introduced so far must be massless.

As already discussed, the gauge fields must be massless, since direct mass terms of the form $m^2 A^2$ violate gauge invariance. Applying, for example, the gauge transformation of Eq. (B.5) on the B field would result in

$$B_\mu B^\mu \longrightarrow B'_\mu B'^\mu = B_\mu B^\mu + \frac{2}{g} B^\mu \partial_\mu \theta + \frac{1}{g^2}(\partial_\mu \theta)(\partial^\mu \theta) \neq B_\mu B^\mu. \tag{B.30}$$

At the same time the fermions also cannot have a mass term. The reason is that such a term for Dirac fermions ψ has the form

$$\mathcal{L}_{\text{Dirac,mass}} = m\bar{\psi}\psi = m\left(\bar{\psi}_R \psi_L + \bar{\psi}_L \psi_R\right). \tag{B.31}$$

As long as left- and right-handed fermions transform in the same way, the respective phase factors would of course compensate; this is the case in QED where the phase transformation acts on all components of the spinors in the same way. Clearly, on the other hand, in the very moment left- and right-handed fermions do have different gauge transformations — as is the case in the SM, manifest for example in Eq. (B.27) — there is no guarantee that this compensation happens. As a consequence, the mass term above is explicitly gauge-violating, as it triggers an uncompensated phase factor stemming from the $SU(2)_L$ transformation acting on the left-handed spinors only.

At the same time, masses for the weak gauge bosons - the $W^\pm$ and Z^0 bosons and for all fermions are well established. This means that either the underlying construction paradigm of the SM, gauge invariance, does not hold true or that the SM in the form presented so far is not complete and needs to be supplemented with a mechanism that allows the generation of mass in a gauge-invariant way. As it turns out, the latter option is realized in nature.

B.1.1.5 The Brout–Englert–Higgs mechanism

The Brout–Englert–Higgs (BEH) mechanism solves the problem of gauge-invariant mass generation in a spectacularly elegant way, by essentially hiding the local phase symmetry through the introduction of a non-symmetric vacuum. Before discussing this in some detail, consider first the question of which of the gauge groups of the SM is the critical one. Following the reasoning above, it is clear that the gauge bosons of the $SU(3)_c$, the gluons, are massless, and that there is also a massless $U(1)$ gauge boson, the photon. On the other hand, the three weak gauge bosons are massive, and fermion masses are disallowed because of the $SU(2)_L$ invariance. It is therefore natural to concentrate on the $SU(2)_L$ part of the Standard Model Lagrangian.

In the BEH mechanism, this proceeds by introducing a complex scalar $\Phi = (\phi^+, \phi^0)^T$, which is coupled to the $SU(2)_L \times U(1)$ part of the SM through a gauge-invariant derivative,

$$D_\mu \Phi_\beta = \left(\partial_\mu \delta_{\alpha\beta} + ig_2 \frac{\sigma^a_{\alpha\beta}}{2} W^a_\mu + ig_1 \frac{Y_W}{2} B_\mu \delta_{\alpha\beta}\right)\Phi_\beta, \tag{B.32}$$

where, again, the α and β label the weak isospin components of the doublet Φ, ϕ^+ and ϕ^0. The relevant quantum numbers of Φ are $T_3 = \pm 1/2$ for the upper and lower components and $Y_W = 1/2$.

Including a potential for the doublet, the original Lagrangian of Eq. (B.28) is supplemented with two new parts, namely

$$\mathcal{L}_{\mathrm{H}} = (D_\mu\Phi)^\dagger(D^\mu\Phi) + \mu^2\Phi^\dagger\Phi - \lambda(\Phi^\dagger\Phi)^2 \tag{B.33}$$

and

$$\mathcal{L}_{\mathrm{HF}} = -f_u^{IJ}\,\bar{Q}_L^{(I)}\tilde{\Phi}u_R^J - f_d^{IJ}\,\bar{Q}_L^{(I)}\Phi d_R^J - f_e^{IJ}\,\bar{L}_L^{(I)}\Phi l_R^J \tag{B.34}$$

for its Yukawa interactions with the fermions, which actually contains both left- and right-handed fermions. Here, μ^2 and λ are real numbers, while the f^{IJ} are arbitrary matrices in generation space. In addition,

$$\tilde{\Phi} = i\sigma^2\,\Phi, \tag{B.35}$$

essentially swapping the position of the ϕ^+ and ϕ^0 components in the Higgs doublet. The new, complete SM Lagrangian is given by

$$\mathcal{L}_{\mathrm{SM}} = \mathcal{L}_{\mathrm{matter}} + \mathcal{L}_{\mathrm{gauge}} + \mathcal{L}_{\mathrm{H}} + \mathcal{L}_{\mathrm{HF}}, \tag{B.36}$$

where all gauge and fermion fields are still massless.

Analysing the form of the Higgs potential in more detail, and determining its ground state through minimization, leads to the condition

$$-\mu^2 + 2\lambda\Phi^\dagger\Phi = 0 \tag{B.37}$$

or

$$\langle\Phi^\dagger\Phi\rangle_0 = \frac{\mu^2}{2\lambda} = \frac{v^2}{2} \tag{B.38}$$

for the expectation value of the fields in the ground state. Identifying the ground state with the physical vacuum gives rise to the notion of the **vacuum expectation value** v of the fields. If μ^2 and λ are both real positive numbers this leads to an infinite number of equivalent vacua, forming a hypersphere with radius $v/\sqrt{2}$ in the space spanned by the doublet. Note that choosing $\mu^2 > 0$ also means that the Φ doublet does not have a physical mass term in the Lagrangian, the corresponding term $\propto \mu^2$ just has the wrong sign.

As the quantization of the fields proceeds by expanding around a single vacuum, for instance by using creation and annihilation operators in canonical quantization, one of these vacua must be picked.[1] Without any loss of generality, it has become customary to define the vacuum state of the Higgs doublet to be

$$\langle\Phi\rangle_0 = \begin{pmatrix} 0 \\ \frac{v}{\sqrt{2}} \end{pmatrix}, \tag{B.39}$$

leading to a reparameterization of the fields as

$$\Phi' = \Phi - \langle\Phi\rangle_0 = \begin{pmatrix} \phi^+ \\ \phi^0 - \frac{v}{\sqrt{2}} \end{pmatrix}. \tag{B.40}$$

---

[1] Once such a vacuum has been picked, the system cannot tunnel out of it as there are infinitely many equivalent states.

Rotating the three W bosons and the corresponding generators by introducing charged states

$$W_\mu^{1,2} \longrightarrow W_\mu^\pm = \frac{1}{\sqrt{2}} \left(W_\mu^1 \mp i W_\mu^2 \right) \tag{B.41}$$

and, correspondingly,

$$\sigma^\pm = \sigma^1 \pm i\sigma^2 = \begin{pmatrix} 0 & 2 \\ 0 & 0 \end{pmatrix}, \begin{pmatrix} 0 & 0 \\ 2 & 0 \end{pmatrix}, \tag{B.42}$$

essentially the ladder operators of $SU(2)$.

This allows to parameterize the Higgs doublet as

$$\Phi(x) = \exp\left[-\frac{i}{v} \sum_{i=\pm,3} \xi_i(x) \tau^i \right] \begin{pmatrix} 0 \\ \frac{v + \eta(x)}{\sqrt{2}} \end{pmatrix} = U^{-1}(\xi) \frac{v + \eta(x)}{\sqrt{2}} \chi \tag{B.43}$$

with $\chi = (0, 1)^T$ and where the spatial dependence of the four real fields $\eta(x)$ and $\xi_i(x)$ has been made explicit. These four fields of course have no vacuum expectation value,

$$\langle v \rangle_0 = \langle \xi_i \rangle_0 = 0. \tag{B.44}$$

A convenient way to see how the breaking of electroweak symmetry (EWSB) proceeds is to chose the **unitary gauge**, fixing the Higgs doublet to have the form

$$\Phi' = \Phi_{\text{unitary}} = U(\xi)\Phi = \frac{v + \eta(x)}{\sqrt{2}} \chi = \begin{pmatrix} 0 \\ \frac{v+\eta(x)}{\sqrt{2}} \end{pmatrix}. \tag{B.45}$$

This essentially means that the Higgs doublet has only one remaining visible field left, η, while the three phase fields ξ_i have been rotated away and must be re-introduced in the various parts of the Lagrangian — which is achieved by the set of transformations in Eq. (B.46),

$$\left(\sum_{i=\pm,3} W_\mu^i \tau^i \right)' = U(\xi) \left[\sum_{i=\pm,3} W_\mu^i \tau^i \right] U^{-1}(\xi) + \frac{i}{g_2} \left(\partial_\mu U(\xi) \right) U^{-1}(\xi)$$

$$B'_\mu = B_\mu \tag{B.46}$$
$$\Psi'_L = U(\xi)\Psi_L$$
$$\Psi'_R = \Psi_R.$$

Taking a closer look at the gauge transformation for the W fields in the first line, it becomes apparent that the three ξ fields that have appear to have vanished as dynamics degrees of freedom from the Higgs sector have resurfaced as parts of the W bosons through the $\partial_\mu U(\xi)$ term. This term in the end results in fields $\partial_\mu \xi$, which together with the derivatives in the kinetic term of the W gauge bosons from a kinetic term for these new fields. It turns out that they indeed are the **massless Goldstone**

bosons, postulted by the celebrated **Goldstone theorem** [589, 590] as a consequence of a broken gauge group. Ultimately, however, they will be "eaten" by the gauge fields, absorbed by another gauge transformation. This mechansim provides the gauge bosons with a third polarization degree of freedom and in turn turns the hitherto massless gauge bosons into massive particles.

Ignoring this part of the Lagrangian, consider the further terms emerging from the transformation above. Starting with the simplest bit, the Higgs potential, the invariance under this kind of transformation is clear — the phases just trivially cancel out, since

$$(\Phi^\dagger \Phi) \longrightarrow (\Phi^\dagger \Phi) = \Phi^\dagger U^\dagger(\xi) U(\xi) \Phi = (\Phi^\dagger \Phi). \tag{B.47}$$

The Higgs potential therefore reads

$$\mathcal{L}_{H,\text{pot}} = -\frac{2\lambda v^2}{2} \eta^2 - \lambda v \eta^3 - \frac{\lambda}{4} \eta^4 + \text{const.} \tag{B.48}$$

with the Higgs field η as the only dynamic member. Also, this field has acquired a mass term with the right sign, leading to a physical mass of

$$m_H = v\sqrt{2\lambda}. \tag{B.49}$$

From the transformations in Eq. (B.46) it is straightforward to see how this works with the kinetic term of the Higgs doublet as well. From

$$(D_\mu \Phi)' = U(\xi) \, D_\mu \Phi \tag{B.50}$$

it is simple to see that again the phases will just cancel out. The $(D_\mu \Phi)^\dagger (D^\mu \Phi)$ term expressed through the transformed fields looks like

$$\mathcal{L}_{H,\text{kin}} = (D'_\mu \Phi')^\dagger (D'^\mu \Phi') = \mathcal{L}_{\eta,\text{kin}} + \mathcal{L}_M + \mathcal{L}_I \tag{B.51}$$

and depends only on the vacuum expectation value v, and the dynamics degrees of freedom given by the Higgs field η and the gauge fields W_μ^i and B_μ, where, for convenience, the primes are omitted. They are given by

$$
\begin{aligned}
\mathcal{L}_{\eta,\text{kin}} &= \frac{1}{2}(\partial_\mu \eta)(\partial^\mu \eta) \\
\mathcal{L}_M &= \frac{v^2 g_2^2}{4} W_\mu^- W^{+,\mu} + \frac{v^2}{8} \begin{pmatrix} B_\mu \\ W_\mu^3 \end{pmatrix}^T \begin{pmatrix} g_1^2 & -g_1 g_2 \\ -g_1 g_2 & g_2^2 \end{pmatrix} \begin{pmatrix} B_\mu \\ W_\mu^3 \end{pmatrix}^T \\
&= m_W^2 \, W_\mu^- W^{+,\mu} + \frac{1}{2} m_Z^2 Z_\mu Z^\mu \\
\mathcal{L}_I &= \frac{m_W^2}{v^2}\left(\eta^2 + 2v\eta\right) W_\mu^- W^{+,\mu} + \frac{m_Z^2}{2v^2}\left(\eta^2 + 2v\eta\right) Z_\mu Z^\mu,
\end{aligned}
\tag{B.52}
$$

with the new fields A_μ and Z_μ — the photon and the Z boson — emerging from the diagonalization of the mass matrix in $\mathcal{L}_M$ as

$$A_\mu = \sin\theta_W W_\mu^3 + \cos\theta_W B_\mu$$
$$Z_\mu = \cos\theta_W W_\mu^3 - \sin\theta_W B_\mu. \tag{B.53}$$

The photon is massless, $m_A = 0$, and the masses of the charged $W^\pm$ and the neutral Z^0 boson are given by

$$m_W = \frac{v g_2}{2} \quad \text{and} \quad m_Z = \frac{v}{2}\sqrt{g_1^2 + g_2^2}, \tag{B.54}$$

fixing also the **weak mixing angle** or **Weinberg angle** as

$$\tan\theta_W = \frac{g_1}{g_2} \quad \text{or} \quad \cos\theta_W = \frac{m_W}{m_Z} = \frac{g_2}{\sqrt{g_1^2 + g_2^2}}. \tag{B.55}$$

It is a tedious but straightforward exercise to show that the kinetic term of the gauge fields as well as their interaction with the fermions is invariant under the gauge transformation of Eq. (B.46).

B.1.1.6 Dealing with the fermions

This leaves only the Yukawa interactions of the fermions with the Higgs doublet. Employing the same transformations as before, Eq. (B.34) expressed through the transformed fields becomes

$$\mathcal{L}_{\mathrm{HF}} = \frac{v + \eta}{\sqrt{2}} \left[f_u^{IJ} \, \bar{u}_L^{(I)} u_R^{(J)} + f_d^{IJ} \, \bar{d}_L^{(I)} d_R^{(J)} + f_\ell^{IJ} \, \bar{\ell}_L^{(I)} \ell_R^{(J)} \right], \tag{B.56}$$

where for convenience again the primes over the fields have been omitted. This however leaves an interesting problem: since the Yukawa matrices $f_{u,d,\ell}$ are arbitrary, there is no reason to assume that they are diagonal, leading to non-diagonal mass terms of the fields. Insisting on well-defined particle masses — which is sensible! — this means that the fields must be rotated in such a way that the mass matrices are diagonal, for example

$$f_u^{IJ} \longrightarrow \frac{\sqrt{2}\, m_u^{(I)}}{v} \delta^{IJ} \tag{B.57}$$

giving rise to in total nine fundamental coupling strengths of the Yukawa couplings of the Higgs fields to the fermions, namely the $m_u^{(I)}$, $m_d^{(I)}$, and $m_\ell^{(I)}$. It is one of the non-trivial predictions of the SM that the Higgs boson couplings to the fermions are directly proportional to their masses.

This sounds very good, but there is a somewhat unwanted consequence of this diagonalization process. Before the diagonalization, the interactions of the fermions were diagonal, obviuosly they cannot stay that way. In other words there are two bases for the fermions, a basis given by the mass eigenstates of the fermions and another one, given by their interaction eigenstates. In order to see how they connect, take a closer look at how this diagonalization proceeds. The starting point are the arbitrary mass matrices $f_{u,d,\ell}^{IJ}$, summarily denoted as M. Such general matrices can be diagonalized by a bi-unitary transformation,

$$M_{\text{diag}} = S^\dagger M T, \tag{B.58}$$

where S and T are unitary and M_{diag} is diagonal with non-zero eigenvalues. In addition, every matrix M can be written as a product of a Hermitian and a unitary matrix, H and U,

$$M = HU, \tag{B.59}$$

and in general $M^\dagger M$ by construction is Hermitian and positive. As a consequence,

$$S^\dagger (M^\dagger M) S = (M^2)_{\text{diag}} \longrightarrow \begin{pmatrix} m_1^2 & 0 & 0 \\ 0 & m_2^2 & 0 \\ 0 & 0 & m_3^2 \end{pmatrix}. \tag{B.60}$$

Up to an arbitrary phase in the diagonal elements, S is unique, so that also

$$S^\dagger F^\dagger (M^\dagger M) F S = (M^2)_{\text{diag}}, \tag{B.61}$$

where

$$F = \begin{pmatrix} e^{i\phi_1} & 0 & 0 \\ 0 & e^{i\phi_2} & 0 \\ 0 & 0 & e^{i\phi_3} \end{pmatrix}. \tag{B.62}$$

These phases will come back in the context of CP **violation** but here this freedom indeed guarantees that $m_i^2 \geq 0$. The Hermitian part H of the decomposition in Eq. (B.59) can be identified with

$$H = S M_{\text{diag}} S^\dagger, \tag{B.63}$$

which fixes U as

$$U = H^{-1} M \quad \text{and} \quad U^\dagger = M^\dagger H^{-1}. \tag{B.64}$$

The hermiticity of H and the unitarity of U are simple to confirm. Using Eq. (B.63) this means that

$$M_{\text{diag}} = S^\dagger H S = S^\dagger M U^\dagger S = S^\dagger M T \tag{B.65}$$

which also defines the matrix T in Eq. (B.58), $T = U^\dagger S$. With this reasoning, the mass terms are diagonalized through

$$\bar{\psi}_L M \psi_R = (\bar{\psi}_L S)(S^\dagger M T)(T^\dagger \psi_R) = \bar{\psi}'_L M_{\text{diag}} \psi'_R. \tag{B.66}$$

In other words, the left-handed fields will transform with S, while the right-handed fermions will transform with T. Looking at the interactions with the gauge bosons effectively leads to three different structures. First of all, there is the interaction of the right-handed fermions, which will assume the form

$$\bar{\psi}_R^I \gamma^\mu \psi_R^I = \bar{\psi}_R'^K \gamma^\mu T_{KI} T_{IL}^\dagger \psi_R'^L = \bar{\psi}_R'^K \gamma^\mu \delta_{KL} \psi_R'^L = \bar{\psi}_R'^K \gamma^\mu \psi_R'^K. \tag{B.67}$$

This means that the right-handed quarks can be rotated to their mass eigenstate without any obvious consequence for their dynamics. The same reasoning also applies for neutral interactions of the left-handed fermions, for their interactions with the gluons and the B and W^3 or, equivalently, the photon and the Z boson. The only difference is to replace the unitary matrix T with the unitary matrix S. However, for their charged interactions with the $W^\pm$ this is not true anymore. There, the fermion current becomes

$$\bar{u}_L^I \gamma^\mu d_L^I = \bar{u}_L'^K \gamma^\mu S_{u,KI}^\dagger S_{d,IL} d_L'^L = \bar{u}_L'^K \gamma^\mu V_{KL}^{(\mathrm{CKM})} d_L'^L, \tag{B.68}$$

where the **Cabibbo-Kobayashi-Maskawa matrix** (CKM matrix)

$$V_{KL}^{(\mathrm{CKM})} = S_{u,KI}^\dagger S_{d,IL} \tag{B.69}$$

has been introduced. This matrix mixes the quark generations in interactions, where a W boson couples to an up-type and a down-type quark, and ultimately allows the quarks of the second and third generation to decay weakly into the first generation.

As a product of two unitary matrices, the CKM matrix itself is unitary, in principle with $n^2 = 1$ free parameters. Using the arbitrary phases in the matrices S, it becomes clear that $(2n - 1)$ phases can be removed by redefining the quark states. In total therefore the CKM matrix has $n^2 - (2n - 1) = (n - 1)^2 = 4$ free paramters, 3 free angles and 1 phase. This gives rise to the **Wolfenstein parameterization**, where the Cabibbo angle $\lambda \approx 0.22$ is the small evolution parameter, and, up to third order in λ

$$V^{(\mathrm{CKM})} = \begin{pmatrix} V_{ud} & V_{us} & V_{ub} \\ V_{cd} & V_{cs} & V_{cb} \\ V_{td} & V_{ts} & V_{tb} \end{pmatrix} = \begin{pmatrix} 1 - \frac{\lambda^2}{2} & \lambda & A\lambda^3(\rho - i\eta) \\ -\lambda & 1 - \frac{\lambda^2}{2} & A\lambda^2 \\ A\lambda^3(1 - \rho - i\eta) & -A\lambda^2 & 1 \end{pmatrix}. \tag{B.70}$$

The further parameters of this parameterization are given by

$$A \approx 0.8, \quad \rho \approx 0.135, \text{ and } \quad \eta \approx 0.35. \tag{B.71}$$

It is this phase η, which actually introduces CP violation into the SM.

B.2   Feynman rules of the Standard Model

The Feynman rules of the SM are obtained from the Lagrangian introduced above via the usual techniques of quantum field theory, that may be found in one of the many standard texts. In short, one considers the **action** of the theory which is related to the Lagrangian by

$$\mathcal{S} = i \int \mathrm{d}^4 x \, \mathcal{L}(x). \tag{B.72}$$

In the free (non-interacting) theory this action leads to two-point functions whose inverses represent particle propagators. The interaction terms in the Lagrangian combine particle fields of different types that are represented by vertices.

B.2.1 Propagators

Fermion propagators arise in the SM through the two-point functions represented by terms in the Lagrangian such as those shown in Eq. (B.7). Recalling the momentum-space replacement $\partial_\mu \to -ip_\mu$ leads to the corresponding propagator factor,

$$i\,(\not{p} - m)^{-1} = \frac{i(\not{p} + m)}{p^2 - m^2}, \tag{B.73}$$

where the inverse is easily obtained.

The bosonic propagators are slightly more complicated to obtain except for the Higgs boson, for which the scalar propagator is trivial. Consider first the W and Z propagators that correspond to the gauge terms in the Lagrangian ($\mathcal{L}_{\text{gauge}}$) of Eq. (B.28), together with the mass terms generated by the BEH mechanism, $\mathcal{L}_M$ of Eq. (B.52). These generate two-point functions, for instance between Z fields, of the form,

$$Z_\mu \left[(p^2 - m_Z^2)g^{\mu\nu} - p^\mu p^\nu \right] Z_\nu. \tag{B.74}$$

The inverse of the factor in square brackets yields a propagator of the form

$$\frac{1}{p^2 - m_Z^2} \left(g^{\mu\nu} - \frac{p^\mu p^\nu}{m_Z^2} \right). \tag{B.75}$$

In the limit $m_Z \to 0$, corresponding to the case of the photon and gluon, the tensor in Eq. (B.74) is not invertible. This necessitates the introduction of additional gauge-fixing terms to the Lagrangian. For instance, for the photon one can add the term

$$\mathcal{L}_{\text{g.f.}} = -\frac{1}{2\xi} \left(\partial^\mu A_\mu \right)^2, \tag{B.76}$$

where ξ is an arbitrary parameter. This leads to the Feynman rules shown. Note that the choice $\xi = 1$, called the **Feynman gauge**, is often the simplest choice since it leads to fewer terms at intermediate stages. Although this is the end of the story for the photon, in a non-Abelian theory the gauge-fixing term introduces unphysical degrees of freedom that must be cancelled by ghost contributions. These contributions are not discussed further here.

The Feynman rules for all of the propagators of the SM are shown in Fig. B.1.

B.2.2 Interactions of gauge bosons with fermions

The interactions of the gauge bosons with the fermions of the SM originate from Eqs. (B.28) and (B.56). The interactions that do not involve the Higgs boson correspond to the covariant derivatives appearing in Eq. (B.27), after accounting for the effects of the BEH mechansim and the modifications due to the CKM matrix indicated in Eq. (B.68). The Yukawa interactions of the fermions with the Higgs boson are manifest already in Eq. (B.56) and can be simplified slightly by identifying the value of the vacuum expectation value through $v = 2m_W/g_W$.

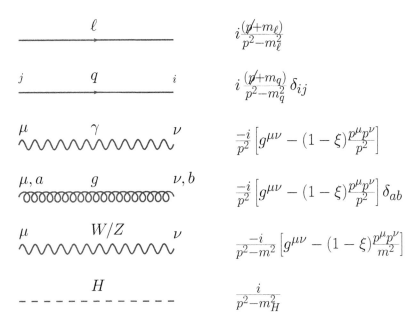

Fig. B.1 Propagator Feynman rules in the Standard Model.

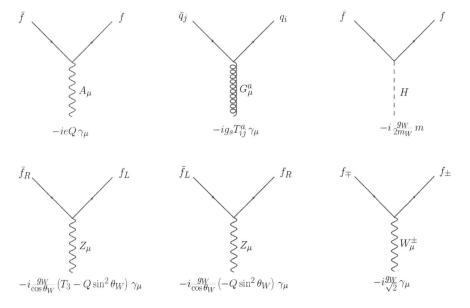

Fig. B.2 Feynman rules for fermion-boson interactions in the Standard Model.

In order to see an example of this in practice, consider the expression for the covariant derivative acting on the left-handed lepton doublet, $D_\mu L_{L,\alpha}^{(I)}$ of Eq. (B.27). Rewriting the Pauli matrices σ^a in terms of the generators τ and the W_1, W_2 fields in terms of $W^\pm$, *cf.* Eq. (B.41), one obtains

$$D_\mu L_{L,\alpha}^{(I)} = \left(\partial_\mu + ig_2\left[\frac{1}{\sqrt{2}}\tau_{\alpha\beta}^+ W_\mu^+ + \frac{1}{\sqrt{2}}\tau_{\alpha\beta}^- W_\mu^- + \tau_{\alpha\beta}^3 W_\mu^3 +\right] + ig_1 \frac{Y_W}{2} B_\mu \delta_{\alpha\beta}\right) L_{L,\beta}^{(I)},$$
(B.77)

where $\tau^\pm$ are defined analogously to $\sigma^\pm$, *cf.* Eq. (B.42). Expressing the fields W_μ^3 and B_μ in terms of the photon and Z-boson fields, *cf.* Eq. (B.53), and using the weak-hypercharge relation in Eq. (B.26) reduces all the fields to physical ones,

$$D_\mu L_{L,\alpha}^{(I)} = \left(\partial_\mu + \frac{ig_2}{\sqrt{2}}\left[\tau_{\alpha\beta}^+ W_\mu^+ + \tau_{\alpha\beta}^- W_\mu^-\right] + i\left[g_2 T_3 \sin\theta_W + g_1(Q - T_3)\cos\theta_W\right] A_\mu \right.$$
$$\left. + i\left[g_2 T_3 \cos\theta_W - g_1(Q - T_3)\sin\theta_W\right] Z_\mu \delta_{\alpha\beta}\right) L_{L,\beta}^{(I)},$$
(B.78)

The final simplification is obtained by relating the couplings g_1 and g_2 through the weak mixing angle, *cf.* Eq. (B.55), and identifying the electromagnetic and weak couplings, $g_2 \to g_W$ and $e = g_W \sin\theta_W$,

$$D_\mu L_{L,\alpha}^{(I)} = \left(\partial_\mu + \frac{ig_W}{\sqrt{2}}\left[\tau_{\alpha\beta}^+ W_\mu^+ + \tau_{\alpha\beta}^- W_\mu^-\right] + ieQ A_\mu \right.$$
$$\left. + \frac{ig_W}{\cos\theta_W}\left[T_3 - Q\sin^2\theta_W\right] Z_\mu \delta_{\alpha\beta}\right) L_{L,\beta}^{(I)}.$$
(B.79)

From this expression it is straightforward to read off the Feynman rules that are shown in Fig. B.2. Similar manipulations of the other covariant derivatives can be performed to reproduce the remaining Feynman rules shown.

B.2.3 Self-interactions of gauge bosons

The self-interactions of the gauge bosons of the SM are generated by the non-Abelian contributions to the term $\mathcal{L}_{\text{gauge}}$ of Eq. (B.28). The corresponding Feynman rules can be derived in a straightforward manner by substitution of the corresponding expressions for the field-strength tensors, *cf.* Eq. (B.29), accounting for identical-particle factors of $1/n!$ where appropriate.

In QCD these terms lead to the three- and four-point vertices shown in Fig. B.3. Note that the sign of the three-point vertex is sensitive to the direction of flow of the momentum; the rule in the figure corresponds to all momenta outgoing (signified by the outward-pointing arrows).

Turning to the electroweak sector, the interactions are again obtained by rewriting the field-strength tensor in terms of the basis of physical fields, $W_\mu^\pm$, Z_μ and A_μ. Thus,

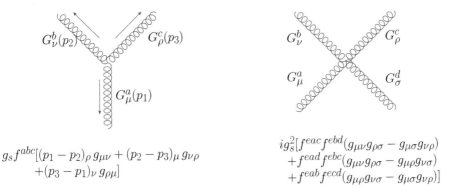

$$g_s f^{abc}[(p_1 - p_2)_\rho \, g_{\mu\nu} + (p_2 - p_3)_\mu \, g_{\nu\rho} + (p_3 - p_1)_\nu \, g_{\rho\mu}]$$

$$ig_s^2[f^{eac} f^{ebd}(g_{\mu\nu}g_{\rho\sigma} - g_{\mu\sigma}g_{\nu\rho}) + f^{ead} f^{ebc}(g_{\mu\nu}g_{\rho\sigma} - g_{\mu\rho}g_{\nu\sigma}) + f^{eab} f^{ecd}(g_{\mu\rho}g_{\nu\sigma} - g_{\mu\sigma}g_{\nu\rho})]$$

Fig. B.3 Feynman rules for boson self-interactions in QCD.

$$
\begin{aligned}
\mathcal{L}_{\text{gauge,EW}} &= \frac{1}{4} W^a_{\mu\nu} W^{a,\mu\nu} \\
&= \left[\frac{1}{4} W^+_\mu W^+_\nu W^-_\rho W^-_\sigma \right. \\
&\quad \left. - \frac{1}{2} W^-_\mu W^+_\nu \left(Z_\rho Z_\sigma g_W^2 \cos\theta_W^2 + A_\rho A_\sigma e^2 + 2 A_\rho Z_\sigma e g_W \cos\theta_W \right) \right] \\
&\quad \times \left(2 g_{\mu\nu} g_{\rho\sigma} - g_{\mu\rho} g_{\nu\sigma} - g_{\mu\sigma} g_{\nu\rho} \right) \\
&\quad + i W^-_\mu W^+_\nu \left(A_\rho e + Z_\rho g_W \cos\theta_W \right) \\
&\quad \times \left((p^{W^-}_\rho - p^{W^+}_\rho) g_{\mu\nu} + (p^{W^+}_\mu - p^V_\mu) g_{\nu\rho} + (p^V_\nu - p^{W^-}_\nu) g_{\rho\mu} \right) \\
&\quad + \text{kinetic terms}
\end{aligned}
\tag{B.80}
$$

This leads to the Feynman rules for self-interactions of electroweak bosons that are shown in Fig. B.4. The universal Lorentz structure of the interactions can be seen immediately from Eq. (B.80), so that the figure summarizes several interactions at once using the definitions,

$$c_\gamma = e, \quad c_Z = g_W \cos\theta_W, \tag{B.81}$$

The cubic and quartic interactions that involve the Higgs and other electroweak bosons can be read off from the contribution $\mathcal{L}_I$ shown in Eq. (B.52). The self-interactions of the Higgs boson are a result of the potential term introduced in Eq. (B.48). The strength of the interactions can be simplified using the implicit expressions for v and λ in terms of the electroweak coupling and boson masses given in Eqs. (B.49) and (B.54).

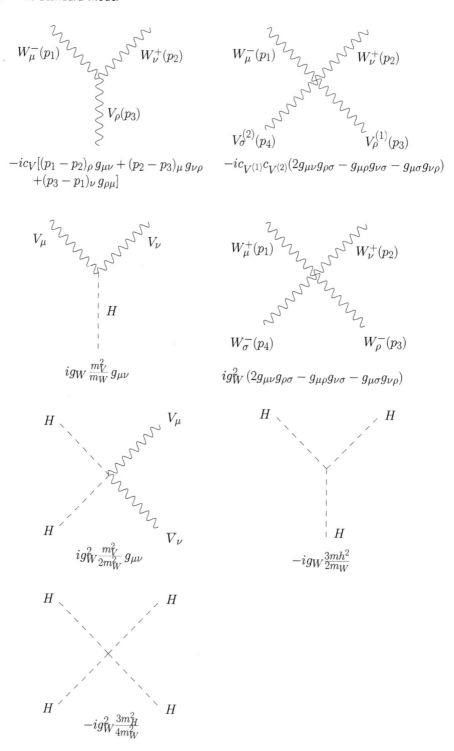

Fig. B.4 Feynman rules for self-interactions of electroweak bosons. In this figure $V^{(i)}$ represent either a photon or a Z boson, with the couplings defined by $c_\gamma = e$ and $c_Z = g_W \cos\theta_W$. In the interactions involving Higgs bosons, V is either a Z or W boson of the appropriate charge.

Appendix C
Catani–Seymour subtraction

C.1 Catani–Seymour subtraction for NLO calculations

In the following, kinematic maps for the construction of real subtraction terms, the corresponding splitting kernels and their integral will be listed. In all cases, massless partons only will be considered.

C.1.1 Final-state splitter — final-state spectator

The kinematic maps $p_i + p_j + p_k \rightarrow \tilde{p}_{ij} + \tilde{p}_k$ such that

$$p_i + p_j + p_k = \tilde{p}_{ij} + \tilde{p}_k \tag{C.1}$$

and

$$
\begin{aligned}
\tilde{p}_{ij} &= p_i + p_j - \frac{y_{ij,k}}{1 - y_{ij,k}} p_k \\
\tilde{p}_k &= \frac{1}{1 - y_{ij,k}} p_k,
\end{aligned}
\tag{C.2}
$$

where the recoil parameter $y_{ij,k}$ and the splitting variable $\tilde{z}_i$ are given by

$$
\begin{aligned}
y_{ij,k} &= \frac{p_i p_j}{p_i p_j + p_j p_k + p_k p_i} \\
\tilde{z}_i &= \frac{p_i p_k}{(p_i + p_j) p_k} = \frac{p_i \tilde{p}_k}{\tilde{p}_{ij} \tilde{p}_k} \quad \text{and} \quad \tilde{z}_j = 1 - \tilde{z}_i.
\end{aligned}
\tag{C.3}
$$

With their dependence on these two parameters understood, the splitting kernels read (CS-5.7–CS-5.9)

$$
\begin{aligned}
\langle s | V_{q_i g_j; k} | s' \rangle &= 8\pi \mu^{2\varepsilon} C_F \alpha_{\mathrm{s}} \left[\frac{2}{1 - \tilde{z}_i (1 - y_{ij,k})} - (1 + \tilde{z}_i) - \varepsilon (1 - \tilde{z}_i) \right] \delta_{ss'} \\
\langle \mu | V_{q_i \bar{q}_j; k} | \nu \rangle &= 8\pi \mu^{2\varepsilon} T_R \alpha_{\mathrm{s}} \left[-g^{\mu\nu} - \frac{2}{p_i p_j} (\tilde{z}_i p_i - \tilde{z}_j p_j)^\mu (\tilde{z}_i p_i - \tilde{z}_j p_j)^\nu \right] \\
\langle \mu | V_{g_i g_j; k} | \nu \rangle &= 16\pi \mu^{2\varepsilon} C_A \alpha_{\mathrm{s}} \left[-g^{\mu\nu} \left(\frac{1}{1 - \tilde{z}_i (1 - y_{ij,k})} + \frac{1}{1 - \tilde{z}_j (1 - y_{ij,k})} - 2 \right) \right. \\
&\quad \left. + \frac{1 - \varepsilon}{p_i p_j} (\tilde{z}_i p_i - \tilde{z}_j p_j)^\mu (\tilde{z}_i p_i - \tilde{z}_j p_j)^\nu \right].
\end{aligned}
\tag{C.4}
$$

Note that here the flavours of the particles resulting from the splitting are indicated as subscripts in the kernels V. The flavour of the spectator does not matter, while the spin states of the splitter do. They are therefore given as arguments in the $\langle\rangle$ brackets.

The spin-averaged kernels are denoted by $\langle V \rangle$ and they are in this case given by

$$\frac{\langle V_{q_i g_j, k} \rangle}{8\pi \alpha_s \mu^{2\varepsilon}} = C_F \left[\frac{2}{1 - \tilde{z}_i(1 - y_{ij,k})} - (1 + \tilde{z}_i) - \varepsilon(1 - \tilde{z}_i) \right]$$

$$\frac{\langle V_{q_i \bar{q}_j; k} \rangle}{8\pi \mu^{2\varepsilon} \alpha_s} = T_R \left[1 - \frac{2\tilde{z}_i(1 - \tilde{z}_i)}{1 - \varepsilon} \right]$$

$$\frac{\langle V_{g_i g_j; k} \rangle}{8\pi \mu^{2\varepsilon} \alpha_s} = 2C_A \left[\frac{1}{1 - \tilde{z}_i(1 - y_{ij,k})} + \frac{1}{1 - (1 - \tilde{z}_i)(1 - y_{ij,k})} - 2 + \tilde{z}_i(1 - \tilde{z}_i) \right].$$

$$(C.5)$$

These are the terms that form the basis of splitting kernels in the construction of parton showers, see Section 5.3.2.

Integrating the spin-averaged kernels over the phase-space $y = y_{ij,k}$ and $z = \tilde{z}_i$ yields

$$\mathcal{V}_{ij}(\varepsilon) = \int_0^1 dz[z(1-z)]^{-\varepsilon} \int_0^1 dy(1-2y)^{1-2\varepsilon} y^{-\varepsilon} \frac{\langle V_{ij,k}(z, y) \rangle}{8\pi \alpha_s \mu^{2\varepsilon}}, \qquad (C.6)$$

where the extra terms in z and y stem from the phase-space integral over momenta rewritten in these quantities, *cf.* (CS-5.16–CS-5.21).

The resulting integrated dipoles are given by

$$\mathcal{V}_{qg}(\varepsilon) = C_F \left[\frac{1}{\varepsilon^2} + \frac{3}{2\varepsilon} + 5 - \frac{\pi^2}{2} + \mathcal{O}(\varepsilon) \right]$$

$$\mathcal{V}_{q\bar{q}}(\varepsilon) = T_R \left[-\frac{2}{3\varepsilon} - \frac{16}{9} + \mathcal{O}(\varepsilon) \right] \qquad (C.7)$$

$$\mathcal{V}_{gg}(\varepsilon) = 2C_A \left[\frac{1}{\varepsilon^2} + \frac{11}{6\varepsilon} + \frac{50}{9} - \frac{\pi^2}{2} + \mathcal{O}(\varepsilon) \right].$$

In general they can be written as

$$\mathcal{V}_i(\varepsilon) = T_i^2 \left(\frac{1}{\varepsilon^2} - \frac{\pi^2}{3} \right) + \gamma_i^{(1)} \left(\frac{1}{\varepsilon} + 1 \right) + K_i + \mathcal{O}(\varepsilon), \qquad (C.8)$$

where the γ_i are the usual first-order anomalous dimensions of the splitting functions Eq. (2.33), and the K_i read

$$K_q = C_F \left(\frac{7}{2} - \frac{\pi^2}{6} \right)$$

$$K_g = C_A \left(\frac{67}{18} - \frac{\pi^2}{6} \right) - T_R n_f \frac{10}{9}. \qquad (C.9)$$

The last term is the usual soft gluon correction term.

C.1.2 Final-state splitter — initial-state spectator

The kinematic maps $p_i + p_j + p_a \to \tilde{p}_{ij} + \tilde{p}_a$ such that

$$p_i + p_j - p_a = \tilde{p}_{ij} - \tilde{p}_a \tag{C.10}$$

and

$$\begin{aligned}
\tilde{p}_{ij} &= p_i + p_j - (1 - x_{ij,a})\, p_a \\
\tilde{p}_a &= (1 - x_{ij,a})\, p_a,
\end{aligned} \tag{C.11}$$

where the recoil parameter $x_{ij,a}$ and the splitting variable $\tilde{z}_i$ are given by

$$\begin{aligned}
x_{ij,a} &= \frac{p_i p_a + p_j p_a - p_i p_j}{(p_i + p_j) p_a} \\
\tilde{z}_i &= \frac{p_i p_a}{(p_i + p_j) p_a} = \frac{p_i \tilde{p}_a}{\tilde{p}_{ij} \tilde{p}_a} \quad \text{and} \quad \tilde{z}_j = 1 - \tilde{z}_i.
\end{aligned} \tag{C.12}$$

With their dependence on these two parameters again understood, the splitting kernels read (CS-5.39–CS-5.41)

$$\begin{aligned}
\langle s | V^a_{q_i g_j} | s' \rangle &= 8\pi\mu^{2\varepsilon} C_F \alpha_{\rm s} \left[\frac{2}{1 - \tilde{z}_i(1 - x_{ij,a})} - (1 + \tilde{z}_i) - \varepsilon(1 - \tilde{z}_i) \right] \delta_{ss'} \\
\langle \mu | V^a_{q_i \bar{q}_j} | \nu \rangle &= 8\pi\mu^{2\varepsilon} T_R \alpha_{\rm s} \left[-g^{\mu\nu} - \frac{2}{p_i p_j} (\tilde{z}_i p_i - \tilde{z}_j p_j)^\mu (\tilde{z}_i p_i - \tilde{z}_j p_j)^\nu \right] \\
\langle \mu | V^a_{g_i g_j} | \nu \rangle &= 16\pi\mu^{2\varepsilon} C_A \alpha_{\rm s} \left[-g^{\mu\nu} \left(\frac{1}{1 - \tilde{z}_i(1 - x_{ij,a})} + \frac{1}{1 - \tilde{z}_j(1 - x_{ij,a})} - 2 \right) \right. \\
&\qquad\qquad \left. + \frac{1 - \varepsilon}{p_i p_j} (\tilde{z}_i p_i - \tilde{z}_j p_j)^\mu (\tilde{z}_i p_i - \tilde{z}_j p_j)^\nu \right].
\end{aligned} \tag{C.13}$$

Apart from the replacement $y_{ij,k} \to x_{ij,a}$ these kernels are identical to the ones in Eq. (C.4), and also the splitting parameter $\tilde{z}_i$ can be related to the one in the case of a final-state splitter with a final-state spectator, with the obvious replacement of $p_k \to p_a$. It is therefore not a surprise that the spin-averaged splitting kernels $\langle V \rangle$ for the FI case here can be obtained from their FF counterparts in Eq. (C.5) by replacing $y_{ij,k}$ with $x_{ij,a}$.

The integration of these terms over the emission phase-space with the replacement $x = x_{ij,a}$ and $z = \tilde{z}_i$, yields, *cf.* (CS-5.50)

$$\mathcal{V}_{ij}(x, \varepsilon) = \Theta(x_{ij,a})\Theta(1 - x_{ij,a}) \left(\frac{1}{1 - x_{ij,a}} \right)^{1+\varepsilon} \int\limits_0^1 \mathrm{d}z [z(1 - z)]^{-\varepsilon} \frac{\langle V_{ij,k}(z, y_{ij,k}) \rangle}{8\pi\alpha_{\rm s}\mu^{2\varepsilon}}, \tag{C.14}$$

and here the first big difference with respect to the case of both splitter and spectator being in the final-state becomes apparent. As the recoil parameter $x_{ij,a}$ is not being integrated over, for $\varepsilon \to 0$, the term $\propto 1/(1 - x_{ij,a})$ before the integral diverges at $x_{ij,a}$, a singularity, which cannot be lifted without taking care in how the two limits, $\varepsilon \to 0$ and $x_{ij,a} \to 1$ are approached together. The usual solution for this kind of problem is to invoke a "+" function such that

$$\mathcal{V}_{ij}(x_{ij,a}, \varepsilon) = [\mathcal{V}_{ij}(x_{ij,a}, \varepsilon)]_+ + \delta(1 - x_{ij,a}) \int_0^1 d\tilde{x} \mathcal{V}_{ij}(\tilde{x}, \varepsilon), \tag{C.15}$$

where

$$[\mathcal{V}_{ij}(x_{ij,a}, \varepsilon)]_+ = [\mathcal{V}_{ij}(x_{ij,a}, 0)]_+ + \mathcal{O}(\varepsilon). \tag{C.16}$$

Therefore, *cf.* (CS-5.57–CS-5.59),

$$\mathcal{V}_{qg}(x, \varepsilon) = C_F \left[\left(\frac{2}{1-x} \log \frac{1}{1-x} \right)_+ - \frac{3}{2} \left(\frac{1}{1-x} \right)_+ + \frac{2}{1-x} \log(2-x) \right]$$
$$+ \delta(1-x) \left[\mathcal{V}_{qg}(\varepsilon) - \frac{3C_F}{2} \right] + \mathcal{O}(\varepsilon)$$

$$\mathcal{V}_{q\bar{q}}(x, \varepsilon) = \frac{2}{3} T_R \left(\frac{1}{1-x} \right)_+ + \delta(1-x) \left[\mathcal{V}_{q\bar{q}}(\varepsilon) + \frac{2T_R}{3} \right] + \mathcal{O}(\varepsilon)$$

$$\mathcal{V}_{gg}(x, \varepsilon) = 2C_A \left[\left(\frac{2}{1-x} \log \frac{1}{1-x} \right)_+ - \frac{11}{6} \left(\frac{1}{1-x} \right)_+ + \frac{2}{1-x} \log(2-x) \right]$$
$$+ \delta(1-x) \left[\mathcal{V}_{gg}(\varepsilon) - \frac{11C_A}{3} \right] + \mathcal{O}(\varepsilon), \tag{C.17}$$

with the $\mathcal{V}_{ij}$ given in Eq. (C.7).

C.1.3 Initial-state splitter — final-state spectator

The kinematic maps $p_a + p_i + p_k \to \tilde{p}_{ai} + \tilde{p}_k$ such that

$$p_i + p_j - p_a = \tilde{p}_{ij} - \tilde{p}_a \tag{C.18}$$

and

$$\begin{aligned} \tilde{p}_{ai} &= x_{ij,a} \, p_a \\ \tilde{p}_k &= p_k + p_i - (1 - x_{ij,a}) \, p_a. \end{aligned} \tag{C.19}$$

It is worth stressing that in this configuration the initial-state splitter actually keeps its direction along the beam axis, with a momentum reduced by $x_{ij,a}$, which now plays the role of a splitting parameter. At the same time, the transverse momentum recoil is transferred to the spectator which, consequently, changes its direction. The splitting variable $x_{ik,a}$ parameterizing this behaviour is given by

$$x_{ik,a} = \frac{p_k p_a + p_i p_a - p_i p_k}{(p_i + p_k)p_a}. \tag{C.20}$$

There is, however, another parameter, u_i, which is being used to decompose the eikonals into two splitter–spectator dipoles:

$$u_i = \frac{p_i p_a}{(p_i + p_k)p_a}. \tag{C.21}$$

With their dependence on these two parameters again understood, the splitting kernels read (CS-5.65–CS-5.68)

$$\langle s|V_k^{q_a g_i}|s'\rangle = 8\pi\mu^{2\varepsilon}C_F\alpha_{\mathrm{s}}\left[\frac{2}{1 - x_{ik,a} + u_i} - (1 + x_{ik,a}) - \varepsilon(1 - x_{ik,a})\right]\delta_{ss'}$$

$$\langle s|V_k^{g_a q_i}|s'\rangle = 8\pi\mu^{2\varepsilon}T_R\alpha_{\mathrm{s}}\left[1 - \varepsilon - 2x_{ik,a}(1 - x_{ik,a})\right]\delta_{ss'}$$

$$\langle\mu|V_k^{q_a\bar{q}_j}|\nu\rangle = 8\pi\mu^{2\varepsilon}C_F\alpha_{\mathrm{s}}\left[-g^{\mu\nu}x_{ik,a} + \frac{1 - x_{ik,a}}{x_{ik,a}}\frac{2u_i(1 - u_i)}{p_i p_k}q_{ik}^{\mu}q_{ik}^{\nu}\right] \tag{C.22}$$

$$\langle\mu|V_k^{g_i g_j}|\nu\rangle = 16\pi\mu^{2\varepsilon}C_A\alpha_{\mathrm{s}}\left[-g^{\mu\nu}\left(\frac{1}{1 - x_{ik,a} + u_i} - 1 + x_{ik,a}(1 - x_{ik,a})\right)\right.$$
$$\left. + (1 - \varepsilon)\frac{1 - x_{ik,a}}{x_{ik,a}}\frac{2u_i(1 - u_i)}{p_i p_k}q_{ik}^{\mu}q_{ik}^{\nu}\right],$$

where

$$q_{ik}^{\mu} = \frac{p_i^{\mu}}{u_i} - \frac{p_k^{\mu}}{1 - u_i}. \tag{C.23}$$

Indicating the number of polarizations of particle a with $n_s(a)$, the spin-averaged kernels in a suitable normalization read

$$\frac{n_s(\tilde{q})}{n_s(q)}\frac{\langle V_k^{qg}\rangle}{8\pi\alpha_{\mathrm{s}}\mu^{2\varepsilon}} = C_F\left[\frac{2}{1 - x_{ik,a} + u_i} - (1 + x_{ik,a}) - \varepsilon(1 - x_{ik,a})\right]$$

$$\frac{n_s(\tilde{q})}{n_s(g)}\frac{\langle V_k^{g\bar{q}}\rangle}{8\pi\alpha_{\mathrm{s}}\mu^{2\varepsilon}} = T_R\left[1 - \frac{2x_{ik,a}(1 - x_{ik,a})}{1 - \varepsilon}\right]$$

$$\frac{n_s(\tilde{g})}{n_s(q}\frac{\langle V_k^{qq}\rangle}{8\pi\alpha_{\mathrm{s}}\mu^{2\varepsilon}} = C_F\left[(1 - \varepsilon)x_{ik,a} + 2\frac{1 - x_{ik,a}}{x_{ik,a}}\right]$$

$$\frac{n_s(\tilde{g})}{n_s(g)}\frac{\langle V_k^{gg}\rangle}{8\pi\alpha_{\mathrm{s}}\mu^{2\varepsilon}} = 2C_A\left[[\frac{1}{1 - x_{ik,a} + u_i} + \frac{1 - x_{ik,a}}{x_{ik,a}} - 1 + x_{ik,a}(1 - x_{ik,a})\right], \tag{C.24}$$

cf. (CS-5.77–CS-5.80).

As in the FI case, the integration over the emission phase-space does not include the parameter $x_{ik,a}$, governing the kinematic map for the initial state. Therefore, this time the integrated dipoles emerge from an integral over u_i only, *cf.* (CS-5.74)

$$\mathcal{V}^{ai,a}(x,\varepsilon) = \Theta(x)\Theta(1-x)\left(\frac{1}{1-x}\right)^{\varepsilon}\int_0^1 du_i[u_i(1-u_i)]^{-\varepsilon}\frac{n_s(\tilde{ai})}{n_s(a)}\frac{\langle V_{ij,k}(\tilde{z}_i,\,y_{ij,k})\rangle}{8\pi\alpha_s\mu^{2\varepsilon}}.$$

(C.25)

As before poles emerge which this time are of the type $f(x)/\varepsilon$ with $f(x)$ is either integrable over x or proportional to $1/(1-x)$. Therefore the integrated dipoles are written as (CS-5.75)

$$\mathcal{V}^{ai,a}(x,\varepsilon) = \frac{1}{\varepsilon}\left\{\frac{1}{x}\left[\varepsilon x\,\mathcal{V}^{ai,a}(x,\varepsilon)\right]_+ + \varepsilon\delta(1-x)\int_0^1 d\tilde{x}\,\tilde{x}\,\mathcal{V}^{ai,a}(\tilde{x},\varepsilon)\right\},$$

(C.26)

which upon expansion in ε ultimately leads to a structure of the form

$$\mathcal{V}^{ai,a}(x,\varepsilon) \sim p^{a,\tilde{ai}}(x) + \propto \left(\frac{1}{\varepsilon}+\mathcal{O}(1)\right) + \mathcal{O}(\varepsilon),$$

(C.27)

where the p^{ij} are related to the Altarelli–Parisi splitting functions from Eq. (2.33) through

$$\begin{aligned}
p^{qg}(x) &= P_{gq}(x)\\
p^{gq}(x) &= P_{qg}(x)\\
p^{qq}(x) &= [P_{qq}(x)]_+\\
p^{gg}(x) &= [P_{gg}(x)]_+ - 2C_A + \delta(1-x)\gamma_g^{(1)},
\end{aligned}$$

(C.28)

cf. (CS-5.85–CS-5.88). This yields the integrated dipoles for the IF case,

$$\mathcal{V}^{qg}(x,\varepsilon) = \left[-\frac{1}{\varepsilon}+\log(1-x)\right]p^{qg}(x) + C_F x + \mathcal{O}(\varepsilon)$$

$$\mathcal{V}^{gq}(x,\varepsilon) = \left[-\frac{1}{\varepsilon}+\log(1-x)\right]p^{gq}(x) + 2T_R x(1-x) + \mathcal{O}(\varepsilon)$$

$$\mathcal{V}^{qq}(x,\varepsilon) = -\frac{1}{\varepsilon}p^{qq}(x) + \delta(1-x)\left[\mathcal{V}_{qg}(\varepsilon) + C_F\left(\frac{2\pi^2}{3}-5\right)\right]$$
$$+ C_F\left[-\left(\frac{4}{1-x}\log\frac{1}{1-x}\right)_+ - \frac{2}{1-x}\log(2-x)\right.$$
$$\left. + (1-x) - (1+x)\log(1-x)\right]$$

$$\mathcal{V}^{gg}(x,\varepsilon) = -\frac{1}{\varepsilon}p^{gg}(x) + \delta(1-x)\left[\frac{1}{2}\mathcal{V}_{gg}(\varepsilon) + n_f\mathcal{V}_{q\bar{q}}(\varepsilon) + C_A\left(\frac{2\pi^2}{3}-\frac{50}{9}\right) + \frac{16}{9}n_f T_R\right]$$

$$+ C_A \left[-\left(\frac{4}{1-x} \log \frac{1}{1-x} \right)_+ - \frac{2}{1-x} \log(2-x) \right.$$
$$\left. + 2\left(-1 + x(1-x) + \frac{1-x}{x} \right) \log(1-x) \right]. \tag{C.29}$$

The first terms $\propto 1/\varepsilon$ in each dipole are related to the collinear divergence in the initial-state, while the terms $\propto \mathcal{V}_{ij}(\varepsilon)$ in $\mathcal{V}^{qq}(x,\varepsilon)$ and $\mathcal{V}^{gg}(x,\varepsilon)$ capture the soft divergences stemming from the eikonals and are related to the emission of soft gluons.

C.1.4 Initial-state splitter — initial-state spectator

In the case of both the splitter and the spectator in the initial-state an interesting problem emerges. In Catani–Seymour subtraction, there are two paradigm for the construction of the kinematics maps between the $(n+1)$–particle and the n–particle configurations, namely

(a) to keep the spectator direction and to just stretch it; and

(b) to keep initial-state particles on the beam axis.

This works very well for splitter partons in the final-state. In particular, in the case of FI dipoles, this was natural — the spectator momentum keeps its direction and merely is stretched to compensate for recoil in its longitudinal direction. In so doing of course the initial-state particle stayed oriented along the beam axis. For IF dipoles both paradigms above already could not be maintained anymore. In fact there this conflict was solved in such a way that the spectator accounts for the balance of transverse momenta, This becomes even more aggrevated in the case of II dipoles, where both splitter and spectator are initial-state partons. In this case, the solution is to let the complete final-state, apart from the emitted parton i, of course, capture the transverse momentum. This is achieved by defining a boost for all final–state momenta $k_j \neq p_i$ in such a way that four-momentum conservation in the $(n+1)$-particle configuration,

$$p_a^\mu + p_b^\mu - p_i^\mu - \sum_{j \neq i} k_j^\mu = 0, \tag{C.30}$$

is maintained also for the n-particle configuration:

$$\tilde{p}_{ai}^\mu + p_b^\mu - \sum_{j \neq i} \tilde{k}_j^\mu = 0. \tag{C.31}$$

Here,

$$\tilde{p}_{ai} = x_{i,ab}\, p_a$$
$$x_{i,ab} = \frac{p_a p_b - p_i(p_a + p_b)}{p_a p_b}. \tag{C.32}$$

As can be seen, the momentum of the spectator parton p_b is not altered in this map, but instead all final-state momenta are modified to read

$$\tilde{k}_j^\mu = k_j^\mu - \frac{2k_j(K+\tilde{K})}{(K+\tilde{K})^2}(K+\tilde{K})^\mu - \frac{2k_jK}{K^2}\tilde{K}^\mu, \tag{C.33}$$

where the momenta K and $\tilde{K}$ are the total momenta of the dipole before and after the mapping onto Born-level kinematics has taken place:

$$\begin{aligned} K^\mu &= p_a^\mu + p_b^\mu - p_i^\mu \\ \tilde{K}^\mu &= \tilde{p}_{ai}^\mu + p_b^\mu. \end{aligned} \tag{C.34}$$

The splitting kernels are given by

$$\begin{aligned} \langle s|V^{q_a g_i,b}|s'\rangle &= 8\pi\mu^{2\varepsilon}C_F\alpha_s\left[\frac{2}{1-x_{i,ab}} - (1+x_{i,ab}) - \varepsilon(1-x_{i,ab})\right]\delta_{ss'} \\ \langle s|V^{g_a q_i,b}|s'\rangle &= 8\pi\mu^{2\varepsilon}T_R\alpha_s\left[1 - \varepsilon - 2x_{i,ab}(1-x_{i,ab})\right]\delta_{ss'} \\ \langle\mu|V^{q_a\bar{q}_j,b}|\nu\rangle &= 8\pi\mu^{2\varepsilon}C_F\alpha_s\left[-g^{\mu\nu}x_{i,ab} + \frac{1-x_{i,ab}}{x_{i,ab}}\frac{2p_ap_b}{p_ip_a\,p_ip_b}q^\mu q^\nu\right] \\ \langle\mu|V^{g_i g_j,b}|\nu\rangle &= 16\pi\mu^{2\varepsilon}C_A\alpha_s\left[-g^{\mu\nu}\left(\frac{1}{1-x_{i,ab}} + x_{i,ab}(1-x_{i,ab})\right)\right. \\ &\qquad\left. + (1-\varepsilon)\frac{1-x_{i,ab}}{x_{i,ab}}\frac{2p_ap_b}{p_ip_a\,p_ip_b}q^\mu q^\nu\right], \end{aligned} \tag{C.35}$$

where

$$q^\mu = p_i^\mu - \frac{p_ip_a}{p_bp_a}p_b^\mu. \tag{C.36}$$

With the notation concerning incident spins already used in the IF case, the phase-space integration, *cf.* (CS-5.153),

$$\mathcal{V}^{\tilde{a},ai}(x,\varepsilon) = -\frac{1}{\varepsilon}\frac{\Gamma^2(1-\varepsilon)}{\Gamma(1-2\varepsilon)}\Theta(x)\Theta(1-x)(1-x)^{-2\varepsilon}\frac{n_s(\tilde{ai})}{n_s(a)}\frac{\langle V^{ai,b}\rangle}{8\pi\alpha_s\mu^{2\varepsilon}} \tag{C.37}$$

of the spin-averaged splitting kernels yields integrated dipoles of the form

$$\mathcal{V}^{\widetilde{ab}}(x,\varepsilon) = \mathcal{V}^{ab}(x,\varepsilon) + \delta^{ab}T_a^2\left[\left(\frac{2}{1-x}\log\frac{1}{1-x}\right)_+ + \frac{2}{1-x}\log(2-x)\right] + \tilde{K}^{ab} \tag{C.38}$$

up to $\mathcal{O}(\varepsilon)$, see also (CS-5.155). Here, the $\mathcal{V}^{ab}(x,\varepsilon)$ are the integrated dipoles in the IF case, *cf.* Eq. (C.29). Not surprisingly, in the II case thus additional terms emerge. The $\tilde{K}^{ab}$ read

$$\tilde{K}^{ab}(x) = P_{ab}^{(\mathrm{reg})}(x)\log(1-x) + \delta^{ab}T_a^2\left[\left(\frac{2}{1-x}\log(1-x)\right)_+ - \frac{\pi^2}{3}\delta(1-x)\right]. \tag{C.39}$$

The regular bits of the splitting functions, $P_{ab}^{(\text{reg})}(x)$ are the remainders after the poles for $x \to 1$ have been subtracted, namely

$$P_{ab}^{(\text{reg})}(x) = \mathcal{P}_{ab}(x) - \delta^{ab} \left[2T_a^2 \left(\frac{1}{1-x} \right)_+ + \gamma_a^{(1)} \delta(1-x) \right], \tag{C.40}$$

where both the splitting functions and their anomalous dimensions γ are given by Eq. (2.33). In particular they read

$$
\begin{aligned}
P_{ab}^{(\text{reg})}(x) &= \mathcal{P}_{ab}(x) \quad \text{for } a \neq b \\
P_{qq}^{(\text{reg})}(x) &= -C_F \left(1 + x \right) \\
P_{gg}^{(\text{reg})}(x) &= 2C_A \left[\frac{1-x}{x} - 1 + x(1-x) \right].
\end{aligned}
\tag{C.41}
$$

Note that the spin-averaged splitting kernels of course effectively are the Altarelli–Parisi splitting functions P in D dimensions,

$$\frac{n_s(\tilde{ai})}{n_s(a)} \frac{\langle V^{ai,b}(x) \rangle}{8\pi\alpha_s\mu^{2\varepsilon}} = P_{a,\tilde{ai}}(x) \tag{C.42}$$

from Eq. (2.33).

C.1.5 Final formulae

After rearranging the various integrated dipole terms and adding the collinear counter-terms the part of the cross-section to be added to the virtual part reads

$$
\begin{aligned}
d\sigma^{(I+C)} = {}& d\sigma_{ab}^{(\text{Born})}(p_a, p_b) \otimes \mathbf{I}(\varepsilon) \\
&+ \sum_{a'} \int_0^1 dx d\sigma_{a'b}^{(\text{Born})}(xp_a, p_b) \otimes \left[\mathbf{K}^{aa'}(x) + \mathbf{P}^{aa'}(xp_a, x;, \mu_F^2) \right] \\
&+ \sum_{b'} \int_0^1 dx d\sigma_{ab'}^{(\text{Born})}(p_a, xp_b) \otimes \left[\mathbf{K}^{bb'}(x) + \mathbf{P}^{bb'}(xp_b, x;, \mu_F^2) \right],
\end{aligned}
\tag{C.43}
$$

see also Eq. (3.202) and (CS-10.27)-CS(10.30). The integrated dipole terms are

$$
\begin{aligned}
\mathbf{I}(\varepsilon) = {}& -\frac{\alpha_s}{2\pi\Gamma(1-\varepsilon)} \\
& \left\{ \sum_i \frac{\mathcal{V}_i(\varepsilon)}{T_i^2} \left[\sum_{k \neq i} T_i \cdot T_k \left(\frac{4\pi\mu^2}{2p_ip_k} \right)^\varepsilon + T_i \cdot T_a \left(\frac{4\pi\mu^2}{2p_ip_a} \right)^\varepsilon + T_i \cdot T_b \left(\frac{4\pi\mu^2}{2p_ip_b} \right)^\varepsilon \right] \right. \\
& \left. + \frac{\mathcal{V}_a(\varepsilon)}{T_a^2} \left[\sum_k T_a \cdot T_k \left(\frac{4\pi\mu^2}{2p_ap_k} \right)^\varepsilon + T_a \cdot T_b \left(\frac{4\pi\mu^2}{2p_ap_b} \right)^\varepsilon \right] \right.
\end{aligned}
$$

$$+ \frac{\mathcal{V}_b(\varepsilon)}{T_b^2} \left[\sum_k T_b \cdot T_k \left(\frac{4\pi\mu^2}{2p_b p_k} \right)^\varepsilon + T_b \cdot T_a \left(\frac{4\pi\mu^2}{2p_b p_a} \right)^\varepsilon \right] \Bigg\}, \tag{C.44}$$

while the collinear counterterms are given by

$$K^{aa'}(x) = \frac{\alpha_\mathrm{s}}{2\pi} \Bigg\{ \overline{K}^{aa'}(x) - K_{(\mathrm{F.S.})}^{aa'}(x) - \sum_i \frac{T_i \cdot T_a}{T_i^2} \tilde{K}^{aa'}(x)$$

$$+ \delta^{aa'} \sum_i \frac{T_i \cdot T_a}{T_i^2} \gamma_i \left[\left(\frac{1}{1-x} \right)_+ + \delta(1-x) \right] \Bigg\}$$

$$P^{aa'}(xp_a, x; \mu_F^2) = \frac{\alpha_\mathrm{s}}{2\pi} \mathcal{P}_{a'a}^{(1)}(x) \left[\sum_i \frac{T_i \cdot T_{a'}}{T_{a'}^2} \log \frac{\mu_F^2}{2xp_a p_i} + \frac{T_b \cdot T_{a'}}{T_{a'}^2} \log \frac{\mu_F^2}{2xp_a p_b} \right], \tag{C.45}$$

where

$$\overline{K}^{qq} = C_F \left[\left(\frac{1+x^2}{1-x} \log \frac{1-x}{x} \right)_+ + (1-x) - \delta(1-x)\left(5 - \pi^2 \right) \right]$$

$$\overline{K}^{gg} = 2C_A \left[\left(\frac{1}{1-x} \log \frac{1-x}{x} \right)_+ + \left(\frac{1-x}{x} - 1 + x(1-x) \right) \log \frac{1-x}{x} \right]$$

$$- \delta(1-x) \left[C_A \left(\frac{50}{9} - \pi^2 \right) - \frac{16}{9} T_R n_f \right] \tag{C.46}$$

$$\overline{K}^{qg} = P_{qg}^{(1)} \log \frac{1-x}{x} + C_F x$$

$$\overline{K}^{gq} = P_{gq}^{(1)} \log \frac{1-x}{x} + 2T_R x(1-x)$$

The terms $\tilde{K}$ contain the regular part of the splitting functions $P_{ba}^{(\mathrm{reg})}$ defined in Eq. (C.40) and listed in Eq. (C.41),

$$\tilde{K}^{ab}(x) = P_{ba}^{(\mathrm{reg})}(x) \log(1-x) + \delta^{ab} T_a^2 \left[\left(\frac{2\log(1-x)}{1-x} \right)_+ - \frac{\pi^2}{3} \delta(1-x) \right]. \tag{C.47}$$

C.2   Catani–Seymour subtraction for parton showers

In the following, the various building blocks for a parton (dipole) shower based on Catani-Seymour ("CS shower") splitting kernels for massless partons will be reviewed. In principle there are four cases to consider, namely final- (initial-) state splittings with spectator partons in the final- (initial-) state, leading to FF, FI, IF, and II cases, respectively. There are typically four ingredients:

- kinematical variables, parameterizing recoil, and splitting;
- their relation to the transverse momentum, $k_\perp$, the default ordering parameter of the dipole shower, and the corresponding Jacobean J;

- the splitting kernels;
- the phase-space map between four momenta before and after the splitting, denoted by $\tilde{p}$ and p.

These ingredients in the original proposal of [778] and in the implementation of [839] will be listed here. It should be noted, though, that in some cases the actual implementation of the CS shower in SHERPA version 2 differs from the one described here, mainly due to the resummation property described in the main test, in the relevant part of Section 5.3.2.

1. FF (final-state splitter with final-state spectator):

    - kinematical parameters:

    $$y_{ij;k} = \frac{p_i p_j}{p_i p_j + p_i p_k + p_j p_k} \quad \text{and} \quad z_i = \frac{p_i p_k}{p_i p_k + p_j p_k} = 1 - z_j ; \quad (C.48)$$

    - transverse momentum:

    $$k_\perp^2 = z_i(1 - z_i) y_{ij;k} Q^2, \qquad (C.49)$$

    where $Q^2 = s_{ijk} = (p_i + p_j + p_k)^2 = (\tilde{p}_{ij} + \tilde{p}_k)^2$, and

    $$J^{(FF)} = 1 - y_{ij;k}. \qquad (C.50)$$

    Therefore, the relevant phase-space element (including the propagator-like denominator) is given by

    $$d\Phi_1 = \frac{1}{16\pi^2} \frac{dk_\perp^2}{k_\perp^2} \frac{dz_i}{z_i(1-z_i)} \frac{d\phi}{2\pi} \left(1 - \frac{k_\perp^2}{z(1-z)Q^2}\right) ; \qquad (C.51)$$

    - splitting kernels:

    $$\mathcal{K}_{qg,k}^{(FF)} = C_F \left[\frac{2}{1 - z_i(1 - y_{ij;k})} - (1 + z_i)\right]$$

    $$\mathcal{K}_{gg,k}^{(FF)} = 2C_A \left[\frac{1}{1 - z_i(1 + y_{ij;k})} + \frac{1}{1 - (1 - z_i)(1 + y_{ij;k})} - 2 + z_i(1 - z_i)\right]$$

    $$\mathcal{K}_{q\bar{q},k}^{(FF)} = T_R \left[1 - 2z_i(1 - z_i)\right] ; $$

    $$(C.52)$$

    - phase-space map (in the emitter-spectator c.m. frame):

    $$p_i = z_i \tilde{p}_{ij} + (1 - z_i) y_{ij;k} \tilde{p}_k + \vec{k}_\perp$$

    $$p_j = (1 - z_i) \tilde{p}_{ij} + z_i y_{ij;k} \tilde{p}_k - \vec{k}_\perp \qquad (C.53)$$

    $$p_k = (1 - y_{ij;k}) \tilde{p}_k.$$

2. FI (final-state splitter with initial-state spectator):

- kinematical parameters:

$$x_{ij;a} = \frac{p_i p_a + p_j p_a - p_i p_j}{p_i p_a + p_j p_a} \quad \text{and} \quad z_i = \frac{p_i p_a}{p_i p_a + p_j p_a} = 1 - z_j; \quad \text{(C.54)}$$

- transverse momentum:

$$k_\perp^2 = z_i(1 - z_i)\frac{1 - x_{ij;a}}{x_{ij;a}} Q^2, \quad \text{(C.55)}$$

where $Q^2 = (p_i + p_j + p_a)^2 = (\tilde{p}_{ij} + \tilde{p}_a)^2$, and

$$J^{(FI)} = \frac{f_{a/h}\left(\frac{\eta_a}{x_{ij;a}}, \mu_F^2\right)}{f_{a/h}\left(\eta_a, \mu_F^2\right)}, \quad \text{(C.56)}$$

where the PDF ratios typical for backwards evolution with η_a the momentum fraction of a are made explict and where the transformation from an integral over $x_{ij;a}$ to one over $k_\perp^2$ cancels the $1/(1 - x_{ij;a})$ term stemming from the one-particle phase-space integral written as a function of $x_{ij;a}$;

- splitting kernels:

$$K_{qg,k}^{(FI)} = C_F\left[\frac{2}{1 - z_i + (1 - x_{ij;a})} - (1 + z_i)\right]$$

$$K_{gg,k}^{(FI)} = 2C_A\left[\frac{1}{1 - z_i + (1 + x_{ij;a})} + \frac{1}{z_i + (1 - x_{ij;a})} - 2 + z_i(1 - z_i)\right]$$

$$K_{q\bar{q},k}^{(FI)} = T_R\left[1 - 2z_i(1 - z_i)\right];$$

$$\text{(C.57)}$$

- phase-space map (in the emitter-spectator Breit frame):

$$p_i = z_i\tilde{p}_{ij} + \frac{(1 - z_i)(1 - x_{ij;a})}{x_{ij;a}} \tilde{p}_k + \vec{k}_\perp$$

$$p_j = (1 - z_i)\tilde{p}_{ij} + \frac{z_i(1 - x_{ij;a})}{x_{ij;a}} \tilde{p}_k - \vec{k}_\perp \quad \text{(C.58)}$$

$$p_k = \frac{1}{x_{ij;a}}\tilde{p}_k .$$

3. IF (initial-state splitter with final-state spectator):

- kinematical parameters:

$$x_{aj;k} = \frac{p_a p_j + p_a p_k - p_j p_k}{p_a p_j + p_a p_k} \quad \text{and} \quad u_a = \frac{p_a p_j}{p_a p_j + p_a p_k} = 1 - u_k \, ; \quad \text{(C.59)}$$

- transverse momentum:

$$k_\perp^2 = u_i(1 - u_i)\frac{1 - x_{aj;k}}{x_{aj;k}} Q^2, \quad \text{(C.60)}$$

where $Q^2 = (p_i + p_j + p_a)^2 = (\tilde{p}_{aj} + \tilde{p}_k)^2$, and

$$J^{(IF)} = \frac{1}{x_{ij;a}} \frac{1 - u_a}{1 - 2u_a} \frac{f_{A/a}\left(\frac{\eta_a}{x_{ij;a}}, \mu_F^2\right)}{f_{A/a}\left(\eta_a, \mu_F^2\right)}, \quad \text{(C.61)}$$

where as before suitable PDF factors have been included;
- splitting kernels:

$$
\begin{aligned}
\mathcal{K}_{qg,k}^{(IF)} &= C_F\left[\frac{2}{1 - x_{aj;k} + u_a} - (1 + x_{aj;k})\right] \\
\mathcal{K}_{qq,k}^{(IF)} &= C_F\left[2\frac{1 - x_{aj;k}}{x_{aj;k}} + x_{aj;k}\right] \\
\mathcal{K}_{gg,k}^{(IF)} &= 2C_A\left[\frac{2}{1 - x_{aj;k} + u_a} + \frac{1 - x_{aj;k}}{x_{aj;k}} - 1 + x_{aj;k}(1 - x_{aj;k})\right] \\
\mathcal{K}_{q\bar{q},k}^{(IF)} &= T_R\left[1 - 2x_{aj;k}(1 - x_{aj;k})\right] \, ;
\end{aligned}
\quad \text{(C.62)}
$$

- phase-space map (in the emitter-spectator Breit frame):

$$
\begin{aligned}
p_a &= \frac{1}{x_{aj;k}} \tilde{p}_{aj} \\
p_j &= (1 - u_a)\frac{1 - x_{aj;k}}{x_{aj;k}} \tilde{p}_{aj} + u_a\tilde{p}_k + \vec{k}_\perp \\
p_k &= u_a\frac{1 - x_{aj;k}}{x_{aj;k}} \tilde{p}_{aj} + (1 - u_a)\tilde{p}_k - \vec{k}_\perp.
\end{aligned}
\quad \text{(C.63)}
$$

This map stemming from a trivial inversion of the original Catani–Seymour phase-space map, however, is not optimally suited for parton shower simulations, since here the transverse momenta act on the colourful spectator alone and not on the total final-state. This is at odds with standard resummation ideas as discussed in Section 5.3.2. Therefore, other phase-space maps have been proposed in [626, 807], which remedy this situation. Essentially, these maps transmit the $k_\perp$ onto the initial-state splitter, which in turn results in a Lorentz transformation that puts the splitter back onto the beam axis at the expense of the complete final-state being subjected to some transverse momentum "kick". Following the more specific [807], this map is given by first constructing the four-momenta according to

$$
p_a = \frac{1 - u_a}{x_{aj;k} - u_a} \tilde{p}_{aj} + \frac{u}{x} \frac{1 - x_{aj;k}}{x_{aj;k} - u_a} \tilde{p}_k + \frac{1}{u_a - x_{aj;k}} \vec{k}_\perp
$$

$$
p_j = \frac{1 - x_{aj;k}}{x_{aj;k} - u_a} \tilde{p}_{aj} + \frac{u}{x} \frac{1 - u_a}{x_{aj;k} - u_a} \tilde{p}_k + \frac{1}{u_a - x_{aj;k}} \vec{k}_\perp \qquad \text{(C.64)}
$$

$$
p_k = \frac{x_{aj;k} - u_a}{x_{aj;k}} \tilde{p}_k \;.
$$

In fact, these two phase-space maps are related to each other by the Lorentz transformation $\Lambda^\mu_\nu(K)$,

$$
\Lambda^\mu_\nu(K) = g^\mu_\nu + \frac{x_{aj;k}}{(1 - u_a)(1 - x_{aj;k})} \frac{k^\mu_\perp k_{\perp\nu}}{\tilde{p}_{aj}\tilde{p}_k} + \frac{u_a(1 - x_{aj;k})}{x_{aj;k} - u_a} \frac{K^\mu K_\nu}{\tilde{p}_{aj}\tilde{p}_k}
$$
$$
+ \frac{x_{aj;k}}{x_{aj;k} - u_a} \frac{k^\mu_\perp K_\nu - K^\mu k_{\perp\nu}}{\tilde{p}_{aj}\tilde{p}_k}
$$

$$
\text{(C.65)}
$$

with $K = \tilde{p}_{aj} + \tilde{p}_k$.

4. II (initial-state splitter with initial-state spectator):
This is, in some sense, a special case, since the spectator momentum is preserved and, consequently, any recoils are captured by the final state. In particular, this leads to a phase-space mapping for the emitter of

$$
\tilde{p}_{aj} = x_{aj;b}\, p_a \quad \text{and} \quad \tilde{p}_b = p_b, \qquad \text{(C.66)}
$$

where

$$
x_{aj;b} = \frac{p_a p_b - p_a p_j - p_b p_j}{p_a p_b}. \qquad \text{(C.67)}
$$

Then, further building blocks are given by

- transverse momentum:

$$
k^2_\perp = \frac{1 - x_{aj;b} - v_j}{x_{aj;b}} v_j\, Q^2, \qquad \text{(C.68)}
$$

where $Q^2 = (p_a + p_j + p_b)^2 = (\tilde{p}_{aj} + \tilde{p}_b)^2$, and

$$
v_j = \frac{p_a p_j}{p_a p_b}. \qquad \text{(C.69)}
$$

The Jacobean reads:

$$
J^{(II)} = \frac{1}{x_{ij;a}} \frac{1 - x_{ij;a} - v_j}{1 - x_{ij;a} - 2v_j} \frac{f_{A/a}\left(\frac{\eta_a}{x_{ij;a}}, \mu^2_F\right)}{f_{A/a}\left(\eta_a, \mu^2_F\right)}; \qquad \text{(C.70)}
$$

- splitting kernels:

$$\mathcal{K}_{qg,k}^{(II)} = C_F \left[\frac{2}{1 - x_{aj;b}} - (1 + x_{aj;b}) \right]$$

$$\mathcal{K}_{qq,k}^{(II)} = C_F \left[\frac{2(1 - x_{aj;b})}{x_{aj;b}} - x_{aj;b} \right]$$

$$\mathcal{K}_{gg,k}^{(II)} = 2C_A \left[\frac{1}{1 - x_{aj;b}} + \frac{1 - x_{aj;b}}{x_{aj;b}} - 1 + x_{aj;b}(1 - x_{aj;b}) \right] \tag{C.71}$$

$$\mathcal{K}_{q\bar{q},k}^{(II)} = T_R \left[1 - 2x_{aj;b}(1 - x_{aj;b}) \right] \ ;$$

- phase-space map (in the emitter-spectator Breit frame):

$$p_a = \frac{1}{x_{aj;k}} \tilde{p}_{aj}$$

$$p_j = \frac{1 - x_{aj;k} - v_j}{x_{aj;k}} \tilde{p}_{aj} + v_j \tilde{p}_b + \vec{k}_\perp \tag{C.72}$$

$$p_b = \tilde{p}_b \ .$$

and all other (FS) momenta k_j are suitably boosted and rotated by a Lorentz transformation Λ

$$k_j = \Lambda(\tilde{p}_{aj} + p_b, \, p_a + p_b - p_j) \, \tilde{k}_j \tag{C.73}$$

given by

$$\Lambda_\nu^\mu(\tilde{K}, \, K) = g_\nu^\mu - 2\frac{(\tilde{K} + K)^\mu (\tilde{K} + K)_\nu}{(\tilde{K} + K)^2} + 2\frac{K^\mu \tilde{K}_\nu}{\tilde{K}^2} . \tag{C.74}$$

References

[1] Aaboud, Morad et al. (2016). Measurement of the total cross section from elastic scattering in pp collisions at $\sqrt{s} = 8$ TeV with the ATLAS detector. *Phys. Lett.*, **B761**, 158–178. doi:10.1016/j.physletb.2016.08.020.

[2] Aad, G. et al. (2011*a*). Charged-particle multiplicities in pp interactions measured with the ATLAS detector at the LHC. *New J. Phys.*, **13**, 053033. doi:10.1088/1367-2630/13/5/053033.

[3] Aad, Georges et al. (2011*b*). Measurement of dijet production with a veto on additional central jet activity in pp collisions at $\sqrt{s} = 7$ TeV using the ATLAS detector. *JHEP*, **09**, 053. doi:10.1007/JHEP09(2011)053.

[4] Aad, Georges et al. (2011*c*). Measurement of multi-jet cross sections in proton-proton collisions at a 7 TeV center-of-mass energy. *Eur. Phys. J.*, **C71**, 1763. doi:10.1140/epjc/s10052-011-1763-6.

[5] Aad, Georges et al. (2011*d*). Measurement of the inelastic proton-proton cross-section at $\sqrt{s} = 7$ TeV with the ATLAS detector. *Nature Commun.*, **2**, 463. doi:10.1038/ncomms1472.

[6] Aad, Georges et al. (2011*e*). Measurement of underlying event characteristics using charged particles in pp collisions at $\sqrt{s} = 900 GeV$ and 7 TeV with the ATLAS detector. *Phys. Rev.*, **D83**, 112001. doi:10.1103/PhysRevD.83.112001.

[7] Aad, Georges et al. (2012*a*). Evidence for the associated production of a W boson and a top quark in ATLAS at $\sqrt{s} = 7$ TeV. *Phys. Lett.*, **B716**, 142–159. doi:10.1016/j.physletb.2012.08.011.

[8] Aad, Georges et al. (2012*b*). Jet mass and substructure of inclusive jets in $\sqrt{s} = 7$ TeV pp collisions with the ATLAS experiment. *JHEP*, **1205**, 128. doi:10.1007/JHEP05(2012)128.

[9] Aad, Georges et al. (2012*c*). Measurement of inclusive jet and dijet production in pp collisions at $\sqrt{s} = 7$ TeV using the ATLAS detector. *Phys. Rev.*, **D86**, 014022. doi:10.1103/PhysRevD.86.014022.

[10] Aad, Georges et al. (2012*d*). Measurement of the charge asymmetry in top quark pair production in pp collisions at $\sqrt{s} = 7$ TeV using the ATLAS detector. *Eur. Phys. J.*, **C72**, 2039. doi:10.1140/epjc/s10052-012-2039-5.

[11] Aad, Georges et al. (2012*e*). Measurement of the cross section for the production of a W boson in association with b^- jets in pp collisions at $\sqrt{s} = 7$ TeV with the ATLAS detector. *Phys. Lett.*, **B707**, 418–437. doi:10.1016/j.physletb.2011.12.046.

[12] Aad, Georges et al. (2012*f*). Measurement of the inclusive $W^{\pm}$ and Z/gamma cross sections in the electron and muon decay channels in pp collisions at $\sqrt{s} = 7$ TeV with the ATLAS detector. *Phys. Rev.*, **D85**, 072004. doi:10.1103/PhysRevD.85.072004.

[13] Aad, Georges et al. (2012*g*). Measurement of the production cross section of an isolated photon associated with jets in proton-proton collisions at $\sqrt{s} = 7$ TeV with the ATLAS detector. *Phys. Rev.*, **D85**, 092014. doi:10.1103/PhysRevD.85.092014.

[14] Aad, Georges et al. (2012*h*). Measurement of the *t*-channel single top-quark production cross section in *pp* collisions at $\sqrt{s} = 7$ TeV with the ATLAS detector. *Phys. Lett.*, **B717**, 330–350. doi:10.1016/j.physletb.2012.09.031.

[15] Aad, Georges et al. (2012*i*). Observation of a new particle in the search for the Standard Model Higgs boson with the ATLAS detector at the LHC. *Phys. Lett.*, **B716**, 1–29. doi:10.1016/j.physletb.2012.08.020.

[16] Aad, Georges et al. (2012*j*). Rapidity gap cross sections measured with the ATLAS detector in *pp* collisions at $\sqrt{s} = 7$ TeV. *Eur. Phys. J.*, **C72**, 1926. doi:10.1140/epjc/s10052-012-1926-0.

[17] Aad, Georges et al. (2013*a*). Measurement of hard double-parton interactions in $W(\to l\nu)+$ 2 jet events at $\sqrt{s}$=7 TeV with the ATLAS detector. *New J. Phys.*, **15**, 033038. doi:10.1088/1367-2630/15/3/033038.

[18] Aad, Georges et al. (2013*b*). Measurement of isolated-photon pair production in *pp* collisions at $\sqrt{s} = 7$ TeV with the ATLAS detector. *JHEP*, **1301**, 086. doi:10.1007/JHEP01(2013)086.

[19] Aad, Georges et al. (2013*c*). Measurement of the cross-section for W boson production in association with b-jets in pp collisions at $\sqrt{s} = 7$ TeV with the ATLAS detector. *JHEP*, **1306**, 084. doi:10.1007/JHEP06(2013)084.

[20] Aad, Georges et al. (2013*d*). Measurement of the production cross section of jets in association with a Z boson in pp collisions at $\sqrt{s} = 7$ TeV with the ATLAS detector. *JHEP*, **1307**, 032. doi:10.1007/JHEP07(2013)032.

[21] Aad, Georges et al. (2013*e*). Measurement of W^+W^- production in pp collisions at $\sqrt{s}$=7 TeV with the ATLAS detector and limits on anomalous WWZ and WWW couplings. *Phys. Rev.*, **D87**(11), 112001. doi:10.1103/PhysRevD.87.112001, 10.1103/PhysRevD.88.079906.

[22] Aad, Georges et al. (2013*f*). Measurement of ZZ production in *pp* collisions at $\sqrt{s} = 7$ TeV and limits on anomalous ZZZ and $ZZ\gamma$ couplings with the ATLAS detector. *JHEP*, **1303**, 128. doi:10.1007/JHEP03(2013)128.

[23] Aad, Georges et al. (2013*g*). Search for light top squark pair production in final states with leptons and b^- jets with the ATLAS detector in $\sqrt{s} = 7$ TeV proton-proton collisions. *Phys. Lett.*, **B720**, 13–31. doi:10.1016/j.physletb.2013.01.049.

[24] Aad, Georges et al. (2014*a*). Comprehensive measurements of *t*-channel single top-quark production cross sections at $\sqrt{s} = 7$ TeV with the ATLAS detector. *Phys. Rev.*, **D90**(11), 112006. doi:10.1103/PhysRevD.90.112006.

[25] Aad, Georges et al. (2014*b*). Evidence for electroweak production of $W^{\pm}W^{\pm}jj$ in *pp* collisions at $\sqrt{s} = 8$ TeV with the ATLAS detector. *Phys. Rev. Lett.*, **113**(14), 141803. doi:10.1103/PhysRevLett.113.141803.

[26] Aad, Georges et al. (2014*c*). Measurement of differential production cross-sections for a Z boson in association with *b*-jets in 7 TeV proton-proton collisions with the ATLAS detector. *JHEP*, **1410**, 141. doi:10.1007/JHEP10(2014)141.

[27] Aad, Georges et al. (2014*d*). Measurement of dijet cross sections in *pp* collisions at 7 TeV centre-of-mass energy using the ATLAS detector. *JHEP*, **1405**, 059. doi:10.1007/JHEP05(2014)059.

[28] Aad, Georges et al. (2014*e*). Measurement of distributions sensitive to the underlying event in inclusive Z-boson production in *pp* collisions at $\sqrt{s} = 7$ TeV with the ATLAS detector. *Eur. Phys. J.*, **C74**(12), 3195. doi:10.1140/epjc/s10052-014-3195-6.

[29] Aad, Georges et al. (2014*f*). Measurement of Higgs boson production in the diphoton decay channel in pp collisions at center-of-mass energies of 7 and 8 TeV with the ATLAS detector. *Phys. Rev.*, **D90**(11), 112015. doi:10.1103/PhysRevD.90.112015.

[30] Aad, Georges et al. (2014*g*). Measurement of the inclusive isolated prompt photons cross section in pp collisions at $\sqrt{s} = 7$ TeV with the ATLAS detector using 4.6 fb^{-1}. *Phys. Rev.*, **D89**(5), 052004. doi:10.1103/PhysRevD.89.052004.

[31] Aad, Georges et al. (2014*h*). Measurement of the inclusive jet cross-section in proton-proton collisions at $\sqrt{s} = 7$ TeV using 4.5 fb^{-1} of data with the ATLAS detector.

[32] Aad, Georges et al. (2014*i*). Measurement of the top quark pair production charge asymmetry in proton-proton collisions at $\sqrt{s} = 7$ TeV using the ATLAS detector. *JHEP*, **1402**, 107. doi:10.1007/JHEP02(2014)107.

[33] Aad, Georges et al. (2014*j*). Measurement of the total cross section from elastic scattering in pp collisions at $\sqrt{s} = 7$ TeV with the ATLAS detector. *Nucl. Phys.*, **B889**, 486–548. doi:10.1016/j.nuclphysb.2014.10.019.

[34] Aad, Georges et al. (2014*k*). Measurement of the Z/γ^* boson transverse momentum distribution in *pp* collisions at $\sqrt{s} = 7$ TeV with the ATLAS detector. *JHEP*, **1409**, 145. doi:10.1007/JHEP09(2014)145.

[35] Aad, Georges et al. (2014*l*). Measurements of fiducial and differential cross sections for Higgs boson production in the diphoton decay channel at $\sqrt{s} = 8$ TeV with ATLAS. *JHEP*, **09**, 112. doi:10.1007/JHEP09(2014)112.

[36] Aad, Georges et al. (2015*a*). Combined measurement of the Higgs boson mass in *pp* collisions at $\sqrt{s} = 7$ and 8 TeV with the ATLAS and CMS experiments. *Phys. Rev. Lett.*, **114**, 191803. doi:10.1103/PhysRevLett.114.191803.

[37] Aad, Georges et al. (2015*b*). Constraints on the off-shell Higgs boson signal strength in the high-mass ZZ and WW final states with the ATLAS detector. *Eur. Phys. J.*, **C75**(7), 335. doi:10.1140/epjc/s10052-015-3542-2.

[38] Aad, Georges et al. (2015*c*). Evidence for the Higgs-boson Yukawa coupling to tau leptons with the ATLAS detector. *JHEP*, **04**, 117. doi:10.1007/JHEP04(2015)117.

[39] Aad, Georges et al. (2015*d*). Jet energy measurement and its systematic uncertainty in proton-proton collisions at $\sqrt{s} = 7$ TeV with the ATLAS detector. *Eur. Phys. J.*, **C75**, 17. doi:10.1140/epjc/s10052-014-3190-y.

[40] Aad, Georges et al. (2015*e*). Measurement of the $t\bar{t}$ production cross-section as a function of jet multiplicity and jet transverse momentum in 7 TeV proton-proton collisions with the ATLAS detector. *JHEP*, **01**, 020. doi:10.1007/JHEP01(2015)020.

[41] Aad, Georges et al. (2015*f*). Measurement of the $WW + WZ$ cross section and limits on anomalous triple gauge couplings using final states with one lepton, missing transverse momentum, and two jets with the ATLAS detector at $\sqrt{s} = 7$ TeV. *JHEP*, **01**, 049. doi:10.1007/JHEP01(2015)049.

[42] Aad, Georges et al. (2015*g*). Measurements of Higgs boson production and couplings in the four-lepton channel in pp collisions at center-of-mass energies of 7 and 8TeV with the ATLAS detector. *Phys. Rev.*, **D91**(1), 012006. doi:10.1103/PhysRevD.91.012006.

[43] Aad, Georges et al. (2015*h*). Measurements of the total and differential Higgs boson production cross sections combining the $H \to \gamma\gamma$ and $H \to ZZ^* \to 4l$ decay channels at $\sqrt{s}$=8TeV with the ATLAS detector. *Phys. Rev. Lett.*, **115**(9), 091801. doi:10.1103/PhysRevLett.115.091801.

[44] Aad, Georges et al. (2015*i*). Measurements of the W production cross sections in association with jets with the ATLAS detector. *Eur. Phys. J.*, **C75**(2), 82. doi:10.1140/epjc/s10052-015-3262-7.

[45] Aad, Georges et al. (2015*j*). Observation and measurement of Higgs boson decays to WW^* with the ATLAS detector. *Phys. Rev.*, **D92**(1), 012006. doi:10.1103/PhysRevD.92.012006.

[46] Aad, Georges et al. (2015*k*). Observation of top-quark pair production in association with a photon and measurement of the $t\bar{t}\gamma$ production cross section in pp collisions at $\sqrt{s} = 7$ TeV using the ATLAS detector. *Phys. Rev.*, **D91**(7), 072007. doi:10.1103/PhysRevD.91.072007.

[47] Aad, Georges et al. (2015*l*). Search for s-channel single top-quark production in proton-proton collisions at $\sqrt{s} = 8$ TeV with the ATLAS detector. *Phys. Lett.*, **B740**, 118–136. doi:10.1016/j.physletb.2014.11.042.

[48] Aad, Georges et al. (2015*m*). Search for the $b\bar{b}$ decay of the Standard Model Higgs boson in associated $(W/Z)H$ production with the ATLAS detector. *JHEP*, **01**, 069. doi:10.1007/JHEP01(2015)069.

[49] Aad, Georges et al. (2015*n*). Study of the spin and parity of the Higgs boson in diboson decays with the ATLAS detector. *Eur. Phys. J.*, **C75**(10), 476. [Erratum: Eur. Phys. J.C76,no.3,152(2016)]. doi:10.1140/epjc/s10052-015-3685-1, 10.1140/epjc/s10052-016-3934-y.

[50] Aad, Georges et al. (2016*a*). Measurement of the charge asymmetry in top-quark pair production in the lepton-plus-jets final state in pp collision data at $\sqrt{s} = 8$ TeV with the ATLAS detector. *Eur. Phys. J.*, **C76**(2), 87. doi:10.1140/epjc/s10052-016-3910-6.

[51] Aad, Georges et al. (2016*b*). Measurement of the inclusive isolated prompt photon cross section in pp collisions at $\sqrt{s} = 8$ TeV with the ATLAS detector. *JHEP*, **08**, 005. doi:10.1007/JHEP08(2016)005.

[52] Aad, Georges et al. (2016*c*). Measurement of the transverse momentum and ϕ_η^* distributions of DrellYan lepton pairs in protonproton collisions at $\sqrt{s} = 8$ TeV with the ATLAS detector. *Eur. Phys. J.*, **C76**(5), 291. doi:10.1140/epjc/s10052-016-4070-4.

[53] Aad, Georges et al. (2016*d*). Measurement of total and differential W^+W^- production cross sections in proton-proton collisions at $\sqrt{s} = 8$ TeV with the ATLAS detector and limits on anomalous triple-gauge-boson couplings. *JHEP*, **09**, 029. doi:10.1007/JHEP09(2016)029.

[54] Aad, Georges et al. (2016*e*). Measurements of the Higgs boson production and decay rates and constraints on its couplings from a combined ATLAS and CMS analysis of the LHC pp collision data at $\sqrt{s} = 7$ and 8 TeV. *JHEP*, **08**, 045. doi:10.1007/JHEP08(2016)045.

[55] Aad, Georges et al. (2016*f*). Measurements of the Higgs boson production and decay rates and coupling strengths using pp collision data at $\sqrt{s} = 7$ and 8 TeV in the ATLAS experiment. *Eur. Phys. J.*, **C76**(1), 6. doi:10.1140/epjc/s10052-015-3769-y.

[56] Aad, Georges et al. (2016*g*). Measurements of top-quark pair differential cross-sections in the lepton+jets channel in *pp* collisions at $\sqrt{s} = 8$ TeV using the ATLAS detector. *Eur. Phys. J.*, **C76**(10), 538. doi:10.1140/epjc/s10052-016-4366-4.

[57] Aaij, Roel et al. (2015). Measurement of the forward Z boson production cross-section in *pp* collisions at $\sqrt{s} = 7$ TeV. *JHEP*, **08**, 039. doi:10.1007/JHEP08(2015)039.

[58] Aaltonen, T. (2009*a*). Measurement of the inclusive isolated prompt photon cross section in $p\bar{p}$ collisions at $\sqrt{s} = 1.96$ TeV using the CDF detector. *Phys. Rev.*, **D80**, 111106. doi:10.1103/PhysRevD.80.111106.

[59] Aaltonen, T. et al. (2008*a*). Measurement of inclusive jet cross-sections in $Z/\gamma^*(\to e^+e^-)$ + jets production in $p\bar{p}$ collisions at $\sqrt{s} = 1.96$-TeV. *Phys. Rev. Lett.*, **100**, 102001. doi:10.1103/PhysRevLett.100.102001.

[60] Aaltonen, T. et al. (2008*b*). Measurement of the cross section for W^- boson production in association with jets in $p\bar{p}$ collisions at $\sqrt{s} = 1.96$-TeV. *Phys. Rev.*, **D77**, 011108. doi:10.1103/PhysRevD.77.011108.

[61] Aaltonen, T. et al. (2008*c*). Measurement of the inclusive jet cross section at the Fermilab Tevatron p anti-p collider using a cone-based jet algorithm. *Phys. Rev.*, **D78**, 052006. doi:10.1103/PhysRevD.79.119902, 10.1103/PhysRevD.78.052006.

[62] Aaltonen, T. et al. (2010). First measurement of the b-jet cross section in events with a W Boson in $p\bar{p}$ collisions at $\sqrt{s} = 1.96$. *Phys. Rev. Lett.*, **104**, 131801. doi:10.1103/PhysRevLett.104.131801.

[63] Aaltonen, T. et al. (2011). Measurement of the cross section for prompt isolated diphoton production in $p\bar{p}$ collisions at $\sqrt{s} = 1.96$ TeV. *Phys. Rev.*, **D84**, 052006. doi:10.1103/PhysRevD.84.052006.

[64] Aaltonen, T. et al. (2012*a*). Study of substructure of high transverse momentum jets produced in proton-antiproton collisions at $\sqrt{s} = 1.96$ TeV. *Phys. Rev.*, **D85**, 091101. doi:10.1103/PhysRevD.85.091101.

[65] Aaltonen, T. et al. (2012*b*). Transverse momentum cross section of e^+e^- pairs in the Z-boson region from $p\bar{p}$ collisions at $\sqrt{s} = 1.96$ TeV. *Phys. Rev.*, **D86**, 052010. doi:10.1103/PhysRevD.86.052010.

[66] Aaltonen, T. et al. (2013a). Higgs boson studies at the Tevatron. *Phys. Rev.*, **D88**(5), 052014. doi:10.1103/PhysRevD.88.052014.

[67] Aaltonen, T. et al. (2013b). Measurement of the cross section for prompt isolated diphoton production using the full CDF Run II data sample. *Phys. Rev. Lett.*, **110**(10), 101801. doi:10.1103/PhysRevLett.110.101801.

[68] Aaltonen, T. et al. (2013c). Measurement of the top quark forward-backward production asymmetry and its dependence on event kinematic properties. *Phys. Rev.*, **D87**, 092002. doi:10.1103/PhysRevD.87.092002.

[69] Aaltonen, T. et al. (2013d). Search for resonant top-antitop production in the lepton plus jets decay mode using the full CDF data set. *Phys. Rev. Lett.*, **110**(12), 121802. doi:10.1103/PhysRevLett.110.121802.

[70] Aaltonen, Timo Antero et al. (2014a). Combination of measurements of the top-quark pair production cross section from the Tevatron Collider. *Phys. Rev.*, **D89**, 072001. doi:10.1103/PhysRevD.89.072001.

[71] Aaltonen, Timo Antero et al. (2014b). Measurement of the inclusive leptonic asymmetry in top-quark pairs that decay to two charged leptons at CDF. *Phys. Rev. Lett.*, **113**, 042001. [Erratum: Phys. Rev. Lett.117,no.19,199901(2016)]. doi:10.1103/PhysRevLett.113.042001, 10.1103/PhysRevLett.117.199901.

[72] Aaltonen, Timo Antero et al. (2015a). Study of the energy dependence of the underlying event in proton-antiproton collisions. *Phys. Rev.*, **D92**(9), 092009. doi:10.1103/PhysRevD.92.092009.

[73] Aaltonen, Timo Antero et al. (2015b). Tevatron combination of single-top-quark cross sections and determination of the magnitude of the Cabibbo-Kobayashi-Maskawa matrix element V_{tb}. *Phys. Rev. Lett.*, **115**(15), 152003. doi:10.1103/PhysRevLett.115.152003.

[74] Aaltonen, T. et al. (2009b, Jun). Measurement of particle production and inclusive differential cross sections in $p\bar{p}$ collisions at $\sqrt{s} = 1.96$ TeV. *Phys. Rev. D*, **79**, 112005. doi:10.1103/PhysRevD.79.112005.

[75] Aaltonen, T. et al. (2010, Dec). Erratum: Measurement of particle production and inclusive differential cross sections in $p\bar{p}$ collisions at $\sqrt{s} = 1.96$ TeV [phys. rev. d **79** , 112005 (2009)]. *Phys. Rev. D*, **82**, 119903. doi:10.1103/PhysRevD.82.119903.

[76] Aaltonen, T. et al. (2015, Jan). Measurement of differential production cross sections for z/γ^* bosons in association with jets in $p\bar{p}$ collisions at $\sqrt{s} = 1.96$ TeV. *Phys. Rev. D*, **91**, 012002. doi:10.1103/PhysRevD.91.012002.

[77] Aaron, F.D. et al. (2010a). Combined measurement and QCD analysis of the inclusive e+- p scattering cross sections at HERA. *JHEP*, **1001**, 109. doi:10.1007/JHEP01(2010)109.

[78] Aaron, F.D. et al. (2010b). Jet production in ep collisions at low Q^2 and determination of α_s. *Eur. Phys. J.*, **C67**, 1–24. doi:10.1140/epjc/s10052-010-1282-x.

[79] Aaron, F. D. et al. (2009). Jet production in ep collisions at high Q^2 and determination of α_s. *Eur. Phys. J.*, **C65**, 363–383. DESY-09-032. doi:10.1140/epjc/s10052-009-1208-7.

[80] Ababekri, Mamut, Dulat, Sayipjamal, Isaacson, Joshua, Schmidt, Carl, and Yuan, C. P. (2016). Implication of CMS data on photon PDFs (arXiv:1603.04874).

[81] Abachi, S. et al. (1995). Observation of the top quark. *Phys. Rev. Lett.*, **74**, 2632–2637. doi:10.1103/PhysRevLett.74.2632.

[82] Abazov, V.M. et al. (2001). The ratio of the isolated photon cross sections at $\sqrt{s}$ = 630 GeV and 1800 GeV. *Phys. Rev. Lett.*, **87**, 251805. doi:10.1103/PhysRevLett.87.251805.

[83] Abazov, V.M. et al. (2008*a*). First measurement of the forward-backward charge asymmetry in top quark pair production. *Phys. Rev. Lett.*, **100**, 142002. doi:10.1103/PhysRevLett.100.142002.

[84] Abazov, V.M. et al. (2008*b*). First study of the radiation-amplitude zero in $W\gamma$ production and limits on anomalous $WW\gamma$ couplings at $\sqrt{s}$ = 1.96- TeV. *Phys. Rev. Lett.*, **100**, 241805. doi:10.1103/PhysRevLett.100.241805.

[85] Abazov, V.M. et al. (2008*c*). Measurement of differential Z/γ^* + jet + X cross sections in $p\bar{p}$ collisions at $\sqrt{s}$ = 1.96-TeV. *Phys. Lett.*, **B669**, 278–286. doi:10.1016/j.physletb.2008.09.060.

[86] Abazov, V.M. et al. (2008*d*). Measurement of the inclusive jet cross-section in $p\bar{p}$ collisions at $s^{(1/2)}$ =1.96-TeV. *Phys. Rev. Lett.*, **101**, 062001. doi:10.1103/PhysRevLett.101.062001.

[87] Abazov, V.M. et al. (2009*a*). Combination of $t\bar{t}$ cross section measurements and constraints on the mass of the top quark and its decays into charged Higgs bosons. *Phys. Rev.*, **D80**, 071102. doi:10.1103/PhysRevD.80.071102.

[88] Abazov, V.M. et al. (2009*b*). Measurements of differential cross sections of Z/γ^* + jets + X events in proton anti-proton collisions at $\sqrt{s}$ = 1.96-TeV. *Phys. Lett.*, **B678**, 45–54. doi:10.1016/j.physletb.2009.05.058.

[89] Abazov, V.M. et al. (2010*a*). Double parton interactions in photon+3 jet events in $p\bar{p}$ collisions $\sqrt{s}$ = 1.96 TeV. *Phys. Rev.*, **D81**, 052012. doi:10.1103/PhysRevD.81.052012.

[90] Abazov, V. et al. (2010*b*). Measurement of direct photon pair production cross sections in $p\bar{p}$ collisions at $\sqrt{s}$ = 1.96 TeV. *Phys. Lett.*, **B690**, 108–117. doi:10.1016/j.physletb.2010.05.017.

[91] Abazov, V.M. et al. (2013*a*). Measurement of the differential cross sections for isolated direct photon pair production in $p\bar{p}$ collisions at $\sqrt{s}$ = 1.96 TeV. *Phys. Lett.*, **B725**, 6–14. doi:10.1016/j.physletb.2013.06.036.

[92] Abazov, V.M. et al. (2013*b*). Measurement of the $p\bar{p} \rightarrow W + b + X$ production cross section at $\sqrt{s}$ = 1.96 TeV. *Phys. Lett.*, **B718**, 1314–1320. doi:10.1016/j.physletb.2012.12.044.

[93] Abazov, V. M. et al. (2006). Measurement of the isolated photon cross section in $p\bar{p}$ collisions at $\sqrt{s}$ = 1.96 TeV. *Phys. Lett.*, **B639**, 151–158. doi:10.1016/j.physletb.2006.04.048.

[94] Abazov, V. M. et al. (2007). Measurement of the shape of the boson rapidity distribution for $p\bar{p} \rightarrow Z/\gamma^* \rightarrow e^+e^-$ + X events produced at $\sqrt{s}$ of 1.96-TeV. *Phys. Rev.*, **D76**, 012003. doi:10.1103/PhysRevD.76.012003.

[95] Abazov, Victor Mukhamedovich et al. (2009*c*). Measurement of the W boson mass. *Phys. Rev. Lett.*, **103**, 141801. doi:10.1103/PhysRevLett.103.141801.

[96] Abazov, Victor Mukhamedovich et al. (2011*a*). Measurements of inclusive W+jets production rates as a function of jet transverse momentum in $p\bar{p}$ collisions at $\sqrt{s} = 1.96$ TeV. *Phys. Lett.*, **B705**, 200–207. doi:10.1016/j.physletb.2011.10.011.

[97] Abazov, Victor Mukhamedovich et al. (2011*b*). $W\gamma$ production and limits on anomalous $WW\gamma$ couplings in $p\bar{p}$ collisions. *Phys. Rev. Lett.*, **107**, 241803. doi:10.1103/PhysRevLett.107.241803.

[98] Abazov, Victor Mukhamedovich et al. (2012). Measurement of the inclusive jet cross section in $p\bar{p}$ collisions at $\sqrt{s} = 1.96$ TeV. *Phys. Rev.*, **D85**, 052006. doi:10.1103/PhysRevD.85.052006.

[99] Abazov, Victor Mukhamedovich et al. (2013*c*). Studies of W boson plus jets production in $p\bar{p}$ collisions at $\sqrt{s} = 1.96$ TeV. *Phys. Rev.*, **D88**(9), 092001. doi:10.1103/PhysRevD.88.092001.

[100] Abazov, Victor Mukhamedovich et al. (2014). Measurement of the forward-backward asymmetry in top quark-antiquark production in pp-bar collisions using the lepton+jets channel. *Phys. Rev.*, **D90**, 072011. doi:10.1103/PhysRevD.90.072011.

[101] Abbiendi, G. et al. (2000). A Study of parton fragmentation in hadronic Z0 decays using Lambda anti-Lambda correlations. *Eur. Phys. J.*, **C13**, 185–195. doi:10.1007/s100520050685.

[102] Abbott, B. et al. (2000). The isolated photon cross-section in $p\bar{p}$ collisions at $\sqrt{s} = 1.8$ TeV. *Phys. Rev. Lett.*, **84**, 2786–2791. doi:10.1103/PhysRevLett.84.2786.

[103] Abbott, B. et al. (2001). Inclusive jet production in $p\bar{p}$ collisions. *Phys. Rev. Lett.*, **86**, 1707–1712. doi:10.1103/PhysRevLett.86.1707.

[104] Abdallah, J. et al. (2006). Evidence for an excess of soft photons in hadronic decays of Z0. *Eur. Phys. J.*, **C47**, 273–294. doi:10.1140/epjc/s2006-02568-8.

[105] Abdallah, J. et al. (2010). Study of the dependence of direct soft photon production on the jet characteristics in hadronic Z^0 decays. *Eur. Phys. J.*, **C67**, 343–366. doi:10.1140/epjc/s10052-010-1315-5.

[106] Abdesselam, A., Kuutmann, E. Bergeaas, Bitenc, U., Brooijmans, G., Butterworth, J. et al. (2011). Boosted objects: A probe of beyond the Standard Model physics. *Eur. Phys. J.*, **C71**, 1661. doi:10.1140/epjc/s10052-011-1661-y.

[107] Abe, F. et al. (1992*a*). Search for squarks and gluinos from $\bar{p}p$ collisions at $\sqrt{s} = 1.8$ TeV. *Phys. Rev. Lett.*, **69**, 3439–3443. doi:10.1103/PhysRevLett.69.3439.

[108] Abe, F. et al. (1992*b*, Mar). Topology of three-jet events in $\bar{p}p$ collisions at $\sqrt{s} = 1.8$ tev. *Phys. Rev. D*, **45**, 1448–1458. doi:10.1103/PhysRevD.45.1448.

[109] Abe, F. et al. (1993). Study of four jet events and evidence for double parton interactions in $p\bar{p}$ collisions at $\sqrt{s} = 1.8$ TeV. *Phys. Rev.*, **D47**, 4857–4871. doi:10.1103/PhysRevD.47.4857.

[110] Abe, F. et al. (1994*a*). A Precision measurement of the prompt photon cross-section in $p\bar{p}$ collisions at $\sqrt{s} = 1.8$ TeV. *Phys. Rev. Lett.*, **73**, 2662–2666. doi:10.1103/PhysRevLett.73.2662.

[111] Abe, F. et al. (1994*b*). Evidence for color coherence in $p\bar{p}$ collisions at $\sqrt{s} = 1.8$ TeV. *Phys. Rev.*, **D50**, 5562–5579. doi:10.1103/PhysRevD.50.5562.

[112] Abe, F. et al. (1995). Observation of top quark production in $\bar{p}p$ collisions. *Phys. Rev. Lett.*, **74**, 2626–2631. doi:10.1103/PhysRevLett.74.2626.

[113] Abe, F. et al. (1997). Measurement of double parton scattering in $\bar{p}p$ collisions at $\sqrt{s} = 1.8$ TeV. *Phys. Rev. Lett.*, **79**, 584–589.

[114] Abelev, Betty et al. (2013). Measurement of inelastic, single- and double-diffraction cross sections in proton-proton collisions at the LHC with ALICE. *Eur. Phys. J.*, **C73**, 2456. doi:10.1140/epjc/s10052-013-2456-0.

[115] Abramowicz, H. et al. (2015). Combination of measurements of inclusive deep inelastic $e^{\pm}p$ scattering cross sections and QCD analysis of HERA data. *Eur. Phys. J.*, **C75**(12), 580. doi:10.1140/epjc/s10052-015-3710-4.

[116] Abulencia, A. et al. (2006a). Measurement of the inclusive jet cross section in $p\bar{p}$ interactions at $\sqrt{s} = 1.96$-TeV using a cone-based jet algorithm. *Phys. Rev.*, **D74**, 071103. doi:10.1103/PhysRevD.74.071103.

[117] Abulencia, A. et al. (2006b). Measurement of the inclusive jet cross section using the k(t) algorithm in p anti-p collisions at $\sqrt{s} = 1.96$-TeV. *Phys. Rev. Lett.*, **96**, 122001. doi:10.1103/PhysRevLett.96.122001.

[118] Abulencia, A. et al. (2006c). Measurement of the $t\bar{t}$ Production Cross Section in $p\bar{p}$ Collisions at $\sqrt{s} = 1.96$ TeV. *Phys. Rev. Lett.*, **97**, 082004. doi:10.1103/PhysRevLett.97.082004.

[119] Abulencia, A. et al. (2007). Measurement of the inclusive jet cross section using the k_T algorithm in $p\bar{p}$ collisions at $\sqrt{s} = 1.96$ TeV with the CDF II detector. *Phys. Rev.*, **D75**, 092006. doi:10.1103/PhysRevD.75.092006.

[120] Acosta, D. et al. (2002). Comparison of the isolated direct photon cross sections in $p\bar{p}$ collisions at $\sqrt{s} = 1.8$-TeV and $\sqrt{s} = 0.63$-TeV. *Phys. Rev.*, **D65**, 112003. doi:10.1103/PhysRevD.65.112003.

[121] Acosta, D. et al. (2004). The underlying event in hard interactions at the Tevatron $\bar{p}p$ collider. *Phys. Rev.*, **D70**, 072002. doi:10.1103/PhysRevD.70.072002.

[122] Acosta, D. et al. (2005). Study of jet shapes in inclusive jet production in $p\bar{p}$ collisions at $\sqrt{s} = 1.96$ TeV. *Phys. Rev.*, **D71**, 112002. doi:10.1103/PhysRevD.71.112002.

[123] Actis, Stefano, Passarino, Giampiero, Sturm, Christian, and Uccirati, Sandro (2008). NLO electroweak corrections to Higgs boson production at hadron colliders. *Phys. Lett.*, **B670**, 12–17. doi:10.1016/j.physletb.2008.10.018.

[124] Adams, D. et al. (2015). Towards an Understanding of the Correlations in Jet Substructure. *Eur. Phys. J.*, **C75**(9), 409. doi:10.1140/epjc/s10052-015-3587-2.

[125] Ade, P. A. R. et al. (2014). Planck 2013 results. XVI. Cosmological parameters. *Astron. Astrophys.*, **571**, A16. doi:10.1051/0004-6361/201321591.

[126] Affolder, T. et al. (2001). Measurement of the inclusive jet cross section in $\bar{p}p$ collisions at $\sqrt{s} = 1.8$ TeV. *Phys. Rev.*, **D64**, 032001. doi:10.1103/PhysRevD.65.039903, 10.1103/PhysRevD.64.032001.

[127] Affolder, T. et al. (2002). Charged jet evolution and the underlying event in proton-antiproton collisions at 1.8 TeV. *Phys. Rev.*, **D65**, 092002.

[128] Aguilar-Saavedra, J.A., Juste, A., and Rubbo, F. (2012). Boosting the t-tbar charge asymmetry. *Phys. Lett.*, **B707**, 92–98. doi:10.1016/j.physletb.2011.12.007.

[129] Ahmed, Taushif, Mandal, M. K., Rana, Narayan, and Ravindran, V. (2014). Rapidity distributions in Drell-Yan and Higgs productions at threshold to third order in QCD. *Phys. Rev. Lett.*, **113**, 212003. doi:10.1103/PhysRevLett.113.212003.

[130] Ahrens, Valentin, Becher, Thomas, Neubert, Matthias, and Yang, Li Lin (2009). Origin of the large perturbative corrections to Higgs production at hadron colliders. *Phys. Rev.*, **D79**, 033013. doi:10.1103/PhysRevD.79.033013.

[131] Aivazis, M. A. G., Collins, John C., Olness, Fredrick I., and Tung, Wu-Ki (1994). Leptoproduction of heavy quarks. II. A unified QCD formulation of charged and neutral current processes from fixed-target to collider energies. *Phys. Rev.*, **D50**, 3102–3118. doi:10.1103/PhysRevD.50.3102.

[132] Akrawy, M.Z. et al. (1990). A Study of coherence of soft gluons in hadron jets. *Phys. Lett.*, **B247**, 617–628. doi:10.1016/0370-2693(90)91911-T.

[133] Albrow, Michael G. et al. (2006). Tevatron-for-LHC Report of the QCD Working Group (hep-ph/0610012).

[134] Alcaraz Maestre, J. et al. (2012). The SM and NLO Multileg and SM MC Working Groups: Summary Report (arXiv:1203.6803).

[135] Alekhin, Sergey et al. (2011). The PDF4LHC Working Group interim report.

[136] Alekhin, S., Blumlein, J., and Moch, S. (2014). The ABM parton distributions tuned to LHC data. *Phys. Rev.*, **D89**(5), 054028. doi:10.1103/PhysRevD.89.054028.

[137] Ali, Ahmed, Pietarinen, E., Kramer, G., and Willrodt, J. (1980). A QCD analysis of the high-energy e^+e^- data from PETRA. *Phys. Lett.*, **B93**, 155. doi:10.1016/0370-2693(80)90116-1.

[138] Alioli, S., Badger, S., Bellm, J., Biedermann, B., Boudjema, F. et al. (2014). Update of the Binoth Les Houches Accord for a standard interface between Monte Carlo tools and one-loop programs. *Comput. Phys. Commun.*, **185**, 560–571. doi:10.1016/j.cpc.2013.10.020.

[139] Alioli, Simone, Nason, Paolo, Oleari, Carlo, and Re, Emanuele (2009). NLO Higgs boson production via gluon fusion matched with shower in POWHEG. *JHEP*, **04**, 002. doi:10.1088/1126-6708/2009/04/002.

[140] Alitti, J. et al. (1991). A Determination of the strong coupling constant α_s from W production at the CERN p anti-p collider. *Phys. Lett.*, **B263**, 563–572. doi:10.1016/0370-2693(91)90505-K.

[141] Almeida, Leandro G., Sterman, George F., and Vogelsang, Werner (2008). Threshold resummation for the top quark charge asymmetry. *Phys. Rev.*, **D78**, 014008. doi:10.1103/PhysRevD.78.014008.

[142] Altarelli, Guido, Ellis, R. Keith, and Martinelli, G. (1979). Large perturbative corrections to the Drell-Yan process in QCD. *Nucl. Phys.*, **B157**, 461. doi:10.1016/0550-3213(79)90116-0.

[143] Altarelli, Guido and Parisi, G. (1977). Asymptotic freedom in parton language. *Nucl. Phys.*, **B126**, 298–318.

[144] Altheimer, A., Arora, S., Asquith, L., Brooijmans, G., Butterworth, J. et al. (2012). Jet substructure at the Tevatron and LHC: New results, new tools, new benchmarks. *J. Phys.*, **G39**, 063001. doi:10.1088/0954-3899/39/6/063001.

[145] Alwall, J. et al. (2007*a*). A standard format for Les Houches Event Files. *Comput. Phys. Commun.*, **176**, 300–304.

[146] Alwall, Johan et al. (2007*b*). MadGraph/MadEvent v4: The new web generation. *JHEP*, **09**, 028.

[147] Alwall, J. et al. (2008). Comparative study of various algorithms for the merging of parton showers and matrix elements in hadronic collisions. *Eur. Phys. J.*, **C53**, 473–500.

[148] Alwall, J., Frederix, R., Frixione, S., Hirschi, V., Maltoni, F., Mattelaer, O., Shao, H. S., Stelzer, T., Torrielli, P., and Zaro, M. (2014). The automated computation of tree-level and next-to-leading order differential cross sections, and their matching to parton shower simulations. *JHEP*, **07**, 079. doi:10.1007/JHEP07(2014)079.

[149] Alwall, Johan, Frederix, R., Gerard, J. M., Giammanco, A., Herquet, M., Kalinin, S., Kou, E., Lemaitre, V., and Maltoni, F. (2007*c*). Is V(t_b) $\simeq$ 1? *Eur. Phys. J.*, **C49**, 791–801. doi:10.1140/epjc/s10052-006-0137-y.

[150] Alwall, Johan, Herquet, Michel, Maltoni, Fabio, Mattelaer, Olivier, and Stelzer, Tim (2011). MadGraph 5 : Going Beyond. *JHEP*, **1106**, 128. doi:10.1007/JHEP06(2011)128.

[151] Amati, D. and Veneziano, G. (1979). Preconfinement as a property of perturbative QCD. *Phys. Lett.*, **B83**, 87. doi:10.1016/0370-2693(79)90896-7.

[152] Ametller, L., Gava, E., Paver, N., and Treleani, D. (1985). Role of the QCD induced gluon - gluon coupling to gauge boson pairs in the multi - TeV region. *Phys. Rev.*, **D32**, 1699. doi:10.1103/PhysRevD.32.1699.

[153] Ametller, L., Paver, N., and Treleani, D. (1986). Possible signature of multiple parton interactions in collider four jet events. *Phys. Lett.*, **B169**, 289. doi:10.1016/0370-2693(86)90668-4.

[154] Anastasiou, Charalampos, Dixon, Lance J., Melnikov, Kirill, and Petriello, Frank (2004*a*). High precision QCD at hadron colliders: Electroweak gauge boson rapidity distributions at NNLO. *Phys. Rev.*, **D69**, 094008. doi:10.1103/PhysRevD.69.094008.

[155] Anastasiou, Charalampos, Duhr, Claude, Dulat, Falko, Furlan, Elisabetta, Gehrmann, Thomas et al. (2014). Higgs boson gluon fusion production at threshold in N^3LO QCD. *Phys. Lett.*, **B737**, 325–328. doi:10.1016/j.physletb.2014.08.067.

[156] Anastasiou, Charalampos, Duhr, Claude, Dulat, Falko, Furlan, Elisabetta, Gehrmann, Thomas, Herzog, Franz, Lazopoulos, Achilleas, and Mistlberger, Bernhard (2016). High precision determination of the gluon fusion Higgs boson cross-section at the LHC. *JHEP*, **05**, 058. doi:10.1007/JHEP05(2016)058.

[157] Anastasiou, Charalampos, Duhr, Claude, Dulat, Falko, Herzog, Franz, and Mistlberger, Bernhard (2015). Higgs boson gluon-fusion production in QCD at three loops. *Phys. Rev. Lett.*, **114**(21), 212001. doi:10.1103/PhysRevLett.114.212001.

[158] Anastasiou, Charalampos and Melnikov, Kirill (2002). Higgs boson production at hadron colliders in NNLO QCD. *Nucl. Phys.*, **B646**, 220–256. doi:10.1016/S0550-3213(02)00837-4.

[159] Anastasiou, Charalampos, Melnikov, Kirill, and Petriello, Frank (2004*b*). A new method for real radiation at NNLO. *Phys. Rev.*, **D69**, 076010.

doi:10.1103/PhysRevD.69.076010.

[160] Anastasiou, Charalampos, Melnikov, Kirill, and Petriello, Frank (2004c). Higgs boson production at hadron colliders: Differential cross sections through next–to–next–to–leading order. *Phys. Rev. Lett.*, **93**, 262002. doi:10.1103/PhysRevLett.93.262002.

[161] Andersen, J. R. et al. (2016). Les Houches 2015: Physics at TeV Colliders Standard Model Working Group Report. In *9th Les Houches Workshop on Physics at TeV Colliders (PhysTeV 2015) Les Houches, France, June 1-19, 2015*.

[162] Andersson, Bo (1997). *The Lund model.* Volume 7. Camb. Monogr. Part. Phys. Nucl. Phys. Cosmol.

[163] Andersson, Bo, Gustafson, G., Ingelman, G., and Sjöstrand, T. (1983). Parton fragmentation and string dynamics. *Phys. Rept.*, **97**, 31–145. doi:10.1016/0370-1573(83)90080-7.

[164] Andersson, Bo, Gustafson, Gosta, Lönnblad, Leif, and Pettersson, Ulf (1989). Coherence effects in deep–inclastic scattering. *Z. Phys.*, **C43**, 625.

[165] Andersson, Bo, Gustafson, G., and Sjöstrand, T. (1985). Baryon production in jet fragmentation and Upsilon-decay. *Phys. Scripta*, **32**, 574. doi:10.1088/0031-8949/32/6/003.

[166] Antchev, G. et al. (2013a). Luminosity-independent measurement of the proton-proton total cross section at $\sqrt{s} = 8$ TeV. *Phys. Rev. Lett.*, **111**(1), 012001. doi:10.1103/PhysRevLett.111.012001.

[167] Antchev, G. et al. (2013b). Luminosity-independent measurements of total, elastic and inelastic cross-sections at $\sqrt{s} = 7$ TeV. *Europhys. Lett.*, **101**, 21004. doi:10.1209/0295-5075/101/21004.

[168] Appell, David, Sterman, George F., and Mackenzie, Paul B. (1988). Soft gluons and the normalization of the Drell-Yan cross-section. *Nucl. Phys.*, **B309**, 259. doi:10.1016/0550-3213(88)90082-X.

[169] Arneodo, M. et al. (1997a). Accurate measurement of F2(d) / F2(p) and $R^d - R^p$. *Nucl. Phys.*, **B487**, 3–26. doi:10.1016/S0550-3213(96)00673-6.

[170] Arneodo, M. et al. (1997b). Measurement of the proton and deuteron structure functions, F2(p) and F2(d), and of the ratio sigma-L / sigma-T. *Nucl. Phys.*, **B483**, 3–43. doi:10.1016/S0550-3213(96)00538-X.

[171] Arnold, K., Bahr, M., Bozzi, Giuseppe, Campanario, F., Englert, C. et al. (2009). VBFNLO: A parton level Monte Carlo for processes with electroweak bosons. *Comput. Phys. Commun.*, **180**, 1661–1670. doi:10.1016/j.cpc.2009.03.006.

[172] Artru, X. and Mennessier, G. (1974). String model and multiproduction. *Nucl. Phys.*, **B70**, 93–115.

[173] ATLAS (2012). Measurement of the inclusive jet cross section in pp collisions at $\sqrt{s} = 2.76$TeV and comparison to the inclusive jet cross section at $\sqrt{s} = 7$TeV using the ATLAS detector (ATLAS-CONF-2012-128, ATLAS-COM-CONF-2012-173).

[174] ATLAS (2013). Measurement of multi-jet cross-section ratios and determination of the strong coupling constant in proton-proton collisions at $\sqrt{s}$=7 TeV with the ATLAS detector (ATLAS-CONF-2013-041).

[175] ATLAS (2015, Jul). Measurement of the $t\bar{t}W$ and $t\bar{t}Z$ production cross sections in pp collisions at $\sqrt{s} = 8$ TeV with the ATLAS detector. Technical Report ATLAS-CONF-2015-032, CERN, Geneva.

[176] Azimov, Yakov I., Dokshitzer, Yuri L., Khoze, Valery A., and Troian, S.I. (1985a). The string effect and QCD coherence. *Phys. Lett.*, **B165**, 147–150. doi:10.1016/0370-2693(85)90709-9.

[177] Azimov, Yakov I., Dokshitzer, Yuri L., Khoze, Valery A., and Troyan, S.I. (1985b). Similarity of parton and hadron Spectra in QCD jets. *Z. Phys.*, **C27**, 65–72. doi:10.1007/BF01642482.

[178] Azimov, Yakov I., Dokshitzer, Yuri L., Khoze, Valery A., and Troyan, S.I. (1986). Hump–backed QCD plateau in hadron spectra. *Z. Phys.*, **C31**, 213. doi:10.1007/BF01479529.

[179] Baak, M., Cth, J., Haller, J., Hoecker, A., Kogler, R., Mnig, K., Schott, M., and Stelzer, J. (2014). The global electroweak fit at NNLO and prospects for the LHC and ILC. *Eur. Phys. J.*, **C74**, 3046. doi:10.1140/epjc/s10052-014-3046-5.

[180] Badger, Simon, Biedermann, Benedikt, Uwer, Peter, and Yundin, Valery (2013a). NLO QCD corrections to multi-jet production at the LHC with a centre-of-mass energy of $\sqrt{s} = 8$ TeV. *Phys. Lett.*, **B718**, 965–978. doi:10.1016/j.physletb.2012.11.029.

[181] Badger, Simon, Biedermann, Benedikt, Uwer, Peter, and Yundin, Valery (2013b). Numerical evaluation of virtual corrections to multi-jet production in massless QCD. *Comput. Phys. Commun.*, **184**, 1981–1998. doi:10.1016/j.cpc.2013.03.018.

[182] Badger, Simon, Biedermann, Benedikt, Uwer, Peter, and Yundin, Valery (2014a). Computation of multi-leg amplitudes with NJet. *J. Phys. Conf. Ser.*, **523**, 012057. doi:10.1088/1742-6596/523/1/012057.

[183] Badger, Simon, Biedermann, Benedikt, Uwer, Peter, and Yundin, Valery (2014b). Next-to-leading order QCD corrections to five jet production at the LHC. *Phys. Rev.*, **D89**, 034019. doi:10.1103/PhysRevD.89.034019.

[184] Badger, S.D. and Glover, E.W. Nigel (2004). Two loop splitting functions in QCD. *JHEP*, **0407**, 040. doi:10.1088/1126-6708/2004/07/040.

[185] Badger, Simon, Guffanti, Alberto, and Yundin, Valery (2014c). Next-to-leading order QCD corrections to di-photon production in association with up to three jets at the Large Hadron Collider. *JHEP*, **1403**, 122. doi:10.1007/JHEP03(2014)122.

[186] Badger, Simon, Sattler, Ralf, and Yundin, Valery (2011). One-loop helicity amplitudes for $t\bar{t}$ production at hadron colliders. *Phys. Rev.*, **D83**, 074020. doi:10.1103/PhysRevD.83.074020.

[187] Baglio, J., Bellm, J., Campanario, F., Feigl, B., Frank, J. et al. (2014). Release Note - VBFNLO 2.7.0 (arXiv:1404.3940).

[188] Bähr, M. et al. (2008). Herwig++ Physics and Manual. *Eur. Phys. J.*, **C58**, 639–707. doi:10.1140/epjc/s10052-008-0798-9.

[189] Balitsky, I. I. and Lipatov, L. N. (1978). The Pomeranchuk singularity in quantum chromodynamics. *Sov. J. Nucl. Phys.*, **28**, 822–829.

[190] Ball, Richard D. et al. (2009). A determination of parton distributions with faithful uncertainty estimation. *Nucl. Phys.*, **B809**, 1–63.

doi:10.1016/j.nuclphysb.2008.09.037.

[191] Ball, Richard D. et al. (2012). Unbiased global determination of parton distributions and their uncertainties at NNLO, NLO, and at LO. *Nucl. Phys.*, **B855**, 153–221. doi:10.1016/j.nuclphysb.2011.09.024.

[192] Ball, Richard D. et al. (2013*a*). Parton distributions with LHC data. *Nucl. Phys.*, **B867**, 244–289. doi:10.1016/j.nuclphysb.2012.10.003.

[193] Ball, Richard D. et al. (2013*b*). Parton distributions with QED corrections. *Nucl. Phys.*, **B877**, 290–320. doi:10.1016/j.nuclphysb.2013.10.010.

[194] Ball, Richard D. et al. (2015). Parton distributions for the LHC Run II. *JHEP*, **04**, 040. doi:10.1007/JHEP04(2015)040.

[195] Ball, Richard D., Bonvini, Marco, Forte, Stefano, Marzani, Simone, and Ridolfi, Giovanni (2013*c*). Higgs production in gluon fusion beyond NNLO. *Nucl. Phys.*, **B874**, 746–772. doi:10.1016/j.nuclphysb.2013.06.012.

[196] Ball, Richard D., Carrazza, Stefano, Del Debbio, Luigi, Forte, Stefano, Gao, Jun et al. (2013*d*). Parton distribution benchmarking with LHC data. *JHEP*, **1304**, 125. doi:10.1007/JHEP04(2013)125.

[197] Ballestrero, Alessandro, Maina, Ezio, and Moretti, Stefano (1994). Heavy quarks and leptons at e^+e^- colliders. *Nucl. Phys.*, **B415**, 265–292. doi:10.1016/0550-3213(94)90112-0.

[198] Banfi, Andrea, Caola, Fabrizio, Dreyer, Frdric A., Monni, Pier F., Salam, Gavin P., Zanderighi, Giulia, and Dulat, Falko (2016). Jet-vetoed Higgs cross section in gluon fusion at $N^3LO+NNLL$ with small-R resummation. *JHEP*, **04**, 049. doi:10.1007/JHEP04(2016)049.

[199] Banfi, Andrea, Monni, Pier Francesco, Salam, Gavin P., and Zanderighi, Giulia (2012). Higgs and Z-boson production with a jet veto. *Phys. Rev. Lett.*, **109**, 202001. doi:10.1103/PhysRevLett.109.202001.

[200] Banfi, Andrea, Salam, Gavin P., and Zanderighi, Giulia (2007). Accurate QCD predictions for heavy-quark jets at the Tevatron and LHC. *JHEP*, **07**, 026. doi:10.1088/1126-6708/2007/07/026.

[201] Bardeen, John, Cooper, L. N., and Schrieffer, J. R. (1957*a*). Microscopic theory of superconductivity. *Phys. Rev.*, **106**, 162. doi:10.1103/PhysRev.106.162.

[202] Bardeen, John, Cooper, L. N., and Schrieffer, J. R. (1957*b*). Theory of superconductivity. *Phys. Rev.*, **108**, 1175–1204. doi:10.1103/PhysRev.108.1175.

[203] Bartel, W. et al. (1983). Particle distribution in three jet events produced by e+ e- annihilation. *Z. Phys.*, **C21**, 37. doi:10.1007/BF01648774.

[204] Bartel, W. et al. (1986). Experimental studies on multijet production in e^+ e^- annihilation at PETRA energies. *Z. Phys.*, **C33**, 23. doi:10.1007/BF01410449.

[205] Bauer, Christian W. and Lange, Bjorn O. (2009). Scale setting and resummation of logarithms in $pp \rightarrow V+$ jets (arXiv:0905.4739).

[206] Baur, U. (2007). Weak boson emission in hadron collider processes. *Phys. Rev.*, **D75**, 013005. doi:10.1103/PhysRevD.75.013005.

[207] Becattini, F. and Passaleva, G. (2002). Statistical hadronization model and transverse momentum spectra of hadrons in high-energy collisions. *Eur. Phys. J.*, **C23**, 551–583. doi:10.1007/s100520100869.

[208] Becher, Thomas, Lorentzen, Christian, and Schwartz, Matthew D. (2012). Precision direct photon and W-boson spectra at high p_T and comparison to LHC data. *Phys. Rev.*, **D86**, 054026. doi:10.1103/PhysRevD.86.054026.

[209] Beenakker, W., Dittmaier, S., Kramer, M., Plumper, B., Spira, M. et al. (2003). NLO QCD corrections to $t\bar{t}H$ production in hadron collisions. *Nucl. Phys.*, **B653**, 151–203. doi:10.1016/S0550-3213(03)00044-0.

[210] Belyaev, Alexander, Christensen, Neil D., and Pukhov, Alexander (2013). CalcHEP 3.4 for collider physics within and beyond the Standard Model. *Comput. Phys. Commun.*, **184**, 1729–1769. doi:10.1016/j.cpc.2013.01.014.

[211] Bena, Iosif, Bern, Zvi, and Kosower, David A. (2005). Twistor-space recursive formulation of gauge-theory amplitudes. *Phys. Rev.*, **D71**, 045008.

[212] Beneke, M., Buchalla, G., Neubert, M., and Sachrajda, Christopher T. (1999). QCD factorization for $B \to \pi\pi$ decays: Strong phases and CP violation in the heavy quark limit. *Phys. Rev. Lett.*, **83**, 1914–1917. doi:10.1103/PhysRevLett.83.1914.

[213] Beneke, M., Buchalla, G., Neubert, M., and Sachrajda, Christopher T. (2000). QCD factorization for exclusive, nonleptonic B meson decays: General arguments and the case of heavy light final states. *Nucl. Phys.*, **B591**, 313–418. doi:10.1016/S0550-3213(00)00559-9.

[214] Bengtsson, Mats and Sjostrand, Torbjorn (1987). A comparative study of coherent and non-coherent parton shower evolution. *Nucl. Phys.*, **B289**, 810–846. doi:10.1016/0550-3213(87)90407-X.

[215] Bengtsson, Mats and Sjöstrand, Torbjorn (1987). Coherent parton showers versus matrix elements: Implications of PETRA - PEP data. *Phys. Lett.*, **B185**, 435. doi:10.1016/0370-2693(87)91031-8.

[216] Bengtsson, Mats, Sjöstrand, Torbjörn, and van Zijl, Maria (1986). Initial state radiation effects on W and jet production. *Z. Phys.*, **C32**, 67. doi:10.1007/BF01441353.

[217] Benvenuti, A.C. et al. (1989). A High statistics measurement of the proton structure functions $F_2(x, Q^2$ and R from deep inelastic muon scattering at High Q^2. *Phys. Lett.*, **B223**, 485. doi:10.1016/0370-2693(89)91637-7.

[218] Berends, Frits A., Gaemers, K. J. F., and Gastmans, R. (1973). α^3 Contribution to the angular asymmetry in $e^+e^- \to \mu^+\mu^-$. *Nucl. Phys.*, **B63**, 381–397. doi:10.1016/0550-3213(73)90153-3.

[219] Berends, Frits A. and Giele, W. (1987). The six gluon process as an example of Weyl-Van Der Waerden spinor calculus. *Nucl. Phys.*, **B294**, 700–732. doi:10.1016/0550-3213(87)90604-3.

[220] Berends, Frits A. and Giele, W.T. (1988). Recursive calculations for processes with n gluons. *Nucl. Phys.*, **B306**, 759. doi:10.1016/0550-3213(88)90442-7.

[221] Berends, Frits A., Giele, W.T., and Kuijf, H. (1989a). Exact expressions for processes involving a vector boson and up to five partons. *Nucl. Phys.*, **B321**, 39. doi:10.1016/0550-3213(89)90242-3.

[222] Berends, Frits A., Giele, W. T., and Kuijf, H. (1989b). On six-jet production at hadron colliders. *Phys. Lett.*, **B232**, 266–270. doi:10.1016/0370-2693(89)91699-7.

[223] Berends, Frits A., Giele, W. T., Kuijf, H., Kleiss, R., and Stirling, W. James (1989c). Multi-jet production in W, Z events at $p\bar{p}$ colliders. *Phys. Lett.*, **B224**, 237. doi:10.1016/0370-2693(89)91081-2.

[224] Berends, Frits A., Kuijf, H., Tausk, B., and Giele, W.T. (1991). On the production of a W and jets at hadron colliders. *Nucl. Phys.*, **B357**, 32–64. doi:10.1016/0550-3213(91)90458-A.

[225] Berger, C.F., Bern, Z., Dixon, Lance J., Febres Cordero, Fernando, Forde, D. et al. (2009). Next-to-leading order QCD predictions for W+3-jet distributions at hadron colliders. *Phys. Rev.*, **D80**, 074036. doi:10.1103/PhysRevD.80.074036.

[226] Berger, C. F. et al. (2008). Automated implementation of on-shell methods for one-loop amplitudes. *Phys. Rev.*, **D78**, 036003. doi:10.1103/PhysRevD.78.036003.

[227] Berger, C. F. et al. (2010). Next-to-leading order QCD predictions for Z, γ^*+3-Jet distributions at the Tevatron. *Phys. Rev.*, **D82**, 074002. doi:10.1103/PhysRevD.82.074002.

[228] Berger, Edmond L. and Campbell, John M. (2004). Higgs boson production in weak boson fusion at next-to-leading order. *Phys. Rev.*, **D70**, 073011. doi:10.1103/PhysRevD.70.073011.

[229] Beringer, J. et al. (2012). Review of Particle Physics (RPP). *Phys. Rev.*, **D86**, 010001. doi:10.1103/PhysRevD.86.010001.

[230] Bern, Z., De Freitas, A., and Dixon, Lance J. (2001). Two loop amplitudes for gluon fusion into two photons. *JHEP*, **09**, 037. doi:10.1088/1126-6708/2001/09/037.

[231] Bern, Z., Diana, G., Dixon, L.J., Febres Cordero, F., Hoeche, S. et al. (2012). Four-jet production at the Large Hadron Collider at next-to-leading order in QCD. *Phys. Rev. Lett.*, **109**, 042001. doi:10.1103/PhysRevLett.109.042001.

[232] Bern, Z., Dixon, L.J., Febres Cordero, F., Hoeche, S., Ita, H. et al. (2013). Next-to-leading order W + 5-jet production at the LHC. *Phys. Rev.*, **D88**, 014025. doi:10.1103/PhysRevD.88.014025.

[233] Bern, Z., Dixon, L.J., Febres Cordero, F., Hoeche, S., Ita, H. et al. (2014). Ntuples for NLO Events at hadron colliders. *Comput. Phys. Commun.*, **185**, 1443–1460. doi:10.1016/j.cpc.2014.01.011.

[234] Bern, Zvi, Dixon, Lance J., and Kosower, David A. (1998). One-loop amplitudes for $e^+e^- \to$ four partons. *Nucl. Phys.*, **B513**, 3–86. doi:10.1016/S0550-3213(97)00703-7.

[235] Bern, Zvi, Dixon, Lance J., and Schmidt, Carl (2002). Isolating a light Higgs boson from the diphoton background at the CERN LHC. *Phys. Rev.*, **D66**, 074018. doi:10.1103/PhysRevD.66.074018.

[236] Bernreuther, W., Brandenburg, A., Si, Z.G., and Uwer, P. (2004). Top quark pair production and decay at hadron colliders. *Nucl. Phys.*, **B690**, 81–137. doi:10.1016/j.nuclphysb.2004.04.019.

[237] Bernreuther, Werner and Si, Zong-Guo (2010). Distributions and correlations for top quark pair production and decay at the Tevatron and LHC. *Nucl. Phys.*, **B837**, 90–121. doi:10.1016/j.nuclphysb.2010.05.001.

[238] Bertone, Valerio, Carrazza, Stefano, and Rojo, Juan (2014). APFEL: A PDF evolution library with QED corrections. *Comput. Phys. Commun.*, **185**, 1647–1668. doi:10.1016/j.cpc.2014.03.007.

[239] Bethke, Siegfried (2013). World summary of α_s (2012). *Nucl. Phys.Proc.Suppl.*, **234**, 229–234. doi:10.1016/j.nuclphysbps.2012.12.020.

[240] Bethke, Siegfried, Dissertori, Gunther, and Salam, Gavin P. (2016). World Summary of α_s (2015). *EPJ Web Conf.*, **120**, 07005. doi:10.1051/epjconf/201612007005.

[241] Bevilacqua, G., Czakon, M., Garzelli, M.V., van Hameren, A., Kardos, A. et al. (2013). HELAC-NLO. *Comput. Phys. Commun.*, **184**, 986–997. doi:10.1016/j.cpc.2012.10.033.

[242] Bevilacqua, G., Czakon, M., Papadopoulos, C.G., Pittau, R., and Worek, M. (2009). Assault on the NLO Wishlist: $pp \to t\bar{t}b\bar{b}$. *JHEP*, **0909**, 109. doi:10.1088/1126-6708/2009/09/109.

[243] Bevilacqua, G., Czakon, M., Papadopoulos, C.G., and Worek, M. (2010). Dominant QCD backgrounds in Higgs boson analyses at the LHC: A Study of $pp \to t\bar{t} + 2$ jets at next-to-leading order. *Phys. Rev. Lett.*, **104**, 162002. doi:10.1103/PhysRevLett.104.162002.

[244] Bevilacqua, G., Czakon, M., Papadopoulos, C.G., and Worek, M. (2011*a*). Hadronic top-quark pair production in association with two jets at next-to-leading order QCD. *Phys. Rev.*, **D84**, 114017. doi:10.1103/PhysRevD.84.114017.

[245] Bevilacqua, Giuseppe, Czakon, Michal, van Hameren, Andreas, Papadopoulos, Costas G., and Worek, Malgorzata (2011*b*). Complete off-shell effects in top quark pair hadroproduction with leptonic decay at next-to-leading order. *JHEP*, **1102**, 083. doi:10.1007/JHEP02(2011)083.

[246] Bevilacqua, G. and Worek, M. (2012). Constraining BSM Physics at the LHC: Four top final states with NLO accuracy in perturbative QCD. *JHEP*, **07**, 111. doi:10.1007/JHEP07(2012)111.

[247] Bigi, Ikaros I.Y., Shifman, Mikhail A., Uraltsev, N.G., and Vainshtein, Arkady I. (1993). QCD predictions for lepton spectra in inclusive heavy flavor decays. *Phys. Rev. Lett.*, **71**, 496–499. doi:10.1103/PhysRevLett.71.496.

[248] Bigi, Ikaros I.Y., Uraltsev, N.G., and Vainshtein, A.I. (1992). Non-perturbative corrections to inclusive beauty and charm decays: QCD versus phenomenological models. *Phys. Lett.*, **B293**, 430–436. doi:10.1016/0370-2693(92)90908-M.

[249] Binosi, D., Collins, J., Kaufhold, C., and Theussl, L. (2009). Jaxodraw: A graphical user interface for drawing feynman diagrams. version 2.0 release notes. *Computer Physics Communications*, **180**(9), 1709 – 1715. doi:http://dx.doi.org/10.1016/j.cpc.2009.02.020.

[250] Binoth, T. et al. (2010*a*). A proposal for a standard interface between Monte Carlo tools and one-loop programs. *Comput. Phys. Commun.*, **181**, 1612–1622. doi:10.1016/j.cpc.2010.05.016.

[251] Binoth, T et al. (2010*b*). The SM and NLO multileg working group: Summary report. Proceedings of the Workshop "Physics at TeV Colliders", Les Houches, France, 8-26 June, 2009.

[252] Binoth, T., Gleisberg, T., Karg, S., Kauer, N., and Sanguinetti, G. (2010c). NLO QCD corrections to $ZZ+$ jet production at hadron colliders. *Phys. Lett.*, **B683**, 154–159. doi:10.1016/j.physletb.2009.12.013.

[253] Binoth, T., Greiner, N., Guffanti, A., Reuter, J., Guillet, J.-Ph. et al. (2010d). Next-to-leading order QCD corrections to $pp \to b\bar{b}b\bar{b} + X$ at the LHC: the quark induced case. *Phys. Lett.*, **B685**, 293–296. doi:10.1016/j.physletb.2010.02.010.

[254] Binoth, T., Guillet, J.P., Pilon, E., and Werlen, M. (2000). A Full next-to-leading order study of direct photon pair production in hadronic collisions. *Eur. Phys. J.*, **C16**, 311–330. doi:10.1007/s100520050024.

[255] Binoth, T., Guillet, J. Ph., Pilon, E., and Werlen, M. (2001). Beyond leading order effects in photon pair production at the Tevatron. *Phys. Rev.*, **D63**, 114016. doi:10.1103/PhysRevD.63.114016.

[256] Biswas, Sandip, Melnikov, Kirill, and Schulze, Markus (2010). Next-to-leading order QCD effects and the top quark mass measurements at the LHC. *JHEP*, **1008**, 048. doi:10.1007/JHEP08(2010)048.

[257] Bloch, F. and Nordsieck, A. (1937). Note on the radiation field of the electron. *Phys. Rev.*, **52**, 54–59. doi:10.1103/PhysRev.52.54.

[258] Blok, B., Dokshitser, Yu., Frankfurt, L., and Strikman, M. (2012). pQCD physics of multiparton interactions. *Eur. Phys. J.*, **C72**, 1963. doi:10.1140/epjc/s10052-012-1963-8.

[259] Blok, B., Dokshitzer, Yu., Frankfurt, L., and Strikman, M. (2011). The Four jet production at LHC and Tevatron in QCD. *Phys. Rev.*, **D83**, 071501. doi:10.1103/PhysRevD.83.071501.

[260] Blok, B., Dokshitzer, Yu., Frankfurt, L., and Strikman, M. (2014). Perturbative QCD correlations in multi-parton collisions. *Eur. Phys. J.*, **C74**, 2926. doi:10.1140/epjc/s10052-014-2926-z.

[261] Blok, B., Koyrakh, L., Shifman, Mikhail A., and Vainshtein, A.I. (1994). Differential distributions in semileptonic decays of the heavy flavors in QCD. *Phys. Rev.*, **D49**, 3356. doi:10.1103/PhysRevD.50.3572, 10.1103/PhysRevD.49.3356.

[262] Blumlein, J., Riemersma, S., Botje, M., Pascaud, C., Zomer, F. et al. (1996). A Detailed comparison of NLO QCD evolution codes.

[263] Bolzoni, Paolo, Maltoni, Fabio, Moch, Sven-Olaf, and Zaro, Marco (2012). Vector boson fusion at NNLO in QCD: SM Higgs and beyond. *Phys. Rev.*, **D85**, 035002. doi:10.1103/PhysRevD.85.035002.

[264] Bolzoni, Paolo, Zaro, Marco, Maltoni, Fabio, and Moch, Sven-Olaf (2010). Higgs production at NNLO in QCD: The VBF channel. *Nucl. Phys.Proc.Suppl.*, **205-206**, 314–319. doi:10.1016/j.nuclphysbps.2010.09.012.

[265] Bonciani, Roberto, Catani, Stefano, Grazzini, Massimiliano, Sargsyan, Hayk, and Torre, Alessandro (2015). The q_T subtraction method for top quark production at hadron colliders. *Eur. Phys. J.*, **C75**(12), 581. doi:10.1140/epjc/s10052-015-3793-y.

[266] Boos, E. et al. (2004). CompHEP 4.4 - automatic computations from Lagrangians to events. *Nucl. Instrum. Meth.*, **A534**, 250–259.

doi:10.1016/j.nima.2004.07.096.

[267] Bothmann, Enrico, Ferrarese, Piero, Krauss, Frank, Kuttimalai, Silvan, Schumann, Steffen, and Thompson, Jennifer (2016). Aspects of perturbative QCD at a 100 TeV future hadron collider. *Phys. Rev.*, **D94**(3), 034007. doi:10.1103/PhysRevD.94.034007.

[268] Botje, Michiel et al. (2011). The PDF4LHC Working Group Interim Recommendations (arXiv:1101.0538).

[269] Boughezal, Radja, Campbell, John M., Ellis, R. Keith, Focke, Christfried, Giele, Walter T., Liu, Xiaohui, and Petriello, Frank (2016*a*). Z-boson production in association with a jet at next-to-next-to-leading order in perturbative QCD. *Phys. Rev. Lett.*, **116**(15), 152001. doi:10.1103/PhysRevLett.116.152001.

[270] Boughezal, Radja, Caola, Fabrizio, Melnikov, Kirill, Petriello, Frank, and Schulze, Markus (2013). Higgs boson production in association with a jet at next-to-next-to-leading order in perturbative QCD. *JHEP*, **06**, 072. doi:10.1007/JHEP06(2013)072.

[271] Boughezal, Radja, Caola, Fabrizio, Melnikov, Kirill, Petriello, Frank, and Schulze, Markus (2015*a*). Higgs boson production in association with a jet at next-to-next-to-leading order. *Phys. Rev. Lett.*, **115**(8), 082003. doi:10.1103/PhysRevLett.115.082003.

[272] Boughezal, Radja, Focke, Christfried, Giele, Walter, Liu, Xiaohui, and Petriello, Frank (2015*b*). Higgs boson production in association with a jet at NNLO using jettiness subtraction. *Phys. Lett.*, **B748**, 5–8. doi:10.1016/j.physletb.2015.06.055.

[273] Boughezal, Radja, Focke, Christfried, and Liu, Xiaohui (2015*c*). Jet vetoes versus giant K-factors in the exclusive Z+1-jet cross section. *Phys. Rev.*, **D92**(9), 094002. doi:10.1103/PhysRevD.92.094002.

[274] Boughezal, Radja, Focke, Christfried, Liu, Xiaohui, and Petriello, Frank (2015*d*). W-boson production in association with a jet at next-to-next-to-leading order in perturbative QCD. *Phys. Rev. Lett.*, **115**(6), 062002. doi:10.1103/PhysRevLett.115.062002.

[275] Boughezal, Radja, Li, Ye, and Petriello, Frank (2014). Disentangling radiative corrections using high-mass Drell-Yan at the LHC. *Phys. Rev.*, **D89**, 034030. doi:10.1103/PhysRevD.89.034030.

[276] Boughezal, Radja, Liu, Xiaohui, and Petriello, Frank (2016*b*). A comparison of NNLO QCD predictions with 7 TeV ATLAS and CMS data for V+jet processes. *Phys. Lett.*, **B760**, 6–13. doi:10.1016/j.physletb.2016.06.032.

[277] Boughezal, Radja, Liu, Xiaohui, and Petriello, Frank (2016*c*). Phenomenology of the Z-boson plus jet process at NNLO. *Phys. Rev.*, **D94**(7), 074015. doi:10.1103/PhysRevD.94.074015.

[278] Boughezal, Radja, Liu, Xiaohui, and Petriello, Frank (2016*d*). W-boson plus jet differential distributions at NNLO in QCD. *Phys. Rev.*, **D94**(11), 113009. doi:10.1103/PhysRevD.94.113009.

[279] Bourhis, L., Fontannaz, M., Guillet, J.P., and Werlen, M. (2001). Next-to-leading order determination of fragmentation functions. *Eur. Phys. J.*, **C19**, 89–98. doi:10.1007/s100520100579.

[280] Bourilkov, D, Group, R C, and Whalley, M R (2006). LHAPDF: PDF use from the Tevatron to the LHC. hep-ph/0605240.

[281] Bowler, M. G. (1981). e^+e^- production of heavy quarks in the string model. *Z. Phys.*, **C11**, 169. doi:10.1007/BF01574001.

[282] Bozzi, Giuseppe, Catani, Stefano, de Florian, Daniel, and Grazzini, Massimiliano (2006). Transverse-momentum resummation and the spectrum of the Higgs boson at the LHC. *Nucl. Phys.*, **B737**, 73–120. doi:10.1016/j.nuclphysb.2005.12.022.

[283] Braunschweig, W. et al. (1990). Global jet properties at 14-GeV to 44-GeV center-of-mass energy in e^+e^- annihilation. *Z. Phys.*, **C47**, 187–198. doi:10.1007/BF01552339.

[284] Bredenstein, A., Denner, A., Dittmaier, S., and Pozzorini, S. (2009). NLO QCD corrections to $pp \to t\bar{t}b\bar{b} + X$ at the LHC. *Phys. Rev. Lett.*, **103**, 012002. doi:10.1103/PhysRevLett.103.012002.

[285] Bredenstein, A., Denner, A., Dittmaier, S., and Pozzorini, S. (2010). NLO QCD Corrections to $t\bar{t}b\bar{b}$ Production at the LHC: 2. full hadronic results. *JHEP*, **1003**, 021. doi:10.1007/JHEP03(2010)021.

[286] Britto, Ruth, Cachazo, Freddy, and Feng, Bo (2005*a*). New recursion relations for tree amplitudes of gluons. *Nucl. Phys.*, **B715**, 499–522. doi:10.1016/j.nuclphysb.2005.02.030.

[287] Britto, Ruth, Cachazo, Freddy, Feng, Bo, and Witten, Edward (2005*b*). Direct proof of the tree-level scattering amplitude recursion relation in Yang-Mills theory. *Phys. Rev. Lett.*, **94**, 181602.

[288] Brucherseifer, Mathias, Caola, Fabrizio, and Melnikov, Kirill (2014). On the NNLO QCD corrections to single-top production at the LHC. *Phys. Lett.*, **B736**, 58–63. doi:10.1016/j.physletb.2014.06.075.

[289] Buchalla, Gerhard, Buras, Andrzej J., and Lautenbacher, Markus E. (1996). Weak decays beyond leading logarithms. *Rev.Mod.Phys.*, **68**, 1125–1144. doi:10.1103/RevModPhys.68.1125.

[290] Buckley, Andy et al. (2011). General-purpose event generators for LHC physics. *Phys. Rept.*, **504**, 145–233. doi:10.1016/j.physrep.2011.03.005.

[291] Buckley, Andy, Butterworth, Jonathan, Lonnblad, Leif, Grellscheid, David, Hoeth, Hendrik, Monk, James, Schulz, Holger, and Siegert, Frank (2013). Rivet user manual. *Comput. Phys. Commun.*, **184**, 2803–2819. doi:10.1016/j.cpc.2013.05.021.

[292] Budnev, V. M., Ginzburg, I. F., Meledin, G. V., and Serbo, V. G. (1974). The two photon particle production mechanism. Physical problems. Applications. Equivalent photon approximation. *Phys. Rept.*, **15**, 181–281.

[293] Burdman, Gustavo and Donoghue, John F. (1992). Union of chiral and heavy quark symmetries. *Phys. Lett.*, **B280**, 287–291. doi:10.1016/0370-2693(92)90068-F.

[294] Buttar, C., Dittmaier, S., Drollinger, V., Frixione, S., Nikitenko, A. et al. (2006). Les Houches Physics at TeV colliders 2005, Standard Model and Higgs working group: Summary report (hep-ph/0604120).

[295] Butterworth, Jon et al. (2016). PDF4LHC recommendations for LHC Run II. *J. Phys.*, **G43**, 023001. doi:10.1088/0954-3899/43/2/023001.

[296] Butterworth, J., Dissertori, G., Dittmaier, S., de Florian, D., Glover, N. et al. (2014). Les Houches 2013: Physics at TeV Colliders: Standard Model Working Group Report (arXiv:1405.1067).

[297] Butterworth, Jonathan M., Davison, Adam R., Rubin, Mathieu, and Salam, Gavin P. (2008). Jet substructure as a new Higgs search channel at the LHC. *Phys. Rev. Lett.*, **100**, 242001. doi:10.1103/PhysRevLett.100.242001.

[298] Butterworth, John M., Forshaw, Jeffrey R., and Seymour, Mike H. (1996). Multiparton interactions in photoproduction at HERA. *Z. Phys.*, **C72**, 637–646.

[299] Byckling, E. and Kajantie, K. (1969). N-particle phase space in terms of invariant momentum transfers. *Nucl. Phys.*, **B9**, 568–576. doi:10.1016/0550-3213(69)90271-5.

[300] C. Balazs, G. Ladinsky, C.-P. Yuan (Fortran); P. Nadolsky (C++); CTEQ Collaboration (CTEQ libraries). Resummation program resbos (http://hep.pa.msu.edu/www/legacy/).

[301] Cacciari, Matteo, Dreyer, Frdric A., Karlberg, Alexander, Salam, Gavin P., and Zanderighi, Giulia (2015). Fully differential VBF Higgs production at NNLO. *Phys. Rev. Lett.*, **115**, 082002. doi:10.1103/PhysRevLett.115.082002.

[302] Cacciari, Matteo, Greco, Mario, and Nason, Paolo (1998). The P(T) spectrum in heavy flavor hadroproduction. *JHEP*, **9805**, 007. doi:10.1088/1126-6708/1998/05/007.

[303] Cacciari, Matteo and Salam, Gavin P. (2008). Pileup subtraction using jet areas. *Phys. Lett.*, **B659**, 119–126. doi:10.1016/j.physletb.2007.09.077.

[304] Cacciari, Matteo, Salam, Gavin P., and Soyez, Gregory (2008). The Anti-kT jet clustering algorithm. *JHEP*, **0804**, 063. doi:10.1088/1126-6708/2008/04/063.

[305] Cacciari, Matteo, Salam, Gavin P., and Soyez, Gregory (2012). FastJet user manual. *Eur. Phys. J.*, **C72**, 1896. doi:10.1140/epjc/s10052-012-1896-2.

[306] Cachazo, Freddy and Svrček, Peter (2005). Lectures on twistor strings and perturbative Yang-Mills theory. *PoS*, **RTN2005**, 004.

[307] Cachazo, Freddy, Svrcek, Peter, and Witten, Edward (2004). MHV vertices and tree amplitudes in gauge theory. *JHEP*, **09**, 006. doi:10.1088/1126-6708/2004/09/006.

[308] Cafarella, Alessandro, Papadopoulos, Costas G., and Worek, Malgorzata (2009). HELAC-PHEGAS: A generator for all parton level processes. *Comput. Phys. Commun.*, **180**, 1941–1955. doi:10.1016/j.cpc.2009.04.023.

[309] Campbell, John and Ellis, R. Keith. MCFM – Monte Carlo for FeMtobarn processes (mcfm.fnal.gov).

[310] Campbell, J.M., Hatakeyama, K., Huston, J., Petriello, F., Andersen, Jeppe R. et al. (2013). Working Group Report: Quantum Chromodynamics.

[311] Campbell, John M. and Ellis, R. Keith (1999). Update on vector boson pair production at hadron colliders. *Phys. Rev.*, **D60**, 113006. doi:10.1103/PhysRevD.60.113006.

[312] Campbell, John M. and Ellis, R. Keith (2015). Top-quark processes at NLO in production and decay. *J. Phys.*, **G42**(1), 015005. doi:10.1088/0954-3899/42/1/015005.

[313] Campbell, John M., Ellis, R. Keith, Li, Ye, and Williams, Ciaran (2016). Predictions for diphoton production at the LHC through NNLO in QCD. *JHEP*, **07**, 148. doi:10.1007/JHEP07(2016)148.

[314] Campbell, John M., Ellis, R. Keith, and Williams, Ciaran (2011). Vector boson pair production at the LHC. *JHEP*, **1107**, 018. doi:10.1007/JHEP07(2011)018.

[315] Campbell, John M., Ellis, R. Keith, and Williams, Ciaran (2014). Bounding the Higgs width at the LHC using full analytic results for $gg- > e^-e^+\mu^-\mu^+$. *JHEP*, **04**, 060. doi:10.1007/JHEP04(2014)060.

[316] Campbell, John M., Ellis, R. Keith, and Williams, Ciaran (2017). Direct photon production at next-tonext-to-leading order. *Phys. Rev. Lett.*, **118**(22), 222001. doi:10.1103/PhysRevLett.118.222001.

[317] Campbell, John M., Frederix, Rikkert, Maltoni, Fabio, and Tramontano, Francesco (2009). Next-to-Leading-Order Predictions for t-Channel Single-Top Production at Hadron Colliders. *Phys. Rev. Lett.*, **102**, 182003. doi:10.1103/PhysRevLett.102.182003.

[318] Campbell, John M., Giele, Walter T., and Williams, Ciaran (2012). The Matrix element method at next-to-leading order. *JHEP*, **1211**, 043. doi:10.1007/JHEP11(2012)043.

[319] Campbell, John M. and Glover, E.W. Nigel (1998). Double unresolved approximations to multiparton scattering amplitudes. *Nucl. Phys.*, **B527**, 264–288. doi:10.1016/S0550-3213(98)00295-8.

[320] Campbell, John M., Huston, J.W., and Stirling, W.J. (2007a). Hard interactions of quarks and gluons: A Primer for LHC physics. *Rept.Prog.Phys.*, **70**, 89. doi:10.1088/0034-4885/70/1/R02.

[321] Campbell, John M., Maltoni, F., and Tramontano, F. (2007b). QCD corrections to J/ψ and Υ production at hadron colliders. *Phys. Rev. Lett.*, **98**, 252002. doi:10.1103/PhysRevLett.98.252002.

[322] Caola, Fabrizio and Melnikov, Kirill (2013). Constraining the Higgs boson width with ZZ production at the LHC. *Phys. Rev.*, **D88**, 054024. doi:10.1103/PhysRevD.88.054024.

[323] Caola, Fabrizio, Melnikov, Kirill, Rontsch, Raoul, and Tancredi, Lorenzo (2015a). QCD corrections to ZZ production in gluon fusion at the LHC. *Phys. Rev.*, **D92**(9), 094028. doi:10.1103/PhysRevD.92.094028.

[324] Caola, Fabrizio, Melnikov, Kirill, Rontsch, Raoul, and Tancredi, Lorenzo (2016). QCD corrections to W^+W^- production through gluon fusion. *Phys. Lett.*, **B754**, 275–280. doi:10.1016/j.physletb.2016.01.046.

[325] Caola, Fabrizio, Melnikov, Kirill, and Schulze, Markus (2015b). Fiducial cross sections for Higgs boson production in association with a jet at next-to-next-to-leading order in QCD. *Phys. Rev.*, **D92**(7), 074032. doi:10.1103/PhysRevD.92.074032.

[326] Caravaglios, F., Mangano, Michelangelo L., Moretti, M., and Pittau, R. (1999). A New approach to multijet calculations in hadron collisions. *Nucl. Phys.*, **B539**, 215–232. doi:10.1016/S0550-3213(98)00739-1.

[327] Caravaglios, Francesco and Moretti, Mauro (1995). An algorithm to compute Born scattering amplitudes without Feynman graphs. *Phys. Lett.*, **B358**, 332–338. doi:10.1016/0370-2693(95)00971-M.

[328] Caravaglios, Francesco and Moretti, M. (1997). e+ e- into four fermions + gamma with ALPHA. *Z. Phys.*, **C74**, 291–296. doi:10.1007/s002880050390.

[329] Carli, Tancredi et al. (2010*a*). A posteriori inclusion of parton density functions in NLO QCD final-state calculations at hadron colliders: The APPLGRID Project. *Eur. Phys. J.*, **C**, 503–524. doi:10.1140/epjc/s10052-010-1255-0.

[330] Carli, Tancredi, Gehrmann, Thomas, and Höche, Stefan (2010*b*). Hadronic final states in deep-inelastic scattering with SHERPA. *Eur. Phys. J.*, **C67**, 73. doi:10.1140/epjc/s10052-010-1261-2.

[331] Carrazza, Stefano, Ferrara, Alfio, Palazzo, Daniele, and Rojo, Juan (2015*a*). APFEL Web. *J. Phys.*, **G42**(5), 057001. doi:10.1088/0954-3899/42/5/057001.

[332] Carrazza, Stefano, Forte, Stefano, Kassabov, Zahari, Latorre, Jose Ignacio, and Rojo, Juan (2015*b*). An unbiased Hessian representation for Monte Carlo PDFs. *Eur. Phys. J.*, **C75**(8), 369. doi:10.1140/epjc/s10052-015-3590-7.

[333] Carrazza, Stefano, Forte, Stefano, Kassabov, Zahari, and Rojo, Juan (2016). Specialized minimal PDFs for optimized LHC calculations. *Eur. Phys. J.*, **C76**(4), 205. doi:10.1140/epjc/s10052-016-4042-8.

[334] Carrazza, Stefano, Latorre, Jos I., Rojo, Juan, and Watt, Graeme (2015*c*). A compression algorithm for the combination of PDF sets. *Eur. Phys. J.*, **C75**, 474. doi:10.1140/epjc/s10052-015-3703-3.

[335] Carrazza, Stefano and Pires, Joao (2014). Perturbative QCD description of jet data from LHC Run-I and Tevatron Run-II. *JHEP*, **1410**, 145. doi:10.1007/JHEP10(2014)145.

[336] Cascioli, F., Gehrmann, T., Grazzini, M., Kallweit, S., Maierhoefer, P., von Manteuffel, A., Pozzorini, S., Rathlev, D., Tancredi, L., and Weihs, E. (2014*a*). ZZ production at hadron colliders in NNLO QCD. *Phys. Lett.*, **B735**, 311–313. doi:10.1016/j.physletb.2014.06.056.

[337] Cascioli, F., Hoeche, S., Krauss, F., Maierhoefer, P., Pozzorini, S., and Siegert, F. (2014*b*). Precise Higgs-background predictions: merging NLO QCD and squared quark-loop corrections to four-lepton + 0,1 jet production. *JHEP*, **01**, 046. doi:10.1007/JHEP01(2014)046.

[338] Cascioli, Fabio, Maierhofer, Philipp, and Pozzorini, Stefano (2012). Scattering amplitudes with open loops. *Phys. Rev. Lett.*, **108**, 111601. doi:10.1103/PhysRevLett.108.111601.

[339] Catani, S. and Ciafaloni, M. (1984). Many-gluon correlations and the quark form factor in QCD. *Nucl. Phys.*, **B236**, 61. doi:10.1016/0550-3213(84)90525-X.

[340] Catani, Stefano, Cieri, Leandro, de Florian, Daniel, Ferrera, Giancarlo, and Grazzini, Massimiliano (2012*a*). Diphoton production at hadron colliders: a

fully-differential QCD calculation at NNLO. *Phys. Rev. Lett.*, **108**, 072001. doi:10.1103/PhysRevLett.108.072001.

[341] Catani, Stefano, Cieri, Leandro, Ferrera, Giancarlo, de Florian, Daniel, and Grazzini, Massimiliano (2009). Vector boson production at hadron colliders: a fully exclusive QCD calculation at NNLO. *Phys. Rev. Lett.*, **103**, 082001. doi:10.1103/PhysRevLett.103.082001.

[342] Catani, Stefano, de Florian, Daniel, and Grazzini, Massimiliano (2001a). Universality of non-leading logarithmic contributions in transverse momentum distributions. *Nucl. Phys.*, **B596**, 299–312. doi:10.1016/S0550-3213(00)00617-9.

[343] Catani, Stefano, de Florian, Daniel, and Rodrigo, German (2012b). Factorization violation in the multiparton collinear limit. *PoS*, **LL2012**, 035.

[344] Catani, Stefano, Dittmaier, Stefan, Seymour, Michael H., and Trocsanyi, Zoltan (2002a). The Dipole formalism for next-to-leading order QCD calculations with massive partons. *Nucl. Phys.*, **B627**, 189–265. doi:10.1016/S0550-3213(02)00098-6.

[345] Catani, S., Dokshitzer, Yuri L., Olsson, M., Turnock, G., and Webber, B. R. (1991a). New clustering algorithm for multi-jet cross-sections in e+ e- annihilation. *Phys. Lett.*, **B269**, 432–438. doi:10.1016/0370-2693(91)90196-W.

[346] Catani, S., Dokshitzer, Yuri L., Seymour, M. H., and Webber, B. R. (1993). Longitudinally invariant K_t clustering algorithms for hadron hadron collisions. *Nucl. Phys.*, **B406**, 187–224. doi:10.1016/0550-3213(93)90166-M.

[347] Catani, S., Dokshitzer, Yuri L., and Webber, B. R. (1992). The $k_\perp$ clustering algorithm for jets in deep-inelastic scattering and hadron collisions. *Phys. Lett.*, **B285**, 291–299. doi:10.1016/0370-2693(92)91467-N.

[348] Catani, S., Fontannaz, M., Guillet, J.P., and Pilon, E. (2002b). Cross-section of isolated prompt photons in hadron hadron collisions. *JHEP*, **0205**, 028. doi:10.1088/1126-6708/2002/05/028.

[349] Catani, Stefano and Grazzini, Massimiliano (2000). Infrared factorization of tree level QCD amplitudes at the next-to-next-to-leading order and beyond. *Nucl. Phys.*, **B570**, 287–325. doi:10.1016/S0550-3213(99)00778-6.

[350] Catani, Stefano and Grazzini, Massimiliano (2007). An NNLO subtraction formalism in hadron collisions and its application to Higgs boson production at the LHC. *Phys. Rev. Lett.*, **98**, 222002. doi:10.1103/PhysRevLett.98.222002.

[351] Catani, S., Krauss, F., Kuhn, R., and Webber, B. R. (2001b). QCD matrix elements + parton showers. *JHEP*, **11**, 063. doi:10.1088/1126-6708/2001/11/063.

[352] Catani, S. and Seymour, M. H. (1996). The dipole formalism for the calculation of QCD jet cross sections at next-to-leading order. *Phys. Lett.*, **B378**, 287–301. doi:10.1016/0370-2693(96)00425-X.

[353] Catani, S. and Seymour, M. H. (1997). A general algorithm for calculating jet cross-sections in NLO QCD. *Nucl. Phys.*, **B485**, 291–419. [Erratum: Nucl. Phys.B510,503(1998)]. doi:10.1016/S0550-3213(96)00589-5, 10.1016/S0550-3213(98)81022-5.

[354] Catani, S. and Trentadue, L. (1989a). Inhibited radiation dynamics in QCD. *Phys. Lett.*, **B217**, 539–544. doi:10.1016/0370-2693(89)90093-2.

[355] Catani, S. and Trentadue, L. (1989*b*). Resummation of the QCD perturbative series for hard processes. *Nucl. Phys.*, **B327**, 323–352. doi:10.1016/0550-3213(89)90273-3.

[356] Catani, S., Webber, B. R., and Marchesini, G. (1991*b*). QCD coherent branching and semi-inclusive processes at large *x*. *Nucl. Phys.*, **B349**, 635–654. doi:10.1016/0550-3213(91)90390-J.

[357] Chanowitz, Michael S. (2009). Bounding CKM Mixing with a fourth family. *Phys. Rev.*, **D79**, 113008. doi:10.1103/PhysRevD.79.113008.

[358] Charles, J. et al. (2005). CP violation and the CKM matrix: Assessing the impact of the asymmetric *B* factories. *Eur. Phys. J.*, **C41**, 1–131. doi:10.1140/epjc/s2005-02169-1.

[359] Chatrchyan, Serguei et al. (2011*a*). Measurement of the differential cross section for isolated prompt photon production in pp collisions at 7 TeV. *Phys. Rev.*, **D84**, 052011. doi:10.1103/PhysRevD.84.052011.

[360] Chatrchyan, Serguei et al. (2011*b*). Measurement of the inclusive jet cross section in *pp* collisions at $\sqrt{s}$ = 7 TeV. *Phys. Rev. Lett.*, **107**, 132001. doi:10.1103/PhysRevLett.107.132001.

[361] Chatrchyan, Serguei et al. (2011*c*). Measurement of the inclusive *W* and *Z* production cross sections in *pp* collisions at $\sqrt{s}$ = 7 TeV. *JHEP*, **1110**, 132. doi:10.1007/JHEP10(2011)132.

[362] Chatrchyan, Serguei et al. (2011*d*). Measurement of the underlying event activity at the LHC with $\sqrt{s}$ = 7 TeV and comparison with $\sqrt{s}$ = 0.9 TeV. *JHEP*, **1109**, 109. doi:10.1007/JHEP09(2011)109.

[363] Chatrchyan, Serguei et al. (2012*a*). Inclusive and differential measurements of the $t\bar{t}$ charge asymmetry in proton-proton collisions at 7 TeV. *Phys. Lett.*, **B717**, 129–150. doi:10.1016/j.physletb.2012.09.028.

[364] Chatrchyan, Serguei et al. (2012*b*). Jet production rates in association with *W* and *Z* bosons in *pp* collisions at $\sqrt{s}$ = 7 TeV. *JHEP*, **01**, 010. doi:10.1007/JHEP01(2012)010.

[365] Chatrchyan, Serguei et al. (2012*c*). Measurement of the charge asymmetry in top-quark pair production in proton-proton collisions at $\sqrt{s}$ = 7 TeV. *Phys. Lett.*, **B709**, 28–49. doi:10.1016/j.physletb.2012.01.078.

[366] Chatrchyan, Serguei et al. (2012*d*). Measurement of the rapidity and transverse momentum distributions of *Z* Bosons in *pp* collisions at $\sqrt{s}$ = 7 TeV. *Phys. Rev.*, **D85**, 032002. doi:10.1103/PhysRevD.85.032002.

[367] Chatrchyan, Serguei et al. (2012*e*). Measurement of the Z/γ^*+b-jet cross section in pp collisions at 7 TeV. *JHEP*, **1206**, 126. doi:10.1007/JHEP06(2012)126.

[368] Chatrchyan, Serguei et al. (2012*f*). Observation of a new boson at a mass of 125 GeV with the CMS experiment at the LHC. *Phys. Lett.*, **B716**, 30–61. doi:10.1016/j.physletb.2012.08.021.

[369] Chatrchyan, Serguei et al. (2012*g*). Observation of a new boson at a mass of 125 GeV with the CMS experiment at the LHC. *Phys. Lett.*, **B716**, 30–61. doi:10.1016/j.physletb.2012.08.021.

[370] Chatrchyan, Serguei et al. (2012*h*). Observation of a new boson at a mass of 125 GeV with the CMS experiment at the LHC. *Phys. Lett.*, **B716**, 30–61. doi:10.1016/j.physletb.2012.08.021.

[371] Chatrchyan, Serguei et al. (2013*a*). Measurement of differential top-quark pair production cross sections in *pp* colisions at $\sqrt{s} = 7$ TeV. *Eur. Phys. J.*, **C73**(3), 2339. doi:10.1140/epjc/s10052-013-2339-4.

[372] Chatrchyan, Serguei et al. (2013*b*). Measurement of the cross section and angular correlations for associated production of a Z boson with b hadrons in pp collisions at $\sqrt{s} = 7$ TeV. *JHEP*, **1312**, 039. doi:10.1007/JHEP12(2013)039.

[373] Chatrchyan, Serguei et al. (2013*c*). Measurement of the differential and double-differential Drell-Yan cross sections in proton-proton collisions at $\sqrt{s} = 7$ TeV. *JHEP*, **12**, 030. doi:10.1007/JHEP12(2013)030.

[374] Chatrchyan, Serguei et al. (2013*d*). Measurement of the inelastic proton-proton cross section at $\sqrt{s} = 7$ TeV. *Phys. Lett.*, **B722**, 5–27. doi:10.1016/j.physletb.2013.03.024.

[375] Chatrchyan, Serguei et al. (2013*e*). Measurement of W^+W^- and ZZ production cross sections in pp collisions at $\sqrt{s} = 8$ TeV. *Phys. Lett.*, **B721**, 190–211. doi:10.1016/j.physletb.2013.03.027.

[376] Chatrchyan, Serguei et al. (2013*f*). Measurements of differential jet cross sections in proton-proton collisions at $\sqrt{s} = 7$ TeV with the CMS detector. *Phys. Rev.*, **D87**(11), 112002. [Erratum: Phys. Rev.D87,no.11,119902(2013)]. doi:10.1103/PhysRevD.87.112002, 10.1103/PhysRevD.87.119902.

[377] Chatrchyan, Serguei et al. (2013*g*). Studies of jet mass in dijet and W/Z + jet events. *JHEP*, **1305**, 090. doi:10.1007/JHEP05(2013)090.

[378] Chatrchyan, Serguei et al. (2014*a*). Measurement of differential cross sections for the production of a pair of isolated photons in pp collisions at $\sqrt{s} = 7$ TeV. *Eur. Phys. J.*, **C74**(11), 3129. doi:10.1140/epjc/s10052-014-3129-3.

[379] Chatrchyan, Serguei et al. (2014*b*). Measurement of four-jet production in proton-proton collisions at $\sqrt{s} = 7$ TeV. *Phys. Rev.*, **D89**(9), 092010. doi:10.1103/PhysRevD.89.092010.

[380] Chatrchyan, Serguei et al. (2014*c*). Measurement of inclusive W and Z boson production cross sections in pp collisions at $\sqrt{s} = 8$ TeV. *Phys. Rev. Lett.*, **112**, 191802. doi:10.1103/PhysRevLett.112.191802.

[381] Chatrchyan, Serguei et al. (2014*d*). Measurement of the production cross section for a W boson and two b jets in pp collisions at $\sqrt{s}=7$ TeV. *Phys. Lett.*, **B735**, 204–225. doi:10.1016/j.physletb.2014.06.041.

[382] Chatrchyan, Serguei et al. (2014*e*). Measurement of the production cross sections for a Z boson and one or more b jets in pp collisions at sqrt(s) = 7 TeV. *JHEP*, **1406**, 120. doi:10.1007/JHEP06(2014)120.

[383] Chatrchyan, Serguei et al. (2014*f*). Measurement of the ratio of inclusive jet cross sections using the anti-k_T algorithm with radius parameters R=0.5 and 0.7 in pp collisions at $\sqrt{s} = 7$ TeV. *Phys. Rev.*, **D90**(7), 072006. doi:10.1103/PhysRevD.90.072006.

[384] Chatrchyan, Serguei et al. (2014g). Measurement of the triple-differential cross section for photon+jets production in proton-proton collisions at $\sqrt{s}$=7 TeV. *JHEP*, **1406**, 009. doi:10.1007/JHEP06(2014)009.

[385] Chatrchyan, Serguei et al. (2014h). Observation of the associated production of a single top quark and a W boson in pp collisions at $\sqrt{s}$ =8 TeV. *Phys. Rev. Lett.*, **112**(23), 231802. doi:10.1103/PhysRevLett.112.231802.

[386] Chatrchyan, Serguei et al. (2014i). Search for new physics in the multijet and missing transverse momentum final state in proton-proton collisions at $\sqrt{s}$= 8 TeV. *JHEP*, **1406**, 055. doi:10.1007/JHEP06(2014)055.

[387] Chatrchyan, Serguei et al. (2014j). Study of double parton scattering using W + 2-jet events in proton-proton collisions at $\sqrt{s}$ = 7 TeV. *JHEP*, **1403**, 032. doi:10.1007/JHEP03(2014)032.

[388] Chay, Junegone, Georgi, Howard, and Grinstein, Benjamin (1990). Lepton energy distributions in heavy meson decays from QCD. *Phys. Lett.*, **B247**, 399–405. doi:10.1016/0370-2693(90)90916-T.

[389] Chekanov, S. et al. (2002). Inclusive jet cross-sections in the Breit frame in neutral current deep inelastic scattering at HERA and determination of α_s. *Phys. Lett.*, **B547**, 164–180. doi:10.1016/S0370-2693(02)02763-6.

[390] Chekanov, S. et al. (2007). Inclusive-jet and dijet cross sections in deep-inelastic scattering at HERA. *Nucl. Phys.*, **B765**, 1–30. doi:10.1016/j.nuclphysb.2006.09.018.

[391] Chekanov, S. et al. (2010). Measurement of isolated photon production in deep inelastic ep scattering. *Phys. Lett.*, **B687**, 16–25. doi:10.1016/j.physletb.2010.02.045.

[392] Chen, X., Gehrmann, T., Glover, E. W. N., and Jaquier, M. (2015). Precise QCD predictions for the production of Higgs + jet final states. *Phys. Lett.*, **B740**, 147–150. doi:10.1016/j.physletb.2014.11.021.

[393] Chetyrkin, K.G. and Tkachov, F.V. (1981). Integration by parts: The algorithm to calculate beta functions in 4 loops. *Nucl. Phys.*, **B192**, 159–204. doi:10.1016/0550-3213(81)90199-1.

[394] Chiappetta, P. and Perrottet, M. (1991). Possible bounds on compositeness from inclusive one jet production in large hadron colliders. *Phys. Lett.*, **B253**, 489–493. doi:10.1016/0370-2693(91)91757-M.

[395] Chiesa, Mauro, Montagna, Guido, Barz, Luca, Moretti, Mauro, Nicrosini, Oreste et al. (2013). Electroweak Sudakov corrections to new physics searches at the LHC. *Phys. Rev. Lett.*, **111**(12), 121801. doi:10.1103/PhysRevLett.111.121801.

[396] Christiansen, Jesper Roy and Sjostrand, Torbjorn (2014). Weak gauge boson radiation in parton showers. *JHEP*, **1404**, 115. doi:10.1007/JHEP04(2014)115.

[397] Chudakov, A. E. (1955). *Ser. Fiz., Izv. Akad. Nauk SSSR*, **19**, 650.

[398] Ciccolini, Mariano, Denner, Ansgar, and Dittmaier, Stefan (2008). Electroweak and QCD corrections to Higgs production via vector-boson fusion at the LHC. *Phys. Rev.*, **D77**, 013002. doi:10.1103/PhysRevD.77.013002.

[399] Cieri, Leandro (2012). Diphoton production at next-to-next-to-leading-order. [*Nucl. Phys. Proc. Suppl.*234,29(2013)]. doi:10.1016/j.nuclphysbps.2012.11.007.

[400] CMS (2009). Particle-flow event reconstruction in CMS and performance for jets, taus, and MET (CMS-PAS-PFT-09-001).

[401] CMS (2013*a*). Projected performance of an upgraded CMS detector at the LHC and HL-lHC: Contribution to the Snowmass Process (arXiv:1307.7135).

[402] CMS (2013*b*). Properties of the observed Higgs-like resonance using the diphoton channel. Technical Report CMS-PAS-HIG-13-016, CERN, Geneva.

[403] CMS (2014*a*). Measurement of the inclusive top-quark pair + photon production cross section in the muon + jets channel in pp collisions at 8 TeV. Technical Report CMS-PAS-TOP-13-011, CERN, Geneva.

[404] CMS (2014*b*). Study of topological distributions of inclusive three- and four-jet events at the LHC. Technical Report CMS-PAS-QCD-11-006, CERN, Geneva.

[405] CMS (2015*a*). Measurement of the double-differential inclusive jet cross section at sqrt(s) = 8 TeV. Technical Report CMS-PAS-SMP-14-001, CERN, Geneva.

[406] CMS (2015*b*). Measurement of top quark pair production in association with a W or Z boson using event reconstruction techniques. Technical Report CMS-PAS-TOP-14-021, CERN, Geneva.

[407] Colangelo, G. and Nason, P. (1992). A theoretical study of the c and b fragmentation function from e^+e^- annihilation. *Phys. Lett.*, **B285**, 167–171. doi:10.1016/0370-2693(92)91317-3.

[408] collaboration, LHCb (2014, Aug). Measurement of the forward W boson cross-section in pp collisions at $\sqrt{s} = 7$ TeV. Technical Report arXiv:1408.4354. LHCB-PAPER-2014-033. CERN-PH-EP-2014-175, CERN, Geneva.

[409] Collins, John C., Soper, Davison E., and Sterman, George F. (1985*a*). Factorization for short-distance hadron–hadron scattering. *Nucl. Phys.*, **B261**, 104. doi:10.1016/0550-3213(85)90565-6.

[410] Collins, John C., Soper, Davison E., and Sterman, George F. (1985*b*). Transverse momentum distribution in Drell-Yan pair and W and Z boson production. *Nucl. Phys.*, **B250**, 199. doi:10.1016/0550-3213(85)90479-1.

[411] Collins, P.D.B. and Spiller, T.P. (1985). The fragmentation of heavy quarks. *J. Phys.*, **G11**, 1289. doi:10.1088/0305-4616/11/12/006.

[412] Cooper, Leon N. (1956). Bound electron pairs in a degenerate Fermi gas. *Phys. Rev.*, **104**, 1189–1190. doi:10.1103/PhysRev.104.1189.

[413] Cooper-Sarkar, A.M. (2011). PDF fits at HERA. *PoS*, **EPS-HEP2011**, 320.

[414] Corcella, G. et al. (2002). HERWIG 6.5 Release Note.

[415] Corcella, G., Knowles, I. G., Marchesini, G., Moretti, S., Odagiri, K., Richardson, P., Seymour, M. H., and Webber, B. R. (2001). HERWIG 6: An Event generator for hadron emission reactions with interfering gluons (including supersymmetric processes). *JHEP*, **01**, 010. doi:10.1088/1126-6708/2001/01/010.

[416] Corcella, G. and Seymour, M. H. (1998). Matrix element corrections to parton shower simulations of heavy quark decay. *Phys. Lett.*, **B442**, 417–426. doi:10.1016/S0370-2693(98)01251-9.

[417] Corcella, Gennaro and Seymour, Michael H. (2000). Initial state radiation in simulations of vector boson production at hadron colliders. *Nucl. Phys.*, **B565**, 227–244. doi:10.1016/S0550-3213(99)00672-0.

[418] Corke, Richard and Sjöstrand, Torbjörn (2009). Multiparton interactions and rescattering. *JHEP*, **01**, 035. doi:10.1007/JHEP01(2010)035.

[419] Corke, Richard and Sjostrand, Torbjorn (2011). Multiparton Interactions with an x-dependent Proton Size. *JHEP*, **1105**, 009. doi:10.1007/JHEP05(2011)009.

[420] Cullen, G., van Deurzen, H., Greiner, N., Luisoni, G., Mastrolia, P., Mirabella, E., Ossola, G., Peraro, T., and Tramontano, F. (2013). Next-to-Leading-Order QCD corrections to Higgs boson production plus three jets in gluon fusion. *Phys. Rev. Lett.*, **111**(13), 131801. doi:10.1103/PhysRevLett.111.131801.

[421] Curci, G., Furmanski, W., and Petronzio, R. (1980). Evolution of parton densities beyond leading order: The non-singlet case. *Nucl. Phys.*, **B175**, 27. doi:10.1016/0550-3213(80)90003-6.

[422] Currie, J, Glover, E. W. N., and Pires, J (2017). NNLO QCD predictions for single jet inclusive production at the LHC. *Phys. Rev. Lett.*, **118**(7), 072002. doi:10.1103/PhysRevLett.118.072002.

[423] Curtin, David, Jaiswal, Prerit, and Meade, Patrick (2013). Charginos hiding In plain sight. *Phys. Rev.*, **D87**(3), 031701. doi:10.1103/PhysRevD.87.031701.

[424] Curtin, David, Meade, Patrick, and Tien, Pin-Ju (2014). Natural SUSY in plain sight. *Phys. Rev.*, **D90**(11), 115012. doi:10.1103/PhysRevD.90.115012.

[425] Czakon, M. (2011). Double-real radiation in hadronic top quark pair production as a proof of a certain concept. *Nucl. Phys.*, **B849**, 250–295. doi:10.1016/j.nuclphysb.2011.03.020.

[426] Czakon, Michal, Fiedler, Paul, Heymes, David, and Mitov, Alexander (2016a). NNLO QCD predictions for fully-differential top-quark pair production at the Tevatron. *JHEP*, **05**, 034. doi:10.1007/JHEP05(2016)034.

[427] Czakon, Michal, Fiedler, Paul, and Mitov, Alexander (2013). The total top quark pair production cross-section at hadron colliders through $\mathcal{O}(\alpha_S^4)$. *Phys. Rev. Lett.*, **110**, 252004. doi:10.1103/PhysRevLett.110.252004.

[428] Czakon, Michal, Fiedler, Paul, and Mitov, Alexander (2015). Resolving the Tevatron top quark forward-backward asymmetry puzzle: Fully differential next-to-next-to-leading-order calculation. *Phys. Rev. Lett.*, **115**(5), 052001. doi:10.1103/PhysRevLett.115.052001.

[429] Czakon, Michal, Heymes, David, and Mitov, Alexander (2016b). Dynamical scales for multi-TeV top-pair production at the LHC (arXiv:1606.03350).

[430] Daleo, A., Gehrmann, T., and Maitre, D. (2007). Antenna subtraction with hadronic initial states. *JHEP*, **04**, 016. doi:10.1088/1126-6708/2007/04/016.

[431] Dasgupta, Mrinal, Fregoso, Alessandro, Marzani, Simone, and Salam, Gavin P. (2013). Towards an understanding of jet substructure. *JHEP*, **1309**, 029. doi:10.1007/JHEP09(2013)029.

[432] Dasgupta, Mrinal, Magnea, Lorenzo, and Salam, Gavin P. (2008). Non-perturbative QCD effects in jets at hadron colliders. *JHEP*, **02**, 055. doi:10.1088/1126-6708/2008/02/055.

[433] Dashen, Roger F. (1969). Chiral $SU(3) \otimes SU(3)$ as a symmetry of the strong interactions. *Phys. Rev.*, **183**, 1245–1260. doi:10.1103/PhysRev.183.1245.

[434] Dashen, Roger F. (1971). Some features of chiral symmetry breaking. *Phys. Rev.*, **D3**, 1879–1889. doi:10.1103/PhysRevD.3.1879.

[435] Dashen, Roger F. and Weinstein, M. (1969). Soft pions, chiral symmetry, and phenomenological lagrangians. *Phys. Rev.*, **183**, 1261–1291. doi:10.1103/PhysRev.183.1261.

[436] Davies, C.T.H. and Stirling, W. James (1984). Non-leading corrections to the Drell-Yan cross section at small transverse momentum. *Nucl. Phys.*, **B244**, 337. doi:10.1016/0550-3213(84)90316-X.

[437] Davies, C.T.H., Webber, B.R., and Stirling, W. James (1985). Drell-Yan cross sections at small transverse momentum. *Nucl. Phys.*, **B256**, 413. doi:10.1016/0550-3213(85)90402-X.

[438] Dawson, S. (1991). Radiative corrections to Higgs boson production. *Nucl. Phys.*, **B359**, 283–300. doi:10.1016/0550-3213(91)90061-2.

[439] Dawson, Sally, Gritsan, Andrei, Logan, Heather, Qian, Jianming, Tully, Chris et al. (2013). Working Group Report: Higgs Boson (arXiv:1310.8361).

[440] Dawson, S., Jackson, C., Orr, L.H., Reina, L., and Wackeroth, D. (2003). Associated Higgs production with top quarks at the large hadron collider: NLO QCD corrections. *Phys. Rev.*, **D68**, 034022. doi:10.1103/PhysRevD.68.034022.

[441] de Florian, D. et al. (2016). Handbook of LHC Higgs Cross Sections: 4. Deciphering the Nature of the Higgs Sector (arXiv:1610.07922).

[442] de Florian, Daniel, Ferrera, Giancarlo, Grazzini, Massimiliano, and Tommasini, Damiano (2011). Transverse-momentum resummation: Higgs boson production at the Tevatron and the LHC. *JHEP*, **11**, 064. doi:10.1007/JHEP11(2011)064.

[443] de Florian, Daniel and Grazzini, Massimiliano (2000). Next–to–next–to–leading logarithmic corrections at small transverse momentum in hadronic collisions. *Phys. Rev. Lett.*, **85**, 4678–4681. doi:10.1103/PhysRevLett.85.4678.

[444] de Florian, Daniel and Grazzini, Massimiliano (2001). The Structure of large logarithmic corrections at small transverse momentum in hadronic collisions. *Nucl. Phys.*, **B616**, 247–285. doi:10.1016/S0550-3213(01)00460-6.

[445] de Florian, Daniel, Hinderer, Patriz, Mukherjee, Asmita, Ringer, Felix, and Vogelsang, Werner (2014). Approximate next-to-next-to-leading order corrections to hadronic jet production. *Phys. Rev. Lett.*, **112**, 082001. doi:10.1103/PhysRevLett.112.082001.

[446] de Florian, Daniel, Sassot, R., Epele, Manuel, Hernandez-Pinto, Roger J., and Stratmann, Marco (2015). Parton-to-pion fragmentation reloaded. *Phys. Rev.*, **D91**(1), 014035. doi:10.1103/PhysRevD.91.014035.

[447] de Florian, Daniel, Sassot, Rodolfo, and Stratmann, Marco (2007*a*). Global analysis of fragmentation functions for pions and kaons and their uncertainties. *Phys. Rev.*, **D75**, 114010. doi:10.1103/PhysRevD.75.114010.

[448] de Florian, Daniel, Sassot, Rodolfo, and Stratmann, Marco (2007*b*). Global analysis of fragmentation functions for protons and charged hadrons. *Phys. Rev.*, **D76**, 074033. doi:10.1103/PhysRevD.76.074033.

[449] De Florian, D. and Signer, A. (2000). $W\gamma$ and $Z\gamma$ production at hadron colliders. *Eur. Phys. J.*, **C16**, 105–114. doi:10.1007/s100520050007.

[450] Del Duca, Vittorio (1993). Parke-Taylor amplitudes in the multi - Regge kinematics. *Phys. Rev.*, **D48**, 5133–5139. doi:10.1103/PhysRevD.48.5133.

[451] Del Duca, Vittorio (1995). An introduction to the perturbative QCD pomeron and to jet physics at large rapidities (hep-ph/9503226).

[452] Del Duca, Vittorio, Dixon, Lance J., and Maltoni, Fabio (2000). New color decompositions for gauge amplitudes at tree and loop level. *Nucl. Phys.*, **B571**, 51–70. doi:10.1016/S0550-3213(99)00809-3.

[453] del Duca, Vittorio, Frizzo, Alberto, and Maltoni, Fabio (2000). Factorization of tree QCD amplitudes in the high-energy limit and in the collinear limit. *Nucl. Phys.*, **B568**, 211–262.

[454] Del Duca, V., Kilgore, W., Oleari, C., Schmidt, C., and Zeppenfeld, D. (2001). Gluon fusion contributions to $H + 2$ jet production. *Nucl. Phys.*, **B616**, 367–399. doi:10.1016/S0550-3213(01)00446-1.

[455] Del Duca, Vittorio, Kilgore, William B., and Maltoni, Fabio (2000). Multi-photon amplitudes for next-to-leading order QCD. *Nucl. Phys.*, **B566**, 252–274. doi:10.1016/S0550-3213(99)00663-X.

[456] Denner, Ansgar and Dittmaier, S. (2006). Reduction schemes for one-loop tensor integrals. *Nucl. Phys.*, **B734**, 62–115. doi:10.1016/j.nuclphysb.2005.11.007.

[457] Denner, Ansgar, Dittmaier, Stefan, and Hofer, Lars (2014). COLLIER - A fortran-library for one-loop integrals. *PoS*, **LL2014**, 071.

[458] Denner, Ansgar, Dittmaier, Stefan, Kallweit, Stefan, and Muck, Alexander (2012). Electroweak corrections to Higgs-strahlung off W/Z bosons at the Tevatron and the LHC with HAWK. *JHEP*, **1203**, 075. doi:10.1007/JHEP03(2012)075.

[459] Denner, A., Dittmaier, S., Kallweit, S., and Pozzorini, S. (2011*a*). NLO QCD corrections to $W^{+}W^{-}b\bar{b}$ production at hadron colliders. *Phys. Rev. Lett.*, **106**, 052001. doi:10.1103/PhysRevLett.106.052001.

[460] Denner, A., Heinemeyer, S., Puljak, I., Rebuzzi, D., and Spira, M. (2011*b*). Standard Model Higgs-Boson branching ratios with uncertainties. *Eur. Phys. J.*, **C71**, 1753. doi:10.1140/epjc/s10052-011-1753-8.

[461] Denner, Ansgar and Pozzorini, Stefano (2001). One loop leading logarithms in electroweak radiative corrections. 1. Results. *Eur. Phys. J.*, **C18**, 461–480. doi:10.1007/s100520100551.

[462] d'Enterria, David and Rojo, Juan (2012). Quantitative constraints on the gluon distribution function in the proton from collider isolated-photon data. *Nucl. Phys.*, **B860**, 311–338. doi:10.1016/j.nuclphysb.2012.03.003.

[463] Dicus, Duane A. and Willenbrock, Scott S.D. (1988). Photon pair production and the intermediate mass Higgs boson. *Phys. Rev.*, **D37**, 1801. doi:10.1103/PhysRevD.37.1801.

[464] Diehl, Markus, Ostermeier, Daniel, and Schafer, Andreas (2012). Elements of a theory for multi-parton interactions in QCD. *JHEP*, **1203**, 089. doi:10.1007/JHEP03(2012)089.

[465] Diehl, Markus and Schafer, Andreas (2011). Theoretical considerations on multi-parton interactions in QCD. *Phys. Lett.*, **B698**, 389–402. doi:10.1016/j.physletb.2011.03.024.

[466] Dinsdale, Michael, Ternick, Marko, and Weinzierl, Stefan (2007). Parton showers from the dipole formalism. *Phys. Rev.*, **D76**, 094003. doi:10.1103/PhysRevD.76.094003.

[467] Dissertori, G., Knowles, I.G., and Schmelling, M. (2003). Quantum Chromodynamics: High energy experiments and theory.

[468] Dittmaier, S. et al. (2011). Handbook of LHC Higgs Cross Sections: 1. Inclusive Observables (arXiv:1101.0593).

[469] Dittmaier, Stefan, Huss, Alexander, and Speckner, Christian (2012). Weak radiative corrections to dijet production at hadron colliders. *JHEP*, **11**, 095. doi:10.1007/JHEP11(2012)095.

[470] Dittmar, M., Forte, S., Glazov, A., Moch, S., Altarelli, G. et al. (2009). Parton Distributions (arXiv:0901.2504).

[471] Dixon, Lance J., Kunszt, Z., and Signer, A. (1998). Helicity amplitudes for $\mathcal{O}(\alpha_s)$ production of W^+W^-, $W^\pm Z$, ZZ, $W^\pm\gamma$, or $Z\gamma$ pairs at hadron colliders. *Nucl. Phys.*, **B531**, 3–23. doi:10.1016/S0550-3213(98)00421-0.

[472] Dixon, Lance J., Kunszt, Z., and Signer, A. (1999). Vector boson pair production in hadronic collisions at $\mathcal{O}(\alpha_s)$: Lepton correlations and anomalous couplings. *Phys. Rev.*, **D60**, 114037. doi:10.1103/PhysRevD.60.114037.

[473] Dixon, Lance J. and Li, Ye (2013). Bounding the Higgs boson width through interferometry. *Phys. Rev. Lett.*, **111**, 111802. doi:10.1103/PhysRevLett.111.111802.

[474] Djouadi, A., Spira, M., and Zerwas, P.M. (1991). Production of Higgs bosons in proton colliders: QCD corrections. *Phys. Lett.*, **B264**, 440–446. doi:10.1016/0370-2693(91)90375-Z.

[475] Dobrescu, Bogdan A. and Lykken, Joseph D. (2013). Coupling spans of the Higgs-like boson. *JHEP*, **1302**, 073. doi:10.1007/JHEP02(2013)073.

[476] Dokshitzer, Yuri L., Diakonov, Dmitri, and Troian, S.I. (1978). On the transverse momentum distribution of massive lepton pairs. *Phys. Lett.*, **B79**, 269–272. doi:10.1016/0370-2693(78)90240-X.

[477] Dokshitzer, Yuri L., Khoze, Valery A., Mueller, Alfred H., and Troyan, S. I. (1991*a*). *Basics of perturbative QCD*. Gif-sur-Yvette, France: Ed. Frontieres.

[478] Dokshitzer, Yuri L., Khoze, Valery A., and Sjöstrand, T. (1992). Rapidity gaps in Higgs production. *Phys. Lett.*, **B274**, 116–121. doi:10.1016/0370-2693(92)90312-R.

[479] Dokshitzer, Yuri L., Khoze, Valery A., and Troian, S.I. (1991*b*). On specific QCD properties of heavy quark fragmentation ('dead cone'). *J. Phys.*, **G17**, 1602–1604. doi:10.1088/0954-3899/17/10/023.

[480] Dokshitzer, Yuri L., Leder, G. D., Moretti, S., and Webber, B. R. (1997). Better jet clustering algorithms. *JHEP*, **08**, 001. doi:10.1088/1126-6708/1997/08/001.

[481] Dokshitzer, Yuri L., Marchesini, G., and Webber, B. R. (1996). Dispersive approach to power-behaved contributions in QCD hard processes. *Nucl. Phys.*, **B469**, 93–142. doi:10.1016/0550-3213(96)00155-1.

[482] Dokshitzer, Yuri L., Troian, S.I., and Khoze, Valery A. (1987). Collective QCD effects in the structure of final multi–hadron states. (in Russian). *Sov. J. Nucl. Phys.*, **46**, 712–719.

[483] Dokshitzer, Yuri L. and Webber, B. R. (1995). Calculation of power corrections to hadronic event shapes. *Phys. Lett.*, **B352**, 451–455. doi:10.1016/0370-2693(95)00548-Y.

[484] Donnachie, A. and Landshoff, P.V. (2004). Does the hard pomeron obey Regge factorization? *Phys. Lett.*, **B595**, 393–399. doi:10.1016/j.physletb.2004.05.068.

[485] Donnachie, A. and Landshoff, P. V. (1992). Total cross-sections. *Phys. Lett.*, **B296**, 227–232. doi:10.1016/0370-2693(92)90832-O.

[486] Draggiotis, Petros D., Kleiss, Ronald H. P., and Papadopoulos, Costas G. (2002). Multi-jet production in hadron collisions. *Eur. Phys. J.*, **C24**, 447–458. doi:10.1007/s10052-002-0955-5.

[487] Draggiotis, Petros D., van Hameren, Andre, and Kleiss, Ronald (2000). SARGE: An Algorithm for generating QCD antennas. *Phys. Lett.*, **B483**, 124–130. doi:10.1016/S0370-2693(00)00532-3.

[488] Dreyer, Frdric A. and Karlberg, Alexander (2016). Vector-Boson fusion Higgs production at three loops in QCD. *Phys. Rev. Lett.*, **117**(7), 072001. doi:10.1103/PhysRevLett.117.072001.

[489] Dulat, Sayipjamal, Hou, Tie-Jiun, Gao, Jun, Guzzi, Marco, Huston, Joey, Nadolsky, Pavel, Pumplin, Jon, Schmidt, Carl, Stump, Daniel, and Yuan, C. P. (2016). New parton distribution functions from a global analysis of quantum chromodynamics. *Phys. Rev.*, **D93**(3), 033006. doi:10.1103/PhysRevD.93.033006.

[490] Dulat, Sayipjamal, Hou, Tie-Jiun, Gao, Jun, Huston, Joey, Nadolsky, Pavel, Pumplin, Jon, Schmidt, Carl, Stump, Daniel, and Yuan, C. P. (2014). Higgs Boson cross section from CTEQ-TEA global analysis. *Phys. Rev.*, **D89**(11), 113002. doi:10.1103/PhysRevD.89.113002.

[491] Dumm, D. Gomez, Roig, P., Pich, A., and Portoles, J. (2010*a*). Hadron structure in $\tau \to KK\pi\nu_\tau$ decays. *Phys. Rev.*, **D81**, 034031. doi:10.1103/PhysRevD.81.034031.

[492] Dumm, D. Gomez, Roig, P., Pich, A., and Portoles, J. (2010*b*). $\tau \to \pi\pi\pi\nu_\tau$ decays and the $a(1)(1260)$ off-shell width revisited. *Phys. Lett.*, **B685**, 158–164. doi:10.1016/j.physletb.2010.01.059.

[493] Dyson, F. J. (1949, Jun). The s matrix in quantum electrodynamics. *Phys. Rev.*, **75**, 1736–1755. doi:10.1103/PhysRev.75.1736.

[494] Ecker, G., Gasser, J., Leutwyler, H., Pich, A., and de Rafael, E. (1989*a*). Chiral Lagrangians for massive spin-1 fields. *Phys. Lett.*, **B223**, 425. doi:10.1016/0370-2693(89)91627-4.

[495] Ecker, G., Gasser, J., Pich, A., and de Rafael, E. (1989*b*). The role of resonances in chiral perturbation theory. *Nucl. Phys.*, **B321**, 311. doi:10.1016/0550-3213(89)90346-5.

[496] Eichten, E., Gottfried, K., Kinoshita, T., Lane, K. D., and Yan, Tung-Mow (1978). Charmonium: The model. *Phys. Rev.*, **D17**, 3090. [Erratum: Phys. Rev.D21,313(1980)]. doi:10.1103/PhysRevD.17.3090, 10.1103/PhysRevD.21.313.

[497] Eichten, Estia and Hill, Brian Russell (1990). An effective field theory for the calculation of matrix elements involving heavy quarks. *Phys. Lett.*, **B234**, 511. doi:10.1016/0370-2693(90)92049-O.

[498] Ellis, John R., Gaillard, Mary K., and Ross, Graham G. (1976). Search for gluons in e^+e^- annihilation. *Nucl. Phys.*, **B111**, 253. doi:10.1016/0550-3213(76)90542-3.

[499] Ellis, R.Keith and Zanderighi, Giulia (2008). Scalar one-loop integrals for QCD. *JHEP*, **0802**, 002. doi:10.1088/1126-6708/2008/02/002.

[500] Ellis, R. Keith, Hinchliffe, I., Soldate, M., and van der Bij, J.J. (1988). Higgs decay to tau+ tau-: A Possible signature of intermediate mass Higgs bosons at the SSC. *Nucl. Phys.*, **B297**, 221. doi:10.1016/0550-3213(88)90019-3.

[501] Ellis, R. Keith, Kunszt, Zoltan, Melnikov, Kirill, and Zanderighi, Giulia (2012). One-loop calculations in quantum field theory: from Feynman diagrams to unitarity cuts. *Phys.Rept.*, **518**, 141–250. doi:10.1016/j.physrep.2012.01.008.

[502] Ellis, R. Keith, Martinelli, G., and Petronzio, R. (1983). Lepton-pair production at large transverse momentum in second order QCD. *Nucl. Phys.*, **B211**, 106. doi:10.1016/0550-3213(83)90188-8.

[503] Ellis, R. Keith and Sexton, J.C. (1986). QCD radiative corrections to parton parton scattering. *Nucl. Phys.*, **B269**, 445. doi:10.1016/0550-3213(86)90232-4.

[504] Ellis, R. Keith, Stirling, W. James, and Webber, Bryan R. (1996). *QCD and collider physics* (1 edn). Volume 8. Cambridge Monogr. Part. Phys. Nucl. Phys. Cosmol.

[505] Ellis, R. Keith and Veseli, Sinisa (1998). *W* and *Z* transverse momentum distributions: Resummation in q_T space. *Nucl. Phys.*, **B511**, 649–669. doi:10.1016/S0550-3213(97)00655-X.

[506] Ellis, Stephen. private communication.

[507] Ellis, S.D., Huston, J., and Tonnesmann, M. (2001). On building better cone jet algorithms. *eConf*, **C010630**, P513.

[508] Ellis, S. D., Huston, J., Hatakeyama, K., Loch, P., and Tonnesmann, M. (2008). Jets in hadron-hadron collisions. *Prog. Part. Nucl. Phys.*, **60**, 484–551. doi:10.1016/j.ppnp.2007.12.002.

[509] Ellis, Stephen D., Kunszt, Zoltan, and Soper, Davison E. (1990). The one-jet inclusive cross section at order α_s^3, quarks and gluons. *Phys. Rev. Lett.*, **64**, 2121. doi:10.1103/PhysRevLett.64.2121.

[510] Ellis, Stephen D., Kunszt, Zoltan, and Soper, Davison E. (1992). Jets at hadron colliders at order α_s^3: A Look inside. *Phys. Rev. Lett.*, **69**, 3615–3618. doi:10.1103/PhysRevLett.69.3615.

[511] Ellis, Stephen D. and Soper, Davison E. (1993). Successive combination jet algorithm for hadron collisions. *Phys. Rev.*, **D48**, 3160–3166. doi:10.1103/PhysRevD.48.3160.

[512] Ellis, Stephen D., Vermilion, Christopher K., and Walsh, Jonathan R. (2009). Techniques for improved heavy particle searches with jet substructure. *Phys. Rev.*, **D80**, 051501. doi:10.1103/PhysRevD.80.051501.

[513] Ellis, Stephen D., Vermilion, Christopher K., and Walsh, Jonathan R. (2010). Recombination algorithms and jet substructure: Pruning as a tool for heavy particle searches. *Phys. Rev.*, **D81**, 094023. doi:10.1103/PhysRevD.81.094023.

[514] Engel, R. (1995). Photoproduction within the two component dual parton model. 1. Amplitudes and cross-sections. *Z. Phys.*, **C66**, 203–214.

doi:10.1007/BF01496594.

[515] Englert, C., Freitas, A., Mhlleitner, M. M., Plehn, T., Rauch, M., Spira, M., and Walz, K. (2014). Precision measurements of Higgs couplings: Implications for new physics scales. *J. Phys.*, **G41**, 113001. doi:10.1088/0954-3899/41/11/113001.

[516] Englert, F. and Brout, R. (1964). Broken symmetry and the mass of gauge vector mesons. *Phys. Rev. Lett.*, **13**, 321–323. doi:10.1103/PhysRevLett.13.321.

[517] Fadin, Victor S., Kuraev, E.A., and Lipatov, L.N. (1975). On the Pomeranchuk singularity in asymptotically free theories. *Phys. Lett.*, **B60**, 50–52. doi:10.1016/0370-2693(75)90524-9.

[518] Fadin, Victor S. and Lipatov, L. N. (1998). BFKL pomeron in the next-to-leading approximation. *Phys. Lett.*, **B429**, 127–134. doi:10.1016/S0370-2693(98)00473-0.

[519] Falk, Adam F., Ligeti, Zoltan, Neubert, Matthias, and Nir, Yosef (1994). Heavy quark expansion for the inclusive decay $\bar{B} \to \tau\bar{\nu} + X$. *Phys. Lett.*, **B326**, 145–153. doi:10.1016/0370-2693(94)91206-8.

[520] Febres Cordero, F., Reina, L., and Wackeroth, D. (2006). NLO QCD corrections to W boson production with a massive b-quark jet pair at the Tevatron $p\bar{p}$ collider. *Phys. Rev.*, **D74**, 034007. doi:10.1103/PhysRevD.74.034007.

[521] Febres Cordero, Fernando, Reina, L., and Wackeroth, D. (2009). W- and Z-boson production with a massive bottom-quark pair at the Large Hadron Collider. *Phys. Rev.*, **D80**, 034015. doi:10.1103/PhysRevD.80.034015.

[522] Feigl, Bastian, Rzehak, Heidi, and Zeppenfeld, Dieter (2012). New Physics Backgrounds to the H→WW Search at the LHC? *Phys. Lett.*, **B717**, 390–395. doi:10.1016/j.physletb.2012.09.033.

[523] Fermi, E. (1924). On the theory of the impact between atoms and electrically charged particles. *Z. Phys.*, **29**, 315–327. doi:10.1007/BF03184853.

[524] Ferrera, Giancarlo, Grazzini, Massimiliano, and Tramontano, Francesco (2011). Associated WH production at hadron colliders: a fully exclusive QCD calculation at NNLO. *Phys. Rev. Lett.*, **107**, 152003. doi:10.1103/PhysRevLett.107.152003.

[525] Field, Richard D. (1989). *Applications of perturbative QCD*. Addison-Wesley, Redwood City, USA. Frontiers in physics, 77.

[526] Field, Rick D. (2001). The Underlying event in hard scattering processes. *eConf*, **C010630**, P501.

[527] Field, R. D. and Feynman, R. P. (1978). A parametrization of the properties of quark jets. *Nucl. Phys.*, **B136**, 1. doi:10.1016/0550-3213(78)90015-9.

[528] Field, Richard D. and Wolfram, Stephen (1983). A QCD Model for e^+e^- Annihilation. *Nucl. Phys.*, **B213**, 65. doi:10.1016/0550-3213(83)90175-X.

[529] Finkemeier, Markus and Mirkes, Erwin (1996). The scalar contribution to $\tau \to K\pi\nu_\tau$. *Z. Phys.*, **C72**, 619–626. doi:10.1007/s002880050284.

[530] Foldy, Leslie L. and Peierls, Ronald F. (1963). Isotopic spin of exchanged systems. *Phys. Rev.*, **130**, 1585–1589. doi:10.1103/PhysRev.130.1585.

[531] Forte, Stefano, Isgr, Andrea, and Vita, Gherardo (2014). Do we need N³LO Parton Distributions? *Phys. Lett.*, **B731**, 136–140. doi:10.1016/j.physletb.2014.02.027.

[532] Forte, Stefano, Laenen, Eric, Nason, Paolo, and Rojo, Juan (2010). Heavy quarks in deep-inelastic scattering. *Nucl. Phys.*, **B834**, 116–162. doi:10.1016/j.nuclphysb.2010.03.014.

[533] Fox, Geoffrey C. and Wolfram, Stephen (1980). A model for parton showers in QCD. *Nucl. Phys.*, **B168**, 285. doi:10.1016/0550-3213(80)90111-X.

[534] Frederix, Rikkert (2014). Top quark induced backgrounds to Higgs production in the $WW^{(*)} \to ll\nu\nu$ decay channel at next-to-leading-order in QCD. *Phys. Rev. Lett.*, **112**(8), 082002. doi:10.1103/PhysRevLett.112.082002.

[535] Frederix, Rikkert and Frixione, Stefano (2012). Merging meets matching in MC@NLO. *JHEP*, **1212**, 061. doi:10.1007/JHEP12(2012)061.

[536] Frederix, Rikkert, Frixione, Stefano, Maltoni, Fabio, and Stelzer, Tim (2009). Automation of next–to–leading order computations in QCD: The FKS subtraction. *JHEP*, **0910**, 003. doi:10.1088/1126-6708/2009/10/003.

[537] Frederix, Rikkert, Gehrmann, Thomas, and Greiner, Nicolas (2008). Automation of the dipole subtraction method in MadGraph/MadEvent. *JHEP*, **0809**, 122. doi:10.1088/1126-6708/2008/09/122.

[538] Frederix, R., Gehrmann, T., and Greiner, N. (2010). Integrated dipoles with MadDipole in the MadGraph framework. *JHEP*, **06**, 086. doi:10.1007/JHEP06(2010)086.

[539] Frixione, S. (1997). A general approach to jet cross sections in QCD. *Nucl. Phys.*, **B507**, 295–314. doi:10.1016/S0550-3213(97)00574-9.

[540] Frixione, Stefano (1998). Isolated photons in perturbative QCD. *Phys. Lett.*, **B429**, 369–374. doi:10.1016/S0370-2693(98)00454-7.

[541] Frixione, S., Hirschi, V., Pagani, D., Shao, H. S., and Zaro, M. (2015). Electroweak and QCD corrections to top-pair hadroproduction in association with heavy bosons. *JHEP*, **06**, 184. doi:10.1007/JHEP06(2015)184.

[542] Frixione, S., Kunszt, Z., and Signer, A. (1996). Three-jet cross-sections to next–to–leading order. *Nucl. Phys.*, **B467**, 399–442. doi:10.1016/0550-3213(96)00110-1.

[543] Frixione, Stefano, Nason, Paolo, and Oleari, Carlo (2007). Matching NLO QCD computations with parton shower simulations: the POWHEG method. *JHEP*, **11**, 070. doi:10.1088/1126-6708/2007/11/070.

[544] Frixione, Stefano, Nason, Paolo, and Webber, Bryan R. (2003). Matching NLO QCD and parton showers in heavy flavor production. *JHEP*, **08**, 007. doi:10.1088/1126-6708/2003/08/007.

[545] Frixione, Stefano and Ridolfi, Giovanni (1997). Jet photoproduction at HERA. *Nucl. Phys.*, **B507**, 315–333. doi:10.1016/S0550-3213(97)00575-0.

[546] Frixione, Stefano and Webber, Bryan R. (2002). Matching NLO QCD computations and parton shower simulations. *JHEP*, **06**, 029. doi:10.1088/1126-6708/2002/06/029.

[547] Froissart, Marcel (1961). Asymptotic behavior and subtractions in the Mandelstam representation. *Phys. Rev.*, **123**, 1053–1057. doi:10.1103/PhysRev.123.1053.

[548] Froissart, M. and Omnes, R. (1965). Introduction to the theory of strong interactions. pp. 89–186.

[549] Furmanski, W. and Petronzio, R. (1980). Singlet parton densities beyond leading order. *Phys. Lett.*, **B97**, 437. doi:10.1016/0370-2693(80)90636-X.

[550] Furmanski, W. and Petronzio, R. (1982). Lepton - Hadron Processes Beyond Leading Order in Quantum Chromodynamics. *Z. Phys.*, **C11**, 293. doi:10.1007/BF01578280.

[551] Gao, Jun, Guzzi, Marco, Huston, Joey, Lai, Hung-Liang, Li, Zhao et al. (2014). The CT10 NNLO Global Analysis of QCD. *Phys. Rev.*, **D89**, 033009. doi:10.1103/PhysRevD.89.033009.

[552] Gao, Jun and Nadolsky, Pavel (2014). A meta-analysis of parton distribution functions. *JHEP*, **07**, 035. doi:10.1007/JHEP07(2014)035.

[553] Gasser, J. and Leutwyler, H. (1983). On the low-energy structure of QCD. *Phys. Lett.*, **B125**, 321. doi:10.1016/0370-2693(83)91293-5.

[554] Gasser, J. and Leutwyler, H. (1984). Chiral perturbation theory to one loop. *Ann. Phys.*, **158**, 142. doi:10.1016/0003-4916(84)90242-2.

[555] Gaunt, Jonathan, Stahlhofen, Maximilian, Tackmann, Frank J., and Walsh, Jonathan R. (2015). N-jettiness subtractions for NNLO QCD calculations. *JHEP*, **09**, 058. doi:10.1007/JHEP09(2015)058.

[556] Gavin, Ryan, Li, Ye, Petriello, Frank, and Quackenbush, Seth (2011). FEWZ 2.0: A code for hadronic Z production at next-to-next-to-leading order. *Comput. Phys. Commun.*, **182**, 2388–2403. doi:10.1016/j.cpc.2011.06.008.

[557] Gehrmann, T., Grazzini, M., Kallweit, S., Maierhfer, P., von Manteuffel, A. et al. (2014). W^+W^- production at hadron colliders in next to next to leading order QCD. *Phys. Rev. Lett.*, **113**(21), 212001. doi:10.1103/PhysRevLett.113.212001.

[558] Gehrmann, Thomas, Hoche, Stefan, Krauss, Frank, Schonherr, Marek, and Siegert, Frank (2013). NLO QCD matrix elements + parton showers in $e^+e^- \to$ hadrons. *JHEP*, **01**, 144. doi:10.1007/JHEP01(2013)144.

[559] Gehrmann, T. and Remiddi, E. (2000). Differential equations for two loop four point functions. *Nucl. Phys.*, **B580**, 485–518. doi:10.1016/S0550-3213(00)00223-6.

[560] Gehrmann, Thomas, von Manteuffel, Andreas, and Tancredi, Lorenzo (2015). The two-loop helicity amplitudes for $q\bar{q}' \to V_1 V_2 \to 4$ leptons. *JHEP*, **09**, 128. doi:10.1007/JHEP09(2015)128.

[561] Gehrmann-De Ridder, A., Gehrmann, T., and Glover, E.W. Nigel (2005*a*). Gluon-gluon antenna functions from Higgs boson decay. *Phys. Lett.*, **B612**, 49–60. doi:10.1016/j.physletb.2005.03.003.

[562] Gehrmann-De Ridder, A., Gehrmann, T., and Glover, E.W. Nigel (2005*b*). Quark-gluon antenna functions from neutralino decay. *Phys. Lett.*, **B612**, 36–48. doi:10.1016/j.physletb.2005.02.039.

[563] Gehrmann-De Ridder, A., Gehrmann, T., and Glover, E. W. Nigel (2005*c*). Antenna subtraction at NNLO. *JHEP*, **09**, 056. doi:10.1088/1126-6708/2005/09/056.

[564] Gehrmann-De Ridder, A., Gehrmann, T., Glover, E. W. N., Huss, A., and Morgan, T. A. (2016*a*). Precise QCD predictions for the production of a Z boson in association with a hadronic jet. *Phys. Rev. Lett.*, **117**(2), 022001. doi:10.1103/PhysRevLett.117.022001.

[565] Gehrmann-De Ridder, Aude, Gehrmann, T., Glover, E. W. N., Huss, A., and Morgan, T. A. (2016*b*). The NNLO QCD corrections to Z boson production at large transverse momentum. *JHEP*, **07**, 133. doi:10.1007/JHEP07(2016)133.

[566] Gehrmann-De Ridder, Aude, Gehrmann, Thomas, Glover, E. W. N., and Pires, Joao (2013). Second order QCD corrections to jet production at hadron colliders: the all-gluon contribution. *Phys. Rev. Lett.*, **110**(16), 162003. doi:10.1103/PhysRevLett.110.162003.

[567] Georgi, Howard (1990). An effective field theory for heavy quarks at low energies. *Phys. Lett.*, **B240**, 447–450. doi:10.1016/0370-2693(90)91128-X.

[568] Georgi, Howard (1991). Comment on heavy baryon weak form-factors. *Nucl. Phys.*, **B348**, 293–296. doi:10.1016/0550-3213(91)90519-4.

[569] Georgi, Howard, Grinstein, Benjamin, and Wise, Mark B. (1990). Λ_b semileptonic decay form-factors for m_c does not equal infinity. *Phys. Lett.*, **B252**, 456–460. doi:10.1016/0370-2693(90)90569-R.

[570] Gerwick, Erik, Plehn, Tilman, and Schumann, Steffen (2012*a*). Understanding jet scaling and jet vetos in Higgs searches. *Phys. Rev. Lett.*, **108**, 032003. doi:10.1103/PhysRevLett.108.032003.

[571] Gerwick, Erik, Plehn, Tilman, Schumann, Steffen, and Schichtel, Peter (2012*b*). Scaling patterns for QCD jets. *JHEP*, **1210**, 162. doi:10.1007/JHEP10(2012)162.

[572] Giele, W., Glover, E.W. Nigel, Hinchliffe, I., Huston, J., Laenen, Eric et al. (2002). The QCD / SM working group: Summary report (hep-ph/0204316). pp. 275–426.

[573] Giele, W.T., Glover, E.W. Nigel, and Kosower, David A. (1993). Higher order corrections to jet cross-sections in hadron colliders. *Nucl. Phys.*, **B403**, 633–670. doi:10.1016/0550-3213(93)90365-V.

[574] Giele, W.T., Glover, E.W. Nigel, and Kosower, David A. (1994). The same side / opposite side two jet ratio. *Phys. Lett.*, **B339**, 181–186. doi:10.1016/0370-2693(94)91152-5.

[575] Giele, W.T. and Zanderighi, G. (2008). On the numerical evaluation of one–loop amplitudes: The gluonic case. *JHEP*, **0806**, 038. doi:10.1088/1126-6708/2008/06/038.

[576] Giele, W. T. and Glover, E. W. Nigel (1992). Higher-order corrections to jet cross sections in e^+e^- annihilation. *Phys. Rev.*, **D46**, 1980–2010. doi:10.1103/PhysRevD.46.1980.

[577] Giele, Walter T., Kosower, David A., and Skands, Peter Z. (2008). A simple shower and matching algorithm. *Phys. Rev.*, **D78**, 014026. doi:10.1103/PhysRevD.78.014026.

[578] Gieseke, Stefan (2005). Uncertainties of Sudakov form-factors. *JHEP*, **01**, 058. doi:10.1088/1126-6708/2005/01/058.

[579] Gieseke, Stefan, Stephens, P., and Webber, Bryan (2003). New formalism for QCD parton showers. *JHEP*, **12**, 045. doi:10.1088/1126-6708/2003/12/045.

[580] Glashow, S.L. (1961). Partial Symmetries of Weak Interactions. *Nucl. Phys.*, **22**, 579–588. doi:10.1016/0029-5582(61)90469-2.

[581] Glashow, S.L., Iliopoulos, J., and Maiani, L. (1970). Weak interactions with lepton-hadron symmetry. *Phys. Rev.*, **D2**, 1285–1292. doi:10.1103/PhysRevD.2.1285.

[582] Gleisberg, Tanju and Höche, Stefan (2008). Comix, a new matrix element generator. *JHEP*, **12**, 039. doi:10.1088/1126-6708/2008/12/039.

[583] Gleisberg, Tanju, Hoeche, Stefan, Krauss, Frank, Schalicke, Andreas, Schumann, Steffen, and Winter, Jan-Christopher (2004). SHERPA 1. alpha: A Proof of concept version. *JHEP*, **02**, 056. doi:10.1088/1126-6708/2004/02/056.

[584] Gleisberg, Tanju and Krauss, Frank (2008). Automating dipole subtraction for QCD NLO calculations. *Eur. Phys. J.*, **C53**, 501–523. doi:10.1140/epjc/s10052-007-0495-0.

[585] Glover, E.W. Nigel (2003). Progress in NNLO calculations for scattering processes. *Nucl. Phys.Proc.Suppl.*, **116**, 3–7. doi:10.1016/S0920-5632(03)80133-0.

[586] Glover, E. W. Nigel and Morgan, A. G. (1994). Measuring the photon fragmentation function at LEP. *Z. Phys.*, **C62**, 311–322. doi:10.1007/BF01560245.

[587] Gluck, M., Pisano, Cristian, and Reya, E. (2002). The Polarized and unpolarized photon content of the nucleon. *Phys. Lett.*, **B540**, 75–80. doi:10.1016/S0370-2693(02)02125-1.

[588] Goebel, C., Halzen, F., and Scott, D.M. (1980). Double Drell-Yan annihilations in hadron collisions: novel tests of the constituent picture. *Phys. Rev.*, **D22**, 2789. doi:10.1103/PhysRevD.22.2789.

[589] Goldstone, J. (1961). Field theories with superconductor solutions. *Nuovo Cim.*, **19**, 154–164. doi:10.1007/BF02812722.

[590] Goldstone, Jeffrey, Salam, Abdus, and Weinberg, Steven (1962). Broken Symmetries. *Phys. Rev.*, **127**, 965–970. doi:10.1103/PhysRev.127.965.

[591] Golonka, Piotr and Was, Zbigniew (2006). PHOTOS Monte Carlo: A Precision tool for QED corrections in Z and W decays. *Eur. Phys. J.*, **C45**, 97–107. doi:10.1140/epjc/s2005-02396-4.

[592] Good, M.L. and Walker, W.D. (1960). Diffraction disssociation of beam particles. *Phys. Rev.*, **120**, 1857–1860. doi:10.1103/PhysRev.120.1857.

[593] Gottschalk, Thomas D. (1983). A realistic model for e+ e- annihilation including parton bremsstrahlung effects. *Nucl. Phys.*, **B214**, 201–222. doi:10.1016/0550-3213(83)90658-2.

[594] Grazzini, Massimiliano (2008). NNLO predictions for the Higgs boson signal in the H → WW → lnu lnu and H → ZZ → 4l decay channels. *JHEP*, **0802**, 043. doi:10.1088/1126-6708/2008/02/043.

[595] Grazzini, Massimiliano, Kallweit, Stefan, Rathlev, Dirk, and Torre, Alessandro (2014). $Z\gamma$ production at hadron colliders in NNLO QCD. *Phys. Lett.*, **B731**, 204–207. doi:10.1016/j.physletb.2014.02.037.

[596] Grazzini, M. *et al. HqT program.* http://theory.fi.infn.it/grazzini/codes.html.

[597] Grellscheid, David and Richardson, Peter (2007). Simulation of τ decays in the Herwig++ event generator (arXiv:0710.1951).

[598] Gribov, V.N. (1968). A Reggeon diagram technique. *Sov.Phys.JETP*, **26**, 414–422.

[599] Group, Tevatron Electroweak Working (2014). Combination of CDF and D0 results on the mass of the top quark using up to 9.7 fb^{-1} at the Tevatron (arXiv:1407.2682).

[600] Grunberg, G. (1980). Renormalization group improved perturbative QCD. *Phys. Lett.*, **B95**, 70. doi:10.1016/0370-2693(80)90402-5.

[601] Guralnik, G.S., Hagen, C.R., and Kibble, T.W.B. (1964). Global conservation laws and massless particles. *Phys. Rev. Lett.*, **13**, 585–587. doi:10.1103/PhysRevLett.13.585.

[602] Gustafson, G. (1986). Dual description of a confined color field. *Phys. Lett.*, **B175**, 453. [,193(1986)]. doi:10.1016/0370-2693(86)90622-2.

[603] Gustafson, Gosta and Pettersson, Ulf (1988). Dipole formulation of QCD cascades. *Nucl. Phys.*, **B306**, 746–758. doi:10.1016/0550-3213(88)90441-5.

[604] Hagiwara, Kaoru and Zeppenfeld, D. (1986). Helicity amplitudes for heavy lepton production in e^+e^- annihilation. *Nucl. Phys.*, **B274**, 1–32. doi:10.1016/0550-3213(86)90615-2.

[605] Hahn, T. and Perez-Victoria, M. (1999). Automatized one loop calculations in four-dimensions and D-dimensions. *Comput. Phys. Commun.*, **118**, 153–165. doi:10.1016/S0010-4655(98)00173-8.

[606] Halzen, Francis and Martin, Alan D. (1984). *Quarks and leptons: An introductory course in modern particle physics*. John Wiley & Sons, New York, USA.

[607] Hamberg, R., van Neerven, W.L., and Matsuura, T. (1991). A complete calculation of the order α_s^2 correction to the Drell-Yan K factor. *Nucl. Phys.*, **B359**, 343–405. doi:10.1016/0550-3213(91)90064-5.

[608] Hamilton, Keith and Nason, Paolo (2010). Improving NLO-parton shower matched simulations with higher order matrix elements. *JHEP*, **06**, 039. doi:10.1007/JHEP06(2010)039.

[609] Hamilton, Keith, Nason, Paolo, Oleari, Carlo, and Zanderighi, Giulia (2013a). Merging H/W/Z + 0 and 1 jet at NLO with no merging scale: a path to parton shower + NNLO matching. *JHEP*, **05**, 082. doi:10.1007/JHEP05(2013)082.

[610] Hamilton, Keith, Nason, Paolo, Re, Emanuele, and Zanderighi, Giulia (2013b). NNLOPS simulation of Higgs boson production. *JHEP*, **1310**, 222. doi:10.1007/JHEP10(2013)222.

[611] Hamilton, Keith, Nason, Paolo, and Zanderighi, Giulia (2012). MINLO: Multi-Scale Improved NLO. *JHEP*, **1210**, 155. doi:10.1007/JHEP10(2012)155.

[612] Hamilton, Keith, Richardson, Peter, and Tully, Jon (2009). A modified CKKW matrix element merging approach to angular-ordered parton showers. *JHEP*, **11**, 038. doi:10.1088/1126-6708/2009/11/038.

[613] Hankele, V., Klamke, G., Zeppenfeld, D., and Figy, T. (2006). Anomalous Higgs boson couplings in vector boson fusion at the CERN LHC. *Phys. Rev.*, **D74**, 095001. doi:10.1103/PhysRevD.74.095001.

[614] Harland-Lang, L. A., Martin, A. D., Motylinski, P., and Thorne, R. S. (2015). Parton distributions in the LHC era: MMHT 2014 PDFs. *Eur. Phys. J.*, **C75**(5), 204. doi:10.1140/epjc/s10052-015-3397-6.

[615] Harlander, Robert V. and Kilgore, William B. (2002). Next–to–next–to–leading order Higgs production at hadron colliders. *Phys. Rev. Lett.*, **88**, 201801. doi:10.1103/PhysRevLett.88.201801.

[616] Harrison, P.F., ed. and Quinn, Helen R., ed. (1998). The BABAR physics book: Physics at an asymmetric *B* factory.

[617] Hasegawa, K., Moch, S., and Uwer, P. (2010). AutoDipole: Automated generation of dipole subtraction terms. *Comput. Phys. Commun.*, **181**, 1802–1817. doi:10.1016/j.cpc.2010.06.044.

[618] Heinemeyer, S et al. (2013). Handbook of LHC Higgs Cross Sections: 3. Higgs Properties (arXiv:1307.1347). doi:10.5170/CERN-2013-004.

[619] Higgs, Peter W. (1964*a*). Broken symmetries and the masses of gauge bosons. *Phys. Rev. Lett.*, **13**, 508–509. doi:10.1103/PhysRevLett.13.508.

[620] Higgs, Peter W. (1964*b*). Broken symmetries, massless particles and gauge fields. *Phys. Lett.*, **12**, 132–133. doi:10.1016/0031-9163(64)91136-9.

[621] Higgs, Peter W. (1966). Spontaneous Symmetry Breakdown without Massless Bosons. *Phys. Rev.*, **145**, 1156–1163. doi:10.1103/PhysRev.145.1156.

[622] Hill, Christopher T. and Simmons, Elizabeth H. (2003). Strong dynamics and electroweak symmetry breaking. *Phys.Rept.*, **381**, 235–402. doi:10.1016/S0370-1573(03)00140-6.

[623] Hinshaw, G. et al. (2013). Nine-Year Wilkinson Microwave Anisotropy Probe (WMAP) Observations: Cosmological Parameter Results. *Astrophys.J.Suppl.*, **208**, 19. doi:10.1088/0067-0049/208/2/19.

[624] Hirai, M., Kumano, S., Nagai, T.-H., and Sudoh, K. (2007). Determination of fragmentation functions and their uncertainties. *Phys. Rev.*, **D75**, 094009. doi:10.1103/PhysRevD.75.094009.

[625] Hirschi, Valentin, Frederix, Rikkert, Frixione, Stefano, Garzelli, Maria Vittoria, Maltoni, Fabio et al. (2011). Automation of one-loop QCD corrections. *JHEP*, **1105**, 044. doi:10.1007/JHEP05(2011)044.

[626] Höche, Stefan, Schumann, Steffen, and Siegert, Frank (2010). Hard photon production and matrix-element parton-shower merging. *Phys. Rev.*, **D81**, 034026. doi:10.1103/PhysRevD.81.034026.

[627] Hoeche, Stefan, Huang, Junwu, Luisoni, Gionata, Schoenherr, Marek, and Winter, Jan (2013*a*). Zero and one jet combined next-to-leading order analysis of the top quark forward-backward asymmetry. *Phys. Rev.*, **D88**(1), 014040. doi:10.1103/PhysRevD.88.014040.

[628] Hoeche, Stefan, Krauss, Frank, and Schonherr, Marek (2014*a*). Uncertainties in MEPS@NLO calculations of h+jets. *Phys. Rev.*, **D90**(1), 014012. doi:10.1103/PhysRevD.90.014012.

[629] Hoeche, Stefan, Krauss, Frank, Schonherr, Marek, and Siegert, Frank (2011). NLO matrix elements and truncated showers. *JHEP*, **1108**, 123. doi:10.1007/JHEP08(2011)123.

[630] Hoeche, Stefan, Krauss, Frank, Schonherr, Marek, and Siegert, Frank (2012). A critical appraisal of NLO+PS matching methods. *JHEP*, **1209**, 049. doi:10.1007/JHEP09(2012)049.

[631] Hoeche, Stefan, Krauss, Frank, Schonherr, Marek, and Siegert, Frank (2013*b*). QCD matrix elements + parton showers: The NLO case. *JHEP*, **1304**, 027. doi:10.1007/JHEP04(2013)027.

[632] Hoeche, Stefan, Krauss, Frank, Schumann, Steffen, and Siegert, Frank (2009). QCD matrix elements and truncated showers. *JHEP*, **05**, 053. doi:10.1088/1126-6708/2009/05/053.

[633] Hoeche, Stefan, Li, Ye, and Prestel, Stefan (2014*b*). Higgs-boson production through gluon fusion at NNLO QCD with parton showers. *Phys. Rev.*, **D90**(5), 054011. doi:10.1103/PhysRevD.90.054011.

[634] Hoeche, Stefan, Li, Ye, and Prestel, Stefan (2015). Drell-Yan lepton pair production at NNLO QCD with parton showers. *Phys. Rev.*, **D91**(7), 074015. doi:10.1103/PhysRevD.91.074015.

[635] Hou, Wei-Shu (1993). Enhanced charged Higgs boson effects in $B^- \to \tau \bar{\nu}_\tau$, $\mu \bar{\nu}_\mu$, and $b \to \tau \bar{\nu}_\tau + X$. *Phys. Rev.*, **D48**, 2342–2344. doi:10.1103/PhysRevD.48.2342.

[636] Hoyer, P., Osland, P., Sander, H. G., Walsh, T. F., and Zerwas, P. M. (1979). Quantum chromodynamics and jets in e^+e^-. *Nucl. Phys.*, **B161**, 349. doi:10.1016/0550-3213(79)90217-7.

[637] Humpert, B. and Odorico, R. (1985). Multiparton scattering and QCD radiation as sources of four-jet events. *Phys. Lett.*, **B154**, 211. doi:10.1016/0370-2693(85)90587-8.

[638] Hunter, J. D. (2007). Matplotlib: A 2d graphics environment. *Computing In Science & Engineering*, **9**(3), 90–95.

[639] Isgur, Nathan, Scora, Daryl, Grinstein, Benjamin, and Wise, Mark B. (1989). Semileptonic B and D decays in the quark model. *Phys. Rev.*, **D39**, 799. doi:10.1103/PhysRevD.39.799.

[640] Isgur, Nathan and Wise, Mark B. (1989). Weak decays of heavy mesons in the static quark approximation. *Phys. Lett.*, **B232**, 113. doi:10.1016/0370-2693(89)90566-2.

[641] Isgur, Nathan and Wise, Mark B. (1990*a*). Influence of the B^* Resonance on $\bar{B} \to \pi e \bar{\nu}_e$. *Phys. Rev.*, **D41**, 151. doi:10.1103/PhysRevD.41.151.

[642] Isgur, Nathan and Wise, Mark B. (1990*b*). Weak transition form factors between heavy mesons. *Phys. Lett.*, **B237**, 527. doi:10.1016/0370-2693(90)91219-2.

[643] Isgur, Nathan and Wise, Mark B. (1991). Heavy baryon weak form-factors. *Nucl. Phys.*, **B348**, 276–292. doi:10.1016/0550-3213(91)90518-3.

[644] Ita, H., Bern, Z., Dixon, L.J., Febres Cordero, Fernando, Kosower, D.A. et al. (2012). Precise predictions for $Z + 4$ lets at hadron colliders. *Phys. Rev.*, **D85**, 031501. doi:10.1103/PhysRevD.85.031501.

[645] Jackson, John David. *Classical electrodynamics* (3rd ed. edn). Wiley, New York, NY.

[646] Jadach, S., Was, Z., Decker, R., and Kuhn, Johann H. (1993). The tau decay library TAUOLA: Version 2.4. *Comput. Phys. Commun.*, **76**, 361–380. doi:10.1016/0010-4655(93)90061-G.

[647] Jaiswal, Prerit, Kopp, Karoline, and Okui, Takemichi (2013). Higgs Production Amidst the LHC Detector. *Phys. Rev.*, **D87**(11), 115017.

doi:10.1103/PhysRevD.87.115017.

[648] James, F. (1980). Monte Carlo theory and practice. *Rept.Prog.Phys.*, **43**, 1145. doi:10.1088/0034-4885/43/9/002.

[649] Jimenez-Delgado, P. and Reya, E. (2009). Dynamical NNLO parton distributions. *Phys. Rev.*, **D79**, 074023. doi:10.1103/PhysRevD.79.074023.

[650] Juste, Aurelio, Mantry, Sonny, Mitov, Alexander, Penin, Alexander, Skands, Peter, Varnes, Erich, Vos, Marcel, and Wimpenny, Stephen (2014). Determination of the top quark mass circa 2013: methods, subtleties, perspectives. *Eur. Phys. J.*, **C74**(10), 3119. doi:10.1140/epjc/s10052-014-3119-5.

[651] Kanaki, Aggeliki and Papadopoulos, Costas G. (2000). HELAC: A Package to compute electroweak helicity amplitudes. *Comput. Phys. Commun.*, **132**, 306–315. doi:10.1016/S0010-4655(00)00151-X.

[652] Karlberg, Alexander, Re, Emanuele, and Zanderighi, Giulia (2014). NNLOPS accurate Drell-Yan production. *JHEP*, **09**, 134. doi:10.1007/JHEP09(2014)134.

[653] Kartvelishvili, V.G., Likhoded, A.K., and Petrov, V.A. (1978). On the fragmentation functions of heavy quarks into hadrons. *Phys. Lett.*, **B78**, 615. doi:10.1016/0370-2693(78)90653-6.

[654] Kauer, Nikolas and Passarino, Giampiero (2012). Inadequacy of zero-width approximation for a light Higgs boson signal. *JHEP*, **08**, 116. doi:10.1007/JHEP08(2012)116.

[655] Khachatryan, Vardan et al. (2011). Strange particle production in pp collisions at $\sqrt{s} = 0.9$ and 7 TeV. *JHEP*, **1105**, 064. doi:10.1007/JHEP05(2011)064.

[656] Khachatryan, Vardan et al. (2014a). Measurement of prompt J/ψ pair production in pp collisions at $\sqrt{s} = 7$ Tev. *JHEP*, **1409**, 094. doi:10.1007/JHEP09(2014)094.

[657] Khachatryan, Vardan et al. (2014b). Measurement of the t-channel single-top-quark production cross section and of the $\mid V_{tb} \mid$ CKM matrix element in pp collisions at $\sqrt{s}$= 8 TeV. *JHEP*, **06**, 090. doi:10.1007/JHEP06(2014)090.

[658] Khachatryan, Vardan et al. (2014c). Observation of the diphoton decay of the Higgs boson and measurement of its properties. *Eur. Phys. J.*, **C74**(10), 3076. doi:10.1140/epjc/s10052-014-3076-z.

[659] Khachatryan, Vardan et al. (2015a). Constraints on the spin-parity and anomalous HVV couplings of the Higgs boson in proton collisions at 7 and 8TeV. *Phys. Rev.*, **D92**(1), 012004. doi:10.1103/PhysRevD.92.012004.

[660] Khachatryan, Vardan et al. (2015b). Differential cross section measurements for the production of a W boson in association with jets in protonproton collisions at $\sqrt{s} = 7$ TeV. *Phys. Lett.*, **B741**, 12–37. doi:10.1016/j.physletb.2014.12.003.

[661] Khachatryan, Vardan et al. (2015c). Limits on the Higgs boson lifetime and width from its decay to four charged leptons. *Phys. Rev.*, **D92**(7), 072010. doi:10.1103/PhysRevD.92.072010.

[662] Khachatryan, Vardan et al. (2015d). Measurement of the differential cross section for top quark pair production in pp collisions at $\sqrt{s} = 8$ TeV. *Eur. Phys. J.*, **C75**(11), 542. doi:10.1140/epjc/s10052-015-3709-x.

[663] Khachatryan, Vardan et al. (2015*e*). Measurement of the inclusive 3-jet production differential cross section in protonproton collisions at 7 TeV and determination of the strong coupling constant in the TeV range. *Eur. Phys. J.*, **C75**(5), 186. doi:10.1140/epjc/s10052-015-3376-y.

[664] Khachatryan, Vardan et al. (2015*f*). Measurement of the $pp \rightarrow ZZ$ production cross section and constraints on anomalous triple gauge couplings in four-lepton final states at $\sqrt{s}$ =8 TeV. *Phys. Lett.*, **B740**, 250–272. [Erratum: Phys. Lett.B757,569(2016)]. doi:10.1016/j.physletb.2016.04.010, 10.1016/j.physletb.2014.11.059.

[665] Khachatryan, Vardan et al. (2015*g*). Measurement of the pp to ZZ production cross section and constraints on anomalous triple gauge couplings in four-lepton final states at sqrt(s) = 8 TeV. *Phys. Lett.*, **B740**, 250. doi:10.1016/j.physletb.2014.11.059.

[666] Khachatryan, Vardan et al. (2015*h*). Measurement of the Z boson differential cross section in transverse momentum and rapidity in protonproton collisions at 8 TeV. *Phys. Lett.*, **B749**, 187–209. doi:10.1016/j.physletb.2015.07.065.

[667] Khachatryan, Vardan et al. (2015*i*). Measurements of differential and double-differential Drell-Yan cross sections in proton-proton collisions at 8 TeV. *Eur. Phys. J.*, **C75**(4), 147. doi:10.1140/epjc/s10052-015-3364-2.

[668] Khachatryan, Vardan et al. (2015*j*). Measurements of jet multiplicity and differential production cross sections of $Z+$ jets events in proton-proton collisions at $\sqrt{s} = 7$ TeV. *Phys. Rev.*, **D91**(5), 052008. doi:10.1103/PhysRevD.91.052008.

[669] Khachatryan, Vardan et al. (2015*k*). Precise determination of the mass of the Higgs boson and tests of compatibility of its couplings with the standard model predictions using proton collisions at 7 and 8 TeV. *Eur. Phys. J.*, **C75**(5), 212. doi:10.1140/epjc/s10052-015-3351-7.

[670] Khachatryan, Vardan et al. (2016*a*). Inclusive and differential measurements of the $t\bar{t}$ charge asymmetry in pp collisions at $\sqrt{s} = 8$TeV. *Phys. Lett.*, **B757**, 154–179. doi:10.1016/j.physletb.2016.03.060.

[671] Khachatryan, Vardan et al. (2016*b*). Measurement of differential cross sections for Higgs boson production in the diphoton decay channel in pp collisions at $\sqrt{s} =$ 8 TeV. *Eur. Phys. J.*, **C76**(1), 13. doi:10.1140/epjc/s10052-015-3853-3.

[672] Khachatryan, Vardan et al. (2016*c*). Measurement of the charge asymmetry in top quark pair production in pp collisions at $\sqrt{(s)} = 8$ TeV using a template method. *Phys. Rev.*, **D93**(3), 034014. doi:10.1103/PhysRevD.93.034014.

[673] Khachatryan, Vardan et al. (2016*d*). Measurement of the W$^+$W$^-$ cross section in pp collisions at $\sqrt{s} = 8$ TeV and limits on anomalous gauge couplings. *Eur. Phys. J.*, **C76**(7), 401. doi:10.1140/epjc/s10052-016-4219-1.

[674] Khachatryan, Vardan et al. (2016*e*). Measurements of $t\bar{t}$ charge asymmetry using dilepton final states in pp collisions at $\sqrt{s} = 8$ TeV. *Phys. Lett.*, **B760**, 365–386. doi:10.1016/j.physletb.2016.07.006.

[675] Kibble, T.W.B. (1967). Symmetry breaking in non-Abelian gauge theories. *Phys. Rev.*, **155**, 1554–1561. doi:10.1103/PhysRev.155.1554.

[676] Kidonakis, Nikolaos and Owens, J.F. (2001). Effects of higher order threshold corrections in high E(T) jet production. *Phys. Rev.*, **D63**, 054019. doi:10.1103/PhysRevD.63.054019.

[677] Kilian, Wolfgang, Ohl, Thorsten, and Reuter, Jurgen (2011). WHIZARD: simulating multi–particle processes at LHC and ILC. *Eur. Phys. J.*, **C71**, 1742. doi:10.1140/epjc/s10052-011-1742-y.

[678] Kinoshita, T. (1962). Mass singularities of Feynman amplitudes. *J. Math. Phys.*, **3**, 650–677. doi:10.1063/1.1724268.

[679] Kirschner, R. (1979). Generalized Lipatov-Altarelli-Parisi equations and jet calculus rules. *Phys. Lett.*, **84B**, 266–270. doi:10.1016/0370-2693(79)90300-9.

[680] Klasen, M. and Kramer, G. (1997). Jet shapes in ep and $p\bar{p}$ collisions in NLO QCD. *Phys. Rev.*, **D56**, 2702–2712. doi:10.1103/PhysRevD.56.2702.

[681] Kleiss, Ronald and Kuijf, Hans (1989). Multi - gluon cross-sections and five jet production at hadron colliders. *Nucl. Phys.*, **B312**, 616–644. doi:10.1016/0550-3213(89)90574-9.

[682] Kleiss, Ronald and Pittau, Roberto (1994). Weight optimization in multichannel Monte Carlo. *Comput. Phys. Commun.*, **83**, 141–146. doi:10.1016/0010-4655(94)90043-4.

[683] Kleiss, R. and Stirling, W.James (1988). Top quark production at hadron colliders: some useful formulae. *Z. Phys.*, **C40**, 419–423. doi:10.1007/BF01548856.

[684] Kleiss, R. and Stirling, W. James (1985). Spinor techniques for calculating p anti-p —¿ W+- / Z0 + jets. *Nucl. Phys.*, **B262**, 235–262. doi:10.1016/0550-3213(85)90285-8.

[685] Kleiss, R., Stirling, W. James, and Ellis, S. D. (1986). A New Monte Carlo treatment of multiparticle phase space at high-energies. *Comput. Phys. Commun.*, **40**, 359. doi:10.1016/0010-4655(86)90119-0.

[686] Kluge, T., Rabbertz, K., and Wobisch, M. (2006). FastNLO: Fast pQCD calculations for PDF fits. hep-ph/0609285. pp. 483–486.

[687] Kniehl, Bernd A., Kramer, G., and Potter, B. (2000). Fragmentation functions for pions, kaons, and protons at next–to–leading order. *Nucl. Phys.*, **B582**, 514–536. doi:10.1016/S0550-3213(00)00303-5.

[688] Koba, Z., Nielsen, Holger Bech, and Olesen, P. (1972). Scaling of multiplicity distributions in high-energy hadron collisions. *Nucl. Phys.*, **B40**, 317–334. doi:10.1016/0550-3213(72)90551-2.

[689] Kodaira, Jiro and Trentadue, Luca (1982). Summing soft emission in QCD. *Phys. Lett.*, **B112**, 66. doi:10.1016/0370-2693(82)90907-8.

[690] Konishi, K., Ukawa, A., and Veneziano, G. (1979). Jet calculus: A Simple algorithm for resolving QCD jets. *Nucl. Phys.*, **B157**, 45–107. doi:10.1016/0550-3213(79)90053-1.

[691] Korchemsky, Gregory P. and Sterman, George F. (1995). Non-perturbative corrections in resummed cross-sections. *Nucl. Phys.*, **B437**, 415–432. doi:10.1016/0550-3213(94)00006-Z.

[692] Kosower, David A. (1990). Light cone recurrence relations for QCD amplitudes. *Nucl. Phys.*, **B335**, 23–44. doi:10.1016/0550-3213(90)90167-C.

[693] Kovarik, K., Schienbein, I., Olness, F.I., Yu, J.Y., Keppel, C. et al. (2011). Nuclear corrections in neutrino-nucleus DIS and their compatibility with global NPDF analyses. *Phys. Rev. Lett.*, **106**, 122301. doi:10.1103/PhysRevLett.106.122301.

[694] Kramer, Michael, 1, Olness, Fredrick I., and Soper, Davison E. (2000). Treatment of heavy quarks in deeply inelastic scattering. *Phys. Rev.*, **D62**, 096007. doi:10.1103/PhysRevD.62.096007.

[695] Krauss, F. (2002). Matrix elements and parton showers in hadronic interactions. *JHEP*, **08**, 015. doi:10.1088/1126-6708/2002/08/015.

[696] Krauss, F., Kuhn, R., and Soff, G. (2002). AMEGIC++ 1.0: A Matrix element generator in C++. *JHEP*, **02**, 044. doi:10.1088/1126-6708/2002/02/044.

[697] Kretzer, S. (2000). Fragmentation functions from flavor inclusive and flavor tagged e^+e^- annihilations. *Phys. Rev.*, **D62**, 054001. doi:10.1103/PhysRevD.62.054001.

[698] Kretzer, S., Lai, H.L., Olness, F.I., and Tung, W.K. (2004). Cteq6 parton distributions with heavy quark mass effects. *Phys. Rev.*, **D69**, 114005. doi:10.1103/PhysRevD.69.114005.

[699] Krohn, David, Thaler, Jesse, and Wang, Lian-Tao (2010). Jet Trimming. *JHEP*, **1002**, 084. doi:10.1007/JHEP02(2010)084.

[700] Kuhn, Johann H. and Rodrigo, German (1998). Charge asymmetry in hadroproduction of heavy quarks. *Phys. Rev. Lett.*, **81**, 49–52. doi:10.1103/PhysRevLett.81.49.

[701] Kuhn, Johann H. and Rodrigo, German (1999). Charge asymmetry of heavy quarks at hadron colliders. *Phys. Rev.*, **D59**, 054017. doi:10.1103/PhysRevD.59.054017.

[702] Kuhn, Johann H. and Rodrigo, German (2012). Charge asymmetries of top quarks at hadron colliders revisited. *JHEP*, **1201**, 063. doi:10.1007/JHEP01(2012)063.

[703] Kuhn, Johann H. and Santamaria, A. (1990). τ decays to pions. *Z. Phys.*, **C48**, 445–452. doi:10.1007/BF01572024.

[704] Kuhn, Johann H., Scharf, A., and Uwer, P. (2006). Electroweak corrections to top-quark pair production in quark-antiquark annihilation. *Eur. Phys. J.*, **C45**, 139–150. doi:10.1140/epjc/s2005-02423-6.

[705] Kuhn, J. H., Scharf, A., and Uwer, P. (2015). Weak interactions in top-quark pair production at hadron colliders: An Update. *Phys. Rev.*, **D91**(1), 014020. doi:10.1103/PhysRevD.91.014020.

[706] Kuhn, Johann H. and Was, Z. (2008). tau decays to five mesons in TAUOLA. *Acta Phys. Polon.*, **B39**, 147–158.

[707] Kulesza, Anna, Sterman, George F., and Vogelsang, Werner (2004). Joint resummation for Higgs production. *Phys. Rev.*, **D69**, 014012. doi:10.1103/PhysRevD.69.014012.

[708] Kuraev, E. A., Lipatov, L. N., and Fadin, Victor S. (1976). Multi-Reggeon processes in the Yang-Mills theory. *Sov.Phys.JETP*, **44**, 443–450.

[709] Kuraev, E. A., Lipatov, L. N., and Fadin, V. S. (1977). The Pomeranchuk singularity in non–Abelian gauge theories. *Sov. Phys. JETP*, **45**, 199–204.

[710] Kuzmin, V.A., Rubakov, V.A., and Shaposhnikov, M.E. (1985). On the anomalous electroweak baryon number nonconservation in the early universe. *Phys. Lett.*, **B155**, 36. doi:10.1016/0370-2693(85)91028-7.

[711] Ladinsky, G.A. and Yuan, C.P. (1994). The non–perturbative regime in QCD resummation for gauge boson production at hadron colliders. *Phys. Rev.*, **D50**, 4239. doi:10.1103/PhysRevD.50.R4239.

[712] Laenen, Eric, Sterman, George F., and Vogelsang, Werner (2000). Higher order QCD corrections in prompt photon production. *Phys. Rev. Lett.*, **84**, 4296–4299. doi:10.1103/PhysRevLett.84.4296.

[713] Lai, Hung-liang, Guzzi, Marco, Huston, Joey, Li, Zhao, Nadolsky, Pavel M. et al. (2010a). New parton distributions for collider physics. *Phys. Rev.*, **D82**, 074024. doi:10.1103/PhysRevD.82.074024.

[714] Lai, Hung-Liang, Huston, Joey, Li, Zhao, Nadolsky, Pavel, Pumplin, Jon et al. (2010b). Uncertainty induced by QCD coupling in the CTEQ global analysis of parton distributions. *Phys. Rev.*, **D82**, 054021. doi:10.1103/PhysRevD.82.054021.

[715] Lai, Hung-Liang, Huston, Joey, Mrenna, Stephen, Nadolsky, Pavel, Stump, Daniel et al. (2010c). Parton distributions for event generators. *JHEP*, **1004**, 035. doi:10.1007/JHEP04(2010)035.

[716] Landry, F., Brock, R., Nadolsky, Pavel M., and Yuan, C.P. (2003). Tevatron Run-1 Z boson data and Collins-Soper-Sterman resummation formalism. *Phys. Rev.*, **D67**, 073016. doi:10.1103/PhysRevD.67.073016.

[717] Landshoff, P.V. and Polkinghorne, J.C. (1971). The dual quark-parton model and high energy hadronic processes. *Nucl. Phys.*, **B32**, 541–556. doi:10.1016/0550-3213(71)90493-7.

[718] Lane, Kenneth D. (1996). Electroweak and flavor dynamics at hadron colliders. hep-ph/9605257.

[719] Lange, Bjorn O., Neubert, Matthias, and Paz, Gil (2005). Theory of charmless inclusive B decays and the extraction of V_{ub}. *Phys. Rev.*, **D72**, 073006. doi:10.1103/PhysRevD.72.073006.

[720] Langenfeld, U., Moch, S., and Uwer, P. (2009). Measuring the running top-quark mass. *Phys. Rev.*, **D80**, 054009. doi:10.1103/PhysRevD.80.054009.

[721] Laporta, S. (2000). High precision calculation of multiloop Feynman integrals by difference equations. *Int.J.Mod.Phys.*, **A15**, 5087–5159. doi:10.1016/S0217-751X(00)00215-7.

[722] Larkoski, Andrew J., Marzani, Simone, Soyez, Gregory, and Thaler, Jesse (2014). Soft Drop. *JHEP*, **05**, 146. doi:10.1007/JHEP05(2014)146.

[723] Lee, Benjamin W., Quigg, C., and Thacker, H.B. (1977). Weak interactions at very high-energies: The Role of the Higgs Boson mass. *Phys. Rev.*, **D16**, 1519. doi:10.1103/PhysRevD.16.1519.

[724] Lee, T.D. and Nauenberg, M. (1964). Degenerate systems and mass singularities. *Phys. Rev.*, **133**, B1549–B1562. doi:10.1103/PhysRev.133.B1549.

[725] Lepage, G. Peter (1980). VEGAS - an Adaptive multi-dimensional integration program. CLNS-80/447.

[726] Lepage, G. Peter and Brodsky, Stanley J. (1980). Exclusive processes in perturbative quantum chromodynamics. *Phys. Rev.*, **D22**, 2157. doi:10.1103/PhysRevD.22.2157.

[727] Lichtenberg, D.B., Namgung, W., Wills, J.G., and Predazzi, E. (1983). Light and Heavy Hadron Masses in a Relativistic Quark Potential Model With Diquark Clustering. *Z. Phys.*, **C19**, 19. doi:10.1007/BF01572332.

[728] Lipatov, L. N. (1976). Reggeization of the vector meson and the vacuum singularity in non–Abelian gauge theories. *Sov. J. Nucl. Phys.*, **23**, 338–345.

[729] Lonnblad, Leif (1992). ARIADNE version 4: A Program for simulation of QCD cascades implementing the color dipole model. *Comput. Phys. Commun.*, **71**, 15–31. doi:10.1016/0010-4655(92)90068-A.

[730] Lonnblad, Leif (2002). Correcting the color dipole cascade model with fixed order matrix elements. *JHEP*, **05**, 046. doi:10.1088/1126-6708/2002/05/046.

[731] Lonnblad, Leif and Pettersson, Ulf (1988). ARIADNE 2: A Monte Carlo for QCD cascades in the color dipole formulation: an update. LU-TP-88-15.

[732] Lonnblad, Leif and Prestel, Stefan (2012). Matching tree–level matrix elements with interleaved showers. *JHEP*, **1203**, 019. doi:10.1007/JHEP03(2012)019.

[733] Lonnblad, Leif and Prestel, Stefan (2013). Unitarising matrix element + parton shower merging. *JHEP*, **02**, 094. doi:10.1007/JHEP02(2013)094.

[734] Low, F.E. (1958). Bremsstrahlung of very low-energy quanta in elementary particle collisions. *Phys. Rev.*, **110**, 974–977. doi:10.1103/PhysRev.110.974.

[735] Low, F.E. (1975). A Model of the Bare Pomeron. *Phys. Rev.*, **D12**, 163–173. doi:10.1103/PhysRevD.12.163.

[736] Luke, Michael E. (1990). Effects of subleading operators in the heavy quark effective theory. *Phys. Lett.*, **B252**, 447–455. doi:10.1016/0370-2693(90)90568-Q.

[737] Lnnblad, Leif and Prestel, Stefan (2013). Merging multi-leg NLO matrix elements with parton showers. *JHEP*, **03**, 166. doi:10.1007/JHEP03(2013)166.

[738] Maltoni, F., Mangano, M. L., Tsinikos, I., and Zaro, M. (2014). Top-quark charge asymmetry and polarization in $t\bar{t}W$ production at the LHC. *Phys. Lett.*, **B736**, 252–260. doi:10.1016/j.physletb.2014.07.033.

[739] Maltoni, F., Paul, K., Stelzer, T., and Willenbrock, S. (2003). Color flow decomposition of QCD amplitudes. *Phys. Rev.*, **D67**, 014026. doi:10.1103/PhysRevD.67.014026.

[740] Maltoni, Fabio and Stelzer, Tim (2003). MadEvent: Automatic event generation with MadGraph. *JHEP*, **02**, 027. doi:10.1088/1126-6708/2003/02/027.

[741] Mana, C. and Martinez, M. (1987). On the radiative Bhabha scattering for the single photon configuration. *Nucl. Phys.*, **B287**, 601. doi:10.1016/0550-3213(87)90120-9.

[742] Mangano, Michelangelo L. (1989). Four-jet production at the Tevatron collider. *Z. Phys.*, **C42**, 331. doi:10.1007/BF01555875.

[743] Mangano, Michelangelo L., Moretti, Mauro, Piccinini, Fulvio, Pittau, Roberto, and Polosa, Antonio D. (2003). ALPGEN, a generator for hard multiparton processes in hadronic collisions. *JHEP*, **07**, 001. doi:10.1088/1126-6708/2003/07/001.

[744] Mangano, Michelangelo L., Moretti, Mauro, and Pittau, Roberto (2002). Multijet matrix elements and shower evolution in hadronic collisions: $Wb\bar{b} + n$ jets as a case study. *Nucl. Phys.*, **B632**, 343–362. doi:10.1016/S0550-3213(02)00249-3.

[745] Mannel, Thomas, Roberts, Winston, and Ryzak, Zbigniew (1991). Baryons in the heavy quark effective theory. *Nucl. Phys.*, **B355**, 38–53. doi:10.1016/0550-3213(91)90301-D.

[746] Manohar, Aneesh, Nason, Paolo, Salam, Gavin P., and Zanderighi, Giulia (2016). How bright is the proton? A precise determination of the photon parton distribution function. *Phys. Rev. Lett.*, **117**(24), 242002. doi:10.1103/PhysRevLett.117.242002.

[747] Manohar, Aneesh V. and Trott, Michael (2012). Electroweak Sudakov Corrections and the Top Quark Forward-Backward Asymmetry. *Phys. Lett.*, **B711**, 313–316. doi:10.1016/j.physletb.2012.04.013.

[748] Manohar, Aneesh V. and Waalewijn, Wouter J. (2012). A QCD Analysis of Double Parton Scattering: Color Correlations, Interference Effects and Evolution. *Phys. Rev.*, **D85**, 114009. doi:10.1103/PhysRevD.85.114009.

[749] Marchesini, G., Trentadue, L., and Veneziano, G. (1981). Space-time description of color screening via Jet calculus techniques. *Nucl. Phys.*, **B181**, 335. doi:10.1016/0550-3213(81)90357-6.

[750] Marchesini, G. and Webber, B. R. (1984). Simulation of QCD jets including soft gluon interference. *Nucl. Phys.*, **B238**, 1. doi:10.1016/0550-3213(84)90463-2.

[751] Marchesini, G. and Webber, B. R. (1990). Simulation of QCD Coherence in Heavy Quark Production and Decay. *Nucl. Phys.*, **B330**, 261–283. doi:10.1016/0550-3213(90)90310-A.

[752] Marquard, Peter, Smirnov, Alexander V., Smirnov, Vladimir A., and Steinhauser, Matthias (2015). Quark mass relations to four-loop order in perturbative QCD. *Phys. Rev. Lett.*, **114**(14), 142002. doi:10.1103/PhysRevLett.114.142002.

[753] Martin, A.D., Roberts, R.G., Stirling, W.J., and Thorne, R.S. (2003). Uncertainties of predictions from parton distributions. 1: Experimental errors. *Eur. Phys. J.*, **C28**, 455–473. doi:10.1140/epjc/s2003-01196-2.

[754] Martin, A. D., Roberts, R. G., Stirling, W. J., and Thorne, R. S. (2005). Parton distributions incorporating QED contributions. *Eur. Phys. J.*, **C39**, 155–161. doi:10.1140/epjc/s2004-02088-7.

[755] Martin, A. D., Stirling, W. J., Thorne, R. S., and Watt, G. (2009). Parton distributions for the LHC. *Eur. Phys. J.*, **C63**, 189–295. doi:10.1140/epjc/s10052-009-1072-5.

[756] Mastrolia, P., Ossola, G., Reiter, T., and Tramontano, F. (2010). Scattering amplitudes from unitarity-based reduction algorithm at the integrand-level. *JHEP*, **1008**, 080. doi:10.1007/JHEP08(2010)080.

[757] Matre, Daniel and Sapeta, Sebastian (2013). Simulated NNLO for high-p_T observables in vector boson + jets production at the LHC. *Eur. Phys. J.*, **C73**(12), 2663. doi:10.1140/epjc/s10052-013-2663-8.

[758] Meissner, W. and Ochsenfeld, R. (1933). Ein neuer effekt bei eintritt der supraleitfhigkeit. *Die Naturwissenschaften*, **21** (**44**), 787–788.

doi:10.1007/BF01504252.

[759] Mekhfi, M. (1985). Multiparton processes: an application to double Drell-Yan. *Phys. Rev.*, **D32**, 2371. doi:10.1103/PhysRevD.32.2371.

[760] Melia, Tom, Nason, Paolo, Rontsch, Raoul, and Zanderighi, Giulia (2011). W^+W^-, WZ and ZZ production in the POWHEG BOX. *JHEP*, **1111**, 078. doi:10.1007/JHEP11(2011)078.

[761] Melnikov, Kirill and Petriello, Frank (2006). Electroweak gauge boson production at hadron colliders through O(alpha(s)**2). *Phys. Rev.*, **D74**, 114017. doi:10.1103/PhysRevD.74.114017.

[762] Melnikov, Kirill and Schulze, Markus (2009). NLO QCD corrections to top quark pair production and decay at hadron colliders. *JHEP*, **0908**, 049. doi:10.1088/1126-6708/2009/08/049.

[763] Melnikov, Kirill and Schulze, Markus (2010). NLO QCD corrections to top quark pair production in association with one hard jet at hadron colliders. *Nucl. Phys.*, **B840**, 129–159. doi:10.1016/j.nuclphysb.2010.07.003.

[764] Mikaelian, K.O., Samuel, M.A., and Sahdev, D. (1979). The magnetic moment of weak bosons produced in pp and $p\bar{p}$ collisions. *Phys. Rev. Lett.*, **43**, 746. doi:10.1103/PhysRevLett.43.746.

[765] Mishra, Kalanand, Becher, Thomas, Barze, Luca, Chiesa, Mauro, Dittmaier, Stefan et al. (2013). Electroweak corrections at high energies. arXiv:1308.1430.

[766] Miu, Gabriela and Sjöstrand, Torbjörn (1999). W production in an improved parton-shower approach. *Phys. Lett.*, **B449**, 313–320. doi:10.1016/S0370-2693(99)00068-4.

[767] Moch, S., Vermaseren, J. A. M., and Vogt, A. (2004). The three-loop splitting functions in QCD: The non-singlet case. *Nucl. Phys.*, **B688**, 101–134. doi:10.1016/j.nuclphysb.2004.03.030.

[768] Monni, Pier Francesco and Zanderighi, Giulia (2015). On the excess in the inclusive $W^+W^- \to l^+l^-\nu\bar{\nu}$ cross section. *JHEP*, **05**, 013. doi:10.1007/JHEP05(2015)013.

[769] Moreno, G., Brown, C.N., Cooper, W.E., Finley, D., Hsiung, Y.B. et al. (1991). Dimuon production in proton - copper collisions at $\sqrt{s} = 38.8$-GeV. *Phys. Rev.*, **D43**, 2815–2836. doi:10.1103/PhysRevD.43.2815.

[770] Moretti, Mauro, Ohl, Thorsten, and Reuter, Jurgen (2001). O'Mega: An optimizing matrix element generator. hep-ph/0102195.

[771] Mueller, Alfred H. and Navelet, H. (1987). An inclusive minijet cross-section and the bare pomeron in QCD. *Nucl. Phys.*, **B282**, 727. doi:10.1016/0550-3213(87)90705-X.

[772] Mukherjee, A. and Pisano, Cristian (2003). Manifestly covariant analysis of the QED Compton process in e p → e gamma p and e p → e gamma X. *Eur. Phys. J.*, **C30**, 477–486. doi:10.1140/epjc/s2003-01308-0.

[773] Murayama, H., Watanabe, I., and Hagiwara, Kaoru (1992). HELAS: HELicity amplitude subroutines for Feynman diagram evaluations. KEK-91-11.

[774] Nadolsky, Pavel M. et al. (2008). Implications of CTEQ global analysis for collider observables. *Phys. Rev.*, **D78**, 013004. doi:10.1103/PhysRevD.78.013004.

[775] Nadolsky, Pavel M., Balazs, C., Berger, Edmond L., and Yuan, C.-P. (2007). Gluon-gluon contributions to the production of continuum diphoton pairs at hadron colliders. *Phys. Rev.*, **D76**, 013008. doi:10.1103/PhysRevD.76.013008.

[776] Nadolsky, Pavel M. and Sullivan, Z. (2001). PDF uncertainties in WH production at Tevatron. *eConf*, **C010630**, P510.

[777] Nagy, Zoltan (2003). Next-to-leading order calculation of three jet observables in hadron hadron collision. *Phys. Rev.*, **D68**, 094002. doi:10.1103/PhysRevD.68.094002.

[778] Nagy, Zoltan and Soper, Davison E. (2006). A New parton shower algorithm: Shower evolution, matching at leading and next-to-leading order level. In *Proceedings, Ringberg Workshop on New Trends in HERA Physics 2005: Ringberg Castle, Tegernsee, Germany, October 2-7, 2005*, pp. 101–123.

[779] Nagy, Zoltan and Soper, Davison E. (2008). Parton showers with quantum interference: leading color, with spin. *JHEP*, **07**, 025. doi:10.1088/1126-6708/2008/07/025.

[780] Nagy, Zoltan and Soper, Davison E. (2010). On the transverse momentum in Z-boson production in a virtuality ordered parton shower. *JHEP*, **03**, 097. doi:10.1007/JHEP03(2010)097.

[781] Nagy, Zoltan and Trocsanyi, Zoltan (1999). Next-to-leading order calculation of four jet observables in electron positron annihilation. *Phys. Rev.*, **D59**, 014020. doi:10.1103/PhysRevD.62.099902, 10.1103/PhysRevD.59.014020.

[782] Nason, Paolo (2004). A new method for combining NLO QCD with shower Monte Carlo algorithms. *JHEP*, **11**, 040. doi:10.1088/1126-6708/2004/11/040.

[783] Nason, P. and Webber, B.R. (1994). Scaling violation in e^+e^- fragmentation functions: QCD evolution, hadronization and heavy quark mass effects. *Nucl. Phys.*, **B421**, 473–517. doi:10.1016/0550-3213(94)90513-4.

[784] Neubert, Matthias (1994). Heavy quark symmetry. *Phys. Rept.*, **245**, 259–396. doi:10.1016/0370-1573(94)90091-4.

[785] Nielsen, Holger Bech and Olesen, P. (1973). Vortex line models for dual strings. *Nucl. Phys.*, **B61**, 45–61. doi:10.1016/0550-3213(73)90350-7.

[786] Norrbin, E. and Sjostrand, T. (2001). QCD radiation off heavy particles. *Nucl. Phys.*, **B603**, 297–342. doi:10.1016/S0550-3213(01)00099-2.

[787] Nussinov, S. (1975). Colored quark version of some hadronic puzzles. *Phys. Rev. Lett.*, **34**, 1286–1289. doi:10.1103/PhysRevLett.34.1286.

[788] Odorico, R. (1980). Exclusive calculations for QCD jets in a Monte Carlo approach. *Nucl. Phys.*, **B172**, 157. doi:10.1016/0550-3213(80)90165-0.

[789] Ohl, Thorsten (1999). Vegas revisited: Adaptive Monte Carlo integration beyond factorization. *Comput. Phys. Commun.*, **120**, 13–19. doi:10.1016/S0010-4655(99)00209-X.

[790] Okun, L. B. Okun and Pomeranchuk, I. I. (1956). . *Sov. Phys. JETP*, **3**, 307.

[791] Olive, K. A. et al. (2014). Review of Particle Physics. *Chin. Phys.*, **C38**, 090001. doi:10.1088/1674-1137/38/9/090001.

[792] Ossola, Giovanni, Papadopoulos, Costas G., and Pittau, Roberto (2007). Reducing full one-loop amplitudes to scalar integrals at the integrand level. *Nucl.*

Phys., **B763**, 147–169. doi:10.1016/j.nuclphysb.2006.11.012.

[793] Ossola, Giovanni, Papadopoulos, Costas G., and Pittau, Roberto (2008). Cut-Tools: A Program implementing the OPP reduction method to compute one-loop amplitudes. *JHEP*, **0803**, 042. doi:10.1088/1126-6708/2008/03/042.

[794] Pagels, Heinz (1975). Departures from chiral symmetry: A review. *Phys.Rept.*, **16**, 219. doi:10.1016/0370-1573(75)90039-3.

[795] Papadopoulos, Costas G. (2001). PHEGAS: A Phase space generator for automatic cross-section computation. *Comput. Phys. Commun.*, **137**, 247–254. doi:10.1016/S0010-4655(01)00163-1.

[796] Parisi, G. and Petronzio, R. (1979). Small transverse momentum distributions in hard processes. *Nucl. Phys.*, **B154**, 427. doi:10.1016/0550-3213(79)90040-3.

[797] Parke, Stephen J. and Taylor, T. R. (1986). An Amplitude for n gluon scattering. *Phys. Rev. Lett.*, **56**, 2459. doi:10.1103/PhysRevLett.56.2459.

[798] Passarino, G. and Veltman, M.J.G. (1979). One–loop corrections for e^+e^- annihilation into $\mu^+\mu^-$ in the Weinberg model. *Nucl. Phys.*, **B160**, 151. doi:10.1016/0550-3213(79)90234-7.

[799] Patrignani, C. et al. (2016). Review of Particle Physics. *Chin. Phys.*, **C40**(10), 100001. doi:10.1088/1674-1137/40/10/100001.

[800] Paver, N. and Treleani, D. (1982). Multi-quark scattering and large $p_\perp$ jet production in hadronic collisions. *Nuovo Cim.*, **A70**, 215. doi:10.1007/BF02814035.

[801] Paver, N. and Treleani, D. (1985). Multiple parton processes on the TeV region. *Z. Phys.*, **C28**, 187. doi:10.1007/BF01575722.

[802] Peraro, Tiziano (2014). Ninja: Automated Integrand Reduction via Laurent Expansion for One-Loop Amplitudes. *Comput. Phys. Commun.*, **185**, 2771–2797. doi:10.1016/j.cpc.2014.06.017.

[803] Peskin, Michael E. and Schroeder, Daniel V. (1995). An introduction to quantum field theory.

[804] Peterson, C., Schlatter, D., Schmitt, I., and Zerwas, Peter M. (1983). Scaling violations in inclusive e^+e^- annihilation spectra. *Phys. Rev.*, **D27**, 105. doi:10.1103/PhysRevD.27.105.

[805] Pettersson, Ulf (1988). Ariadne: A Monte Carlo for QCD cascades in the color dipole formulation. LU-TP-88-5.

[806] Platzer, Simon (2013). Controlling inclusive cross sections in parton shower + matrix element merging. *JHEP*, **1308**, 114. doi:10.1007/JHEP08(2013)114.

[807] Platzer, Simon and Gieseke, Stefan (2011). Coherent parton showers with local recoils. *JHEP*, **1101**, 024. doi:10.1007/JHEP01(2011)024.

[808] Platzer, Simon and Gieseke, Stefan (2012). Dipole showers and automated NLO matching in Herwig++. *Eur. Phys. J.*, **C72**, 2187. doi:10.1140/epjc/s10052-012-2187-7.

[809] Platzer, Simon and Sjodahl, Malin (2012). Subleading-$N-c$ improved Parton Showers. In *Proceedings, 20th International Workshop on Deep-Inelastic Scattering and Related Subjects (DIS 2012): Bonn, Germany, March 26-30, 2012*, pp. 709–712. doi:10.3204/DESY-PROC-2012-02/234.

[810] Plehn, T., Rainwater, David L., and Zeppenfeld, D. (2000). A method for identifying $H \to \tau^+\tau^- \to e^\pm \mu^\mp p_T$ at the CERN LHC. *Phys. Rev.*, **D61**, 093005. doi:10.1103/PhysRevD.61.093005.

[811] Plehn, Tilman, Rainwater, David L., and Zeppenfeld, Dieter (2002). Determining the structure of Higgs couplings at the LHC. *Phys. Rev. Lett.*, **88**, 051801. doi:10.1103/PhysRevLett.88.051801.

[812] Plothow-Besch, H. (1995). The Parton distribution function library. *Int.J.Mod.Phys.*, **A10**, 2901–2920. doi:10.1142/S0217751X9500139X.

[813] Politzer, H. David and Wise, Mark B. (1988). Effective field theory approach to processes involving both light and heavy fields. *Phys. Lett.*, **B208**, 504. doi:10.1016/0370-2693(88)90656-9.

[814] Pomeranchuk, I. I. (1956). . *Sov. Phys. JETP*, **3**, 306.

[815] Pomeranchuk, I. I. (1958). Equality of the nucleon and antinucleon total interaction cross section at high energies. *Sov. Phys. JETP*, **7**, 499.

[816] Pumplin, Jon (2009). Data set diagonalization in a global fit. *Phys. Rev.*, **D80**, 034002. doi:10.1103/PhysRevD.80.034002.

[817] Pumplin, J., Stump, D., Brock, R., Casey, D., Huston, J. et al. (2001). Uncertainties of predictions from parton distribution functions. 2. The Hessian method. *Phys. Rev.*, **D65**, 014013. doi:10.1103/PhysRevD.65.014013.

[818] Pumplin, J., Stump, D.R., Huston, J., Lai, H.L., Nadolsky, Pavel M. et al. (2002). New generation of parton distributions with uncertainties from global QCD analysis. *JHEP*, **0207**, 012. doi:10.1088/1126-6708/2002/07/012.

[819] Radescu, Voica (2010). Combination and QCD analysis of the HERA inclusive cross sections. *PoS*, **ICHEP2010**, 168.

[820] Rainwater, David L. and Zeppenfeld, D. (1997). Searching for $H \to \gamma\gamma$ in weak boson fusion at the LHC. *JHEP*, **12**, 005. doi:10.1088/1126-6708/1997/12/005.

[821] Rainwater, David L., Zeppenfeld, D., and Hagiwara, Kaoru (1998). Searching for $H \to \tau^+\tau^-$ in weak boson fusion at the CERN LHC. *Phys. Rev.*, **D59**, 014037. doi:10.1103/PhysRevD.59.014037.

[822] Ramond, Pierre (1981). Field theory. A modern primer. *Front.Phys.*, **51**, 1–397.

[823] Regge, T. (1959). Introduction to complex orbital momenta. *Nuovo Cim.*, **14**, 951. doi:10.1007/BF02728177.

[824] Regge, T. (1960). Bound states, shadow states and Mandelstam representation. *Nuovo Cim.*, **18**, 947–956. doi:10.1007/BF02733035.

[825] Richardson, Peter (2001). Spin correlations in Monte Carlo simulations. *JHEP*, **11**, 029. doi:10.1088/1126-6708/2001/11/029.

[826] Ritzmann, M., Kosower, D.A., and Skands, P. (2013). Antenna showers with hadronic initial states. *Phys. Lett.*, **B718**, 1345–1350. doi:10.1016/j.physletb.2012.12.003.

[827] Rojo, Juan, Accardi, Alberto, Ball, Richard D., Cooper-Sarkar, Amanda, de Roeck, Albert et al. (2015). The PDF4LHC report on PDFs and LHC data: Results from Run I and preparation for Run II (arXiv:1507.00556).

[828] Rolbiecki, Krzysztof and Sakurai, Kazuki (2013). Light stops emerging in WW cross section measurements? *JHEP*, **1309**, 004. doi:10.1007/JHEP09(2013)004.

[829] Rontsch, Raoul and Schulze, Markus (2014). Constraining couplings of top quarks to the Z boson in $t\bar{t}$ + Z production at the LHC. *JHEP*, **07**, 091. doi:10.1007/JHEP07(2014)091.

[830] Rubin, Mathieu, Salam, Gavin P., and Sapeta, Sebastian (2010). Giant QCD K-factors beyond NLO. *JHEP*, **09**, 084. doi:10.1007/JHEP09(2010)084.

[831] Rubin, V. C., Thonnard, N., and Ford, Jr., W. K. (1980). Rotational properties of 21 SC galaxies with a large range of luminosities and radii, from NGC 4605 /R = 4kpc/ to UGC 2885 /R = 122 kpc/. *Astrophys. J.*, **238**, 471. doi:10.1086/158003.

[832] Ryskin, M.G. and Snigirev, A.M. (2011). A fresh look at double parton scattering. *Phys. Rev.*, **D83**, 114047. doi:10.1103/PhysRevD.83.114047.

[833] Ryskin, M.G. and Snigirev, A.M. (2012). Double parton scattering in double logarithm approximation of perturbative QCD. *Phys. Rev.*, **D86**, 014018. doi:10.1103/PhysRevD.86.014018.

[834] Sakharov, A.D. (1967). Violation of CP invariance, c asymmetry, and baryon asymmetry of the universe. *Pisma Zh.Eksp.Teor.Fiz.*, **5**, 32–35. doi:10.1070/PU1991v034n05ABEH002497.

[835] Salam, Abdus (1968). Weak and electromagnetic interactions. *Proceedings of the eighth Nobel symposium,* Elementary particle physics: relativistic groups and analyticity, N. Svartholm, ed., Almqvist & Wiskell.

[836] Salam, Gavin P. and Soyez, Gregory (2007). A practical seedless infrared-safe cone jet algorithm. *JHEP*, **0705**, 086. doi:10.1088/1126-6708/2007/05/086.

[837] Schmidt, Carl, Pumplin, Jon, Stump, Daniel, and Yuan, C. P. (2016). CT14QED parton distribution functions from isolated photon production in deep inelastic scattering. *Phys. Rev.*, **D93**(11), 114015. doi:10.1103/PhysRevD.93.114015.

[838] Schönherr, M. and Krauss, F. (2008). Soft photon radiation in particle decays in SHERPA. *JHEP*, **12**, 018. doi:10.1088/1126-6708/2008/12/018.

[839] Schumann, Steffen and Krauss, Frank (2008). A Parton shower algorithm based on Catani-Seymour dipole factorisation. *JHEP*, **03**, 038. doi:10.1088/1126-6708/2008/03/038.

[840] Schwinger, Julian (1951*a*). On the greens functions of quantized fields. i. *Proceedings of the National Academy of Sciences*, **37**(7), 452–455. doi:10.1073/pnas.37.7.452.

[841] Schwinger, Julian S. (1951*b*). On gauge invariance and vacuum polarization. *Phys. Rev.*, **82**, 664–679. doi:10.1103/PhysRev.82.664.

[842] Seymour, Michael H. (1994). Searches for new particles using cone and cluster jet algorithms: A comparative study. *Z. Phys.*, **C62**, 127–138. doi:10.1007/BF01559532.

[843] Seymour, Michael H. (1995). Matrix element corrections to parton shower algorithms. *Comput. Phys. Commun.*, **90**, 95–101. doi:10.1016/0010-4655(95)00064-M.

[844] Shekhovtsova, O., Przedzinski, T., Roig, P., and Was, Z. (2012). Resonance chiral Lagrangian currents and τ decay Monte Carlo. *Phys. Rev.*, **D86**, 113008. doi:10.1103/PhysRevD.86.113008.

[845] Shelest, V.P., Snigirev, A.M., and Zinovev, G.M. (1982). The Multiparton distribution equations in QCD. *Phys. Lett.*, **B113**, 325. doi:10.1016/0370-2693(82)90049-1.

[846] Sherstnev, A. and Thorne, R.S. (2008). Parton distributions for LO generators. *Eur. Phys. J.*, **C55**, 553–575. doi:10.1140/epjc/s10052-008-0610-x.

[847] Shifman, Mikhail A. and Voloshin, M.B. (1987). On annihilation of mesons built from heavy and light quark and $\bar{B}_0 \leftrightarrow B_0$ oscillations. *Sov. J. Nucl. Phys.*, **45**, 292.

[848] Shifman, Mikhail A. and Voloshin, M.B. (1988). On production of D and D^* mesons in B meson decays. *Sov. J. Nucl. Phys.*, **47**, 511.

[849] Sjöstrand, Torbjörn (1984). Jet fragmentation of nearby partons. *Nucl. Phys.*, **B248**, 469. doi:10.1016/0550-3213(84)90607-2.

[850] Sjostrand, Torbjorn (1984). The merging of jets. *Phys. Lett.*, **B142**, 420–424. doi:10.1016/0370-2693(84)91354-6.

[851] Sjostrand, Torbjorn (1985). A Model for initial state parton showers. *Phys. Lett.*, **157B**, 321–325. doi:10.1016/0370-2693(85)90674-4.

[852] Sjostrand, Torbjorn, Eden, Patrik, Friberg, Christer, Lonnblad, Leif, Miu, Gabriela et al. (2001). High-energy physics event generation with PYTHIA 6.1. *Comput. Phys. Commun.*, **135**, 238–259. doi:10.1016/S0010-4655(00)00236-8.

[853] Sjostrand, Torbjorn, Mrenna, Stephen, and Skands, Peter Z. (2006). PYTHIA 6.4 Physics and Manual. *JHEP*, **05**, 026. doi:10.1088/1126-6708/2006/05/026.

[854] Sjostrand, T. and Skands, Peter Z. (2003). Baryon number violation and string topologies. *Nucl. Phys.*, **B659**, 243. doi:10.1016/S0550-3213(03)00193-7.

[855] Sjostrand, T. and Skands, Peter Z. (2004). Multiple interactions and the structure of beam remnants. *JHEP*, **03**, 053. doi:10.1088/1126-6708/2004/03/053.

[856] Sjostrand, T. and Skands, Peter Z. (2005). Transverse-momentum-ordered showers and interleaved multiple interactions. *Eur. Phys. J.*, **C39**, 129–154. doi:10.1140/epjc/s2004-02084-y.

[857] Sjostrand, Torbjorn and van Zijl, Maria (1987). A Multiple interaction model for the event structure in hadron collisions. *Phys. Rev.*, **D36**, 2019. doi:10.1103/PhysRevD.36.2019.

[858] Skands, Peter, Plehn, Tilman, and Rainwater, David (2005). QCD radiation in the production of high–$\hat{s}$ final states. *ECONF*, **C0508141**, ALCPG0417.

[859] Skands, Peter, Webber, Bryan, and Winter, Jan (2012). QCD coherence and the top quark asymmetry. *JHEP*, **1207**, 151. doi:10.1007/JHEP07(2012)151.

[860] Skands, Peter Z. and Wicke, Daniel (2007). Non-perturbative QCD effects and the top mass at the Tevatron. *Eur. Phys. J.*, **C52**, 133–140. doi:10.1140/epjc/s10052-007-0352-1.

[861] Snigirev, A.M. (2003). Double parton distributions in the leading logarithm approximation of perturbative QCD. *Phys. Rev.*, **D68**, 114012. doi:10.1103/PhysRevD.68.114012.

[862] Spergel, D.N. et al. (2003). First year Wilkinson Microwave Anisotropy Probe (WMAP) observations: Determination of cosmological parameters. *Astrophys.J.Suppl.*, **148**, 175–194. doi:10.1086/377226.

[863] Sterman, George F. (1987). Summation of large corrections to short-distance hadronic cross sections. *Nucl. Phys.*, **B281**, 310. doi:10.1016/0550-3213(87)90258-6.

[864] Stevenson, Paul M. (1981). Optimized perturbation theory. *Phys. Rev.*, **D23**, 2916. doi:10.1103/PhysRevD.23.2916.

[865] Stewart, Iain W., Tackmann, Frank J., Walsh, Jonathan R., and Zuberi, Saba (2014). Jet p_T resummation in Higgs production at $NNLL' + NNLO$. *Phys. Rev.*, **D89**(5), 054001. doi:10.1103/PhysRevD.89.054001.

[866] Stirling, W.J. and Vryonidou, E. (2013). Electroweak corrections and Bloch-Nordsieck violations in 2-to-2 processes at the LHC. *JHEP*, **1304**, 155. doi:10.1007/JHEP04(2013)155.

[867] Stump, Daniel, Huston, Joey, Pumplin, Jon, Tung, Wu-Ki, Lai, H. L., Kuhlmann, Steve, and Owens, J. F. (2003). Inclusive jet production, parton distributions, and the search for new physics. *JHEP*, **10**, 046. doi:10.1088/1126-6708/2003/10/046.

[868] Stump, D., Pumplin, J., Brock, R., Casey, D., Huston, J. et al. (2001). Uncertainties of predictions from parton distribution functions. 1. The Lagrange multiplier method. *Phys. Rev.*, **D65**, 014012. doi:10.1103/PhysRevD.65.014012.

[869] 't Hooft, Gerard (1974). A Planar diagram theory for strong interactions. *Nucl. Phys.*, **B72**, 461. doi:10.1016/0550-3213(74)90154-0.

[870] Tait, Timothy M. P. and Yuan, C. P. (2000). Single top quark production as a window to physics beyond the standard model. *Phys. Rev.*, **D63**, 014018. doi:10.1103/PhysRevD.63.014018.

[871] The H1 and ZEUS Collaborations (2009). Combined measurement and QCD analysis of the inclusive $e^{\pm}p$ scattering cross sections at HERA. *JHEP*, **01**, 109. doi:10.1007/JHEP01(2010)109.

[872] Thorne, R.S. and Roberts, R.G. (1998a). A Practical procedure for evolving heavy flavor structure functions. *Phys. Lett.*, **B421**, 303–311. doi:10.1016/S0370-2693(97)01580-3.

[873] Thorne, R.S. and Roberts, R.G. (1998b). An Ordered analysis of heavy flavor production in deep inelastic scattering. *Phys. Rev.*, **D57**, 6871–6898. doi:10.1103/PhysRevD.57.6871.

[874] Towell, R.S. et al. (2001). Improved measurement of the anti-d / anti-u asymmetry in the nucleon sea. *Phys. Rev.*, **D64**, 052002. doi:10.1103/PhysRevD.64.052002.

[875] Tung, Wu-Ki, Kretzer, Stefan, and Schmidt, Carl (2002). Open heavy flavor production in QCD: Conceptual framework and implementation issues. *J. Phys.*, **G28**, 983–996. doi:10.1088/0954-3899/28/5/321.

[876] Tung, W. K., Lai, H. L., Belyaev, A., Pumplin, J., Stump, D., and Yuan, C. P. (2007). Heavy quark mass effects in deep inelastic scattering and global QCD analysis. *JHEP*, **02**, 053. doi:10.1088/1126-6708/2007/02/053.

[877] Van der Meer, S. (1968). Calibration of the effective beam height in the ISR. *Internal CERN report, ISR-PO/68-31*.

[878] van der Waerden, B. L. (1974). *Group theory and quantum mechanics*. Springer, Berlin, Germany. Die Grundlehren der math. Wissenschaften.

[879] van Hameren, A. (2011). OneLOop: For the evaluation of one-loop scalar functions. *Comput. Phys. Commun.*, **182**, 2427–2438. doi:10.1016/j.cpc.2011.06.011.

[880] van Hameren, Andre and Papadopoulos, Costas G. (2002). A hierarchical phase space generator for QCD antenna structures. *Eur. Phys. J.*, **C25**, 563–574. doi:10.1007/s10052-002-1000-4.

[881] van Oldenborgh, G.J. (1991). FF: A Package to evaluate one loop Feynman diagrams. *Comput. Phys. Commun.*, **66**, 1–15. doi:10.1016/0010-4655(91)90002-3.

[882] Virkus, T. et al. (2008). Direct measurement of the Chudakov effect. *Phys. Rev. Lett.*, **100**, 164802. doi:10.1103/PhysRevLett.100.164802.

[883] Vogt, A., Moch, S., and Vermaseren, J. A. M. (2004). The three–loop splitting functions in QCD: The singlet case. *Nucl. Phys.*, **B691**, 129–181. doi:10.1016/j.nuclphysb.2004.04.024.

[884] von Weizsacker, C.F. (1934). Radiation emitted in collisions of very fast electrons. *Z. Phys.*, **88**, 612–625. doi:10.1007/BF01333110.

[885] Watt, G. and Thorne, R. S. (2012). Study of Monte Carlo approach to experimental uncertainty propagation with MSTW 2008 PDFs. *JHEP*, **08**, 052. doi:10.1007/JHEP08(2012)052.

[886] Webber, Bryan R. (1984). A QCD Model for jet fragmentation including soft gluon interference. *Nucl. Phys.*, **B238**, 492–528. doi:10.1016/0550-3213(84)90333-X.

[887] Webber, Bryan R. (1994). Estimation of power corrections to hadronic event shapes. *Phys. Lett.*, **B339**, 148–150. doi:10.1016/0370-2693(94)91147-9.

[888] Weinberg, Steven (1967). A Model of leptons. *Phys. Rev. Lett.*, **19**, 1264–1266. doi:10.1103/PhysRevLett.19.1264.

[889] Werner, Klaus, Liu, Fu-Ming, and Pierog, Tanguy (2006). Parton ladder splitting and the rapidity dependence of transverse momentum spectra in deuteron-gold collisions at RHIC. *Phys. Rev.*, **C74**, 044902. doi:10.1103/PhysRevC.74.044902.

[890] Weyl, Hermann (1931). *The theory of groups and quantum mechanics*. Dover, New York, USA.

[891] Whalley, M. R., Bourilkov, D., and Group, R. C. (2005). The Les Houches accord PDFs (LHAPDF) and LHAGLUE. In *HERA and the LHC: A Workshop on the implications of HERA for LHC physics. Proceedings, Part B*, pp. 575–581.

[892] Williams, E.J. (1934). Nature of the high-energy particles of penetrating radiation and status of ionization and radiation formulae. *Phys. Rev.*, **45**, 729–730. doi:10.1103/PhysRev.45.729.

[893] Wilson, K.G. and Zimmermann, W. (1972). Operator product expansions and composite field operators in the general framework of quantum field theory. *Commun.Math.Phys.*, **24**, 87–106. doi:10.1007/BF01878448.

[894] Wilson, Kenneth G. (1974). Confinement of quarks. *Phys. Rev.*, **D10**, 2445–2459. doi:10.1103/PhysRevD.10.2445.

[895] Winter, Jan-Christopher and Krauss, Frank (2008). Initial-state showering based on colour dipoles connected to incoming parton lines. *JHEP*, **07**, 040. doi:10.1088/1126-6708/2008/07/040.

[896] Winter, Jan-Christopher, Krauss, Frank, and Soff, Gerhard (2004). A Modified cluster hadronization model. *Eur. Phys. J.*, **C36**, 381–395. doi:10.1140/epjc/s2004-01960-8.

[897] Witten, Edward (2004). Perturbative gauge theory as a string theory in twistor space. *Commun. Math. Phys.*, **252**, 189–258. doi:10.1007/s00220-004-1187-3.

[898] Wobisch, M. and Wengler, T. (1998). Hadronization corrections to jet cross-sections in deep inelastic scattering. hep-ph/9907280.

[899] Wolfram, Stephen (1980). Parton and hadron Production in e^+e^- Annihilation. In *Elementary constituents and hadronic structure. Proceedings, 15th Rencontres de Moriond, Les Arcs, France, March 9-21, 1980. vol. 1*, pp. 549–588.

[900] Yang, Un-Ki et al. (2001). Measurements of F_2 and $xF_3^\nu - xF_3^{\bar\nu}$ from CCFR ν_μ–Fe and $\bar\nu_\mu$–Fe data in a physics model independent way. *Phys. Rev. Lett.*, **86**, 2742–2745. doi:10.1103/PhysRevLett.86.2742.

[901] Yennie, D. R., Frautschi, Steven C., and Suura, H. (1961). The infrared divergence phenomena and high-energy processes. *Annals Phys.*, **13**, 379–452. doi:10.1016/0003-4916(61)90151-8.

[902] Zeppenfeld, D., Kinnunen, R., Nikitenko, A., and Richter-Was, E. (2000). Measuring Higgs boson couplings at the CERN LHC. *Phys. Rev.*, **D62**, 013009. doi:10.1103/PhysRevD.62.013009.

[903] Zinovev, G.m., Snigirev, A.m., and Shelest, V.p. (1982). Equations for many-parton distributions in Quantum Chromodynamics. *Theor. Math. Phys.*, **51**, 523–528. doi:10.1007/BF01017270.

Index